Novel Biomaterials for Tissue Engineering 2018

Novel Biomaterials for Tissue Engineering 2018

Special Issue Editor

Emmanuel Stratakis

MDPI • Basel • Beijing • Wuhan • Barcelona • Belgrade

MDPI

Special Issue Editor
Emmanuel Stratakis
Laser and Applications division of the Institute of Electronic Structure and Laser (IESL),
Foundation of Research and Technology-Hellas (FORTH),
Vassilika Vouton Heraklion Crete,
Greece

Editorial Office
MDPI
St. Alban-Anlage 66
4052 Basel, Switzerland

This is a reprint of articles from the Special Issue published online in the open access journal *International Journal of Molecular Sciences* (ISSN 1422-0067) from 2017 to 2018 (available at: https: //www.mdpi.com/journal/ijms/special_issues/tis_engineering_2017)

For citation purposes, cite each article independently as indicated on the article page online and as indicated below:

LastName, A.A.; LastName, B.B.; LastName, C.C. Article Title. *Journal Name* **Year**, *Article Number*, Page Range.

ISBN 978-3-03897-543-4 (Pbk)
ISBN 978-3-03897-544-1 (PDF)

Cover image courtesy of Emmanuel Stratakis.

Contents

About the Special Issue Editor

Emmanuel Stratakis is a Research Director at the Institute of Electronic Structure and Laser (IESL) (www.iesl.forth.gr) of the Foundation for Research and Technology—Hellas (FORTH) (www.forth.gr). He received his Ph.D. in Physics from the University of Crete in 2001. After graduating, he became a visiting Researcher at IESL-FORTH, working on the ultrafast laser engineering of materials, and worked as an Adjunct Professor at the Department of Materials Science and Technology, University of Crete. In the fall semesters of 2006 and 2008 he was appointed as a visiting Researcher at the Department of Mechanical Engineering of the University of California, Berkeley. In 2007, he was elected as a Researcher at IESL-FORTH where he is leading the "Ultrafast Laser Micro- and Nano-processing" laboratory (http://stratakislab.iesl.forth.gr) which comprises a team of 35 postdocs, PhD students, technicians and administrative personnel. His research interests are in the fields of ultrafast laser interactions with materials for: (a) biomimetic micro- and nano-structuring; (b) advanced photonic processes for photovoltaics and energy storage; and (c) biomaterials processing for tissue engineering applications. He has delivered more than 40 invited and keynote lectures and has been the organizer and chair in major international scientific conferences. He has over 170 SCI publications and more than 6000 citations and has coordinated many national and EU grants. Since 2015, he is the Director of the European Nanoscience Facility of FORTH, part of the NFFA-Europe EU Infrastructure, where he is a member of the General Assembly. He is a national representative to the High-Level Group of EU on Nanosciences, Nanotechnology and Advanced Materials and a national expert for the Horizon 2020 committee configurations on: nanotechnologies, advanced materials, biotechnology, advanced manufacturing and processing. He is a member of the Scientific Committee of COST, of the Physical Sciences Sectoral Scientific Council of the National Council for Research and Innovation of Greece and the national delegate of the OECD Working Party on Bio-, Nano- and Converging Tech (BNCT).

International Journal of
Molecular Sciences

MDPI

Editorial

Novel Biomaterials for Tissue Engineering 2018

Emmanuel Stratakis

Institute of Electronic Structure and Laser, Foundation for Research and Technology Hellas, Nikolaou Plastira 100, Heraklion, Crete, GR-70013, Greece; stratak@iesl.forth.gr

Received: 7 December 2018; Accepted: 7 December 2018; Published: 9 December 2018

The concept of regenerating tissues, with properties and functions that mimic natural tissues, has attracted significant attention in recent years. It provides potential solutions for many diseases treatment and other healthcare problems. To fully realize the potential of the approach, it is crucial to have a rational biomaterial design to create novel scaffolds and other material systems suitable for tissue engineering, repair, and regeneration. Research advances in the topic include the design of new biomaterials and their composites, the scaffold fabrication via subtractive and additive manufacturing approaches, the development of implantable scaffolds for disease monitoring, diagnostics, and treatment, as well as the understanding of cell–biomaterial scaffolds interaction.

In the Special Issue "Novel Biomaterials for Tissue Engineering 2018", promising findings on different approaches to designing and developing new biomaterials, biomaterial systems, and methods for tissue engineering, are presented and discussed.

In particular, Rahali et al. report on the synthesis and characterization of novel gelatin methacrylate hydrogels functionalized with nanoliposomes and nanodroplets [1]. This nanofunctionalization approach enables control over the design of the biomaterial, via tailoring the type of incorporated nanoparticle for the specific application. Furthermore, hydroxyapatite (HA) films with minute amounts (ca. 1 weight %) of (rhenium-doped) fullerene-like MoS_2 nanoparticles (IF) were deposited through an electrophoretic process [2]. Tribological tests revealed that the nanoparticles endow the HA film very low friction and wear characteristics. As a consequence, HA-IF films could be of interest for various medical technologies. Al-Kattan et al. [3] reviewed the application of bare, ligand-free, laser-synthesized nanoparticles of Au and Si as functional modules (additives) in scaffold platforms intended for tissue engineering purposes. In addition, a new biodegradable medical adhesive material was recently developed by blending Poly(lactic acid) (PLA) with Poly(trimethylene carbonate) (PTMC) in ethyl acetate [4]. It is shown that in addition to having a positive effect on hemostasis and no sensibility to wounds, PLA-PTMC can efficiently prevent infections. Moreover, Babaliari et al., demonstrated the use of ultrafast laser-fabricated microstructured culture substrates on different materials, including Si, Polyethylene terephthalate (PET), and Poly(lactide-*co*-glycolide) (PLGA), as a mean to control cellular adhesion and orientation [5]. This property is potentially useful in the field of neural tissue engineering and for microenvironment systems that simulate in vivo conditions. The recent achievements in the field of non-apatitic calcium phosphate materials (CaPs) substituted with various ions were reviewed by Laskus and Kolmas [6]. The authors focused particularly on tricalcium phosphates (TCP) and "additives" such as magnesium, zinc, strontium, and silicate ions, all of which have been widely investigated thanks to their important biological role. The review also highlights some of the potential biomedical applications of non-apatitic substituted CaPs. Besides this, Khan and Tanaka [7] discussed the prospect of using functional biomaterials, which respond to external stimulus, for the development of smart 3D biomimetic scaffolds. The authors elaborated on how smart biomaterials are designed to interact with biological systems, for a wide range of biomedical applications, including the delivery of bioactive molecules, cell adhesion mediators, and cellular functioning for the engineering of functional tissues to treat diseases.

Human mesenchymal stem cells (MSCs) have been widely studied for therapeutic development in tissue engineering and regenerative medicine. However, directing the differentiation of MSCs still remains challenging for tissue engineering applications. To address this issue, Balikov et al. [8] developed a low-cost, scalable, and effective copolymer film to direct angiogenic differentiation of MSCs. hMSCs were cultured over several passages without the loss of reactive oxygen species handling or differentiation capacity. In another approach, Lee et al. [9] developed in situ cross-linkable gelatin–hydroxyphenyl propionic acid hydrogels that can direct endothelial differentiation of MSCs, thereby promoting vascularization of scaffolds towards tissue engineering and regenerative medicine applications in humans. Besides this, the development of techniques and devices for the development of new cellular microenvironments (i.e, niche), which is poorly represented by the typical plastic substrate used for the two-dimensional growth of MSCs in a tissue culture flask, is of critical importance. Aubert et al. [10] presented a collagen-based medical device as a mimetic niche for MSCs with the ability to preserve human MSC stemness in vitro. Nativel et al. [11] reported on the application of droplet millifluidics to encapsulate and support viable hMSCs in a polysaccharide hydrogel.

This Special Issue also presents recent advances in bone tissue engineering and regeneration as well as in osteogenic differentiation. Specifically, resorbable bacterial cellulose membranes, treated by electron beam irradiation, have been reported to be excellent biomaterials for guided bone regeneration [12]. Moreover, Hum et al. [13] developed highly porous bioactive glass-based scaffolds, fabricated by the foam replica technique and coated with collagen. The combination of bioactivity, mechanical competence, and cellular response makes this novel scaffold system attractive for bone tissue engineering. In another approach, Hsieh et al. [14] explored the development of solid biomimetic scaffolds of Poly(ε-Caprolactone)/Hydroxyapatite and Glycidyl-Methacrylate-Modified Hyaluronic Acid. In vivo experiments on the healing of osteochondral defects, performed on the knees of miniature pigs, concluded that the structural design of the scaffold should be reconsidered to match the regeneration process of both cartilage and subchondral bone. Besides this, different approaches to osteogenic differentiation are reported. In particular, the osteogenic differentiation effect of the FN Type 10-Peptide Amphiphile (FNIII10-PA) on Polycaprolactone fibers has been investigated [15]. It is shown that the FNIII10-PA-induced the osteogenic differentiation of MC3T3-E1 cells, indicating its potential as a new biomaterial for bone tissue engineering applications. Sobreiro-Almeida et al. investigated the hMSCs osteogenic differentiation on piezoelectric Poly(vinylidene fluoride) microsphere substrates [16]. It is concluded that such microspheres are a suitable support for bone tissue engineering purposes, as hMSCs can proliferate, be viable, and undergo osteogenic differentiation when chemically stimulated. Finally, Paun et al. developed 3D magnetic structures, fabricated by direct laser writing via two-photon polymerization and coated with a thin layer of collagen-chitosan-hydroxyapatite-magnetic nanoparticles composite [17]. In vitro experiments showed that such scaffolds stimulate the osteogenesis in the static magnetic field, via promotion of the MG-63 osteoblast-like cells proliferation and differentiation.

Tissue engineering methods to address skin regeneration and fertility restoration have been additionally reported. Specifically, Pang et al. [18] evaluated the effects of total flavonoids from *Blumea balsamifera* (L.) DC. on skin excisional wounds on the backs of Sprague-Dawley rats and revealed its chemical constitution, as well as its action mechanism. The study concluded that flavonoids were the main active constituents that contribute to excisional wound healing. Del Vento et al., reviewed the tissue engineering approaches to the improvement of immature testicular tissue and cell transplantation outcomes [19]. It is concluded that such bioengineering techniques may be a step closer to fertility restoration for prepubertal boys exposed to gonadotoxic treatments.

Finally, this Special Issue includes recent advances in biofabrication techniques for tissue engineering purposes. In particular, Zhang et al., provided an overview of the application of the layer-by-layer (LbL) self-assembly technology for the surface design and control of biomaterial scaffolds to mimic the unique features of native extracellular matrices [20]. It is concluded that LbL self-assembly not only provides advances for molecular deposition but also opens avenues for the design and development of innovative biomaterials for tissue engineering. Another emerging biofabrication tool is 3D printing, which has been recently applied for the development of an artificial trachea [21]. It is shown that epithelial cells in the 3D bioprinted artificial trachea were effective in respiratory epithelium regeneration. Furthermore, chondrogenic-differentiated bone marrow-derived MSCs had more neo-cartilage formation potential in a short period, although in a localized area. Furthermore, Park et al. [22] demonstrated the generation of oriented ligamentous architectures, driven by a 3D-printed microgroove pattern. The results of this study demonstrate that 3D-printed topographical approaches can regulate spatiotemporal cell organizations that offer strong potential for adaptation to complex tissue defects to regenerate ligament-bone complexes. In another study, a new bone substitute developed from 3D-printed structures of Polylactide (PLA) loaded with collagen I, have been demonstrated [23]. The results obtained from in vitro cultures of various cell types, including osteoblasts, osteoblast-like, fibroblasts, and endothelial, indicate the potential use of 3D-printed PLA scaffolds in bone tissue engineering. Being a promising biofabrication technique, electrospinning has been widely used for the fabrication of extracellular matrix (ECM)-mimicking fibrous scaffolds for several decades. In this context, Jun et al. [24] summarized the fundamental principles of electrospinning processes for generating complex fibrous scaffold geometries that are similar in structural complexity to the ECM of living tissues. Qasim et al. [25] reviewed the research progress on the electrospinning of chitosan and its composite formulations for creating fibers in combination with other natural polymers to be employed in tissue engineering. The review shows that evidence exists in support of the favorable properties and biocompatibility of chitosan electrospun composite biomaterials for a range of applications. It is concluded, however, that further research and in vivo studies are required to translate these materials from the laboratory to clinical applications.

References

1. Rahali, K.; Ben Messaoud, G.; Kahn, C.J.; Sanchez-Gonzalez, L.; Kaci, M.; Cleymand, F.; Fleutot, S.; Linder, M.; Desobry, S.; Arab-Tehrany, E. Synthesis and Characterization of Nanofunctionalized Gelatin Methacrylate Hydrogels. *Int. J. Mol. Sci.* **2017**, *18*, 2675. [CrossRef] [PubMed]
2. Shalom, H.; Feldman, Y.; Rosentsveig, R.; Pinkas, I.; Kaplan-Ashiri, I.; Moshkovich, A.; Perfilyev, V.; Rapoport, L.; Tenne, R. Electrophoretic Deposition of Hydroxyapatite Film Containing Re-Doped MoS$_2$ Nanoparticles. *Int. J. Mol. Sci.* **2018**, *19*, 657. [CrossRef]
3. Al-Kattan, A.; Nirwan, V.P.; Popov, A.; Ryabchikov, Y.V.; Tselikov, G.; Sentis, M.; Fahmi, A.; Kabashin, A.V. Recent Advances in Laser-Ablative Synthesis of Bare Au and Si Nanoparticles and Assessment of Their Prospects for Tissue Engineering Applications. *Int. J. Mol. Sci.* **2018**, *19*, 1563. [CrossRef] [PubMed]
4. Zhang, S.; Li, H.; Yuan, M.; Yuan, M.; Chen, H. Poly(Lactic Acid) Blends with Poly(Trimethylene Carbonate) as Biodegradable Medical Adhesive Material. *Int. J. Mol. Sci.* **2017**, *18*, 2041. [CrossRef] [PubMed]
5. Babaliari, E.; Kavatzikidou, P.; Angelaki, D.; Chaniotaki, L.; Manousaki, A.; Siakouli-Galanopoulou, A.; Ranella, A.; Stratakis, E. Engineering Cell Adhesion and Orientation via Ultrafast Laser Fabricated Microstructured Substrates. *Int. J. Mol. Sci.* **2018**, *19*, 2053. [CrossRef] [PubMed]
6. Laskus, A.; Kolmas, J. Ionic Substitutions in Non-Apatitic Calcium Phosphates. *Int. J. Mol. Sci.* **2017**, *18*, 2542. [CrossRef] [PubMed]
7. Khan, F.; Tanaka, M. Designing Smart Biomaterials for Tissue Engineering. *Int. J. Mol. Sci.* **2018**, *19*, 17. [CrossRef] [PubMed]
8. Balikov, D.A.; Crowder, S.W.; Lee, J.B.; Lee, Y.; Ko, U.H.; Kang, M.-L.; Kim, W.S.; Shin, J.H.; Sung, H.-J. Aging Donor-Derived Human Mesenchymal Stem Cells Exhibit Reduced Reactive Oxygen Species Loads and Increased Differentiation Potential Following Serial Expansion on a PEG-PCL Copolymer Substrate. *Int. J. Mol. Sci.* **2018**, *19*, 359. [CrossRef]
9. Lee, Y.; Balikov, D.A.; Lee, J.B.; Lee, S.H.; Lee, S.H.; Lee, J.H.; Park, K.D.; Sung, H.-J. In Situ Forming Gelatin Hydrogels-Directed Angiogenic Differentiation and Activity of Patient-Derived Human Mesenchymal Stem Cells. *Int. J. Mol. Sci.* **2017**, *18*, 1705. [CrossRef]

10. Aubert, L.; Dubus, M.; Rammal, H.; Bour, C.; Mongaret, C.; Boulagnon-Rombi, C.; Garnotel, R.; Schneider, C.; Rahouadj, R.; Laurent, C.; et al. Collagen-Based Medical Device as a Stem Cell Carrier for Regenerative Medicine. *Int. J. Mol. Sci.* **2017**, *18*, 2210. [CrossRef]
11. Nativel, F.; Renard, D.; Hached, F.; Pinta, P.-G.; D'Arros, C.; Weiss, P.; Le Visage, C.; Guicheux, J.; Billon-Chabaud, A.; Grimandi, G. Application of Millifluidics to Encapsulate and Support Viable Human Mesenchymal Stem Cells in a Polysaccharide Hydrogel. *Int. J. Mol. Sci.* **2018**, *19*, 1952. [CrossRef] [PubMed]
12. An, S.-J.; Lee, S.-H.; Huh, J.-B.; Jeong, S.I.; Park, J.-S.; Gwon, H.-J.; Kang, E.-S.; Jeong, C.-M.; Lim, Y.-M. Preparation and Characterization of Resorbable Bacterial Cellulose Membranes Treated by Electron Beam Irradiation for Guided Bone Regeneration. *Int. J. Mol. Sci.* **2017**, *18*, 2236. [CrossRef] [PubMed]
13. Hum, J.; Boccaccini, A.R. Collagen as Coating Material for 45S5 Bioactive Glass-Based Scaffolds for Bone Tissue Engineering. *Int. J. Mol. Sci.* **2018**, *19*, 1807. [CrossRef] [PubMed]
14. Hsieh, Y.-H.; Shen, B.-Y.; Wang, Y.-H.; Lin, B.; Lee, H.-M.; Hsieh, M.-F. Healing of Osteochondral Defects Implanted with Biomimetic Scaffolds of Poly(ε-Caprolactone)/Hydroxyapatite and Glycidyl-Methacrylate-Modified Hyaluronic Acid in a Minipig. *Int. J. Mol. Sci.* **2018**, *19*, 1125. [CrossRef] [PubMed]
15. Yun, Y.-R.; Kim, H.-W.; Jang, J.-H. The Osteogenic Differentiation Effect of the FN Type 10-Peptide Amphiphile on PCL Fiber. *Int. J. Mol. Sci.* **2018**, *19*, 153. [CrossRef] [PubMed]
16. Sobreiro-Almeida, R.; Tamaño-Machiavello, M.N.; Carvalho, E.O.; Cordón, L.; Doria, S.; Senent, L.; Correia, D.M.; Ribeiro, C.; Lanceros-Méndez, S.; Sabater i Serra, R.; et al. Human Mesenchymal Stem Cells Growth and Osteogenic Differentiation on Piezoelectric Poly(vinylidene fluoride) Microsphere Substrates. *Int. J. Mol. Sci.* **2017**, *18*, 2391. [CrossRef] [PubMed]
17. Paun, I.A.; Popescu, R.C.; Calin, B.S.; Mustaciosu, C.C.; Dinescu, M.; Luculescu, C.R. 3D Biomimetic Magnetic Structures for Static Magnetic Field Stimulation of Osteogenesis. *Int. J. Mol. Sci.* **2018**, *19*, 495. [CrossRef] [PubMed]
18. Pang, Y.; Zhang, Y.; Huang, L.; Xu, L.; Wang, K.; Wang, D.; Guan, L.; Zhang, Y.; Yu, F.; Chen, Z.; et al. Effects and Mechanisms of Total Flavonoids from *Blumea balsamifera* (L.) DC. on Skin Wound in Rats. *Int. J. Mol. Sci.* **2017**, *18*, 2766. [CrossRef]
19. Del Vento, F.; Vermeulen, M.; de Michele, F.; Giudice, M.G.; Poels, J.; des Rieux, A.; Wyns, C. Tissue Engineering to Improve Immature Testicular Tissue and Cell Transplantation Outcomes: One Step Closer to Fertility Restoration for Prepubertal Boys Exposed to Gonadotoxic Treatments. *Int. J. Mol. Sci.* **2018**, *19*, 286. [CrossRef] [PubMed]
20. Zhang, S.; Xing, M.; Li, B. Biomimetic Layer-by-Layer Self-Assembly of Nanofilms, Nanocoatings, and 3D Scaffolds for Tissue Engineering. *Int. J. Mol. Sci.* **2018**, *19*, 1641. [CrossRef] [PubMed]
21. Bae, S.-W.; Lee, K.-W.; Park, J.-H.; Lee, J.; Jung, C.-R.; Yu, J.; Kim, H.-Y.; Kim, D.-H. 3D Bioprinted Artificial Trachea with Epithelial Cells and Chondrogenic-Differentiated Bone Marrow-Derived Mesenchymal Stem Cells. *Int. J. Mol. Sci.* **2018**, *19*, 1624. [CrossRef]
22. Park, C.H.; Kim, K.-H.; Lee, Y.-M.; Giannobile, W.V.; Seol, Y.-J. 3D Printed, Microgroove Pattern-Driven Generation of Oriented Ligamentous Architectures. *Int. J. Mol. Sci.* **2017**, *18*, 1927. [CrossRef] [PubMed]
23. Ritz, U.; Gerke, R.; Götz, H.; Stein, S.; Rommens, P.M. A New Bone Substitute Developed from 3D-Prints of Polylactide (PLA) Loaded with Collagen I: An In Vitro Study. *Int. J. Mol. Sci.* **2017**, *18*, 2569. [CrossRef] [PubMed]
24. Jun, I.; Han, H.-S.; Edwards, J.R.; Jeon, H. Electrospun Fibrous Scaffolds for Tissue Engineering: Viewpoints on Architecture and Fabrication. *Int. J. Mol. Sci.* **2018**, *19*, 745. [CrossRef] [PubMed]
25. Qasim, S.B.; Zafar, M.S.; Najeeb, S.; Khurshid, Z.; Shah, A.H.; Husain, S.; Rehman, I.U. Electrospinning of Chitosan-Based Solutions for Tissue Engineering and Regenerative Medicine. *Int. J. Mol. Sci.* **2018**, *19*, 407. [CrossRef] [PubMed]

International Journal of
Molecular Sciences

MDPI

Article

Synthesis and Characterization of Nanofunctionalized Gelatin Methacrylate Hydrogels

Kamel Rahali [1,†], Ghazi Ben Messaoud [1,†], Cyril J.F. Kahn [1], Laura Sanchez-Gonzalez [1], Mouna Kaci [1], Franck Cleymand [2], Solenne Fleutot [2], Michel Linder [1], Stéphane Desobry [1] and Elmira Arab-Tehrany [1,*]

[1] Laboratoire d'Ingénierie des Biomolécules (LIBio), Université de Lorraine, 2 Avenue de la Forêt de Haye–BP 20163, 54505 Vandoeuvre-lès-Nancy, France; kamel.rahali.iut@gmail.com (K.R.); ghazi.benmessaoud@hotmail.fr (G.B.M.); cyril.kahn@univ-lorraine.fr (C.J.F.K.); laura.sanchez-gonzalez@univ-lorraine.fr (L.S.-G.); mouna.kaci@gmail.com (M.K.); michel.linder@univ-lorraine.fr (M.L.); stephane.desobry@univ-lorraine.fr (S.D.)
[2] Institut Jean Lamour (UMR CNRS 7198), Université de Lorraine, Parc de Saurupt, CS 50840, 54011 Nancy CEDEX, France; franck.cleymand@univ-lorraine.fr (F.C.); solenne.fleutot@univ-lorraine.fr (S.F.)
* Correspondence: elmira.arab-tehrany@univ-lorraine.fr; Tel.: +33-3-72-74-41-05
† These authors contributed equally to this work.

Received: 8 November 2017; Accepted: 28 November 2017; Published: 10 December 2017

Abstract: Given the importance of the extracellular medium during tissue formation, it was wise to develop an artificial structure that mimics the extracellular matrix while having improved physico-chemical properties. That is why the choice was focused on gelatin methacryloyl (GelMA), an inexpensive biocompatible hydrogel. Physicochemical and mechanical properties were improved by the incorporation of nanoparticles developed from two innovative fabrication processes: High shear fluid and low frequencies/high frequencies ultrasounds. Both rapeseed nanoliposomes and nanodroplets were successfully incorporated in the GelMA networks during the photo polymerization process. The impact on polymer microstructure was investigated by Fourier-transform infrared spectroscopy (FTIR), scanning electron microscopy (SEM), and enzymatic degradation investigations. Mechanical stability and viscoelastic tests were conducted to demonstrate the beneficial effect of the functionalization on GelMA hydrogels. Adding nanoparticles to GelMA improved the surface properties (porosity), tuned swelling, and degradability properties. In addition, we observed that nanoemulsion didn't change significantly the mechanical properties to shear and compression solicitations, whereas nanoliposome addition decreased Young's modulus under compression solicitations. Thus, these ways of functionalization allow controlling the design of the material by choosing the type of nanoparticle (nanoliposome or nanoemulsion) in function of the application.

Keywords: GelMA; functionalized hydrogel; nanoliposome; nanoemulsion; LF/HF ultrasounds; mechanical properties; tissue engineering

---

1. Introduction

Hydrogels are polymeric networks made of hydrophilic groups or domains, which allowed the material to absorb and retain large amounts of water. Thus, to avoid dissolution of the hydrogel into the aqueous phase, crosslinks have to be present [1].

Hydrogels were first synthesized by Wichterle and Lim in 1960 [2]. Since then, they have been developed for a widely range of applications such as food additives, biomedical implants, pharmaceutical & diagnostics products, biosensors, drug carriers, and wound dressing materials [3–7]. Today, hydrogels are extensively studied as biomaterials due to their properties (mechanical, physicochemical, etc.) and functional resemblance with the extracellular matrix (ECM) [1].

Thus, their characteristics and biocompatibility make them an interesting research material in the fields of tissue engineering, cell encapsulation, and drug delivery applications [8].

Hydrogels contain hydrophilic groups that, in an aqueous medium, absorb a large amount of water into its pores and develop a superabsorbent material [9–11].

The water retention capacity of these type of materials can be considered as one of their most important characteristic property where they have the ability to imbibe water up to 20 times more than its original molecular weight [12] and becoming soft with a degree of flexibility similar to living tissues [13,14].

Hydrogel scaffolds can be formed by natural and/or synthetic crosslinked polymer chains [15] Synthetic polymers allow a control of their composition, polymerization, and mechanical properties and their microstructure, thereby their degradability. The chemical properties of hydrogels to create an optimized cellular environment that improve the cellular performance and activities [8,16–18]. In addition to their physicochemical characteristics, hydrogels can be functionalized by the incorporation of molecules of interest or vectors for the delivery of active molecules in a native functional state.

Among the various hydrogels, gelatin is one of these natural polymers that are extensively used in biomedical application [19,20]. It is a natural cytocompatible protein [21] derived from either an acid or an alkaline hydrolysis of a natural product called collagen [22,23].

It is an interesting polymer due to the presence of many bioactive sequences such as arginine-glycine-aspartic acid (RGD), providing cell attachment and growth of different type of active molecules [24–26]. However, to enhance its final physicochemical properties and avoid the thermo reversibility due to its relatively low melting point, a chemical modification of this polymer is necessary. For this, gelatin methacryloyl (GelMA), a semi synthetic derived hydrogel, was chosen [19].

GelMA is produced by the substitution of the free amine groups of the gelatin with methacrylate anhydride while preserving the arginine-glycine-aspartic acid (RGD) sequences that promote cell attachment. Conveniently, introduction of methacryloyl substituent groups and photoinitiator confers to gelatin the property of photocrosslinking by exposure to UV radiation [27]. This polymerization does not need to be done at specific conditions (room temperature, neutral pH, in aqueous environments, etc.), and allows the control of the reaction in terms of time and space [26]. This enables microfabrication of the hydrogels to create unique patterns, morphologies, and 3D structures for applications such as platforms to control cellular behaviors in order to study cell-biomaterial interactions, cell-laden microtissues and microfluidic devices [28].

Properties of GelMA (swelling, strength, porosity, etc.) can be optimized further by incorporating nanoparticles within the hydrogel. Previous works [29,30] showed that nanoliposomes improve mechanical properties and porosity of the GelMA hydrogels. In order to compare the various properties of GelMA functionalized with lipidic nanoparticles differing by their external surface hydrophilicity and their structure. Thus, using the nanoemulsion as soft nanoparticle [31] can be another possibility to improve the physico-chemical properties of GelMA compared to GelMA-Nanoliposome functionalization.

2. Results and Discussion

2.1. Solutions Characterization

Size, polydispersity index, and zeta potential of nanoliposomes solution and nanoemulsion were measured immediately after preparation (Table 1). The average size of rapeseed nanoliposomes was 170 nm, and it was consistent with the results previously obtained with the sonication method [32] with a better polydispersity index. Rapeseed nanodroplets average size was 238 nm with a polydispersity index extremely close to that of nanoliposomes. This index corresponds to a good distribution of the particles and a good homogeneity of the solution. Zeta potentials were substantially similar with −43.7 mV and −48.8 mV, respectively, for liposomes and oil nanodroplets. It is due to the negative electrophoretic mobility of the phospholipids in nanoliposomes [30,33] and the hydroxide ions in the oil nanodroplets [34]. Small particle size, low polydispersity index, and negative charges are important to prevent an aggregation of the liposomes or a coalescence of the droplets during functionalization of the hydrogels.

Table 1. Mean particle size (nm), polydispersity index (PDI), and zeta potential (mV) of the rapeseed nanoliposomes and nanodroplets of rapeseed oil emulsion.

Solutions	Particle Size (nm)	Polydispersity Index	Zeta Potential (mV)
Nanoliposomes	169.7 ± 2	0.360 ± 0.03	-43.7 ± 1.6
Nanoemulsion	237.9 ± 7	0.393 ± 0.01	-48.8 ± 0.6

2.2. Hydrogels Morphology

After photopolymerization of the hydrogels under UV irradiation, the rinsed hydrogel discs had smooth and homogenous surfaces: pure 15% GelMA disc was completely transparent, while functionalized GelMA discs were translucent for GelMA-Nanoemulsion and opaque for GelMA-Nanoliposome. GelMA-Nanoliposomes hydrogel was a yellowish colored opaque and GelMA-Emulsion hydrogel was translucent white. Coloration of the hydrogels indicates that the incorporation of nanoliposomes from rapeseed lecithin and nanodroplets from rapeseed emulsion are uniform (Figure 1).

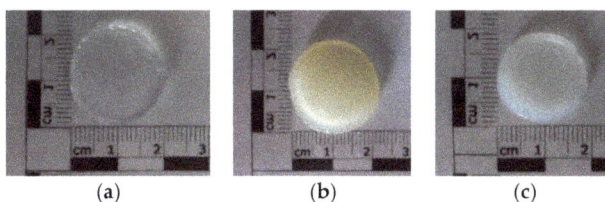

(a)　　　　　　　　　(b)　　　　　　　　　(c)

Figure 1. GelMA scaffolds after photopolymerization (**a**) 15% GelMA; (**b**) GelMA-Nanoliposomes 15%:1.5% (*w/w*); (**c**) GelMA-Emulsion 15%:1.5% (*w/w*).

2.3. Fourier-Transform Infrared Spectroscopy (FTIR)

Material excited by infrared sources can be characterized by the spectra of molecular absorption and transmission peaks obtained from vibration frequencies between the bounds of atoms. Fourier-transform infrared spectroscopy (FTIR) is used essentially to characterize the presence of specific chemical groups in the hydrogels and to study the interaction between the blended polymers and the effect of the added nanoliposomes or nanodroplets in the polymers. The results show the spectrum of nanoliposomes, emulsion, pure GelMA hydrogel, and GelMA functionalized with nanoliposomes or nanoemulsion (Figure 2).

The FTIR spectrum of pure rapeseed oil (emulsion) presents the 22 specific regions [35,36]. It consists of a major quantity of triglycerides that the main bands appearing in spectrum due to the asymmetric and symmetric stretching vibration of CH at 2924 and 2854 cm^{-1} and stretching of C=O at 1744 cm^{-1}. Weaker bands appear at 1464, 1377, 1242, 1161, 1119 and 723 cm^{-1} that correspond to scissoring of CH (CH_2), bending of CH (CH_3), CH (cis), CH_2, CO stretching, and rocking of $-(CH_2)_n-$, respectively. The spectrum of nanoliposomes displays the main characteristic bands of phospholipids presented in liposomes: maximum of peaks at 2854 and 2924 cm^{-1}, corresponding to the symmetric and antisymmetric stretching in the CH_2 groups of alkyl chains, respectively. The broad band from 3750 to 3050 cm^{-1} represents OH band. The band at 1732 cm^{-1} corresponds to the stretching vibrations of the ester carbonyl groups of phospholipids, and the relatively strong band centered at 1651 cm^{-1}

corresponds to the stretching vibrations of alkene carbon–carbon double bond –C=C–. The scissoring vibrations of the CH$_2$ groups are represented by the band at 1456 cm^{-1}, and the band at 1406 cm^{-1} corresponds to (=C–H) bending (rocking) vibrations. While the relatively weak band at 1394 cm^{-1} represents the umbrella deformation vibrations of the CH$_3$ groups of alkyl chains. In addition, the spectral bands at 1086 and 1224 cm^{-1} represent the symmetric and antisymmetric PO$_2$– stretching vibration of phospholipids, and the band representing the antisymmetric N$^+$/CH$_3$ stretching vibrations was detected at 970 cm^{-1} [37].

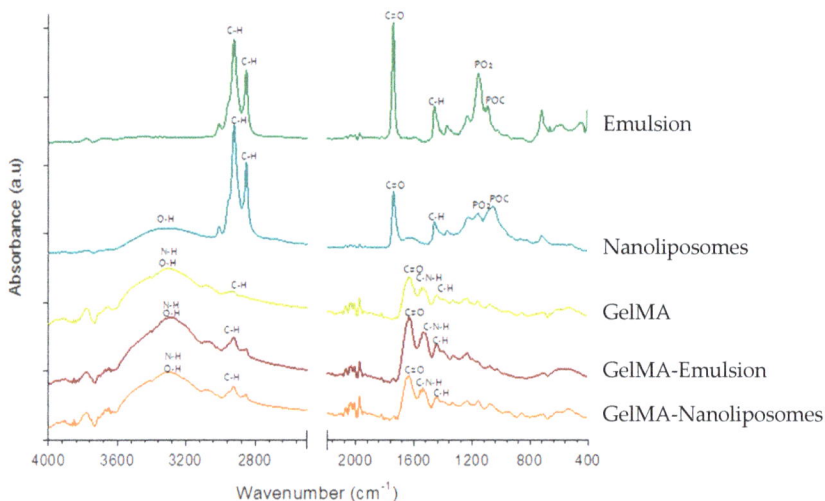

Figure 2. Fourier-transform infrared spectroscopy (FTIR) spectra of Emulsion, Nanoliposomes, GelMA, GelMA-Emulsion, and GelMA-Nanoliposomes. a.u., arbitrary unit.

All the hydrogels spectra (GelMA, GelMA-Emulsion, and GelMA-Nanoliposomes) showed a broad peak with a peak position at 3290 cm^{-1} associated with the stretching of the hydrogen bonded hydroxyl groups. The spectrum of GelMA hydrogel derived from the modification of the gelatin with the methacrylate anhydride. A strong peak appears at 1650 cm^{-1} related to amide I primarily C=O stretching groups. The band at 1500–1570 cm^{-1} corresponds to C–N–H bending while the band at 3200–3400 cm^{-1} indicates the presence of peptide bonds (mainly N–H stretching). The peak at 3062 cm^{-1} represents the C–H stretching groups [38,39]. The peak at 1640 cm^{-1} corresponds to carbon double bond in GelMA that presents the interaction between gelatin and methacrylate anhydride. The spectra of the hydrogel with the soft nanoparticles (GelMA-Nanoliposomes) does not show the peaks corresponding to the liposomes spectra but it presents an increasing of the intensity of certain peaks specially around 3300 cm^{-1}. The spectra of hydrogel with emulsion (GelMA-Emulsion) presents a similar increasing in the same region corresponding to the presence of a small amount of fatty acids. These results show an interaction between the GelMA and the nanoliposomes and nanoemulsion.

2.4. Scanning Electron Microscopy (SEM)

The morphological properties of pure GelMA and functionalized hydrogels were investigated by scanning electron microscopy (SEM). Upon examination with SEM, we observed the porosity of GelMA hydrogels with different diameters on the microscale (Figure 3).

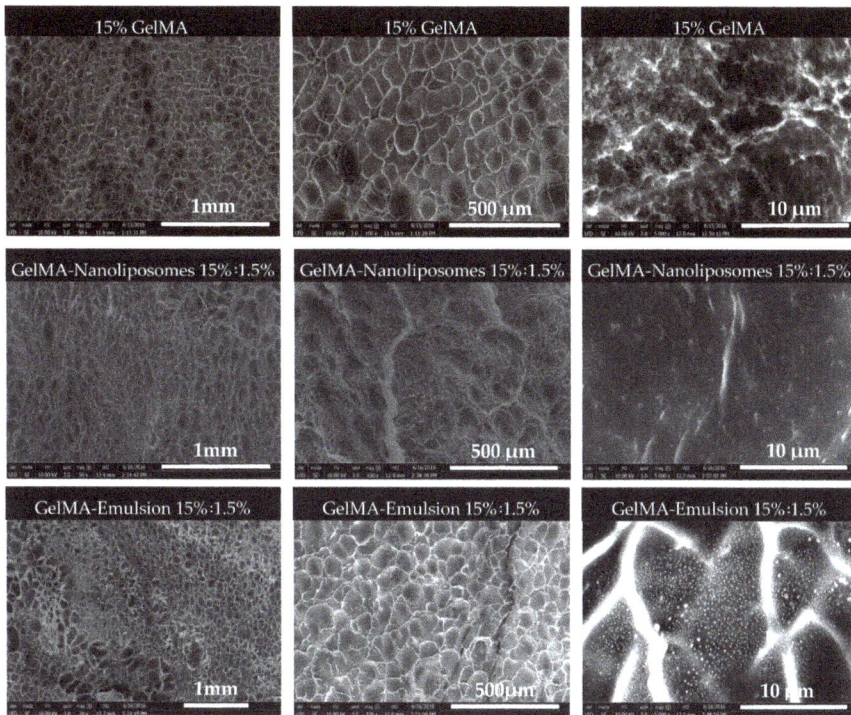

Figure 3. Scanning electron microscopy (SEM) images of 3D hydrogels (GelMA, GelMA-Nanoliposomes, and GelMA-Emulsion).

Although freeze drying procedure can alter material porosity, all samples were dried by the same process and there was no difference in the pore distribution. Pores were shaped pockets like which do not seem interconnected and are separated by thin walls. They are different size sand were formed during the photopolymerization, the syneresis phenomenon acts by squeezing locally excess of water out of the polymer in an attempt to reach its equilibrium swelling concentration (by a thermodynamically favored state of the polymer) and prevent the system from forming a fully homogenous material [40]. Higher magnification (Figure 3, ×5000; scale bar 10 µm) showed that nanoliposomes were totally assimilated (a homogenous surface morphology) in the network, while nanodroplets were adsorbed at the surface of GelMA after nanofunctionalization.

2.5. Degradability

For biomedical applications, biodegradability of hydrogel materials must be investigated. Indeed, in order to develop and obtain a mature tissue, cells must be able to degrade and remodel their hydrogel environment. GelMA has already been tested, and like gelatin, in its native state, maintains its susceptibility to enzymatic degradation. Thus, the influence of functionalized by nanoliposome or nanoemulsion hydrogel must investigate toward enzymatic activity. GelMA hydrogel and functionalized hydrogels were added to collagenase type II dissolved in phosphate buffer solution (PBS) at an enzyme concentration of 2 µg/mL. Enzymatic degradation of the GelMA hydrogels was examined to ensure that the inlay of nanoparticles did not decrease the enzymatic degradability.

The degradation profiles of GelMA, GelMA-Emulsion, and GelMA-Nanoliposomes (Figure 4) showed that at a concentration of 2 µg/mL after 2 h of incubation the degradation rate was the same in the three systems, but after 4 h of incubation, type II collagenase had a tendency to degrade the

GelMA-Nanoliposomes most rapidly. This tendency was confirmed after 8 h of incubation, and the rate of mass loss was approximately double (39.2% in GelMA-Nanoliposomes hydrogels against 21.0% and 18.3% in pure GelMA and GelMA-Emulsion hydrogels, respectively). This increase can be due to a physical phenomenon, as nanoliposomes incorporation increases pores size and enhances their distribution [29], which releases a way of privileged access to the collagenase, multiplying by twice the area of access sequences Arg-Gly-Asp (RGD), which are considered matrix metalloproteinase 2 (MMP-2) binding sites. These enzymes can degrade and remodel the extracellular matrix for cell spreading and migration.

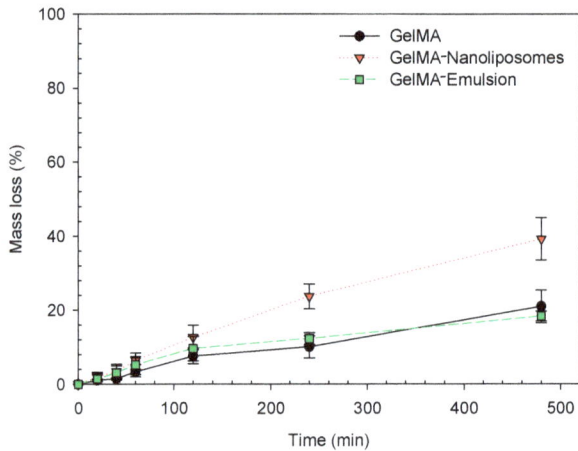

Figure 4. GelMA degradability GelMA (●), GelMA-Nanoliposomes (▼), and GelMA-Emulsion (■) hydrogels of uniform size were exposed to 2 μg/mL exogenous collagenase. Mass losses (%) were measured during 8 h. Error bars represent standard deviation.

2.6. Swelling Behavior Study

To study of swelling characteristics of polymers allows us to evidence the surface properties as well as to explain the mechanical characteristics and the diffusion process. The pore size of a polymer network, the interaction between the polymer and the solvent, the methacrylation degree, and the amount of photoinitiator will determine its degree of swelling. That is why we investigated the mass swelling ratio of 15% GelMA polymers (pure and functionalized) in two different solvents (water and PBS solution, pH = 7.4).

The calculated swelling ratio in water (Figure 5a) demonstrated that the functionalization of the GelMA did not affect the water retention capacity, and there was no significant difference between the hydrogels, they have absorbed in mean from four to six times their weight of water. In PBS solution (Figure 5b), the three hydrogels swallowed with the same proportions from four to five times their weight of water. However, a difference in the equilibrium time was observed, in deionized water the equilibrium time was reached after 2 h only while otherwise in PBS solution, it took four times longer. This difference can be explained by a capillary process associated with a higher osmosis pressure due to the deionized water which accelerate filling of interconnected pores.

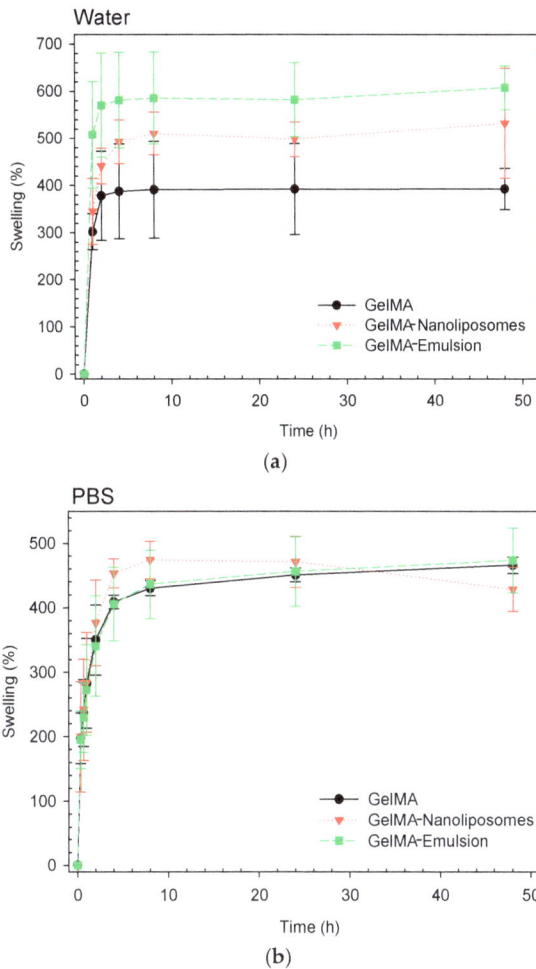

Figure 5. Degree of swelling of based hydrogels (GelMA (●), GelMA-Nanoliposomes (▼), and GelMA-Emulsion (■)) in two different solvents (**a**: deionized water and **b**: PBS solution), at 37 °C. Error bars represent standard deviation.

2.7. Mechanical Stability

The compression curves of the hydrogels (Figure 6) showed the same force–displacement behavior for GelMA and GelMA-Emulsion hydrogels. In the case of these two systems, it is observed that the deformation requires a force superior to the limit of the device. The earlier increase can be related to the stiffness of the crosslinked GelMA (Young's modulus of 15% GelMA = 30 kPa [40]), adding nanodroplets do not seem to influence negatively on the stiffness of the network of GelMA, that is why deformation profile are relatively the same for both hydrogels. In the case of GelMA-Nanoliposomes, we observed that the deformation is much higher at equal strength showing a softer property. It is noted that the force required for a displacement of about 0.1 mm is equal to 13 N, this deformation can be due to the mechanical action of the nanoliposomes on the GelMA network. Nanoliposomes are soft particles anchored in the GelMA scaffold, which explains the same deformation of GelMA-Nanoliposomes needs less external force.

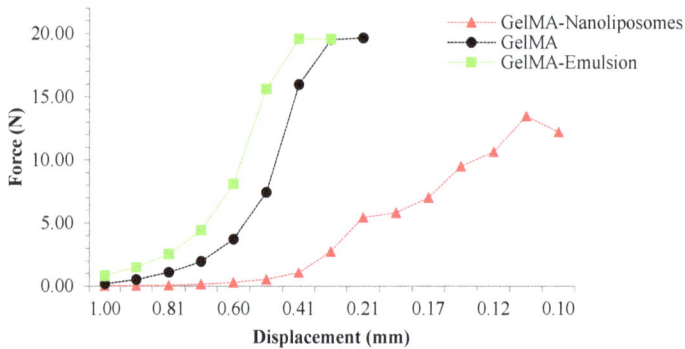

Figure 6. Typical force–displacement curves of GelMA hydrogels: pure GelMA (●) and Functionalized GelMA (▲: GelMA-Nanoliposomes and ■: GelMA-Emulsion).

2.8. Viscoelastic Measurements

Dynamic shear oscillation measurement at small strain was used to characterize the viscoelastic properties and related differences in the molecular structure of functionalized hydrogels based on GelMA.

2.8.1. Amplitude Sweep Test

To determine the linear viscoelastic region (LVER), an amplitude sweep test was performed over strain range (from 0.01 to 100%). The graph (Figure 7) gives an indication of stability. While sample structure is maintained (between 0.1% and 1%), the complex modulus is constant. When the applied stress becomes too high (beyond 1%), decomposition of the internal structure occurs, and the modulus decreases. All the samples have the same LVER. The length of the LVER is a measure of stability that is why frequency sweep test must be performed in the LVER. The amplitude sweeps of the hydrogels also showed that all of them had a $G'/G'' \gg 1$, typical of a stronger hydrogel character.

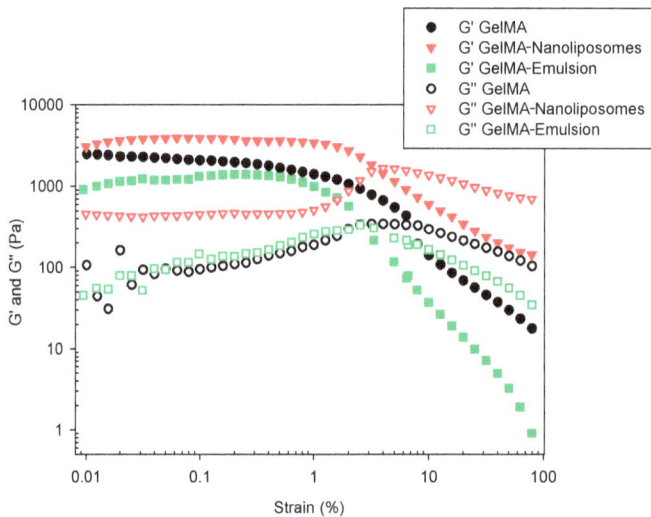

Figure 7. Amplitude sweeps showing storage moduli (G') and loss moduli (G'') of the hydrogels performed over the whole strain range at a frequency of 1 Hz.

2.8.2. Frequency Sweeps Test

Rheological frequency sweep tests were performed on three-dimensional (discs) GelMA gel. The elastic (G′) and viscous (G″) moduli of hydrogels were investigated by dynamic mechanical analyses. The frequency dependence of these moduli was reported in Figure 8.

It is known that physical gelation results from a thermo-reversible conformation change from the triple helix to individual polypeptide coils at above approximately 40 °C. Upon cooling below 35 °C, the random coils join locally and associate into helix, which grows interconnect and forms larger domains until the whole volume is percolated [41]. Chemical cross-linking results through photo-polymerization of vinyl groups after UV initiation [27].

The first observation was that in all hydrogels, elastic modulus values G′ exhibits a pronounced plateau in the frequency range investigated. G′ values were superior to viscous modulus values G″, confirming that the pure GelMA hydrogel and functionalized GelMA hydrogels have a predominantly elastic rather than viscous character. This criterion discriminates viscoelastic solids behavior such as hydrogels from viscous liquids. Thus, the deformation energy is recovered in the elastic stretching of chemical bonds [42]. The constant values of functionalized hydrogels' storage modulus with frequency sweep indicated the absence of relaxation processes, which may be explained by a stability of the intermolecular junctions. There has been no release of nanoparticles that would have resulted in a change in G′ value.

The high degree of methacrylation and the presence of strong chemical cross-links allow an elevated stability despite gel oscillatory in the frequency range (0.01 to 10 Hz). In this case, the contribution of physical cross-links cannot explain the high values because the experiment was done at 37 °C.

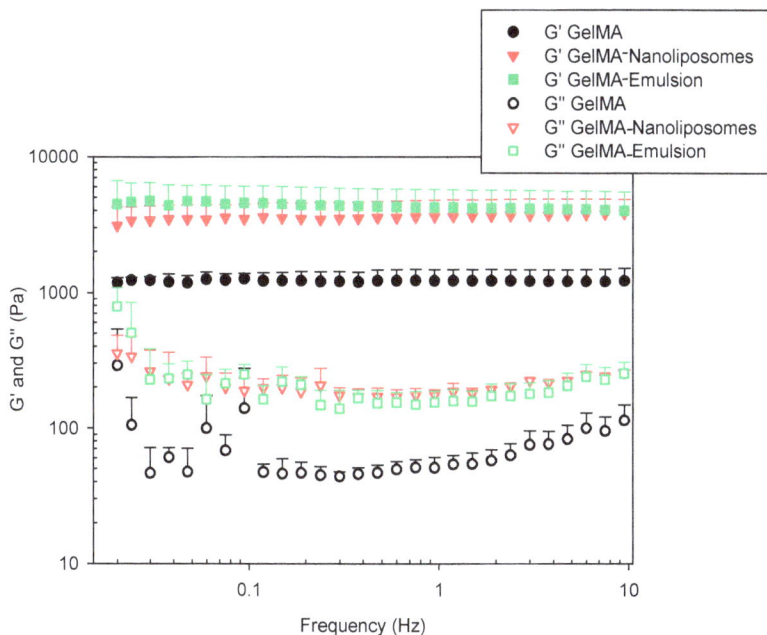

Figure 8. Frequency sweep test of hydrogels. (■, ●, and ▼) elastic modulus (G′) and (□, ○ and ▽) viscous modulus (G″) at a shear strain of 0.1%.

The incorporation of nanoliposomes and nanodroplets results in a significant increase of both G′ and G″. Considering that GelMA provides a big part of the hydrogel mechanical stability, the incorporation of nanoliposomes in the network walls and the adsorption of droplets on GelMA scaffold improve the elastic component of the resulting hydrogels.

3. Materials and Methods

3.1. Material

Gelatin (type A, 300 bloom from porcine skin), methacrylic anhydride (MA), photoinitiator (PI), 2-hydroxy-4'-(2-hydroxyethoxy)-2-methylpropiophenone and phosphate buffered saline tablets (PBS) were purchased from Sigma-Aldrich (Chemie, Stuttgart, Germany). Rapeseed lecithin was acquired from Solae Europe SA society (Geneva, Switzerland). Rapeseed oil (vegetable oil, Lesieur, Asnière-sur-Seine, France) was purchased from a local supermarket (Nancy, France).

3.2. Methacrylated Gelatin Synthesis

Methacrylated gelatin was synthesized as described previously [26]. Briefly, type A porcine skin gelatin was mixed at 10% (w/v) into phosphate buffered saline solution at 60 °C and stirred until fully dissolved. Then, 8 mL of MA was added very slowly and dropwise under stirring into the first solution. After 3 h, the reaction was stopped following a 1:5 dilution using warm phosphate buffered saline at 50 °C and allowed to react for 1 h. The mixture was then dialyzed for one week at 40 °C against distilled water using a dialysis membrane (Spectro/Por molecular porous membrane tubing, MWCO 12–14,000, SpectrumLabs, Inc., Rancho Dominguez, CA, USA) to remove salts and methacrylic acid. The solution was finally lyophilized for one week to generate white porous foam and stored at −20 °C until further use.

3.3. Nanoliposomes Preparation

We prepared 5% nanoliposomes solution by adding 5 g of rapeseed lecithin to 95 mL of water. The suspension was mixed for 5–6 h under agitation at inert atmosphere (nitrogen) and a temperature of 40 °C. Then, the lecithin solution was passed one time through high pressure homogenizer microfluidizer (Microfluidics LM 20, Newton, MA, USA). The lecithin solution was fed into the microfluidizer through a 300 mL glass reservoir (pressurized by an intensifier pump) and was passed through the homogenization unit for one pass at a constant pressure of 500 bar (50 MPa). Liposome samples were stored in glass bottle in the dark at 4 °C after preparation.

3.4. Nanoemulsion Preparation

A stable rapeseed oil in water emulsion was prepared according to GENIALIS patent (N149668A1) by a low frequency/high frequency (LF/HF) ultrasound method. 10% (w/w) oil/water dispersion was homogenized 3 min at 12,000 rpm then sonicated at low frequency at 20 kHz for 3 min (30 s on, 30 s off, amplitude 40%) in an ice bath. The dispersion was placed in a device composed of a thermostated reactor (4 °C) in which was located three piezoelectric transducers, a peristaltic pump, and a stirring modulus for 4.

The oil ratio in emulsion was calculated at the end of emulsification process by measuring the difference between oil initially put in emulsion and non-emulsified oil remaining at emulsion surface as follows:

$$EC = \frac{(IV - FV)}{EV} \tag{1}$$

where EC is the emulsification capacity, IV the initial oil volume, FV the floating oil volume and EV the emulsion volume.

3.5. Hydrogels Preparation

3.5.1. GelMA Hydrogel

Freeze-dried GelMA macromer was mixed at 15% (w/v) into PBS solution containing 1% (w/v) 2-hydroxy-1-(4-(hydroxyethoxy) phenyl)-2-methyl-1-propanone as a photoinitiator at 80 °C until fully dissolved.

One mL of 15% GelMA solution was poured on a specific silicone mold with the controlled dimensions (ø = 2 cm, h = 0.2 cm) and then exposed to UV light (360–480 nm) for 240 s. The PI absorbs the UV light and transforms the solution onto gel. The obtained gel was then washed with PBS before use.

3.5.2. Nanoliposomes/GelMA Hydrogel

One mL of GelMA/nanoliposomes mixture (0.5 mL 30% GelMA + 0.5 mL 3% nanoliposomes + 1% PI) was poured on specific silicone mold and exposed to UV light (360–480 nm) for 240 s. The solution was transformed to gel by the action of PI reaction. The obtained gel (15% GelMA:1.5% nanoliposomes:0.5% PI) was then washed with PBS before use.

3.5.3. Nanoemulsion/GelMA Hydrogel

One mL of GelMA/nanoemulsion mixture (0.5 mL 30% GelMA + 0.5 mL 3% emulsion + 1% PI) was poured on specific silicone mold and exposed to UV light for 240 s. The obtained gel (15% GelMA:1.5% nanoemulsions:0.5% PI) was then washed with PBS before use.

3.6. Size Measurement

The size distribution (mean diameter and polydispersity index) of the liposome dispersions was measured by dynamic light scattering (DLS) using a Malvern ZetasizerNano ZS (Malvern Instruments Ltd., Malvern, UK). Prior to measuring size, the samples were diluted (nanoliposomes: 1/400 and nanoemulsion 1/100) into a distilled water ultra-filtrate which measures the mass distribution of particle size. Measurements were made at 37 °C with a fixed angle of 173°. Sizes quoted are the z-average mean (dz) for the liposomal hydrodynamic diameter (nm). Measurements were made in triplicate.

3.7. Zeta Potential Measurements

Nanoliposomes and nanoemulsions Zeta potentials were measured with ZetasizerNano ZS (Malvern Instruments Ltd., Malvern, UK) using dynamic light scattering (DLS). The determined potential is an important parameter to analyze the effect of the nanoparticles in suspension. The samples were diluted (as previously) and introduced into disposable capillary cells equipped with gold electrodes designed to afford maximum zeta potential measurement capability. All measurements were carried out at 37 °C.

3.8. Scanning Electron Microscopy

The surface morphology of the different hydrogels was characterized by Quanta 200 high-resolution scanning electron microscope low vacuum mode (FEI-Japan, Tokyo, Japan). The use of "low vacuum" mode presents powerful tools for the observation of the surface topography of biological materials without sputter-coated. It also preserves the delicate samples from the electron beam damaging. The maximal resolution attained, employing an electron beam spot size of 3, could be lower than 5 nanometers. The analysis was executed by the use of a large field detector (LFD). The squared shaped samples with dimensions 9 × 9 mm^2 were inserted and maintained in a holder inside the SEM chamber and the tests were performed at laboratory temperature of 25 °C with a relative humidity of 50%. A partial vacuum was created within the chamber, and the air was evacuated using a pump, which provides a regular pressure of 60 mbar. The images were taken from a distance of 10.0 mm with an acceleration voltage of 10.00 kV. The pictures were provided utilizing software "xT microscope server."

3.9. Fourier-Transform Infrared Spectroscopy

Fourier-transform infrared (FTIR) spectra of freeze-dried nanoliposomes, emulsion and hydrogels were recorded with a Tensor 27 mid-FTIR Bruker spectrometer (Bruker, Karlsruhe, Germany) equipped with a diamond ATR (Attenuated Total Reflectance) module and a DTGS (Deuterated-Triglycine Sulfate) detector. Between 4000 and 400 cm^{-1} at 4 cm^{-1} resolutions, 128 scans were used for both reference and samples. Spectral manipulations were then performed using OPUS software (Bruker, Karlsruhe, Germany). Raw absorbance spectra were smoothed using a nine-points smoothing function. After elastic baseline correction using 200 points, H_2O/CO_2 correction was then applied. Then, spectra were centered and normalized. All tests were run in triplicate.

3.10. Swelling Behavior Study

For determining the water retention capability of the prepared hydrogels, immediately following hydrogel formation three replicas (discs of ø = 2 cm, h = 0.2 cm) of each hydrogel were dipped in deionized water or phosphate buffer saline solution (PBS–pH 7.4) at 37 °C. The swelling ratios of hydrogels were measured after 20 min, 40 min, 60 min, 2 h, 4 h, 8 h, 24 h and 48 h (discs were dried with absorbent paper to remove the residual liquid before each weighing and dipped in water or PBS after).

The swelling ratio (*SR*) was calculated using the following equation:

$$SR = \frac{\left(W_f - W_i\right)}{W_i} \tag{2}$$

where W_f is the final weight and W_i is the initial weight.

3.11. Characterization of Hydrogel Degradation with Collagenase

PBS pre-incubated discs of GelMA, GelMA-Nanoliposomes, and GelMA-Emulsion hydrogels were added to collagenase type II (crude collagenase from Clostridium histolyticum, reference number C-6885, Sigma-Aldrich) in PBS containing 2 µg/mL (0.5 U/mL) of enzyme. Samples were then placed in an oven at 37 °C. At the indicated time points, hydrogel samples were removed, wiped, and weighed. Mass loss was determined by normalizing sample weight measurements at time zero.

3.12. Mechanical Stability

To investigate mechanical stability of nanofunctionalized GelMA hydrogels, the three different systems were compressed individually between two parallel plates. A rotational rheometer Malvern kinexux (Malvern Instruments Ltd., Malvern, UK) with a plate-and-plate (20 mm) geometry was used. A force gap test used to compress the discs from 1.50 to 0.10 mm with a linear compression speed of 10 µm/s [43]. The gap and the normal force being imposed were measured simultaneously at the upper plate. At least three replicates were considered for each type of hydrogel.

3.13. Viscoelastic Measurements

Dynamic viscoelastic measurements were carried out using a Kinexus pro rheometer (Malvern Instruments Ltd., Malvern, UK) equipped with a plate-and-plate geometry (20 mm). A dynamic frequency sweep test from 0.02 to 30 Hz was performed to determine the dynamic storage (G′) and viscous (G″) modulus, at a strain rate of 0.1% confirmed to be in the linear viscoelastic range (LVER) for each type of hydrogel by a prior strain amplitude sweep (strain: from 0.01 to 100% at a frequency of 1.00 Hz).

During the rheological experiments, the temperature was maintained at 37 °C and the measuring system was covered with a humidity chamber to minimize water evaporation. Three different hydrogel disks were tested for each type of hydrogel with the same experimental settings and average values are presented.

4. Conclusions

In this report, we developed two approaches to synthesize functionalized GelMA hydrogels for cosmetic applications. Physicochemical and mechanical proprieties were characterized.

Both high shear fluid and LF/HF sonication processes have enabled us to make stable negatively charged nanoparticles, which can be used as nanovehicles to encapsulate and transport active molecules. FTIR results indicated the presence of a small amount of nanoparticles that does not disturb the polymer organization. No interactions have been revealed between the nanoparticles and GelMA scaffold.

SEM images demonstrated that functionalized hydrogels were porous and that they could serve as niches where eukaryotic cells can grow, interact, and benefit from a regular intake of active substances conveyed by nanoliposomes.

The mechanical properties of functionalized hydrogels investigated by small amplitude oscillatory shear rheology showed greater resistance to shear and twist and therefore better stability after functionalization.

Acknowledgments: This research was supported by Fonds Unique Interministériel (FUI).

Author Contributions: Kamel Rahali, Ghazi Ben Messaoud, Mouna Kaci, Franck Cleymand and Solenne Fleutot, designed and performed the experiments. Cyril J.F. Kahn, Laura Sanchez-Gonzalez, Michel Linder, Stephane Desobry, and Elmira Arab-Tehrany wrote and corrected the paper manuscript.

Conflicts of Interest: The authors declare no conflict of interest.

References

1. Serafim, A.; Tucureanu, C.; Petre, D.-G.; Dragusin, D.-M.; Salageanu, A.; Van Vlierberghe, S.; Dubruel, P.; Stancu, I.-C. One-pot synthesis of superabsorbent hybrid hydrogels based on methacrylamide gelatin and polyacrylamide. Effortless control of hydrogel properties through composition design. *New J. Chem.* **2014**, *38*, 3112–3126. [CrossRef]

2. Wichterle, O.; LíM, D. Hydrophilic Gels for Biological Use. *Nature* **1960**, *185*, 117–118. [CrossRef]

3. Corkhill, P.H.; Hamilton, C.J.; Tighe, B.J. Synthetic hydrogels VI. Hydrogel composites as wound dressings and implant materials. *Biomaterials* **1989**, *10*, 3–10. [CrossRef]

4. Kashyap, N.; Kumar, N.; Kumar, M.N.V.R. Hydrogels for Pharmaceutical and Biomedical Applications. *Crit. Rev. Ther. Drug Carr. Syst.* **2005**, *22*, 107–150. [CrossRef]

5. Lee, K.Y.; Mooney, D.J. Hydrogels for Tissue Engineering. *Chem. Rev.* **2001**, *101*, 1869–1880. [CrossRef] [PubMed]

6. Zhu, J.; Marchant, R.E. Design properties of hydrogel tissue-engineering scaffolds. *Expert Rev. Med. Devices* **2011**, *8*, 607–626. [CrossRef] [PubMed]

7. Khademhosseini, A.; Langer, R. Microengineered hydrogels for tissue engineering. *Biomaterials* **2007**, *28*, 5087–5092. [CrossRef] [PubMed]

8. Annabi, N.; Tamayol, A.; Uquillas, J.A.; Akbari, M.; Bertassoni, L.E.; Cha, C.; Camci-Unal, G.; Dokmeci, M.R.; Peppas, N.A.; Khademhosseini, A. 25th Anniversary Article: Rational Design and Applications of Hydrogels in Regenerative Medicine. *Adv. Mater.* **2014**, *26*, 85–124. [CrossRef] [PubMed]

9. Peppas, N.A.; Huang, Y.; Torres-Lugo, M.; Ward, J.H.; Zhang, J. Physicochemical Foundations and Structural Design of Hydrogels in Medicine and Biology. *Annu. Rev. Biomed. Eng.* **2000**, *2*, 9–29. [CrossRef] [PubMed]

10. Hoffman, A.S. Hydrogels for biomedical applications. *Adv. Drug Deliv. Rev.* **2002**, *54*, 3–12. [CrossRef]

11. McClements, D.J. Encapsulation, protection, and release of hydrophilic active components: Potential and limitations of colloidal delivery systems. *Adv. Colloid Interface Sci.* **2015**, *219*, 27–53. [CrossRef] [PubMed]

12. Elzoghby, A.O. Gelatin-based nanoparticles as drug and gene delivery systems: Reviewing three decades of research. *J. Control. Release* **2013**, *172*, 1075–1091. [CrossRef] [PubMed]

13. Rosiak, J.M.; Yoshii, F. Hydrogels and their medical applications. *Nucl. Instrum. Methods Phys. Res. Sect. B* **1999**, *151*, 56–64. [CrossRef]

14. Kalshetti, P.P.; Rajendra, V.B.; Dixit, D.N.; Parekh, P.P. Hydrogels as a drug delivery system and applications: A review. *Int. Pharm. Pharm. Sci.* **2012**, *4*, 1–7.

15. Samchenko, Y.; Ulberg, Z.; Korotych, O. Multipurpose smart hydrogel systems. *Adv. Colloid Interface Sci.* **2011**, *168*, 247–262. [CrossRef] [PubMed]

16. Lin, C.-C.; Metters, A.T. Hydrogels in controlled release formulations: Network design and mathematical modeling. *Adv. Drug Deliv. Rev.* **2006**, *58*, 1379–1408. [CrossRef] [PubMed]

17. Gunatillake, P.; Mayadunne, R.; Adhikari, R. Recent developments in biodegradable synthetic polymers. In *Biotechnology Annual Review*; Elsevier: Amsterdam, The Netherlands, 2006; Volume 12, pp. 301–347, ISBN 978-0-444-52724-0.

18. Drury, J.L.; Mooney, D.J. Hydrogels for tissue engineering: Scaffold design variables and applications. *Biomaterials* **2003**, *24*, 4337–4351. [CrossRef]

19. Yue, K.; Trujillo-de Santiago, G.; Alvarez, M.M.; Tamayol, A.; Annabi, N.; Khademhosseini, A. Synthesis, properties, and biomedical applications of gelatin methacryloyl (GelMA) hydrogels. *Biomaterials* **2015**, *73*, 254–271. [CrossRef] [PubMed]

20. Knopf-Marques, H.; Barthes, J.; Wolfova, L.; Vidal, B.; Koenig, G.; Bacharouche, J.; Francius, G.; Sadam, H.; Liivas, U.; Lavalle, P.; et al. Auxiliary Biomembranes as a Directional Delivery System to Control Biological Events in Cell-Laden Tissue-Engineering Scaffolds. *ACS Omega* **2017**, *2*, 918–929. [CrossRef]

21. Barbetta, A.; Dentini, M.; Zannoni, E.M.; De Stefano, M.E. Tailoring the Porosity and Morphology of Gelatin-Methacrylate PolyHIPE Scaffolds for Tissue Engineering Applications. *Langmuir* **2005**, *21*, 12333–12341. [CrossRef] [PubMed]

22. Ward, A.G.; Courts, A. (Eds.) *The Science and Technology of Gelatin*; Food Science and Technology; Academic Press: London, UK; New York, NY, USA, 1977; ISBN 978-0-12-735050-9.

23. Veis, A. *The Macromolecular Chemistry of Gelatin*; Molecular Biology; Academic Press: Cambridge, MA, USA, 1964.

24. Galis, Z.S.; Khatri, J.J. Matrix metalloproteinases in vascular remodeling and atherogenesis: The good, the bad, and the ugly. *Circ. Res.* **2002**, *90*, 251–262. [PubMed]

25. Van den Steen, P.E.; Dubois, B.; Nelissen, I.; Rudd, P.M.; Dwek, R.A.; Opdenakker, G. Biochemistry and Molecular Biology of Gelatinase B or Matrix Metalloproteinase-9 (MMP-9). *Crit. Rev. Biochem. Mol. Biol.* **2002**, *37*, 375–536. [CrossRef] [PubMed]

26. Nichol, J.W.; Koshy, S.T.; Bae, H.; Hwang, C.M.; Yamanlar, S.; Khademhosseini, A. Cell-laden microengineered gelatin methacrylate hydrogels. *Biomaterials* **2010**, *31*, 5536–5544. [CrossRef] [PubMed]

27. Van Den Bulcke, A.I.; Bogdanov, B.; De Rooze, N.; Schacht, E.H.; Cornelissen, M.; Berghmans, H. Structural and rheological properties of methacrylamide modified gelatin hydrogels. *Biomacromolecules* **2000**, *1*, 31–38. [CrossRef] [PubMed]

28. Aubin, H.; Nichol, J.W.; Hutson, C.B.; Bae, H.; Sieminski, A.L.; Cropek, D.M.; Akhyari, P.; Khademhosseini, A. Directed 3D cell alignment and elongation in microengineered hydrogels. *Biomaterials* **2010**, *31*, 6941–6951. [CrossRef] [PubMed]

29. Kadri, R.; Ben Messaoud, G.; Tamayol, A.; Aliakbarian, B.; Zhang, H.Y.; Hasan, M.; Sánchez-González, L.; Arab-Tehrany, E. Preparation and characterization of nanofunctionalized alginate/methacrylated gelatin hybrid hydrogels. *RSC Adv.* **2016**, *6*, 27879–27884. [CrossRef]

30. Hasan, M.; Belhaj, N.; Benachour, H.; Barberi-Heyob, M.; Kahn, C.J.F.; Jabbari, E.; Linder, M.; Arab-Tehrany, E. Liposome encapsulation of curcumin: Physico-chemical characterizations and effects on MCF7 cancer cell proliferation. *Int. J. Pharm.* **2014**, *461*, 519–528. [CrossRef] [PubMed]

31. Kaci, M.; Meziani, S.; Arab-Tehrany, E.; Gillet, G.; Desjardins-Lavisse, I.; Desobry, S. Emulsification by high frequency ultrasound using piezoelectric transducer: Formation and stability of emulsifier free emulsion. *Ultrason. Sonochem.* **2014**, *21*, 1010–1017. [CrossRef] [PubMed]

32. Arab Tehrany, E.; Kahn, C.J.F.; Baravian, C.; Maherani, B.; Belhaj, N.; Wang, X.; Linder, M. Elaboration and characterization of nanoliposome made of soya; rapeseed and salmon lecithins: Application to cell culture. *Colloids Surf. B Biointerfaces* **2012**, *95*, 75–81. [CrossRef] [PubMed]

33. Chansiri, G.; Lyons, R.T.; Patel, M.V.; Hem, S.L. Effect of surface charge on the stability of oil/water emulsions during steam sterilization. *J. Pharm. Sci.* **1999**, *88*, 454–458. [CrossRef] [PubMed]

34. Dickinson, W. The effect of pH upon the electrophoretic mobility of emulsions of certain hydrocarbons and aliphatic halides. *Trans. Faraday Soc.* **1941**, *37*, 140. [CrossRef]

35. Guillén, M.D.; Cabo, N. Infrared spectroscopy in the study of edible oils and fats. *J. Sci. Food Agric.* **1997**, *75*, 1–11. [CrossRef]

36. Zhang, Q.; Liu, C.; Sun, Z.; Hu, X.; Shen, Q.; Wu, J. Authentication of edible vegetable oils adulterated with used frying oil by Fourier Transform Infrared Spectroscopy. *Food Chem.* **2012**, *132*, 1607–1613. [CrossRef]

37. Hasan, M.; Ben Messaoud, G.; Michaux, F.; Tamayol, A.; Kahn, C.J.F.; Belhaj, N.; Linder, M.; Arab-Tehrany, E. Chitosan-coated liposomes encapsulating curcumin: Study of lipid-polysaccharide interactions and nanovesicle behavior. *RSC Adv.* **2016**, *6*, 45290–45304. [CrossRef]

38. Hermanto, S.; Surmalin, L.O.; Fatimah, W. Differentiation of Bovine and Porcine Gelatin Based on Spectroscopic and Electrophoretic Analysis. *J. Food Pharm. Sci.* **2012**, *1*, 68–73.

39. Sadeghi, M.; Heidari, B. Crosslinked Graft Copolymer of Methacrylic Acid and Gelatin as a Novel Hydrogel with pH-Responsiveness Properties. *Materials* **2011**, *4*, 543–552. [CrossRef] [PubMed]

40. MacQueen, L.; Chebotarev, O.; Chen, M.; Usprech, J.; Sun, Y.; Simmons, C.A. Three-Dimensional mechanical compression of biomaterials in a microfabricated bioreactor with on-chip strain sensors. In Proceedings of the 16th International Conference on Miniaturized Systems for Chemistry and Life Sciences, Okinawa, Japan, 28 October–1 November 2012; pp. 1141–1143.

41. Joly-Duhamel, C.; Hellio, D.; Djabourov, M. All Gelatin Networks: 1. Biodiversity and Physical Chemistry[†]. *Langmuir* **2002**, *18*, 7208–7217. [CrossRef]

42. Stendahl, J.C.; Rao, M.S.; Guler, M.O.; Stupp, S.I. Intermolecular Forces in the Self—Assembly of Peptide Amphiphile Nanofibers. *Adv. Funct. Mater.* **2006**, *16*, 499–508. [CrossRef]

43. Leick, S.; Kott, M.; Degen, P.; Henning, S.; Päsler, T.; Suter, D.; Rehage, H. Mechanical properties of liquid-filled shellac composite capsules. *Phys. Chem. Chem. Phys.* **2011**, *13*, 2765–2773. [CrossRef] [PubMed]

International Journal of
Molecular Sciences

MDPI

Article

Electrophoretic Deposition of Hydroxyapatite Film Containing Re-Doped MoS$_2$ Nanoparticles

Hila Shalom [1], Yishay Feldman [2], Rita Rosentsveig [1], Iddo Pinkas [2], Ifat Kaplan-Ashiri [2],
Alexey Moshkovich [3], Vladislav Perfilyev [3], Lev Rapoport [3] and Reshef Tenne [1,*]

[1] Department of Materials and Interfaces, Weizmann Institute, Rehovot 76100, Israel;
hila.shalom@weizmann.ac.il (H.S.); rita.rosentsveig@weizmann.ac.il (R.R.)

[2] Department of Chemical Research Support, Weizmann Institute, Rehovot 76100, Israel;
Isai.Feldman@weizmann.ac.il (Y.F.); iddo.pinkas@weizmann.ac.il (I.P.);
ifat.kaplan-ashiri@weizmann.ac.il (I.K.-A.)

[3] Department of Science, Holon Institute of Technology, P.O. Box 305, 52 Golomb St., Holon 58102, Israel;
alexeym@hit.ac.il (A.M.); vladper@hit.ac.il (V.P.); rapoport@hit.ac.il (L.R.)

* Correspondence: reshef.tenne@weizmann.ac.il; Tel.: +972-8-934-2394

Received: 6 November 2017; Accepted: 12 January 2018; Published: 26 February 2018

Abstract: Films combining hydroxyapatite (HA) with minute amounts (ca. 1 weight %) of (rhenium doped) fullerene-like MoS$_2$ (IF) nanoparticles were deposited onto porous titanium substrate through electrophoretic process (EPD). The films were analyzed by scanning electron microscopy (SEM), X-ray diffraction and Raman spectroscopy. The SEM analysis showed relatively uniform coatings of the HA + IF on the titanium substrate. Chemical composition analysis using energy dispersive X-ray spectroscopy (EDS) of the coatings revealed the presence of calcium phosphate minerals like hydroxyapatite, as a majority phase. Tribological tests were undertaken showing that the IF nanoparticles endow the HA film very low friction and wear characteristics. Such films could be of interest for various medical technologies. Means for improving the adhesion of the film to the underlying substrate and its fracture toughness, without compromising its biocompatibility are discussed at the end.

Keywords: hydroxyapatite; electrophoretic deposition; nanoparticles; inorganic fullerene-like; tribology

1. Introduction

Self-lubricating solid-state films are used for a variety of applications where fluid lubricants can-not support the excessive load or are prohibitive. Two examples for the use of such films are, under vacuum or low temperature conditions, where lubrications by fluids are not relevant. Other uses include the automotive, medical devices, power generation, machining, shipping, aerospace industries as well as many others [1]. Often such films are in fact a nanocomposite made of hard matrix containing a minority phase of a soft metal like copper or silver, or impregnated nanoparticles with good tribological performance [2–6]. More recently, self-lubricating films containing carbon nanotubes [7], MoS$_2$ [8] and WS$_2$ [9] nanoparticles have been described.

Hydroxyapatite (HA, Ca$_{10}$(PO$_4$)$_6$(OH)$_2$) is used as a bone replacement material in a variety of orthopedic implants and artificial prostheses [10,11]. Given the fact that already 15% of the population is above 65 and increasing, artificial orthopedic implants have become a major health issue. However, this material suffers from high wear and poor fracture toughness, which can be improved by composing it with a toughening phase. To alleviate these problems various methods were conceived including incorporation of nanoparticles (NP) into the HA films. In particular, HA films containing carbon [12] and boron nitride nanotubes [13] were prepared by spark plasma sintering technique.

Among the different methods to prepare HA films on metallic substrates, electrophoretic deposition (EPD) is well established and documented in the literature [14–16].

Frequently, HA phase also contains associated minerals and materials, including brushite and portlandite. Brushite—(CaH(PO$_4$)·2H$_2$O) is a metastable compound in physiological conditions and therefore it transforms into hydroxyapatite after implantation of a prostheses [17]. In one report, HA was synthesized in a hydrothermal reaction of CaO and monetite (CaHPO$_4$). High concentration of calcium oxide in the reaction led to the formation of excess portlandite—Ca(OH)$_2$, while low concentration of calcium oxide resulted in hydroxyapatite [18]. Biphasic calcium phosphate (BCP) is an intimate mixture of two phases of HA and β-TCP (Ca$_3$(PO$_4$)$_2$) in variety of ratios, which appears after annealing of HA above 700 °C [19,20].

Nanoslabs (graphene-like) of MoS$_2$ and numerous other layered materials are currently studied intensively for variety of optoelectronic as well as for energy harvesting and energy-storage devices [21]. WS$_2$ and MoS$_2$ nanoparticles with fullerene-like (IF) structure were also synthesized and were found to perform well as solid lubricants [22,23]. They are presently used in various commercial products, mostly as additives to lubricating fluids, greases, metal working fluids and in high performance bearings (available online at: www.nisusacorp.com and www.utausa.com). Self-lubricating films containing these NP were obtained by co-deposition of the IF NP and various metallic films [24–26]. Recently, doping of IF-MoS$_2$ nanoparticles with minute amounts (<200 ppm) of rhenium atoms (Re:IF-MoS$_2$) was demonstrated [27–30]. Figure 1a shows high-resolution scanning electron microscope (HRSEM) micrograph of the Re-doped IF NP powder. The oblate shape of the nanoparticles with smooth surfaces is clearly delineated. The size range of the nanoparticles is 70–170 nm with a minor content (<10%) of NP larger than 200 nm. Figure 1b shows high-resolution transmission electron microscopy (HRTEM) image of one such nanoparticle made of some 20 closed and nested layers of MoS$_2$. The crystalline perfection and atomically smooth (sulfur-terminated) surface of the IF NP contributes to their excellent mechanical [31] and tribological performance [27,30]. The synthesized nanoparticles are highly agglomerated (see Figure S1) and must be deagglomerated before use. However, in general the agglomerates are weakly bound, and hence only light sonication suffices to disperse them well in aqueous or ethanolic suspensions, which is particularly true for the Re-doped IF-MoS$_2$ nanoparticles.

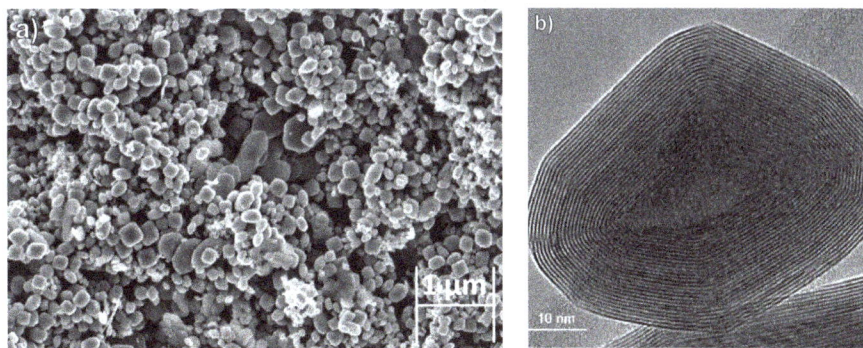

Figure 1. (**a**) High-resolution scanning electron microscopy (HRSEM) image of Re:IF-MoS$_2$ nanoparticles powder in In-lens detector 5 kV; (**b**) high-resolution transmission electron microcopy (HRTEM) image of Re:IF-MoS$_2$ nanoparticle (see experimental section for details).

Zeta potential (ZP) analysis, showed that the IF NP have an excess negative charge on their surface [28,29]. The negative surface charge was attributed to the presence of intrinsic defects (sulfur vacancies) as well as the extra charge induced by the ionized doping (Re) atoms. Furthermore, the negative surface charge, particularly in the Re:IF-MoS$_2$ NP was shown to induce self-repulsion and formation of stable suspensions of the colloidal nanoparticles in different fluids. Consequently, when added to lubricating fluids and medical gels the Re:IF-MoS$_2$ NP produced exceedingly small friction and wear, compared to the undoped nanoparticles and microcrystalline WS$_2$ and MoS$_2$ [29].

One of the most critical aspects of the usage of nanomaterials is their toxicity and biocompatibility. Several studies deliberated on this issue for MoS$_2$ and WS$_2$ NP and their IF structure, in particular [32–34]. In contrast to various other nanoparticles, the IF NP were found to be non-toxic in general, up to a very high dosage (>100 µg/mL). These encouraging findings could be beneficial for the development of medical technologies based on such nanoparticles [30].

In the present work, HA films impregnated with Re-doped IF-MoS$_2$ (Re:IF-MoS$_2$) NP were prepared by electrophoretic deposition on a porous TiO$_2$ substrate, obtained via anodization in fluoride solution. The films were characterized by a number of techniques and their tribological performance was evaluated. The addition of small amounts of the above nanoparticles to the HA films led to substantial improvement in their tribological behavior. Future research into ameliorating the mechanical properties of the films and their biocompatibility is discussed in brief. Such films could, potentially be useful in the future for orthopedic implants, which in general suffer from poor wear resistance.

2. Experimental

2.1. Sample Preparation

2.1.1. Surface Treatment

Anodization of titanium surface produces a highly-textured surface comprising an organized array of TiO$_2$ nanotubes [35–37]. Prior to the anodization, each titanium electrode (30 × 5 × 0.3 mm, 97 wt % purity) was polished with silicon carbide paper to a mirror finish. It was subsequently cleaned by sonicating in a series of solvents, i.e., acetone, ethanol, methanol, isopropanol and finally distilled water, then dried under a nitrogen stream.

2.1.2. Titanium Anodization

An electrochemical cell containing two-electrodes, i.e., platinum (cathode) and titanium (anode) was used. The electrolyte solution contained 1 M (NH$_4$)$_2$SO$_4$ and 0.5 wt % NH$_4$F. All electrolytes were prepared from reagent grade chemicals and deionized water. The electrochemical treatment was conducted with a DC power source operated at 2.5 V and 1.5 A, at room temperature for 2.5 h. After the electrochemical treatment, the samples were rinsed with deionized water and dried under nitrogen stream.

2.1.3. Electrophoretic Deposition

The detailed synthesis of the Re:IF-MoS$_2$ nanoparticles (Re content < 0.1 at %), which were added to the coating processes, was reported before [38]. Three different chemical baths were used for electrophoretic deposition of HA + IF NP on the porous titanium substrate [39]. Titanium samples were used as the working electrode (cathode), while a platinum plate served as the anode. The final volume of all three electrolyte solutions containing 1 mg of the IF NP was 50 mL.

Solution A: The electrolyte solution consisted of 42 mM Ca(NO$_3$)$_2$ and 25 mM NH$_4$H$_2$PO$_4$, 1 mg Re:IF-MoS$_2$ sonicated in 3 mM cetyl trimethylammonium bromide (CTAB). Ethyl alcohol was added into the above solution in a 1:1 ratio in order to reduce the hydrogen evolution on the titanium electrode [40]. The initial pH of the electrolyte solution was 4.5. The coating process was carried out at 40 °C with a DC power supply at 20 V bias and 0.11 A for 3 h. The samples were washed with deionized water and dried for 24 h at 100 °C.

Solution B: The electrolyte solution consisted of 5.25 mM Ca(NO$_3$)$_2$, 10.5 mM NH$_4$H$_2$PO$_4$, and 150 mM NaCl. The initial pH of the solution was adjusted to 5.30 by adding NaOH. 1 mg Re:IF-MoS$_2$ was sonicated in distilled water for 15 min and added to the electrolyte solution. The coating process was conducted with a DC power source operated at 2.5 V and 0.11 A at room temperature for 3 h.

Solution C: The electrolyte solution consisted of 3 mM $Ca(NO_3)_2$ and 1.8 mM KH_2PO_4, 1 mg Re:IF-MoS_2 sonicated in 3 mM CTAB. The initial pH of the electrolyte solution was 5. The coating process was conducted with a DC power source operated at 6 V and 1 A at room temperature for 1 h. The resulting samples, after coating, were washed with deionized water and dried in room temperature.

The bath showing the most uniform coating and good adhesion (solution A) was then further studied by changing the deposition time to 2, 3 and 4 h and subsequent annealing at 700 °C for 1 h. Obviously this process involved a lot of trial and error, using different parameters for the EPD process and annealing. Ultimately, the optimized coating procedure exhibited also the best tribological performance.

2.2. Characterization

2.2.1. High-Resolution Scanning Electron Microscopy (HRSEM) and High-Resolution Transmission Electron Microcopy (HRTEM)

The surface morphology of the titanium samples was analyzed by (HRSEM) (Zeiss Ultra 55) after each step. For topographical information, the secondary electrons were recorded using the SE2 and In-lens detectors. For atomic number contrast the backscattering electron (BSE) detector was used. In order to avoid the sample charging during the analysis, the imaging was done under relatively low accelerating voltage (2–5 kV) and low current. Energy dispersive spectroscopy (EDS) analysis (EDS Bruker XFlash/60mm) of the samples was undertaken as well. The reported results of the EDS were based on standard-less analysis and hence is semi-quantitative in nature.

TEM was performed with a JEOL 2100 microscope (JEOL Ltd., Tokyo, Japan) operating at 200 kV, equipped with a Thermo Fisher EDS analyzer. High-resolution TEM (HRTEM) images were recorded with a Tecnai F30 UT (FEI) microscope (FEI, Eindhoven, the Netherlands) operating a 300 kV. The TEM grids were prepared by dripping an ethanolic solution of the nanoparticles onto a collodion-coated Cu grids.

2.2.2. X-ray Diffraction (XRD)

The film was removed from the Ti substrate and carefully crushed into a powder. The powder was analyzed by X-ray powder diffraction (XRD) using TTRAX III (Rigaku, Tokyo, Japan) theta-theta diffractometer equipped with a rotating copper anode X-ray tube operating at 50 kV/200 mA. A scintillation detector aligned at the diffracted beam was used after a bent Graphite monochromator. The samples were scanned in specular diffraction mode ($\theta/2\theta$ scans) from 10 to 80 degrees (2θ) with step size of 0.025 degrees and scan rate of 0.5 degree per minute. Phase identification and quantitative analysis were performed using the Jade 2010 software (MDI) (available online: http://ksanalytical.com/jade-2010/) and PDF-4+ (2016) database (available online: http://www.icdd.com/products/pdf4.htm).

2.2.3. Raman Spectroscopy

Raman spectra of the powders ground from the films (see Section 2.2.2) were obtained with Horiba-Jobin Yivon (Lille, France) LabRAM HR Evolution set-up using solid state laser with a wavelength of 532 nm. The instrument was equipped with Olympus objectives MPlan N 100 × NA 0.9. The measurements were conducted using a 600 grooves/mm grating. Each spectrum was acquired for 20 s and the spectra were averaged 100 times, which enabled using low excitation power thereby preserving the sample integrity. The spectral ranges collected were from 100 to 1800 cm^{-1}.

2.2.4. Zeta Potential Measurements

The surface charge of the HA suspension with and without the nanoparticles was determined by zeta potential (ZP) measurements using ZetaSizer Nano ZS (Malvern Instruments Inc., Malvern, UK) with a He-Ne light source (632 nm). To prepare the samples for these measurements, IF (0.6 mg) NP were deagglomerated in 20 mL purified water by sonicating for 5–10 min using an ultrasonic bath (see Figure S1 for a typical SEM image of such an agglomerate). Subsequently, 0.2 mL of the

IF suspension was added to 1.5 mL aqueous solutions with pH varying from 1 to 12 and sonicated for an extra 5 min. Before the addition of the IF NP, the pH of each solution was adjusted using concentrated NaOH or HCl. The final concentration of the IF NP was 0.004 mg/mL. The ZP of the solutions was measured in a folded capillary cell (DTS1060) made from polycarbonate with gold plated beryllium/copper electrodes.

2.2.5. Tribological Testing

A home-made ball-on-flat rig was used for the tribological tests. The tests were carried-out at room temperature and humidity of ~40%. Each test was repeated 5-times. Tribological tests were performed on the titanium samples at every step of the experimental procedure. The tribological testing was done under dry friction conditions. This testing method utilizes flat lower samples and a ball-shaped upper specimen, which slides against the flat specimen. The two surfaces move relative to each other in a linear, back and forth sliding motion, under a prescribed set of conditions. In this testing method, the load is applied vertically downwards through the ball against the horizontally mounted flat specimen. Two measurements procedures were used in these series of tests. Sliding speed of 0.3 mm/s was common to both series. In one series of measurements the load was 10 g; the diameter of the ball (hard steel—AISI 301) was 10 mm and consequently a Hertzian pressure of 150 MPa was applied on the film (20 cycles). In another series, the load was 20 g, the diameter of the ball 2 mm, i.e., a Hertzian pressure of 600 MPa was applied, and the number of cycles was 100.

3. Results and Discussion

3.1. SEM Analysis

The surface morphology of the titanium before the pretreatment preceding the anodization is presented in Figure S2a,b in the supporting information—SI. Visibly, the fresh surface was heavily contaminated with a dense network of scratches. After treatment of the titanium (Section 2.1.1), a smooth surface with low density of scratches and clean from contaminants was obtained (Figure S2c,d). The smooth surface was imperative for achieving reproducible tribological measurements.

The surface of the titanium after anodization is displayed in Figure S3. Visibly the anodized titanium surface consists of a dense array of (TiO_2) nanotubes with the range of pore diameters between 50–130 nm, which form a highly organized, roughly hexagonal, pattern on the Ti surface [35–37].

The formal molar Ca/P ratio in HA is 5:3 (1.67). The Ca/P ratio in each coating was calculated based on semi-quantitative EDS analysis. For solution A, the ratio was found to be 2.6. The higher abundance of calcium in this coating could be attributed to the presence of portlandite ($Ca(OH)_2$)—see XRD analysis (Section 3.3). The Ca/P ratio of the coating obtained from solution B, which was highly crystalline and discontinuous was 1.5, which agrees well with the HA formula (1.66). The ratio is 1 for the coating obtained from solution C, which can be ascribed to the presence of calcium pyrophosphate phase ($Ca_2(P_2O_7)$) in the coating—see XRD analysis (Section 3.3).

It is clear that the surface morphology of the HA film prepared via solutions A (Figure 2) and C was more homogeneous and could be successfully combined with the Re:IF-MoS$_2$ NP in the films as opposed to the film obtained from solution B, which was highly crystalline but non-uniform. The surface morphology of the film obtained from solutions B and C are shown in Figures S4 and S5, respectively.

In the next step, the experimental parameters of the Ti-substrate anodization and the electrophoretic deposition from solution A were varied in order to obtain uniform coatings having optimized tribological performance.

The SEM images of the surface of the HA films with Re:IF-MoS$_2$ nanoparticles obtained from solution A for different deposition periods are shown in Figure 3. The surface of the coated film shows defects, including the presence of cracks and pores with circular shape. Such pores can be probably attributed to the formation of $H_2(g)$ bubbles during the coating process.

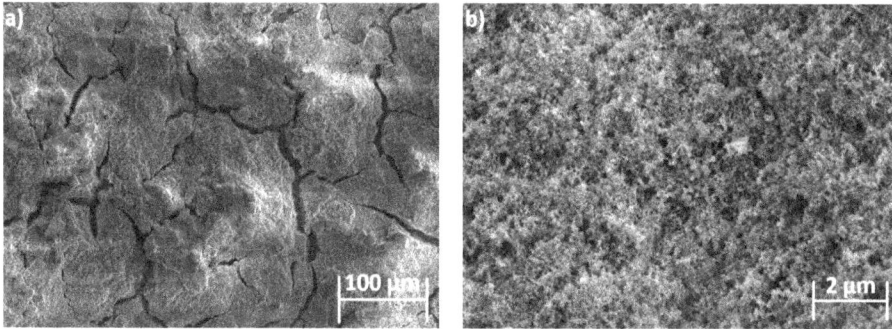

Figure 2. HRSEM pictures of HA with Re:IF-MoS$_2$ nanoparticles coating obtained from solution A on porous titanium substrate in two magnifications: (**a**) 100 μm ; (**b**) 2 μm. The film is continuous but visibly is heavily cracked.

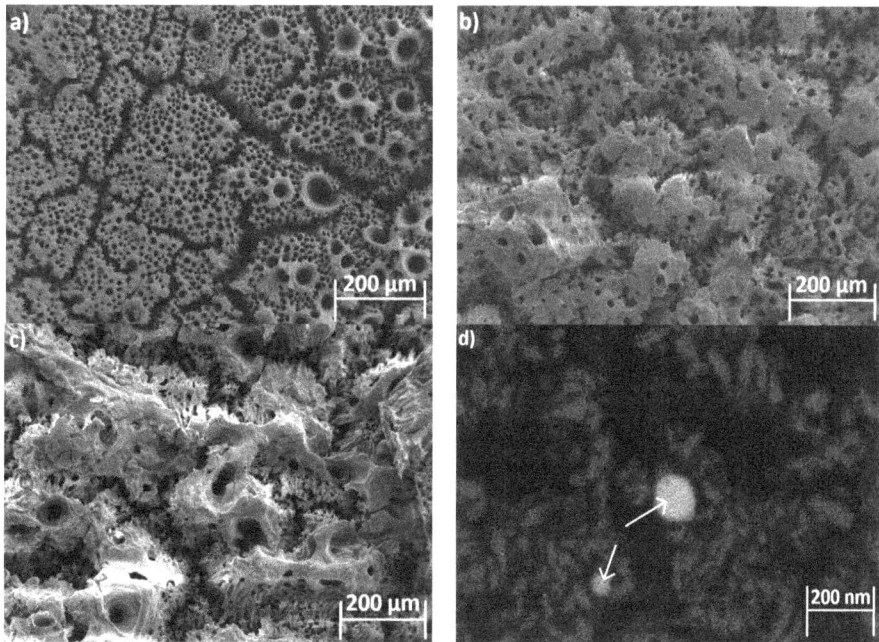

Figure 3. HRSEM images of the HA film with Re:IF-MoS$_2$ obtained from solution A after 2 (**a**), 3 (**b**), and 4 h (**c**) deposition. The Re:IF-MoS$_2$ nanoparticles in the film (**c**) are observed in the backscattering electron (BSE) mode (**d**). The arrows in Figure 3d point on the Re:IF-MoS$_2$ nanoparticles occluded in the HA film.

Interestingly, the bias applied during EPD for solution B (and C) was appreciably smaller (2.5 V) compared to solution A (20 V). On the other hand, the film obtained by EPD from solution A was quasi-continuous. It was highly crystalline but less uniform in the case of solution B, i.e., the apparent current density was higher than that calculated on the basis of the formal electrode surface. The higher voltage used for the EPD from solution A implied a much higher rate of hydrogen production, which could explain the porous structure of this film. The density of the pores and their sizes could be possibly tuned by the bias applied on the cathode during the electrophoretic deposition. Furthermore, addition of surface active agents, like CTAB and others, could reduce the size of the pores. This optimization process is reserved for a future study. In any event, the large cracks are diminished, and the pore-size decreased as the coating time was prolonged. The Re:IF-MoS$_2$ NP cannot be easily discerned from the HA, due possibly to charging of the film during SEM analysis. Furthermore, the thickness of the coating was a few microns, therefore the nanoparticles could have been buried under the film surface and even be closer to the titanium substrate. Using low energy beam (2 keV) in the BSE mode, the IF NP could be nevertheless observed—see Figure 3d.

3.2. Zeta Potential Measurements

Figure 4 shows the results of the Zeta potential (ZP) measurements performed with the three solutions containing Re:IF-MoS$_2$ nanoparticles as a function of pH—up to pH7. The ZP of all the solutions containing the nanoparticles is positive for pH below 6.5. Beyond that pH—the ZP of solution B becomes negative, while that of solutions A and C remain positive. This difference can be attributed to the addition of the CTAB, which is a cationic surfactant, to solutions A and C. The (positive) ZP of the natural solutions used for EPD is marked on Figure 4 for all three solutions.

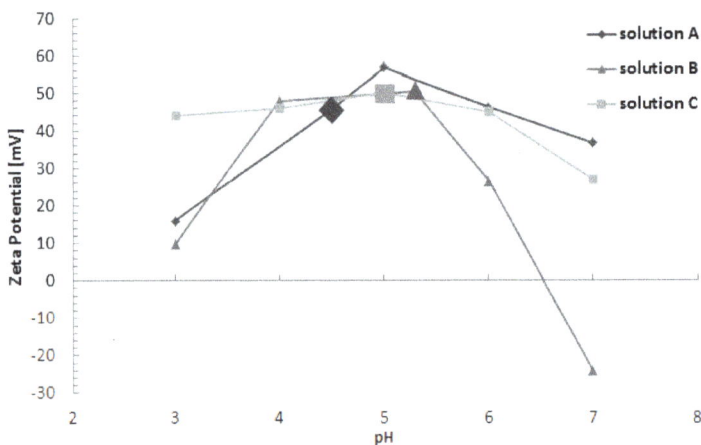

Figure 4. Zeta-potential vs. pH for Re:IF-MoS$_2$ nanoparticles. The (positive) ZP of the genuine solutions used for EPD of the HA + IF film are marked by enlarged symbols.

The ZP measurements show that the species in the HA solution containing the IF NP are positively charged and consequently, the HA film could be deposited on the negative electrode (Ti) during the EPD process. The ZP of the IF NP in pure water, ethanol solution, CTAB in water, and the three solutions used for the EPD (included also in Figure 4) are summarized in Figure S6 of the supporting information, the errors of the ZP measurements is about 2%.

3.3. X-ray Diffraction (XRD)

The results of the XRD analyses are summarized in Figure 5 and in Table 1. The XRD patterns of the different coatings obtained from solutions A, B and C are shown in Figure 5a. The major phase obtained by EPD of these solutions is HA. Nonetheless, the coating obtained from solution A contained appreciable amounts (25 wt %) of portlandite (Ca(OH)$_2$). Solution B, on the other hand, contained, in addition to the HA, also significant amounts of brushite—(CaH(PO$_4$)·2H$_2$O). The film obtained from solution C contained calcium pyrophosphate—(Ca$_2$(P$_2$O$_7$)). The presence of the Re:IF-MoS$_2$ nanoparticles in the coatings is confirmed by the tiny peak at 14.3°. The content of the IF NP is calculated as 0.2 wt % for solutions A, 1.5 wt % for solution B and 1.4 wt % for solution C. This amount is rather small but could nevertheless lead to major improvements of the tribological properties of the film without compromising its mechanical robustness.

Following the annealing of the film obtained from solution A (Figure 5b), the HA became biphasic calcium phosphate (BCP) [19], i.e., intimate mixture of two phases: HA (73.6 wt %) and β-TCP (5.9 wt %), and 0.1 wt % Re:IF-MoS$_2$ NP.

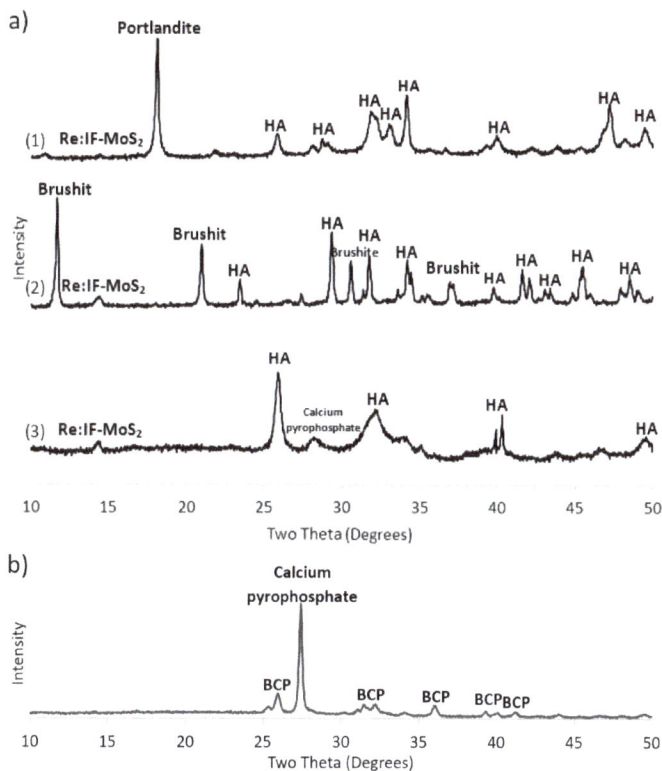

Figure 5. (**a**) XRD patterns of the HA films incorporating Re:IF-MoS$_2$ nanoparticles: film obtained from solution A (1), solution B (2) and solution C (3); (**b**) shows the XRD pattern of the film obtained from solution A (3 h) after annealing (700 °C for 1 h). Here, a strong crystalline peak associated with calcium pyrophosphate phase (Ca$_2$(P$_2$O$_7$)) is observed. This phase is obtained through water evaporation from the HA (Ca$_{10}$(PO$_4$)$_6$(OH)$_2$) film. The presence of the Re:IF-MoS$_2$ nanoparticles did not change appreciably upon annealing, suggesting that these NP are thermally stable at the annealing conditions.

Table 1. Composition of the films deposited from different solutions determined from the XRD analysis.

EPD Films	HA	Portlandite	Brushite	Calcium Pyrophosphate	β-TCP	Re:IF-MoS$_2$
Film obtained from solution A	74.8 wt %	25 wt %				0.2 wt %
Film obtained from solution B	17.2 wt %		81.3 wt %			1.5 wt %
Film obtained from solution C	81.1 wt %			17.5 wt %		1.4 wt %
Film obtained from solution A after annealing	73.6 wt %			20.4 wt %	5.9 wt %	0.1 wt %

The XRD patterns of the films obtained from solution A without the NP (a) and with the IF NP for different deposition times (b–c) is shown in Figure 6. The percentages of the compounds in each film is presented in Table 2. The major phase in the films is hydroxyapatite. However, one can conclude from Figure 6 that the relative amount of the portlandite in the film increases with extending deposition times. The relative amount of the calcium oxide does not seem to vary with the deposition time which is also true for the relative content of the IF NP. Although the signal of the IF NP is non-visible in Figure 6, their presence is confirmed through both electron microscopy (Figure 3) and the Raman measurements (see below Figure 7).

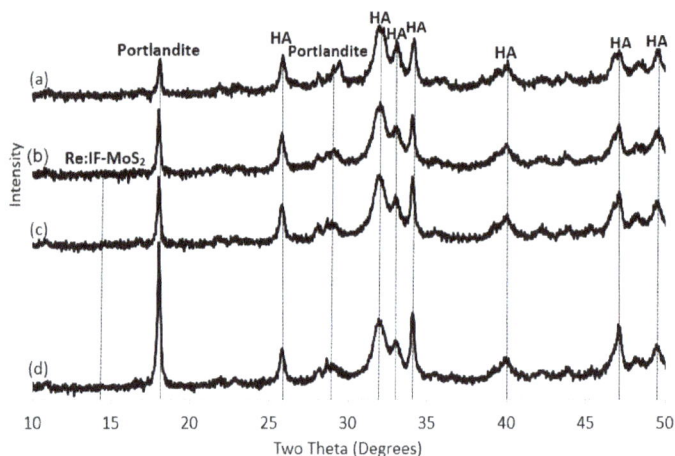

Figure 6. XRD patterns: Films obtained from solution A without the Re:IF-MoS$_2$ NP (**a**) and (with the IF NP) for different deposition periods: after 2 h (**b**), 3 h (**c**) and 4 h (**d**).

Table 2. Composition of the film determined via XRD analysis for different deposition times (from solution A).

EPD films	HA	Portlandite	Calcium Oxide	Re:IF-MoS$_2$
Film obtained from solution A without Re:IF-MoS$_2$ (3 h)	87.8 wt %	4.6 wt %	7.6 wt %	
Film obtained from solution A (2 h)	82.6 wt %	7.4 wt %	9.1 wt %	0.3 wt %
Film obtained from solution A (3 h)	80.4 wt %	11.3 wt %	8.0 wt %	0.3 wt %
Film obtained from solution A (4 h)	77.8 wt %	13.6 wt %	8.3 wt %	0.3 wt %

3.4. Raman Spectroscopy

The Raman spectra of HA + IF films prepared from solution A for different deposition times (2, 3 and 4 h) are shown in Figure 7. The spectra show the characteristic vibration bands of calcium hydroxide (wide peak at 1600 cm^{-1}) and poorly crystalline phosphoric moieties, especially phosphate PO$_4$$^{-3}$ bands at 469 (v_2), 562–603 (v_4), 962 (v_1) and 1000–1104 cm^{-1} (v_3) [39,41]. These bands are typical of HA. The Raman spectra showed also the typical MoS$_2$ modes at 383 (E$_{2g}$) and 408 cm^{-1} (A$_{1g}$) [42,43]. Interestingly, in contrast to the XRD pattern (Figure 6), the Raman bands of the IF NP in the HA film are easily discerned here.

3.5. Tribological Testing

Table 3 summarizes the data for the friction coefficient and surface roughness of the different samples under dry conditions. In general, the friction coefficient was found to go down along with the stages of the experimental procedure of preparing the film. The low friction coefficient of the HA film obtained from solution A can be attributed to the IF nanoparticle structure. The nanoparticles exhibit facile rolling when released from the film surface. In addition, gradual peeling/crushing of the NP and material transfer from the film surface to the counter surface of the ball contributed to the facile shearing of the mating surfaces and low friction coefficients. Interestingly, the friction coefficient of the HA film obtained from solution A was maintained also after 700 °C annealing.

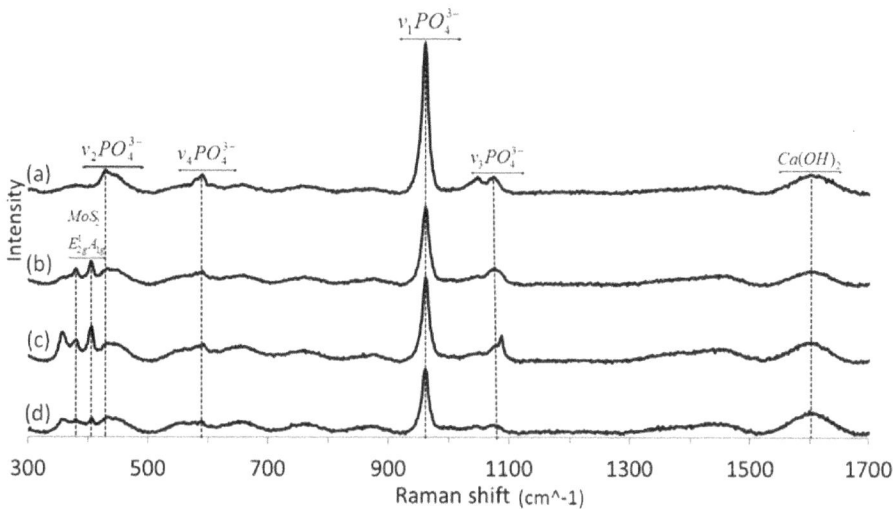

Figure 7. Raman spectra of HA powder film without (**a**) and with the Re:IF-MoS$_2$ nanoparticles obtained from solution A for different EPD periods: after 2 (**b**), 3 (**c**) and 4 h (**d**).

Table 3. Summary of the initial and final friction coefficients and the initial roughness for different stages of preparation of the composite HA + IF film. Measurement conditions: diameter of the test ball 10 mm; load = 10 g (Hertzian pressure—P = 150 MPa).

Tested Film	Initial Coefficient of Friction	Final Coefficient of Friction (after 20 Cycles)	Initial Roughness (μm)
Titanium after surface treatment	0.50 ± 0.01	0. 60 ± 0.02	0.23 ± 0.03
Titanium after anodization	0.15 ± 0.01	0.23 ± 0.03	0.50 ± 0.05
Film of HA with Re:IF-MoS$_2$ NP obtained from solution A on anodized titanium	0.11 ± 0.01	0.13 ± 0.01	0.45 ± 0.4
Film of HA with Re:IF-MoS$_2$ NP obtained from solution B on anodized titanium	0.21 ± 0.02	0.43 ± 0.08	0.37 ± 0.03
Film of HA with Re:IF-MoS$_2$ NP obtained from solution C on anodized titanium	0.37 ± 0.23	0.30 ± 0.18	0.52 ± 0.02
Film of HA with Re:IF-MoS$_2$ NP obtained from solution A on anodized titanium after annealing	0.12 ± 0.01	0.11 ± 0.02	0.49 ± 0.7

Table 4 shows the dry friction coefficient of the coatings obtained from solutions A without (3 h) and with the NP after 2, 3 and 4 h of deposition time on the anodized titanium substrate. Note that, in this series of measurements a higher Hertzian pressure (600 MPa) was used for the tribological test. The dry friction coefficient was reduced with increasing coating-time of the film.

It should be noted that, relative to the previous measurement (see Table 3), there was an increase in the friction coefficient for the HA coating with Re:IF-MoS$_2$ nanoparticles obtained from solution A for 3 h deposition time. This result can be accounted for by the weaker adhesion of the coating to the titanium samples under the high pressure (600 MPa) applied onto the film. Nonetheless, following the 4 h deposition time the friction coefficient was very low (0.12) attesting to the quality of the composite film.

Therefore, it is clear that the extended deposition of the composite film resulted in lower friction under very high load. However, the mechanical stability of the film might have been partially compromised. The surface roughness of the films was in the sub-micron range for all the films containing the NP.

Figure 8 shows optical micrographs of the wear of the ball and the wear track on the film (inset) after different periods of EPD (600 MPa) and 100 cycles. In analogy to the friction coefficient, the visible wear scar on the ball and the wear track on the film were markedly reduced with the deposition time of the HA + IF NP film.

Table 4. The initial and final friction coefficients and the initial roughness of the coating on titanium substrate obtained from solution A for different periods of deposition. Measurement conditions: diameter of the test ball 2 mm; load 20 g and Hertzian pressure of P = 600 MPa.

Tested Film	Initial Coefficient of Friction	Final Coefficient of Friction (after 100 Cycles)	Initial Roughness (μm)
Pure HA film obtained from solution A without NP after 3 h deposition	0.66 ± 0.08	0.78 ± 0.04	1.59 ± 0.28
HA film with Re:IF-MoS$_2$ NP obtained from solution A after 2 h	0.75 ± 0.05	0.63 ± 0.03	0.49 ± 0.05
HA film with Re:IF-MoS$_2$ NP obtained from solution A after 3 h	0.53 ± 0.03	0.55 ± 0.04	0.57 ± 0.17
HA film with Re:IF-MoS$_2$ NP obtained from solution A after 4 h	0.13 ± 0.01	0.12 ± 0.02	0.48 ± 0.02

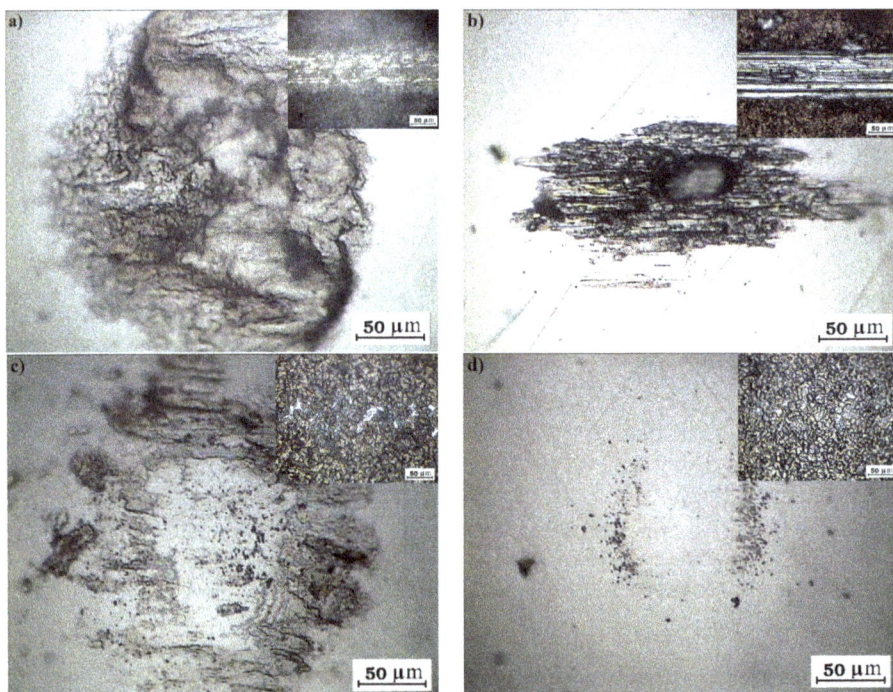

Figure 8. Optical image of wear on the ball and inside the track of HA film without (**a**) and with the Re:IF-MoS$_2$ nanoparticles obtained from solution A for different periods: after 2 h (**b**), 3 h (**c**) and 4 h (**d**) on anodized titanium.

Int. J. Mol. Sci. **2018**, *19*, 657

4. Conclusions

Given the fact that the aging population of the world is approaching one billion people, inabilities due to orthopedic failures and consequently artificial bone implants became a major health concern. Hydroxyapatite (a form of calcium phosphate) is the main constituency of the bone. Therefore, calcium-phosphate coatings occupy important part of modern research in medical technology. Calcium phosphate coatings containing up to 1.5 wt % Re:IF-MoS$_2$ nanoparticles were deposited on porous titanium substrate by electrophoretic deposition using DC bias. Three different solutions were used for the deposition. Solution A is based on ethanol-water mixture as solvent, solution B contained sodium chloride and solution C without (ethanol or NaCl) additive. The major phase in each coating was hydroxyapatite which successfully incorporated small amounts of Re:IF-MoS$_2$ nanoparticles. The electrophoretic deposition from solution B was found to be highly crystalline and discontinuous, i.e., it did not fully cover the porous titanium substrate. The film obtained from solution C was found to have high friction coefficient. The tribology test performed on the coatings showed lower friction coefficient and wear as the time of the deposition increased beyond 3 h period. The low friction coefficient was maintained also after annealing of the sample (solution A). The good tribological performance of the film indicates also that the film is robust and suffers no adhesion problems during the tests in dry conditions.

The present films must be improved in order to increase the adhesion of the film to the underlying substrate and its fracture toughness, while maintaining its biocompatibility, especially under wet conditions [44]. Future studies will attempt to use a third component, possibly in the form of a biocompatible polymer, which could be suitable as a binder in this case. Remarkably though, the present composite films show very low friction and reasonable adhesion to the underlying rough substrate even under very high load (600 MPa).

Supplementary Materials: Supplementary materials can be found at http://www.mdpi.com/1422-0067/19/3/657/s1.

Acknowledgments: We thank the H. Perlman Foundation; the Irving and Azelle Waltcher Foundations in honor of M. Levy; and the Irving and Cherna Moskowitz Center for Nano and Bio-Nano Imaging are acknowledged.

Author Contributions: Hila Shalom did much of the film deposition and analysis; Yishay Feldman did the XRD analysis; Rita Rosentsveig synthesized the IF nanoparticles; Iddo Pinkas did the Raman analysis; Ifat Kaplan-Ashiri did the SEM analysis; Alexey Moshkovich, Vladislav Perfilyev and Lev Rapoport were responsible for the tribological testing; Reshef Tenne led the project and supervised the daily work of Hila Shalom and disseminated the article.

Conflicts of Interest: The authors declare no conflict of interest.

References

1. Clauses, F.J. *Solid Lubricants and Self-Lubricating Solids*; Academic Press: New York, NY, USA, 1972.
2. Basnyat, P.; Luster, B.; Kertzman, Z.; Stadler, S.; Kohli, P.; Aouadi, S.; Xu, J.; Mishra, S.R.; Eryilmaz, O.L.; Erdemir, A. Mechanical and tribological properties of CrAlN-Ag self-lubricating films. *Surf. Coat. Technol.* **2007**, *202*, 1011–1016. [CrossRef]
3. Fusaro, R.L. Self-lubricating polymer composites and polymer transfer film lubrication for space applications. *Tribol. Int.* **1990**, *23*, 105–122. [CrossRef]
4. Li, J.L.; Xiong, D.S. Tribological properties of nickel-based self-lubricating composite at elevated temperature and counterface material selection. *Wear* **2008**, *265*, 533–539. [CrossRef]
5. Skeldon, P.; Wang, H.W.; Thompson, G.E. Formation and characterization of self-lubricating MoS$_2$ precursor films on anodized aluminum. *Wear* **1997**, *206*, 187–196. [CrossRef]
6. Erdemir, A. A crystal chemical approach to the formulation of self-lubricating nanocomposite coatings. *Surf. Coat. Technol.* **2005**, *200*, 1792–1796. [CrossRef]
7. Moghadam, A.D.; Omrani, E.; Menezes, P.L.; Rohatgi, P.K. Mechanical and tribological properties of self-lubricating metal matrix nanocomposites reinforced by carbon nanotubes (CNTs) and grapheme—A review. *Compos. Part B* **2015**, *77*, 402–420. [CrossRef]

8. Liu, E.Y.; Wang, W.Z.; Gao, Y.M.; Jia, J.H. Tribological properties of Ni-based self-lubricating composites with addition of silver and molybdenum disulfide. *Tribol. Int.* **2013**, *57*, 235–241. [CrossRef]

9. Lian, Y.; Deng, J.; Li, S.; Yan, G.; Lei, S. Friction and wear behavior of WS2/Zr self-lubricating soft coatings in dry sliding against 40Cr-hardened steel balls. *Tribol. Lett.* **2014**, *53*, 237–246. [CrossRef]

10. Petit, R. The use of hydroxyapatite in orthopaedic surgery: A ten-year review. *Eur. J. Orthop. Surg. Traumatol.* **1999**, *9*, 71–74. [CrossRef]

11. Mucalo, M. *Hydroxyapatite (HAp) for Biomedical Applications*; Woodhead Publishing Series in Biomaterials; Elsevier: Cambridge, MA, USA, 2015; ISBN 978-1-78242-041-5.

12. Lahiri, D.; Singh, V.; Keshri, A.K.; Seal, S.; Agarwal, A. Carbon nanotube toughened hydroxyapatite by spark plasma sintering: Microstructural evolution and multiscale tribological properties. *Carbon* **2010**, *48*, 3103–3120. [CrossRef]

13. Lahiri, D.; Singh, V.; Benaduce, A.P.; Seal, S.; Kos, L.; Agarwal, A. Boron nitride nanotube reinforced hydroxyapatite composite: Mechanical and tribological performance and in-vitro biocompatibility to osteoblasts. *J. Mech. Behav. Biomed.* **2011**, *4*, 44–56. [CrossRef] [PubMed]

14. Zhitomirsky, I.; Gal-Or, L. Electrophoretic deposition of hydroxyapatite. *J. Mater. Sci. Mater. Med.* **1997**, *8*, 213–219. [CrossRef] [PubMed]

15. Kwok, C.T.; Wonga, P.K.; Cheng, F.T.; Man, H.C. Characterization and corrosion behavior of hydroxyapatite coatings on Ti6Al4V fabricated by electrophoretic deposition. *Appl. Surf. Sci.* **2009**, *255*, 6736–6744. [CrossRef]

16. Sridhar, T.M.; Eliaz, N.; Kamachi Mudali, U.; Raj, B. Electrophoretic deposition of hydroxyapatite coatings and corrosion aspects of metallic implants. *Corros. Rev.* **2002**, *20*, 255–293. [CrossRef]

17. Theiss, F.; Apelt, D.; Brand, B.; Kutter, A.; Zlinszky, K.; Bohner, M. Biocompatibility and resorption of a brushite calcium phosphate cement. *Biomaterials* **2005**, *26*, 4383–4394. [CrossRef] [PubMed]

18. Rodriguez-Lugo, V.; Angeles-Chavez, C.; Mondragon, G.; Recillas-Gispert, S.; Castano, V.M. Synthesis and structural characterization of hydroxyapatite obtained from CaO and CaHPO4 by a hydrothermal method. *Mater. Res. Innov.* **2005**, *9*, 20–22. [CrossRef]

19. Kuo, M.C.; Yen, S.K. The process of electrochemical deposited hydroxyapatite coatings on biomedical titanium at room temperature. *Mater. Sci. Eng. C* **2002**, *20*, 153–160. [CrossRef]

20. Legeros, R.Z.; Lin, S.; Rohanizadeh, R.; Mijares, D.; Legeros, J.P. Biphasic calcium phosphate bioceramics: Preparation, properties and applications. *J. Mater. Sci. Mater. Med.* **2003**, *14*, 201–209. [CrossRef] [PubMed]

21. Manzeli, S.; Ovchinnikov, D.; Pasquier, D.; Yazyev, O.V.; Kis, A. 2D transition metal dichalcogenides. *Nat. Rev. Mater.* **2017**, *44*, 16399–16404. [CrossRef]

22. Rapoport, L.; Bilik, Y.; Feldman, Y.; Homyonfer, M.; Cohen, S.R.; Tenne, R. Hollow nanoparticles of WS2 as potential solid-state lubricants. *Nature* **1997**, *387*, 791–793. [CrossRef]

23. Rosentsveig, R.; Tenne, R.; Gorodnev, A.; Feuerstein, N.; Friedman, H.; Fleischer, N.; Tannous, J.; Dassenoy, F. Fullerene-like MoS2 nanoparticles and their tribological behavior. *Tribol. Lett.* **2009**, *36*, 175–182. [CrossRef]

24. André, B.; Gustavsson, F.; Svahn, F.; Jacobson, S. Performance and tribofilm formation of a low-friction coating incorporating inorganic fullerene like nano-particles. *Surf. Coat. Technol.* **2012**, *206*, 2325–2329. [CrossRef]

25. Alberdi, A.; Hatto, P.; Díaz, B.; Csillag, S. Tribological behavior of nanocomposite coatings based on fullerene-like structures. *Vacuum* **2011**, *85*, 1087–1092. [CrossRef]

26. Polcar, T.; Mohan, D.B.; Silviu Sandu, C.; Radnoczi, G.; Cavaleiro, A. Properties of nanocomposite film combining hard TiN matrix with embedded fullerene-like WS2 nanoclusters. *Thin Solid Films* **2011**, *519*, 3191–3195. [CrossRef]

27. Yadgarov, L.; Petrone, V.; Rosentsveig, R.; Feldman, Y.; Tenne, R.; Senatore, A. Tribological studies of rhenium doped fullerene-like MoS2 nanoparticles in boundary, mixed and elasto-hydrodynamic lubrication conditions. *Wear* **2013**, *297*, 1103–1110. [CrossRef]

28. Chhetri, M.; Gupta, U.; Yadgarov, L.; Rosentsveig, R.; Tenne, R.; Rao, C.N.R. Beneficial effect of Re doping on the electrochemical HER activity of MoS2 fullerenes. *Dalton Trans.* **2015**, *44*, 16399–16404. [CrossRef] [PubMed]

29. Yadgarov, L.; Rosentsveig, R.; Leitus, G.; Albu-Yaron, A.; Moshkovich, A.; Perfilyev, V.; Vasic, R.; Frenkel, A.I.; Enyashin, A.N.; Seifert, G.; et al. Controlled doping of MS2 (M = W, Mo) nanotubes and fullerene-like nanoparticles. *Angew. Chem. Int. Ed.* **2012**, *51*, 1148–1151. [CrossRef] [PubMed]

30. Sedova, A.; Ron, R.; Goldbart, O.; Elianov, O.; Yadgarov, L.; Kampf, N.; Rosentsveig, R.; Shumalinsky, D.; Lobik, L.; Shay, B.; et al. Re-doped fullerene-like MoS$_2$ nanoparticles in relationship with soft lubrication. *Nanomater. Energy* **2014**, *4*, 30–38. [CrossRef]

31. Xu, F.; Kobayashi, T.; Yang, Z.; Sekine, T.; Chang, H.; Wang, N.; Xia, Y.; Zhu, Y. Low cytotoxicity and genotoxicity of two-dimensional MoS$_2$ and WS$_2$. *ACS Nano* **2017**, *11*, 8114–8121. [CrossRef] [PubMed]

32. Appel, J.H.; Li, D.O.; Podlevsky, J.D.; Debnath, A.; Green, A.A.; Wang, Q.H.; Chae, J. Low cytotoxicity and genotoxicity of two-dimensional MoS$_2$ and WS$_2$. *ACS Biomater. Sci. Eng.* **2016**, *2*, 361–367. [CrossRef]

33. Goldman, E.B.; Zak, A.; Tenne, R.; Kartvelishvily, E.; Levin-Zaidman, S.; Neumann, Y.; Stiubea-Cohen, R.; Palmon, A.; Hovav, A.H.; Aframian, D.J. Biocompatibility of Tungsten Disulfide Inorganic Nanotubes and Fullerene-Like Nanoparticles with Salivary Gland Cells. *Tissue Eng. Part A* **2014**, *21*, 1013–1023. [CrossRef] [PubMed]

34. Pardo, M.; Shuster-Meiseles, T.; Levin-Zaidman, S.; Rudich, A.; Rudich, Y. Low Cytotoxicity of Inorganic Nanotubes and Fullerene-Like Nanostructures in Human Bronchial Epithelial Cells: Relation to Inflammatory Gene Induction and Antioxidant Response. *Environ. Sci. Technol.* **2014**, *48*, 3457–3466. [CrossRef] [PubMed]

35. Zwilling, V.; Aucouturier, M.; Darque-Ceretti, E. Anodic oxidation of titanium and TA6V alloy in chromic media. An electrochemical approach. *Electrochim. Acta* **1999**, *45*, 921–929. [CrossRef]

36. Macak, J.M.; Tsuchiya, H.; Schmuki, P. Nanoporous materials high-aspect-ratio TiO$_2$ nanotubes by anodization of titanium. *Angew. Chem. Int. Ed.* **2005**, *44*, 2100–2102. [CrossRef] [PubMed]

37. Gong, D.; Grimes, C.A.; Varghese, O.K.; Hu, W.; Singh, R.S.; Chen, Z.; Dickey, E.C. Titanium oxide nanotube arrays prepared by anodic oxidation. *J. Mater. Res.* **2001**, *16*, 3331–3334. [CrossRef]

38. Yadgarov, L.; Stroppa, D.G.; Rosentsveig, R.; Ron, R.; Enyashin, A.N.; Houben, L.; Tenne, R. Investigation of Rhenium-Doped MoS$_2$ Nanoparticles with Fullerene-Like Structure. *Z. Anorg. Allg. Chem.* **2012**, *638*, 2610–2616. [CrossRef]

39. Rey, C.; Combes, C.; Drouet, C.; Lebugle, A.; Sfihi, H.; Barroug, A. Nanocrystalline apatites in biological systems: Characterisation, structure and properties. *Materialwiss. Werkstofftech.* **2007**, *38*, 996–1002. [CrossRef]

40. Chen, J.S.; Juang, H.Y.; Hon, M.H. Second order Raman spectrum of MoS$_2$. *J. Mater. Sci. Mater. Med.* **1998**, *9*, 297–300. [CrossRef] [PubMed]

41. Kendall, B. *Nanocrystalline Apatite-Based Biomaterials: Synthesis; Processing and Characterization*; Nova Science Publishers: New York, NY, USA, 2008; pp. 93–143.

42. Chen, J.M.; Wang, C.S. Second order Raman spectrum of MoS$_2$. *Solid State Commun.* **1974**, *14*, 857–860. [CrossRef]

43. Sourisseau, C.; Fouassier, M.; Alba, M.; Ghorayeb, A.; Gorochov, O. Resonance Raman, inelastic neutron scattering and lattice dynamics studies of 2H-WS$_2$. *Mater. Sci. Eng. B* **1989**, *3*, 119–123. [CrossRef]

44. Grandfield, K.; Sun, F.; FitzPatrick, M.; Cheong, M.; Zhitomirsky, I. Electrophoretic deposition of polymer-carbon nanotube–hydroxyapatite composites. *Surf. Coat. Technol.* **2009**, *203*, 1481–1487. [CrossRef]

International Journal of
Molecular Sciences

MDPI

Review

Recent Advances in Laser-Ablative Synthesis of Bare Au and Si Nanoparticles and Assessment of Their Prospects for Tissue Engineering Applications

Ahmed Al-Kattan [1,*]**, Viraj P. Nirwan** [1,2]**, Anton Popov** [1]**, Yury V. Ryabchikov** [1,3]**, Gleb Tselikov** [1]**, Marc Sentis** [1,4]**, Amir Fahmi** [2] **and Andrei V. Kabashin** [1,4]

[1] Aix Marseille University, CNRS, LP3, 13288 Marseille, France; Viraj.nirwan@outlook.com (V.P.N.); popov@lp3.univ-mrs.fr (A.P.); ryabchikov@lp3.univ-mrs.fr (Y.V.R.); tselikov@lp3.univ-mrs.fr (G.T.); marc.sentis@univ-amu.fr (M.S.); kabashin@lp3.univ-mrs.fr (A.V.K.)
[2] Faculty of Technology and Bionics, Rhin-waal University of Applied Science, Marie-Curie-Straße 1, 47533 Kleve, Germany; Amir.Fahmi@hochschule-rhein-waal.de
[3] P.N. Lebedev Physical Institute of Russian Academy of Sciences, 53 Leninskii Prospekt, 199991 Moscow, Russia
[4] MEPhI, Institute of Engineering Physics for Biomedicine (PhysBio), 115409 Moscow, Russia
* Correspondence: ahmed.al-kattan@univ-amu.fr; Tel.: +33-(0)4-91-82-92-86

Received: 24 April 2018; Accepted: 18 May 2018; Published: 24 May 2018

Abstract: Driven by surface cleanness and unique physical, optical and chemical properties, bare (ligand-free) laser-synthesized nanoparticles (NPs) are now in the focus of interest as promising materials for the development of advanced biomedical platforms related to biosensing, bioimaging and therapeutic drug delivery. We recently achieved significant progress in the synthesis of bare gold (Au) and silicon (Si) NPs and their testing in biomedical tasks, including cancer imaging and therapy, biofuel cells, etc. We also showed that these nanomaterials can be excellent candidates for tissue engineering applications. This review is aimed at the description of our recent progress in laser synthesis of bare Si and Au NPs and their testing as functional modules (additives) in innovative scaffold platforms intended for tissue engineering tasks.

Keywords: laser ablation in liquid; Electrospinning; Nanoparticles; Nanofibers; Scaffolds; Tissue engineering; Nanotheranostics

---

1. Introduction

Tissue engineering is an important multidisciplinary field, which is focused on the development of biomaterial substitutes capable of replacing, detecting and treating failed or diseased tissues [1–3]. Due to a variety of tissues (e.g., bone, cartilage, nerve, cardiac/skeletal muscles, etc.), and their different functional and structural properties (e.g., stiffness, cell interconnection, etc.), the elaboration of substitutes presents a really challenging task, which has to gather a multitude of physicochemical characteristics (mechanical, electrical, etc.), structural properties (e.g., 3D architecture, surface topography, porosity, etc.) and advanced theranostic functionalities [4–9]. Biocompatibility and biodegradability are other key properties, which should be taken into account to ensure cell adhesion, growth and differentiation of surrounding tissues, preventing any rejection or toxicity issues [10–12].

Based on recent advances of nanotechnology, most efforts are now applied on the fabrication of nanostructured scaffolds, which could mimic the mesoporosity and nanoscale morphology of natural extracellular matrices (ECM) [13–15] and provide advanced functional properties for diagnostics/treatment of diseases or the restoration of biological functions [15,16]. A variety of scaffolds (hydrogel, nanofibers, etc.) made from ceramic, synthetic/natural polymers or composites are now explored for these tasks. Due to their chemical and structural similarity to natural mineral tissues, calcium phosphate ceramics present a class of tunable bioactive scaffolds, which are currently largely exploited in bone regeneration and orthopedic surgery [17]. Due to their capability of inducing osteoblastic differentiation, a plethora of coatings based on calcium phosphate compositions are employed as bioactive interfaces for implants [18,19]. However, clinical applications of these materials are still quite challenging because of brittleness, difficulty of shaping for implantation and uncontrolled degradation rate [20–22]. As another example, synthetic polymers (e.g., polystyrene) are also actively exploited as scaffolds to offer tunable architectures and degradability option [23]. Nevertheless, the interaction of many synthetic polymers with biological tissues is controversial because of their low bioactivity and side effects [23]. Owing to outstanding biocompatibility and biodegradability, natural polymers (chitosan, collagen, etc.) have appeared as very promising scaffolds, which can naturally promote cell adhesion and growth [24]. However, due to weak dissolution the shaping of natural polymers is challenging, while some of their physical properties including mechanical and conductivity characteristics must be improved [25,26].

One of promising ways to improve intrinsic properties of scaffolds and/or acquire additional functionalities consists in the incorporation of nanoparticles (NPs) as additives [27–30]. Considerable progress in this field has contributed to the fabrication of advanced multifunctional NPs for biomedical tasks including drug delivery, imaging and cell labeling. Applications of such NPs in tissue engineering could dramatically enhance physicochemical properties of scaffolds and contribute to their proper integration into tissue-specific microenvironments. Silver (Ag) NPs present a prominent example of non-exhaustive antibacterial objects, which are intensively exploited in tissue engineering tasks using a variety of polymeric scaffolds [31,32]. Iron-oxide NPs present another example of nanomaterials exhibiting antimicrobial and superparamagnetic properties, which can potentially be used in hyperthermia, gene therapy and bioimaging [33,34]. Owing to prominent optical and physical properties and chemical reactivity, gold (Au) NPs can offer many functionalities for biosensing, bioimaging and theranostics [28,29]. Due to their large surface area, porosity and biocompatibility, mesoporous silica (SiO_2) NPs are also considered as relevant additives for drug delivery [35]. Finally, crystalline silicon (Si) looks as one of most promising candidates for tissue engineering, as it is ideally biocompatible [36] and biodegradable [37], as well can offer a large range of imaging and therapeutic functionalities, including room temperature photoluminescence for bioimaging [36–40], light-induced generation of singlet oxygen for photodynamic cancer therapy [41], and infrared radiation [42], radio frequency radiation [43] and ultrasound-induced [44] hyperthermia for cancer therapy. However, almost all currently employed NPs are synthesized by conventional chemical or electrochemical routes, which involve hazardous products (e.g., HF, nitrate salts, chloride, citrate, etc.) and various ligands. The presence of these products generally leads to surface contamination by residual toxic products, which is not consistent with targeted biomedical applications [45–47]. In addition, the elaboration of these nanomaterials takes place under extreme thermal, pH-metric and pressure conditions, which requires a rigorous control of the synthesis procedure. These reactions also frequently require different organic solvents (e.g., ethanol, THF, etc.) and many switching steps between organic/aqueous solutions, which complicates the fabrication and purification procedures [48,49].

Emerged as a new "green" nanosynthesis route, pulsed laser ablation gathers numerous benefits over chemical methods in the elaboration of ultraclean NPs [50]. This physical method implies the ablation of a solid target by pulsed laser radiation, which gives rise to a natural formation of nanoclusters [51,52]. If created in gaseous ambience, the nanoclusters can be deposited on a substrate forming a thin nanostructured film [53–57]. When created in liquid ambience, the nanoclusters can be released into the liquid forming nanoparticle colloidal solutions [58–67]. In both cases, the ablation can be performed in ultraclean environment (helium or argon gas, deionized water), excluding any contamination [50]. Here, the employment of ultra-short laser pulses can lead to the fabrication of extremely stable colloidal solutions of low-size-dispersed crystalline NPs even in the absence of any protective ligands [62,63,65]. In the case of Si, such procedure can be modified by the ablation (fragmentation) from microcolloids, prepared preliminarily by mechanical milling of a Si wafer [43,

66,67]. Such a fragmentation method makes possible the fabrication of concentrated solutions of low size-dispersed Si-NPs with a variable mean size [66]. It was also shown that "bare" (ligand-free) surface can exhibit unusual reactivity and physicochemical properties [68–71]. Finally, bare laser-synthesized NPs are relevant nanotheranostics tools for biomedical applications [68,72]. We believe that due to the existence of such unique properties, the use of these nanomaterials as additives in tissue engineering looks very promising opening up opportunities for future innovative developments.

This article reviews our recent achievements in the elaboration of bare inorganic and metallic NPs based on silicon and gold for biomedical applications, as well as highlights prospectives of such nanomaterials for tissue engineering applications. Here, we first remind the principle and the advantages of laser synthesis compared to conventional chemical synthesis routes. Second, the physicochemical characteristics of bare NPs and their interaction with biological matrices in vitro and in vivo are reviewed. We finally present results of our tests on the incorporation of Si and Au NPs as functional additives in electrospun nanofibers based on natural polymer chistosan-poly(ethylene oxide)(chitosan (PEO)) [73]. This multidisciplinary work aims to describe potential advantages of using bare NPs in tissue engineering applications.

2. Pulsed Laser Ablation in Liquids (PLAL) for the Synthesis of Colloidal Nanomaterials

Driven by its flexibility, simplicity and rapidity, PLAL appears to be particularity important in the elaboration of ultra-clean, bare (ligand-free) NPs for a variety of applications, including electronics, energy production and biomedicine (for review, see [50,74]. Based on a combination of top-down and bottom-up approaches, this method profits from laser-target interaction to ablate material of the target and thus naturally form nanoclusters. The nanoclusters then coalesce in liquid medium to produce a colloidal NPs solution. Figure 1 illustrates a basic experimental set-up, which offers an easy handling to produce colloidal solutions of nanomaterials. In conventional ablation geometry, the set-up consists of a pulsed laser and focusing optics to ablate a target placed on the bottom of a cuvette filled with a liquid solution (Figure 1a). In this case, the platform with the target should be constantly moved to avoid the ablation from the same area on the target. In an alternative fragmentation setup, material is ablated from liquid-dispersed micro/nano particles, while the liquid is constantly mixed by a magnetic stirrer (Figure 1b).

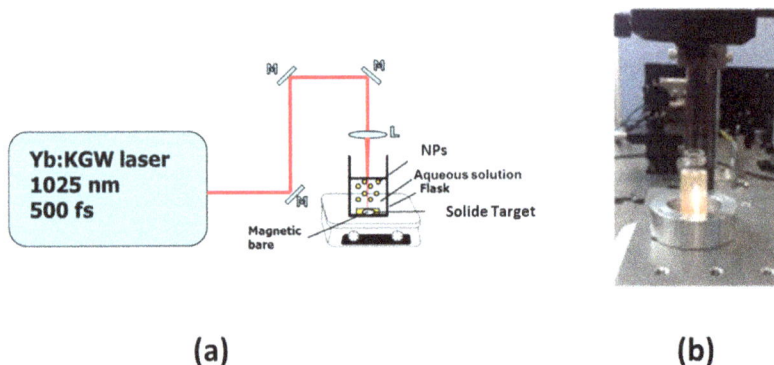

(a) **(b)**

Figure 1. (a) Typical PLAL setup; (b) Illustrative image of colloidal Si NPs solution prepared by femtosecond (fs) laser fragmentation.

PLAL has numerous advantages over conventional chemical methods such as the possibility for NPs synthesis in ultrapure non-contaminated solutions (e.g., deionized water) without surfactants or ligands. NPs formed under these conditions can exhibit unique "bare" surface, which is characterized by different reactivity compared to chemically-synthesized NPs [69–73]. It is also important that such ultrapure surface does not require any additional purification, as it often takes place in the case of chemically-synthesized NPs. Here, to condition an appropriate chemical composition ranging from elemental to hydroxide, water can be replaced by organic solvents, polymer or saline solutions. Moreover, by varying the liquid composition (e.g., oils, superfluids, etc.) one can monitor the NPs shape [74]. As another advantage, laser-ablative synthesis makes possible (bio)functionalization of formed nanomaterials in situ [74].

It should be noted that laser ablation mechanism is a complex process involving extreme phenomena (shock wave, plasma plume and cavitation bubble, etc.) under severe physical and thermodynamic conditions (thousands of kelvins and hundreds of pascals). Furthermore, all processes take place simultaneously during a very short time period (a few ns) and depend on laser parameters (pulses length, wavelength, fluence, etc.) [75]. Many experimental methods (spectroscopic analysis, acoustic measurements, x-ray imaging techniques, acoustic measurements, CCD cameras observations) and theoretical modeling investigations were devoted to clarifying the ablation mechanism [75]. In most cases, laser ablation tends to provide a broadened size dispersion of NPs (from several tens to hundreds of nm) with polymodal size populations. As a debated scenario, a sequence of different ablation mechanisms occurs during plasma plume expansion (cooling) and the explosion of a cavitation bubble. Such phenomena are especially important for a "long" ns laser ablation regime. The addition of reactive chemical products during the ablation process makes possible efficient "chemical" control of NPs size, but in this case the cleanness of NPs can be compromised. On the other hand, laser ablation at ultrashort regime (fs) looks like the most promising "fine" tool, which is now accepted by the whole laser processing community. Due its specific conditions, much less radiation is transferred to the cavitation bubble, which limits cavitation phenomena and thus makes possible much more rigorous control of NPs size and size dispersion [50,75].

It is important that initial colloidal NPs solutions obtained by laser ablation (Figure 2a) can be subjected to the second laser "fragmentation" step (Figure 2b) [65]. The fragmentation mechanism still not fully understood, but "photothermal evaporation" and "Coulomb explosion" are considered as the main mechanisms responsible for ablation. Moreover, preliminary colloids can be directly obtained from micropowder suspensions [66,68]. Here, fs laser ablation from micro/nano colloids manifested itself a very efficient method to achieve desired controllable size characteristics of formed NPs. In general, PLAL looks as a reliable tool for the fabrication of a variety of nanomaterials, which enables one to control NPs characteristics by adjusting laser parameters (focusing point position, fluence, repetition rate, fragmentation duration, etc.) and physicochemical conditions (e.g., concentration of raw material).

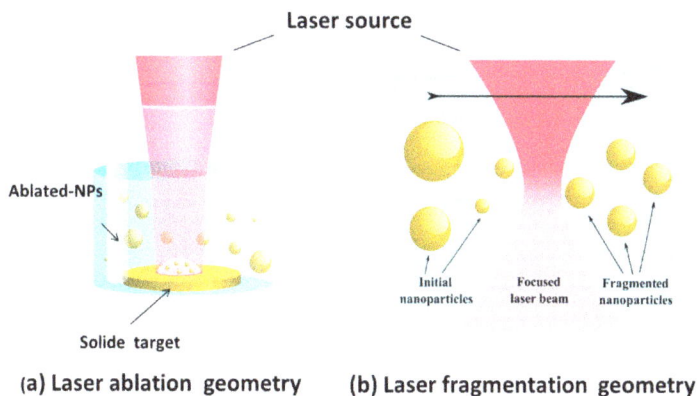

Figure 2. Schematic presentation of laser ablation (**a**) and laser fragmentation (**b**) geometries.

3. PLAL Synthesis of Bare Nanomaterials for Biomedical Applications

3.1. Bare Laser-Synthesized Si Nanoparticles

Silicon (Si) is a group IV semiconductor, which participates in many biochemical processes, including bone mineralization (e.g., osteoblastogenesis), connective tissue metabolism, signal transduction [76–78]. Moreover, Si improves the adsorption of crucial minerals such as magnesium and copper, which are involved in the proliferation of lymphocyte cells and their immune response. In addition, Si nanostructures are water-dissolvable and biodegradable, as in biological environment they convert into orthosilicic acid $Si(OH)_4$, which is naturally excreted with urine [37]. Finally, Si NPs offer a large panel of applications in biomedicine including photoluminescence-based imaging [39,40,79], photodynamic therapy and hyperthermia-based therapy for cancer treatment [41–44]. Si NPs can be synthesized using chemical reduction methods, microemulsion techniques, electrochemical synthesis, etc., which typically require numerous purification steps to clean the NPs surface [37,47,80–82]. Here, specific installations (hood, vacuum box, etc.) and operating skills are required. We recently introduced ultrashort (femtosecond) laser ablation in aqueous solutions as a novel approach to fabricate ultrapure Si NPs for biomedical applications [66,67]. In a typical laser-synthesized protocol, Si NPs are prepared from Si microparticles powder dispersed at 0.35 g·L^{-1} in deionized water by sonication step for 30 min. The microparticles are fragmented by focused femtosecond laser irradiation (Yb:KGW laser, Amplitude systems, 1025 nm, 480 fs, 1 kHz) for 1 h (for more details see ref. [67]). Physicochemical characterization showed that so formed Si NPs have a tunable mean side between 10 and 100 nm under narrow size dispersion. Structural and analytical measurements showed that the NPs are surrounded by a thin oxidized layer of SiO_x ($1 \leq x \leq 2$) with a ζ-potential of -45 ± 1.5 mV preventing thus any aggregation phenomenon between the Si NPs (Figure 3b).

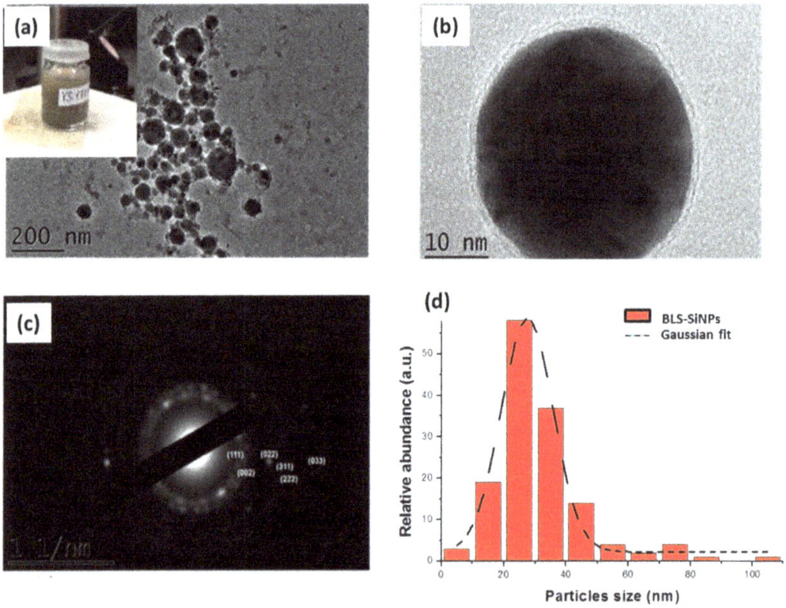

Figure 3. (a) HR-TEM image of Si NPs obtained by laser fragmentation at 0.35 g·L^{-1} initial concentration of microcolloids (Inset, typical image of Si NPs solution). (b) Single laser-synthesized Si nanoparticle. Characteristic electron diffraction pattern of Si NPs (c) and corresponding size distributions (d). Adapted from ref. [67].

Moreover, we established that by varying the amount of dissolved oxygen in water, one can control the oxidation state and potentially create silicon oxide defects inside Si NPs crystals, which can lead to much accelerated dissolution of NPs in aqueous solutions (Figure 4) [67]. Other advantages of PLAL approach are related to the possibility of controlling mean NPs size by varying the initial Si microparticles concentration [66]. Such approach can provide "calibrated" additives with monitored structural properties.

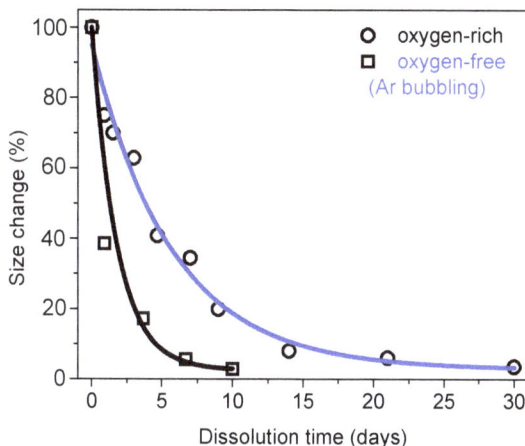

Figure 4. Size evolution (in percent, relative to the initial size of Si NPs prepared under oxygen-rich (black) and oxygen-free (blue, Ar bubbling) conditions as a function of dialysis duration in deionized water. Adapted from ref. [67].

The interaction of Si NPs with biological matrices was investigated in vitro and in vivo in our earlier studies [67,83]. No obvious cytotoxicity effect on human cells (HMEC) was observed up to 100 $\mu g \cdot mL^{-1}$ with cell survival rate around 80% (Figure 5a). In addition, TEM examination revealed that Si NPs are readily uptaken by cells via classical endocytosis mechanism without damage of cell compartments (Figure 5b). In vitro study was also completed by in vivo tests performed in a nude mouse model at different incubation time (from 3 h to 7 days) with a single dose (20 mg/kg) of intravenous administration [83]. Based on the examination of behavior of mice and their growth, we concluded that all animals showed normal physiological activities without lethargy or apathy. The biodistribution and the fate of Si NPs were followed by a control of a panel of biochemical parameters and the examination of organ tissues. This study revealed that Si NPs were completely safe. Furthermore, these NPs were cleared from biological matrices within one week, while similar porous Si-based nanoformulations prepared by electrochemical routes require 4–6 weeks for the excretion [37]. The functionality of laser-synthesized Si NPs as sensitizers of radiofrequency (RF)-induced hyperthermia was tested on Lewis lung carcinoma in vivo and compared to porous silicon-based nanostructures (Figure 5c–e) [43]. Here, we observed efficient tumor inhibition without any side effects, while laser-synthesized NPs demonstrated much stronger therapeutic outcome. We believe that such sensitizing properties of laser-synthesized Si NPs can be used as a novel functionality in the development of tissue engineering platforms.

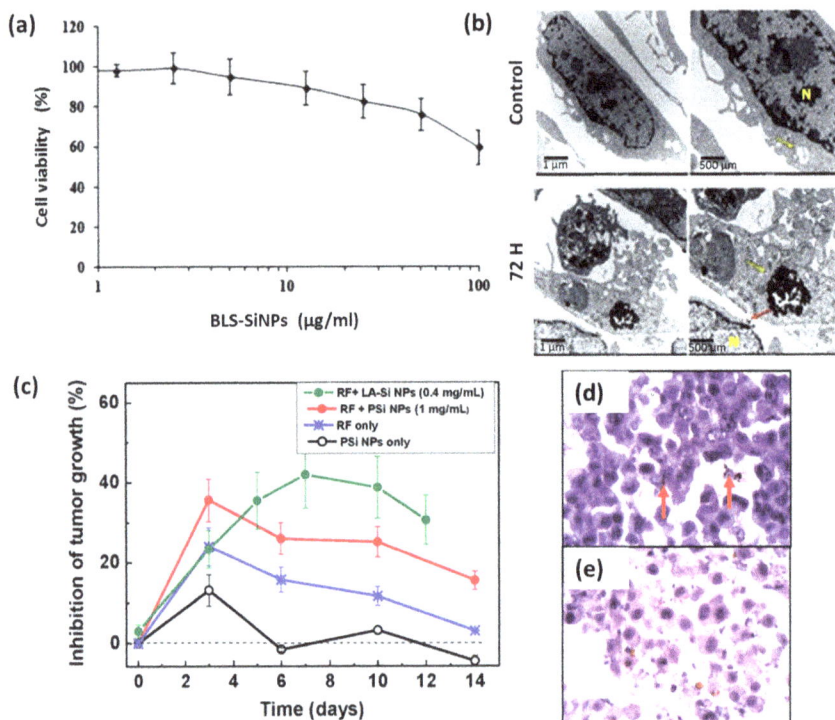

Figure 5. (**a**) MTT assays of HMEC cells viability following their exposure to different concentration of Si NPs (1.25–100 µg/mL) for 72 h. (**b**) TEM images of HMEC cells showing kinetics of Si NPs internalization 72 h after incubation time with 50 µg/mL of NPs. (**c**) Inhibition of the tumor growth after the following treatments: the injection of Si NPs suspension without RF irradiation (black curve); 2 min treatment of tumor area by RF irradiation with the intensity of 2 W/cm^2 (blue); injection of a suspension of porous Si NPs (PSi NPs) (0.5 mL, 1 mg/mL) followed by 2 min RF irradiation treatment (red); injection of a suspension of laser-synthesized Si NPs (LA-Si NPs) (0.2 mL, 0.4 mg/mL) followed by 2 min RF irradiation treatment (green). (**d**,**e**) are histology images of a tumor area 1 h and 3 days after the PSi NP injection and RF-based treatment using PSi NPs as nanosensitizers, respectively. Cancer cells are visible as dark blue spots. Examples of agglomerations of PSi NPs in the cells are indicated by red arrows. Adapted from refs. [43,83].

3.2. Bare Laser-Synthesized Au Nanoparticles

Nanostructured gold (Au) has attracted a considerable attention of biomedical community due to their unique physical, optical and chemical properties [84,85]. Owing to optical excitations of free electron oscillations (plasmons), electric field is strongly enhanced in the vicinity of metal surface, which can be used in various applications, including biosensing [86–89], imaging [90,91], photothermal therapy [92–94], gene and drug delivery. Numerous methods have been reported to fabricate a wide variety of Au NPs shapes (nanospheres, nanorods, nanoplates, nanoshells, etc.) opening a wide avenue for applications in energy, biomedicine, material science and tissue engineering. Here, the surface of Au NPs can be functionalized by polymers (e.g., PEG) [95,96], functional groups (e.g., Amine and Carboxyl) [97,98], as well as by biomolecules including DNA [99] and peptides [100]. Such functionalizations can help to enhance specificity and efficacy of Au NPs toward specific cell types and organelles such as nucleus and mitochondria. However, in general such NPs are fabricated by chemical routes, involving stabilizing molecules or ligands, which are not always biocompatible [101].

First, the presence of stabilizing agent on Au NPs surface can potentially hinder their direct interactions with biological environment and can compromise their future functionalization [102,103]. Second, these molecules can interfere with plasmonic properties of Au NPs [104].

To overcome such limitations, we recently elaborated PLAL technique to synthesize bare Au NPs in aqueous solutions in the absence of any stabilizing molecules (Figure 6) [62,63,65,68]. In a typical procedure a solid Au target (99.99%, GoodFellow, France) was placed at the bottom of the glace vessel and filled with 7 mL of deionized water. The target was then irradiated with femtosecond laser (Yb:KGW laser, Amplitude systems, 1025 nm, 480 fs, 1 kHz) for 15 min. In order to reduce size dispersion, Au NPs produced by laser ablation step were then subjected to the second "fragmentation" step for 30 min (for more details see ref. [72]). Structural and microscopic observations revealed that Au NPs were spherical in shape and free from any residual contaminants, enabling high chemical [69–71] and catalytic [68] activity. In addition, due to the partial oxidation of surface (Au–O$^-$/Au–OH$^-$), the Au NPs exhibit a negative surface charge (-23 ± 2.3 mV) conferring thus a great stability and limiting any agglomeration effect. The interaction of such NPs with biological matrices was assessed during in vitro tests under relatively high concentration (10 mg/L^{-1}) of NPs up to 72 h. TEM analyses demonstrated biological safety characteristics of Au NPs without any side effects on morphology and cytoskeleton cells [72]. Au NPs were internalized by a classical endocytosis mechanism without penetration into the nucleus cell. In addition, the analysis of protein corona on NPs surface revealed interactions with abundant proteins such as Albumine and Apos, which are known to play crucial roles in intracellular trafficking [72]. Due to their unique structures and excellent biocompatibility, bare Au NPs can be considered alternative candidates as additives for tissue engineering.

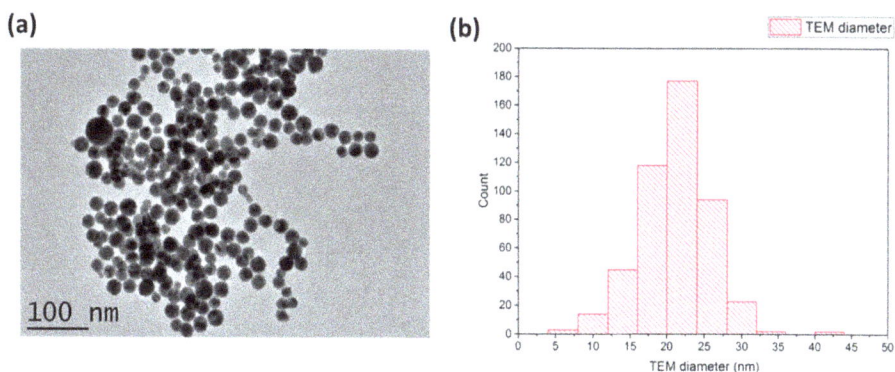

Figure 6. Typical HR-TEM image of Au NPs prepared by PLAL (**a**) and corresponding size distribution (**b**).

4. Potential Applications of BLS-NPs in Tissue Engineering

Biological tissue presents a complex environment with specific structural, biological, chemical and physical characteristics. Therefore, the creation of artificial functional tissue structures (scaffolds) has to gather variety of properties such as a good cell adhesion, high porosity, adequate pore size for cell seeding and diffusion, structural rigidity, biocompatibility and biodegradability. Many studies are now devoted to the elaboration of functional scaffolds, which could mimic the ECM. Scaffolds can be fabricated by variety conventional techniques (e.g., solvent casting/particle leaching [105]), but electrospinning looks as the most promising approach to fabricate biocompatible/biodegradable nanofibrous scaffolds. Based on the application of electrical field in polymer solutions, this process offers plenty of advantages such as the possibility of working with a variety of materials including natural/synthetic polymers and their composites, generation of micro- to nano-scale nanofibers, cost effectiveness and easy scaling-up [106,107]. However, despite these benefits and approach flexibility, a number of problems need to be solved such as the fabrication of uniform nanofibers with desired diameter, morphology, mechanical strength, conductivity and chemistry. On the other hand, the elaboration of electrospun nanofibers with multi-functionalities (biological and therapeutic characteristics) is still required.

We recently carried out tests to explore the potential of using bare laser-synthesized Si and au NPs as functional additives in order to (i) improve/optimize intrinsic properties of nanofibers; (ii) enable advanced biomedical/biological properties. As a first approach, we recently functionalized biologically-derived polymer based on nanofibers chitosan (PEO) by bare Si and Au NPs [73]. At optimized chitosan:PEO ratio, the NPs were directly introduced at increased concentration in the polymer solution before electrospinning. Numerous analyses were then conducted on obtained nanofibers based on microscopic, thermal and analytical methods. First, it appeared that the NPs were properly attached via electrostatic interaction and homogenously dispersed on the nanofiber surface, while the presence of NPs did not affect the morphology of fiber networks and their chemical properties. Second, we observed a reduction of the fiber diameter by a factor 2 when the fibers are co-electrospun with Si NPs. In addition, functionalized nanofibers exhibited better thermal stability at higher temperature and this effect was especially prominent for Si NPs. Safety properties of the hybrid scaffold were also assessed by preliminary MTT tests and did not show ant toxicity.

These first tests confirmed the possibility of using NPs as functional additives in the elaboration of innovative scaffolds for tissue engineering (Figure 7). In particular, the presence of NPs on the nanofibers surface can be exploited as additional anchoring site interacting with cells. Here, the bare surface of laser-synthesized NPs looks very important as it can be tuned with specific biomolecules and growth factors to increase the nanofibers bioactivity toward cells. On the other hand, NPs can be used as sensitive probes to track variety of biomolecules (DNA, RNA, protein, etc.) and other materials including metal ions. Furthermore, the reduction of nanofibers diameter can potentially lead to higher bioactivity characteristics as it was noted in literature [108]. Besides, the incorporation of BLS-NPs into nanofibers improves thermal stability of the fiber matrix, which can be exploited for therapeutic applications and extended to other physicochemical parameters such as pH control. Despite encouraging results, the development of NPs as functional additives is in its early stage and many issues should still be clarified. Here, many physical properties including mechanical and electrical characteristics have to be assessed. In addition, the scaffold fate (e.g., dissolution of nanofibers, release of NPs from scaffold, etc.) has to be monitored and evaluated. Other parameter such as the size of NPs has to be varied to highlight their effect on the physicochemical properties of the fibers.

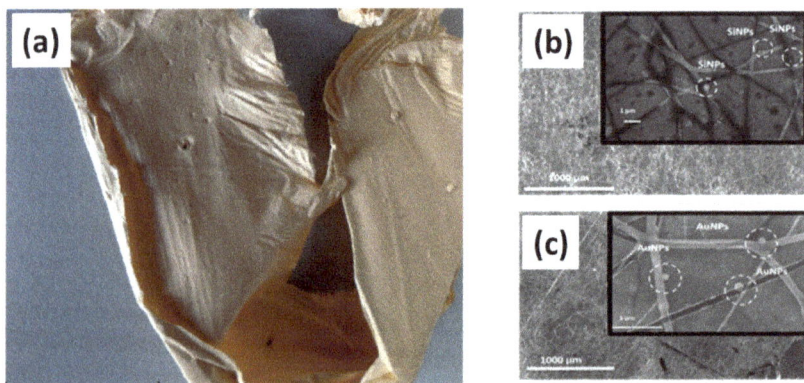

Figure 7. (**a**) Illustrative image of electrospun chitosan(PEO) nanofibers functionalized with bare Si NPs. (**b**) SEM of hybrid chitosan (PEO) nanofibers functionalized with bare Si NPs at 30 wt. %. (**c**) SEM of hybrid chitosan (PEO) nanofibers functionalized with bare Au NPs at 30 wt. %. Adapted from ref. [73].

Int. J. Mol. Sci. **2018**, *19*, 1563

5. Conclusions

In conclusion, bare laser-synthesized NPs open a wide range of opportunities toward the elaboration of functional scaffolds for tissue engineering enabling advanced biomedical modalities. First, PLAL method enables to fabricate NPs exempt of any contaminants, while the NPs surface can exhibit high reactivity and much better biocompatibility compared to chemically-synthesized Si and Au counterparts. Second, structural properties of laser-synthesized NPs can be easily designed to control their size and dissolution behavior. Third, one can use Si and Au NPs to enable a variety of therapy modalities [109], as well as imaging modalities including fluorescence imaging [40], SERS [110], SEIRAS [111]. As first preliminary work, we highlighted the possibility of incorporating of Si and Au NPs as functional additives for hybrid-electrospun nanofibers based on chitosan (PEO), without any effect on nanofiber compositions. Here, we observed a drastic decrease of nanofiber diameter promising a much improved bioactivity of nanofibers, while thermal analysis revealed a better stability of nanofibers at higher temperatures which can be exploited for advanced therapeutic tasks. Finally, the presence of NPs on the nanofibers promises its additional reactive surface toward biological tissue. Si and Au NPs can offer the opportunity to fabricate innovative scaffolds systems, which are capable of treating specific information related to surrounding tissues. The employment of NPs as functional modules for tissue engineering is still in a very early stage. Intensive research is still required to assess all potential benefits from such nano-engineered systems.

Author Contributions: All authors participated in the experimental study related to testing of laser-synthesized Si and Au NPs as additives in tissue engineering matrices based on chitosan (PEO) nanofibers. A.A.-K. and A.V.K. conceived the manuscript and wrote main parts. V.P.N., A.P., Y.V.R., G.T., M.S., A.F. reviewed the text and added their comments and corrections.

Acknowledgments: The authors express their thanks to DAAD program (ID:50015559), EHYBIOMED project, The International Associated Laboratory (LIA) MINOS project, LASERNANNOCANCER project, GRAVITY project of the ITMO "Plan Cancer 2014–2019" INSERM program, YVR acknowledges a support from COST project (ECOST-STSM-BM1205-120416-072252) and from Center for Research Strategy of Free University of Berlin (0503121810) for performing experiments. AVK and MS acknowledge the Competitiveness Program of MEPHI.

Conflicts of Interest: The authors declare no conflict of interest.

Abbreviations

NPs   Nanoparticles
PLAL   Pulsed Laser Ablation in Liquid

References

1. Amini, A.R.; Laurencin, C.T.; Nukavarapu, S.P. Bone Tissue Engineering: Recent Advances and Challenges. *Crit. Rev. Biomed. Eng.* **2012**, *40*, 363–408. [CrossRef] [PubMed]
2. Langer, R.; Vacanti, J. Advances in Tissue Engineering. *J. Pediatr. Surg.* **2018**, *51*, 8–12. [CrossRef] [PubMed]
3. Wobma, H.; Vunjak-Novakovic, G. Tissue Engineering and Regenerative Medicine 2015: A Year in Review. *Tissue Eng. Part B. Rev.* **2016**, *22*, 101–113. [CrossRef] [PubMed]
4. Salgado, A.J.; Oliveira, J.M.; Martins, A.; Teixeira, F.G.; Silva, N.A.; Neves, N.M.; Sousa, N.; Reis, R.L. Chapter One—Tissue Engineering and Regenerative Medicine: Past, Present, and Future. In *Tissue Engineering of the Peripheral Nerve: Stem Cells and Regeneration Promoting Factors*; International Review of Neurobiology; Geuna, S., Perroteau, I., Tos, P., Battiston, B., Eds.; Academic Press: Cambridge, MA, USA, 2013; Volume 108, pp. 1–33.
5. Zhang, L.; Webster, T.J. Nanotechnology and Nanomaterials: Promises for Improved Tissue Regeneration. *Nano Today* **2009**, *4*, 66–80. [CrossRef]
6. Bouten, C.V.C.; Driessen-Mol, A.; Baaijens, F.P.T. In Situ Heart Valve Tissue Engineering: Simple Devices, Smart Materials, Complex Knowledge. *Expert Rev. Med. Devices* **2012**, *9*, 453–455. [CrossRef] [PubMed]
7. Andreas, K.; Sittinger, M.; Ringe, J. Toward in Situ Tissue Engineering: Chemokine-Guided Stem Cell Recruitment. *Trends Biotechnol.* **2018**, *32*, 483–492. [CrossRef] [PubMed]

8. Khan, Y.; Yaszemski, M.J.; Mikos, A.G.; Laurencin, C.T. Tissue Engineering of Bone: Material and Matrix Considerations. *J. Bone Jt. Surg. Am.* **2008**, *90* (Suppl. 1), 36–42. [CrossRef] [PubMed]
9. Skoulas, E.; Manousaki, A.; Fotakis, C.; Stratakis, E. Biomimetic Surface Structuring Using Cylindrical Vector Femtosecond Laser Beams. *Sci. Rep.* **2017**, *7*, 45114. [CrossRef] [PubMed]
10. Wan, Y.; Wu, H.; Yu, A.; Wen, D. Biodegradable Polylactide/chitosan Blend Membranes. *Biomacromolecules* **2006**, *7*, 1362–1372. [CrossRef] [PubMed]
11. Rezwan, K.; Chen, Q.Z.; Blaker, J.J.; Boccaccini, A.R. Biodegradable and Bioactive Porous Polymer/inorganic Composite Scaffolds for Bone Tissue Engineering. *Biomaterials* **2006**, *27*, 3413–3431. [CrossRef] [PubMed]
12. Ma, L.; Gao, C.; Mao, Z.; Zhou, J.; Shen, J.; Hu, X.; Han, C. Collagen/chitosan Porous Scaffolds with Improved Biostability for Skin Tissue Engineering. *Biomaterials* **2003**, *24*, 4833–4841. [CrossRef]
13. Karuppuswamy, P.; Venugopal, J.R.; Navaneethan, B.; Laiva, A.L.; Sridhar, S.; Ramakrishna, S. Functionalized Hybrid Nanofibers to Mimic Native ECM for Tissue Engineering Applications. *Appl. Surface Sci.* **2014**, *322*, 162–168. [CrossRef]
14. Rosellini, E.; Zhang, Y.S.; Migliori, B.; Barbani, N.; Lazzeri, L.; Shin, S.R.; Dokmeci, M.R.; Cascone, M.G. Protein/polysaccharide-Based Scaffolds Mimicking Native Extracellular Matrix for Cardiac Tissue Engineering Applications. *J. Biomed. Mater. Res. A* **2018**, *106*, 769–781. [CrossRef] [PubMed]
15. Webber, M.J.; Khan, O.F.; Sydlik, S.A.; Tang, B.C.; Langer, R. A Perspective on the Clinical Translation of Scaffolds for Tissue Engineering. *Ann. Biomed. Eng.* **2015**, *43*, 641–656. [CrossRef] [PubMed]
16. Billström, G.H.; Blom, A.W.; Larsson, S.; Beswick, A.D. Application of Scaffolds for Bone Regeneration Strategies: Current Trends and Future Directions. *Injury* **2018**, *44*, S28–S33. [CrossRef]
17. Shepherd, J.H.; Best, S.M. Calcium Phosphate Scaffolds for Bone Repair. *JOM* **2011**, *63*, 83–92. [CrossRef]
18. El-Bassyouni, G.T.; Beherei, H.H.; Mohamed, K.R.; Kenawy, S.H. Fabrication and Bioactivity Behavior of HA/bioactive Glass Composites in the Presence of Calcium Hexaboride. *Mater. Chem. Phys.* **2016**, *175*, 92–99. [CrossRef]
19. Ning, C.Q.; Zhou, Y. In Vitro Bioactivity of a Biocomposite Fabricated from HA and Ti Powders by Powder Metallurgy Method. *Biomaterials* **2002**, *23*, 2909–2915. [CrossRef]
20. Wang, M. Developing Bioactive Composite Materials for Tissue Replacement. *Biomaterials* **2003**, *24*, 2133–2151. [CrossRef]
21. Valle, S.D.; Miño, N.; Muñoz, F.; González, A.; Planell, J.A.; Ginebra, M.-P. In Vivo Evaluation of an Injectable Macroporous Calcium Phosphate Cement. *J. Mater. Sci. Mater. Med.* **2007**, *18*, 353–361. [CrossRef] [PubMed]
22. Tancred, D.C.; Carr, A.J.; McCormack, B.A.O. Development of a New Synthetic Bone Graft. *J. Mater. Sci. Mater. Med.* **1998**, *9*, 819–823. [CrossRef] [PubMed]
23. Place, E.S.; George, J.H.; Williams, C.K.; Stevens, M.M. Synthetic Polymer Scaffolds for Tissue Engineering. *Chem. Soc. Rev.* **2009**, *38*, 1139–1151. [CrossRef] [PubMed]
24. Stratton, S.; Shelke, N.B.; Hoshino, K.; Rudraiah, S.; Kumbar, S.G. Bioactive Polymeric Scaffolds for Tissue Engineering. *Bioact. Mater.* **2016**, *1*, 93–108. [CrossRef] [PubMed]
25. Pakravan, M.; Heuzey, M.-C.; Ajji, A. A Fundamental Study of chitosan/PEO Electrospinning. *Polymer Guildf.* **2011**, *52*, 4813–4824. [CrossRef]
26. Deeken, C.R.; Fox, D.B.; Bachman, S.L.; Ramshaw, B.J.; Grant, S.A. Characterization of Bionanocomposite Scaffolds Comprised of Amine-Functionalized Gold Nanoparticles and Silicon Carbide Nanowires Crosslinked to an Acellular Porcine Tendon. *J. Biomed. Mater. Res. Part B Appl. Biomater.* **2011**, *97*, 334–344. [CrossRef] [PubMed]
27. Vieira, S.; Vial, S.; Reis, R.L.; Oliveira, J.M. Nanoparticles for Bone Tissue Engineering. *Biotechnol. Prog.* **2017**, *33*, 590–611. [CrossRef] [PubMed]
28. Kim, E.Y.; Kumar, D.; Khang, G.; Lim, D.-K. Recent Advances in Gold Nanoparticle-Based Bioengineering Applications. *J. Mater. Chem. B* **2015**, *3*, 8433–8444. [CrossRef]
29. Vial, S.; Reis, R.L.; Oliveira, J.M. Recent Advances Using Gold Nanoparticles as a Promising Multimodal Tool for Tissue Engineering and Regenerative Medicine. *Curr. Opin. Solid State Mater. Sci.* **2017**, *21*, 92–112. [CrossRef]
30. Kim, E.-S.; Ahn, E.H.; Dvir, T.; Kim, D.-H. Emerging Nanotechnology Approaches in Tissue Engineering and Regenerative Medicine. *Int. J. Nanomed.* **2014**, *9* (Suppl. 1), 1–5. [CrossRef] [PubMed]

31. Annur, D.; Wang, Z.K.; Liao, J.D.; Kuo, C. Plasma-Synthesized Silver Nanoparticles on Electrospun Chitosan Nanofiber Surfaces for Antibacterial Applications. *Biomacromolecules* **2015**, *16*, 3248–3255. [CrossRef] [PubMed]

32. Ali, S.W.; Rajendran, S.; Joshi, M. Synthesis and Characterization of Chitosan and Silver Loaded Chitosan Nanoparticles for Bioactive Polyester. *Carbohydr. Polym.* **2011**, *83*, 438–446. [CrossRef]

33. Colombo, M.; Carregal-Romero, S.; Casula, M.F.; Gutierrez, L.; Morales, M.P.; Bohm, I.B.; Heverhagen, J.T.; Prosperi, D.; Parak, W.J. Biological Applications of Magnetic Nanoparticles. *Chem. Soc. Rev.* **2012**, *41*, 4306–4334. [CrossRef] [PubMed]

34. Sensenig, R.; Sapir, Y.; MacDonald, C.; Cohen, S.; Polyak, B. Magnetic Nanoparticle-Based Approaches to Locally Target Therapy and Enhance Tissue Regeneration in Vivo. *Nanomedicine* **2012**, *7*, 1425–1442. [CrossRef] [PubMed]

35. Rosenholm, J.M.; Zhang, J.; Linden, M.; Sahlgren, C. Mesoporous Silica Nanoparticles in Tissue Engineering—A Perspective. *Nanomedicine* **2016**, *11*, 391–402. [CrossRef] [PubMed]

36. Canham, L.T. Bioactive Silicon Structure Fabrication Through Nanoetching Techniques. *Adv. Mater.* **1995**, *7*, 1033–1037. [CrossRef]

37. Park, J.-H.; Gu, L.; von Maltzahn, G.; Ruoslahti, E.; Bhatia, S.N.; Sailor, M.J. Biodegradable Luminescent Porous Silicon Nanoparticles for in Vivo Applications. *Nat. Mater.* **2009**, *8*, 331–336. [CrossRef] [PubMed]

38. Erogbogbo, F.; Yong, K.-T.; Roy, I.; Xu, G.; Prasad, P.N.; Swihart, M.T. Biocompatible Luminescent Silicon Quantum Dots for Imaging of Cancer Cells. *ACS Nano* **2008**, *2*, 873–878. [CrossRef] [PubMed]

39. Gu, L.; Hall, D.J.; Qin, Z.; Anglin, E.; Joo, J.; Mooney, D.J.; Howell, S.B.; Sailor, M.J. In Vivo Time-Gated Fluorescence Imaging with Biodegradable Luminescent Porous Silicon Nanoparticles. *Nat. Commun.* **2013**, *4*, 2326. [CrossRef] [PubMed]

40. Gongalsky, M.B.; Osminkina, L.A.; Pereira, A.; Manankov, A.A.; Fedorenko, A.A.; Vasiliev, A.N.; Solovyev, V.V.; Kudryavtsev, A.A.; Sentis, M.; Kabashin, A.V.; et al. Laser-Synthesized Oxide-Passivated Bright Si Quantum Dots for Bioimaging. *Sci. Rep.* **2016**, *6*, 24732. [CrossRef] [PubMed]

41. Timoshenko, V.Y.; Kudryavtsev, A.A.; Osminkina, L.A.; Vorontsov, A.S.; Ryabchikov, Y.V.; Belogorokhov, I.A.; Kovalev, D.; Kashkarov, P.K. Silicon Nanocrystals as Photosensitizers of Active Oxygen for Biomedical Applications. *JETP Lett.* **2006**, *83*, 423–426. [CrossRef]

42. Lee, C.; Kim, H.; Hong, C.; Kim, M.; Hong, S.S.; Lee, D.H.; Lee, W.I. Porous Silicon as an Agent for Cancer Thermotherapy Based on near-Infrared Light Irradiation. *J. Mater. Chem.* **2008**, *18*, 4790–4795. [CrossRef]

43. Tamarov, K.P.; Osminkina, L.A.; Zinovyev, S.V.; Maximova, K.A.; Kargina, J.V.; Gongalsky, M.B.; Ryabchikov, Y.; Al-Kattan, A.; Sviridov, A.P.; Sentis, M.; et al. Radio Frequency Radiation-Induced Hyperthermia Using Si Nanoparticle-Based Sensitizers for Mild Cancer Therapy. *Sci. Rep.* **2014**, *4*, 7034. [CrossRef] [PubMed]

44. Sviridov, A.P.; Andreev, V.G.; Ivanova, E.M.; Osminkina, L.A.; Tamarov, K.P.; Timoshenko, V.Y. Porous Silicon Nanoparticles as Sensitizers for Ultrasonic Hyperthermia. *Appl. Phys. Lett.* **2013**, *103*, 193110. [CrossRef]

45. Balasubramanian, S.K.; Yang, L.; Yung, L.-Y.L.; Ong, C.-N.; Ong, W.-Y.; Yu, L.E. Characterization, Purification, and Stability of Gold Nanoparticles. *Biomaterials* **2010**, *31*, 9023–9030. [CrossRef] [PubMed]

46. Goodman, C.M.; McCusker, C.D.; Yilmaz, T.; Rotello, V.M. Toxicity of Gold Nanoparticles Functionalized with Cationic and Anionic Side Chains. *Bioconjug. Chem.* **2004**, *15*, 897–900. [CrossRef] [PubMed]

47. English, D.S.; Pell, L.E.; Yu, Z.; Barbara, P.F.; Korgel, B.A. Size Tunable Visible Luminescence from Individual Organic Monolayer Stabilized Silicon Nanocrystal Quantum Dots. *Nano Lett.* **2002**, *2*, 681–685. [CrossRef]

48. Ono, H.; Takahashi, K. Preparation of Silica Microcapsules by Sol-Gel Method in W/O Emulsion. *J. Chem. Eng. Jpn.* **1998**, *31*, 808–812. [CrossRef]

49. Zou, J.; Baldwin, R.K.; Pettigrew, K.A.; Kauzlarich, S.M. Solution Synthesis of Ultrastable Luminescent Siloxane-Coated Silicon Nanoparticles. *Nano Lett.* **2004**, *4*, 1181–1186. [CrossRef]

50. Kabashin, A.V.; Delaporte, P.; Grojo, D.; Torres, R.; Sarnet, T.; Sentis, M. Nanofabrication with Pulsed Lasers. *Nanoscale Res. Lett.* **2010**, *5*, 454–463. [CrossRef] [PubMed]

51. Marine, W.; Patrone, L.; Luk'yanchuk, B.; Sentis, M. Strategy of Nanocluster and Nanostructure Synthesis by Conventional Pulsed Laser Ablation. *Appl. Surface Sci.* **2000**, *154–155*, 345–352. [CrossRef]

52. Geohegan, D.B.; Puretzky, A.A.; Duscher, G.; Pennycook, S.J. Time-Resolved Imaging of Gas Phase Nanoparticle Synthesis by Laser Ablation. *Appl. Phys. Lett.* **1998**, *72*, 2987–2989. [CrossRef]

53. Patrone, L.; Nelson, D.; Safarov, V.I.; Sentis, M.; Marine, W.; Giorgio, S. Photoluminescence of Silicon Nanoclusters with Reduced Size Dispersion Produced by Laser Ablation. *J. Appl. Phys.* **2000**, *87*, 3829–3837. [CrossRef]

54. Kabashin, A.V.; Meunier, M.; Leonelli, R. Photoluminescence Characterization of Si-Based Nanostructured Films Produced by Pulsed Laser Ablation. *J. Vac. Sci. Technol. B Microelectron. Nanomed. Struct. Process. Meas. Phenom.* **2001**, *19*, 2217–2222. [CrossRef]

55. Kabashin, A.V.; Meunier, M. Visible Photoluminescence from Nanostructured Si-Based Layers Produced by Air Optical Breakdown on Silicon. *Appl. Phys. Lett.* **2003**, *82*, 1619–1621. [CrossRef]

56. Kabashin, A.V.; Meunier, M. Laser-Induced Treatment of Silicon in Air and Formation of Si/SiO$_x$ Photoluminescent Nanostructured Layers. *Mater. Sci. Eng. B* **2003**, *101*, 60–64. [CrossRef]

57. Kabashin, A.V.; Sylvestre, J.-P.; Patskovsky, S.; Meunier, M. Correlation between Photoluminescence Properties and Morphology of Laser-Ablated Si/SiO$_x$ Nanostructured Films. *J. Appl. Phys.* **2002**, *91*, 3248–3254. [CrossRef]

58. Fojtik, A.; Henglein, A. Laser Ablation of Films and Suspended Particles in Solvent-Formation of Cluster and Colloid Solutions. *Ber. Bunsenges. Phys. Chem.* **1993**, *97*, 252.

59. Sibbald, M.S.; Chumanov, G.; Cotton, T.M. Reduction of Cytochrome c by Halide-Modified, Laser-Ablated Silver Colloids. *J. Phys. Chem.* **1996**, *100*, 4672–4678. [CrossRef]

60. Mafuné, F.; Kohno, J.; Takeda, Y.; Kondow, T.; Sawabe, H. Formation of Gold Nanoparticles by Laser Ablation in Aqueous Solution of Surfactant. *J. Phys. Chem. B* **2001**, *105*, 5114–5120. [CrossRef]

61. Dolgaev, S.I.; Sinakin, A.V.; Vornov, V.V.; Shafeev, G.A.; Bozon-Verduaz, F. Nanoparticles Produced by Laser Ablation of Solids in Liquid Environment. *Appl. Surface Sci.* **2002**, *186*, 546–551. [CrossRef]

62. Kabashin, V.K.; Meunier, M. Synthesis of Colloidal Nanoparticles during Femtosecond Laser Ablation of Gold in Water. *J. Appl. Phys.* **2003**, *94*, 7941. [CrossRef]

63. Kabashin, V.K.; Meunier, M. Femtosecond Laser Ablation in Aqueous Solutions: A Novel Method to Synthesize Non-Toxic Metal Colloids with Controllable Size. *J. Phys. Conf. Ser.* **2007**, *59*, 354–359. [CrossRef]

64. Besner, S.; Degorce, J.-Y.; Kabashin, A.V.; Meunier, M. Influence of Ambient Medium on Femtosecond Laser Processing of Silicon. *Appl. Surface Sci.* **2005**, *247*, 163–168. [CrossRef]

65. Maximova, K.; Aristov, A.; Sentis, M.; Kabashin, A.V. Size-Controllable Synthesis of Bare Gold Nanoparticles by Femtosecond Laser Fragmentation in Water. *Nanotechnology* **2015**, *26*, 065601. [CrossRef] [PubMed]

66. Blandin, P.; Maximova, K.A.; Gongalsky, M.B.; Sanchez-Royo, J.F.; Chirvony, V.S.; Sentis, M.; Timoshenko, V.Y.; Kabashin, A.V. Femtosecond Laser Fragmentation from Water-Dispersed Microcolloids: Toward Fast Controllable Growth of Ultrapure Si-Based Nanomaterials for Biological Applications. *J. Mater. Chem. B* **2013**, *1*, 2489–2495. [CrossRef]

67. Al-Kattan, A.; Ryabchikov, Y.V.; Baati, T.; Chirvony, V.; Sánchez-Royo, J.F.; Sentis, M.; Braguer, D.; Timoshenko, V.Y.; Estève, M.-A.; Kabashin, A.V. Ultrapure Laser-Synthesized Si Nanoparticles with Variable Oxidation States for Biomedical Applications. *J. Mater. Chem. B* **2016**, *4*, 7852–7858. [CrossRef]

68. Hebié, S.; Holade, Y.; Maximova, K.; Sentis, M.; Delaporte, P.; Kokoh, K.B.; Napporn, T.W.; Kabashin, A.V. Advanced Electrocatalysts on the Basis of Bare Au Nanomaterials for Biofuel Cell Applications. *ACS Catal.* **2015**, *5*, 6489–6496. [CrossRef]

69. Sylvestre, J.P.; Kabashin, A.V.; Sacher, E.; Meunier, M.; Luong, J.H.T. Stabilization and Size Control of Gold Nanoparticles during Laser Ablation in Aqueous Cyclodextrins. *J. Am. Chem. Soc.* **2004**, *126*, 7176–7177. [CrossRef] [PubMed]

70. Kabashin, A.V.; Meunier, M.; Kingston, C.; Luong, J.H.T. Fabrication and Characterization of Gold Nanoparticles by Femtosecond Laser Ablation in an Aqueous Solution of Cyclodextrins. *J. Chem. Phys. B* **2003**, *107*, 4527–4531. [CrossRef]

71. Sylvestre, J.; Poulin, S.; Kabashin, A.V.; Sacher, E.; Meunier, M.; Luong, J.H.T. Surface Chemistry of Gold Nanoparticles Produced by Laser Ablation in Aqueous Media. *J. Phys. Chem. B* **2004**, *108*, 16864–16869. [CrossRef]

72. Correard, F.; Maximova, K.; Estève, M.-A.; Villard, C.; Roy, M.; Al-Kattan, A.; Sentis, M.; Gingras, M.; Kabashin, A.V.; Braguer, D. Gold Nanoparticles Prepared by Laser Ablation in Aqueous Biocompatible Solutions: Assessment of Safety and Biological Identity for Nanomedicine Applications. *Int. J. Nanomed.* **2014**, *9*, 5415–5430.

Int. J. Mol. Sci. **2018**, *19*, 1563

73. Al-kattan, A.; Nirwan, V.P.; Munnier, E.; Chourpa, I. Toward Multifunctional Hybrid Platforms for Tissue Engineering Based on Chitosan (PEO) Nanofibers Functionalized by Bare Laser-Synthesized Au and Si Nanoparticles. *RSC Adv.* **2017**, *7*, 31759–31766. [CrossRef]

74. Zhang, D.; Gökce, B.; Barcikowski, S. Laser Synthesis and Processing of Colloids: Fundamentals and Applications. *Chem. Rev.* **2017**, *117*, 3990–4103. [CrossRef] [PubMed]

75. Dell'Aglio, M.; Gaudiuso, R.; Pascale, O.D.; Giacomo, A.D. Mechanisms and Processes of Pulsed Laser Ablation in Liquids during Nanoparticle Production. *Appl. Surface Sci.* **2015**, *348*, 4–9. [CrossRef]

76. Carlisle, E.M. Silicon: A Requirement in Bone Formation Independent of Vitamin D1. *Calcif. Tissue Int.* **1981**, *33*, 27–34. [CrossRef] [PubMed]

77. Jugdaohsingh, R. Silicon and bone health. *J. Nutr. Health Aging* **2007**, *11*, 99–110. [PubMed]

78. Emerick, R.J.; Kayongo-Male, H. Interactive Effects of Dietary Silicon, Copper, and Zinc in the Rat. *J. Nutr. Biochem.* **1990**, *1*, 35–40. [CrossRef]

79. Neiner, D.; Chiu, H.W.; Kauzlarich, S.M. Low-Temperature Solution Route to Macroscopic Amounts of Hydrogen Terminated Silicon Nanoparticles. *J. Am. Chem. Soc.* **2006**, *128*, 11016–11017. [CrossRef] [PubMed]

80. Tilley, R.D.; Yamamoto, K. The Microemulsion Synthesis of Hydrophobic and Hydrophilic Silicon Nanocrystals. *Adv. Mater.* **2006**, *18*, 2053–2056. [CrossRef]

81. Cabrera, Z.Y.; Aceves-Mijares, M.; Cabrera, M.A.I. Single Electron Charging and Transport in Silicon Rich Oxide. *Nanotechnology* **2006**, *17*, 3962.

82. Arul Dhas, N.; Raj, C.P.; Gedanken, A. Preparation of Luminescent Silicon Nanoparticles: A Novel Sonochemical Approach. *Chem. Mater.* **1998**, *10*, 3278–3281. [CrossRef]

83. Baati, T.; Al-kattan, A.; Esteve, M.; Njim, L.; Ryabchikov, Y.; Chaspoul, F.; Hammami, M.; Sentis, M.; Kabashin, A.V.; Braguer, D. Ultrapure Laser-Synthesized Si- Based Nanomaterials for Biomedical Applications: In Vivo Assessment of Safety and Biodistribution. *Sci. Rep.* **2016**, *6*, 1–13. [CrossRef] [PubMed]

84. Chen, H.; Kou, X.; Yang, Z.; Ni, W.; Wang, J. Shape- and Size-Dependent Refractive Index Sensitivity of Gold Nanoparticles. *Langmuir* **2008**, *24*, 5233–5237. [CrossRef] [PubMed]

85. Lee, K.-S.; El-Sayed, M.A. Gold and Silver Nanoparticles in Sensing and Imaging: Sensitivity of Plasmon Response to Size, Shape, and Metal Composition. *J. Phys. Chem. B* **2006**, *110*, 19220–19225. [CrossRef] [PubMed]

86. Liedberg, B.; Nylander, C.; Lundström, I. Biosensing with surface plasmon resonance—How it all started. *Biosens. Bioelectron.* **1995**, *10*, 1–9. [CrossRef]

87. Anker, J.N.; Hall, W.P.; Lyandres, O.; Shah, N.C.; Zhao, J.; Van Duyne, R.P. Biosensing with plasmonic nanosensors. *Nat. Mater.* **2008**, *7*, 442–453. [CrossRef] [PubMed]

88. Patskovsky, S.; Kabashin, A.V.; Meunier, M.; Luong, J.H.T. Silicon-based surface plasmon resonance sensing with two surface plasmon polariton modes. *Appl. Opt.* **2003**, *42*, 6905–6909. [CrossRef] [PubMed]

89. Nemova, G.; Kabashin, A.V.; Kashyap, R. Surface plasmon-polariton Mach-Zehnder refractive index sensor. *J. Opt. Soc. Am. B* **2008**, *25*, 1673–1677. [CrossRef]

90. Wang, Y.; Xie, X.; Wang, X.; Ku, G.; Gill, K.L.; O'Neal, D.P.; Stoica, G.; Wang, L.V. Photoacoustic Tomography of a Nanoshell Contrast Agent in the in Vivo Rat Brain. *Nano Lett.* **2004**, *4*, 1689–1692. [CrossRef]

91. Yguerabide, J.; Yguerabide, E.E. Light-Scattering Submicroscopic Particles as Highly Fluorescent Analogs and Their Use as Tracer Labels in Clinical and Biological Applications. *Anal. Biochem.* **1998**, *262*, 137–156. [CrossRef] [PubMed]

92. Hirsch, L.R.; Stafford, R.J.; Bankson, J.A.; Sershen, S.R.; Rivera, B.; Price, R.E.; Hazle, J.D.; Halas, N.J.; West, J.L. Nanoshell-mediated near-infrared thermal therapy of tumors under magnetic resonance guidance. *Proc. Natl. Acad. Sci. USA* **2003**, *100*, 13549–13554. [CrossRef] [PubMed]

93. Loo, C.; Lowery, A.; Halas, N.J.; West, J.; Drezek, R. Immunotargeted nanoshells for integrated cancer imaging and therapy. *Nano Lett.* **2005**, *5*, 709–711. [CrossRef] [PubMed]

94. Huang, X.; El-Sayed, I.H.; Qian, W.; El-Sayed, M.A. Cancer cell imaging and photothermal therapy in near-infrared region by using gold nanorods. *J. Am. Chem. Soc.* **2006**, *128*, 2115–2120. [CrossRef] [PubMed]

95. Lipka, J.; Semmler-Behnke, M.; Sperling, R.A.; Wenk, A.; Takenaka, S.; Schleh, C.; Kissel, T.; Parak, W.J.; Kreyling, W.G. Biodistribution of PEG-Modified Gold Nanoparticles Following Intratracheal Instillation and Intravenous Injection. *Biomaterials* **2010**, *31*, 6574–6581. [CrossRef] [PubMed]

96. Cho, W.-S.; Cho, M.; Jeong, J.; Choi, M.; Han, B.S.; Shin, H.-S.; Hong, J.; Chung, B.H.; Jeong, J.; Cho, M.-H. Size-Dependent Tissue Kinetics of PEG-Coated Gold Nanoparticles. *Toxicol. Appl. Pharmacol.* **2010**, *245*, 116–123. [CrossRef] [PubMed]

97. Lee, S.H.; Bae, K.H.; Kim, S.H.; Lee, K.R.; Park, T.G. Amine-Functionalized Gold Nanoparticles as Non-Cytotoxic and Efficient Intracellular siRNA Delivery Carriers. *Int. J. Pharm.* **2008**, *364*, 94–101. [CrossRef] [PubMed]

98. Wangoo, N.; Bhasin, K.K.; Mehta, S.K.; Suri, C.R. Synthesis and Capping of Water-Dispersed Gold Nanoparticles by an Amino Acid: Bioconjugation and Binding Studies. *J. Colloid Interface Sci.* **2008**, *323*, 247–254. [CrossRef] [PubMed]

99. Javier, D.J.; Nitin, N.; Levy, M.; Ellington, A.; Richards-Kortum, R. Aptamer-Targeted Gold Nanoparticles As Molecular-Specific Contrast Agents for Reflectance Imaging. *Bioconjug. Chem.* **2008**, *19*, 1309–1312. [CrossRef] [PubMed]

100. Sun, L.; Liu, D.; Wang, Z. Functional Gold Nanoparticle−Peptide Complexes as Cell-Targeting Agents. *Langmuir* **2008**, *24*, 10293–10297. [CrossRef] [PubMed]

101. Munday, R. Toxicity of Thiols and Disulphides: Involvement of Free-Radical Species. *Free Radic. Biol. Med.* **1989**, *7*, 659–673. [CrossRef]

102. Pernodet, N.; Fang, X.; Sun, Y.; Bakhtina, A.; Ramakrishnan, A.; Sokolov, J.; Ulman, A.; Rafailovich, M. Adverse Effects of Citrate/gold Nanoparticles on Human Dermal Fibroblasts. *Small* **2006**, *2*, 766–773. [CrossRef] [PubMed]

103. Aubin-Tam, M.-E.; Hamad-Schifferli, K. Gold Nanoparticle-Cytochrome C Complexes: The Effect of Nanoparticle Ligand Charge on Protein Structure. *Langmuir* **2005**, *21*, 12080–12084. [CrossRef] [PubMed]

104. Etchegoin, P.G.; Le Ru, E.C. A Perspective on Single Molecule SERS: Current Status and Future Challenges. *Phys. Chem. Chem. Phys.* **2008**, *10*, 6079–6089. [CrossRef] [PubMed]

105. Deyao, K.; Tao, P.; Goosen, M.F.A.; Min, J.M.; He, Y.Y. Ph-Sensitivity of Hydrogels Based on Complex-Forming Chitosan—Polyether Interpenetrating Polymer Network. *J. Appl. Polym. Sci.* **1993**, *48*, 343–354.

106. Ma, Z.; Kotaki, M.; Yong, T.; He, W.; Ramakrishna, S. Surface Engineering of Electrospun Polyethylene Terephthalate (PET) Nanofibers towards Development of a New Material for Blood Vessel Engineering. *Biomaterials* **2005**, *26*, 2527–2536. [CrossRef] [PubMed]

107. Agarwal, S.; Wendorff, J.H.; Greiner, A. Use of Electrospinning Technique for Biomedical Applications. *Polymer* **2008**, *49*, 5603–5621. [CrossRef]

108. Chen, M.; Patra, P.K.; Warner, S.B.; Bhowmick, S. Role of Fiber Diameter in Adhesion and Proliferation of NIH 3T3 Fibroblast on Electrospun Polycaprolactone Scaffolds. *Tissue Eng.* **2007**, *13*, 579–587. [CrossRef] [PubMed]

109. Kabashin, A.V.; Timoshenko, V.Y. What theranostic applications could ultrapure laser-synthesized Si nanoparticles have in cancer? *Nanomedicine* **2016**, *11*, 2247–2250. [CrossRef] [PubMed]

110. Kögler, M.; Ryabchikov, Y.V.; Uusitalo, S.; Popov, A.; Popov, A.; Tselikov, G.; Välimaa, A.-L.; Al-Kattan, A.; Hiltunen, J.; Laitinen, R.; et al. Bare Laser-Synthesized Au-Based Nanoparticles as Non-Disturbing SERS Probes for Bacteria Identification. *J. Biophotonics* **2018**. [CrossRef] [PubMed]

111. Bibikova, O.; Haas, J.; Lopez-Lorente, A.; Popov, A.; Yury, R.; Kinnunen, M.; Kabashin, A.V.; Meglinski, I.; Mizaikoff, B. Surface enhanced infrared absorption spectroscopy based on gold nanostars and spherical nanoparticles. *Anal. Chimica Acta* **2017**, *990*, 141–149. [CrossRef] [PubMed]

International Journal of
Molecular Sciences

MDPI

Article

Poly(Lactic Acid) Blends with Poly(Trimethylene Carbonate) as Biodegradable Medical Adhesive Material

Shuang Zhang [†], Hongli Li [†], Mingwei Yuan, Minglong Yuan * and Haiyun Chen *

Engineering Research Center of Biopolymer Functional Materials of Yunnan, Yunnan Minzu University, Kunming 650500, China; zhshu@deakin.edu.au (S.Z.); honglili_1982@163.com (H.L.); yuanmingwei@163.com (M.Y.)
* Correspondence: yml@ynni.edu.cn (M.Y.); chenhy1960@163.com (H.C.); Tel.: +86-871-6591-4825 (M.Y. & H.C.)
† These authors contributed equally to this work.

Received: 16 August 2017; Accepted: 13 September 2017; Published: 28 September 2017

Abstract: A novel medical adhesive was prepared by blending poly(lactic acid) (PLA) with poly(trimethylene carbonate) (PTMC) in ethyl acetate, and the two materials were proven to be biodegradable and biocompatible. The medical adhesive was characterized by ^{1}HNMR nuclear magnetic resonance (^{1}HNMR), gel permeation chromatography (GPC), scanning electron microscopy (SEM) and differential scanning calorimetry (DSC). The water vapor transmission rate (WVTR) of this material was measured to be 7.13 g·cm^{-2}·24 h^{-1}. Its degree of comfortability was confirmed by the extensibility (E) and the permanent set (PS), which were approximately 7.83 N·cm^{-2} and 18.83%, respectively. In vivo tests regarding rabbit immunoglobulin M (IgM), rabbit immunoglobulin G (IgG), rabbit bone alkaline phosphatase (BALP), rabbit interleukin 6 (IL-6), rabbit interleukin 10 (IL-10), rabbit tumor necrosis factor α(TNFα), glutamic-oxaloacetic transaminase (AST/GOT), glutamic-pyruvic transaminase (ALT/GPT), alkaline phosphatase (AKP), blood urea nitrogen (BUN) and creatinine (Cr) indicated that the PLA-PTMC medical adhesive was not harmful to the liver and kidneys. Finally, pathological sections indicated that PLA-PTMC was more effective than the control group. These data suggest that in addition to having a positive effect on hemostasis and no sensibility to wounds, PLA-PTMC can efficiently prevent infections and has great potential as a medical adhesive.

Keywords: Poly(Lactic Acid)-Poly(Trimethylene Carbonate); medical adhesive; avoid infection; hemostasis

1. Introduction

Each year, millions of people suffer from many types of wounds, including traumatic or surgical wounds, which require a proper closure. To develop better materials for the treatment of wounds, the translation of modern biomaterials from the lab to the clinical hospital will permit an in-depth understanding of how biomaterials interact with biological systems at both cellular and molecular levels, with the ultimate purpose of creating more suitable materials and products [1–7].

Previously, there have been many techniques for curing a wide range of wounds, such as sutures and staples. However, although sutures can provide great tensile strength and show relatively low failure rates [8,9], the disadvantages of suturing are that it is time consuming, is not always technically possible, thus requiring anesthesia, and induces undesirable scar formation [9,10]. Staples easily damage surrounding tissues while evoking an inflammatory response and causing scar tissue formation. Most importantly, the use of staples also results in a significant failure rate [8,9,11]. Robust adhesion and cohesive integrity are valuable characteristics of a medical adhesive, especially for applications requiring a long-term performance of the material. However, the limitations of currently-approved synthetic adhesives include poor adhesion in the presence of biological fluids, sensitization, an allergic response and inflammation. They can always cause problematic infections, which can cause many types of diseases. Thus, development of new biomaterials with a low cost, low toxicity and less infection is quite essential.

Poly(lactic acid) (PLA), apart from being derived from renewable resources (e.g., corn, whey [12,13], wheat and rice), is biodegradable, recyclable and compostable [14,15]. Its production also consumes carbon dioxide [16]. Due to its ability to be degraded and assimilated inside the human body within a few months, its first applications were in the biomedical field [17–20]. Moreover, PLA degradation products are non-toxic at a lower composition, making it a natural choice for biomedical applications [21,22]. Furthermore, PLA requires 25–55% less energy to produce than do petroleum-based polymers, and estimations indicate that this can be reduced further to less than 10% in the future [23]. Although it is an ideal biomaterial with biocompatibility and biodegradability, PLA is a very brittle material with less than 10% elongation at break [24,25], which largely limits its practical use in the medical field.

Blending of PLA with other high strength materials is one of the most extensively-used methods to improve the mechanical properties of the material. To overcome the disadvantages mentioned above, PLA has been blended with some biodegradable polymers such as poly(para-dioxanone) [26], poly(propylene carbonate) [27], poly(butylene succinate) and derivatives [28,29] to improve its mechanical properties, especially toughness. Among these, PTMC is an amorphous polymer with a glass transition at 12 °C with low crystallinity and high mechanical properties [30], and it has good biocompatibility and biodegradability [31,32]. High-molecular-weight PTMC maintains elastic properties at ambient temperature [33–35], and the mechanical performance of PLA can be improved efficiently.

Therefore, the aim of this study was to produce and characterize PLA-based, high-performance biocompatible composite materials for biomaterial applications such as medical adhesives. This novel synthesized composite contains different percentages of PTMC to enhance the properties of the polymer. Medical adhesives with different ratios of PLA and PTMC were first produced and characterized by ^{1}H nuclear magnetic resonance (^{1}HNMR), gel permeation chromatography (GPC), scanning electron microscopy (SEM) and differential scanning calorimetry (DSC). The film-forming time, water vapor transmission rate, degree of comfortability and water contact angle were measured afterwards.

A ratio of 7:3 of PLA:PTMC was found to have the most suitable properties, and animal experiments were conducted to ensure the performance of this product. In the experiment, 10 rabbits with similar weights were separated into two groups: the PLA-PTMC group and the positive control group. Then, a wound model was created and observed, and the wounds' status was recorded until the wounds recovered. During this period, glutamic-oxaloacetic transaminase (AST/GOT), glutamic-pyruvic transaminase (ALT/GPT), alkaline phosphatase (AKP), blood urea nitrogen (BUN) and creatinine (Cr) were measured to determine the influence of the medical adhesives on hepatic and renal tissues, and enzyme-linked immunosorbent assay (ELISA) assays were performed to determine the content of immunoglobulin M (IgM), rabbit immunoglobulin G (IgG), rabbit bone alkaline phosphatase (BALP), rabbit interleukin 6 (IL-6), rabbit interleukin 10 (IL-10) and rabbit tumor necrosis factor α(TNFα). The results suggests that this type of biomaterial can efficiently reduce infections because it can form a thin and ventilated film quickly on the surface of the skin and promote wound repair, with less harm to other organs.

2. Results and Discussion

2.1. Characterization of Medical Adhesive Films

To characterize the properties of the medical adhesive and choose the most suitable ratio of PLA/PTMC, poly(lactic acid) (PLA, Mw = 280 kDa, Mw/Mn = 1.98) was modified with poly(trimethylene carbonate) (PTMC, Mw = 100 kDa, Mw/Mn = 1.70) in different ratios of 9:1,

8:2 and 7:3 to facilitate their use in biomaterial studies. A solvent evaporation technique with 50 mL of chloroform was used to prepare thin films to be characterized by ^{1}HNMR, GPC and SEM (from Figures 1–3 and Table 1). From the GPC data, when the content of PTMC increased, the polydispersity index (PDI) of the blends decreased (see Table 1). The raw polymer and blends exhibited high molar masses ranging from 98,972–77,253, and the PDI of the blends ranged from 2.05–1.85. The chemical compositions of the blends were determined by ^{1}HNMR analysis. As shown in Figure 1, Peak 1 and Peak 4 are assigned to lactyl CH and CH$_3$ groups, respectively, and Peak 2 and Peak 3 are assigned to the trimethylene carbonate (TMC) CH$_2$ group. The different ratios of PLA and PTMC are presented with the different peak intensities. SEM was used to track changes in the film surface morphology among different blends. Figure 2 shows the SEM photographs of A, B, C and D; with an increased scale of PLA-PTMC, the composite looked smoother than pure PLA; namely, the blends became better distributed when the ratio of PLA:PTMC was 7:3 [36]. Figure 3 shows the DSC thermograms of PLA and PLA-PTMC blends. The glass transition (Tg) of the PLA was observed at approximately 58 °C; the same result was achieved in Martin's work [37]. This is consistent with PLA having a Tg between 50 and 80 °C [38–40]. With increasing PTMC content of the polyurethane, the Tg of the polymers decreased. When the ratio of PLA:PTMC was 7:3, the Tg was measured as 38.8 °C, which is lower compared with pure PLA, due to the distribution of PTMC in the film. Because the glass transition of PTMC is around −20 °C [32,41,42], the addition of PTMC made the glass transition of PLA decrease.

Table 1. Molecular characteristics of poly(lactic acid) (PLA) and PLA/PTMC (poly(lactic acid)/poly(trimethylene carbonate)) blends with different ratios.

Groups	PLA/PTMC [a]	Solvent	Temperature (°C)/Time (h)	Mn [b]	PDI [b]
A	10:0	ethyl acetate	25/6	98,972	2.052
B	9:1	ethyl acetate	25/6	93,193	1.957
C	8:2	ethyl acetate	25/6	89,888	1.871
D	7:3	ethyl acetate	25/6	77,253	1.846

[a] Calculated from ^{1}HNMR data using CDCl$_3$ as the solvent; [b] determined from GPC data. PDI: polydispersity index.

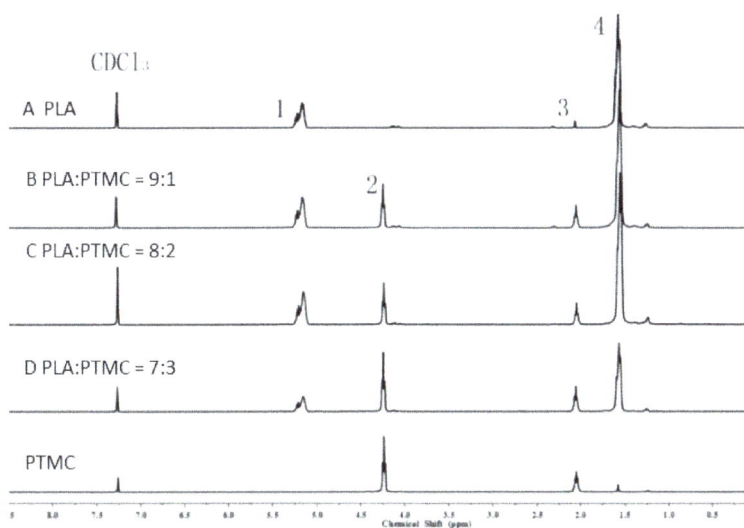

Figure 1. [1]H nuclear magnetic resonance (^{1}HNMR) spectra of poly(lactic acid) (PLA) and poly(trimethylene carbonate) (PTMC) and their blends.

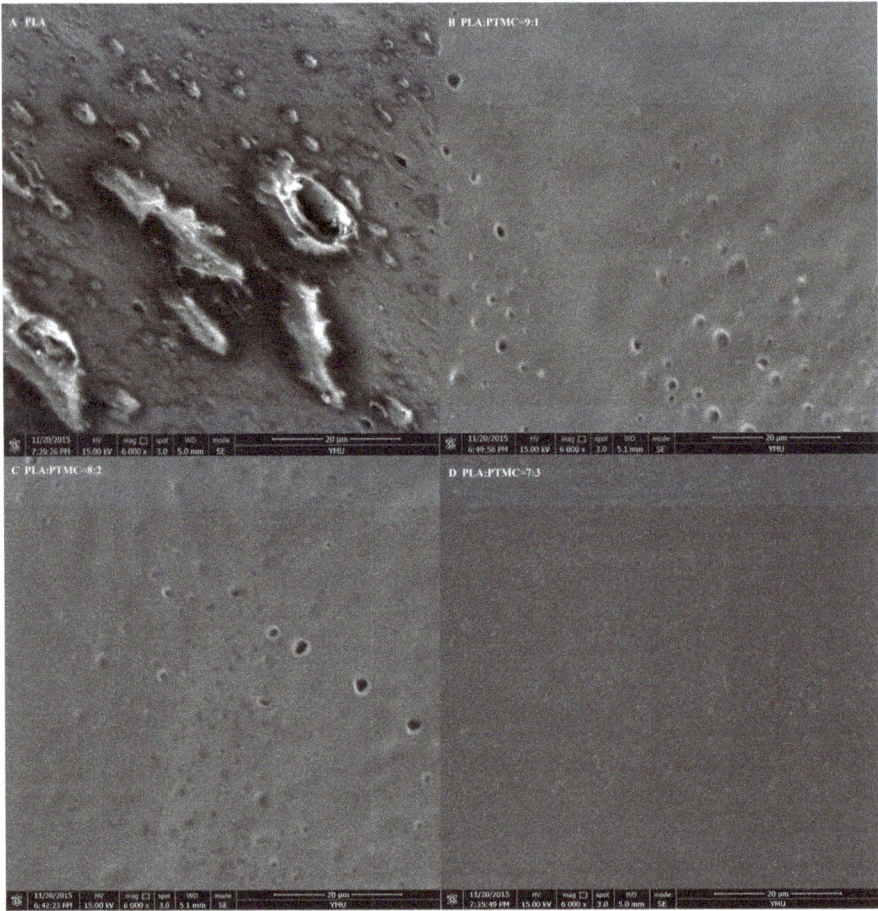

Figure 2. SEM micrographs of medical adhesive with different ratios of PLA-PTMC. (**A**) PLA; (**B**) PLA:PTMC = 9:1; (**C**) PLA:PTMC = 8:2; (**D**) PLA:PTMC = 7:3. Scale bar: 20 μm.

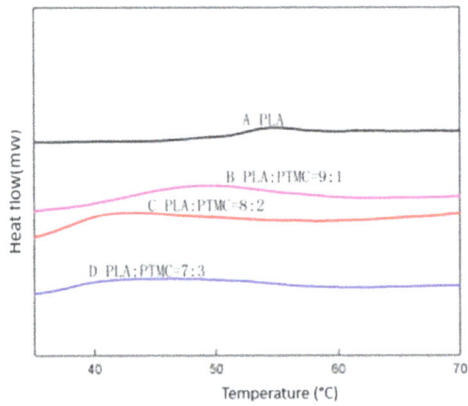

Figure 3. Differential scanning calorimetry (DSC) of different ratios of PLA-PTMC.

2.2. Evaluation of Physical Properties

The physical properties of blends of different ratios of PLA/PTMC were evaluated using film formation time, water vapor transmission rate, the comfortability degree and the contact angle. With the addition of PTMC, the film formation time increased sharply; however, when the ratio of PLA-TMC decreased from 8:2–7:3, the film formation time increased slightly, from 60.33–62.33 min. Figure 4 shows that the water vapor transmission rate of the sample rose sharply, from 4.59–7.13 g·cm^{-2}·24 h^{-1} with a ratio change from 8:2–7:3. Additionally, the water vapor permeability of the films is better than the PLA film reported in the literature [36,43]. The result that air permeability increased may be caused by the increasing content of PTMC in the composite. Additionally, based on the extensibility (E) and permanent set (PS), the comfortability of the material was evaluated. Just as Figure 5 presented, with the increasing content of PTMC in the composite, the comfortability of the medical adhesive was improved. Additionally, when the ratio of PLA:PTMC went up to 7:3, the comfortability of the material rose to the highest, and the extensibility and permanent set reached 7.83 N·cm^{-2} and 18.83%, respectively. This indicates that a ratio of PLA: PTMC of 7:3 was the optimal choice for a medical adhesive. The contact angle indicates that when the ratio of PLA-PTMC reached 7:3, the hydrophobicity was the highest, which is desirable for preventing wound infections. This is presented in Table 2.

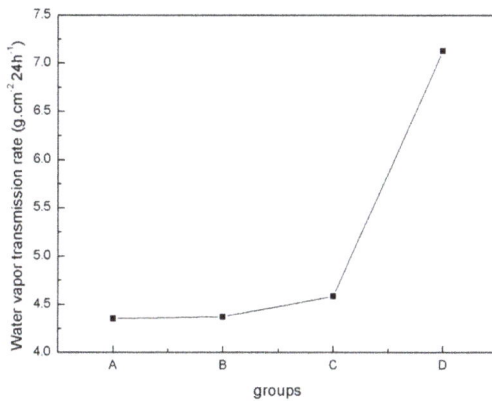

Figure 4. Water vapor transmission rate determined using the cup method.

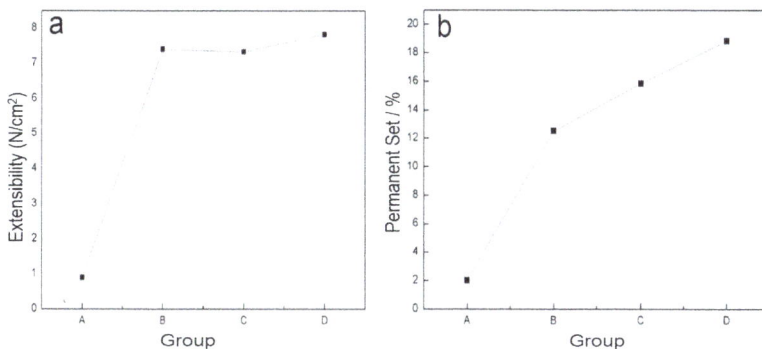

Figure 5. The comfortability of the blends was assigned through extensibility (**a**) and permanent set (**b**).

Table 2. Contact angle measurement.

Group	Contact Angle (Degree)
PLA	72.45
PLA-PTMC 9:1	74.96
PLA-PTMC 8:2	80.75
PLA-PTMC 7:3	87.38

2.3. Biological Performance

After testing the physical and chemical properties, the optimal sample was chosen for animal experiments. The sterility test of PLA/PTMC at a ratio of 7:3 demonstrated no obvious colony formations after 36 h, and the cytotoxicity of the medical adhesive was evaluated using 3-(4,5-dimethylthiazol-2-yl)-2,5-diphenyl tetrazolium bromide (MTT). The IC_{50} of a PLA/PTMC medical adhesive with a ratio of 7:3 was determined to be 6.24 µg/mL, according to linear regression analysis of the information provided in Figure 6. In some research, the IC_{50} of some polymers was smaller than 10 µg/mL [44,45]. When comparing with the positive group, the IC_{50} value was smaller. That is to say, the growth rate was higher in the experiment group than the positive group as presented in Figure 6. Kulkarni reported that PLA was non-toxic, and Schakenraad presented that PLA was tissue compatible [23,46]. Furthermore, after the ears were covered with the medical adhesive for 2, 4, 8, 12 and 24 h, the ears of the rabbits did not display any aberrant phenomenon. These preliminary results suggest that this type of medical adhesive can be used as a biomaterial.

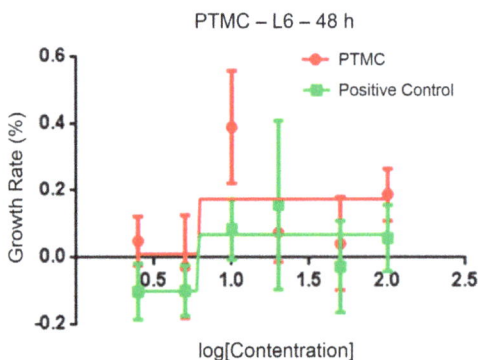

Figure 6. Cytotoxicity of the medical adhesive determined by 3-(4,5-dimethylthiazol-2-yl)-2,5-diphenyl tetrazolium bromide (MTT).

2.4. The Result of Animal Experiments

The result of the skin irritability tests suggests that PLA-PTMC as a medical adhesive does not trigger allergies. According to the data (from the first day of establishing the wound model to the last day of wound recovery), the weight of the rabbits showed an increasing trend during the experiments, as shown in Figure 7a, particularly in the PLA-PTMC group, with a 0.16 kg average. Figure 7b shows that one fever occurred during the experimental process with PLA-PTMC and the positive control group. However, this phenomenon lasted only one day.

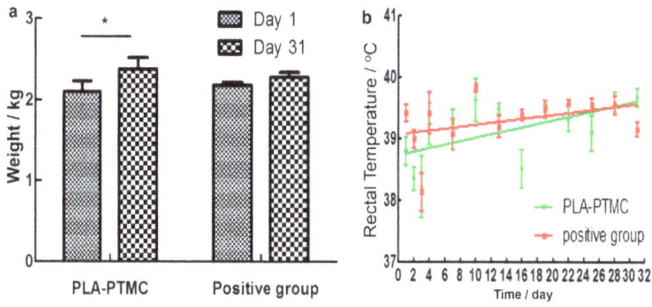

Figure 7. Weight and temperature changes of the rabbits during the experiments. (**a**) Presents the change of weight during the experiment; (**b**) shows the rectal temperature of the rabbits. "*" present there exists statistic difference.

Granulation tissue was observed to grow on the sixteenth day after the operation. The wounds in the PLA-PTMC group were completely healed on the 30th day, but the positive control group had residual wounds of 0.4 cm × 0.3 cm that had not healed fully. Figure 8a–c are images of the wounds that were covered by PLA-PTMC after 1, 20 and 31 days after the operation, respectively, and Figure 8d–f are images of the positive group. Figure 9 shows the wound healing rate from the 13th–31st day with PLA-PTMC and the positive group. The wound healing rate for the PLA-PTMC group was significantly higher than that of the positive group (n = 5).

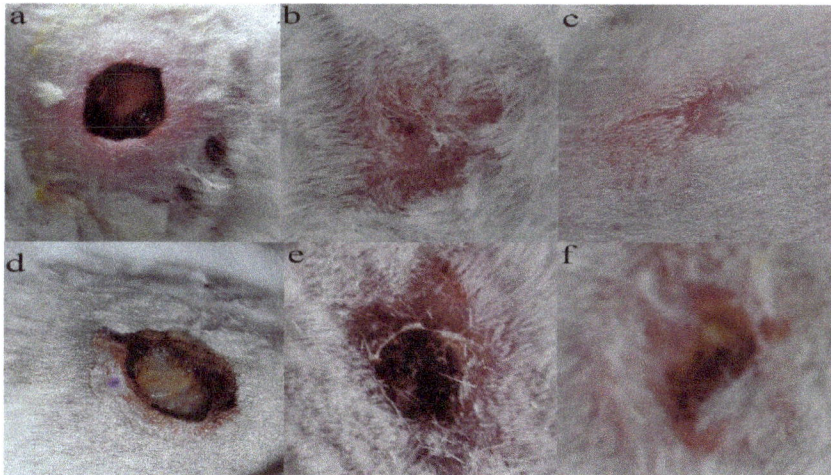

Figure 8. The records of the wound during different periods after being covered by PLA-PTMC and the positive group medical adhesive. (**a–c**) represent the state of the wound covered by PLA-PTMC medical adhesive after 1, 20 and 31 days, and (**d–f**) show the situation of the positive group at the same time, respectively.

The influence on hepatic functions of rabbits in the PLA-PTMC and the positive control groups was evaluated based on the serum levels of ALT/GP, AST/GOT, AKP, BNU and creatinine, which were determined using the appropriate kits. The statistics suggested that AKP activity and BNU concentrations were significantly different for the PLA-PTMC medical adhesive, whereas the positive medical adhesive and other indicators produced no obvious significant difference.

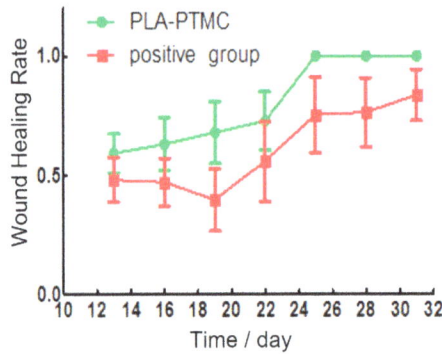

Figure 9. The wound healing rate of the PLA-PTMC group and the positive control group.

Figure 10a shows the AKP activity using PLA-PTMC and a positive medical adhesive, and Figure 10b presents the BUN content. BUN is used as a surrogate marker of neurohormonal activation, and AKP has been employed as one indicator in the very sensitive enzyme-linked immunosorbent assay (ELISA) [46]; it is one of the most commonly-assayed enzymes in clinical practice to diagnose different types of diseases [47]. Figure 10a shows that the activity of AKP in the experiment was lower than the positive group, around 30 U/100 mL. From Figure 10b, the BUN content of the experimental group, around 10 mmol/L, is smaller than the positive group, which is about 12 mmol/L, in a normal range 5–15 mmol/L [48]. In each case, the PLA-PTMC results were lower than those of the positive group.

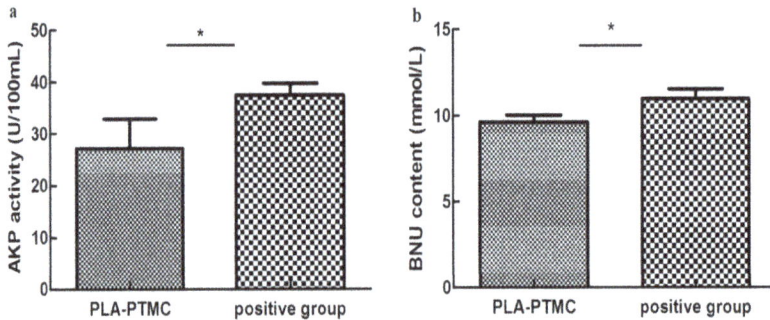

Figure 10. The AKP (alkaline phosphatase) activity (**a**) and the blood urea nitrogen (BUN) content (**b**) of PLA-PTMC and the positive group. Values are the means ± S.E.M. ($n = 5$). * $p < 0.05$ compared with the control group; * $p < 0.01$ compared with the control group; * $p > 0.05$ was considered not to be statistically different. These indicators were detected using different kits that were obtained from the Nanjing Jiancheng biological technology limited company (Nanjing, China).

The immunogenic specificity of the materials was measured by ELISA as described previously [49,50]. The amounts of the cytokines TNF-α, IL-6 and IL-10 and IgG, IgM and BALP in the rabbit sera were determined using double-antibody sandwich ELISAs, and only IL-10 and IgG appeared to be statistically different between the PLA-PTMC and positive groups. IgG concentration reached to around 60 pg/mL, and IL-10 concentration was about 160 pg/mL. IL-10 is a pleiotropic cytokine that plays an important role in the development of inflammation and immune response and disease. Figure 11 shows that the expression of IL-10 and IgG in the PLA-PTMC group was higher than that in the positive control group [51]. To some degree, the experimental group was better than the positive group in terms of avoiding infection.

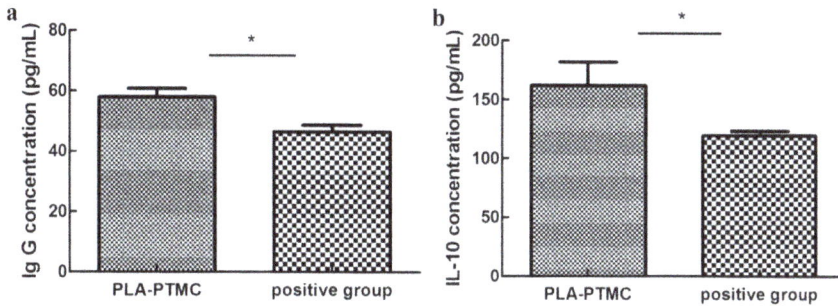

Figure 11. The results of ELISA testing. (**a**) is the difference of rabbit immunoglobulin G (IgG), and (**b**) is the statistical difference of rabbit interleukin 10 (IL-10). Values are the means $\pm$ S.E.M. ($n = 5$). * $p < 0.05$ compared with the control group; * $p > 0.05$ was considered not to be statistically different. These indicators were detected using kits that were obtained from CUSABIO (CUSABIO, Wuhan, China).

During the experiments, three of the five samples in the positive group produced wound infections, which attracted our attention. After completing the initial tests, skin pathological section analyses were conducted. The tissues were excised immediately, and adhering tissues were trimmed and fixed with 4% paraformaldehyde for 24 h for the histological studies. The specimens were embedded in paraffin, and the sections were stained with hematoxylin-eosin (HE) for light microscopic observations. The sections were examined and photographed using an Olympus CX-31 Microscope (Olympus Corporation, Shinjuku, Tokyo, Japan). Three areas on each slide were chosen randomly for microscopic examination. The slides were further examined and evaluated blindly by two investigators. The images in Figure 12 indicate that there was apparent re-epithelialization that extended sufficiently to cover the wound in the PLA-PTMC group.

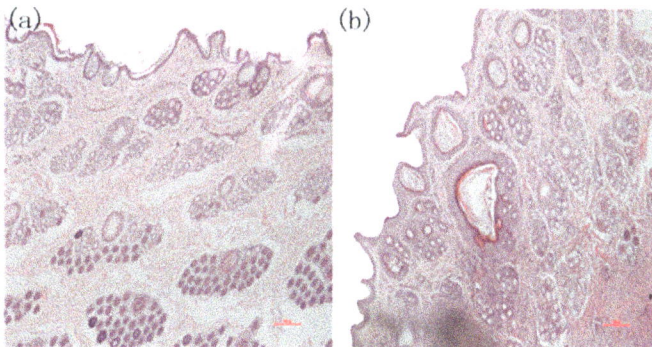

Figure 12. The microscopic state of the recovered wound tissue. (**a**) Skin pathological section from the PLA-PTMC group; (**b**) skin pathological section from the positive group.

Figure 12 is the photomicrograph of the PLA-PTMC medical adhesive and control sites 31 days after surgery ($\times$100). Figure 12a shows that the intact cellular structure and the fibroblastic activity of the wound in the PLA-PTMC group had both recovered 31 days after the surgery. Obviously, the positive control group had more viral factors than did the PLA-PTMC group. This was be confirmed by Majola, who suggested that PLA has no inflammation or foreign body reaction through experiments on rats [52,53], and Von Schroeder, who indicated that PLA was well tolerated with minimal inflammatory response through experiments on dogs in 1991 [54]. Thus, during the experiment, the wounds were infected more seriously in the positive control group than in the PLA-PTMC group, and inflammatory cells were more numerous than in the PLA-PTMC group. This entire process was conducted at the Yunnan University of Traditional Chinese-Medicine.

3. Materials and Methods

3.1. Synthesis of PLA-PTMC Medical Adhesive

Different ratios of poly(lactic acid) (PLA, Mw = 280 kDa, Mw/Mn = 1.98) purchased from NatureWorks® LLC (Minnetonka, NE, USA) and poly(trimethylene carbonate) (PTMC, Mw = 100 kDa, Mw/Mn = 1.70), which was prepared in the laboratory of the Engineering Research Center of Biopolymer Functional Materials of Yunnan, Yunnan Minzu University (Kunming, China), were weighed and added to 40.0 g of ethyl acetate (>99.5%, KESHI, Chengdu, China). Dissolution required approximately 6 h in a closed environment at room temperature (approximately 25 °C).

3.2. Preparation of Films

PLA-PTMC medical adhesive membranes were fabricated using a solvent evaporation technique. Approximately 10.0 g of PLA-PTMC medical adhesive were weighed and dissolved in 50 mL of chloroform (>99.0%) using a batch mixer. After vigorous mixing, the film-forming solution was applied to a polytetrafluoroethylene (PTFE) plate. The solvent was allowed to evaporate at room temperature under the previous conditions to produce a PLA-PTMC medical adhesive membrane with a thickness of 0.6 mm.

The time for the pouring of the solution and for the film formation were recorded as Time 1 and Time 2, respectively. The film formation time for the PLA-PTMC medical adhesive was Time 2 minus Time 1. The composite films were then cut into 100 mm × 25.4 mm sections for investigating the film properties.

3.3. Water Vapor Transmission Rate

The water vapor transmission rate was determined gravimetrically using a water vapor transmission measuring cup at 35 °C and at 50% relative humidity (RH) in accordance with the ASTM E96-95 standard method [55]. Film samples were mounted over the acrylic cups and sealed with paraffin and rubber. The covered cups were placed in a constant temperature and RH-controlled chamber using the same conditions required for film equilibration. The weight loss of the measuring cup was measured as a function of time for 12 h [56]. Every sample was tested at least 8 times. The result was expressed as the average of the measurements.

3.4. Comfortability Degree of PLA-PTMC Medical Adhesives

According to the YY/T0471.4-2004 test methods for primary wound dressing, Part 4, comfortability, the sample was cut to a width of 25 cm, and the sample was ensured to relax freely for at least 300 s. Two parallel marks were made on the sample at a distance of 100 cm ± 0.5 mm and were recorded as *L1*, and the range of the two marks to both ends was kept equal. Finally, a Universal tensile machine (CMT4104; Power Supply 220 V; Max Force 40 KN; Accuracy 1 level; MTS SYSTEMS (CHINA) CO. LTC, Sichuan University, Chengdu, China) was used to stretch the sample at a speed of 300 mm/min, and the maximal load named (ML) was recorded when the sample was stretched by 20%. The sample was then relaxed for 1–2 s in this state, and after 5 min, the distance between the marks was recorded as *L2*. The extensibility (E) was calculated according to Formula (1), and the permanent set (PS) was calculated according to Formula (2). An average of at least five test values was obtained for each sample.

$$E = \frac{ML}{2.5} \tag{1}$$

$$PS\% = \left[(L2 - L1) \times \frac{100}{L1} \right] \tag{2}$$

3.5. Contact Angle

According to ISO 15989:2004 (Plastic film and sheet corona processing thin film of water contact Angle measurement) the contact angles for the film samples were measured using a Kruss Tensiometer K100 (Hamburg, Germany) at 25 °C, and the wetting characteristics of the polymer surfaces were quantified by following the Wilhelmy method. The films were doused with distilled water, dried at 25 °C and then cut into 20 × 30 mm pieces. The measurements were carried out in water at a rate of 25 $\mu m \cdot s^{-1}$. Three measurements were conducted for each sample.

3.6. Gel Permeation Chromatography

Gel permeation chromatography (GPC) measurements were performed on a Waters 515 system equipped with a refractive index (RI) detector using tetrahydrofuran (THF) as the solvent at a flow rate of 1.0 mL/min, and 60 μL of a 1.0 $w/v\%$ solution were injected for each analysis. Calibration was accomplished using polystyrene standards (Polysciences, Warrington, PA, USA).

3.7. ^1H Nuclear Magnetic Resonance (^1HNMR)

^{1}H nuclear magnetic resonance (^{1}HNMR) spectra were recorded using a Bruker AVANCE400 spectrometer (Bruker Corporation, Switzerland) operating at 400 MHz using deuterated chloroform ($CDCl_3$) as the solvent. Chemical shifts (d) were obtained in ppm with respect to tetramethylsilane (TMS).

3.8. Differential Scanning Calorimetry

DSC analysis of PLA-PTMC medical adhesives was conducted using a DSC214 instrument (Netzsch, Selb, German) with dry nitrogen gas at a flow rate of 60 mL/min, Approximately 6 mg of each sample were placed in a small crucible with a sealed surrounding and then heated from 20–210 °C at 10 °C·min^{-1} to identify possible changes in the crystallization and in the melting transition. Subsequently, the sample was cooled to room temperature at a cooling rate of 10 °C·min^{-1} and then further heated to 220 °C at 10 °C·min^{-1}.

3.9. Scanning Electron Microscopy

A section of the sample was sheared before performing the test. The sheared surfaces of the composite films were observed via scanning electron microscopy (SEM) under high vacuum using an SEM instrument (Nova 450, FEI Corporation, Brno, Czech Republic) in liquid nitrogen to observe the interior of the unstressed composites. To optimize the SEM examination, the sheared surfaces of samples were gold-sprayed to produce a thin conductive gold layer 5 nm thick on the exterior of the sheared surfaces.

3.10. Animal Experiments

The trial was approved by the laboratory animals Ethics Committee of Yunnan Minzu University (29 December 2016) and was registered on the Kunming Science and Technology Bureau (SYXK(Yunnan)K2017-0001, 16 January 2017).

3.10.1. Cytotoxicity

Cell viability was determined according to ISO-10993-5 standard tests using 3-(4,5-dimethylthiazol-2-yl)-2,5-diphenyl tetrazolium bromide (MTT). Briefly, RAEC cells were seeded at 5.0×10^4 cells/mL in 198 μL of appropriate culture medium containing 10% serum and 1% antibiotics in a 96-well plate and incubated for 48 h with 5% CO_2, at 37 °C. When the cells were completely adherent, 2 μL of PLA/PTMC (D) medical adhesive at different concentrations (2.5, 5, 10, 20, 50 and 100 μg/mL) were added to the cells. At the same time, 2 μL of Dimethyl Sulphoxide (DMSO) were added to the positive control group cells. Three parallel holes were used, and the cells were incubated for 24 and 48 h. Then, 20 μL of 5 mg/mL MTT were added to the cells, and after the cells were co-incubated for 4 h, the medium was aspirated from the cells. Then, 200 μL of DMSO were added to the holes, and the plates were incubated at room temperature in the dark for 30 min and then homogenized by shaking for approximately 20 min. The formazan absorbance was measured at 570 nm using a microplate reader. A standard curve was obtained using different drug concentrations, and the growth inhibition for L6 cells was analyzed using Prism 6.0 (Nanjing Jiancheng biological technology limited company, Nanjing, China). The cytotoxicity of the PLA/PTMC (D) medical adhesive was recorded as the IC_{50}.

3.10.2. Test of Sterility

The same batch of PLA/PTMC (D) medical adhesive was coated evenly on the surface with the medium and then incubated in a thermostatic incubator for 36 h.

3.10.3. Skin Irritability

Ten male and female rabbits were included in this experiment. The PLA/PTMC (D) medical adhesive and positive reference substance were coated on the inner ears of the rabbits and observed after 2, 4, 8, 12 and 24 h.

3.10.4. Creation of the Wound Model

Animals and Tissue Preparation

Male and female rabbits (2.0–2.5 kg) were obtained from the Laboratory Animal Unit of Kunming Medical University (Kunming, China). All experiments performed in this study were approved by the Committee on the Use of Live Animals in Teaching and Research of Yunnan Minzu University (18 July 2016–16 November 2016). After injecting air into the ear vein of the rabbits, the wound granulation tissue and the surrounding skin on the back were excised, and the tissues were fixed immediately with 4% paraformaldehyde for 24 h.

Establishment of the Wound Model

Ten rabbits were randomly divided into 2 groups, with 5 rabbits in the experimental group and another 5 in the positive control group. The hair on the back of the rabbits was removed on the day before the surgery. The rabbits were anesthetized with 2 mL/kg of 10% chloral hydrate; then, a 2 cm × 2 cm square full-thickness skin wound was produced on the left side of the back, and a full-thickness skin scratch of a length of 2 cm was created on the right side of the back. The surgical site and scratches were treated with the PTMC medical adhesives (0.5 mm-thick membranes) and the positive control drug, respectively. After surgery, the food intake, wound granulation tissue growth and inflammation of the rabbits were observed, and the temperature was measured.

Wound healing rate was one of the direct indicators of wound healing. In our work, the wound healing rate was evaluated as in the work of Nagelschmidt's and Pulok K. Mukherjee's group [56,57]. The changes in healing of the wound, namely the measurement of wound area on graph paper, was expressed as units of mm^2. The healing rate of the wounds was expressed as Formula (3).

$$\text{Healing rate of wound} = \frac{\text{original area} - \text{unhealed wound area}}{\text{original area}} \tag{3}$$

Weight and Temperature Transform

On the day of operation, the weights of the rabbits that were used in the wound experiments were obtained and recorded. The weights were also recorded on the 31st day. In addition, rectal

Int. J. Mol. Sci. **2017**, *18*, 2041

temperatures were recorded once per day after the operation and, after four days, once per three days until the wound recovered completely.

Hepatic and Renal Function

On the 31st day after the wounds were created, 15 mL of blood gained was collected from the heart and prepared for next further study. Serum was prepared by centrifugation for 5 min at 37 °C at 4500 rpm. The hepatic indexes, including AST/GOT, ALT/GPT, AKP, BUN and Cr, were determined using special kits (Nanjing Jiancheng biological technology limited company, Nanjing, China), according to the manufacturer's instructions.

ELISA Assay

IgM, IgG, BALP, IL-6, IL-10 and TNFα were determined in the rabbit serum using double-antibody sandwich ELISAs (CUSABIO), according to the manufacturer's instructions.

Skin Pathology Analyses

Skin pathological section analyses were performed after the other tests were completed. The tissues were excised, and adhering tissues were trimmed and then fixed with 4% paraformaldehyde for 24 h for histological studies. The specimens were embedded in paraffin and sectioned, and the sections were stained with hematoxylin-eosin (HE) for light microscopic observations. The sections were examined and photographed using an Olympus CX-31 Microscope (Olympus Corporation, Shinjuku, Tokyo, Japan).

3.11. Statistical Analysis

All data were expressed as the means ± S.E.M. Statistical analyses were performed using Prism 6.0. Comparisons between two groups were performed using unpaired tests. Comparisons among three or more groups were performed using a one-way ANOVA. $p < 0.05$ was considered statistically significant, and $p > 0.05$ was considered to have no statistically-significant difference.

4. Conclusions

In this study, an adhesive that reduces infections effectively because of its breathability was prepared from PLA modified with PTMC via blending at room temperature. Through a series of characterizations of the performance of different ratios of PLA and PTMC, an optimal ratio was determined, and the effectiveness of the best ratio of PLA and PTMC was demonstrated through animal experiments. The results suggest that a ratio of PLA to PTMC of 7:3 produces a material with suitable gas permeability that can reduce the incidence of wound infections. Furthermore, it produces no harm to the liver and kidneys. It also promotes the proliferation of L6 cells, fibroblasts and epidermal stem cells within the skin wound. Therefore, a blend of PLA and PTMC at a ratio of 7:3 has promise as a medical adhesive.

Acknowledgments: This work was supported by the National Natural Science Foundation of China (Project Nos. 31360417, 31460247, 81460542), the Biodegradable Materials Innovative Research Team (in Science and Technology) at the University of Yunnan Province and Innovation Team Based on Research and Application of Biological Functional Materials of Yunnan Minzu University (2017HC034).

Author Contributions: Minglong Yuan and Haiyun Chen conceived and designed the experiments; Shuang Zhang and Mingwei Yuan performed the experiments; Shuang Zhang and Hongli Li analyzed the data; Hongli Li and Mingwei Yuan contributed reagents/materials/analysis tools; Shuang Zhang and Hongli Li wrote the paper.

Conflicts of Interest: The authors declare no conflict of interest.

References

1. Langer, R.; Tirrell, D.A. Designing materials for biology and medicine. *Nature* **2004**, *428*, 487. [CrossRef] [PubMed]
2. Vacanti, J.P.; Langer, R. Tissue engineering: The design and fabrication of living replacement devices for surgical reconstruction and transplantation. *Lancet* **1999**, *354*, S32–S34. [CrossRef]
3. Somorjai, G.A.; Frei, H.; Park, J.Y. Advancing the frontiers in nanocatalysis, biointerfaces, and renewable energy conversion by innovations of surface techniques. *J. Am. Chem. Soc.* **2009**, *131*, 16589–16605. [CrossRef] [PubMed]
4. Place, E.S.; Evans, N.D.; Stevens, M.M. Complexity in biomaterials for tissue engineering. *Nat. Mater.* **2009**, *8*, 457. [CrossRef] [PubMed]
5. Wang, X.; Gan, H.; Zhang, M.; Sun, T. Modulating cell behaviors on chiral polymer brush films with different hydrophobic side groups. *Langmuir* **2012**, *28*, 2791–2798. [CrossRef] [PubMed]
6. Nel, A.E.; Mädler, L.; Velegol, D.; Xia, T.; Hoek, E.M.; Somasundaran, P.; Klaessig, F.; Castranova, V.; Thompson, M. Understanding biophysicochemical interactions at the nano-bio interface. *Nat. Mater.* **2009**, *8*, 543. [CrossRef] [PubMed]
7. Peppas, N.A.; Langer, R. New challenges in biomaterials. *Science* **1994**, *263*, 1715–1719. [CrossRef] [PubMed]
8. Bouten, P.J.; Zonjee, M.; Bender, J.; Yauw, S.T.; van Goor, H.; van Hest, J.C.; Hoogenboom, R. The chemistry of tissue adhesive materials. *Prog. Polym. Sci.* **2014**, *39*, 1375–1405. [CrossRef]
9. Lloyd, J.D.; Marque, M.J.; Kacprowicz, R.F. Closure techniques. *Emerg. Med. Clin. N. Am.* **2007**, *25*, 73–81. [CrossRef] [PubMed]
10. Duarte, A.; Coelho, J.; Bordado, J.; Cidade, M.; Gil, M. Surgical adhesives: Systematic review of the main types and development forecast. *Prog. Polym. Sci.* **2012**, *37*, 1031–1050. [CrossRef]
11. Tajirian, A.L.; Goldberg, D.J. A review of sutures and other skin closure materials. *J. Cosmet. Laser Ther.* **2010**, *12*, 296–302. [CrossRef] [PubMed]
12. Holm, V.K.; Ndoni, S.; Risbo, J. The stability of poly(lactic acid) packaging films as influenced by humidity and temperature. *J. Food Sci.* **2006**, *71*, E40–E44. [CrossRef]
13. Cava, D.; Gavara, R.; Lagaron, J.; Voelkel, A. Surface characterization of poly(lactic acid) and polycaprolactone by inverse gas chromatography. *J. Chromatogr. A* **2007**, *1148*, 86–91. [CrossRef] [PubMed]
14. Sawyer, D.J. Bioprocessing–No Longer a Field of Dreams. In *Macromolecular Symposia*; Wiley Online Library: New York, NY, USA, 2003; pp. 271–282.
15. Drumright, R.E.; Gruber, P.R.; Henton, D.E. Polylactic acid technology. *Adv. Mater.* **2000**, *12*, 1841–1846. [CrossRef]
16. Dorgan, J.R.; Lehermeier, H.J.; Palade, L.I.; Cicero, J. Polylactides: Properties and prospects of an environmentally benign plastic from renewable resources. In *Macromolecular Symposia*; Wiley Online Library: New York, NY, USA, 2001; pp. 55–66.
17. Barrows, T. Degradable implant materials: A review of synthetic absorbable polymers and their applications. *Clin. Mater.* **1986**, *1*, 233–257. [CrossRef]
18. Seyednejad, H.; Gawlitta, D.; Kuiper, R.V.; de Bruin, A.; van Nostrum, C.F.; Vermonden, T.; Dhert, W.J.; Hennink, W.E. In vivo biocompatibility and biodegradation of 3D-printed porous scaffolds based on a hydroxyl-functionalized poly(ε-caprolactone). *Biomaterials* **2012**, *33*, 4309–4318. [CrossRef] [PubMed]
19. Mainil-Varlet, P.; Curtis, R.; Gogolewski, S. Effect of in vivo and in vitro degradation on molecular and mechanical properties of various low-molecular-weight polylactides. *J. Biomed. Mater. Res. A* **1997**, *36*, 360–380. [CrossRef]
20. Tsuji, H.; Sumida, K. Poly(L-lactide): V. Effects of storage in swelling solvents on physical properties and structure of poly(L-lactide). *J. Appl. Polym. Sci.* **2001**, *79*, 1582–1589. [CrossRef]
21. Bleach, N.; Tanner, K.; Kellomäki, M.; Törmälä, P. Effect of filler type on the mechanical properties of self-reinforced polylactide–calcium phosphate composites. *J. Mater. Sci.* **2001**, *12*, 911–915.
22. Eling, B.; Gogolewski, S.; Pennings, A. Biodegradable materials of poly(L-lactic acid): 1. Melt-spun and solution-spun fibres. *Polymer* **1982**, *23*, 1587–1593. [CrossRef]
23. Athanasiou, K.A.; Niederauer, G.G.; Agrawal, C.M. Sterilization, toxicity, biocompatibility and clinical applications of polylactic acid/polyglycolic acid copolymers. *Biomaterials* **1996**, *17*, 93–102. [CrossRef]

24. Janorkar, A.V.; Metters, A.T.; Hirt, D.E. Modification of poly(lactic acid) films: Enhanced wettability from surface-confined photografting and increased degradation rate due to an artifact of the photografting process. *Macromolecules* **2004**, *37*, 9151–9159. [CrossRef]

25. Vink, E.T.; Rabago, K.R.; Glassner, D.A.; Gruber, P.R. Applications of life cycle assessment to natureworks™ polylactide (PLA) production. *Polym. Degrad. Stab.* **2003**, *80*, 403–419. [CrossRef]

26. Pezzin, A.; Ekenstein, V.; Alberda, G.; Zavaglia, C.; Ten Brinke, G.; Duek, E. Poly(para-dioxanone) and poly(L-lactic acid) blends: Thermal, mechanical, and morphological properties. *J. Appl. Polym. Sci.* **2003**, *88*, 2744–2755. [CrossRef]

27. Ma, X.; Yu, J.; Wang, N. Compatibility characterization of poly(lactic acid)/poly(propylene carbonate) blends. *J. Polym. Sci. B* **2006**, *44*, 94–101. [CrossRef]

28. Shibata, M.; Inoue, Y.; Miyoshi, M. Mechanical properties, morphology, and crystallization behavior of blends of poly(L-lactide) with poly(butylene succinate-*co*-L-lactate) and poly(butylene succinate). *Polymer* **2006**, *47*, 3557–3564. [CrossRef]

29. Chen, G.-X.; Kim, H.-S.; Kim, E.-S.; Yoon, J.-S. Compatibilization-like effect of reactive organoclay on the poly(L-lactide)/poly(butylene succinate) blends. *Polymer* **2005**, *46*, 11829–11836. [CrossRef]

30. Pego, A.; Van Luyn, M.; Brouwer, L.; Van Wachem, P.; Poot, A.A.; Grijpma, D.W.; Feijen, J. In vivo behavior of poly(1,3-trimethylene carbonate) and copolymers of 1,3-trimethylene carbonate with D,L-lactide or ε-caprolactone: Degradation and tissue response. *J. Biomed. Mater. Res. A* **2003**, *67*, 1044–1054. [CrossRef] [PubMed]

31. Albertsson, A.C.; Eklund, M. Influence of molecular structure on the degradation mechanism of degradable polymers: In vitro degradation of poly(trimethylene carbonate), poly(trimethylene carbonate-co-caprolactone), and poly(adipic anhydride). *J. Appl. Polym. Sci.* **1995**, *57*, 87–103. [CrossRef]

32. Nederberg, F.; Bowden, T.; Hilborn, J. Induced surface migration of biodegradable phosphoryl choline functional poly(trimethylene carbonate). *Polym. Adv. Technol.* **2005**, *16*, 108–112. [CrossRef]

33. Márquez, Y.; Franco, L.; Puiggalí, J. Thermal degradation studies of poly(trimethylene carbonate) blends with either polylactide or polycaprolactone. *Thermochim. Acta* **2012**, *550*, 65–75. [CrossRef]

34. Rocha, D.N.; Brites, P.; Fonseca, C.; Pêgo, A.P. Poly(trimethylene carbonate-*co*-ε-caprolactone) promotes axonal growth. *PLoS ONE* **2014**, *9*, e88593. [CrossRef] [PubMed]

35. Rhim, J.-W. Effect of pla lamination on performance characteristics of agar/κ-carrageenan/clay bio-nanocomposite film. *Food Res. Int.* **2013**, *51*, 714–722. [CrossRef]

36. Li, H.; Chang, J.; Qin, Y.; Wu, Y.; Yuan, M.; Zhang, Y. Poly(lactide-*co*-trimethylene carbonate) and polylactide/polytrimethylene carbonate blown films. *Int. J. Mol. Sci.* **2014**, *15*, 2608–2621. [CrossRef] [PubMed]

37. Martin, O.; Averous, L. Poly(lactic acid): Plasticization and properties of biodegradable multiphase systems. *Polymer* **2001**, *42*, 6209–6219. [CrossRef]

38. Nampoothiri, K.M.; Nair, N.R.; John, R.P. An overview of the recent developments in polylactide (PLA) research. *Bioresour. Technol.* **2010**, *101*, 8493–8501. [CrossRef] [PubMed]

39. Södergård, A.; Stolt, M. Properties of lactic acid based polymers and their correlation with composition. *Prog. Polym. Sci.* **2002**, *27*, 1123–1163. [CrossRef]

40. Taylor, M.; Daniels, A.; Andriano, K.; Heller, J. Six bioabsorbable polymers: In vitro acute toxicity of accumulated degradation products. *J. Appl. Biomater.* **1994**, *5*, 151–157. [CrossRef] [PubMed]

41. Qin, Y.; Yuan, M.; Li, L.; Guo, S.; Yuan, M.; Li, W.; Xue, J. Use of polylactic acid/polytrimethylene carbonate blends membrane to prevent postoperative adhesions. *J. Biomed. Mater. Res. B* **2006**, *79*, 312–319. [CrossRef] [PubMed]

42. Wang, H.; Dong, J.H.; Qiu, K.Y. Synthesis and characterization of ABA-type block copolymer of poly(trimethylene carbonate) with poly(ethylene glycol): Bioerodible copolymer. *J. Polym. Sci. A* **1998**, *36*, 695–702. [CrossRef]

43. Żenkiewicz, M.; Richert, J.; Różański, A. Effect of blow moulding ratio on barrier properties of polylactide nanocomposite films. *Polym. Test.* **2010**, *29*, 251–257. [CrossRef]

44. Bockenstedt, P.; Greenberg, J.; Handin, R. Structural basis of von willebrand factor binding to platelet glycoprotein Ib and collagen. Effects of disulfide reduction and limited proteolysis of polymeric von willebrand factor. *J. Clin. Investig.* **1986**, *77*, 743. [CrossRef] [PubMed]

45. Duangjai, A.; Luo, K.; Zhou, Y.; Yang, J.; Kopeček, J. Combination cytotoxicity of backbone degradable hpma copolymer gemcitabine and platinum conjugates toward human ovarian carcinoma cells. *Eur. J. Pharm. Biopharm.* **2014**, *87*, 187–196. [CrossRef] [PubMed]

46. Blake, M.S.; Johnston, K.H.; Russell-Jones, G.J.; Gotschlich, E.C. A rapid, sensitive method for detection of alkaline phosphatase-conjugated anti-antibody on western blots. *Anal. Biochem.* **1984**, *136*, 175–179. [CrossRef]

47. Xiong, L.-H.; He, X.; Xia, J.; Ma, H.; Yang, F.; Zhang, Q.; Huang, D.; Chen, L.; Wu, C.; Zhang, X. Highly sensitive naked-eye assay for enterovirus 71 detection based on catalytic nanoparticle aggregation and immunomagnetic amplification. *ACS Appl. Mater. Interfaces* **2017**, *9*, 14691–14699. [CrossRef] [PubMed]

48. Zhang, C.; Wang, X.; Tang, J. Evaluation of acute kidney injury using contrast ultrasonography in a rabbit model of crush syndrome. *Ultrasound Med. Biol.* **2017**, *43*, 494–499. [CrossRef] [PubMed]

49. Goding, J.W. Use of staphylococcal protein a as an immunological reagent. *J. Immunol. Methods* **1978**, *20*, 241–253. [CrossRef]

50. Habib, S.; Ali, R. Acquired antigenicity of DNA after modification with peroxynitrite. *Int. J. Biol. Macromol.* **2005**, *35*, 221–225. [CrossRef] [PubMed]

51. Lee, C.G.; Homer, R.J.; Cohn, L.; Link, H.; Jung, S.; Craft, J.E.; Graham, B.S.; Johnson, T.R.; Elias, J.A. Transgenic overexpression of interleukin (IL)-10 in the lung causes mucus metaplasia, tissue inflammation, and airway remodeling via IL-13-dependent and-independent pathways. *J. Biol. Chem.* **2002**, *277*, 35466–35474. [CrossRef] [PubMed]

52. Klompmaker, J.; Jansen, H.; Veth, R.; de Groot, J.; Nijenhuis, A.; Pennings, A. Porous polymer implant for repair of meniscal lesions: A preliminary study in dogs. *Biomaterials* **1991**, *12*, 810–816. [CrossRef]

53. Majola, A.; Vainionpää, S.; Vihtonen, K.; Mero, M.; Vasenius, J.; Törmälä, P.; Rokkanen, P. Absorption, biocompatibility, and fixation properties of polylactic acid in bone tissue: An experimental study in rats. *Clin. Orthop. Relat. Res.* **1991**, *268*, 260–269.

54. Von Schroeder, H.P.; Kwan, M.; Amiel, D.; Coutts, R.D. The use of polylactic acid matrix and periosteal grafts for the reconstruction of rabbit knee articular defects. *J. Biomed. Mater. Res. A* **1991**, *25*, 329–339. [CrossRef] [PubMed]

55. Vásconez, M.B.; Flores, S.K.; Campos, C.A.; Alvarado, J.; Gerschenson, L.N. Antimicrobial activity and physical properties of chitosan–tapioca starch based edible films and coatings. *Food Res. Int.* **2009**, *42*, 762–769. [CrossRef]

56. Nagelschmidt, M.; Becker, D.; Bonninghoff, N.; Engelhardt, G.H. Effect of fibronectin therapy and fibronectin deficiency on wound healing: A study in rats. *J. Trauma Acute Care Surg.* **1987**, *27*, 1267–1271. [CrossRef]

57. Mukherjee, P.K.; Verpoorte, R.; Suresh, B. Evaluation of in-vivo wound healing activity of hypericum patulum (family: Hypericaceae) leaf extract on different wound model in rats. *J. Ethnopharmacol.* **2000**, *70*, 315–321. [CrossRef]

International Journal of
Molecular Sciences

MDPI

Article

Engineering Cell Adhesion and Orientation via Ultrafast Laser Fabricated Microstructured Substrates

Eleftheria Babaliari [1,2], Paraskevi Kavatzikidou [1], Despoina Angelaki [1,3], Lefki Chaniotaki [2], Alexandra Manousaki [1], Alexandra Siakouli-Galanopoulou [4], Anthi Ranella [1,*] and Emmanuel Stratakis [1,2,*]

[1] Foundation for Research and Technology—Hellas (F.O.R.T.H.), Institute of Electronic Structure and Laser (I.E.S.L.), Vassilika Vouton, 711 10 Heraklion, Greece; ebabaliari@iesl.forth.gr (E.B.); ekavatzi@iesl.forth.gr (P.K.); angelaki@iesl.forth.gr (D.A.); manousa@iesl.forth.gr (A.M.)
[2] Department of Materials Science and Technology, University of Crete, 70013 Crete, Greece; mst1095@edu.materials.uoc.gr
[3] Department of Physics, University of Crete, 70013 Crete, Greece
[4] Department of Biology, University of Crete, 70013 Crete, Greece; siakouli@biology.uoc.gr
* Correspondence: ranthi@iesl.forth.gr (A.R.); stratak@iesl.forth.gr (E.S.); Tel.: +30-2810-391319 (A.R.); +30-2810-391274 (E.S.)

Received: 24 June 2018; Accepted: 10 July 2018; Published: 14 July 2018

Abstract: Cell responses depend on the stimuli received by the surrounding extracellular environment, which provides the cues required for adhesion, orientation, proliferation, and differentiation at the micro and the nano scales. In this study, discontinuous microcones on silicon (Si) and continuous microgrooves on polyethylene terephthalate (PET) substrates were fabricated via ultrashort pulsed laser irradiation at various fluences, resulting in microstructures with different magnitudes of roughness and varying geometrical characteristics. The topographical models attained were specifically developed to imitate the guidance and alignment of Schwann cells for the oriented axonal regrowth that occurs in nerve regeneration. At the same time, positive replicas of the silicon microstructures were successfully reproduced via soft lithography on the biodegradable polymer poly(lactide-*co*-glycolide) (PLGA). The anisotropic continuous (PET) and discontinuous (PLGA replicas) microstructured polymeric substrates were assessed in terms of their influence on Schwann cell responses. It is shown that the micropatterned substrates enable control over cellular adhesion, proliferation, and orientation, and are thus useful to engineer cell alignment in vitro. This property is potentially useful in the fields of neural tissue engineering and for dynamic microenvironment systems that simulate in vivo conditions.

Keywords: cell adhesion; cell orientation; Schwann cells; topography; laser fabrication; soft lithography; polymeric materials

1. Introduction

Cell behavior in vivo is influenced by a variety of extracellular signals. It is currently clear that many cellular aspects, including adhesion, migration, spreading, proliferation, survival, apoptosis, and gene expression, are modulated by interdependent signaling cascades of soluble signals, shear stresses, other supportive cells, and the nature of the extracellular matrix (ECM) [1]. Thus, the main challenge in and goal of tissue engineering is to mimic the features of the ECM and the surrounding environment of cells sufficiently so that cells function in the artificial medium as they would in vivo [2]. Furthermore, individual cells recognize structures that have comparable dimensions to the those at the cellular level, which is at the micro scale. Consequently, control over micro/nanotopography is desirable. At the cell–material interface, all the cellular processes are governed by the physical and chemical stimuli of substrate stiffness (or rigidity), topography, and chemistry, respectively, while at the intracellular level, focal adhesions are key molecular complexes for sensing the environmental conditions as significant mechanosensitive players [3–6]. Indeed, many studies confirmed that the surface topography influences the adhesion, migration, polarization, proliferation, and differentiation of cells [7–12]. These parameters are of high significance for the design and development of advanced biomaterials in regenerative medicine and tissue engineering. Therefore, a considerable amount of research is devoted to the modification of materials' surfaces for use as platforms to study cell viability, differentiation, motility, and apoptosis [13,14].

Generally, there are various materials and fabrication techniques that aim to reconstruct the ECM architecture in vitro with very specific compositions, ligand presentations, mechanical properties, and organization that vary between different tissues [13–15]. Indeed, previous studies have detailed the major fabrication techniques, the produced types of micro/nanostructured substrates, and advantages/disadvantages of the techniques [16,17]. Among various techniques that have been developed for surface modification, laser irradiation has proved to be important in the enhancement of material biocompatibility, particularly via the creation of new functional groups and the precise topography formation at the cellular and subcellular scale [10,18–21]. In particular, microstructuring via ultrafast lasers provides unique control over the uniformity and regularity of micron and submicron features [22].

The in vitro guiding of neurite outgrowth is important in tissue regeneration and for the development of neuronal interfaces with useful characteristics. To date, this has been achieved with micro- and nanofabrication techniques that give rise to various anisotropic continuous or discontinuous geometries [13,14]. Previous studies have demonstrated that anisotropic continuous electrospun polymeric fibers can influence neurite growth, alignment, and differentiation [23–25]. It has been also reported that in photolithographically fabricated continuous grooved substrates, axons grew on top of the ridges [25–27]. Moreover, studies have contributed significant insight into the impact of the disordered/anisotropic nanotopographical features on neuron differentiation and maturation by mechanotransduction pathways in PC12 cells [28–30]. It has also been reported that laser-microstructured discontinuous Si substrates not only support cellular adhesion and viability, but also significantly affect cell morphology, growth, orientation, and differentiation in a surface-dependent manner [10,20,21]. Furthermore, it was reported that both Schwann cells and axons of sympathetic neurons were parallel oriented on microcone patterns of elliptical cross-sections, while they exhibited a random orientation on the microcones exhibiting arbitrarily shaped cross-sections. As a result, it is suggested that an anisotropic continuous and discontinuous topographical patterns could promote Schwann cell and axonal alignment, provided that the pattern presents anisotropic geometrical features, even though their sizes are at a subcellular scale [20]. The same topographical model was used to study PC12 differentiation after treatment with nerve growth factor (NGF). It was shown that, unlike surfaces with low and medium roughness, those that are highly rough and exhibit large distances between microcones did not support PC12 cell differentiation, although cells had been stimulated with NGF [21]. Such substrates were also shown to support macrophage adherence and antigen presentation process in vitro, and to induce specific antibody production upon implantation in vivo [31].

Soft lithography is used to produce substrates with distinct surface topographies at the nano- and micrometer scale. It has been successfully used to transfer well-defined microsized patterns from silicon or stainless-steel masters to surfaces of soft biomaterials [32,33], allowing the replication of controlled microenvironments and in-depth study of the influence of surface properties on cell behavior [34].

As mentioned in our previous studies, we have thoroughly characterized discontinuous Si surfaces as cell substrates, and we have extensively investigated cell-specific responses of various neuronal cell types to these surfaces. In this study, we aim to demonstrate the reproducibility (or not) of Schwann cell behavior—focusing on growth, adhesion, and orientation—on laser-patterned polymeric microstructures, including those made from polyethylene terephthalate (PET) and poly(lactide-*co*-glycolide) (PLGA), compared with the Si substrates. PET has been widely used for cell culturing, surgical suture material, and prosthetic vascular grafts due to its biocompatibility and its excellent mechanical strength and resistance [35,36]. Moreover, PLGA is a biocompatible and biodegradable synthetic polymer that is used in various microfabrication techniques to create patterned substrates for various applications in tissue engineering and regenerative medicine [37–40]. In particular, microstructured substrates with different continuous microgroove (MG) and discontinuous microspike (MS) topographies were fabricated via either ultrafast laser direct writing of 2D planar PET substrates [41], or through soft lithography of PLGA replicas from microstructured Si substrates, respectively. The morphological, topographical, wetting, and optical properties of these substrates were investigated, and then their interactions with Schwann cells (SW10)—a murine glia cell line—in terms of adhesion, orientation, and proliferation were determined.

2. Results and Discussion

2.1. Scanning Electron Microscopy (SEM) Images of Laser-Microstructured Substrates on PET (PET-MG) and PLGA-MS (1:10) Replicas (from Laser-Microstructured Si Substrates with Three Different Laser Fluences)

Figure 1 depicts the SEM images of PET coverslips that were ablated by the femtosecond laser at a constant fluence of 11.9 J/cm^2, scan velocity of 7 mm/s, and an x_{step} (distance between two consecutive scan lines) of 50 μm fabricated using a linear Gaussian beam. Thus, using these parameters, we fabricated microstructured substrates with a continuous microgroove geometry (PET-MG substrates). Such a surface morphology occurs due to the overlap between adjacent spots during the scanning process [42]. Figures 2 and 3 respectively represent the different stages of the soft lithography process; from the Si master with microspikes, through the poly(dimethylsiloxane) (PDMS) negative mold, to the PLGA replicas with the three different topographies (Si-MS and PLGA-MS replicas). We have successfully reproduced (irradiated Si topography and PLGA replicas) patterned substrates exhibiting 2D–3D surface characteristics, resulting in an additional parameter to control cell growth and network formation. As shown in Figure 3, each culture substrate consisted of these three microstructured areas, irradiated using 0.42 J/cm^2 (25 mW, low roughness), 0.58 J/cm^2 (40 mW, medium roughness), and 0.72 J/cm^2 (65 mW, high roughness). The main difference between the three PLGA-MS replicas is the distance between the spikes, that is, the spike size (aspect ratio). The higher the laser fluence, the higher the spike size, characterized as high roughness or topography.

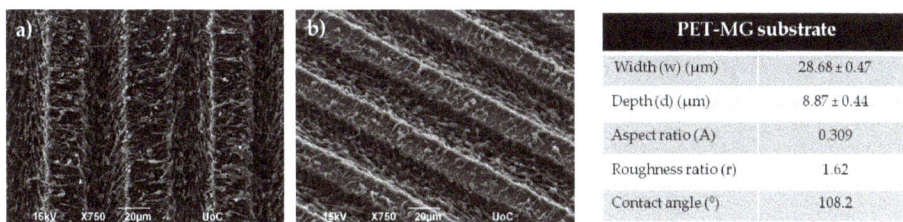

PET-MG substrate	
Width (w) (μm)	28.68 ± 0.47
Depth (d) (μm)	8.87 ± 0.44
Aspect ratio (A)	0.309
Roughness ratio (r)	1.62
Contact angle (°)	108.2

Figure 1. Scanning electron microscopy (SEM) images (top (**a**) and tilted (**b**) view) of polyethylene terephthalate microgroove (PET-MG) substrates. Measurements of the geometrical parameters of the surface of the PET-MG substrates in the tilted view of the SEM images were processed by Fiji ImageJ. A series of measurements were obtained for the surface characterization, such as width (w) and depth (d) of microgrooves, aspect ratio (A = d/w) and roughness ratio (r = 1 + 2d/w). Measurements of the contact angles were performed with the use of a tensiometer.

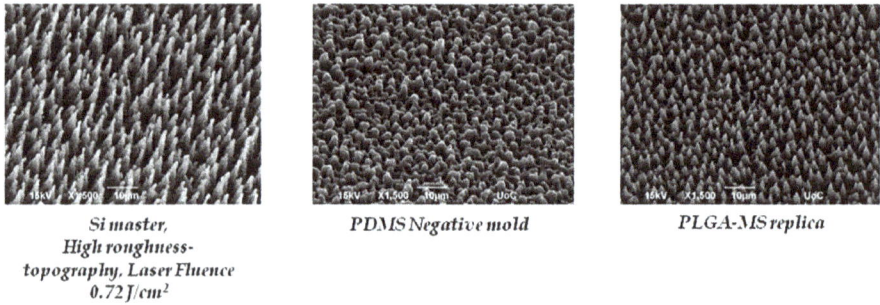

*Si master,
High roughness-
topography, Laser Fluence
0.72 J/cm²*

PDMS Negative mold

PLGA-MS replica

Figure 2. Scanning electron microscopy (SEM) images (tilted view) of a laser-microstructured Si-microspike (MS) substrate (high roughness/topography) mold, poly(dimethylsiloxane) (PDMS) negative mold, and poly(lactide-*co*-glycolide) (PLGA)-MS replica.

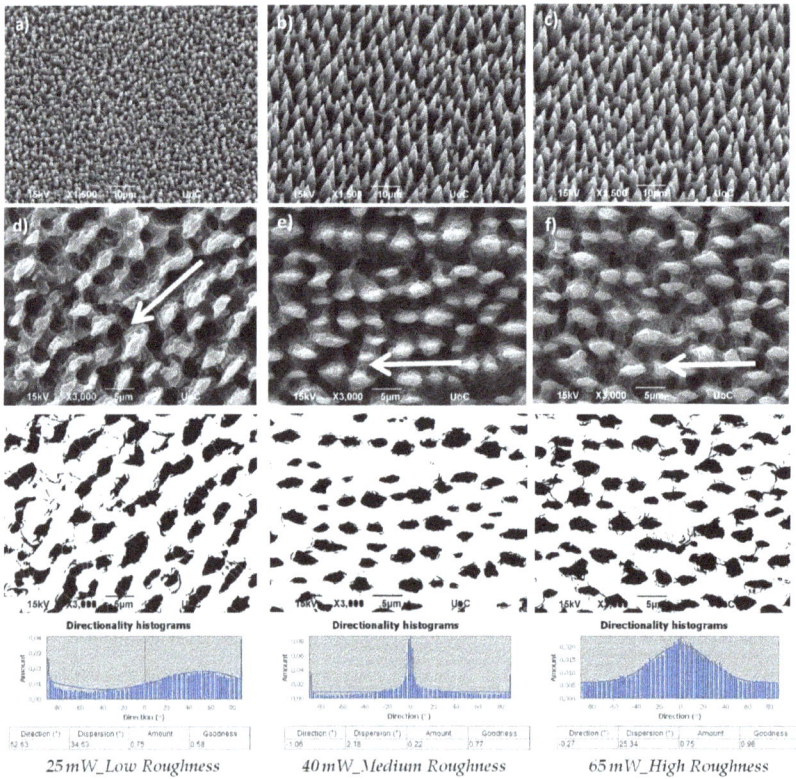

25 mW_Low Roughness

40 mW_Medium Roughness

65 mW_High Roughness

Figure 3. *Cont.*

Groups PLGA-MS	Density [10(4) / mm(2)]	Height (h) (µm)	Width (b) (µm)	Aspect ratio (A)	Roughness ratio (r)	Contact Angle (º)
25mW_Low Roughness	7.18 ± 1.30	3.06 ± 0.40	2.93 ± 0.30	1.044	3.1	120
40mW_Medium Roughness	5.35 ± 0.25	4.34 ± 0.36	2.16 ± 0.31	2.005	5.0	124
65mW_High Roughness	4.69 ± 0.19	10.55 ± 1.10	4.68 ± 0.41	2.252	5.5	133

Figure 3. Scanning electron microscopy (SEM) images (tilted (**a**–**c**) and top (**d**–**f**) view) of PLGA-MS replicas of the three different topographies (25 mW_Low Roughness, 40 mW_Medium Roughness, and 65 mW_High Roughness). The white arrows represent the spikes' direction. Under the images, directionality histograms and tables with statistics are presented, which were generated using the Fiji ImageJ plug-in "Directionality" [43]. Above the histogram, the plug-in generates statistics for the highest peak found. The highest peak is fitted by a Gaussian function, taking into account the periodic nature of the histogram. In the tables, the "Direction (°)" column reports the center of the Gaussian; the "Dispersion (°)" column reports the standard deviation of the Gaussian; the "Amount" column is the sum of the histogram from center-std to center+std, divided by the total sum of the histogram; the "Goodness" column reports the goodness of the fit, where 1 is good, 0 is bad. Measurements of the geometrical parameters of the surface of the PLGA-MS (10%) replicas at the three different topographies (25 mW_Low Roughness, 40 mW_Medium Roughness, and 65 mW_High Roughness) on the SEM images (top and tilted view) of the PLGA-MS replicas were processed using an image-processing algorithm (Fiji ImageJ) to determine the topological characteristics of the MSs. Measurements include the height (h), width (d), aspect ratio (A), and roughness ratio (r) from the top (highest magnification) and tilted-view SEM images. The aspect ratio was calculated by dividing the height by the radius of the spike's base. For the PLGA-MS replicas, a surface plot of each image was produced by Fiji ImageJ, and the height and spike's base were measured. From each image, at least 10 measurements were performed. The roughness ratio, r, was calculated by dividing the actual, unfolded, surface area of spikes by the total irradiated area (r = 1 + 2h/b, where b is the width of spikes). The mean value was calculated from four individual surfaces in each case. Measurements of the contact angles were performed with the use of a tensiometer.

The measurements of the geometrical parameters of the PET-MG substrates, as calculated from SEM images, are summarized in Figure 1. The width of the microgrooves was 28.68 ± 0.47 µm, the depth 8.87 ± 0.44 µm, the aspect ratio 0.309, and the roughness ratio 1.62. The geometrical characteristics of the spikes on the Si substrates have been previously determined [20,21]. Here, in Figure 3, we also show the measurements of the geometrical parameters of the surface of the PLGA-MS replicas for the three different topographies. As calculated from SEM images, the spike height varied from 3.06 ± 0.40 µm in the low-roughness structures to 10.55 ± 1.10 µm in the high-roughness structures (Figure 3). While spike density was the lowest in the high-roughness structures, the spikes' height and roughness, thus aspect ratio, increased. These findings demonstrate the anisotropic nature of the PLGA-MS substrates. Furthermore, it is clear from Figure 3, and specifically from the directionality histograms, that there is a varied orientation between the replicas. The medium- and high-roughness PLGA-MS substrates showed a directionality at the area of zero degrees, while the low roughness substrate showed a lower directionality at the area close to 52 degrees.

2.2. Measurements of Wettability of Irradiated PET (PET-MG), Non-Irradiated PET (PET-Flat), and PLGA-MS (1:10) Replicas (from Irradiated Si Substrates)

The contact angle measured on the irradiated PET (PET-MG substrate) is presented in Figure 1. Specifically, the contact angle of the non-irradiated PET (PET-Flat) was ~77.8°, which is in agreement with previous studies [35,44], while the contact angle of the irradiated (PET-MG) is 108.2°. We observed a decrease in the hydrophilicity of the PET-MG substrate, which is attributed to the increased roughness of the surfaces after irradiation with the femtosecond laser [45]. Figure 3 shows the measured contact angles of PLGA-MS (1:10) replicas with the three topographies. Increasing the roughness of the PLGA-MS replica's surface decreased the hydrophilicity. According to the literature, lactide is more hydrophobic than glycolide, therefore, PLGA copolymers rich in lactide (the PLGA in this study) are less hydrophilic and absorb less water, leading to a slower degradation of the polymer chains [46]. Therefore the topography enhances the degradation rate of this PLGA copolymer.

2.3. UV–Vis Measurements of Irradiated PET (PET-MG), Non-Irradiated PET (PET-Flat), and PLGA-MS (1:10) Replicas (from Irradiated Si Substrates)

In order to determine changes to the surface chemistry of the microstructured substrates, ultraviolet–visible (UV–Vis) spectroscopy was used. We observed an increase of the absorption in the irradiated PET (PET-MG) due to the structuring process. Moreover, we noticed the development of an absorption band in the region of 300–500 nm in PET-MG, likely due to the presence of aromatic hydroxylated species produced during the photooxidation of PET (Figure 4a), which is in agreement with previous studies [47–49]. Specifically, previous work [47–49] has demonstrated the development of an absorption band at around 340 nm in the UV range. The relevant absorbance (a.u.) exhibits an increase in the PLGA topographies compared with the flat PLGA and the glass substrate, as shown in Figure 4b. By increasing the topography or laser fluence of the PLGA-MS replicas, the absorbance (A) was increased (A_25 mW_Low Roughness < A_40 mW Medium Roughness < A_65 mW High Roughness). All absorption bands (obtained here by UV–Vis and unpublished data using ATR-FTIR on these replicas) found in the spectra agree with those given in the literature for PLGA copolymers [46,50]. There was a slight difference in the relevant absorbance of the microstructured replicas compared with the flat PLGA, but it was negligible.

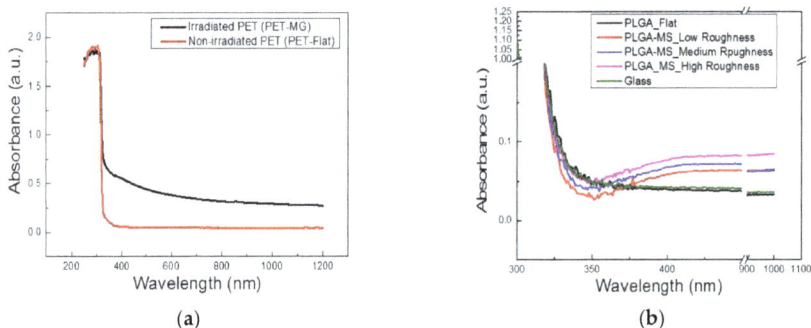

Figure 4. (**a**) UV–Vis measurements of irradiated PET (PET-MG) and non-irradiated PET (PET-Flat); (**b**) UV–Vis measurements of all the PLGA-MS replicas, as well as the PLGA flat and glass substrates.

2.4. Cell Seeding of Laser-Microstructured Substrates on PET (PET-MG) and PLGA-MS Replicas with Schwann Cells

In Figure 5, we show the morphology of Schwann cells for two different time points (4 days and 6 days) cultured on the PET substrates (PET-MG and PET-Flat). The anisotropic continuous microgrooves had a width of 28.68 ± 0.47 μm and a depth of 8.87 ± 0.44 μm. We noticed that the cells exhibited a branched shape and flattened morphology with long cellular extensions, which indicates good adhesion and growth of the cells on the microgrooves. Moreover, we noticed that the cells appeared to be oriented along the direction of the microgrooves for 4 and 6 days of culture, while they showed a random orientation on the flat PET. This is also demonstrated by the directionality histograms in Figure 5, which show that the amount is higher in the domain parallel to the microgrooves (±90 degrees). It is obvious from the SEM images in Figure 5 that, although surface roughness did not affect the proliferation of the cells (cells were equally grown on flat PET and PET-MG substrates), surface morphology significantly controlled the outgrowth of the cells. Consequently, cells could sense continuous directional topographical cues, with sizes at the subcellular scale.

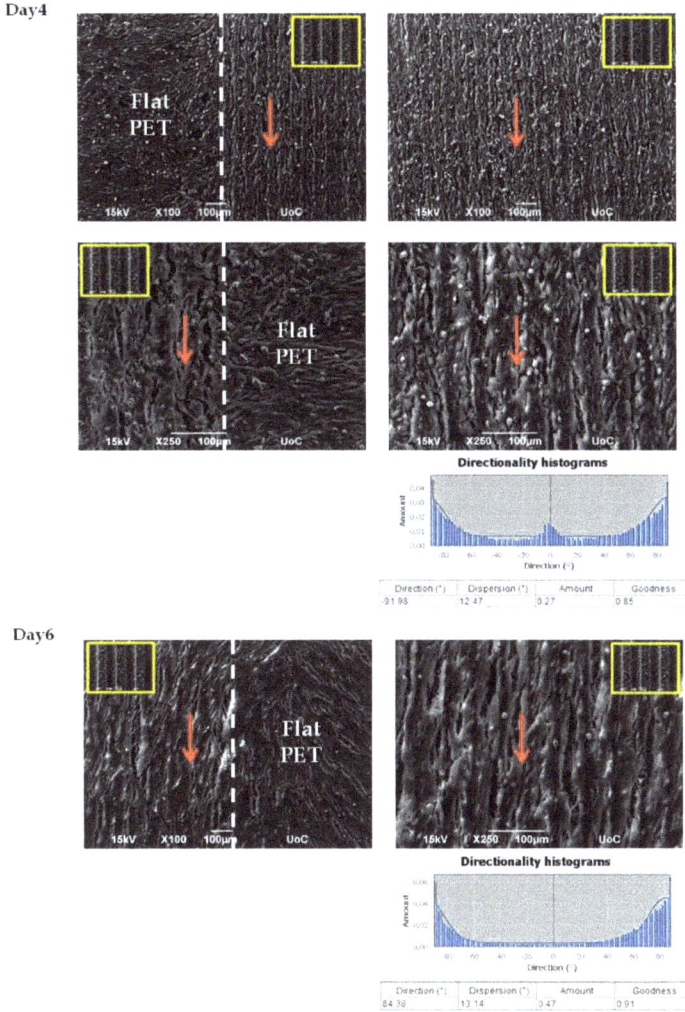

Figure 5. Scanning electron microscopy (SEM) images of Schwann cells cultured on the PET substrates (PET-MG and PET-Flat) for 4 and 6 days. The red arrows represent the directionality of Schwann cells, which are oriented according to the direction of the microgrooves. The inset SEM images, indicated by the yellow box, show the geometry of microgrooves. Under the SEM images, directionality histograms and the tables with statistics are presented, which were generated using the Fiji ImageJ plug-in "Directionality" [43]. Above the histogram, the plug-in generates statistics for the highest peak found. The highest peak is fitted by a Gaussian function, taking into account the periodic nature of the histogram. In the tables, the "Direction (°)" column reports the center of the Gaussian; the "Dispersion (°)" column reports the standard deviation of the Gaussian; the "Amount" column is the sum of the histogram from center-std to center+std, divided by the total sum of the histogram; the "Goodness" column reports the goodness of the fit, where 1 is good, 0 is bad.

According to Figure 6, there is an apparent finding that all three discontinuous topographies on the PLGA-MS replicas equally support Schwann cells' growth. It is demonstrated that a growth pattern/profile of the Schwann cells is observed mainly on the medium- and high-roughness PLGA-MS replicas, compared with the low-roughness PLGA-MS replica and flat PLGA substrate, for 3 days of culture. The cells adhered and aligned on the ridge of the spikes of surfaces with medium and high topography. On the contrary, on the flat substrate and on the low-roughness PLGA-MS replica, there is arbitrary cell growth occurring. Elongated cells are present in greater numbers on the high-roughness and the medium-roughness PLGA-MS replicas (as demonstrated from the directionality histograms in Figure 6, where there is a clear concentration of the quantity at the area of zero degrees) compared to low-roughness PLGA-MS replicas and flat PLGA. In this study, the topography (due to the laser irradiation process) of the polymeric substrates ranges at both the micro and nano scale. Our previous studies demonstrated that the adhesion, alignment, proliferation, and differentiation of different types of neural cells depend on the topography [12–15]. Specifically, it has been proved that there is directional cell outgrowth dictated by substrates of medium and high roughness [20]. In the present study, all the three topographies showed cell growth according to the spikes' orientation (red arrows on Figure 6), indicating that the cells could sense the discontinuous directional topographical features at subcellular scales. To date, there are no widely accepted hypotheses regarding the mechanism for the effects of topography substrates on cell adhesion, orientation, and proliferation. Moreover, a study demonstrated that in microgrooved features, the ridge width is commonly larger than or equal to the size of a single cell, permissive for cell attachment and migration, as well as cell alignment following the geometrical guidance. In contrast, nanogrooved features are similar to the ECM architecture and are typically much smaller than a single cell, thus inducing cell alignment in a more fundamental way, such as mimicking or signaling the cell membrane receptors [51].

2.5. Fluorescent Images of Schwann Cells Seeded on Laser-Microstructured Substrates on PET (PET-MG) (Immunostaining) and PLGA-MS Replicas (Immunostaining)

In Figure 7, we present the fluorescent images of Schwann cells cultured on the PET substrates (PET-MG and PET-Flat) for 4 and 6 days. The actin filament of cytoskeleton is visualized with red color, while the nuclei is indicated by blue color. We noticed that the cytoskeleton of the cells was elongated along the direction of the microgrooves, whereas a random orientation was observed on the flat PET. It is important to mention here that the width of the microgrooves is a critical parameter for the alignment of Schwann cells. The width of the Schwann cells ranges from 5 to 10 µm. It has been shown [14] that pattern widths or spacings varying from 2 to 30 µm are optimal for the alignment of Schwann cells. Indeed, when we used a topography of microgrooves with a width of 28.68 ± 0.47 µm, cells appeared to be oriented along the direction of microgrooves (as it is shown in Figure 7), while, when the width of the microgrooves was 168.12 ± 1.38 µm, a random orientation of cells was observed We found that PLGA-MS replicas seeded with Schwann cells resulted in the presence of elongated and round cells at the proliferation stage and signs of orientation according to the spikes (3 days of culture). Specifically, Schwann cells grew more randomly and in an isotropic manner on low-roughness PLGA-MS, comparable with the growth on flat PLGA. On medium- and high-roughness PLGA-MS replicas, the cells exhibited a directional growth. At the fifth day, the presence of elongated cells at different layers was observed, and there was full coverage of the surface. There was no difference between the three different topographies and the control substrate at this time point (Figure 8).

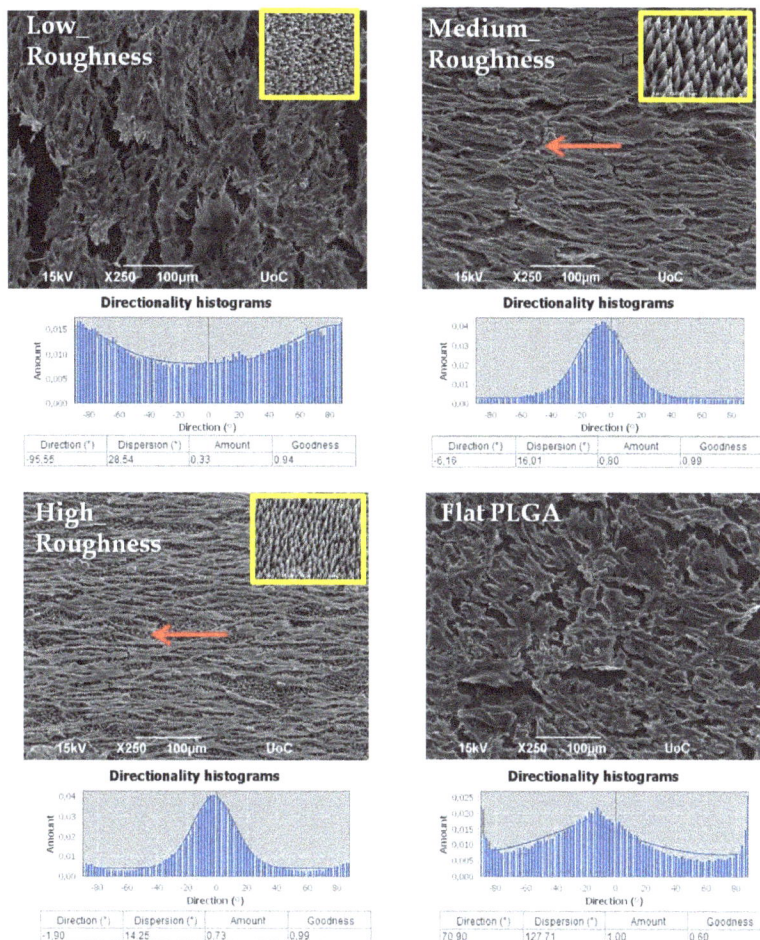

Figure 6. Scanning electron microscopy (SEM) images of Schwann cells cultured on the PLGA-MS replicas (three topographies) and on flat PLGA for 3 days. The red arrows represent the directionality of Schwann cells, which are oriented according to the topography of the PLGA-MS replica (inset SEM image, indicated by the yellow box, on the right side of each group). Under the SEM images, directionality histograms and tables with statistics are presented, which were generated using the Fiji ImageJ plug-in "Directionality" [43]. Above the histogram, the plug-in generates statistics for the highest peak found. The highest peak is fitted by a Gaussian function, taking into account the periodic nature of the histogram. In the tables, the "Direction (°)" column reports the center of the Gaussian; the "Dispersion (°)" column reports the standard deviation of the Gaussian; the "Amount" column is the sum of the histogram from center-std to center+std, divided by the total sum of the histogram; the "Goodness" column reports the goodness of the fit, where 1 is good, 0 is bad.

The outgrowth of Schwann cells (number of cells/mm^2) on the PET-MG substrate and on flat PET was evaluated by counting cell nuclei stained with DAPI (Figure 7). Nuclei number was assessed with Fiji ImageJ analysis. Figure 9 depicts the mean cell number on the PET-MG substrate and flat PET for 4 and 6 days of culture. The cell outgrowth was improved on PET-MG substrate compared to the flat PET, in agreement with the SEM and fluorescent images, with a significant difference between 4 days and 6 days at PET-MG. According to Figure 10, all three topographies of the PLGA-MS replicas support Schwann cells' growth (results that are also confirmed from the SEM and fluorescent images) and proliferation for up to 5 days. The high-roughness PLGA-MS replicas had the highest cell number at both time points, followed by the medium-roughness PLGA-MS replica, while the lowest cell proliferation was observed for the low-roughness PLGA-MS replica. Taking into consideration Figures 8 and 10, it is clear that these findings are in agreement with our previous work on Si microstructured substrates, where the surface roughness did not influence the Schwann cell growth, but the surface morphology (discontinuous pattern) played a key role in cell response [20]. Schwann cells seemed to be aligned with the orientation of the spikes' topographical features and, specifically, this preference was more pronounced as the roughness increased. The key geometrical characteristics (height, width, and aspect ratio) of the substrates leading to the anisotropic nature of the spikes and their parallel orientation varied between the three topographies and significantly affected the degree of cell alignment. The previous findings of the group are also demonstrated in the present study, with cell growth on the low-roughness substrates having an isotropic manner similar to flat and control materials (shown clearly in Figure 8), and cell growth on medium- and high-roughness substrates exhibiting a more pronounced anisotropic growth (shown in Figures 6 and 8). It should be noted here that the height of the spikes and the interspike distance cannot be controlled by the microfabrication techniques used in this study, and, since there is the step of replication of the topography (from the Si master mold, through the PDMS negative mold, to the final PLGA-MS replica), there is definitely a slight difference between the fabricated topography and the replicated topography in terms of the height of the spikes. These results demonstrate that the micro- and nanostructures favor the cell outgrowth.

In this study, we demonstrated that ultrafast pulsed laser irradiation is a simple and effective method to fabricate micro- and nanostructures with controlled geometry and pattern regularity. Two different synthetic polymers—the fabricated PET-MG substrates and the produced PLGA-MS replicas at a range of laser fluences, resulting in different levels of roughness, and geometrical characteristics were investigated for their selective cellular adhesion, proliferation, and orientation. In this context, we studied the effects of an anisotropic continuous topography and three anisotropic discontinuous topographies on cellular response.

The morphological characterization of the PET-MG substrates and the PLGA-MS replicas (SEM images) indicated a topography with microgrooves (anisotropic continuous) for the PET substrates and microspikes (anisotropic discontinuous) for the PLGA replicas. This is due to the different fabrication processes used; PET substrates were laser-irradiated directly, and the PLGA-MS replicas were produced by soft lithography of laser-irradiated Si substrates. Thus, although the same laser irradiation process was used, the different materials formed a range of topographies, as shown in Figure 11. The composition and the mechanical properties of the material play a significant role in the topography [52]. The wetting and absorbance (related to optical properties) were assessed by the contact angle and the UV–Vis system, respectively. These properties were mainly affected by the topography of the material. Schwann cells attached strongly and proliferated on all the substrates. The cell adhesion/orientation engineering profile was mainly affected by the topography, while the cell proliferation was influenced by the topography.

The specific cell patterning model involving anisotropic continuous microgrooves (PET-MG) and anisotropic discontinuous microspikes with parallel orientations (PLGA-MS replicas) were developed in an attempt to imitate native nerve regeneration support structures, particularly imitating the guidance/alignment and growth of Schwann cells. It is known that primary Schwann cells transiently proliferate and form longitudinal bands of Bürger (boB) [53]. Aligned Schwann cells and their extracellular matrix are indispensable pathways for oriented axonal regrowth. The boB formation from a molecular point of view is unknown. A potential mechanism could be the polarized expression of adhesion proteins along the proximal–distal cell axis [53]. It was reported that placement of dissimilar adhesion characteristics in separate Schwann cell surface domains could aid longitudinal cell alignment.

From a physical point of view, the basal lamina tube (enwrapping Schwann cells and myelinated axons) is the guiding cue for axonal regrowth [53].

Two different "axonal guidance"' models were studied here. By using the same microfabrication techniques, two models were fabricated with different topographical (anisotropic continuous vs. discontinuous) geometries. The same cell type was tested. Schwann cells adhered, grew, equally aligned, and proliferated in both the models. Both models feature topographical cues (pattern) with a combination of nano- and microcharacteristics and are proposed to overcome the weaknesses of the existing and well-studied horizontal (grooves and ridges) or vertical (pillars, pores) cell patterning models.

The ability of this micropatterning strategy to control cellular adhesion and growth, and thus to engineer cell alignment in vitro, could be potentially useful in a wide range of neuroscience subfields, including basic research to understand cell interactions and network behavior; dynamic microenvironment systems that would better simulate the desired in vivo conditions; and, finally, neural tissue engineering, with the creation of implantable scaffolds for nerve tissue regeneration.

Day4

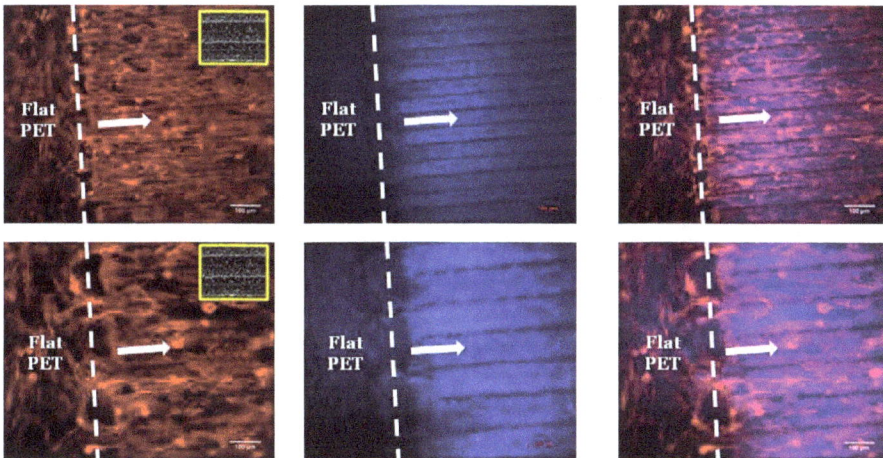

Day6

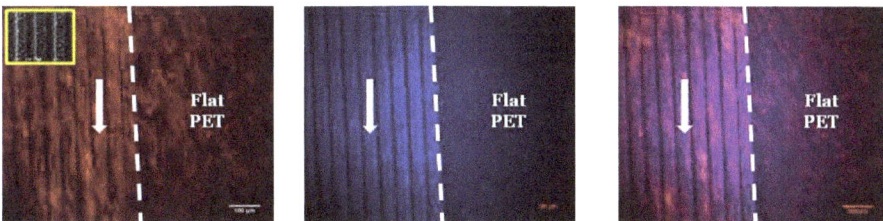

Figure 7. Fluorescent images of Schwann cells cultured on the PET substrates (PET-MG and PET-Flat) for 4 and 6 days. The cytoskeleton of the cells is visualized with red color (Alexa Fluor® 568 Phalloidin), while the nuclei are indicated with blue color (DAPI). The white arrows represent the directionality of Schwann cytoskeleton, which is according to the direction of the microgrooves. The inset SEM images, indicated by the yellow box, show the geometry of microgrooves.

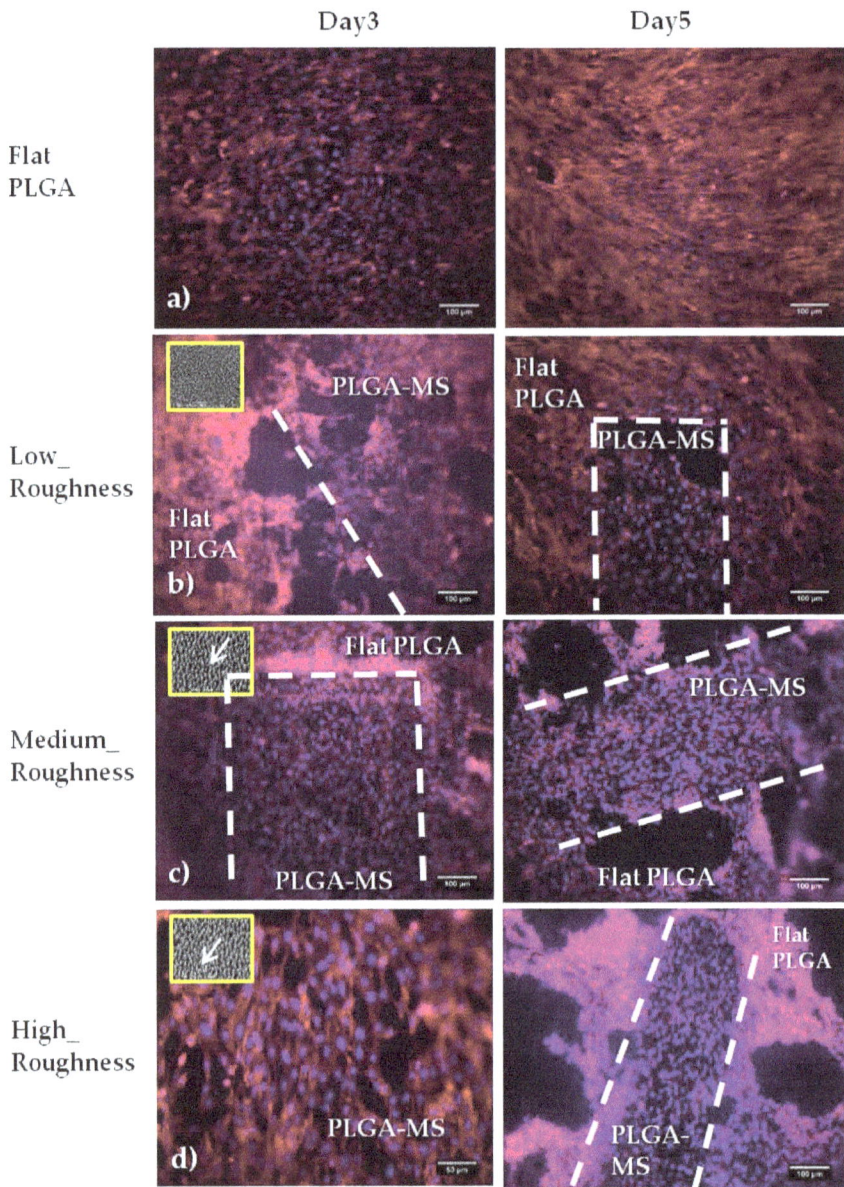

Figure 8. Fluorescent images of Schwann cells cultured on the (**a**) flat PLGA and PLGA-MS replicas (**b–d**) for 3 and 5 days. Each replica is defined by low (**b**), medium (**c**), and high (**d**) roughness. The cytoskeleton of the cells is visualized with red color (Alexa Fluor® 568 Phalloidin). The white interrupted lines represent the PLGA-MS area, and thus the area with the spikes. It should be noted that for this specific study using the actin/DAPI assay, the PLGA-MS replicas (spike's area) were 0.3–0.5 mm (width) and 1.5 mm (length). The inset SEM images on the left side, indicated by the yellow box, show the topography of the PLGA-MS replicas, and the white arrows represent the directionality of the spikes.

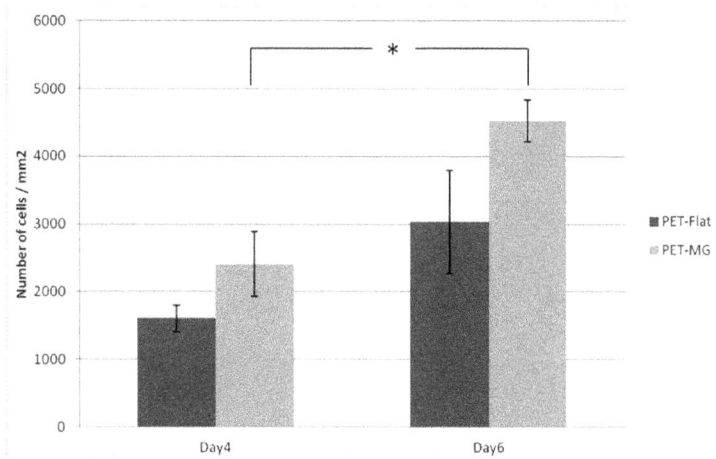

Figure 9. Proliferation of Schwann cells (number of cells/mm^2) cultured on the PET substrates (PET-MG and PET-Flat) (via DAPI) for 4 and 6 days. The data were subjected to ANOVA with post hoc Tukey HSD test. A significant difference (* $p < 0.0.5$) was observed between 4 days and 6 days for the PET-MG substrate.

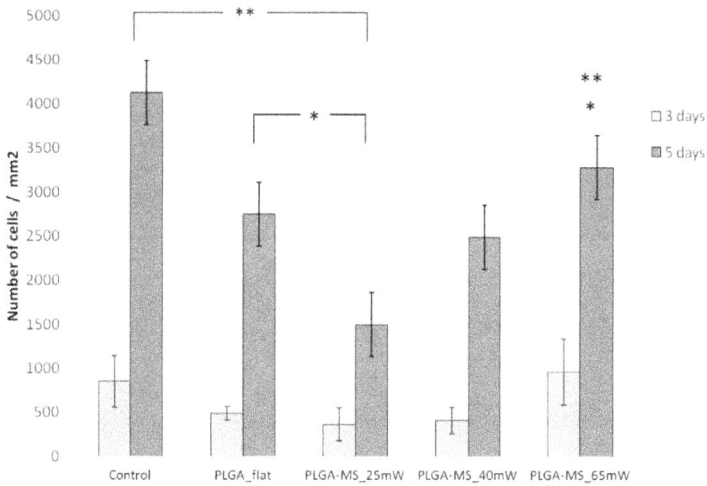

Figure 10. Proliferation of Schwann cells (number of cells /mm^2) cultured on the PLGA-MS replicas, and PLGA flat and control samples (via live/dead assay) for 3 and 5 days. The data were subjected to ANOVA with post hoc Tukey HSD test for multiple comparisons between the groups. At 3 days, the p value > 0.05; therefore, the treatments (groups) were not significantly different for that level of significance. However, at 5 days, we observed some significant differences, strongly suggesting that one or more pairs of treatments (groups) are significantly different. In particular, the control group is significantly different from PLGA flat and PLGA-MS replicas of 25 mW_Low Roughness and 65 mW_High Roughness (** $p < 0.001$); PLGA-MS replica 25 mW_Low Roughness is significantly different from PLGA flat and the replica 65 mW-High Roughness (* $p < 0.05$).

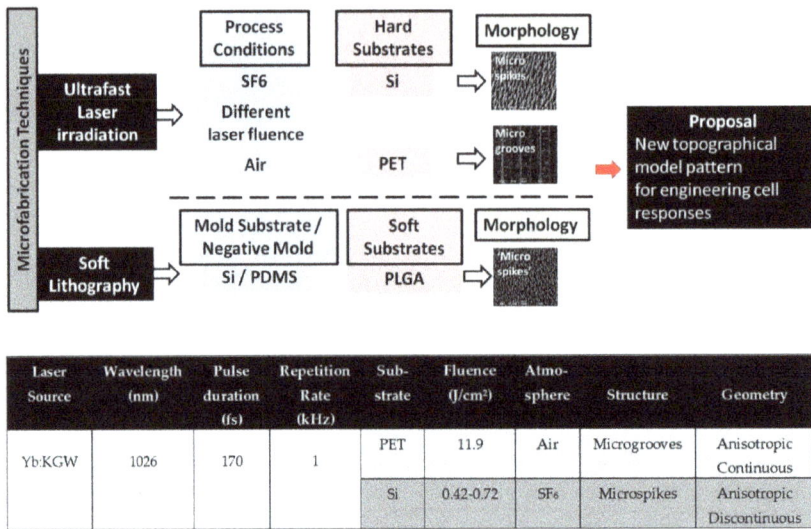

Figure 11. Comparison of the microfabricating techniques used in this study to fabricate the laser-microstructured substrates; the table demonstrates the conditions of the ultrafast laser irradiation process.

3. Materials and Methods

3.1. Experimental Setup Used for the Fabrication of Laser-Microstructured Substrates

The microstructured substrates were prepared by ultrafast laser structuring, which is a simple but effective method to fabricate micro/nanostructures with different geometries [41]. The specially treated PET (polyethylene terephthalate) coverslips for cell culturing were subjected to laser irradiation. A Yb:KGW laser was used with a pulse duration equal to 170 fs, 1 kHz repetition rate, and 1026 nm wavelength. The beam propagated through a half waveplate and a linear polarizer (which were used to vary the values of power), to a shutter (that was used to control the exposure time and thus the number of pulses receptive to the sample), then to a convex lens of 10 mm focal length, and, finally, to the sample (Figure 12). The microstructured substrates were fabricated at a constant fluence of 11.9 J/cm^2, scan velocity of 7 mm/s, and an x_{step} (distance between two consecutive scan lines) of 50 μm. The overall patterned area was 4 mm × 4 mm.

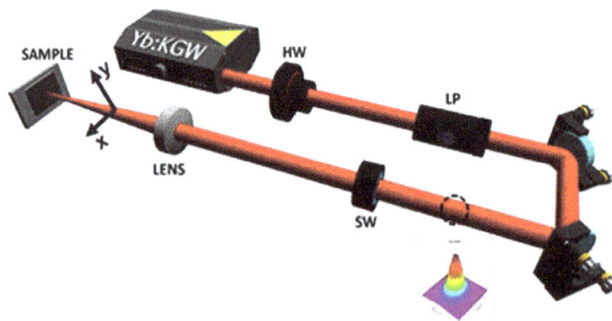

Figure 12. Experimental setup used for the fabrication of laser-microstructured substrates [42].

The same laser setup (Figure 12) was used to fabricate the Si substrates as described above. Single-crystal n-type silicon (1 0 0) wafers were subjected to laser irradiation in a vacuum chamber evacuated down to a residual pressure of 10^{-2} mbar. A constant sulfur hexafluoride (SF6) pressure of 650 mbar was maintained during the process through a precision microvalve system. A Yb:KGW laser was used with a pulse duration equal to 170 fs, 1 kHz repetition rate, and 1026 nm wavelength. The sample was mounted on a high-precision X–Y translation stage normal to the incident laser beam. The laser fluence used in these experiments was in the range 0.42–0.72 J/cm^2, thus creating three different topographies, defined as low, medium, and high topography [20,21]. The overall spike area was 5 mm × 5 mm. After laser irradiation, microstructured surfaces were morphologically characterized by scanning electron microscopy (SEM). The top SEM images (Figure 3d–f) revealed an arbitrarily shaped cross-section of the microstructures at low fluences that became almost elliptical as the laser fluence increased.

The laser-fabricated Si substrate is characterized as the "master" substrate. Negative replicas of the three master Si substrates were produced on elastomeric PDMS (SYLGARD 184, Dow Corning). In particular, liquid PDMS pre-polymer consisting of a "base" and "curing agent", typically mixed in a 10:1 *w:w* ratio, was poured onto each substrate [54]. Then, the PDMS-coated Si substrates were placed into a vacuum chamber to remove residual air bubbles, thus providing for better penetration of the polymer into the laser microstructures. After heating at 80 °C for 2 h, a mold, which holds the negative of the original pattern, was peeled off of each Si substrate. An adequate number of PDMS negative molds was produced. Using the PDMS negative mold (negative spikes morphology), replicas of the initial morphology can be made out of several polymeric materials. In this study, we demonstrated the successful reproduction of the initial Si morphologies by producing PLGA replicas. A PLGA (lactide:glycolide 65:35, MW 40–75 k) polymeric solution of 1:10 (Code No: P2066, Sigma Aldrich, St. Louis, MO, USA) in dichloromethane (DCM) was carefully prepared. The PLGA solution was magnetically stirred for 2 h at room temperature (RT). One droplet of the PLGA solution was poured onto each PDMS negative mold and slowly finger-pressed with a glass disk. Following the evaporation of the solvent (24–48 h in −20 °C), the PLGA-coated PDMS mold was placed in 4 °C for 2 h. Then, the PLGA replica was peeled from the PDMS negative mold with a pair of tweezers.

3.2. Characterization of Laser-Microstructured Substrates

3.2.1. Scanning Electron Microscopy (SEM)

The laser-microstructured substrates were morphologically characterized by scanning electron microscopy (SEM) (JEOL JSM-6390 LV, Jeol USA Inc, Peabody, MA, USA). Specifically, the substrates were sputter-coated with a 15 nm layer of gold (Baltec SCD 050, BAL-TEC AG, Balzers, Liechtenstein) and observed under the microscope with an acceleration voltage of 15 kV. Fiji ImageJ, an image processing software, was used to perform the analysis of the geometrical characteristics of the microgrooves and microspikes on the three topographies, as described in [20,21]. Briefly, the aspect ratio of the microgrooves/microspikes, A, was calculated by dividing the depth/height of the microgrooves/microspikes by the width. The roughness ratio, r, was calculated by dividing the actual, unfolded surface area of microgrooves/microspikes by the total irradiated area.

For the determination of the directionality of the PLGA-MS replicas, the "Local gradient orientation" for directionality was performed using the Fiji ImageJ plug-in "Directionality" [43].

3.2.2. Wettability Measurements of Laser-Microstructured Substrates

The contact angles of the laser-microstructured substrates were calculated via an automated tensiometer, using the sessile drop method. A droplet of distilled, deionized Millipore water with a volume of 4 μL was positioned on the surface of the substrates using a microsyringe, and images were taken to measure the angle formed at the liquid–solid interface.

3.2.3. UV–Vis Measurements of Laser-Microstructured Substrates

The UV–Vis absorption spectra of the laser-microstructured substrates were measured with a LAMBDA 950 UV/VIS/NIR spectrophotometer from Perkin Elmer with spectral range from 250 nm to 1200 nm. Laser-microstructured substrates and their relevant flat substrates were used for these measurements.

3.3. Cell Culture

The Schwann (SW10) mouse cell line is an established adherent neuronal Schwann cell line; it has been immortalized with SV40 large T antigen. SW10 cells were obtained from ATCC® (Code: CRL-2766™). Schwann cells were grown in cell culture flasks using Dulbecco's modified Eagle's medium (DMEM (Invitrogen, Grand Island, NY, USA) supplemented with 10% fetal bovine serum (Biosera, Sussex, UK) in a 5% CO_2 incubator (Thermo Scientific, OH, USA) at 33 °C. Laser-microstructured substrates were UV sterilized and transferred into sterile wells of 24-well plates (Sarstedt; Numbrecht, Germany). Then, 3×10^4 cells in culture medium were seeded on the substrates and were cultured for a series of different time periods depending on the substrates, ranging from 3 to 6 days. The cell orientation and proliferation were better assessed at 4 and 6 days for the PET substrates, while for the PLGA replicas, the optimized time points were 3 and 5 days. The control samples in all the experiments were PET (polyethylene terephthalate) coverslips for cell culture.

3.3.1. Morphology of Schwann (SW10) Cells by Scanning Electron Microscopy (SEM)

The laser-microstructured substrates seeded with the SW10 cells were removed from the incubator, washed twice with 0.1 M sodium cacodylate buffer (SCB), and fixed with 2% glutaraldehyde (GDA) and 2% paraformaldehyde (PFA) in 0.1 M SCB for 30 min. Thereafter, they were washed twice with 0.1 M SCB and dehydrated in increasing concentrations (from 30–100%) of ethanol. Finally, they were dried in a critical point drier (Baltec CPD 030, , BAL-TEC AG, Balzers, Liechtenstein), sputter-coated with a 15 nm layer of gold (Baltec SCD 050, BAL-TEC AG, Balzers, Liechtenstein), and observed under a scanning electron microscope (JEOL JSM-6390 LV, Jeol USA Inc, Peabody, MA, USA) at an accelerating voltage of 15 kV. For the PLGA replicas, CPD cannot be used since it deforms the polymer, so, after an optimization process, a hexamethyldisilizane (HDMS) protocol was established. After the dehydration steps with ethanol (EtOH), EtOH:HDMS solutions (50:50) were used for specific time points for all the replicas, and then the same procedure was repeated with HDMS solutions. Finally, the replicas were left to dry at room temperature overnight.

To investigate changes in the directional orientation of Schwann cells on the microstructured substrates, the "Local gradient orientation" for directionality was performed using the Fiji ImageJ plug-in "Directionality" [43].

3.3.2. Immunocytochemical Assay

SW10 cells were stained for F-actin. Specifically, after 4 and 6 days of culture, the samples were fixed with 4% PFA for 15 min and permeabilized with 0.1% Triton X-100 in PBS for 5 min. The non-specific binding sites were blocked with 2% BSA in PBS for 30 min. Then, the samples were incubated for 2 h at room temperature with Alexa Fluor® 568 Phalloidin (Invitrogen, Thermo Fisher Scientific) (1:250 in PBS–BSA 1%) for F-actin staining. Finally, the samples were washed with PBS and put on coverslips with DAPI (Molecular Probes by Life Technologies, Carlsbad, CA, USA) for nuclei staining. Cell imaging was performed using an epifluorescence microscope coupled to a high-resolution Carl Zeiss Axiocam color camera. The objectives of ×10 and ×20 were used. The number of SW10 cells that were grown on the microstructured substrates were determined by counting cell nuclei stained with DAPI. Nuclei number was assessed with ImageJ (cell counter plugin). The results represent the means of three different experiments (n = 10 field-of-view images for each substrate and time point).

3.3.3. Live/Dead Assay

SW10 cells were seeded onto a PLGA replica to a density of 3×10^4 cells/well. After 3 and 5 days of incubation under standardized culture conditions, medium was removed and replaced by a live/dead viability/cytotoxicity solution. The LIVE/DEAD™ Viability/Cytotoxicity Kit for mammalian cells (L3224, Thermo Scientific) was used for evaluating cell viability and proliferation. The cell-adhered replicas were washed twice with PBS. A live/dead solution was prepared by adding 20 µL of the supplied 2 mM ethidium homodimer-1 (EthD-1) stock solution to 10 mL of sterile PBS (thus reaching the desired concentration of 4 µM EthD-1 solution) and, after mixing thoroughly, 5 µL of the supplied 4 mM calcein AM stock solution was added to the 10 mL EthD-1 solution (thus reaching the desired concentration of 2 µM calcein AM solution). The solution was directly added to the replicas in order to cover the whole sample and was left for 45 min at room temperature. Finally, the cells were washed once with PBS, and fluorescent images were obtained by fluorescent microscope (images not shown here). The number of SW10 cells that were grown on the microstructured substrates were determined by counting cell nuclei stained with calcein. Nuclei number was assessed with ImageJ (cell counter plugin). The results represent the means of three different experiments (n = 10 field-of-view images for each substrate and time point).

3.4. Statistical Analysis

The data were subjected to ANOVA with post hoc Tukey HSD test to compare the significance levels ($p < 0.05$) between multiple groups.

4. Conclusions

Successful fabrication of micropatterned substrates was accomplished via ultrafast laser irradiation and soft lithography. Ultrafast pulsed laser irradiation is a simple and effective method to fabricate micro- and nanostructures with controlled geometry and pattern regularity. The anisotropic continuous (PET-MG) and discontinuous (PLGA-MS replicas) microstructured polymeric substrates were assessed in terms of their geometrical and topographical parameters (aspect ratio, roughness, and directionality), their influence on Schwann cell responses, and the reproducibility of these responses. The cells attached strongly and proliferated well on the substrates. Surface topography affected Schwann cells. Moreover, cells appeared to be oriented along the direction of the microgrooves and microspikes. This micropatterned strategy to control cellular adhesion and growth, thus engineering cell alignment in vitro, could be potentially useful in the field of neural tissue engineering and for the assessment in dynamic microenvironments by sufficiently simulating in vivo conditions.

Author Contributions: E.B. performed the full study of the PET microstructured substrates (from their fabrication, characterization of their properties, cell studies and analyzing the data); P.K. performed the PLGA replicas and their assessment (from their replication, characterization, cell studies and analyzing the data); D.A. performed the fabrication of the Si microstructured substrates used for the PLGA replication; L.C. performed the characterization of the PLGA replicas in terms of morphology, contact angle measurements and optical properties; A.M. performed all the SEM imaging; A.S.-G. optimized the HDMS protocol for the successful preparation of the PLGA replicas for SEM imaging and performed some of the SEM imaging on PLGA replicas with SW10 cells; A.R. contributed to the study objectives, supervision and writing of paper; E.S. contributed with funding this study, supervision, and writing of paper.

Acknowledgments: This work was supported from funding by NFFA (EU H2020 framework programme) under grant agreement n. 654360 from 1/9/2015 to 31/8/2019, State Scholarship Foundation (IKY) within the framework of the Action "Doctoral Research Support" (MIS 5000432), ESPA 2014-2020 Program, CN: 2016-ESPA-050-0502-5321 and Onassis Foundation through the G ZM 039-1/2016-2017 scholarship grant. We acknowledge also support of this work by the project "Advanced Research Activities in Biomedical and Agro alimentary Technologies" (MIS 5002469) which is implemented under the "Action for the Strategic Development on the Research and Technological Sector", funded by the Operational Programme "Competitiveness, Entrepreneurship and Innovation" (NSRF 2014-2020) and co-financed by Greece and the European Union (European Regional Development Fund).

Conflicts of Interest: The authors declare no conflict of interest.

Abbreviations

2D	Two-dimensional
A	Absorbance
ATR-FTIR	Attenuated Total Reflection-Fourier Transform Infrared (spectroscopy)
BSA	Bovine Serum Albumen
CO_2	Carbon dioxide
CPD	Critical Point Dryer
DAPI	4′,6-Diamidino-2-Phenylindole
DCM	Dichloromethane
DMEM	Dulbecco's Modified Eagles Medium
ECM	Extracellular matrix
EthD-1	Ethidium Homodimer-1
EtOH	Ethanol
GDA	Glutaraldehyde
HDMS	Hexamethyldisilizane
IESL	Institute of Electronic Structure and Laser
MG	Microgrooves
MS	Microspikes
NGF	Nerve Growth Factor
PBS	Phosphate-buffered saline
PC12	Pheochromocytoma
PDMS	Poly(dimethylsiloxane)
PET	Polyethylene terephthalate
PFA	Paraformaldehyde
PLGA	Poly(lactide-co-glycolide)
RT	Room Temperature
SCB	Sodium Cacodylate buffer
SEM	Scanning Electron Microscopy
SF_6	Sulfur hexafluoride
Si	Silicon
SW10	Schwann cells
UV	Ultraviolet
UV-Vis	Ultraviolet-Visible
UV/VIS/NIR	Ultraviolet/Visible/Near Infrared
Yb:KGW	Ytterbium-doped potassium gadolinium tungstate

References

1. Wu, R.-X.; Yin, Y.; He, X.-T.; Li, X.; Chen, F.-M. Engineering a Cell Home for Stem Cell Homing and Accommodation. *Adv. Biosyst.* **2017**, *1*, 1700004, doi:10.1002/adbi.201700004. [CrossRef]
2. Stevens, M.M.; George, J.H. Exploring and Engineering the Cell Surface Interface. *Science* **2005**, *310*, 1135–1138. [CrossRef] [PubMed]
3. Pasapera, A.M.; Schneider, I.C.; Rericha, E.; Schlaepfer, D.D.; Waterman, C.M. Myosin II activity regulates vinculin recruitment to focal adhesions through FAK-mediated paxillin phosphorylation. *J. Cell Biol.* **2010**, *188*, 877–890. [CrossRef] [PubMed]
4. Crowder, S.W.; Leonardo, V.; Whittaker, T.; Papathanasiou, P.; Stevens, M.M. Material Cues as Potent Regulators of Epigenetics and Stem Cell Function. *Cell Stem Cell* **2016**, *18*, 39–52. [CrossRef] [PubMed]
5. Dalby, M.J.; Gadegaard, N.; Oreffo, R.O.C. Harnessing nanotopography and integrin–matrix interactions to influence stem cell fate. *Nat. Mater.* **2014**, *13*, 558–569. [CrossRef] [PubMed]
6. Stukel, J.M.; Willits, R.K. Mechanotransduction of Neural Cells Through Cell–Substrate Interactions. *Tissue Eng. Part B Rev.* **2016**, *22*, 173–182. [CrossRef] [PubMed]

Int. J. Mol. Sci. **2018**, 19, 2053

7. Yiannakou, C.; Simitzi, C.; Manousaki, A.; Fotakis, C.; Ranella, A.; Stratakis, E. Cell patterning via laser micro/nano structured silicon surfaces. *Biofabrication* **2017**, *9*, 25024, doi:10.1088/1758-5090/aa71c6. [CrossRef] [PubMed]

8. Wang, X.; Ohlin, C.A.; Lu, Q.; Hu, J. Cell directional migration and oriented division on three-dimensional laser-induced periodic surface structures on polystyrene. *Biomaterials* **2008**, *29*, 2049–2059. [CrossRef] [PubMed]

9. Rebollar, E.; Frischauf, I.; Olbrich, M.; Peterbauer, T.; Hering, S.; Preiner, J.; Hinterdorfer, P.; Romanin, C.; Heitz, J. Proliferation of aligned mammalian cells on laser-nanostructured polystyrene. *Biomaterials* **2008**, *29*, 1796–1806. [CrossRef] [PubMed]

10. Ranella, A.; Barberoglou, M.; Bakogianni, S.; Fotakis, C.; Stratakis, E. Tuning cell adhesion by controlling the roughness and wettability of 3D micro/nano silicon structures. *Acta Biomater.* **2010**, *6*, 2711–2720. [CrossRef] [PubMed]

11. Orgovan, N.; Peter, B.; Bősze, S.; Ramsden, J.J.; Szabó, B.; Horvath, R. Dependence of cancer cell adhesion kinetics on integrin ligand surface density measured by a high-throughput label-free resonant waveguide grating biosensor. *Sci. Rep.* **2015**, *4*, 4034, doi:/10.1038/srep04034. [CrossRef] [PubMed]

12. Makarona, E.; Peter, B.; Szekacs, I.; Tsamis, C.; Horvath, R. ZnO Nanostructure Templates as a Cost-Efficient Mass-Producible Route for the Development of Cellular Networks. *Materials (Basel)* **2016**, *9*, 256, doi:10.3390/ma9040256. [CrossRef] [PubMed]

13. Simitzi, C.; Karali, K.; Ranella, A.; Stratakis, E. Controlling the Outgrowth and Functions of Neural Stem Cells: The Effect of Surface Topography. *ChemPhysChem* **2018**, doi:10.1002/cphc.201701175. [CrossRef] [PubMed]

14. Simitzi, C.; Ranella, A.; Stratakis, E. Controlling the morphology and outgrowth of nerve and neuroglial cells: The effect of surface topography. *Acta Biomater.* **2017**, *51*, 21–52. [CrossRef] [PubMed]

15. Abdeen, A.A.; Lee, J.; Kilian, K.A. Capturing extracellular matrix properties in vitro: Microengineering materials to decipher cell and tissue level processes. *Exp. Biol. Med.* **2016**, *241*, 930–938. [CrossRef] [PubMed]

16. Chen, W.; Shao, Y.; Li, X.; Zhao, G.; Fu, J. Nanotopographical surfaces for stem cell fate control: Engineering mechanobiology from the bottom. *Nano Today* **2014**, *9*, 759–784. [CrossRef] [PubMed]

17. Griffith, L.G.; Swartz, M.A. Capturing complex 3D tissue physiology in vitro. *Nat. Rev. Mol. Cell Biol.* **2006**, *7*, 211–224. [CrossRef] [PubMed]

18. Mirzadeh, H.; Dadsetan, M. Influence of laser surface modifying of polyethylene terephthalate on fibroblast cell adhesion. *Radiat. Phys. Chem.* **2003**, *67*, 381–385. [CrossRef]

19. Dadsetan, M.; Mirzadeh, H.; Sharifi-Sanjani, N.; Salehian, P. IR Laser Surface Modification of Polyethylene Terephthalate as Biomaterial. In *Processing and Fabrication of Advanced Materials VIII*; World Scientific: Singapore, 2001; pp. 221–229.

20. Simitzi, C.; Efstathopoulos, P.; Kourgiantaki, A.; Ranella, A.; Charalampopoulos, I.; Fotakis, C.; Athanassakis, I.; Stratakis, E.; Gravanis, A. Laser fabricated discontinuous anisotropic microconical substrates as a new model scaffold to control the directionality of neuronal network outgrowth. *Biomaterials* **2015**, *67*, 115–128. [CrossRef] [PubMed]

21. Simitzi, C.; Stratakis, E.; Fotakis, C.; Athanassakis, I.; Ranella, A. Microconical silicon structures influence NGF-induced PC12 cell morphology. *J. Tissue Eng. Regen. Med.* **2015**, *9*, 424–434. [CrossRef] [PubMed]

22. Stratakis, E.; Ranella, A.; Fotakis, C. Biomimetic micro/nanostructured functional surfaces for microfluidic and tissue engineering applications. *Biomicrofluidics* **2011**, *5*, 13411, doi:10.1063/1.3553235. [CrossRef] [PubMed]

23. Chow, W.N.; Simpson, D.G.; Bigbee, J.W.; Colello, R.J. Evaluating neuronal and glial growth on electrospun polarized matrices: Bridging the gap in percussive spinal cord injuries. *Neuron Glia Biol.* **2007**, *3*, 119–126. [CrossRef] [PubMed]

24. Schnell, E.; Klinkhammer, K.; Balzer, S.; Brook, G.; Klee, D.; Dalton, P.; Mey, J. Guidance of glial cell migration and axonal growth on electrospun nanofibers of poly-ε-caprolactone and a collagen/poly-ε-caprolactone blend. *Biomaterials* **2007**, *28*, 3012–3025. [CrossRef] [PubMed]

25. Hoffman-Kim, D.; Mitchel, J.A.; Bellamkonda, R.V. Topography, cell response, and nerve regeneration. *Annu. Rev. Biomed. Eng.* **2010**, *12*, 203–231. [CrossRef] [PubMed]

26. Johansson, F.; Carlberg, P.; Danielsen, N.; Montelius, L.; Kanje, M. Axonal outgrowth on nano-imprinted patterns. *Biomaterials* **2006**, *27*, 1251–1258. [CrossRef] [PubMed]

27. Yao, L.; Wang, S.; Cui, W.; Sherlock, R.; O'connell, C.; Damodaran, G.; Gorman, A.; Windebank, A.; Pandit, A. Effect of functionalized micropatterned PLGA on guided neurite growth. *Acta Biomater.* **2008**, *5*, 580–588. [CrossRef] [PubMed]

28. Schulte, C.; Rodighiero, S.; Cappelluti, M.A.; Puricelli, L.; Maffioli, E.; Borghi, F.; Negri, A.; Sogne, E.; Galluzzi, M.; Piazzoni, C.; et al. Conversion of nanoscale topographical information of cluster-assembled zirconia surfaces into mechanotransductive events promotes neuronal differentiation. *J. Nanobiotechnol.* **2016**, *14*, 18, doi:10.1186/s12951-016-0171-3. [CrossRef] [PubMed]

29. Maffioli, E.; Schulte, C.; Nonnis, S.; Grassi Scalvini, F.; Piazzoni, C.; Lenardi, C.; Negri, A.; Milani, P.; Tedeschi, G. Proteomic Dissection of Nanotopography-Sensitive Mechanotransductive Signaling Hubs that Foster Neuronal Differentiation in PC12 Cells. *Front. Cell. Neurosci.* **2018**, *11*, 417, doi:10.3389/fncel.2017.00417. [CrossRef] [PubMed]

30. Schulte, C.; Ripamonti, M.; Maffioli, E.; Cappelluti, M.A.; Nonnis, S.; Puricelli, L.; Lamanna, J.; Piazzoni, C.; Podestà, A.; Lenardi, C.; et al. Scale Invariant Disordered Nanotopography Promotes Hippocampal Neuron Development and Maturation with Involvement of Mechanotransductive Pathways. *Front. Cell. Neurosci.* **2016**, *10*, 267. [CrossRef] [PubMed]

31. Zerva, I.; Simitzi, C.; Siakouli-Galanopoulou, A.; Ranella, A.; Stratakis, E.; Fotakis, C.; Athanassakis, I. Implantable vaccine development using in vitro antigen-pulsed macrophages absorbed on laser micro-structured Si scaffolds. *Vaccine* **2015**, *33*, 3142–3149. [CrossRef] [PubMed]

32. Geissler, M.; Xia, Y. Patterning: Principles and Some New Developments. *Adv. Mater.* **2004**, *16*, 1249–1269. [CrossRef]

33. Whitesides, G.M. The origins and the future of microfluidics. *Nature* **2006**, *442*, 368–373. [CrossRef] [PubMed]

34. Nikkhah, M.; Edalat, F.; Manoucheri, S.; Khademhosseini, A. Engineering microscale topographies to control the cell–substrate interface. *Biomaterials* **2012**, *33*, 5230–5246. [CrossRef] [PubMed]

35. Chollet, C.; Chanseau, C.; Remy, M.; Guignandon, A.; Bareille, R.; Labrugère, C.; Bordenave, L.; Durrieu, M.-C. The effect of RGD density on osteoblast and endothelial cell behavior on RGD-grafted polyethylene terephthalate surfaces. *Biomaterials* **2009**, *30*, 711–720. [CrossRef] [PubMed]

36. Li, Y.; Ma, T.; Yang, S.-T.; Kniss, D.A.; Kniss, D.A. Thermal compression and characterization of three-dimensional nonwoven PET matrices as tissue engineering scaffolds. *Biomaterials* **2001**, *22*, 609–618. [CrossRef]

37. Lima, M.J.; Correlo, V.M.; Reis, R.L. Micro/nano replication and 3D assembling techniques for scaffold fabrication. *Mater. Sci. Eng. C* **2014**, *42*, 615–621. [CrossRef] [PubMed]

38. Qian, L.; Ahmed, A.; Glennon-Alty, L.; Yang, Y.; Murray, P.; Zhang, H. Patterned substrates fabricated by a controlled freezing approach and biocompatibility evaluation by stem cells. *Mater. Sci. Eng. C* **2015**, *49*, 390–399. [CrossRef] [PubMed]

39. Tay, C.; Pal, M.; Yu, H.; Leong, W.; Tan, N.; Ng, K.W.; Venkatraman, S.; Boey, F.; Leong, D.T.; Tan, L.P. Bio-inspired Micropatterned Platform to Steer Stem Cell Differentiation. *Small* **2011**, *7*, 1416–1421. [CrossRef] [PubMed]

40. Mandoli, C.; Pagliari, F.; Pagliari, S.; Forte, G.; Di Nardo, P.; Licoccia, S.; Traversa, E. Stem Cell Aligned Growth Induced by CeO$_2$ Nanoparticles in PLGA Scaffolds with Improved Bioactivity for Regenerative Medicine. *Adv. Funct. Mater.* **2010**, *20*, 1617–1624. [CrossRef]

41. Stratakis, E.; Ranella, A.; Fotakis, C. *Laser-Based Biomimetic Tissue Engineering*; Springer: Berlin/Heidelberg, Germany, 2013; pp. 211–236.

42. Skoulas, E.; Manousaki, A.; Fotakis, C.; Stratakis, E. Biomimetic surface structuring using cylindrical vector femtosecond laser beams. *Sci. Rep.* **2017**, *7*, 45114. [CrossRef] [PubMed]

43. Schindelin, J.; Arganda-Carreras, I.; Frise, E.; Kaynig, V.; Longair, M.; Pietzsch, T.; Preibisch, S.; Rueden, C.; Saalfeld, S.; Schmid, B.; et al. Fiji: An open-source platform for biological-image analysis. *Nat. Methods* **2012**, *9*, 676–682. [CrossRef] [PubMed]

44. Donelli, I.; Taddei, P.; Smet, P.F.; Poelman, D.; Nierstrasz, V.A.; Freddi, G. Enzymatic surface modification and functionalization of PET: A water contact angle, FTIR, and fluorescence spectroscopy study. *Biotechnol. Bioeng.* **2009**, *103*, 845–856. [CrossRef] [PubMed]

45. Zorba, V.; Stratakis, E.; Barberoglou, M.; Spanakis, E.; Tzanetakis, P.; Anastasiadis, S.H.; Fotakis, C. Biomimetic Artificial Surfaces Quantitatively Reproduce the Water Repellency of a Lotus Leaf. *Adv. Mater.* **2008**, *20*, 4049–4054. [CrossRef]

46. D'avila, C.; Erbetta, C.; Alves, R.J.; Resende, J.M.; Fernando De Souza Freitas, R.; Geraldo De Sousa, R. Synthesis and Characterization of Poly(D,L-Lactide-co-Glycolide) Copolymer. *J. Biomater. Nanobiotechnol.* **2012**, *3*, 208–225. [CrossRef]

47. Grossetête, T.; Rivaton, A.; Gardette, J.L.; Hoyle, C.E.; Ziemer, M.; Fagerburg, D.R.; Clauberg, H. Photochemical degradation of poly(ethylene terephthalate)-modified copolymer. *Polym. (Guildf)* **2000**, *41*, 3541–3554. [CrossRef]

48. Fechine, G.J.; Rabello, M.S.; Souto Maior, R.M.; Catalani, L.H. Surface characterization of photodegraded poly(ethylene terephthalate). The effect of ultraviolet absorbers. *Polym. (Guildf)* **2004**, *45*, 2303–2308. [CrossRef]

49. Prasad, S.G.; De, A.; De, U. Structural and Optical Investigations of Radiation Damage in Transparent PET Polymer Films. *Int. J. Spectrosc.* **2011**, *2011*, 1–7, doi:10.1155/2011/810936. [CrossRef]

50. Upson, S.J.; Partridge, S.W.; Tcacencu, I.; Fulton, D.A.; Corbett, I.; German, M.J.; Dalgarno, K.W. Development of a methacrylate-terminated PLGA copolymer for potential use in craniomaxillofacial fracture plates. *Mater. Sci. Eng. C* **2016**, *69*, 470–477. [CrossRef] [PubMed]

51. Qi, L.; Li, N.; Huang, R.; Song, Q.; Wang, L.; Zhang, Q.; Su, R.; Kong, T.; Tang, M.; Cheng, G. The Effects of Topographical Patterns and Sizes on Neural Stem Cell Behavior. *PLoS ONE* **2013**, *8*, e59022. [CrossRef] [PubMed]

52. Serrano, M.C.; Chung, E.J.; Ameer, G.A. Advances and Applications of Biodegradable Elastomers in Regenerative Medicine. *Adv. Funct. Mater.* **2010**, *20*, 192–208. [CrossRef]

53. Lietz, M.; Dreesmann, L.; Hoss, M.; Oberhoffner, S.; Schlosshauer, B. Neuro tissue engineering of glial nerve guides and the impact of different cell types. *Biomaterials* **2006**, *27*, 1425–1436. [CrossRef] [PubMed]

54. Koufaki, N.; Ranella, A.; Aifantis, K.E.; Barberoglou, M.; Psycharakis, S.; Fotakis, C.; Stratakis, E. Controlling cell adhesion via replication of laser micro/nano-textured surfaces on polymers. *Biofabrication* **2011**, *3*, 45004, doi:10.1088/1758-5082/3/4/045004. [CrossRef] [PubMed]

International Journal of
Molecular Sciences

MDPI

Review

Ionic Substitutions in Non-Apatitic Calcium Phosphates

Aleksandra Laskus and Joanna Kolmas *

Department of Inorganic and Analytical Chemistry, Faculty of Pharmacy with Laboratory Medicine Division, Medical University of Warsaw, ul. Banacha 1, 02-097 Warsaw, Poland; aleksandralaskus@gmail.com
* Correspondence: joanna.kolmas@wum.edu.pl; Tel.: +48-225-720-784

Received: 9 November 2017; Accepted: 24 November 2017; Published: 27 November 2017

Abstract: Calcium phosphate materials (CaPs) are similar to inorganic part of human mineralized tissues (i.e., bone, enamel, and dentin). Owing to their high biocompatibility, CaPs, mainly hydroxyapatite (HA), have been investigated for their use in various medical applications. One of the most widely used ways to improve the biological and physicochemical properties of HA is ionic substitution with trace ions. Recent developments in bioceramics have already demonstrated that introducing foreign ions is also possible in other CaPs, such as tricalcium phosphates (amorphous as well as α and β crystalline forms) and brushite. The purpose of this paper is to review recent achievements in the field of non-apatitic CaPs substituted with various ions. Particular attention will be focused on tricalcium phosphates (TCP) and "additives" such as magnesium, zinc, strontium, and silicate ions, all of which have been widely investigated thanks to their important biological role. This review also highlights some of the potential biomedical applications of non-apatitic substituted CaPs.

Keywords: calcium phosphates; ionic substitution; brushite; αTCP; βTCP; bioceramics

1. Introduction

Calcium phosphates (CaPs) are commonly used biomaterials in various medical fields, i.e., mineralized tissue surgery, implantology, orthopaedics, and stomatology. Due to their special properties, such as biocompatibility, bioactivity, nontoxicity, and osteoconductivity, they play a crucial role as bone grafts, bone fillers, and coating materials [1,2]. Most commonly, they are applied in a form of porous granules, scaffolds, or hydraulic, ready-to-use, mouldable cements. What is more, they may serve as local drug delivery systems to introduce medicines directly into the mineralized tissue [3–5].

CaPs can be obtained in different crystalline or amorphous phases, depending on the synthesis conditions (see Scheme 1). These materials differ in Ca/P molar ratio and both physicochemical and biological properties, such as solubility, biodegradability, and bioactivity [1]. Moreover, they are stable in various pHs: for example, HA is chemically stable in aqueous solution in pH > 8, whereas octacalcium phosphate (OCP) and dicalcium phosphate dihydrate (DCPD) in neutral and pH < 6, respectively. HA and tetracalcium phosphate (TTCP) are almost insoluble CaPs (solubility at 25 °C is for HA and TTCP 0.0003 and 0.0007 g/dm^3). Regarding the easily soluble materials, amorphous calcium phosphate, monocalcium phosphate monohydrate and anhydrous (ACP, MCPM, and MCPA, respectively) should be mentioned [1].

Amorphous Calcium Phosphate	• (ACP)	• $Ca_xH_y(PO_4)_z \cdot nH_2O$
Monocalcium Phosphate	• Anhydrous (MCPA) • Monohydrate (MCPM)	• $Ca(H_2PO_4)_2$ • $Ca(H_2PO_4)_2 \cdot H_2O$
Dicalcium Phosphate	• Anhydrous (DCPA) • Dihydrate (DCPD)	• $CaHPO_4$ • $CaHPO_4 \cdot 2H_2O$
Octacalcium Phosphate	• (OCP)	• $Ca_8(HPO_4)_2(PO_4)_4 \cdot 5H_2O$
Tricalcium Phosphate	• Alpha (αTCP) • Beta (βTCP)	• $\alpha\text{-}Ca_3(PO_4)_2$ • $\beta\text{-}Ca_3(PO_4)_2$
Hydroxyapatite	• (HA)	• $Ca_{10}(PO_4)_6(OH)_2$
Tetracalcium Phosphate	• (TTCP)	• $Ca_4(PO_4)_2O$

solubility →

Scheme 1. Calcium phosphates of biomedical interest.

Among all CaPs, hydroxyapatite (HA) with the formula $Ca_{10}(PO_4)_6(OH)_2$ is the most thoroughly studied material. Thanks to its similarity to the biological apatite, the main constituent of the inorganic part of human mineralized tissues, HA is an attractive material used alone as a bioceramic or as a component of hybrid composites for biomedical engineering [6–8]. It is thoroughly documented that the chemical and biological properties of HA may be improved by the incorporation of foreign ions into its crystal structure [9,10]. In order to provide a few examples, the introduction of Mn^{2+} favours the activity and proliferation of osteoblasts and has a positive impact on the osteointegration process [11]. Fe ions improve the response of osteoblasts [9,11]. In turn, doping CaPs with selenium may give the material an anticancer potential [9,12]. Moreover, the slight replacement of Ca^{2+} with Ag^+ cations in the HA structure results in gaining additional antibacterial properties by the material [9,13].

Recently, it turned out that other calcium phosphates, i.e., tricalcium phosphates (TCPs) and dicalcium phosphate dihydrate (DCPD), also possess a natural ability to exchange ions within their crystal lattice. Due to the fact that these CaPs are more soluble than HA [1], it gives an opportunity to create a more resorbable material, releasing therapeutic agents with favourable kinetics.

Several reviews with different approaches have been published on the topic of substituted calcium phosphates. Some of them focused on various ionic substitution in HA [9], antibacterial ions introduced into the HA crystal lattice [14], biological activity of substituted HA [15], or ion-substituted calcium phosphate coatings [16].

In turn, the aim of this work is to review the attempts made so far in the field of ionic substitutions in non-apatitic CaPs and to summarize their effects. To begin with, the biological role of the most relevant ionic substitution, followed by a short presentation of the main, non-apatitic CaPs used in biomedical applications will be presented. The final section will report the recent achievements in ionic modification of non-apatitic CaPs.

2. The Most Relevant Ionic Substitution within the CaPs Crystal Structure

Comprehensive studies on the composition of bone and dental tissues have led to the conclusion that biological apatite is not a pure, stoichiometric HA, but is instead, substituted by different ions, including CO_3^{2-}, HPO_4^{2-}, SiO_4^{2-}, Mg^{2+}, K^+, Na^+, Zn^{2+}, Mn^{2+}, F^-, or Cl^- [1,17]. The quantitative and qualitative content of these "impurities" varies according to the age and condition of the tissue. However, it should be concluded that all of them play a crucial role in mineralized tissue metabolism [17]. Therefore, recent developments in ionic substitutions in synthetic CaP materials have mainly considered these ions that naturally occur in biological apatites. In the following sections, the most ubiquitously introduced ions will be presented.

2.1. Magnesium (Mg^{2+})

Magnesium is the fourth most abundant element in the human body. Significant amount of this element is present in bone, dentin, and enamel. As much as 60% of the body's total Mg is deposited in bone, which acts as a natural reservoir for the metal. Skeletal magnesium is located either on the surface of hydroxyapatite or in the hydrated layer around the crystal [13,18,19]. Magnesium contributes to the maintenance of homeostasis in mineralized tissues. An appropriate level of the metal is vital for preventing osteoporosis and other bone tissue impairments. The element's influence on mineralized tissue is both direct and indirect [19]. When it comes to the direct influence, magnesium has a significant impact on bone crystal structure and skeletal cell activity. Magnesium deficiency leads immediately to hypomagnesaemia, which is compensated through the mobilization of bone deposits [20]. This, in turn, leads to the alteration of the biological apatite structure and, as a result, a decrease in mechanical strength. Both in vitro and in vivo studies have confirmed the element's influence on bone cell activity. Magnesium stimulates osteoblast activity [21]. On the other hand, metal deficiency promotes low-grade inflammation and, as a consequence, an increase in osteoclast activity, which contributes to bone mass loss [19–21]. Among its indirect effects, magnesium has an impact on the secretion of parathyroid hormone (PTH) and thereby a secretion of $1.25(OH)_2$ vitamin D. Experimental evidence show that the metal shortages lower the concentration of both PTH and vitamin D in serum, thereby leading to hypocalcaemia [19,20].

2.2. Zinc (Zn^{2+})

Zinc is the second most abundant trace metal in the human body. Approximately 86% of the total amount of this element is localized in skeletal muscles and bone tissue. The highest proportion of Zn^{2+} ions within the bones is localized in the osteoid and non-mineralized matrix [13,22]. Zinc acts as a cofactor for numerous enzymes and, due to this, is involved in DNA and RNA replication, protein synthesis, and bone metabolism. The element promotes both differentiation and proliferation of osteoblasts [2]. The metal's stimulating effect on osteoblastic differentiation has been partially attributed to its ability to augment the expression of runt-related transcription factor 2, which is a key transcription factor in osteoblastogenesis. In vitro studies showed that zinc also stimulates the osteoblastic production of growth factors, i.e., insulin-like growth factor (IGF-I) and transforming growth factor (TGF-β), both of which promote bone formation [22,23]. Experimental evidence show that zinc exhibits a stimulating effect on alkaline phosphatase (ALP), which promotes bone mineralization through the dephosphorylation of organic pyrophosphates. Simultaneously, Zn^{2+} ions suppress osteoclastic activity. In vitro studies revealed that zinc inhibits osteoclastic resorption mediated by PTH and proinflammatory cytokines [22]. Studies in vivo confirmed zinc's specific impact on bone cell activity and its contribution to the normal development of the skeletal system. Experiments conducted with chicks [24] and rats [25] showed that Zn supplementation increases the strength parameters of long bones, decrease extracellular bone resorption markers, and augments the level of bone formation indicators.

2.3. Strontium (Sr^{2+})

Strontium is a trace element present in significant amounts in calcified tissues [13,26]. This particularly concerns the mineral phase of bone, which is characterized by high metabolic turnover. A new developing bone is likely to contain higher concentrations of Sr than a mature one [26]. Low doses of strontium contribute to proper bone formation, while higher amounts are believed to cause osteomalacia [26,27]. It is noteworthy that strontium ranelate is one of the drugs administered to treat postmenopausal osteoporosis. It reduces significantly the risk of vertebral and non-vertebral fractures. Strontium ranelate exhibits a specific, dualistic influence on mineralized tissues. On the one hand, it stimulates bone formation, while on the other, it decreases its resorption. Studies in vitro showed that this particular dual-mode action of the element is caused by its stimulating effect on osteoblastic differentiation and simultaneous inhibition of osteoclastic activity [28]. What is more, experiments conducted on osteoarthritic, subchondral bone osteoblasts revealed that strontium ranelate decreases the expression of key factors affecting bone resorption [29]. An interesting in vivo study with a zebrafish as an animal model was also designed. The outcomes of the experiment demonstrated an increase in vertebral mineralization when compared with the control with lower strontium concentrations. Higher doses of strontium caused inhibited mineral deposition in dose-dependent order [30]. All of the facts mentioned above make strontium ranelate one of the drugs that is still administered as a treatment of osteoporosis [31].

2.4. Silicon (SiO_3^{2-}, SiO_4^{2-})

Silicon is the third most abundant trace element in the human body [32]. It significantly affects bone metabolism and contributes to proper mineralized tissue formation. A particularly high concentration of silicon can be found in metabolically active skeletal cells. A relatively high level of the element is found in the mitochondria of these cells. A pioneer in investigating silicon's influence on bone metabolism was Carlisle [33–36]. In vitro and in vivo studies conducted by Carlisle [33–36] proved that silicon is localized in active bone growth areas. This, in turn, suggested that Si plays a physiological role in the mineralization process. What is more, the scientists observed that silicon deficiency contributes to problems with normal growth and skeletal development abnormalities [32,33,37]. Over the years, Carlisle's observations have been confirmed. Dietary silicon can be absorbed in the form of orthosilicic acid (H_4SiO_4). Studies [32,37,38] have investigated the influence of this acid on bone metabolism. Experiments conducted in osteoblastic cell line MG-63 showed that orthosilicic acid promotes the synthesis of collagen type 1 and augments osteoblastic differentiation markers, i.e., alkaline phosphatase and osteocalcin levels [37]. Silicon does not affect the expression of the collagen type 1 gene, but it does stimulate the prolyl hydrolase involved in collagen synthesis [32]. In vivo studies also confirmed a stimulating effect of the element on collagen type 1, including randomized trials conducted in osteopenic women [38]. Overall, an adequate Si level is crucial for maintaining higher bone mineral density (BMD). It contributes to bone mineralization and accelerates the calcification rate [39–42].

3. Non-Apatitic Calcium Phosphates

Unlike biological apatites, non-apatitic calcium phosphates are not present in normal mineralized tissues. They frequently occur in pathological calcifications such as dental calculi, urinary stones or heart valve calcifications [43]. Non-apatitic CaPs, except crystalline TCP and TTCP, can be easily synthesized using standard wet methods, by adjusting the Ca/P ratio, the pH of the solution and the temperature beforehand [44]. As was mentioned above, the various types of non-apatitic CaPs differ in their solubility, which decreases as follows: ACP < MCPM < MCPA < DCPD < OCP < αTCP < TTCP < βTCP [45]. In general, non-apatitic CaPs are more soluble and less stable than hydroxyapatite, which allows them transform easily into apatites and substituted apatites or other calcium phosphates under physiological conditions through the dissolution-precipitation process [44].

3.1. Amorphous Calcium Phosphate (ACP)

Amorphous calcium phosphate (ACP) is a hydrated, thermodynamically unstable, transient phase that commonly precipitates during the formation of more stable CaPs in aqueous systems. The structure of ACP is still being discussed in the literature [46]. Briefly, it is proposed that the main structural unit has a nearly spherical cluster measuring 9.5 Å in diameter with the basic composition of $Ca_9(PO_4)_6$ [47]. Moreover, the clusters are interspersed with water molecules (in various contents) [48]. The presence of acidic phosphates in the ACP structure is also under investigation. ACP can be obtained by wet precipitation in aqueous medium at low temperature (the wet method) and by using high energy processing and high temperature (the dry method) [48,49]. As the wet method is more ubiquitously applied, only this route will be briefly outlined below. Wet synthesis may be undertaken either in aqueous medium or a water-alcohol solution. Usually, it consists of two steps: rapid mixing of the reagents and precipitate filtration. Low or room temperature, high supersaturation, and a pH close to 10 are required. The reaction consists of a double decomposition of calcium (e.g., $Ca(NO_3)_2 \cdot 4H_2O$) and phosphate (e.g., $(NH_4)_2HPO_4$) salt in the aqueous or water-alcohol medium [48,49]. After rapid mixing and precipitate filtration, the powder is lyophilized. The prepared ACP is stored in a freezer in order to prevent conversion or phase transformation. The ACP's structure and chemical composition depends on the pH and composition of the mother solution. Hence, the Ca/P ratio of the synthesized ACP can range from 1 to 2, or even higher. Having said this, the most commonly known Ca/P ratios are 1.5 and 1.33 [48]. The Pβ diffractogram of ACP is typical for amorphous materials: it only presents a very broad line with no narrow reflections [47].

Due to the significant chemical and structural similarities to calcified tissues, as well as its excellent biocompatibility and bioresorbability, ACP is commonly used as a component of calcium phosphates cements (CPCs) in surgery or dentistry. Owing to the fact that it can easily transform into biological apatites, it could also be a promising material for the artificial bone grafts engineering. According to the literature, several mechanisms of ACP transformation into crystalline apatite are proposed, depending on pH or the presence of other ions [47,48]. In several cases, the intermediate phase is OCP [48]. It should be also noted, that ACP occurs naturally in soft tissue pathological calcifications, i.e., heart valve calcifications [43,49,50].

3.2. Dicalcium Phosphate Dihydrate (DCPD)

Dicalcium phosphate dihydrate, also known as brushite or calcium hydrogen phosphate dihydrate, is a crystalline calcium phosphate that can be described using the formula $CaHPO_4 \cdot 2H_2O$. It crystallizes in the monoclinic *Ia* space group (see Figure 1). The DCPD crystals consist of CaP chains arranged parallel to each other, with water molecules situated between the chains [47].

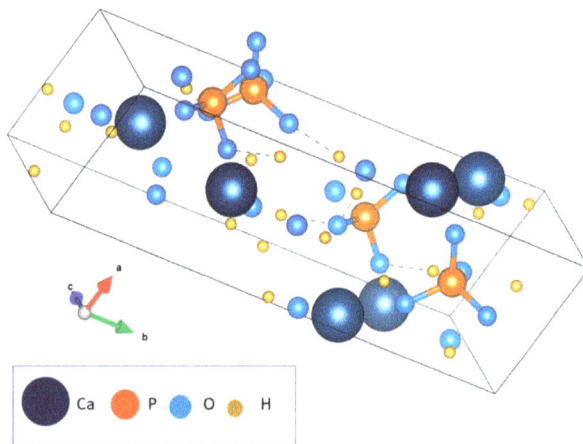

Figure 1. Crystalline structure of dicalcium phosphate dihydrate (DCPD).

DCPD is naturally present in dental calculi, urinary stones and precipitations involved in chondrocalcinosis [43]. In laboratory terms, it crystallizes easily from aqueous solutions and is stable in acidic environments, namely the pH between 4 and 6. Most commonly, it is obtained by mixing calcium chloride $CaCl_2$ with NaH_2PO_4 or $NH_4H_2PO_4$ in a water solution. The synthesis is carried out according to the following chemical reaction [51]:

$$Na_2HPO_4 + CaCl_2 + 2H_2O \rightarrow CaHPO_4 \cdot 2H_2O + 2NaCl$$

Apart from the starting materials listed above, other calcium and phosphate sources can also be employed. Among them, calcium hydroxide (Ca(OH)$_2$), calcium nitrate (Ca(NO)$_3$) or calcium acetate (CH$_3$COOH)$_2$Ca and phosphoric acid (H$_3$PO$_4$) should be mentioned [49,52–54]. As demonstrated above, the reagents are mixed in an equimolar ratio. The proper pH value is usually maintained by the addition of HNO$_3$, HCl, H$_3$PO$_4$, and KOH or NH$_4$OH. The literature provides data showing that brushite forms at the pH of 5 and the temperature of 37 °C. In order to synthesize DCPD at the pH of 6, the temperature should be lowered to 25 °C [55].

Brushite can be also obtained by mixing two different calcium phosphate powders in water [54]. The starting materials are βTCP and monocalcium phosphate hydrate (Ca(H$_2$PO$_4$)$_2$·H$_2$O). Sodium pyrophosphate (Na$_2$H$_2$PO$_4$) is added to the mixture as a setting regulator. The synthesis may be conducted in a sulphuric acid solution or in inorganic (silica) and organic (collagen) gels [55].

It should be noted that brushite is a metastable material. It transforms easily into the anhydrous form monetite (CaHPO$_4$). Dehydration begins already in atmospheric conditions [51]. Under the physiological pH ranging from 7 to 7.5, DCPD transforms into HA. An essential aspect is that it resorbs much faster than apatite and, due to this, is widely applied in surgery as a CPC component. Moreover, in dentistry it is used as an anti-plaque factor in toothpastes [43,55,56].

3.3. Tricalcium Phosphate (TCP)

Tricalcium phosphate, which can be described with the general formula Ca$_3$(PO$_4$)$_2$, exists in two allotropic forms, i.e., α and βTCP, both of which are of great importance from the biological point of view. Schematic models of the crystal structures of these materials are presented in Figure 2. αTCP crystallizes in the monoclinic space group P2$_1$/a, whereas βTCP possesses a rhombohedral structure [47].

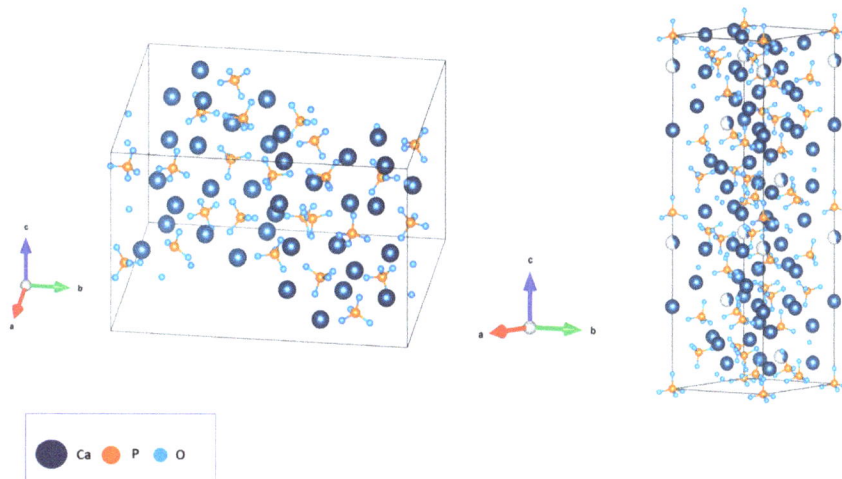

Figure 2. The crystalline structures of tricalcium phosphate (TCP): α form (**left**) and β form (**right**).

As the hydroxyapatite earned its immense popularity in biomedicine due to its great similarity to biological apatite, TCP is still gaining interest thanks to its relatively high solubility. Both HA and TCP form a strong bond with the mineralized tissue and favour bone formation. Due to the poor resorbability of HA, both CaPs are often mixed together in different ratios in order to increase the bioreactivity of the material [57–59].

αTCP is a high temperature allotropic form, which comes into being during the heat treatment of βTCP above 1125 °C [60]. The synthesis of αCa$_3$(PO$_4$)$_2$ brings one major problem, which is the metastability of the obtained material under the room temperature. Its thermal stability ranges from

1430 to 1470 °C. Thus, synthesizing αTCP at a high temperature requires immediate cooling to stabilize the structure after the previous heat treatment [60,61]. αTCP is highly soluble and together with a liquid phase forms a hard and stable material that is used in bone cements. It should be also noted that αTCP transforms into CDHA in water medium according the following reaction [60]:

$$3Ca_3(PO_4)_2 + H_2O \rightarrow Ca_9(PO_4)_5(HPO_4)OH$$

αTCP can be obtained via different methods, i.e., precursor's thermal transformation, solid-state precursors' reaction and self-propagating, high-temperature or combustion synthesis [61]. When it comes to thermal transformation, the decomposition may consider CDHA, ACP or βTCP (see Equations (1)–(3)):

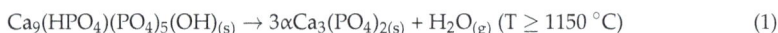

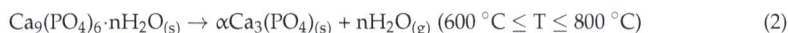

$$Ca_9(HPO_4)(PO_4)_5(OH)_{(s)} \rightarrow 3\alpha Ca_3(PO_4)_{2(s)} + H_2O_{(g)} \ (T \geq 1150\ ^\circ C) \tag{1}$$

$$Ca_9(PO_4)_6 \cdot nH_2O_{(s)} \rightarrow \alpha Ca_3(PO_4)_{(s)} + nH_2O_{(g)} \ (600\ ^\circ C \leq T \leq 800\ ^\circ C) \tag{2}$$

$$\beta Ca_3(PO_4)_{(s)} \rightarrow \alpha Ca_3(PO_4)_{(s)} \ (1125\ ^\circ C \leq T \leq 1430\ ^\circ C \tag{3}$$

Both the decomposition of CDHA (Equation (1)) and ACP (Equation (2)) are commonly applied [57]. Starting materials are usually synthesized using the wet precipitation method. Nonetheless, high temperature, βTCP to αTCP crystalline phase transition is the most direct and the simplest way to obtain αCa$_3$(PO$_4$) [62–66].

Considering obtaining αTCP via solid-state reaction, the most commonly applied synthesis routes are listed below [67–71].

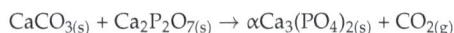

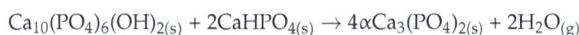

$$CaCO_{3(s)} + 2CaHPO_{4(s)} \rightarrow \alpha Ca_3(PO_4)_{2(s)} + CO_{2(g)} + H_2O_{(g)}$$

$$3CaCO_{3(s)} + 2NH_4H_2PO_{4(s)} \rightarrow \alpha Ca_3(PO_4)_{2(s)} + 3CO_{2(g)} + 3H_2O_{(g)} + 2NH_{3(g)}$$

$$CaCO_{3(s)} + Ca_2P_2O_{7(s)} \rightarrow \alpha Ca_3(PO_4)_{2(s)} + CO_{2(g)}$$

$$Ca_{10}(PO_4)_6(OH)_{2(s)} + 2CaHPO_{4(s)} \rightarrow 4\alpha Ca_3(PO_4)_{2(s)} + 2H_2O_{(g)}$$

In general, solid-state reaction scheme consists of a few stable steps: milling the powder mixture, putting the ground mixture under pressure and finally heating the powder above the transformation temperature. The first two stages serve to reduce the particles size, promote proper homogenization and increase the contact area. When it comes to the heat treatment parameters, the recommended temperature ranges from 1250 to 1500 °C. In turn, the sintering time may vary from 2 to 48 h. To avoid the reversion of the crystal phase, quenching should be employed immediately [61].

Less common ways of obtaining αTCP are self-propagating, high-temperature, and combustion synthesis [72–75]. Briefly, they consist in preparing a mixture of Ca and P sources in a proper ratio, either as a solid pellet or solution, and the heat treatment. In case of the self-propagating, high temperature method, the heat is applied from an external source, while in case of the self-combustion route, the reagent solution contains an organic, inflammable fuel (e.g., urea) that acts as a catalyst for the self-combustion of the mixture [72–75].

Recently, some novel approaches to βTCP synthesis have been made [51,76–78]. Herein, they will be briefly presented, together with more conventional and ubiquitously employed methods. In general, βTCP synthesis methods include:

- Thermal transformation of the precursor;
- Solid-state synthesis;
- Wet chemical precipitation;
- Sol-gel technique;
- Self-combustion method.

The most conventional methods from the above are thermal transformation of the precursor, solid-state synthesis and wet chemical precipitation. Most frequently, thermal transformation of the precursor consists in calcination of previously precipitated CDHA at the temperature above 800 °C [55,77]. When it comes to solid-state synthesis, possible reactions schemes that can be employed are similar to those presented for αTCP. The only differences concern the heat treatment parameters, which should be maintained at the level preventing βTCP to αTCP phase transition. In turn, wet chemical precipitation consists in precipitating CaP at a Ca/P molar ratio of ≈1.5 followed by the product's calcination.

Another interesting method is a sol-gel technique [77]. In this case, the precipitation is carried out in very acidic conditions. The proper pH value is usually achieved by the addition of citric acid or concentrated nitric acid, and varies from 2 to 3. The sol-gel transformation is caused by vigorous mixing of Ca and P sources in a liquid medium during the simultaneous heating up to 80 to 90 °C [76,77].

The self-combustion method has already been described. An alternative to this method is microwave self-combustion synthesis [78].

An essential property of TCP is that, similarly to hydroxyapatite, it forms a strong bond with mineralized tissue and favours bone formation. Due to its relatively high solubility, it is commonly applied as a component of calcium phosphate cements and other bone substitutes [56,57,61].

4. Ionic Substitutions in Non-Apatitic Calcium Phosphates

4.1. Substituted Amorphous Calcium Phosphate (ACP)

As mentioned above, unsubstituted ACP cannot be formed at the physiological pH level. It usually forms at the alkaline pH of about 10. Introducing some specific ions, i.e., Mg^{2+}, Sn^{2+}, Al^{3+}, $P_2O_7^{4-}$, and CO_3^{2-}, stabilizes the ACP and inhibits its transformation into HA [48]. Sometimes, such a modification can also promote hydrolysis to brushite instead of apatite. Lee and Kumta [51] decided to modify ACP with Mg^{2+} ions mainly because of their excellent biocompatibility. They used a precipitation method, with $MgCl_2$ as a source of magnesium. The concentrations of Ca and P in the two conducted experiments corresponded to the composition of HA (Ca/P = 1.67) and TCP (Ca/P = 1.5). The magnesium content was maintained at the level of 30 mol % and 20 mol % for the HA and TCP, respectively. Physicochemical analysis was carried out. The results revealed that Mg ions act as a phase stabilizer. During the heat treatment of ACP, αTCP is known to form at 600 °C. The addition of Mg^{2+} stabilized MgTCP at this temperature and retarded its transformation [51].

The in vitro biological activity of Mg-substituted ACP, amorphous magnesium phosphate (AMP) and pure HA was tested on MC3T3-E1 preosteoblasts [79]. In general, the results for amorphous materials (Mg-substituted ACP and AMP) have demonstrated higher proliferation and differentiation rate as well as higher mineralization of preosteoblast cells than HA samples.

The effect of silicon doping on the transformation of ACP to αTCP has been investigated by Dong et al. [80]. During the synthesis of HA with high silicon concentration, the amorphous phase enriched with Si was easily formed and then transformed to αSiTCP at lower temperatures than SiHA.

Kato et al. [81] have prepared potassium-substituted ACP for potential application in dentin hypersensitivity treatment. It has been shown that potassium release from amorphous material is significantly larger than from potassium-substituted hydroxyapatite. ACP enriched in silver ions was prepared via chemical precipitation method by Yu et al. [82]. The obtained AgACP material, together with slightly acidic compounds were then used to produce calcium phosphate cements (CPCs). The sufficient silver release and high cytotoxic effect toward *Escherichia coli* were demonstrated only in the samples prepared without heat treatment.

4.2. Substituted Dicalcium Phosphate Dihydrate (DCPD)

Modifying brushite with foreign ions is also an object of scientific interest. DCPD is a material commonly used as a component of CPCs, as it transforms easily to HA under physiological conditions. The most ubiquitously applied ionic substitution in case of brushite is magnesium, mainly due to the potential enhancement of the bioactivity of such a modified material [51,83–86]. Some studies also deal with the introduction of silicon [87], strontium [88,89], cuprum [90], zinc [91], and iron [92]. Recently, a study concerning the introduction of nickel has been also conducted [93]. Alkhraisat and Cabrejos-Azama focused on magnesium-modified, brushite-based CPC [83–85]. Various ways of introducing Mg^{2+} ions into brushite cements were investigated and reported, also.

Alkhraisat et al. [83] used TCP and magnesium-substituted TCP as substrates in this process. In their studies, the preparation of brushite cement consisted of the following few steps:

- Preparation of βTCP and βMgTCP powder via solid-state synthesis;
- Addition of MCPM to previously crushed and mixed powders;
- Addition of water to the composed powder;
- Mixing in mortar.

By modifying the substrate (TCP) with different amounts of Mg^{2+}, Alkhraisat et al. [83] managed to replace 3% of the Ca^{2+} with Mg^{2+} ions in the final product. The final setting time of the CPC increased, together with its magnesium content. Moreover, the release rate of the ions was affected. The release of Ca^{2+} decreased as the Mg^{2+} concentration increased, which is a result of magnesium's inhibiting influence on the dissolution of brushite. A similar observation was made by Lee and Kumta in their work [51]. Hence, it could be stated that Mg^{2+} acts as a phase stabilizer of DCPD.

In turn, Correia et al. [94] studied the role of Mg^{2+} ions on the growth of acidic calcium phosphate crystals: OCP and DCPD. It was shown that the presence of Mg^{2+} could affect the morphology of the coatings obtained by electrodeposition. Magnesium inhibited the OCP and DCPD crystal growth due to surface adsorption and doping process, respectively [94].

In the study [84], the biological properties of magnesium-modified brushite cement were examined (see Figures 3 and 4). The investigation involved cell studies and animal model experiments. The evaluation of the in vitro response to the cement was conducted using the osteoblastic cell line MG-63. In turn, in vivo tests were performed using a rabbit model. Both in vitro and in vivo studies proved that there was an increased biological response in case of MgCPC in comparison with unmodified CPC. Specifically, cell proliferation, cell adhesion, and bone formation all increased. No manifestation of inflammation was observed [84].

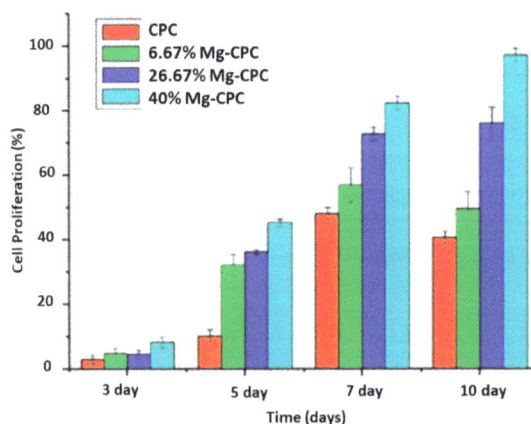

Figure 3. Cell proliferation after seeding on calcium phosphates cement (CPC) and MgCPCs (error bars represent SD). Reprinted from [84] with permission from Elsevier.

Alkhraisat et al. [88] also carried out studies concerning the impact of strontium and pyrophosphates on the physicochemical properties of brushite cement. The results showed that both Sr^{2+} and $P_2O_7^{2-}$ inhibit the cement setting reaction.

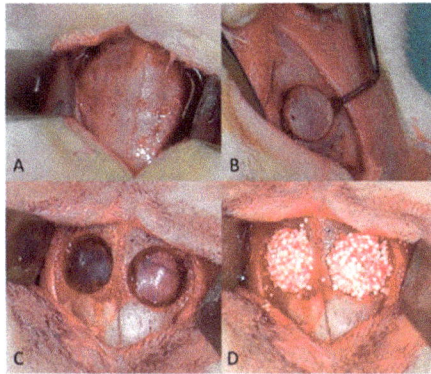

Figure 4. (**A**) Rabbit calvaria exposed; (**B**) removal of bone block; (**C**) bone defects exposed; (**D**) bone defects filled with bone substitute. Reprinted from [84] with permission from Elsevier.

Cabrejos-Azama et al. [85] used magnesium-substituted DCPD cement as a drug carrier. Vancomycin, an antibiotic commonly used against *Staphylococcus aureus*, which is one of the most frequent pathogens associated with mineralized tissue infections, was chosen as a model drug. The cement was loaded with the antibiotic through the adsorption from the solution or introduction into the solid phase of cement. The release profile varied together with Mg content in the TCP used as one of the reagents in the preparation of brushite cement (see Figure 5). The material prepared with βMgTCP containing 66.67% Mg^{2+} released vancomycin with zero-order kinetics. Vancomycin was also used as a drug model in study [87], in which silicon was chosen as an ionic modifier. The investigated material was CPC-silica gel composite. The composite was obtained through the infiltration of the macropores of the CPC by the silica gel. The infiltration process altered both density of the material and the release mode of vancomycin. In comparison with unmodified CPC, 25% of previously introduced drug remained in the SiCPC matrix. The cytocompatibility of obtained composite was also examined. The composites with the highest silicate content performed as high cell proliferation as hydroxyapatite, which was one of the reference materials. However, biological response did not correlate with the release rate of silicon from the material, but seemed to be attributed to the phosphate release and magnesium absorption from the cell culture medium [87].

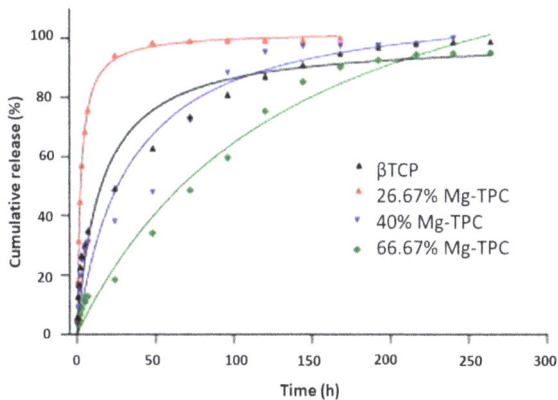

Figure 5. Release of vancomycin from different CPCs. Reprinted from [85] with permission from Elsevier.

The available literature contains also the co-substituted DCPD cements information [95,96]. Torres et al. [95] focused on co-doping DCPD with the combination of Mn^{2+} and Sr^{2+}. The injectable cement contained also an addition of sucrose, which improved its biological and mechanical properties. In turn, Vahabzadeh et al. [96] studied the influence of Si and Zn ions as dopants of DCPD on the physical, biological, and mechanical properties of cements. In vivo results revealed that Si and Zn addition may improve the early stage osseointegration process [96].

4.3. TCP Substituted with Foreign Ions

Ionic substitution in both α and βTCP is currently being investigated. A solid-state reaction, as well as the heat treatment of already precipitated powders, are commonly applied to synthesize these materials [97–146]. It is worth underlining that there are two main directions in ionic modification of α and βTCP. One of them is improving the properties of CaP, while the second one is stabilizing the phase. In the following paragraph the most interesting studies on this topic will be briefly discussed.

4.3.1. αTCP Substituted with Different Ions

Various elements, including Mg, Sr, Si, and Ag, are being successfully introduced into αTCP's crystal lattice. It should be underlined that Si is the most widely applied and thoroughly examined dopant.

αSiTCP samples can be obtained using various wet and solid-state methods [98–103]. Reid et al. [98] prepared Si-substituted αTCP using one of the most ubiquitous ways of synthesis, i.e., by sintering previously precipitated precursor, which was Si-enriched CaP. The applied temperature was 1250 °C, and the single-phase samples contained 0.59–1.14 wt % of Si.

As stated before, pure αTCP is stable only in the range of 1430 to 1470 °C. Thus, it is important to note that the addition of silicon favours the formation of αTCP and helps to stabilize its structure in lower temperature [97]. The chemical reactivity of αSiTCP was also examined by Moitsuke et al. [100]. αTCP is known to convert easily to calcium-deficient hydroxyapatite in water solution. However, the study showed that, in comparison with pure αTCP, the reactivity of αSiTCP was significantly lower. This could be related to the release of silicon and formation of an amorphous Si layer slowing down the dissolution-precipitation reaction.

Interesting conclusions were made by Reid et al. in their work [105]. The study considered the influence of the presence of magnesium impurities in one of the reactants on the phase composition of obtained powder. To synthesize the final material, the sol-gel technique combined with the appropriate heat treatment was applied. Surprisingly, it turned out that an Mg content of barely 250 to the chemical structure of αSiTCP was investigated by Duncan et al. [99], who obtained high-purity αSiTCP through a solid-state reaction in a furnace at 1300 °C. Detailed physicochemical analysis (X-ray diffraction with Rietveld analysis and solid-state nuclear magnetic resonance) allowed to confirm crystal structure and silicon incorporation. Moreover, the obtained results suggested the presence of silicate and disilicate ions that partially substituted phosphate groups in crystallographic sites. Silicon substitution into the αTCP crystals resulted in a significant increase of *b* axis length and the β angle [99].

During the last few years αSiTCP has been also examined in vivo [101–103]. Cylindrical implants, ceramic blocks [102,103] or porous granules based on this material [101] have already been prepared and examined by conducting biological tests on rabbit and rat models. All the materials demonstrated good bioactivity and biocompatibility as well as mechanical properties [102]. What is more, the addition of Si promoted osteogenesis and again, retarded biodegradation of αTCP [101].

Not only silicon has been used as a ionic modifier of αTCP. The study [104] considered applying a hydrothermal method to obtain hydroxyapatite (as a precursor of αTCP) modified with Cu^{2+} and Ag+. The as prepared HA powder was put under the temperature of 1200 °C for two hours in order to examine the influence of dopants on the creation of biphasic HA/αTCP powder. The aim of the study was also to investigate the antimicrobial activity of the prepared material. All of the annealed powders consisted mainly of spherical particles (see Figure 6). The doping of foreign ions had slightly negative influence on the creation of biphasic material. The antimicrobial activity of both modified HA and biphasic powder were evaluated against *Staphylococcus aureus*, *Escherichia coli*, *Pseudomonas aeruginosa*, and *Candida albicans*—the bacterial and yeast strains responsible for common human infections.

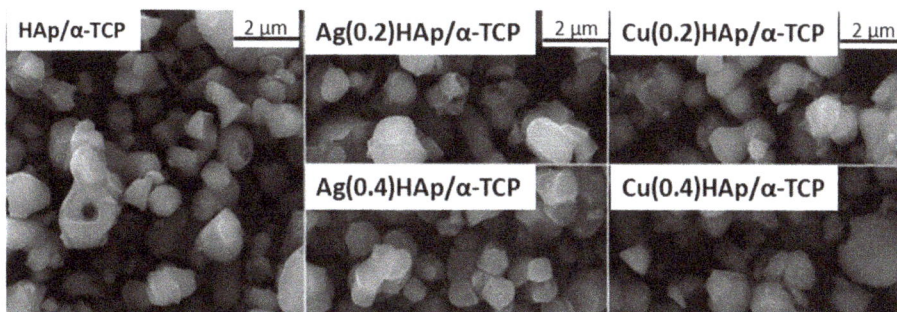

Figure 6. Field emission scanning electron microscopy (FESEM) micrographs of the powders calcified at 1200 °C (CuAg). Scale bar: 2 μm. Reprinted from [104] with permission from Elsevier.

Although the tests showed that more ions were released from the doped HA, the biphasic powder performed more homogenous and, in some cases, better antimicrobial activity than modified HA. The authors believe that this phenomenon can be explained by the adsorption of microorganisms on the surface of the HA/αTCP, which put them in close contact with antimicrobial agents, i.e., Ag^+ and Cu^{2+}, preventing further expansion [104].

In turn, Tong et al. [106] presented an unconventional approach, which examined the introduction of foreign ions into αTCP. Using a solid-state technique, they modified the CaP with Eu^{3+}. The outcomes of the study suggested that, due to the self-reduction, europium could serve as a spectroscopic probe for the detection of $αCa_3(PO_4)_3$ in the phase transition process.

4.3.2. βTCP Substituted with Foreign Ions

Among non-apatitic CaPs, the most thoroughly investigated material considering ionic substitution is βTCP [108–116]. In turn, magnesium is definitely the most widely incorporated dopant into its crystal lattice [108,109,111,112,115–119,121–125]. The most ubiquitous method of obtaining Mg-substituted βTCP is annealing already precipitated CDHA powders modified with Mg^{2+}. Analysing the behaviour of these materials during heat treatment, several main trends may be observed. Introducing Mg^{2+} lowers the thermal stability of HA and therefore facilitates CDHA's transformation into βTCP. In study [118] the process occurred within the temperature range of 600 to 700 °C in case of Mg-substituted CDHA, whereas in the absence of the dopant, the phase transition took place at higher temperatures (700 to 800 °C). Introducing magnesium into $βCa_3(PO_4)_3$ suppresses β to α phase transformation. Marchi et al. proved that the addition of at least 1.5 mol % is enough to retard the above process [119].

The maximum substitution of Mg^{2+} in Ca^{2+} sites has been also examined and varies from 14 to 15 mol % [115,116]. Mg ions occupy Ca(4) and Ca(5) sites in the crystal structure. Successful attempts were made to create interconnected micro and macroporous Mg-substituted βTCP scaffolds.

In study [123], porous structures were obtained as an effect of the in situ foaming of the examined slurry. NH_4HCO_3 was used as a foaming factor. Subsequently, the viscous mass was sintered at 1150 °C for two hours. The field emission scanning electron microscopy (FESEM) images of these prepared materials are presented below (see Figure 7).

Another method to create βTCP—based porous scaffolds was applied in study [122]. Briefly, βMgTCP powder was made into slurry with distilled water and polyvinyl alcohol. Subsequently, polyurethane foams were soaked in the previously prepared slurry (see Figure 2) [122]. The excessive slurry was removed by squeezing. The soaked foams were dried and put under a heat treatment of above 1500 °C. The sintering temperature was higher than the β to αTCP transition temperature. The results of the study showed that the addition of magnesium was crucial to prevent the transformation during heat treatment.

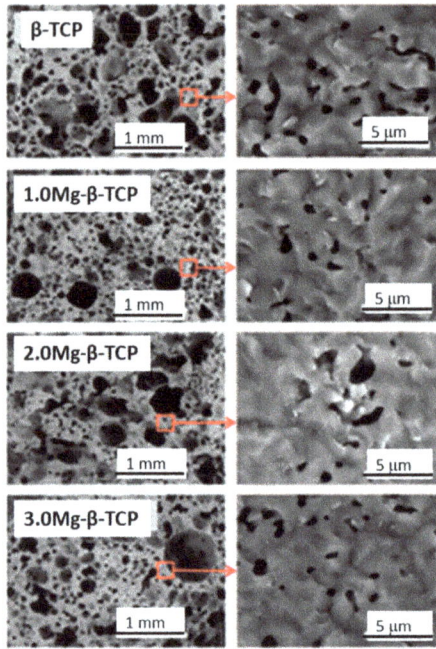

Figure 7. FESEM images Mg-containing micro-macroporous scaffolds fracture surfaces. Reprinted from [123] with permission from Elsevier.

Figure 8. Typical macroscopic structure of: (a) the polyurethane foam template and (b–h) TCP foams prepared with calcium phosphate slurries containing different amounts of MgO. Reprinted from [122] with permission from Elsevier.

It is not only magnesium that is being introduced into the βTCP crystal lattice. In the literature, following examples of other ionic modifications can be found: sodium [127,143], potassium [128], silver [135], manganese [132], silicon [111,133], strontium [107,112–114], copper [136], cobalt [137], aluminium [146], iron [139–141], lantanium [131], and rare earth elements [145].

Divalent cobalt ions (Co^{2+}) were introduced into the βTCP structure due to their potential induction of angiogenesis process [137]. The physicochemical and biological properties of the obtained materials containing 2 and 5 mol % of Co^{2+} were analysed. The results suggested that, similarly to magnesium, the addition of cobalt ions stabilizes the βTCP phase. The samples performed also low toxicity and stimulated significantly the synthesis of the vascular endothelial growth factor (VEGF).

In turn, studies conducted by Kannan et al. [126–128] investigated the influence of sodium, potassium and chlorine on CaP formation. Firstly, aqueous precipitation was applied to obtain CDHA modified with Na^+, K^+ or Cl^-. $NaNO_3$, KNO_3 or NH_4Cl were used as sources of dopants. The prepared powders were calcified at the temperature of 800 °C. Physicochemical studies confirmed successful ionic incorporation in each case. The resulting materials were biphasic mixtures consisting of HA and βTCP modified with the appropriate elements. The main effect of introducing monovalent ions into the HA crystal lattice was an increase in its thermal stability to the temperature of between 1200 and 1300 °C [127].

Recent works also focused on iron-doped βTCP [139–141]. Considering the fact that CPCs substituted with magnetic ions (i.e., Fe^{3+}) exhibit ferromagnetic properties, they may be used as heat mediators in hyperthermia treatment of tumors, drug delivery systems or biomaterials for magnetic resonance imaging (MRI) [141]. The studies provided by Singh et al. [140] indicated that addition of the Fe^{3+} ions preserved the structural stability of βTCP. The limit of iron substitution was found to be 5.02 mol %. It was also shown that only high content of Fe^{3+} ions introduced into βTCP structure may produce pronounced hyperthermia effect. Detailed studies on ferric doping mechanism was presented in [141]. Hyperthermia effect was investigated on Fe^{3+}/Ni^{2+} co-substituted βTCP [139]. High nickel content exhibited significant cytotoxicity. However, iron and nickel co-substitution displayed optimal hyperthermia effect.

The potential antibacterial activity of βTCP substituted with appropriate ions was also investigated. Gokcekaya et al. [135] and Rau et al. [136] provided data considering the activity of AgβTCP and CuβTCP, respectively. Small amounts of silver or copper ions were nontoxic for human cells and effectively lethal for bacteria. Moreover, in one of the studies, Matsumoto et al. [129] presented the effects of co-doping βTCP with monovalent and divalent antibacterial ions. They introduced Ag^+ and Cu^{2+} or Zn^{2+} using a solid-state method. The Ag^+ concentration was constant and maintained at the level of 9.09 mol %, while the amount of Cu^{2+} or Zn^{2+} varied from 0 to 15 mol %. The antibacterial activity of synthesized materials against *S. aureus* and *E. coli* was evaluated. Interestingly, co-doped powders exhibited better antibacterial effect than AgTCP and pure TCP. What is more, the release rates of the ions from CuAgTCP and ZnAgTCP were slower than in case of AgTCP and unsubstituted TCP. Hence, the results of the study strongly suggest that co-doped materials can be used over a long period of time, exhibiting a good antibacterial activity [129].

Not only antibacterial ions are being introduced into βTCP. Scientific research is also focused on loading porous, modified βTCP nanoparticles with drugs, which are commonly used to treat hard tissue infections [114,124]. An interesting method to obtain porous, doxycycline loaded material was applied in study [124]. Mesoporous MgβTCP nanospheres were obtained from amorphous $Mg_2P_2O_7$ using EDTA ions as a nucleating factor. Subsequently, the prepared material was loaded with the antibiotic by immersion. The nanoparticles exhibited sustained drug release and easy cell reuptake [124].

A simultaneous introduction of more than one element into βTCP structure is also being investigated. In studies [142,143], magnesium ions were co-substituted together with strontium [142] or sodium ions [143]. All the structures were determined using detailed X-ray diffractometry and Rietveld analysis. The obtained results showed that Mg^{2+} ions play the crucial role in the significant reduction of the *a* and *c* axis lengths. It was also revealed that Sr^{2+} occupy Ca(1,2,3,4) sites whereas Mg^{2+} is located at the six fold coordinated Ca(5) site [142]. Thermal stability as well as the limit of substitution was examined. An interesting work was presented by Meenambal et al. [144]. A simultaneous substitution of Gd^{3+} and Dy^{3+} into the βTCP crystals was investigated. The substitution limit for co-substituted ions was determined as ca. 2.2 mol % and was sufficient to exhibit paramagnetic effects. Biological tests revealed negligible toxicity of the obtained materials. Therefore, this new material may find potential application as multifunctional bioprobe or a contrast agent in MRI and CT (Computer Tomography) [144]. The simultaneous introduction of more than two ions into the βTCP crystal lattice has been also conducted. Bose et al. [108] applied a solid-state technique to obtain $βCa_3(PO_4)_3$ modified with magnesium, strontium, and silicon ions at the same time. SrO, MgO, and SiO_2 were used as dopants sources.

5. Conclusions

The ideal bioceramics for bone grafting must combine several biological properties such as bioactivity, biocompatibility, osteoconductivity, and non-immunogenicity. Moreover, these materials should possess mechanical resistance together with suitable resorbability, which allows for bone remodelling and new bone formation.

Different strategies can be taken to improve the biological properties of CaP bioceramics. One of them is to enrich the materials with particular ions, which, even in small amounts, have significant impact on the osseous cell activity, mechanical and thermal properties and solubility.

This review provides evidence for the growing interest in substituted, non-apatitic CaPs. Until now, this research has focused on these ions, which are favourable considering the biocompatibility, bioactivity, and osteointegration of synthesized materials. There has not yet been sufficient investigation in the field of ionic substitution providing biological properties i.e., antimicrobial or anticarcinogenic. We think that co-substitution in non-apatitic CaPs may represent a considerable field of scientific research, too.

Acknowledgments: This work was supported by the research programme (UMO-2016/22/E/ST5/00564) of the National Science Center, Poland.

Conflicts of Interest: The authors declare no conflict of interest.

References

1. Dorozhkin, S.V. Nanosized and nanocrystalline calcium orthophosphates. *Acta Biomater.* **2010**, *6*, 715–734. [CrossRef] [PubMed]
2. Boanini, E.; Gazzano, M.; Bigi, A. Ionic substitutions in calcium phosphates synthesized at low temperature. *Acta Biomater.* **2010**, *6*, 1882–1894. [CrossRef] [PubMed]
3. Dubnika, A.; Loca, D.; Rudovica, V.; Parekh, M.B.; Berzina-Cimdina, L. Functionalized silver doped hydroxyapatite scaffolds for controlled simultaneous silver ion and drug delivery. *Ceram. Int.* **2017**, *43*, 3698–3705. [CrossRef]
4. Jafari, S.; Maleki-Dizaji, N.; Barar, J.; Barzegar-Jalali, M.; Rameshrad, M.; Adibkia, K. Methylprednisolone acetate-loaded hydroxyapatite nanoparticles as a potential drug delivery system for treatment of rheumatoid arthritis: In vitro and in vivo evaluations. *Eur. J. Pharm. Sci.* **2016**, *91*, 225–235. [CrossRef] [PubMed]
5. Yu, W.-L.; Sun, T.-W.; Qi, C.; Zhao, H.-K.; Ding, Z.-Y.; Zhang, Z.-W.; Sun, B.-B.; Shen, J.; Chen, F.; Zhu, Y.-J.; et al. Enhanced osteogenesis and angiogenesis by mesoporous hydroxyapatite microspheres derived simvastatin sustained release system for superior bone regeneration. *Sci. Rep.* **2017**, *7*. [CrossRef] [PubMed]
6. Kolanthai, E.; Ganesan, K.; Epple, M.; Kalkura, S.N. Synthesis of nanosized hydroxyapatite/agarose powders for bonefiller and drug delivery application. *Mater. Today Commun.* **2016**, *8*, 31–40. [CrossRef]
7. Terukina, T.; Saito, H.; Tomita, Y.; Hattori, Y.; Otsuka, M. Development and effect of a sustainable and controllable simvastatin releasing device based on PLGA microspheres/carbonate apatite cement composite: In vitro evaluation for use as a drug delivery system from bone-like biomaterial. *J. Drug Deliv. Sci. Technol.* **2017**, *37*, 74–80. [CrossRef]
8. Lei, Y.; Xu, Z.; Ke, Q.; Yin, W.; Chen, Y.; Zhang, C.; Guo, Y. Strontium hydroxyapatite/chitosan nanohybrid scaffolds with enhanced osteoinductivity for bone tissue engineering. *Mater. Sci. Eng. C* **2017**, *72*, 134–142. [CrossRef] [PubMed]
9. Supova, M. Substituted hydroxyapatites for biomedical applications: A review. *Ceram. Int.* **2015**, *41*, 9203–9231. [CrossRef]
10. Shepherd, J.H.; Shepherd, D.V.; Best, S.M. Substituted hydroxyapatites for bone repair. *J. Mater. Sci. Mater. Med.* **2012**, *23*, 2335–2347. [CrossRef] [PubMed]
11. Li, Y.; Widodo, J.; Lim, S.; Ooi, C.P. Synthesis and cytocompatibility of manganese (II) and iron (III) substituted hydroxyapatite nanoparticles. *J. Mater. Sci.* **2012**, *47*, 754–763. [CrossRef]
12. Wang, Y.; Hao, H.; Li, Y.; Zhang, S. Selenium-substituted hydroxyapatite nanoparticles and their in vivo antitumor effect on hepatocellular carcinoma. *Colloids Surf. B Biointerfaces* **2016**, *140*, 297–306. [CrossRef]

13. Mourino, V.; Cattalini, J.P.; Boccaccini, A.R. Metallic ions as therapeutic agents in tissue engineering scaffolds: An overview of their biological applications and strategies for new developments. *J. R. Soc. Interface* **2012**, *9*, 401–419. [CrossRef] [PubMed]

14. Kolmas, J.; Groszyk, E.; Kwiatkowska-Różycka, D. Substituted hydroxyapatites with antibacterial properties. *BioMed Res. Int.* **2014**, *2014*. [CrossRef] [PubMed]

15. Ratnayake, J.T.B.; Mucalo, M.; Dias, G.J. Substituted hydroxyapatites for bone regeneration: A review of current trends. *J. Biomed. Mater. Res. B* **2017**, *105*, 1285–1299. [CrossRef] [PubMed]

16. Graziani, G.; Bianchi, M.; Sassoni, E.; Russo, A.; Marcacci, M. Ion-substituted calcium phosphate coatings deposited by plasma-assisted techniques: A review. *Mater. Sci. Eng. C* **2017**, *74*, 219–229. [CrossRef] [PubMed]

17. Combes, C.; Cazalbou, S.; Rey, C. Apatite biominerals. *Minerals* **2016**, *6*, 34. [CrossRef]

18. Alshemary, Z.; Akrama, M.; Goh, Y.-F.; Tariq, U.; Butt, F.K.; Abdolahi, A.; Hussain, R. Synthesis, characterization, in vitro bioactivity and antimicrobial activity of magnesium and nickel doped silicate hydroxyapatite. *Ceram. Int.* **2015**, *41*, 11886–11898. [CrossRef]

19. Castiglioni, S.; Cazzaniga, A.; Albisetti, W.; Maier, J.A.M. Magnesium and Osteoporosis: Current State of Knowledge and Future Research Directions. *Nutrients* **2013**, *5*, 3022–3033. [CrossRef] [PubMed]

20. Rude, R.K.; Singer, F.R.; Gruber, H.E. Skeletal and Hormonal Effects of Magnesium Deficiency. *J. Am. Coll. Nutr.* **2009**, *28*, 131–141. [CrossRef] [PubMed]

21. He, L.Y.; Zhang, X.M.; Liu, B.; Tian, Y.; Ma, W.H. Effect of magnesium ion on human osteoblast activity. *Braz. J. Med. Biol. Res.* **2016**, *49*, 1414–1431. [CrossRef] [PubMed]

22. Smith, B.J.; Hermann, J. Aging, zinc, and bone health. In *Bioactive Food as Dietary Interventions for the Aging Population*; Watson, R.R., Preedy, V.R., Eds.; Academic Press: San Diego, CA, USA, 2013; pp. 433–443.

23. Shepherd, D. Zinc-substituted hydroxyapatite for the inhibition of osteoporosis. In *Hydroxyapatite (HAp) in Biomedical Applications*; Mucalo, M., Ed.; Woodhead Publishing: Cambridge, UK, 2015; pp. 107–126.

24. Kwiecień, M.; Winiarska-Mieczan, A.; Milczarek, A.; Tomaszewska, E.; Matras, J. Effects of zinc glycine chelate on growth performance, carcass characteristics, bone quality, and mineral content in bone of broiler chicken. *Livest. Sci.* **2016**, *191*, 43–50. [CrossRef]

25. Hadley, K.B.; Newman, S.M.; Hunt, J.R. Dietary zinc reduces osteoclast resorption activities and increases markers of osteoblast differentiation, matrix maturation, and mineralization in the long bones of growing rats. *J. Nutr. Biochem.* **2010**, *21*, 297–303. [CrossRef] [PubMed]

26. Querido, W.; Rossi, A.L.; Farina, M. The effects of strontium on bone mineral: A review on current knowledge and microanalytical approaches. *Micron* **2016**, *80*, 122–134. [CrossRef] [PubMed]

27. Kaygili, O.; Keser, S.; Kom, M.; Eroksuz, Y.; Dorozhkin, S.V.; Ates, T.; Ozercan, I.H.; Tatar, C.; Yakuphanoglu, F. Strontium substituted hydroxyapatites: Synthesis and determination of their structural properties, in vitro and in vivo performance. *Mater. Sci. Eng. C* **2015**, *55*, 538–546. [CrossRef] [PubMed]

28. Bonnelye, E.; Chabadel, A.; Saltel, F.; Jurdic, P. Dual effect of strontium ranelate: Stimulation of osteoblast differentiation and inhibition of osteoclast formation and resorption in vitro. *Bone* **2008**, *42*, 129–138. [CrossRef] [PubMed]

29. Tat, K.S.; Pelletier, J.P.; Mineau, F.; Caron, J.; Martel-Pelletier, J. Strontium ranelate inhibits key factors affecting bone remodeling in human osteoarthritic subchondral bone osteoblasts. *Bone* **2011**, *49*, 559–567. [CrossRef] [PubMed]

30. Pasqualetti, S.; Banfi, G.; Mariotti, M. The effects of strontium on skeletal development in zebrafish embryo. *J. Trace Elem. Med. Biol.* **2013**, *27*, 375–379. [CrossRef] [PubMed]

31. Reginster, J.Y.; Neuprez, A.; Dardenne, N.; Beaudart, C.; Emonts, P.; Bruyere, O. Efficacy and safety of currently marketed antiosteoporosis medications. *Best Pract. Res. Clin. Endocrinol. Metab.* **2014**, *28*, 809–834. [CrossRef] [PubMed]

32. Jurkić, M.L.; Cepanec, I.; Pavelić, S.K.; Pavelić, K. Biological and therapeutic effects of ortho-silicic acid and some ortho-silicic acid-releasing compounds: New perspectives for therapy. *Nutr. Metab.* **2013**, *10*, 1–12. [CrossRef] [PubMed]

33. Carlisle, E.M. Silicon: A Possible Factor in Bone Calcification. *Science* **1970**, *167*, 279–280. [CrossRef] [PubMed]

34. Carlisle, E.M. Silicon: An essential element for the chick. *Science* **1972**, *178*, 619–621. [CrossRef] [PubMed]

35. Carlisle, E.M. Silicon: A requirement in bone formation independent of vitamin D_1. *Calcif. Tissue Int.* **1981**, *33*, 27–34. [CrossRef] [PubMed]

36.   Carlisle, E.M. Silicon as an essential trace element in animal nutrition. *Ciba Found. Symp.* **1986**, *121*, 123–139. [CrossRef] [PubMed]

37.   Reffitt, D.M.; Ogston, N.; Jugdaohsingh, R.; Cheung, H.F.J.; Evans, B.A.J.; Thompson, R.P.H.; Powell, J.J.; Hampson, G.N. Orthosilicic acid stimulates collagen type 1 synthesis and osteoblastic differentiation in human osteoblast-like cells in vitro. *Bone* **2003**, *32*, 127–135. [CrossRef]

38.   Spector, T.D.; Calomme, M.R.; Anderson, S.H.; Clement, G.; Bevan, L.; Demeester, N.; Swaminathan, R.; Jugdaohsingh, R.; Berghe, D.A.; Powell, J.J. Choline-stabilized orthosilicic acid supplementation as an adjunct to Calcium/Vitamin D3 stimulates markers of bone formation in osteopenic females: A randomized, placebo-controlled trial. *BMC Muscoskelet. Dis.* **2008**, *9*, 1–10. [CrossRef] [PubMed]

39.   Camaioni, A.; Cacciotti, I.; Campagnolo, L.; Bianco, A. Silicon-substituted hydroxyapatite for biomedical applications. In *Hydroxyapatite (HAp) in Biomedical Applications*; Mucalo, M., Ed.; Woodhead Publishing: Cambridge, UK, 2015; pp. 333–373.

40.   Jugdaohsingh, R.; Tucker, K.L.; Qiao, N.; Cupples, L.A.; Kiel, D.P.; Powell, J.J. Dietary Silicon Intake is Positively Associated with Bone Mineral Density in Men and Premenopausal Women of the Framingham Offspring Cohort. *J. Bone Miner. Res.* **2004**, *19*, 297–307. [CrossRef] [PubMed]

41.   Judgaohsingh, R. Silicon and Bone Health. *J. Nutr. Health Aging* **2007**, *11*, 99–110.

42.   Henstock, J.R.; Canham, L.T.; Anderson, S.I. Silicon: The evolution of its use in biomaterials. *Acta Biomater.* **2015**, *11*, 17–26. [CrossRef] [PubMed]

43.   Dorozhkin, S.V.; Epple, M. Biological and Medical Significance of Calcium Phosphates. *Angew. Chem. Int. Ed. Engl.* **2002**, *41*, 3130–3146. [CrossRef]

44.   LeGeros, Z.R.; Ito, A.; Ishikawa, K.; Sakae, T.; LeGeros, J.P. Fundamentals of Hydroxyapatite and Related Calcium Phosphates. In *Advanced Biomaterials*; Basu, B., Katti, D.S., Kumar, A., Eds.; Wiley: Newark, NJ, USA, 2009; pp. 19–52.

45.   Chow, L.C. Solubility of Calcium Phosphates. In *Octacalcium Phosphate*; Chow, L.C., Eanes, E.D., Eds.; Karger: Basel, Switzerland, 2001; Volume 18, pp. 94–111.

46.   Du, L.-W.; Bian, S.; Gou, B.-D.; Jiang, Y.; Huang, J.; Gao, Y.-X.; Zhao, Y.-D.; Wen, W.; Zhang, T.-L.; Wang, K. Structure of Clusters and Formation of Amorphous Calcium Phosphate and Hydroxyapatite: From the Perspective of Coordination Chemistry. *Cryst. Growth Des.* **2013**, *13*, 3103–3109. [CrossRef]

47.   Elliott, J.C. *Structure and Chemistry of the Apatites and Other Calcium Orthophosphates*; Elsevier Science: Amsterdam, The Netherlands, 1994; pp. 53–61.

48.   Combes, C.; Rey, C. Amorphous calcium phosphates: Synthesis, properties and uses in biomaterials. *Acta Biomater.* **2010**, *6*, 3362–3378. [CrossRef] [PubMed]

49.   Dorozhkin, S.V. Amorphous calcium (ortho) phosphates. *Acta Biomater.* **2016**, *6*, 4457–4475. [CrossRef] [PubMed]

50.   Habraken, W.; Habibovic, P.; Epple, M.; Bohner, M. Calcium phosphates in biomedical applications: Materials for the future? *Mater. Today* **2016**, *19*, 69–87. [CrossRef]

51.   Lee, D.; Kumta, P.N. Chemical synthesis and stabilization of magnesium substituted brushite. *Mater. Sci. Eng. C* **2010**, *30*, 934–943. [CrossRef]

52.   Ferreira, A.; Oliveira, C.; Rocha, F. The different phases in the precipitation of dicalcium phosphate dehydrate. *J. Cryst. Growth* **2003**, *252*, 599–611. [CrossRef]

53.   Oliveira, A.; Ferreira, A.; Rocha, F. Dicalcium Phosphate Dihydrate: Precipitation Characterization and Crystal Growth. *Chem. Eng. Res. Des.* **2007**, *85*, 1655–1661. [CrossRef]

54.   Chen, G.G.; Luo, G.S.; Yang, L.M.; Xu, J.H.; Sun, Y.; Wang, J.D. Synthesis and size control of $CaHPO_4$ particles in a two-liquid phase micro-mixing process. *J. Cryst. Growth* **2005**, *279*, 501–507. [CrossRef]

55.   Kumta, P.N.; Sfeir, C.; Lee, D.-H.; Olton, D.; Choi, D. Nanostructured calcium phosphates for biomedical applications: Novel synthesis and characterization. *Acta Biomater.* **2005**, *1*, 65–83. [CrossRef] [PubMed]

56.   Ginebra, M.-P.; Canal, C.; Espanol, M.; Pastorino, D.; Montufar, E.B. Calcium phosphate cements as drug delivery materials. *Adv. Drug Deliv. Rev.* **2012**, *64*, 1090–1110. [CrossRef] [PubMed]

57.   Vallet-Regí, M.; González-Calbet, J.M. Calcium phosphates as substitution of bone tissues. *Prog. Solid State Chem.* **2004**, *32*, 1–31. [CrossRef]

58.   Dorozhkin, S.V. Bioceramics of calcium orthophosphates. *Biomaterials* **2010**, *31*, 1465–1485. [CrossRef] [PubMed]

59.   Kwon, S.-H.; Jun, Y.-K.; Hong, S.-H.; Kim, H.-E. Synthesis and dissolution behavior of β-TCP and HA/b-TCP composite powders. *J. Eur. Ceram. Soc.* **2003**, *23*, 1039–1045. [CrossRef]

60. Kolmas, J.; Kaflak, A.; Zima, A.; Slosarczyk, A. Alpha-tricalcium phosphate synthesized by two different routes: Structural and spectroscopic characterization. *Ceram. Int.* **2015**, *41*, 5727–5733. [CrossRef]

61. Carrodeguas, R.G.; De Aza, S. α-Tricalcium phosphate: Synthesis, properties and biomedical applications. *Acta Biomater.* **2011**, *7*, 3536–3546. [CrossRef] [PubMed]

62. Somrani, S.; Rey, C.; Jemal, M. Thermal evolution of amorphous tricalcium phosphate. *J. Mater. Chem.* **2003**, *13*, 888–892. [CrossRef]

63. Camiré, C.L.; Jegou, S.J.S.; Hansen, S.; McCarthy, I.; Lidgren, L. Hydration characteristics of α-tricalcium phosphates: Comparison of preparation routes. *J. Appl. Biomater. Biomech.* **2005**, *3*, 106–111. [PubMed]

64. Jokic, B.; Jankovic-Castvan, I.; Veljovic, D.; Bucevac, D.; Obradovic-Djuricic, K.; Petrovic, R. Synthesis and settings behavior of alpha-TCP from calcium deficient hyroxyapatite obtained by hydrothermal method. *J. Optoelectron. Adv. Mater.* **2007**, *9*, 1904–1910.

65. Bohner, M.; Brunner, T.J.; Doebelin, N.; Tang, R.K.; Stark, W.J. Effect of thermal treatments on the reactivity of nanosized tricalcium phosphate powders. *J. Mater. Chem.* **2008**, *18*, 4460–4467. [CrossRef]

66. Döbelin, N.; Brunner, T.J.; Stark, W.J.; Eggimann, M.; Fisch, M.; Bohne, M. Phase evolution of thermally treated amorphous tricalcium phosphate nanoparticles. *Key Eng. Mater.* **2009**, *396–398*, 595–598. [CrossRef]

67. Famery, R.; Richard, N.; Boch, P. Preparation of α- and β-tricalcium phosphate ceramics, with and without magnesium addition. *Ceram. Int.* **1994**, *20*, 327–336. [CrossRef]

68. Durucan, C.; Brown, P.W. Reactivity of α-tricalcium phosphate. *J. Mater. Sci.* **2002**, *37*, 963–969. [CrossRef]

69. Cai, S.; Wang, Y.W.; Li, J.Y.; Guan, Y.H.; Yao, K.D. Effects of dispersed medium and mineralizer CaF_2 on the synthesis of alpha-tricalcium phosphate. *J. Inorg. Mater.* **2004**, *19*, 852–858.

70. Camiré, C.L.; Gbureck, U.; Hirsiger, W.; Bohner, M. Correlating crystallinity and reactivity in α-tricalcium phosphate. *Biomaterials* **2005**, *26*, 2787–2794. [CrossRef] [PubMed]

71. Wei, X.; Ugurlu, O.; Akinc, A. Hydrolysis of α-tricalcium phosphate in simulated body fluid and dehydration behaviour during the drying process. *J. Am. Ceram. Soc.* **2007**, *90*, 2315–2321. [CrossRef]

72. Burkes, D.E.; Moore, J.J.; Ayers, R.A. Method for Producing Calcium Phosphate Powders Using an Auto-Ignition Combustion Synthesis Reaction. U.S. Patent Application No. 11/848,520, 15 May 2008.

73. Ayers, R.A.; Simske, S.J.; Moore, J.J.; Castillo, M.; Gottoli, G. Manufacture of Porous Net-Shaped Materials Comprising Alpha or Beta Tricalcium Phosphate or Mixtures Thereof. U.S. Patent Application No. 8,545,786, 1 October 2013.

74. Ayers, R.; Nielsen-Preiss, S.; Ferguson, V.; Gotolli, G.; Moore, J.J.; Kleebe, H.-J. Osteoblast-like cell mineralization induced by multiphasic calcium phosphate ceramic. *Mater. Sci. Eng. C* **2006**, *26*, 1333–1337. [CrossRef]

75. Volkmer, T.M.; Bastos, L.L.; Sousa, V.C.; Santos, L.A. Obtainment of alpha-tricalcium phosphate by solution combustion synthesis method using urea ascombustible. *Key Eng. Mater.* **2009**, *396–398*, 591–594. [CrossRef]

76. Ghosh, R.; Sarkar, R. Synthesis and characterization of sintered beta-tricalcium phosphate: A comparative study on the effect of preparation route. *Mater. Sci. Eng. C* **2016**, *67*, 345–352. [CrossRef] [PubMed]

77. Sanosh, K.P.; Chu, M.-C.; Balakrishnan, A.; Kim, T.N.; Cho, S.-J. Sol-gel synthesis of pure nanosized β-tricalcium phosphate crystalline powders. *Curr. Appl. Phys.* **2010**, *10*, 68–71. [CrossRef]

78. Thomas, D.; Su, S.; Qiu, J.; Pantoya, M.L. Microwave synthesis of functionally graded tricalcium phosphate for osteoconduction. *Mater. Today* **2016**, *9*, 47–53. [CrossRef]

79. Nabiyouni, M.; Ren, Y.; Bhaduri, S.B. Magnesium substitution in the structure of orthopaedic nanoparticles: A comparison between amorphous magnesium phosphates, calcium magnesium phosphates and hydroxyapatites. *Mater. Sci. Eng. C* **2015**, *52*, 11–17. [CrossRef] [PubMed]

80. Dong, G.; Zheng, Y.; He, L.; Wu, G.; Deng, C. The effect of silicon doping on the transformation of amorphous calcium phosphate to silicon-substituted α-tricalcium phosphate by heat treatment. *Ceram. Int.* **2016**, *42*, 883–890. [CrossRef]

81. Kato, N.; Hatoko, Y.; Yamamoto, E.; Furuzono, T.; Hontsu, S. Preparation and application of a potassium-substituted calcium phosphate sheet as a dental material for treating dentin hypersensitivity. *Key Eng. Mater.* **2017**, *720*, 102–107. [CrossRef]

82. Yu, T.; Gao, C.; Ye, J.; Zhang, M. Synthesis and characterization of a novel silver-substituted calcium phosphate cement. *J. Mater. Sci. Technol.* **2014**, *30*, 686–691. [CrossRef]

83. Alkhraisat, M.H.; Cabrejos-Azama, J.; Rodríguez, C.R.; Blanco Jerez, L.; López-Cabarcos, E. Magnesium substitution in brushite cements. *Mater. Sci. Eng. C* **2013**, *33*, 475–481. [CrossRef] [PubMed]

84. Cabrejos-Azama, J.; Alkhraisat, M.H.; Rueda, C.; Torres, J.; Blanco, L.; López-Cabarcos, E. Magnesium substitution in brushite cements for enhanced bone tissue regeneration. *Mater. Sci. Eng. C* **2014**, *43*, 403–410. [CrossRef] [PubMed]

85. Cabrejos-Azama, J.; Alkhraisat, M.H.; Rueda, C.; Torres, J.; Pintado, C.; Blanco, L.; López-Cabarcos, E. Magnesium substitution in brushite cements: Efficacy of a new biomaterial loaded with vancomycin for the treatment of *Staphylococcus aureus* infections. *Mater. Sci. Eng. C* **2016**, *61*, 72–78. [CrossRef] [PubMed]

86. Saleh, T.; Ling, L.S.; Hussain, R. Injectable magnesium-doped brushite cement for controlled drug release application. *J. Mater. Sci.* **2016**, *51*, 7427–7439. [CrossRef]

87. Geffers, M.; Barralet, J.E.; Groll, J.; Gbureck, U. Dual-setting brushite—Silica gel cements. *Acta Biomater.* **2015**, *11*, 467–476. [CrossRef] [PubMed]

88. Alkhraisat, M.H.; Mariño, F.T.; Rodríguez, C.R.; Jerez, L.B.; Cabarcos, E.L. Combined effect of strontium and pyrophosphate on the properties of brushite cements. *Acta Biomater.* **2008**, *4*, 664–670. [CrossRef] [PubMed]

89. Alkhraisat, M.H.; Rueda, C.; Cabarcos, E.L. Strontium ion substitution in brushite crystals: The role of strontium chloride. *J. Funct. Biomater.* **2011**, *2*, 31–38. [CrossRef] [PubMed]

90. Ewald, A.; Kappel, C.; Vorndran, E.; Moseke, C.; Gelinsky, M.; Gbureck, U. The effect of Cu(II)-loaded brushite scaffolds on growth and activity of osteoblastic cells. *J. Biomed. Mater. Res. A* **2012**, *100*, 2392–2400. [CrossRef] [PubMed]

91. Nishikawa, H.; Honda, M.; Yokota, T.; Shimizu, Y.; Aizawa, M. Preparation of Spherical Zn-Substituted Tricalcium Phosphate Powder by Ultrasonic Spray-Pyrolysis Technique and Its Characterization. *J. Nanomater.* **2016**, *2016*, 1–8. [CrossRef]

92. El-dek, S.I.; Mansour, S.F.; Ahmed, M.A.; Ahmed, M.K. Microstructural feature of flower like Fe brushite. *Prog. Nat. Sci.* **2017**, *27*, 520–526. [CrossRef]

93. Guerra-López, J.R.; Güida, J.A.; Ramos, G.; Punte, M. The influence of nickel on brushite structure stabilization. *J. Mol. Struct.* **2017**, *1137*, 720–724. [CrossRef]

94. Correia, M.B.; Junior, J.P.G.; Macedo, M.C.S.S.; Resende, C.X.; dos Santos, E.A. Effect of Mg^{2+} on acidic calcium phosphate phases grown by electrodeposition. *J. Cryst. Growth* **2017**, *475*, 328–333. [CrossRef]

95. Torres, P.M.C.; Marote, A.; Cerqueira, A.R.; Calado, A.J.; Abrantes, J.C.C.; Olhero, S.; da Cruz e Silva, O.A.B.; Vieira, S.I.; Ferreira, J.M.F. Injectable MnSr-doped brushite bone cements with improved biological performance. *J. Mater. Chem. B* **2017**, *5*, 2775–2787. [CrossRef]

96. Vahabzadeh, S.; Bandyopadhyay, A.; Bose, S.; Mandal, R.; Nandi, S.K. IGF-loaded silicon and zinc doped brushite cement: Physico-mechanical characterization and in vivo ostogenesis evaluation. *Integr. Biol.* **2015**, *7*, 1561–1573. [CrossRef] [PubMed]

97. Wei, X.; Akinc, M. Crystal structure analysis of Si- and Zn-codoped tricalcium phosphate by neutron powder diffraction. *J. Am. Ceram. Soc.* **2007**, *90*, 2709–2715. [CrossRef]

98. Reid, J.W.; Tuck, L.; Sayer, M.; Fargo, K.; Hendry, J.A. Synthesis and characterization of single-phase silicon-substituted α-tricalcium phosphate. *Biomaterials* **2006**, *27*, 2916–2925. [CrossRef] [PubMed]

99. Duncan, J.; Hayakawa, S.; Osaka, A.; MacDonald, J.F.; Hanna, J.V.; Skakle, J.M.S.; Gibson, I.R. Furthering the understanding of silicate-substitution in α-tricalcium phosphate: An X-ray diffraction, X-ray fluorescence and solid-state nuclear magnetic resonance study. *Acta Biomater.* **2014**, *10*, 1443–1450. [CrossRef] [PubMed]

100. Motisuke, M.; Mestres, G.; Reno, C.O.; Carrodeguas, R.G.; Zavaglia, C.A.C.; Ginebra, M.P. Influence of Si substitution on the reactivity of α-tricalcium phosphate. *Mater. Sci. Eng. C* **2017**, *75*, 816–821. [CrossRef] [PubMed]

101. Kamitakahara, M.; Tatsukawa, E.; Shibata, Y.; Umemoto, S.; Yokoi, T.; Ioku, K.; Ikeda, T. Effect of silicate incorporation on in vivo responses of α-tricalcium phosphate ceramics. *J. Mater. Sci. Mater. Med.* **2016**, *27*, 97. [CrossRef] [PubMed]

102. Mate-Sanchez de Val, J.E.; Calvo-Guirado, J.L.; Delgado-Ruiz, R.A.; Ramirez-Fernandez, M.P.; Negri, B.; Abboud, M.; Martinez, I.M.; de Aza, P.N. Physical properties, mechanical behavior, and electron microscopy study of a new α-TCP block graft with silicon in an animal model. *J. Biomed. Mater. Res. A* **2012**, *100*, 3446–3454. [CrossRef] [PubMed]

103. Mate-Sanchez de Val, J.E.; Calvo-Guirado, J.L.; Delgado-Ruiz, R.A.; Ramirez-Fernandez, M.P.; Martinez, I.M.; Granero-Marin, J.M.; Negri, B.; Chiva-Garcia, F.; Martinez-Gonzalez, J.M.; de Aza, P.N. New block graft of α-TCP with silicon in critical size defects in rabbits: Chemical characterization, histological, histomorphometric and micro-CT study. *Ceram. Int.* **2012**, *38*, 1563–1570. [CrossRef]

104. Radovanović, P.Z.; Jokić, B.; Veljović, D.; Dimitrijević, S.; Kojic, V.; Petrović, R.; Janáčković, A. Antimicrobial activity and biocompatibility of Ag^+- and Cu^{2+}-doped biphasic hydroxyapatite/α-tricalcium phosphate obtained from hydrothermally synthesized Ag^+- and Cu^{2+}-doped hydroxyapatite. *Appl. Surf. Sci.* **2014**, *307*, 513–519. [CrossRef]

105. Reid, J.W.; Fargo, K.; Hendry, J.A.; Sayer, M. The influence of trace magnesium content on the phase composition of silicon-stabilized calcium phosphate powders. *Mater. Lett.* **2007**, *61*, 3851–3854. [CrossRef]

106. Tong, C.; Zhu, Y.; Xu, C.; Li, Y. Preliminary observation of self-reduction of Eu ions in α-$Ca_3(PO_4)_2$ phosphors prepared in air condition. *Phys. B Condens. Matter* **2016**, *500*, 20–23. [CrossRef]

107. Pina, S.; Torres, P.M.; Goetz-Neunhoeffer, F.; Neubauer, J.; Ferreira, J.M.F. Newly developed Sr-substituted α-TCP bone cements. *Acta Biomater.* **2010**, *6*, 928–935. [CrossRef] [PubMed]

108. Bose, S.; Tarafder, S.; Banerjee, S.S.; Davies, N.M.; Bandyopadhyay, A. Understanding in vivo response and mechanical property variation in MgO, SrO and SiO_2 doped β-TCP. *Bone* **2011**, *48*, 1282–1290. [CrossRef] [PubMed]

109. Singh, R.K.; Kannan, S. Synthesis, Structural analysis, Mechanical, antibacterial and Hemolytic activity of Mg^{2+} and Cu^{2+} co-substitutions in β-$Ca_3(PO_4)_2$. *Mater. Sci. Eng. C* **2014**, *45*, 530–538. [CrossRef] [PubMed]

110. Matsumoto, N.; Sato, K.; Yoshida, K.; Hashimoto, K.; Toda, Y. Preparation and characterization of α-tricalcium phosphate co-doped with monovalent and divalent antibacterial metal ions. *Acta Biomater.* **2009**, *5*, 3157–3164. [CrossRef] [PubMed]

111. García-Páez, H.; García Carrodeguas, R.; De Aza, A.H.; Baudín, C.; Pena, P. Effect of Mg and Si co-substitution on microstructure and strength of tricalcium phosphate ceramics. *J. Mech. Behav. Biomed. Mater.* **2014**, *30*, 1–15. [CrossRef] [PubMed]

112. Banerjee, S.S.; Tarafder, S.; Davies, N.M.; Bandyopadhyay, A.; Bose, S. Understanding the influence of MgO and SrO binary doping on the mechanical and biological properties of β-TCP ceramics. *Acta Biomater.* **2010**, *6*, 4167–4174. [CrossRef] [PubMed]

113. Bigi, A.; Foresti, E.; Gandolfi, M.; Gazzano, M.; Roveri, N. Isomorphous Substitutions in β-Tricalcium Phosphate: The Different Effects of Zinc and Strontium. *J. Inorg. Biochem.* **1997**, *66*, 259–265. [CrossRef]

114. Alkhraisat, M.H.; Rueda, C.; Cabrejos-Azama, J.; Lucas-Aparicio, J.; Mariño, F.T.; Torres García-Denche, J.; Jerez, L.B.; Gbureck, U.; Cabarcos, E.L. Loading and release of doxycycline hyclate from strontium-substituted calcium phosphate cement. *Acta Biomater.* **2010**, *6*, 1522–1528. [CrossRef] [PubMed]

115. Araújo, C.; Sader, M.S.; Moreira, E.L.; Moraes, V.C.A.; LeGeros, R.Z.; Soares, G.A. Maximum substitution of magnesium for calcium sites in Mg-β-TCP structure determined by X-ray powder diffraction with the Rietveld refinement. *Mater. Chem. Phys.* **2009**, *118*, 337–340. [CrossRef]

116. Enderle, R.; Götz-Neunhoeffer, F.; Göbbels, M.; Müller, F.A.; Greil, P. Influence of magnesium doping on the phase transformation temperature of β-TCP ceramics examined by Rietveld refinement. *Biomaterials* **2005**, *26*, 3379–3384. [CrossRef] [PubMed]

117. Kannan, S.; Ventura, J.M.; Ferreira, J.M.F. Aqueous precipitation method for the formation of Mg-stabilized β-tricalcium phosphate: An X-ray diffraction study. *Ceram. Int.* **2007**, *33*, 637–641. [CrossRef]

118. Kannan, S.; Lemos, I.A.F.; Rocha, J.H.G.; Ferreira, J.M.F. Synthesis and characterization of magnesium substituted biphasic mixtures of controlled hydroxyapatite/β-tricalcium phosphate ratios. *J. Solid State Chem.* **2005**, *178*, 3190–3196. [CrossRef]

119. Marchi, A.; Dantas, A.C.S.; Greil, P.; Bressiani, J.C.; Bressiani, A.H.A.; Müller, F.A. Influence of Mg-substitution on the physicochemical properties of calcium phosphate powders. *Mater. Res. Bull.* **2007**, *42*, 1040–1050. [CrossRef]

120. Zhou, H.; Hou, S.; Zhang, M.; Chai, H.; Liu, Y.; Bhaduri, S.; Yang, L.; Deng, L. Synthesis of β-TCP and CPP containing biphasic calcium phosphates by a robust technique. *Ceram. Int.* **2016**, *42*, 11032–11038. [CrossRef]

121. Ryu, H.-S.; Hong, K.S.; Lee, J.-K.; Kim, D.J.; Lee, J.H.; Chang, B.-S.; Lee, D.-H.; Lee, C.-K.; Chung, S.-S. Magnesia-doped HA/β-TCP ceramics and evaluation of their biocompatibility. *Biomaterials* **2004**, *25*, 393–401. [CrossRef]

122. Nikaido, T.; Tsuru, K.; Munar, M.; Maruta, M.; Matsuya, S.; Nakamura, S.; Ishikawa, K. Fabrication of β-TCP foam: Effects of magnesium oxide as phase stabilizer on its properties. *Ceram. Int.* **2015**, *41*, 14245–14250. [CrossRef]

123. Salma-Ancane, A.; Stipniece, L.; Putnins, A.; Berzina-Cimdina, L. Development of Mg-containing porous β-tricalcium phosphate scaffolds for bone repair. *Ceram. Int.* **2015**, *41*, 4996–5004. [CrossRef]

124. Wang, S.; Liu, R.; Yao, J.; Wang, Y.; Li, H.; Dao, R.; Guan, J.; Tang, G. Fabrication of mesoporous magnesium substituted β-tricalcium phosphate nanospheres by self-transformation and assembly involving EDTA ions. *Microporous Microporous Mater.* **2013**, *179*, 172–181. [CrossRef]

125. Tavares, D.; Castro, L.O.; Soares, G.D.A.; Alves, G.G.; Granjeiro, J.M. Synthesis and cytotoxicity evaluation of granular magnesium substituted β-tricalcium phosphate. *J. Appl. Oral Sci.* **2013**, *21*, 37–42. [CrossRef]

126. Grigg, T.; Mee, M.; Mallinson, P.M.; Fong, S.K.; Gan, Z.; Dupree, R.; Holland, D. Cation substitution in β-tricalcium phosphate investigated using multi-nuclear, solid-state NMR. *J. Solid State Chem.* **2014**, *212*, 227–236. [CrossRef]

127. Kannan, S.; Ventura, J.M.G.; Lemos, A.F.; Barba, A.; Ferreira, J.M.F. Effect of sodium addition on the preparation of hydroxyapatites and biphasic ceramics. *Ceram. Int.* **2008**, *34*, 7–13. [CrossRef]

128. Kannan, S.; Ventura, J.M.G.; Ferreira, J.M.F. Synthesis and thermal stability of potassium substituted hydroxyapatites and hydroxyapatite/β-tricalcium phosphate mixtures. *Ceram. Int.* **2007**, *33*, 1489–1494. [CrossRef]

129. Kannan, S.; Rebelo, A.; Lemos, A.F.; Barba, A.; Ferreira, J.M.F. Synthesis and mechanical behaviour of chlorapatite and chlorapatite/β-TCP composites. *J. Eur. Ceram. Soc.* **2007**, *27*, 2287–2294. [CrossRef]

130. Matsumoto, A.; Yoshida, K.; Hashimoto, K.; Toda, Y. Thermal stability of β-tricalcium phosphate doped with monovalent metal ions. *Mater. Res. Bull.* **2009**, *44*, 1889–1894. [CrossRef]

131. Meenambal, R.; Singh, R.K.; Kumar, P.N.; Kannan, S. Synthesis, structure, thermal stability, mechanical and antibacterial behaviour of lanthanum (La^{3+}) substitutions in β-tricalcium phosphate. *Mater. Sci. Eng. C* **2014**, *43*, 598–606. [CrossRef] [PubMed]

132. Mayer, A.; Cuisinier, F.J.G.; Gdalya, S.; Popov, I. TEM study of the morphology of Mn^{2+}-doped calcium hydroxyapatite and β-tricalcium phosphate. *J. Inorg. Biochem.* **2008**, *102*, 311–317. [CrossRef] [PubMed]

133. Ananth, P.; Shanmugam, S.; Jose, S.P.; Nathanael, A.J.; Oh, T.H.; Mangalaraj, D.; Ballamurugan, A.M. Structural and chemical analysis of silica-doped β-TCP ceramic coatings on surgical grade 316L SS for possible biomedical application. *J. Asian Ceram. Soc.* **2015**, *3*, 317–324. [CrossRef]

134. Kawabata, A.; Yamamoto, T.; Kitada, A. Substitution mechanism of Zn ions in β-tricalcium phosphate. *Phys. B Condens. Matter* **2011**, *406*, 890–894. [CrossRef]

135. Gokcekaya, O.; Ueda, K.; Ogasawara, K.; Kanetaka, H.; Narushima, T. In vitro evaluation of Ag-incorporated β-tricalcium phosphate. *Mater. Sci. Eng. C* **2017**, *75*, 926–933. [CrossRef] [PubMed]

136. Rau, J.V.; Wu, V.M.; Graziani, V.; Fadeeva, I.V.; Fomin, A.S.; Fosca, M.; Uskoković, V. The bone building blues: Self-hardening copper-doped calcium phosphate cement and its in vitro assessment against mammalian cells and bacteria. *Mater. Sci. Eng. C* **2017**, *79*, 270–279. [CrossRef] [PubMed]

137. Zhang, M.; Wu, C.; Li, H.; Yuen, J.; Chang, J.; Xiao, Y. Preparation, characterization and in vitro angiogenic capacity of cobalt substituted β-tricalcium phosphate ceramics. *J. Mater. Chem.* **2012**, *22*, 21686–21694. [CrossRef]

138. Zhu, H.; Guo, D.; Qi, W.; Xu, K. Development of Sr-incorporated biphasic calcium phosphate bone cement. *Biomed. Mater.* **2017**, *12*, 015016. [CrossRef] [PubMed]

139. Singh, R.K.; Srivastava, M.; Prasad, N.K.; Kannan, S. Structural analysis and hyperthermia effect of Fe^{3+}/Ni^{2+} co-substitutions in $β-Ca_3(PO_4)_2$. *J. Alloys Compd.* **2017**, *725*, 393–402. [CrossRef]

140. Singh, R.K.; Srivastava, M.; Prasad, N.K.; Awasthi, S.; Dhayalan, A.; Kannan, S. Iron doped β-tricalcium phosphate: Synthesis, characterization, hyperthermia effect, biocompatibility and mechanical evaluation. *Mater. Sci. Eng. C* **2017**, *78*, 715–726. [CrossRef] [PubMed]

141. Gomes, S.; Kaur, A.; Greneche, J.-M.; Nedelec, J.-M. Atomic scale modeling of iron-doped biphasic calcium phosphate bioceramics. *Acta Biomater.* **2017**, *50*, 78–88. [CrossRef] [PubMed]

142. Kannan, S.; Goetz-Neunhoeffer, F.; Neubauer, J.; Pina, S.; Torres, P.M.C.; Ferreira, J.M.F. Synthesis and structural characterization of strontium-and magnesium-co-substituted β-tricalcium phosphate. *Acta Biomater.* **2010**, *6*, 571–576. [CrossRef] [PubMed]

143. Matsumoto, N.; Yoshida, K.; Hashimoto, K.; Toda, Y. Preparation of beta-tricalcium phosphate powder substituted with Na/Mg ions by polymerized complex method. *J. Am. Ceram. Soc.* **2010**, *93*, 3663–3670. [CrossRef]

144. Meenambal, R.; Kumar, P.N.; Poojar, P.; Geethanath, S.; Kannan, S. Simultaneous substitutions of Gd^{3+} and Dy^{3+} in $β-Ca_3(PO_4)_2$ as a potential multifunctional bio-probe. *Mater. Des.* **2017**, *120*, 336–344. [CrossRef]

145. El Khouri, A.; Elaatmani, M.; Ventura, G.D.; Sodo, A.; Rizzi, R.; Rossi, M.; Capitelli, A. Synthesis, structure refinement and vibrational spectroscopy of new rare-earth tricalcium phosphates $Ca_9RE(PO_4)_7$ (RE = La, Pr, Nd, Eu, Gd, Dy, Tm, Yb). *Ceram. Int.* **2017**, *43*, 15645–15653. [CrossRef]
146. Goldberg, M.A.; Smirnov, V.V.; Protsenko, P.V.; Antonova, O.S.; Smirnov, S.V.; Fomina, A.A.; Konovalov, A.A.; Leonov, A.V.; Ashmarin, A.A.; Barinov, S.M. Influence of aluminium substitutions on phase composition and morphology of β-tricalcium phosphate nanopowders. *Ceram. Int.* **2017**, *43*, 13881–13884. [CrossRef]

International Journal of
Molecular Sciences

MDPI

Opinion

Designing Smart Biomaterials for Tissue Engineering

Ferdous Khan [1],*,† and Masaru Tanaka [1,2],*

[1]   Soft-Materials Chemistry, Institute for Materials Chemistry and Engineering, Kyushu University,
      744 Motooka Nishi-ku, Fukuoka 819-0395, Japan
[2]   Frontier Center for Organic Materials, Yamagata University, 4-3-16 Jonan, Yonezawa,
      Yamagata 992-8510, Japan
*   Correspondence: ferdous.khan0@gmail.com (F.K.); masaru_tanaka@ms.ifoc.kyushu-u.ac.jp or
      tanaka@yz.yamagata-u.ac.jp (M.T.)
†   Current address: ECOSE-Biopolymer, Knauf Insulation Limited, P.O. Box 10, Stafford Road,
      ST. HELENS WA10 3NS, UK.

Received: 1 November 2017; Accepted: 1 December 2017; Published: 21 December 2017

Abstract: The engineering of human tissues to cure diseases is an interdisciplinary and a very attractive field of research both in academia and the biotechnology industrial sector. Three-dimensional (3D) biomaterial scaffolds can play a critical role in the development of new tissue morphogenesis via interacting with human cells. Although simple polymeric biomaterials can provide mechanical and physical properties required for tissue development, insufficient biomimetic property and lack of interactions with human progenitor cells remain problematic for the promotion of functional tissue formation. Therefore, the developments of advanced functional biomaterials that respond to stimulus could be the next choice to generate smart 3D biomimetic scaffolds, actively interacting with human stem cells and progenitors along with structural integrity to form functional tissue within a short period. To date, smart biomaterials are designed to interact with biological systems for a wide range of biomedical applications, from the delivery of bioactive molecules and cell adhesion mediators to cellular functioning for the engineering of functional tissues to treat diseases.

Keywords: tissue engineering; smart materials; extracellular matrix; stimuli responsive polymer

1. Introduction

Research on polymer biomaterials has been a subject of interest both in academia and industry since 1960 [1]. The variety of polymeric materials has been rationally designed by incorporating distinct functional groups into the molecular chain to control physical, chemical, and biological properties for versatile biomedical arenas, such as the controlled delivery of bioactive molecules and cell-based therapeutic applications [2–4]. However, the nature of the biomedical applications of such material systems depends on their macromolecular structure, their interactions with living cells and cytocompatibility, and how the macromolecules organize by themselves to form three-dimensional (3D) architecture. Polymer-based biomaterials have been developed both from natural and synthetic (man-made) resources. Polysaccharides and proteins are well–established natural polymeric biomaterials, which have found numerous applications in tissue regeneration. For example, chitosan is one of the best polysaccharides with exceptional biocompatibility and biodegradability, consisting of multifunctional groups that make it a potential candidate for future biomaterials development (via processing) for cellular functioning and differentiation towards tissue engineering [5]. Protein-based biomaterials, for example, collagen, metrigel, and hyaluronic acid, have been derived from animal sources and explored in tissue regeneration. Although proteins are highly biocompatible, their very fast degradation and low mechanical strength result in a lack of structural support during tissue development. Polymeric biomaterials are also developed by polymerizing one or more monomers either by homopolymer reaction or copolymer reaction to form non-biodegradable [6,7] and biodegradable materials [8,9]. The structure, molecular chain length, and stereochemistry can be tailored by varying chemical and physical parameters during synthesis. In recent years, a significant demand for biodegradable materials development, by means of synthesis, in addition to their processing and fabrication of 3D structural scaffolds has arisen in tissue regeneration and the delivery of bioactive molecules, due to the demand for cost-effective surgical procedures. Thus, continuous efforts need to be undertaken in this field.

For biomedical applications, biomaterial scientists have devoted continuous efforts to developing methods, fabricating new medical devices, and processing and synthesizing novel biomaterials. Recently, Liu and co-workers [10] developed a alginate-collagen 3D hydrogel composite, which has improved the treatment of corneal diseases. We developed high-throughput technology for cell-compatible biomaterials discovery, as well as the processing and fabrication of 3D scaffolds for potential application in skeletal tissue regeneration [11,12]. To date, a wide range of biomaterials have been used for biomedical applications, many of which do not have biomimetic properties, and require the engineering of these materials in a cost-effective manner for maximum output. Therefore, there is a socioeconomic need to develop new biomaterials for resolving such problems in the biomedical arena. Polymers with hydrophilic properties form 3D crosslinked networks, known as hydrogels, a type of novel smart biomaterial that responds to external stimulus and exhibits potential application as a scaffold for tissue regeneration and the delivery of bioactive molecules [10,13,14].

For tissue engineering (TE), a biomaterial scaffold provides mechanical and porous network structural support, shape, and hierarchy architecture with surface chemistry for cell attachment, cell-cell communication through the porous network, as well as proliferation and differentiation for tissue regeneration. To date, most synthetic biomaterials are derived for TE application, which are synthesized either from lactic acid, caprolactone, or glycolide monomers to form poly(L-Lactide), or poly(ε-caprolactone), or poly(L-glycolic acid), respectively, and/or their combination to form copolymers, or the physical blending of these polymers [11,12]. Natural polymers such as chitosan, alginate, starch, collagen, hyaluronic acid, cellulose, fibrin, silk, and their derivatives are also used [5,15,16]. A variety of techniques and methods has been developed to fabricate biomaterial scaffolds to control shapes, sizes, porosities, and architectures for TE applications [17,18]. In any tissue regeneration, it is important that biomaterial scaffolds not only provide temporary structural integrity but also play a role in the interaction of the cells and biomolecules, cell attachment and growth, as well as in the process of tissue development. However, poly(L-Lactide) and poly(ε-caprolactone) are known to be biodegradable materials, which have been implanted to identify their host and tissue regeneration [19,20]. However, such traditional biomaterials do not have the ability to adapt to living tissues during changing pH and body temperature caused by disease. Therefore, polymer scientists have been trying to create smart polymeric biomaterials [21] that mimic living tissues in the last two decades.

It is well known that the properties of smart polymers change reversibly with the sliding variation of external or internal parameters in the system, such as temperature, pH, ionic factor, biological molecules, and so on. Thus, smart polymeric biomaterials, in particular as delivery systems of bioactive molecules; tissue engineering scaffolds; and cellular attachment and growth have become novel research topics.

2. Tissue Engineering

The aim of TE is to develop new functional tissue and regenerate tissue either in vitro or in vivo to cure diseases when a surgical intervention is needed. In such circumstances, 3D biomaterial scaffolds play a critical role in repairing injury. For optimal functional tissue development, the scaffold should interact with cells without any adverse effect to provide cellular attachment, proliferation, growth, and accumulation of mineral matrix. In addition, it is essential to provide structural support, a design akin to the natural extracellular matrix, and a suitable surface, porosity, and heterogeneous pore sizes to promote cell-cell communication and differentiation, as well as transport nutrients. The 3D scaffolds should comply with sufficient mechanical strength similar to that of native tissue and a crosslinked network structure, as demonstrated in Figure 1a,b.

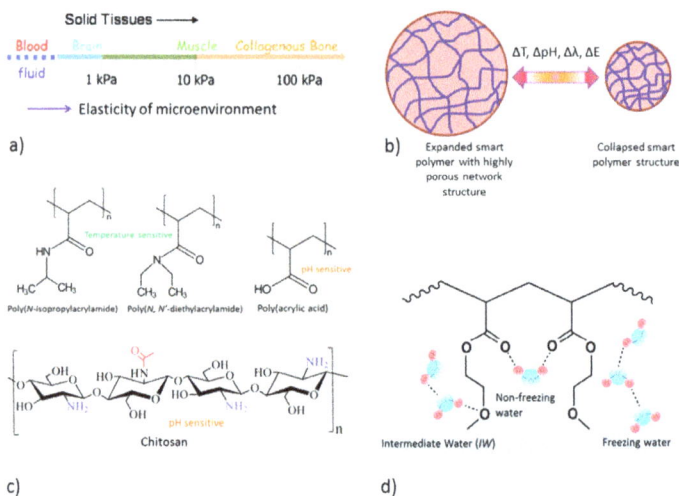

Figure 1. Smart polymer biomaterials for tissue engineering applications: (**a**) Representation of elastic modulus of various tissues in consideration to design a smart scaffold; (**b**) stimuli responsive system (ΔT: variation of temperature change, ΔpH: variation of pH, Δλ: variation of wave length, and ΔE: variation of electric field); (**c**) structures of some synthetic and natural smart polymers; and (**d**) novel scaffold of poly(2-methoxyethyl acrylate) (PMEA)—hydrated PMEA forms intermediate water, which affects the protein adsorption, as well as cell adhesion, proliferation, and differentiation.

The 3D structured biomaterials can be designed utilizing a variety of natural and synthetic polymers (examples are presented in Figure 1c,d), ceramics such as hydroxyl apatite (HAp), and tricalcium phosphate (TCP), as well as their combinations to mimic native tissue while maintaining cell viability and functions. Furthermore, 3D scaffolds should act as a delivery agent for bioactive molecules or drugs and can be encapsulated into the materials during their process and fabrication for faster curing, if needed. The design and selection of smart biomaterials depends on their specific application, some of which are more suitable than others. Biomaterials can have a solid or hydrogel structure before implantation, or be in injectable forms that harden in situ.

Extracellular Matrix (ECM)

The biological ECM provides structural support along with all necessary biological functions during tissue regeneration and maintenance. The scaffolds used in TE are also expected to have the same level of support and functional capability as that of the biological ECM [21]. ECM is a dynamic and complex structure, directly involved in specific gene regulation for a particular tissue development [22,23], and actively interacting with cells to remodel tissue [24]. The structure and components of ECM are continuously changing due to its dynamic nature with the development of tissues, remodeling, and repairing. It is critically important to develop smart and biomimetic 3D structure biomaterials that closely match the characteristic nature of the native ECM [25]. However, understanding complex functions and the structure of ECM in mature and/or during tissue development are extremely difficult. Combining multiple approaches, biomaterial scientists have developed a 3D structure with close approximation of natural ECM, and this has remained an active research area.

Several types of animal-derived proteins (e.g., collagen, laminin, and metrigel) have been utilized to fabricate 3D scaffolds for various TE, such as skin replacement, bone substitutes, artificial blood vessels [26], and cell delivery [27]. Hydrogels have been synthesized by the modification of polysaccharides via crosslinking for both drug release and tissue regeneration [28,29]. Hyaluronic acid (HA) is a polysaccharide consisting of linear glycosaminoglycan, found in the ECM and known to influence cell signaling pathways; it plays a crucial role in functional TE [29]. HA is the Food and Drug Administration (FDA) approved biomaterial for human application, and clinical trials confirmed that HAs are effective for osteoarthritis applications [30–32]. Chitosan (Figure 1c) is an important class of polysaccharide consisting of multiple functional groups, allowing ease of chemical modification, and has been used as a delivery agent, as well as in cell encapsulation, cellular functioning, and differentiation toward TE application [5,16,33,34]. Alginate is another example of a polysaccharide that has been used for cell encapsulation and injectable gel in TE [35].

3. Designing Smart Biomaterials

Designing bio-functional materials is critically important for potential biomedical applications. There are several approaches in the design of hybrid smart biomaterials, composed of synthetic and natural polymers, that considerably enhance the potential applications of such materials. These techniques permit the insertion of bio-recognition moieties into the structure of polymers that influence their self-assembly into precisely defined 3D structure formations. The design of smart polymer-based biomaterials with desired properties and network structure mainly depends on the characteristic nature of functional monomers and their feed ratio, method of polymerization and kinetics, building of molecular architecture during synthesis, and crosslink network formation. The development of advanced functional biomaterials requires control over some physical and chemical parameters in the design, synthesis, processing, and fabrication. The 3D network structure smart material mainly depends on the characteristic nature of the polymers, monomers, oligomers, and the methods of synthesis. There is a wide variety of methods which include radiation-induced crosslinking, chemical and physical crosslinking, for the preparation of 3D network biomaterials, namely hydrogels with different structures and properties. The hydrogel is not soluble in water, due to the presence of a 3D polymer network, and swells at equilibrium. The 3D networks in hydrogels are formed either by covalent bonding between polymer chains or physical interactions such as hydrogen boning and ionic bonding. The crosslinked network provides an equilibrium state between dispersing and cohesive forces between polymer chains, resulting in the insolubility of the gel in water. These materials have versatile applications as biomaterials, scaffolds for functional tissue regeneration, delivery agents, and medical devices.

Key factors that can influence the swelling of the 3D network of smart materials include: (i) the positive change of thermal energy during mixing of a polymer and solvent favors swelling; and (ii) the change of thermal energy is related to the polymer chain conformations, and negative thermal change prevents swelling.

These polymer networks can be classified as to their preparation methods, such as chemical reactions and physical blending. The examples of the chemical reactions of hydrogels are from the graft-copolymer [36], block copolymer [37], and crosslinking reactions of polymers [38]. For the hydrogel synthesis, the reactions can be initiated either by chemical initiation methods [38] or ionizing radiation-induced methods [39]. The chemical initiation methods require the incorporation of a bifunctional reagent and chemical catalyst, thus introducing impurities into the final product. Sometimes it can be extremely difficult to remove impurities from the hydrogel networks. Alternatively, the ionizing radiation-induced methods offer advantages that allow efficient crosslinking and yield a much cleaner product as compared to those of the chemical initiation methods. High-energy (e.g., γ-radiation, electron beam) radiation, plasma-radiation, and UV-radiation induced methods can be adopted to synthesize polymer hydrogels.

The physical blending of different polymers processed either by solution or thermally is an alternative approach and an economically viable route for the development of new structural smart materials [40], although it retains the characteristic properties of each polymer in the blend. There are several advantages to this system, including the ease of controlling the processing and fabrication of smart devices. This approach would allow achieving desired properties such as porosity, pore sizes, swelling ratio, biodegradability, and mechanical property by changing the process parameters, feed ratios of starting polymers, and selection of appropriate solvent. Therapeutic molecules can easily be incorporated into the blended solution or dispersed in micro-phase reservoirs.

Scientists are capable of designing and synthesizing polymeric biomaterials with complex architecture, by copolymerizing multiple chemical functional building blocks and tailoring biocompatibility, biodegradation, mechanical properties, hydrophilicity, and chemical and biological response to external stimuli, in a well-controlled manner. In such biomaterial system, the copolymer chains have the ability to organize by themselves in the water phase through intra- and intermolecular interactions, thus forming structural diversity. There are experimental challenges to designing model copolymer systems that will allow a faster response with slight variation of multiple stimuli. Despite the challenges of fabricating smart materials that respond to multiple stimuli (e.g., temperature, pH, and light), they have biomedical industrial importance. In the future, material scientists will continue to investigate such systems to achieve scientific breakthroughs.

In a broader aspect, smart biomaterials can be designed by incorporating peptide and/or protein into the polymer network, enabling the creation of a 3D scaffold for TE application [41,42]. For example, Ito and co-workers [42] developed smart biomaterials by genetically engineered collagen-type II protein, which was used for 3D scaffold fabrication to tailor chondrocytes proliferation and migration, and promoted artificial cartilage formation [42]. In a separate study, it was reported that cysteine-tagged fibronectin coupled with poly(ethylene glycol) (PEG) derivatives by employing Michael-type addition, then crosslinked with thiol-modified HA [43], formed a 3D hydrogel matrix. They demonstrated that these hydrogels were cytocompatible, fully supported human fibroblasts adhesion, proliferation, and robust migration, and after implantation in porcine cutaneous promoted dermal wound healing [39]. Similarly, a number of elastin-based block copolymers exhibiting a wide range of mechanical properties have been developed for TE application [44]. The signaling molecules in scaffold materials can be chemically linked to enhance cell binding. The functionalization of synthetic polymers by PNIPAm-(Arg-Gly-Asp (RGD)) peptide promotes cells attachment, migration, and functionality, and has been well reviewed [45].

In recent years, self-assembly systems have been developed both by chemically and biologically synthesized peptides for neural tissue engineering [46] and the growth of human dermal fibroblasts with excellent biocompatibility [47,48]. In the development of new biomaterials, self-assembly peptides and proteins are currently and will remain an active field of research, as this system responds to the external environment, namely, pH and temperature [49]. Although the generation of 3D-structure scaffolds by using self-assembly peptides is highly reproducible, it faces a number of challenges such as scale-up the process, production, and purification in a cost-effective manner. Without a doubt, enormous effort is required, focusing on the rational design, discovery, fabrication, and the process development.

With a deeper understanding of the fundamental behavior of biological complexity and the physicochemical nature of native tissue, the rational design of new smart biomaterials and production is made possible by applying multidisciplinary approaches, bringing together polymer chemists, biomaterials scientists, tissue engineers, and medical surgeons.

4. Importance of Smart Biomaterials in TE

There is a socioeconomic need to replace and/or repair tissue with more advanced materials approach, techniques, and methods focusing on functional tissue reconstitution [50] without adverse effect. Although significant advances have been made in understanding the properties (e.g., physical, chemical, and biological) of so-called smart biomaterials, so far a very limited number of such type of materials have met the demand of clinical need. Nevertheless, the European technology platform reported that smart biomaterials could be key to enabling technology for regenerative medicine and therapeutics [51].

In the context of TE, several researchers [52–55] have demonstrated that smart biomaterials have the ability to maintain and control cellular behavior for functional tissue regeneration. Yuan and coworkers [55] demonstrated that porous ceramic scaffolds alone induce substantial ectopic bone formation without the incorporation of molecules and/or cells. Furthermore, authors have clarified that, in a large bone defect in sheep, micro-structured tricalcium phosphate ceramic had the highest osteoinductive potential for bone tissue repair without the addition of cells and/or growth factors, and they defined such ceramics as a class of smart biomaterials for bone TE.

Generally, smart biomaterials respond to one or more environmental variables (e.g., temperature, pH, ionic concentration, light, electric and magnetic fields), which influence cells behavior and functionality as well as tissue modeling. A list of smart biomaterials that relate to cell culture and TE is presented in Table 1 and Figures 1 and 2. A number of research strategies can be adopted for cells attachment, growth, and differentiation towards specific tissue development.

Table 1. Smart biomaterials for cellular/TE applications that response to various types of stimuli. PEO—polyethylene oxide; PPO—polypropylene oxide; DOX—doxorubicin.

Examples of Smart Biomaterials	External Stimuli	Applications
Poly(N-isopropylacrylamide)	Temperature	Patterned cells seeding and co-culture [56].
Pluronics® (poly(ethylene oxide)-poly(propylene oxide)-poly(ethylene oxide))	Temperature	Tissue engineering processes (new cartilage formation [57].
PNIPAm-Arg-Gly-Asp (RGD)	Temperature	Controlling osteoblast adhesion and proliferation [58].
Poly(2-propylacrylic acid)	pH	Protein/DNA intercellular delivery [59].
Chitosan/Polyethyleneimine (CS/PEI) blend	pH	Scaffolds for cellular functioning and cartilage tissue engineering [40].
Self-assembling peptide	Temperature and pH	Neural tissue engineering [46].
Self-assembling peptide	Temperature and pH	Peptide (P_{11}-4) supported primary human dermal fibroblasts growth and proliferation [47].
Azobenzene-containing polymer brushes	Light	Human umbilical vein endothelial cells [60].
Spiropyran-containing polymer brushes/graft copolymer	Light	Cell capture and release [61].
Poly(2-acrylamido-2-methyl-propane sulphonic acid-co-N-butylmethacrylate)	Electric field	Controlled delivery of drug and cells [62].
Poly(N-isopropylacrylamide-acrylamide-chitosan) (PAC)-coated magnetic nanoparticles (MNPs)	Magnetic field, temperature, and pH	Human dermal fibroblasts and normal prostate epithelial cells culture and cancer drug delivery [63].
Poly(6-O-methacryloyl-D-galactopyranose)-SS-poly(γ-benzyl-L-glutamate) (PMAgala-SS-PBLG)	Redox reaction	DOX delivery and human hepatoma cell receptor targeting [64].
Poly(ethylene-glycol)-Poly(acrylate)	Light	Human mesenchymal stem cells growth, proliferation, and chondrogenic differentiation [65].
Gold membrane microchip	Electrochemical	Controlled release in implants [66].
Antibacterial Ti-Ni-Cu shape memory alloys	Temperature	Cellular compatible (e.g., L929 and MG63) [67].

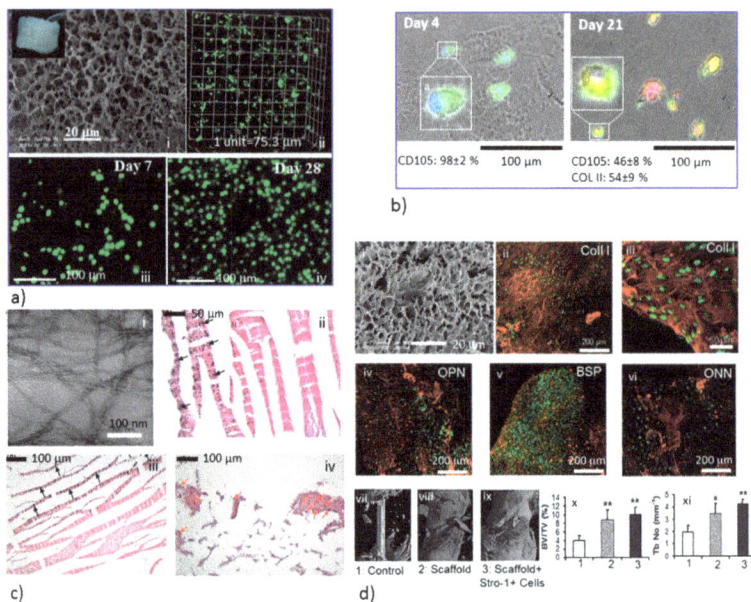

Figure 2. Stimuli-responsive polymeric biomaterials for TE applications: (**a**) CS/PEI, pH responsive hydrogel scaffold, scanning electron microscope image hydrogels frozen in liquid nitrogen and freeze-dried (i), a confocal image of HeLa cells (labeled with CellTracker Green) grown within the hydrogel (day 21) (ii), human fetal skeletal cells (labeled with CellTracker Green) grown within the hydrogel on day 7 (iii) and day 28 (iv); (**b**) influence of photodegradable dynamic microenvironment on chondrogenic differentiation of hMSCs was verified by immunostaining for the hMSC marker CD105 (fluorescein isothiocyanate (FITC), green) and the chondrocyte marker COLII (tetramethyl rhodamine isothiocyanate–labeled, red), cells did not produce COLII on day 4 (left), almost half of the cells with peptide sequence (Arg-Gly-Asp-Ser (RGDS)) strongly expressed CD105, and the other half produced COLII (right) on day 21; (**c**) self-assembly of complementary peptides hydrogels; (i) TEM image of P_{11}–13/P_{11}–14 peptide fibrils and fibers, prepared at pH 7.4, primary human dermal fibroblasts grown within hydrogel and their histological images P_{11}–13/P_{11}–14 hydrogel after 14 days of culture (ii,iii). Black arrows indicate possible cell remnants of black, circular aggregates on some fibers. P_{11}-4 hydrogel with primary human dermal fibroblasts (red arrows) after 14 days of culture (iv) showing Neo-ECM deposition; (**d**) bioresorbable scaffolds fabricated from polymer blend (CS/Polyvinyl acetate (PVAc)/PLLA: 50/25/25) for bone TE; (i) SEM image of scaffold prepared by freeze drying using a solvent-evaporation technique showing 3D porous network structure, immunostaining for osteogenic bone-matrix proteins of STRO-1 + cells cultured on the scaffold (in vitro), cell nuclei are stained with DAPI (green) and each bone matrix protein is stained by the Alexa 594 fluorochrome-conjugated secondary antibody (red). Confocal microscopic images show Collagen type I (ii, iii), osteopontin (OPN) (iv), bone sialoprotein (BSP) (v) and osteonectin (ONN) (vi). (**d**) (vii–xi) Quantitative μ CT analysis for bone tissue regeneration of selected regions of interest within the osteotomy defect after 28 days. Enhanced bone formation is demonstrated in both scaffold groups (without and with STRO-1 + cells, respectively) when compared to the control group. Assessment of new bone regeneration in the defect regions in femora of mice at 28 days following implantation, using indices of bone volume/total volume (BV/TV) (x) and trabecular number (Tb No) (xi). Results are presented as mean ± SD, $n = 4$ per group, $* = p < 0.05$, $** = p < 0.005$. (**a**) reproduced with permission [40]. Copyright 2009, Wiley-VCH Verlag, Germany. (**b**) Reproduced with permission [65]. Copyright 2009, Science. (**c**) Reproduced with permission [47]. Copyright 2012, Wiley-VCH Verlag. (**d**) Reproduced with permission [12]. Copyright 2013, Wiley-VCH Verlag, Germany.

For TE, the biocompatibility—or, no cytotoxicity—is the critical requirement for a smart biomaterial. Successful biocompatibility indicates that such smart materials can be used to develop 3D-specific scaffolds using advanced techniques [17]. Scaffolds can be loaded with specific cell types, proteins, and/or antibiotics, and subsequently implanted for functional tissue development. From a clinical aspect, such smart implants should respond to the variation of their environment in a valuable biological pathway.

For active tissue engineering, smart polymeric biomaterials are being designed for the regulation of stem cell activity and to understand complex cellular processes. Poly(N-isopropyl acrylamide) and poly(N,N'-diacrylamide) (Figure 1b) are common thermo-responsive dynamic polymers with lower critical solution temperature (LCST), which have been explored for cells adhesion, spreading, and release. Light is another stimulus of particular interest, which has been utilized for the degradation of 3D hydrogel networks after stem cell encapsulation to enhance cellular expansion and their chondrogenic differentiation [65], and has proved that the photodegradation of hydrogels networks provide dynamic elastic modulus in a microenvironment very similar to soft tissue, as well as influence the valvular interstitial cell function [68]. Light has also been employed to achieve stronger materials via new crosslinking, resulting in the alteration of material mechanics [69]. In biomaterial scaffolds, dynamic properties can also be introduced by ionic crosslinking such as alginate crosslinking by divalent cations for cells mobility and migration [70], DNA crosslinking for fibroblast remodeling [71], and hydrogen bonding interactions between CS and polyethylenimine for the chondrogenic differentiation of fetal skeletal cells [40].

The control of responsive biomaterials after implantation in the body is more challenging. Hillel and coworkers developed an injectable scaffold for soft tissue replacement using poly(ethylene glycol) (PEG) and hyaluronic acid (HA) that can be crosslinked in situ with light exposure to form a 3D network [72] or modify the characteristic properties of scaffold materials [73]. Ultrasound is another stimulus that has been utilized for ultimate drug delivery in a controlled manner [74]. An example of a non-polymeric trigger, electrochemically activated microchip system, such as a gold membrane, has been developed for controlled release in implanted materials [66].

Shape-memory materials are a subset of stimuli-responsive biomaterials that change their geometry and mechanical properties with the variation of temperature or light [75,76], and may lead to the next generation of dynamic 3D implantable smart biomaterials [77]. Shape-memory alloys (e.g., Nitinol) have already found commercial use in orthodontic, orthopedic vascular, neurosurgical, and surgical fields, and have been well documented [67,78,79]. In the last few decades, a number of polymeric systems have been developed using a variety of functional monomers to generate block, graft, and brush copolymers, along with biodegradable constructs and crosslinked polymers; moreover, some such systems are available for biomedical applications.

Interactions of smart scaffold biomaterials with cells are crucially important, in which cells need to receive signals continuously from biomaterials to regulate cellular function [80]. Smart biomaterials should have the ability to mobilize growth factors in order to modulate cell proliferation and phenotypes (as discussed before). Thus, the surface functionalization of biomaterials can be a key technology to promote the biological performance of smart biomaterials. Smart biomaterials with complex biological functionalities can be achieved either by the physical and/or chemical modification of 3D structural properties to enhance tissue in-growth, vascularization, and nutrient passing [12,81,82]. Furthermore, smart biomaterials could be designed as absorbable scaffolds that are absorbed by the host metabolic activity. However, the design of such absorbable constructs having biodegradation kinetics continuously matching with the progress of new functional tissue development of the host remains challenging. Nevertheless, structural design with sufficient elastic modulus and bioresorption kinetics can be introduced to meet the physical, chemical, and biological properties of implantation that enable minimally-invasive surgery, focusing on repairing that could lead to a more economically viable healthcare system.

Finally, the progress of non-active biomaterials is described. When the TE scaffold comes into contact with cell culture medium or human body fluids, first of all, water molecules adsorb into the TE scaffold, followed by protein adsorptions, deformations, and cell adhesions. The presence of water molecules in TE scaffolds may play a pivotal role in mediating biochemical reactions between cells and scaffold materials. In order to design the appropriate TE scaffold, the water structure and dynamics of the TE scaffold must be considered. There are two types of water structures, namely non-freezing and freezing water, that form the hydrated TE scaffold [83–85] (Figure 1d). It has been reported in the published literature [83–85] that hydrated biomacromolecules such as DNA, RNA, proteins, and polysaccharides also form intermediate water structures in addition to those of non-freezing and freezing water. The intermediate water is also observed in biocompatible/inert/non-fouling synthetic polymers (Figure 1d) [83–85]. The intermediate water dictates the cells behavior such as cellular attachment, proliferation, migration, and differentiation [86–88], and can be considered as a potential parameter for designing biomaterial scaffolds.

The molecular structure and dynamics that affect the water structure can be controlled by interactions between the backbone and side chain of polymeric biomaterials [83,89]. Therefore, in theory, the structure and dynamics of polymer molecules can be changed by altering the chain length, the functional groups in its side chain, or the backbone of the polymer. The molecular structure and dynamics of polymers can dictate the intermediate water contents. Using the concept of intermediate water, a biodegradable and biocompatible polymer consisting of an aliphatic carbonyl group has been designed and synthesized by ring-opening polymerization [90]. In the future, we hope to find a novel route and methodology to generate well-defined smart biomaterials by considering chemical, physical, and biological approaches towards the realization of a combinatorial, high-throughput pathway for specific TE applications.

5. Concluding Remarks

It is clear smart biomaterials will find many applications in the biomedical field. To date, most of the research has focused on pH- and temperature-responsive smart materials systems as carriers for drugs, antibiotics, proteins, and DNA delivery to the cells. Thermal-responsive smart polymers derived from natural polysaccharides, proteins, synthetic origins, and/or their conjugation and their thermal behavior have been described in conjunction with cell behavior and tissue engineering. The design of smart biomaterials with hierarchy architecture along with stimuli response and healing is crucial. To date, advanced chemical technology provides the knowledge and tools for the synthesis, processing, and characterization of smart polymers with integrated bioactive functionality aiming for biomedical applications. Therefore, the performance and function of such biomaterial systems will eventually depend on the method of assembly and interaction with complex biological interfaces. Although it is not necessary to have all biomimetic properties of a native tissue in smart biomaterials [25], an in-depth understanding of complex biology will facilitate rational design for intended tissue application. To date, the most efficient biomaterials have been achieved from various functional monomers, macromolecules, or oligomers by crosslinking along with biodegradable segments. Such systems are able to interact with cells and promote cell-cell communication in response to stimuli, and are expected to be useful for functional tissue generation, and concurrently act as a delivery agent (e.g., for growth factors, antibiotics, drugs) for faster curing. Self-assembly peptides are excellent biomaterials that respond to physico-chemical stimuli, and promising results have been achieved for 3D cell culture and tissue engineering. However, a number of issues need to be addressed, as this technology in its early stage of development.

Biomaterials after implantation experience a tremendously dynamic environment in physiological complexes that demand better techniques and methodology to monitor biomaterials from a molecular perspective, biodegradation to structural integrity changes, and functional tissue formation. Finally, to facilitate future discovery processes, the rational design of smart biomaterials and their creation in an economically viable route will remain an active field.

Acknowledgments: We acknowledge the Funding Program for Next Generation World-Leading Researchers (NEXT Program, LS017, Japan). This work was partially supported by the Cooperative Research Program of Network Joint Research Centre for Materials and Devices, and Dynamic Alliance for Open Innovation Bridging Human, Environment, and Materials.

Conflicts of Interest: The authors declare no conflict of interest.

References

1. Wichterle, O.; Lim, D. Hydrophilic Gels for Biological Use. *Nature* **1960**, *185*, 117–118. [CrossRef]
2. Kwon, I.C.; Bae, Y.H.; Kim, S.W. Electrically Erodible Polymer Gel for Controlled Release of Drugs. *Nature* **1991**, *354*, 291–293. [CrossRef] [PubMed]
3. Hoffman, A.S. Hydrogels for biomedical applications. *Adv. Drug Deliv. Rev.* **2002**, *43*, 3–13. [CrossRef]
4. Shi, D. *Introduction to Biomaterials*; Tsinghua University Press and World Scientific Publishing Co. Pte. Ltd.: Beijing, China, 2006; pp. 143–210. ISBN 981-256-627-9.
5. Khan, F.; Ahmad, S.R. Polysaccharides and Their Derivatives for Versatile Tissue Engineering Application. *Macromol. Biosci.* **2013**, *13*, 395–421. [CrossRef] [PubMed]
6. Tare, R.S.; Khan, F.; Tourniaire, G.; Morgan, S.M.; Bradley, M.; Oreffo, R.O.C. A microarray approach to the identification of polyurethanes for the isolation of human skeletal progenitor cells and augmentation of skeletal cell growth. *Biomaterials* **2009**, *30*, 1045–1055. [CrossRef] [PubMed]
7. Medine, C.N.; Lucendo-Villarin, B.; Storck, C.; Wang, F.; Szkolnicka, D.; Khan, F.; Pernagallo, S.; Black, J.R.; Marriage, H.M.; Ross, J.A.; et al. Developing high-fidelity hepatotoxicity models from pluripotent stem cells. *Stem Cells Trans. Med.* **2013**, *2*, 505–509. [CrossRef] [PubMed]
8. Mystkowska, J.; Mazurek-Budzynska, M.; Piktel, E.; Niemirowicz, K.; Karalus, W.; Deptula, P.; Pogoda, K.; Lysik, D.; Dabrowski, J.R.; Rokicki, G.; et al. Assessment of aliphatic poly(ester-carbonate-urea-urethane)s potential as materials for biomedical application. *J. Polym. Res.* **2017**, *24*, 144. [CrossRef]
9. Khan, F.; Valere, S.; Fuhrmann, S.; Arrighi, V.; Bradley, M. Synthesis and cellular compatibility of multi-block biodegradable poly(ε-caprolactone)-based polyurethanes. *J. Mater. Chem. B Mater. Biol. Med.* **2013**, *1*, 2590–2600. [CrossRef]
10. Liu, W.G.; Griffith, M.; Li, F. Alginate microsphere-collagen composite hydrogels for ocular drug delivery and implantation. *J. Mater. Sci. Mater. Med.* **2008**, *19*, 3365–3371. [CrossRef] [PubMed]
11. Khan, F.; Tare, R.S.; Kanczler, J.M.; Oreffo, R.O.; Bradley, M. Strategies for Cell Manipulation and Skeletal Tissue Engineering Using High-Throughput Polymer Blend Formulation and Microarray Techniques. *Biomaterial* **2010**, *31*, 2216–2228. [CrossRef] [PubMed]
12. Khan, F.; Smith, J.O.; Kanczler, J.M.; Tare, R.S.; Oreffo, R.O.C.; Bradley, M. Discovery and Evaluation of a Functional Ternary Polymer Blend for Bone Repair: Translation from a Microarray to a Clinical Model. *Adv. Funct. Mater.* **2013**, *23*, 2850–2862. [CrossRef]
13. Yang, F.; Wang, Y.; Zhang, Z.; Hsu, B.; Jabs, E.W.; Elisseeff, J.H. The study of abnormal bone development in the Apert syndrome Fgfr2+/S252W mouse using a 3D hydrogel culture model. *Bone* **2008**, *43*, 55–63. [CrossRef] [PubMed]
14. Eljarrat-Binstock, E.; Orucov, F.; Frucht-Pery, J.; Pe'er, J.; Domb, A.J. Methylprednisolone delivery to the back of the eye using hydrogel iontophoresis. *J. Ocul. Pharm. Ther.* **2008**, *24*, 344–350. [CrossRef] [PubMed]
15. Weigel, T.; Schinkel, G.; Lendlein, A. Design and preparation of polymeric scaffolds for tissue engineering. *Expert Rev. Med. Devices* **2006**, *3*, 835–851. [CrossRef] [PubMed]
16. Khan, F.; Ahmad, S.R. Biomimetic Polysaccharides and Derivatives for Cartilage Tissue Regeneration. In *Biomimetics: Advancing Nanobiomaterials and Tissue Engineering*; Ramalingam, M., Wang, X., Chen, G., Ma, P., Cui, F.Z., Eds.; Scrivener Publishing, Wiley: Austin, TX, USA, 2013; pp. 1–22, ISBN 978-1-118-46962-0.
17. Khan, F.; Tanaka, M.; Ahmad, S.R. Fabrication of Polymeric Biomaterials: A Strategy for Tissue Engineering and Medical Devices. *J. Mater. Chem. B Mater. Biol. Med.* **2015**, *3*, 8224–8249. [CrossRef]
18. Tsang, V.L.; Bhatia, S.N. Fabrication of three-dimensional tissues. *Adv. Biochem. Eng. Biotechnol.* **2007**, *103*, 189–205. [PubMed]
19. Vince, D.G.; Hunt, J.A.; Williams, D.F. Quantitative assessment of the tissue response to implanted biomaterials. *Biomaterials* **1991**, *12*, 731–736. [CrossRef]

20. Smith, J.O.; Tayton, E.R.; Khan, F.; Aarvold, A.; Cook, R.B.; Goodship, A.; Bradley, M.; Oreffo, R.O.C. Large animal in vivo evaluation of a binary blend polymer scaffold for skeletal tissue-engineering strategies; translational issues. *J. Tissue Eng. Regen. Med.* **2017**, *11*, 1065–1076. [CrossRef] [PubMed]

21. Rosso, F.; Marino, G.; Giordano, A.; Barbaris, M.; Parmeggiani, D.; Barbarisi, A. Smart materials as scaffolds for tissue engineering. *J. Cell. Physiol.* **2005**, *203*, 465–470. [CrossRef] [PubMed]

22. Jones, P.L.; Schmidhauser, C.; Bissell, M.J. Regulation of gene expression and cell function by extracellular matrix. *Crit. Rev. Eukaryot. Gene Expr.* **1993**, *3*, 137–154. [PubMed]

23. Juliano, R.L.; Haskill, S. Signal transduction from the extracellular matrix. *J. Cell Biol.* **1993**, *120*, 577–585. [CrossRef] [PubMed]

24. Birkedal-Hansen, H. Proteolytic remodeling of extracellular matrix. *Curr. Opin. Cell Biol.* **1995**, *7*, 728–735. [CrossRef]

25. Lutolf, M.P.; Hubbell, J.A. Synthetic biomaterials as instructive extracellular microenvironments for morphogenesis in tissue engineering. *Nat. Biotechnol.* **2005**, *23*, 47–55. [CrossRef] [PubMed]

26. Lee, C.H.; Singla, A.; Lee, Y. Biomedical applications of collagen. *Int. J. Pharm.* **2001**, *221*, 1–22. [CrossRef]

27. Yannas, I.V. Natural materials. In *Biomaterials Science: An Introduction to Materials in Medicine*; Ratner, B.D., Hoffman, A.S., Schoen, F.J., Lemons, J.E., Eds.; Academic Press: New York, NY, USA, 1996; pp. 84–94, ISBN 0-12-582463-7.

28. Coviello, T.; Matricardi, P.; Marianecci, C.; Alhaique, F. Polysaccharide hydrogels for modified release formulations. *J. Control. Release* **2007**, *119*, 5–24. [CrossRef] [PubMed]

29. Spicer, A.P.; Tien, J.Y. Hyaluronan and morphogenesis. *Birth Defects Res. C Embryo Today* **2004**, *72*, 89–108. [CrossRef] [PubMed]

30. Waddell, D.D.; Bricker, D.C. Hylan G-F 20 tolerability with repeat treatment in a large orthopedic practice: A retrospective review. *J. Surg. Orthop. Adv.* **2006**, *15*, 53–59. [PubMed]

31. Waddell, D.D.; Bricker, D.C. Clinical experience with the effectiveness and tolerability of hylan G-F 20 in 1047 patients with osteoarthritis of the knee. *J. Knee Surg.* **2006**, *19*, 19–27. [CrossRef] [PubMed]

32. Kemper, F.; Gebhardt, U.; Meng, T.; Murray, C. Tolerability and shortterm effectiveness of hylan G-F 20 in 4253 patients with osteoarthritis of the knee in clinical practice. *Curr. Med. Res. Opin.* **2005**, *21*, 1261–1269. [CrossRef] [PubMed]

33. Shi, C.; Zhu, Y.; Ran, X.; Wang, M.; Su, Y.; Cheng, T. Therapeutic potential of chitosan and its derivatives in regenerative medicine. *J. Surg. Res.* **2006**, *133*, 185–192. [CrossRef] [PubMed]

34. Khor, E.; Lim, L.Y. Implantable applications of chitin and chitosan. *Biomaterials* **2003**, *24*, 2339–2349. [CrossRef]

35. Gutowska, A.; Jeong, B.; Jasionowski, M. Injectable gels for tissue engineering. *Anat. Rec.* **2001**, *263*, 342–349. [CrossRef] [PubMed]

36. Yoshida, R.; Uchida, K.; Kaneko, Y.; Sakai, K.; Kikuchi, A.; Sakurai, Y.; Okano, T. Comb-type grafted hydrogels with rapid de-swelling response to temperature changes. *Nature* **1995**, *374*, 240–242. [CrossRef]

37. Lee, S.H.; Oh, J.M.; Son, J.S.; Lee, J.W.; Kim, B.S.; Khang, G.; Han, D.K.; Kim, J.H.; Lee, H.B.; Kim, M.S. Controlled preparation of poly(ethylene glycol) and poly(L-lactide) block copolymers in the presence of a monomer activator. *J. Polym. Sci. Part A Polym. Chem.* **2009**, *47*, 5917–5922. [CrossRef]

38. Ehrick, J.D.; Stokes, S.; Bachas-Daunert, S.; Moschou, E.A.; Deo, S.K.; Bachas, L.G.; Daunert, S. Chemically tunable lensing of stimuli-responsive hydrogel microdomes. *Adv. Mater.* **2007**, *19*, 4024–4027. [CrossRef]

39. Satarkar, N.S.; Hilt, J.Z. Hydrogel nanocomposites as remote-controlled biomaterials. *Acta Biomater.* **2008**, *4*, 11–16. [CrossRef] [PubMed]

40. Khan, F.; Tare, R.S.; Oreffo, R.O.C.; Bradley, M. Versatile biocompatible polymer hydrogels: Scaffolds for cell growth. *Angew. Chem. Int. Ed.* **2009**, *48*, 978–982. [CrossRef] [PubMed]

41. Anderson, D.G.; Burdick, J.A.; Langer, R. Materials science: Smart biomaterials. *Science* **2004**, *305*, 1923–1924. [CrossRef] [PubMed]

42. Ito, H.; Steplewski, A.; Alabyeva, T.; Fertala, A. Testing the utility of rationally engineered recombinant collagen-like proteins for applications in tissue engineering. *J. Biomed. Mater. Res. A* **2006**, *76*, 551–560. [CrossRef] [PubMed]

43. Ghosh, K.; Ren, X.D.; Shu, X.Z.; Prestwich, G.D.; Clark, R.F.A. Fibronectin functional domains coupled to hyaluronan stimulate adult human dermal fibroblast responses critical for wound healing. *Tissue Eng.* **2006**, *12*, 601–613. [CrossRef] [PubMed]

44. Nagapudi, K.; Brinkman, W.T.; Thomas, B.S.; Park, J.O.; Srinivasarao, M.; Wright, E.; Conticello, V.P.; Chaikof, E.L. Viscoelastic and mechanical behavior of recombinant protein elastomers. *Biomaterials* **2005**, *26*, 4695–4706. [CrossRef] [PubMed]

45. Hersel, U.; Dahmen, C.; Kessler, H. RGD modified polymers: Biomaterials for stimulated cell adhesion and beyond. *Biomaterials* **2003**, *24*, 4385–4415. [CrossRef]

46. Koss, K.M.; Unsworth, L.D. Neural tissue engineering: Bioresponsive nanoscaffolds using engineered self-assembling peptides. *Acta Biomater.* **2016**, *44*, 2–15. [CrossRef] [PubMed]

47. Kyle, S.; Felton, S.; McPherson, M.J.; Aggeli, A.; Ingham, E. Rational molecular design of complementary self-assembling peptide hydrogels. *Adv. Healthc. Mater.* **2012**, *1*, 640–645. [CrossRef] [PubMed]

48. Kyle, S.; Aggeli, A.; Ingham, E.; McPherson, M.J. Recombinant self-assembling peptides as biomaterials for tissue engineering. *Biomaterials* **2010**, *31*, 9395–9405. [CrossRef] [PubMed]

49. Kyle, S.; Aggeli, A.; Ingham, E.; McPherson, M.J. Production of self-assembling biomaterials for tissue engineering. *Trends Biotechnol.* **2009**, *27*, 423–433. [CrossRef] [PubMed]

50. Hutmacher, D.W. Regenerative medicine will impact, but not replace, the medical device industry. *Expert Rev. Med. Devices* **2006**, *3*, 409–412. [CrossRef] [PubMed]

51. Boisseau, P. *Nanomedicine, European Technology Platform., ETP Nanomedicine, NANOMEDICINE 2020, Contribution of Nanomedicine to Horizon 2020, NANOMED2020*; ETP Nanomedicine: Grenoble, France, May 2013; p. 7.

52. Boyan, B.D.; Schwartz, Z. Regenerative medicine: Are calcium phosphate ceramics 'smart' biomaterials? *Nat. Rev. Rheumatol.* **2011**, *7*, 8–9. [CrossRef] [PubMed]

53. Furth, M.E.; Atala, A.; Van Dyke, M.E. Smart biomaterials design for tissue engineering and regenerative medicine. *Biomaterials* **2007**, *28*, 5068–5073. [CrossRef] [PubMed]

54. Mieszawska, A.J.; Kaplan, D.L. Smart biomaterials-regulating cell behavior through signaling molecules. *BMC Biol.* **2010**, *8*, 59. [CrossRef] [PubMed]

55. Yuan, H.; Fernandes, H.; Habibovic, P.; de Boer, J.; Barradas, A.M.; Walsh, W.R.; van Blitterswijk, C.A.; De Bruijn, J.D. Smart biomaterials and osteoinductivity. *Nat. Rev. Rheumatol.* **2011**, *7*, c1. [CrossRef] [PubMed]

56. Yamato, M.; Konno, C.; Utsumi, M.; Kikuchi, A.; Okano, T. Thermally responsive polymer-grafted surfaces facilitate patterned cell seeding and co-culture. *Biomaterials* **2002**, *23*, 561–567. [CrossRef]

57. Cao, Y.; Ibarra, C.; Vacanti, C. Preparation and use of thermoresponsive polymers. In *Tissue Engineering: Methods and Protocols*; Morgan, J.R., Yarmush, M.L., Eds.; Humana Press: New York, NY, USA, 2006; pp. 75–84, ISBN 978-1-59259-602-7.

58. Stile, R.A.; Shull, K.R.; Healy, K.E. Axisymmetric Adhesion Test to Examine the Interfacial Interactions between Biologically-Modified Networks and Models of the Extracellular Matrix. *Langmuir* **2003**, *19*, 1853–1860. [CrossRef]

59. Stayton, P.S.; El-Sayed, M.E.; Murthy, N.; Bulmus, V.; Lackey, C.; Cheung, C.; Hoffman, A.S. 'Smart' delivery systems for biomolecular therapeutics. *Orthod. Craniofac. Res.* **2005**, *8*, 219–225. [CrossRef] [PubMed]

60. Kollarigowda, R.H.; Fedele, C.; Rianna, C.; Calabuig, A.; Manikas, A.C.; Pagliarulo, V.; Ferraro, P.; Cavalli, S.; Netti, P.A. Light-responsive polymer brushes: Active topographic cues for cell culture applications. *Polym. Chem.* **2017**, *8*, 3271–3278. [CrossRef]

61. Cao, Z.; Bian, Q.; Chen, Y.; Liang, F.; Wang, G. Light-Responsive Janus-Particle-Based Coatings for Cell Capture and Release. *ACS Macro Lett.* **2017**, *6*, 1124–1128. [CrossRef]

62. Zhao, X.; Kim, J.; Cezar, C.A.; Huebsch, N.; Lee, K.; Bouhadir, K. Active scaffolds for on-demand drug and cell delivery. *Proc. Natl. Acad. Sci. USA* **2011**, *108*, 67–72. [CrossRef] [PubMed]

63. Sundaresan, V.; Menon, J.U.; Rahimi, M.; Nguyen, K.T.; Wadajkar, A.S. Dual-responsive polymer-coated iron oxide nanoparticles for drug delivery and imaging applications. *Int. J. Pharm.* **2014**, *466*, 1–7. [CrossRef] [PubMed]

64. Wang, Z.; Sheng, R.; Luo, T.; Sun, J.; Cao, A. Synthesis and self-assembly of diblock glycopolypeptide analogues PMAgala-b-PBLG as multifunctional biomaterials for protein recognition, drug delivery and hepatoma cell targeting. *Polym. Chem.* **2017**, *8*, 472–484. [CrossRef]

65. Kloxin, A.M.; Kasko, A.M.; Salinas, C.N.; Anseth, K.S. Photodegradable hydrogels for dynamic tuning of physical and chemical properties. *Science* **2009**, *324*, 59–63. [CrossRef] [PubMed]

66. Santini, J.T., Jr.; Cima, M.J.; Langer, R. A controlled-release microchip. *Nature* **1999**, *397*, 335–338. [CrossRef] [PubMed]

67. Li, H.F.; Qiu, K.J.; Zhou, F.Y.; Li, L.; Zheng, Y.F. Design and development of novel antibacterial Ti-Ni-Cu shape memory alloys for biomedical application. *Sci. Rep.* **2016**, *6*, 37475. [CrossRef] [PubMed]

68. Kloxin, A.M.; Benton, J.A.; Anseth, K.S. In situ elasticity modulation with dynamic substrates to direct cell phenotype. *Biomaterials* **2010**, *31*, 1–8. [CrossRef] [PubMed]

69. Guvendiren, M.; Burdick, J.A. Stiffening hydrogels to probe short- and long-term cellular responses to dynamic mechanics. *Nat. Commun.* **2012**, *3*, 792. [CrossRef] [PubMed]

70. Gillette, B.M.; Jensen, J.A.; Wang, M.; Tchao, J.; Sia, S.K. Dynamic hydrogels: Switching of 3D microenvironments using two component naturally derived extracellular matrices. *Adv. Mater.* **2010**, *22*, 686–691. [CrossRef] [PubMed]

71. Jiang, F.X.; Yurke, B.; Schloss, R.S.; Firestein, B.L.; Langrana, N.A. The relationship between fibroblast growth and the dynamic stiffnesses of a DNA crosslinked hydrogel. *Biomaterials* **2010**, *31*, 1199–1212. [CrossRef] [PubMed]

72. Hillel, A.T.; Unterman, S.; Nahas, Z.; Reid, B.; Coburn, J.M.; Axelman, J.; Chae, J.J.; Guo, Q.; Trow, R.; Thomas, A.; et al. Photoactivated composite biomaterial for soft tissue restoration in rodents and in humans. *Sci. Transl. Med.* **2011**, *3*, 93ra67. [CrossRef] [PubMed]

73. Lee, T.T.; García, J.R.; Paez, J.I.; Singh, A.; Phelps, F.A.; Weis, S.; Shafiq, Z.; Shekaran, A.; Campo, A.; García, A.J. Light-triggered in vivo activation of adhesive peptides regulates cell adhesion, inflammation and vascularization of biomaterials. *Nat. Mater.* **2015**, *14*, 352–360. [CrossRef] [PubMed]

74. Epstein-Barash, H.; Orbey, G.; Polat, B.E.; Ewoldt, R.H.; Feshitan, J.; Langer, R.; Borden, M.A.; Kohane, D.S. A microcomposite hydrogel for repeated on-demand ultrasound-triggered drug delivery. *Biomaterials* **2010**, *31*, 5208–5217. [CrossRef] [PubMed]

75. Lendlein, A.; Behl, M.; Hiebl, B.; Wischke, C. Shape-memory polymers as a technology platform for biomedical applications. *Expert Rev. Med. Devices* **2010**, *7*, 357–379. [CrossRef] [PubMed]

76. Serrano, M.C.; Ameer, G.A. Recent insights into the biomedical applications of shape-memory polymers. *Macromol. Biosci.* **2012**, *12*, 1156–1171. [CrossRef] [PubMed]

77. Lendlein, A.; Langer, R. Biodegradable, elastic shape memory polymers for potential biomedical applications. *Science* **2002**, *296*, 1673–1676. [CrossRef] [PubMed]

78. Hannula, S.-P.; Soderberg, O.; Jamsa, T.; Lindroos, V.K. Shape memory alloys for biomedical applications. *Adv. Sci. Technol.* **2006**, *49*, 109–118. [CrossRef]

79. Nespoli, A.; Villa, E.; Bergo, L.; Rizzacasa, A.; Passaretti, F. DSC and three-point bending test for the study of the thermo-mechanical history of NiTi and NiTi-based orthodontic archwires-The material point of view. *J. Therm. Anal. Calorimet.* **2015**, *120*, 1129–1138. [CrossRef]

80. Rosso, F.; Giordano, A.; Barbarisi, M.; Barbarisi, A. From cell–ECM interactions to tissue engineering. *J. Cell. Physiol.* **2004**, *199*, 174–180. [CrossRef] [PubMed]

81. Hutmacher, D.W.; Schantz, J.T.; Lam, C.X.; Tan, K.C.; Lim, T.C. State of the art and future directions of scaffold-based bone engineering from a biomaterials perspective. *J. Tissue Eng. Regen. Med.* **2007**, *1*, 245–260. [CrossRef] [PubMed]

82. Karageorgiou, V.; Kaplan, D. Porosity of 3D biomaterial scaffolds and osteogenesis. *Biomaterials* **2005**, *26*, 5474–5491. [CrossRef] [PubMed]

83. Tanaka, M.; Sato, K.; Kitakami, E.; Kobayashi, S.; Hoshiba, T.; Fukushima, K. Design of biocompatible and biodegradable polymers based on intermediate water concept. *Polym. J.* **2015**, *47*, 114–121. [CrossRef]

84. Tanaka, M.; Motomura, T.; Kawada, M.; Anzai, T.; Yuu, K.; Shiroya, T.; Shimura, K.; Onishi, M.; Akira, M. Blood compatible aspects of poly(2-methoxyethylacrylate) (PMEA)—relationship between protein adsorption and platelet adhesion on PMEA surface. *Biomaterials* **2000**, *21*, 1471–1481. [CrossRef]

85. Bag, M.A.; Valenzuela, L.M. Impact of the Hydration States of Polymers on Their Hemocompatibility for Medical Applications: A Review. *Int. J. Mol. Sci.* **2017**, *18*, 1422. [CrossRef] [PubMed]

86. Hoshiba, T.; Nikaido, M.; Tanaka, M. Characterization of the Attachment Mechanisms of Tissue-Derived Cell Lines to Blood-Compatible Polymers. *Adv. Healthc. Mater.* **2014**, *3*, 775–784. [CrossRef] [PubMed]

87. Hoshiba, T.; Nemoto, E.; Sato, K.; Maruyama, H.; Endo, C.; Tanaka, M. Promotion of Adipogenesis of 3T3-L1 Cells on Protein Adsorption-Suppressing Poly(2-methoxyethyl acrylate) Analogous. *Biomacromolecules* **2016**, *17*, 3808–3815. [CrossRef] [PubMed]

88. Hoshiba, T.; Orui, T.; Endo, C.; Sato, K.; Yoshihiro, A.; Minagawa, Y.; Tanaka, M. Adhesion-based simple capture and recovery of circulating tumor cells using a blood-compatible and thermo-responsive polymer-coated substrate. *RSC Adv.* **2016**, *6*, 89103–89112. [CrossRef]

89. Sato, K.; Kobayashi, S.; Sekishita, A.; Wakui, M.; Tanaka, M. Synthesis and Thrombogenicity Evaluation of Poly(3-methoxypropionic acid vinyl ester): A Candidate for Blood-Compatible Polymers. *Biomacromolecules* **2017**, *18*, 1609–1616. [CrossRef] [PubMed]

90. Fukushima, K.; Inoue, Y.; Haga, Y.; Ota, T.; Honda, K.; Sato, C.; Tanaka, M. Monoether-tagged Biodegradable Polycarbonate Preventing Platelet Adhesion and Demonstrating Adhesion of Vascular Cells: A Promising Material for Resorbable Vascular Grafts and Stents. *Biomacromolecules* **2017**, *18*, 3834–3843. [CrossRef] [PubMed]

International Journal of
Molecular Sciences

MDPI

Article

Aging Donor-Derived Human Mesenchymal Stem Cells Exhibit Reduced Reactive Oxygen Species Loads and Increased Differentiation Potential Following Serial Expansion on a PEG-PCL Copolymer Substrate

Daniel A. Balikov [1], Spencer W. Crowder [2], Jung Bok Lee [1,3], Yunki Lee [1,4], Ung Hyun Ko [5], Mi-Lan Kang [3], Won Shik Kim [6], Jennifer H. Shin [5] and Hak-Joon Sung [3,4,*]

[1]   Department of Biomedical Engineering, Vanderbilt University, Nashville, TN 37235, USA;
      daniel.a.balikov@Vanderbilt.Edu (D.A.B.); jungboklee01@gmail.com (J.B.L.); yungi2710@gmail.com (Y.L.)
[2]   Department of Materials and Department of Bioengineering, Imperial College London,
      London SW7 2AZ, UK; spencer.crowder@gmail.com
[3]   Severance Biomedical Science Institute, College of Medicine, Yonsei University, Seoul 03722, Korea;
      milan511@yuhs.ac
[4]   Department of Mechanical Engineering, Georgia Institute of Technology, Atlanta, GA 30332, USA
[5]   Department of Mechanical Engineering, Korea Advanced Institute of Science and Technology,
      Daejeon 34141, Korea; unghyunk@gmail.com (U.H.K.); jennifer.shinpark@gmail.com (J.H.S.)
[6]   Department of Otorhinolaryngology, College of Medicine, Yonsei University, Seoul 03722, Korea;
      wskim78@yuhs.ac
*   Correspondence: hj72sung@yuhs.ac; Tel.: +82-2-2228-0834; Fax: +82-2-2227-8283

Received: 9 January 2018; Accepted: 23 January 2018; Published: 25 January 2018

Abstract: Human mesenchymal stem cells (hMSCs) have been widely studied for therapeutic development in tissue engineering and regenerative medicine. They can be harvested from human donors via tissue biopsies, such as bone marrow aspiration, and cultured to reach clinically relevant cell numbers. However, an unmet issue lies in the fact that the hMSC donors for regenerative therapies are more likely to be of advanced age. Their stem cells are not as potent compared to those of young donors, and continue to lose healthy, stemness-related activities when the hMSCs are serially passaged in tissue culture plates. Here, we have developed a cheap, scalable, and effective copolymer film to culture hMSCs obtained from aged human donors over several passages without loss of reactive oxygen species (ROS) handling or differentiation capacity. Assays of cell morphology, reactive oxygen species load, and differentiation potential demonstrate the effectiveness of copolymer culture on reduction in senescence-related activities of aging donor-derived hMSCs that could hinder the therapeutic potential of autologous stem cell therapies.

Keywords: biomaterial; copolymer; stem cell; regenerative medicine; cell culture

1. Introduction

Human mesenchymal stem cells (hMSCs) offer a potential stem cell source for the translation of tissue engineering strategies to repair or replace damaged tissues. In fact, several proof-of-principle studies of direct stem cell injections into injury sites have resulted in improved function, such as in bone [1], cartilage [2], heart [3], and large blood vessels [4]. However, to effectively translate these studies to human clinical trials, hundreds of millions of hMSCs need to be grown for transplantation at the injury site to be effective, as has been demonstrated in large animal studies [5]. Thus, a few hundred thousand hMSCs that can be isolated from the bone marrow of any typical human donor require expansion and break the Hayflick limit in the process [6]. Originally described in 1961,

Leonard Hayflick and his colleagues observed that cells had decreased proliferation as they were serially passaged in Petri dishes. This could become an obstacle to translating hMSCs for therapeutic applications, because donors who would most often utilize such stem cell therapies are of advanced age, and their hMSCs have likely undergone the process of senescence due to the many cycles of cell division occurring over the donor's lifetime.

hMSC phenotype undergoes senescence-associated changes from serial passaging on tissue culture polystyrene (TCPS) due to cytotoxic insults such as the accumulation of the intracellular reactive oxygen species (ROS) [7]. Seminal work by Wagner and colleagues reported that excessive passaging of hMSCs resulted in a host of biochemical and functional alterations that were detrimental to cell health [8]. Among their findings, surface markers unique to hMSCs diminished or disappeared, coupled with profound alterations in mRNA expression profiles indicative of spontaneous differentiation, apoptosis, cell cycle alterations, and inflammatory regulation. Cell proliferation also arrested and was coupled to decreased differentiation capacity and increased β-galactosidase staining (a well-established positive marker of cellular senescence). These observations were further verified in hematopoietic stem cell progenitors by the same research group in order to demonstrate that senescence-associated changes incurred in vitro were limited to hMSCs [9].

Similarly, Heo et al. conducted serially passaging studies on hMSCs that not only confirmed the original findings by Wagner et al., but also specifically reported the loss of ROS handing proteins as the upstream event that leads to hMSC senescence in vitro [7]. Reduced expression of APE/Ref-1 due to serial passaging yielded increased ROS loads within hMSCs, which in turn accelerated accumulation of β-galactosidase. These effects could be countered if APE/Ref-1 was overexpressed, and thus restored homeostatic ROS levels. Kasper and colleagues further explored this phenomenon by looking at proteomic alterations in young and old rat MSCs [10]. Older rats had fewer MSCs in the bone marrow with concordant susceptibility in cellular senescence due to their inability to process ROS. Proteomic profiling validated that the overabundance of ROS resulting in extensive macromolecular damage could not be overcome with hindered antioxidant mechanisms caused by cellular senescence.

Aged hMSCs exhibit altered differentiation potential: younger cells are able to maintain multipotency (e.g., osteogenic, adipogenic, chondrogenic differentiation capacity), but over multiple passages, older cells are only able to differentiate into osteogenic and adipogenic lineages, and eventually only the osteogenic lineage [11]. We have also reported that serially passaged hMSCs are more susceptible to cancerous transformation, and the probability and degree of cancerous transformation is closely correlated with β-galactosidase staining [12]. This becomes especially concerning for aged donors whose cells have already undergone many divisions and could have hMSCs that harbor ROS-mediated DNA damage that can more easily yield cancerous stem cells.

In order to expedite clinical translation, new, cheap, easily scalable strategies to maintain or reinstate hMSC fitness following expansion must be developed to counteract this inherent decline in cell health [13]. Generation of spheroid aggregates of hMSCs has been considered one of the most effective cultural formats for maintaining hMSC therapeutic potency and avoiding passage-associated senescence [14,15]. In fact, hMSC cell aggregates benefit from having increased functional capacity such as pro-angiogenic [16] and anti-inflammatory properties [15,17–19]. However, generating large quantities of aggregates from bioreactors or cell-repellant substrates can result in varying degrees of success, as evidenced in the literature. Maintaining bioreactor systems has the complexity of interconnected bioreactor components and unique contamination risks (e.g., chemical or biological) that complicate the scaling up of this process [20]. Several groups have noted that optimization of bioreactor yields requires improvement despite the careful monitoring of growth conditions [21], and that precursor hMSC populations from the original in vivo aspirate are modified and/or lost after 20 population doublings [22]. Additionally, controlling the size of the hMSC aggregates during a scale-up of the bioreactor risks the aggregates developing necrotic cores that could result in negative consequences for the recipient tissues receiving the cells [23–25]. Moreover, the ability to handle and break down these aggregates into single cell suspensions for injection is difficult.

Approaches other than aggregates have also been developed for the extensive culture of hMSCs. With respect to culture substrates, many groups have employed inverted hanging drop wells that have cell-repellant surfaces [15,26–28]. In one study by Ng et al., adipose MSCs were expanded in vitro by culturing on extracellular matrix (ECM) protein produced by fetal MSCs [29]. Although the adipose MSCs demonstrated increased functional capacity over several passages, concerns regarding immunogenicity arise when using ECM from another human donor, which could ultimately stimulate a negative immune response by the stem cell recipient. Also, the financial burden of continually generating uniform fetal ECM and harvesting fetal MSCs to generate the ECM could become exponential. Alternatively, other groups like Duffy et al. have developed synthetic polymer culture substrates that reduce harsh passaging techniques to grow hMSCs over many passages [30]. While their enzyme-free substrate did demonstrate marginal improvement in adipose MSC differentiation, the surface marker profiles were not maintained, thereby challenging the efficacy of the system beyond simple differentiation assays and the monolayer appearance of the cells.

Using these studies as inspiration, we set out to demonstrate a cheap, easily-reproducible, and effective culture platform that could maintain stem cell phenotype and functional capacity over serial passaging. In previous work, we found that by using a poly(ethylene glycol) (PEG) and poly(ε-caprolactone) (PCL) copolymer film as a culture substrate, human bone marrow-derived MSCs maintained high stemness, retained key surface protein markers lost during in vitro culture (STRO-1), contained low reactive oxygen species load, and adopted decreased proliferation rates compared to conventional TCPS plates [31]. Morphologically, the moderate repellency of specific PEG-PCL (poly(ethylene glycol)-poly(ε-caprolactone)) film compositions created an optimal interface that allowed hMSCs to bind the amorphous PCL domains while the hydrophilic PEG domains forced the cells to aggregate into spheroids that were morphologically similar to marginating hMSCs in the bone marrow. Therefore, because we had validated that our PEG-PCL copolymers could reproduced in the aforementioned findings in random hMSC donors, we hypothesized that the PEG-PCL copolymer films could be used to serially-passage hMSCs from aging donors (age > 65 years old) as a means to attenuate senescence-associated activities resulting from serial in vitro expansion. We demonstrate that culture on this material maintains the cells in a pro-stemness state throughout expansion, as evidenced by a series of functional assays including flow cytometry detection of reactive oxygen species (ROS), and adipogenic and osteogenic differentiation assays. To our knowledge, this is the first paper to demonstrate that the therapeutic capacity of hMSCs isolated from aging donors can be enhanced through serial passaging on a custom synthetic material. The findings presented here offer insight for designing clinically relevant materials for hMSC-based therapies.

2. Results

2.1. Experimental Design

The synthesis of the PEG-PCL polymer was based on ring opening polymerization of ε-caprolactone onto methoxy-PEG (Figure 1A). X and Y refer to the mole percent fraction of PEG and PCL, respectively. Based on prior work in our group [31], we utilized a 5% PEG–95% PCL copolymer in which the PEG block was 2000 Da in size [32]. This polymer served as a favorable hMSC culture substrate in that in vivo-like cell morphologies were adopted by the cells in addition to reinstatement of an in vivo surface marker, STRO-1, and lowered ROS load compared to hMSCs cultured on TCPS. For the current study, a spin coater was employed in order to create easily reproducible copolymer films that the cells could grow on (Figure 1B). As illustrated, a Pyrex© Petri dish (or coverslip) was placed in the spin coater with a small amount of copolymer solution added in the center. The rotation of the block spread the solution evenly across the surface of the dish (or coverslip) yielding the copolymer film in a culture-ready vessel, and additional spinning at a higher rate allowed for the volatile organic solvent to evaporate, leaving a thin, uniform polymer film on the glass surface.

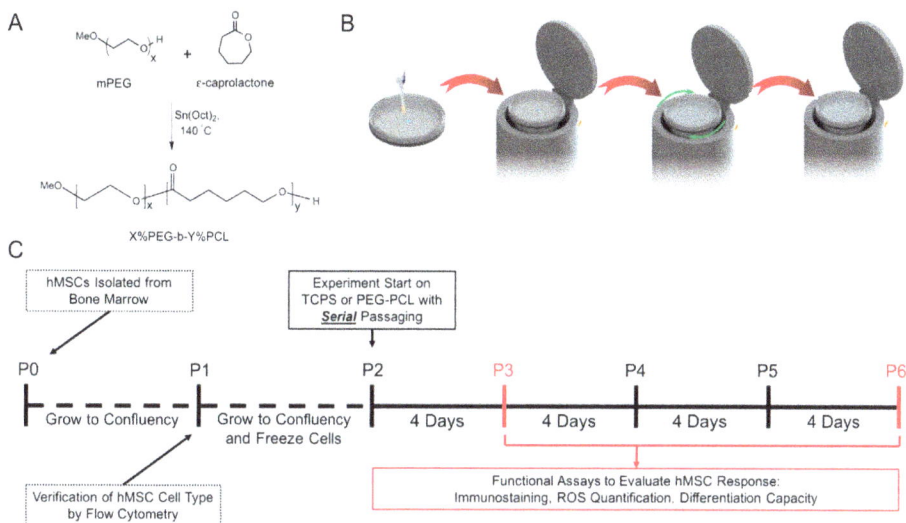

Figure 1. Experimental overview. (**A**) The poly(ethylene glycol)-poly(ε-caprolactone) (PEG-PCL) copolymer was synthesized using methoxy-PEG and ring-opening polymerization of ε–caprolactone; (**B**) polymer films were generated by spin-coating copolymer solution onto coverslips or into Pyrex© petri dishes as shown in the illustration. First, a set volume of 1% *w/v* PEG-PCL solution was dropped onto coverglass or Pyrex© Petri dishes. The coverglass or dishes were placed into the spin coater and a spin program (green arrows) was applied to the substrates. The PEG-PCL solution was evenly spread out on the surface with the solvent evaporating in the process; (**C**) the timeline of experiments with respect to passage number of the donor human mesenchymal stem cells (hMSCs). Red indicates that passage numbers where imaging or functional tests were performed.

For the longitudinal study, patient hMSCs were isolated from donors, expanded two passages on TCPS, and then subsequently cultured to passage 6 on either TCPS or PEG-PCL substrates (Figure 1C). Upon initial collection of bone marrow aspirate, the bone marrow was passed through a 70 μm filter, cultured on Histoplaque, and the mononuclear cells were collected and subsequently plated on TCPS. The hMSCs were the only cells to adhere to TCPS dishes at passage 0. Non-adhesive cells obtained from the filtered bone marrow aspirate (e.g., hematopoietic stem cells) were removed by media aspiration and gentle media washes. The adhesive cells were grown to confluence before being evaluated for appropriate positive and negative MSC markers at passage 1 (refer to Balikov et al. [31]). Cells were frozen following a standard stem cell culture method utilizing 70% complete media, 20% FBS (fetal bovine serum), and 10% DMSO (dimethyl sulfoxide) prior to serial passaging. As indicated in the Figure 1C, cells were passaged every 4 days, the time needed for hMSCs from all three donors to become confluent on TCPS. At day 4, cells were removed from either TCPS or PEG-PCL and then re-plated onto a fresh culture substrate of the same material at the same cell seeding density of 10,000 cells/cm^2. Four days of cell growth was chosen due to TCPS culture reaching nearly 100% confluency at 96 h post-seeding at 10,000 cells/cm^2. At passages 3 and 6, hMSCs were evaluated for functional capacity by evaluating cell morphology, intracellular ROS load, and osteogenic and adipogenic differentiation capacity. The total number of cells collected from PEG-PCL films were nearly double than that initially seeded, while the total number of cells collected from TCPS was nearly triple the amount originally seeded.

2.2. Morphological Change of hMSCs on TCPS and PEG-PCL over Passages

hMSCs from all donors showed markedly altered cellular morphology when passaged on TCPS or PEG-PCL (Figure 2). At passage 3, hMSCs grown on TCPS displayed a flattened, spread shape typical of this cell type. Most cells were oriented along a single major axis, and actin stress fiber organization was clearly visible. However, when hMSCs were cultured on PEG-PCL, distinct cell clusters, reminiscent of hanging drop aggregates, were formed. Cells within the aggregates were round in morphology, with some cells exhibiting spindle-like extensions. Actin stress fibers were only present on the few cells that had spindle-like extensions, while rounded cells within the cell aggregate had minimal polarized actin fiber staining. At passage 6, TCPS hMSCs were highly aligned with strong spindle morphology, forming a cell sheet. Actin stress fibers were still clearly demarcated, coupled with the cells aligning along a major axis. With respect to PEG-PCL substrates, hMSCs continued to form aggregate cell clusters, but both diameter and number of constituent cells increased by visual inspection. Furthermore, the diversity in spheroid morphology can be seen among the donors, in which donor 2 maintained a tight, enlarged spheroid of cells, donor 1 had more cellular spindle projections anchoring to the copolymer surface, and donor 3 contained a spheroid with a broad based of spindle-shaped hMSCs along the copolymer surface like a cell-feeder layer. Of note, passage 6 was not exceeded in this study due to the spheroids becoming so large that they no longer adhered to the surface of the PEG-PCL, thereby rendering the beneficial aspects of the copolymer substrate ineffective.

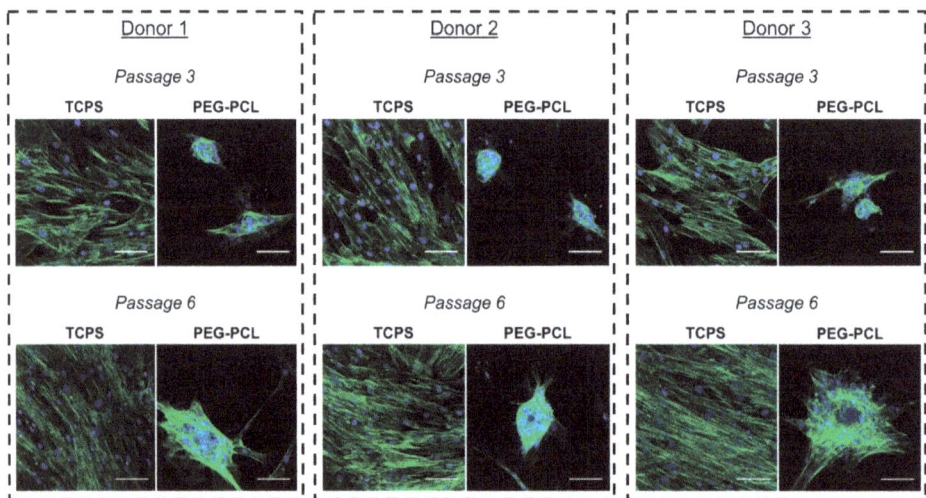

Figure 2. Morphological changes occur over serially passaging human mesenchymal stem cells on their respective substrates. Cells were stained with AlexaFluor-488-conjugated phalloidin (green) and Hoechst nuclear counterstain (blue). Scale bar = 100 μm.

2.3. ROS Load

All donors displayed decreased levels of detected intracellular ROS when grown on the PEG-PCL compared to TCPS (Figure 3). Passage 3 fluorescent signal was decreased by ~1 order of magnitude, and this effect was maintained at passage 6. TCPS curves (blue) had a tight population distribution while PEG-PCL (green) was more heterogenous, as seen by the increased peak width. This could be due to differences in the cells closer to the material interface (likely with higher ROS) compared to the cells within in the cell aggregate (lower ROS).

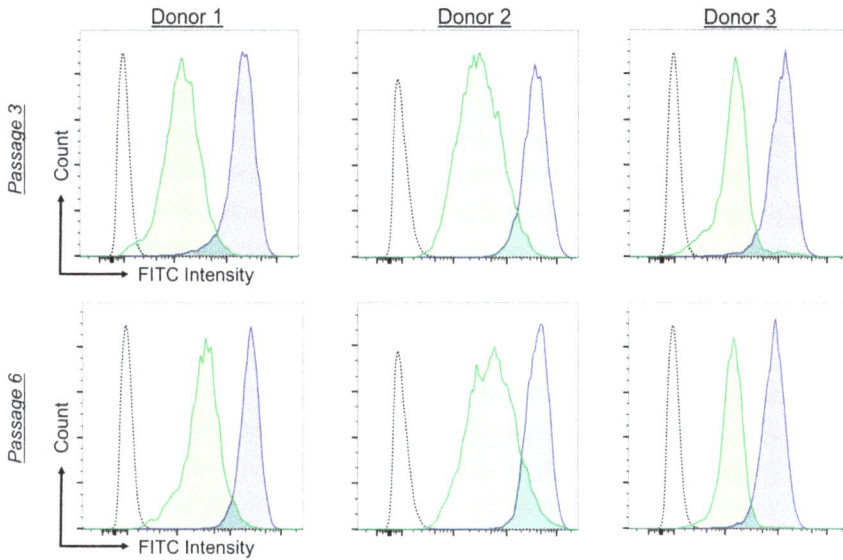

Figure 3. PEG-PCL copolymers reduce intracellular ROS load of donor cells at both passages. Human mesenchymal stem cells (hMSCs) were incubated with DCFDA, and FITC intensity correlated with active ROS species. The graphs shown are representative results from *n* = 3 independent experimental replicates. Blue is TCPS and green is PEG-PCL. All donors for both passages had decreased ROS loads for hMSCs grown on PEG-PCL compared to TCPS.

2.4. Differentiation Capacity

The degree of osteogenic differentiation, as evaluated by image-based quantification of Alizarin Red stain, was maintained when hMSCs were serially passaged on PEG-PCL (Figure 4). Different staining patterns were observed across all donors at passage 3, with enhanced mineralization for donor 2. However, at passage 6, staining intensity was markedly decreased on TCPS, with minimal staining for donor 2. Adipogenic differentiation had mixed results across all donors (Figure 5). Staining patterns of Oil Red O for TCPS and PEG-PCL did not show unique patterning or oil droplet shape, nor were there statistically significant differences in staining intensity between the substrates at either passage.

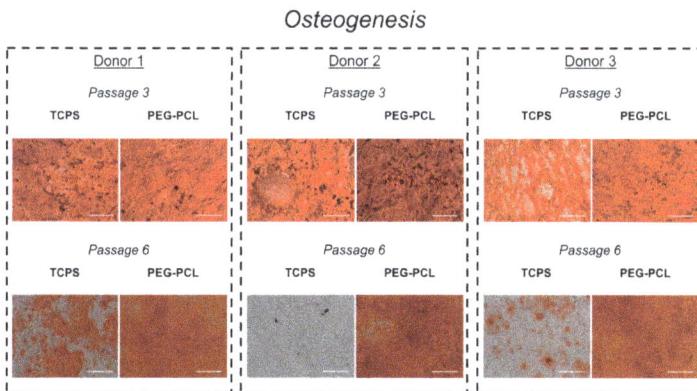

Figure 4. *Cont.*

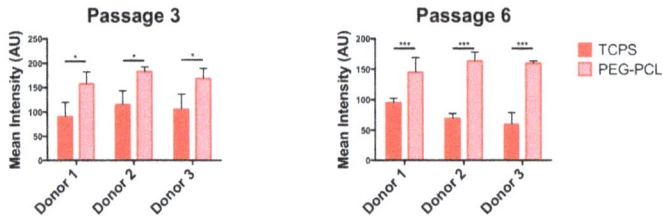

Figure 4. Osteogenic differentiation of hMSCs. Human mesenchymal stem cells (hMSCs) were stained with Alizarin Red after one month of osteogenic differentiation. Increased Alizarin Red staining was observed on PEG-PCL at both passage 3 and passage 6 compared to TCPS. The staining morphology was drastically different at for TCPS at passage 6 relative to TCPS at passage 3 with little staining for donor 2, while donors 1 and 3 had patches of positive staining, indicating a decrease in differentiation potential over serially passaging of the cells. Mean intensity of the stain is plotted for all donors for each substrate with $n = 3$ independent experimental replicates, in which PEG-PCL had increased relative stain intensity compared to TCPS for all donors at both passages. * $p < 0.05$, *** $p < 0.001$. Scale bar = 100 μm.

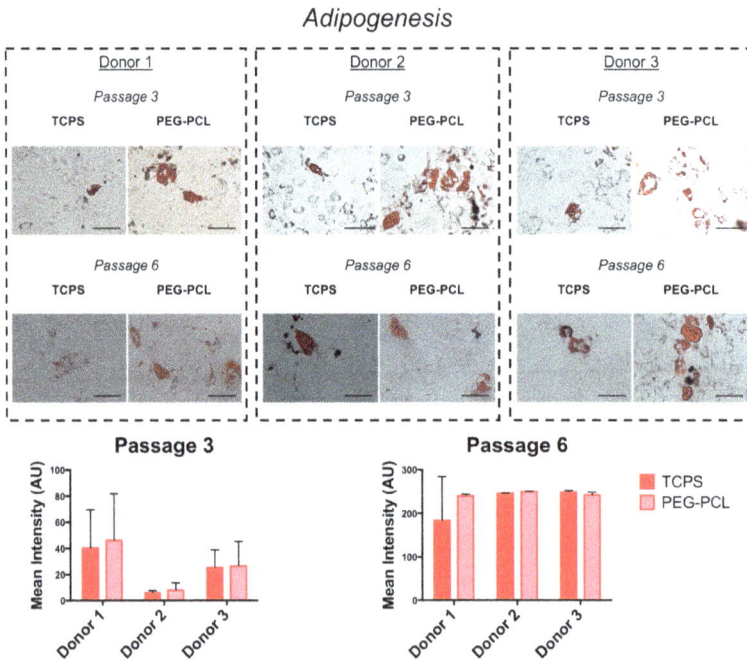

Figure 5. Adipogenic differentiation of hMSCs. hMSCs were stained with Oil Red O after one month of adipogenic differentiation. Oil Red O staining did not appear different across all donors for both passages, and heterogeneity of droplet size can be seen in the images. Mean intensity of the stain is plotted for all donors for each substrate with $n = 3$ independent experimental replicates, and no significant differences between TCPS and PEG-PCL were reported. Scale bar = 100 μm.

3. Discussion

Longitudinal, serial passage of hMSCs for regenerative medicine and tissue engineering-based therapies is undoubtedly a prerequisite that is critical to clinical success. Regardless of allogenic or autologous donor cells, ensuring that a sizable stem cell mass is prepared for host injection enables the greatest chance for engraftment. However, in the process of expanding stem cells, phenotype characteristics associated with senescence could inhibit success rates of future stem cell therapies and potentially raise the risk of harming patients. Therefore, different culture strategies ranging from pro-aggregate culture vessels to minimally-disrupted monolayer substrates have been explored for culturing hMSCs ex vivo [15,20,26–28]. Here, we sought to employ a copolymer composed of PEG-PCL that has been shown in previous work to regulate hMSC function through nanoscale interactions as a culture platform to expand healthy cells from aging donors. We have demonstrated that hMSCs cultured on the PEG-PCL material maintain osteogenic capacity and a lower ROS level than those cultured on TCPS.

Prior work in our lab discovered that carefully tuning of PEG-PCL copolymer composition manipulated the homeostasis of hMSCs towards a more potent and potentially therapeutic phenotype [31]. Of note, this previous work also demonstrated that cells on PEG-PCL materials exhibited a significantly reduced proliferation rate compared to those grown on TCPS, although cells on both materials were proliferative (approximately 40% on TCPS versus 10–20% on PEG-PCL). Similar to the aforementioned study, cells on the copolymer were into a forced aggregation state, but as the passage number progressed, the size of these aggregates increased (Figure 2). The formation of aggregates with a small interfacing cell layer on the material that act like a "feeder layer" has been seen in vivo in the bone marrow [33–35]. Additionally, speculation of hMSCs being related to pericytes, which is still under debate in the field, demonstrates the same balance of cell-cell (pericyte-endothelial cell) and cell-matrix interaction (pericyte-surrounding ECM), and the PEG-PCL appears to cause the hMSCs to behave as if they were in a bone marrow or capillary-like environment [36]. A separate study conducted by our research group observed that the hMSCs culture on PEG-PCL formed a thin monolayer interface with the copolymer film followed by a significant accumulation of cells that created a spheroid cell mass [32]. While these structures were only studied for one passage on the copolymer film, the change in hMSC phenotype over continuous serially passaging on the copolymer films could influence the preference of the hMSCs to engage in stronger and more numerous cell-cell contacts, hence resulting in larger cell aggregates by passage 6. Of additional note, the qualitative differences in spheroid morphologies observed are likely due to inherent differences among the donors, as has been reported in the literature [37,38], although no measurable parameter can be identified as causing these morphological disparities.

Because the functional capacity of hMSCs has direct effects on their clinical usefulness, the first functional test evaluated was ROS load in the hMSCs. It is well known that general cell health decreases as ROS increases in the cell, thereby increasing the likelihood of cancerous or apoptotic-inducing changes occurring [39]. ROS loads for hMSCs cultured on TCPS for both passages were higher than those cultured on PEG-PCL (Figure 3). This aligned with previous literature reporting stem cell niches (exhaustively reviewed by Zhou, Shao, and Spitz [40]). A small, healthy fraction of stem cells that replenish the stem cell population has naturally low amounts of ROS. This is also logical given that ROS can also damage DNA, increasing risk of abnormal cell behavior. Because hMSCs cultured on PEG-PCL maintained a similar level of ROS through passage 6, our substrate appears to keep these hMSCs in a pro-stem cell state and correlates with the growing size of the aggregates (Figure 2). This explanation is further supported by a study by Zhang et al. that showed spheroid aggregates of gingiva-derived hMSCs expressed SOD2 (an antioxidant protein) throughout the aggregate [41]. As seen in Figure 2, the ROS curves for cells on PEG-PCL are wider than those for TCPS, indicating a greater distribution in the ROS load for cells on the copolymer substrate, which could be due to their location within the cell aggregates. That is, cells in direct contact with the surface might have a different ROS state than those within the middle of the cell aggregate.

hMSCs used for future tissue engineering therapies will likely be differentiated into other cell types for healing damaged tissues. Hence, differentiation assays were conducted for osteogenesis and adipogenesis (Figures 4 and 5). In the original publication defining the bone marrow stromal population, researchers confirmed that the stromal cells were multipotent and able to differentiate into multiple mesenchymal stem types including bone, cartilage, and fat [42]. hMSCs have also been differentiated into neurons [43] and cardiac cells [44], but for this study osteogenesis and adipogenesis were employed in order to compared to long-established literature. Osteogenic potential was maintained for hMSCs cultured on PEG-PCL, while the ability for successful mineralization was abrogated by passage 6 for TCPS. Loss of osteogenic potential is known to occur due to senescence-associated changes [45–47], and that donor source does not influence the degree of differentiation but rather the phenotype of the hMSCs prior to differentiation induction [48], thus validating that the pro-stemness phenotype was maintained on PEG-PCL. Of equal note, it has been shown that spheroid morphologies adopted in hMSC culture improve the degree of osteogenesis [49], even going so far as to determine that enhanced epigenetic changes promoting pro-stem cell transcription factors prior to induction drove increased activity in pro-osteogenic proteins like alkaline phosphatase [50]. However, no significant changes were seen in adipogenic differentiation. This result has been reported by Cheng et al. in which bone-marrow derived hMSCs did not exhibit increased efficiency in adipogenic differentiation following culture on chitosan films, while osteogenic differentiation was markedly enhanced along with increased RUNX2 expression [51]. The results could also be due to the age of the donors utilized in this study, thereby reducing their differentiation capacity [11].

4. Materials and Methods

4.1. Polymer Substrate Preparation

5% PEG (Mw = 2000 Da)—95% PCL (PEG-PCL) was synthesized using methods previously described [52]. Briefly, PCL was extended from the free end of methoxy-PEG by ring-opening polymerization of distilled ε-caprolactone at 120 °C for 4 h and then precipitated in ice-cold diethyl ether. The copolymer was then desiccated to remove excess diethyl ether, thereby leaving the final copolymer product behind. Spin-coated polymer films were prepared with a commercial spin-coater (Laurell Technologies, North Wales, PA, USA). 15 mm circular glass cover slips (Fisher Scientific, Hampton, NH, USA) were first cleaned with 100% ethanol (Sigma Aldrich, St. Louis, MO, USA), rinsed with dH$_2$O, and heated to 80 °C for ~20 min to dry. A 1% weight/volume (w/v) solution of the copolymer in tetrahydrofuran (THF, Sigma Aldrich, St. Louis, MO, USA) was spun for 30 s at 3000 RPM atop the clean glass cover slip. For preparation of "large-scale" Petri dish polymer films, Pyrex Petri dishes (Corning Inc., Corning, NY, USA) were cleaned as described above, and 1 mL of a 1% w/v solution of polymer in THF was spun for 2 min at 1500 RPM to coat the surface. All samples were then exposed to constant vacuum for ≥30 min to remove excess solvent and kept in a desiccator until use. Coverslips and dishes were UV sterilized for 60 min before use for cell culture.

4.2. Cell Culture

hMSCs were acquired from three patients at Vanderbilt University Medical Center in cooperation with Dr. Pampee P. Young, according to previously published methods [53]. To briefly summarize, bone marrow isolates were diluted in HBSS and passed through a 70 μm filter, which was subsequently plated onto Histopaque (Sigma Aldrich). The non-adherent cells were washed away and replated on TCPS in complete media. Once the cells were confluent, they were passaged and a small fraction was saved for MSC phenotyping using a Human MSC Phenotyping Kit from Miltenyi Biotec (Auburn, CA, USA). Flow cytometry was performed with this kit to evaluate expression of CD14, CD20, CD34, CD45, CD73, CD90, and CD105. All donors were CD14/20/34/45 negative and CD73/90/105 positive at greater than 99% of the sample population as described in our previous report by Balikov et al. [31]. All patients provided informed consent for use of their bone marrow aspirates for research purposes. Processing and handling of the cells were carried out in accordance to relevant guidelines and regulations established by both Vanderbilt University and the National Institutes of Health, and experimental protocols were reviewed and approved by the Vanderbilt Institutional Review Board. All patients were male and over the age of 65 with no known blood disorders or cancer diagnosis at the time of bone marrow harvest. hMSCs were maintained in complete media (CM) composed of alpha-minimum essential media with nucleosides (αMEM, Life Technologies, Carlsbad, CA, USA) with 16.7% fetal bovine serum (Life Technologies), 1% penicillin/streptomycin (Life Technologies), and 4 μg/mL plasmocin (InvivoGen, San Diego, CA, USA). Cells were kept in a humidified incubator at 37 °C and 5% CO_2, and media was replaced twice each week. For all experiments, hMSCs were seeded at a density of 10,000 viable cells/cm^2, as determined by exclusion of Trypan blue, and cultured for four days before passaging.

4.3. Immunocytochemistry

hMSCs were fixed with 4% paraformaldehyde (PFA, Sigma Aldrich) for 15 min, permeabilized with 0.3% Triton-X (Sigma Aldrich) for 15 min when probing intracellular targets, and blocked with 10% goat serum (Sigma Aldrich) for >2 h, all at room temperature. Cells were incubated with Alexa488-phallodin (1:40 *v*/*v* in Phosphate-Buffered Saline, Life Technologies) for 10 min followed by counterstaining with Hoechst (Sigma Aldrich, 2 μg/mL) for 20 min at room temperature. Imaging was performed with a Zeiss LSM 710 confocal microscope (Carl Zeiss, Oberkochen, Germany), and images were processed with ImageJ (National Institutes of Health, Bethesda, MD, USA).

4.4. Measuring Levels of Intracellular Reactive Oxygen Species (ROS)

hMSCs were incubated with 10 μM 5-(and-6)-chloromethyl-2',7'-dichlorodihydrofluorescein diacetate acetyl ester (DCFDA) (Life Technologies) in serum-free DMEM for 30 min following the manufacturer's instructions. Cells were trypsinized and run on a BD LSR Fortessa (BD Biosciences, Franklin Lakes, NJ, USA) with the appropriate unstained control. *n* = 3 biological replicates were conducted per substrate condition. Data were analyzed by FlowJo software (Tree Star Inc., Ashland, OR, USA).

4.5. Differentiation Assay

hMSCs were grown on TCPS or PEG-PCL at their indicated passage for 4 days before being trypsinized and moved to 24-well TCPS plates. Differentiation assays were performed based on pre-established protocols [54,55]. Adipogenic media using AMEM contained 16.7% FBS, 1% penicillin/streptomycin, 4 μg/mL plasmocin, 0.1 μM dexamethasone, 0.45 mM 3-isobutyl-1-methylxantine, 0.2 mM indomethacin 1 μg/mL insulin, and 1 μM rosiglitazone. Osteogenic media using AMEM contained 16.7% FBS, 1% P/S, 4 μg/mL plasmocin, 10 nM dexamethasone, 5 mM β-glycerophosphate, and 50 μg/mL ascorbate-2-phosphate. All specialized differentiation media reagents were purchased from Sigma-Aldrich. Cells were cultured under induction media for one month and then fixed with 4% PFA. Cells were stained with Oil Red O (ORO) and Alizarin Red S (ARS) for adipogenic and osteogenic staining, respectively. Images were taken with a Nikon Ti inverted microscope (Nikon Instruments Inc., Melville, NY, USA) and processed with ImageJ. Stain quantification was performed on *n* = 3 independent experimental replicates. Images were first converted to a RGB stack followed by setting a threshold range in the green channel to account for variation in background light from brightfield imaging. The images were inverted resulting in a new grayscale image, and mean intensity was measured.

4.6. Statistical Analysis for ROS and Differentiation Assays

Comparisons between substrates for differentiation assays were performed with a Student's unpaired *t*-test. In all cases, $p < 0.05$ is considered statistically significant. Mean ± standard deviation is reported, unless otherwise noted.

5. Conclusions

In this study, we explored the ramifications of serially passaging human bone marrow-derived hMSCs from aged patient donors on a novel PEG-PCL copolymer film to maintain functional capacity and stem cell phenotype for future applications in tissue engineering and regenerative medicine. hMSCs grown on the films illustrated morphologies representative of hMSCs found in vivo and maintained low ROS loads that if unchecked are known to be associated with the progression of senescence-associated changes. Finally, the maintenance of differentiation capacity of PEG-PCL hMSCs demonstrated relevance of using our alternative copolymer film to maintain stem cell functionality for downstream hMSC adoption of target tissue cell types.

Acknowledgments: This research work was funded and supported by the Faculty Research Assistance Program of Yonsei University College of Medicine for 2000 (6-2016-0031) (HJS) and the Basic Science Research Program through the National Research Foundation of Korea (NRF) funded by the Ministry of Science, ICT & Future Planning (2016M3A9E9941743) (HJS). This work was supported by the National Research Foundation of Korea (NRF) (2015M3A9B3028685) (JHS). This study was also supported in part by Vanderbilt CTSA grant UL1 TR000445 from NCATS/NIH (VICTR Resource Request #12676) (DAB).

Author Contributions: Daniel A. Balikov, Spencer W. Crowder, Jung Bok Lee, and Yunki Lee performed the experiments. Daniel A. Balikov, Spencer W. Crowder, Ung Hyun Ko, Jennifer H. Shin, and Hak-Joon Sung analyzed the data. Daniel A. Balikov, Spencer W. Crowder, Jung Bok Lee, Yunki Lee, Ung Hyun Ko, Mi-Lan Kang, Won Shik Kim, Jennifer H. Shin, and Hak-Joon Sung wrote the manuscript.

Conflicts of Interest: The authors declare no conflict of interest.

References

1. Robey, P.G.; Kuznetsov, S.A.; Ren, J.; Klein, H.G.; Sabatino, M.; Stroncek, D.F. Generation of clinical grade human bone marrow stromal cells for use in bone regeneration. *Bone* **2015**, *70*, 87–92. [CrossRef] [PubMed]
2. Fisher, M.B.; Belkin, N.S.; Milby, A.H.; Henning, E.A.; Soegaard, N.; Kim, M.; Pfeifer, C.; Saxena, V.; Dodge, G.R.; Burdick, J.A.; et al. Effects of Mesenchymal Stem Cell and Growth Factor Delivery on Cartilage Repair in a Mini-Pig Model. *Cartilage* **2016**, *7*, 174–184. [CrossRef] [PubMed]
3. Russo, V.; Young, S.; Hamilton, A.; Amsden, B.G.; Flynn, L.E. Mesenchymal stem cell delivery strategies to promote cardiac regeneration following ischemic injury. *Biomaterials* **2014**, *35*, 3956–3974. [CrossRef] [PubMed]
4. Madonna, R.; Delli Pizzi, S.; Tartaro, A.; De Caterina, R. Transplantation of mesenchymal cells improves peripheral limb ischemia in diabetic rats. *Mol. Biotechnol.* **2014**, *56*, 438–448. [CrossRef] [PubMed]
5. Petite, H.; Viateau, V.; Bensaid, W.; Meunier, A.; de Pollak, C.; Bourguignon, M.; Oudina, K.; Sedel, L.; Guillemin, G. Tissue-engineered bone regeneration. *Nat. Biotechnol.* **2000**, *18*, 959–963. [CrossRef] [PubMed]
6. Hayflick, L.; Moorhead, P.S. The serial cultivation of human diploid cell strains. *Exp. Cell Res.* **1961**, *25*, 585–621. [CrossRef]
7. Heo, J.Y.; Jing, K.; Song, K.S.; Seo, K.S.; Park, J.H.; Kim, J.S.; Jung, Y.J.; Hur, G.M.; Jo, D.Y.; Kweon, G.R.; et al. Downregulation of APE1/Ref-1 is involved in the senescence of mesenchymal stem cells. *Stem Cells* **2009**, *27*, 1455–1462. [CrossRef] [PubMed]
8. Wagner, W.; Horn, P.; Castoldi, M.; Diehlmann, A.; Bork, S.; Saffrich, R.; Benes, V.; Blake, J.; Pfister, S.; Eckstein, V.; et al. Replicative senescence of mesenchymal stem cells: A continuous and organized process. *PLoS ONE* **2008**, *3*, e2213. [CrossRef] [PubMed]
9. Wagner, W.; Bork, S.; Horn, P.; Krunic, D.; Walenda, T.; Diehlmann, A.; Benes, V.; Blake, J.; Huber, F.X.; Eckstein, V.; et al. Aging and replicative senescence have related effects on human stem and progenitor cells. *PLoS ONE* **2009**, *4*, e5846. [CrossRef] [PubMed]
10. Kasper, G.; Mao, L.; Geissler, S.; Draycheva, A.; Trippens, J.; Kuhnisch, J.; Tschirschmann, M.; Kaspar, K.; Perka, C.; Duda, G.N.; et al. Insights into mesenchymal stem cell aging: Involvement of antioxidant defense and actin cytoskeleton. *Stem Cells* **2009**, *27*, 1288–1297. [CrossRef] [PubMed]
11. Muraglia, A.; Cancedda, R.; Quarto, R. Clonal mesenchymal progenitors from human bone marrow differentiate in vitro according to a hierarchical model. *J. Cell Sci.* **2000**, *113 Pt 7*, 1161–1166. [PubMed]

12. Crowder, S.W.; Horton, L.W.; Lee, S.H.; McClain, C.M.; Hawkins, O.E.; Palmer, A.M.; Bae, H.; Richmond, A.; Sung, H.J. Passage-dependent cancerous transformation of human mesenchymal stem cells under carcinogenic hypoxia. *FASEB J.* **2013**, *27*, 2788–2798. [CrossRef] [PubMed]

13. Bara, J.J.; Richards, R.G.; Alini, M.; Stoddart, M.J. Concise review: Bone marrow-derived mesenchymal stem cells change phenotype following in vitro culture: Implications for basic research and the clinic. *Stem Cells* **2014**, *32*, 1713–1723. [CrossRef] [PubMed]

14. Frith, J.E.; Thomson, B.; Genever, P.G. Dynamic three-dimensional culture methods enhance mesenchymal stem cell properties and increase therapeutic potential. *Tissue Eng. Part C Methods* **2010**, *16*, 735–749. [CrossRef] [PubMed]

15. Bartosh, T.J.; Ylostalo, J.H.; Mohammadipoor, A.; Bazhanov, N.; Coble, K.; Claypool, K.; Lee, R.H.; Choi, H.; Prockop, D.J. Aggregation of human mesenchymal stromal cells (MSCs) into 3D spheroids enhances their antiinflammatory properties. *Proc. Natl. Acad. Sci. USA* **2010**, *107*, 13724–13729. [CrossRef] [PubMed]

16. Potapova, I.A.; Gaudette, G.R.; Brink, P.R.; Robinson, R.B.; Rosen, M.R.; Cohen, I.S.; Doronin, S.V. Mesenchymal stem cells support migration, extracellular matrix invasion, proliferation, and survival of endothelial cells in vitro. *Stem Cells* **2007**, *25*, 1761–1768. [CrossRef] [PubMed]

17. Bartosh, T.J.; Ylostalo, J.H. Preparation of anti-inflammatory mesenchymal stem/precursor cells (MSCs) through sphere formation using hanging-drop culture technique. *Curr. Protoc. Stem Cell Biol.* **2014**, *6*, 28.

18. Bartosh, T.J.; Ylostalo, J.H.; Bazhanov, N.; Kuhlman, J.; Prockop, D.J. Dynamic compaction of human mesenchymal stem/precursor cells into spheres self-activates caspase-dependent IL1 signaling to enhance secretion of modulators of inflammation and immunity (PGE2, TSG6, and STC1). *Stem Cells* **2013**, *31*, 2443–2456. [CrossRef] [PubMed]

19. Zimmermann, J.A.; McDevitt, T.C. Pre-conditioning mesenchymal stromal cell spheroids for immunomodulatory paracrine factor secretion. *Cytotherapy* **2014**, *16*, 331–345. [CrossRef] [PubMed]

20. Elseberg, C.L.; Salzig, D.; Czermak, P. Bioreactor expansion of human mesenchymal stem cells according to GMP requirements. *Methods Mol. Biol.* **2015**, *1283*, 199–218. [PubMed]

21. Tozetti, P.A.; Caruso, S.R.; Mizukami, A.; Fernandes, T.R.; da Silva, F.B.; Traina, F.; Covas, D.T.; Orellana, M.D.; Swiech, K. Expansion strategies for human mesenchymal stromal cells culture under xeno-free conditions. *Biotechnol. Prog.* **2017**. [CrossRef] [PubMed]

22. Larson, B.L.; Ylostalo, J.; Lee, R.H.; Gregory, C.; Prockop, D.J. Sox11 is expressed in early progenitor human multipotent stromal cells and decreases with extensive expansion of the cells. *Tissue Eng. Part A* **2010**, *16*, 3385–3394. [CrossRef] [PubMed]

23. Tsai, A.C.; Liu, Y.; Yuan, X.; Ma, T. Compaction, fusion, and functional activation of three-dimensional human mesenchymal stem cell aggregate. *Tissue Eng. Part A* **2015**, *21*, 1705–1719. [CrossRef] [PubMed]

24. Mueller-Klieser, W. Three-dimensional cell cultures: From molecular mechanisms to clinical applications. *Am. J. Physiol.* **1997**, *273*, C1109–1123. [CrossRef] [PubMed]

25. Mueller-Klieser, W. Multicellular spheroids. A review on cellular aggregates in cancer research. *J. Cancer Res. Clin. Oncol.* **1987**, *113*, 101–122. [CrossRef] [PubMed]

26. Lin, R.Z.; Chang, H.Y. Recent advances in three-dimensional multicellular spheroid culture for biomedical research. *Biotechnol. J.* **2008**, *3*, 1172–1184. [CrossRef] [PubMed]

27. Leight, J.L.; Liu, W.F.; Chaturvedi, R.R.; Chen, S.; Yang, M.T.; Raghavan, S.; Chen, C.S. Manipulation of 3D Cluster Size and Geometry by Release from 2D Micropatterns. *Cell. Mol. Bioeng.* **2012**, *5*, 299–306. [CrossRef] [PubMed]

28. Baraniak, P.R.; McDevitt, T.C. Scaffold-free culture of mesenchymal stem cell spheroids in suspension preserves multilineage potential. *Cell Tissue Res.* **2012**, *347*, 701–711. [CrossRef] [PubMed]

29. Ng, C.P.; Sharif, A.R.; Heath, D.E.; Chow, J.W.; Zhang, C.B.; Chan-Park, M.B.; Hammond, P.T.; Chan, J.K.; Griffith, L.G. Enhanced ex vivo expansion of adult mesenchymal stem cells by fetal mesenchymal stem cell ECM. *Biomaterials* **2014**, *35*, 4046–4057. [CrossRef] [PubMed]

30. Duffy, C.R.; Zhang, R.; How, S.E.; Lilienkampf, A.; De Sousa, P.A.; Bradley, M. Long term mesenchymal stem cell culture on a defined synthetic substrate with enzyme free passaging. *Biomaterials* **2014**, *35*, 5998–6005. [CrossRef] [PubMed]

31. Balikov, D.A.; Crowder, S.W.; Boire, T.C.; Lee, J.B.; Gupta, M.K.; Fenix, A.M.; Lewis, H.N.; Ambrose, C.M.; Short, P.A.; Kim, C.S.; et al. Tunable Surface Repellency Maintains Stemness and Redox Capacity of Human Mesenchymal Stem Cells. *ACS Appl. Mater. Interfaces* **2017**. [CrossRef] [PubMed]

32. Crowder, S.W.; Balikov, D.A.; Boire, T.C.; McCormack, D.; Lee, J.B.; Gupta, M.K.; Skala, M.C.; Sung, H.-J. Copolymer-Mediated Cell Aggregation Promotes a Proangiogenic Stem Cell Phenotype In Vitro and In Vivo. *Adv. Healthcare Mater.* **2016**. [CrossRef] [PubMed]
33. Mendez-Ferrer, S.; Michurina, T.V.; Ferraro, F.; Mazloom, A.R.; Macarthur, B.D.; Lira, S.A.; Scadden, D.T.; Ma'ayan, A.; Enikolopov, G.N.; Frenette, P.S. Mesenchymal and haematopoietic stem cells form a unique bone marrow niche. *Nature* **2010**, *466*, 829–834. [CrossRef] [PubMed]
34. Morikawa, S.; Mabuchi, Y.; Kubota, Y.; Nagai, Y.; Niibe, K.; Hiratsu, E.; Suzuki, S.; Miyauchi-Hara, C.; Nagoshi, N.; Sunabori, T.; et al. Prospective identification, isolation, and systemic transplantation of multipotent mesenchymal stem cells in murine bone marrow. *J. Exp. Med.* **2009**, *206*, 2483–2496. [CrossRef] [PubMed]
35. Sacchetti, B.; Funari, A.; Michienzi, S.; Di Cesare, S.; Piersanti, S.; Saggio, I.; Tagliafico, E.; Ferrari, S.; Robey, P.G.; Riminucci, M.; et al. Self-renewing osteoprogenitors in bone marrow sinusoids can organize a hematopoietic microenvironment. *Cell* **2007**, *131*, 324–336. [CrossRef] [PubMed]
36. Crisan, M.; Yap, S.; Casteilla, L.; Chen, C.W.; Corselli, M.; Park, T.S.; Andriolo, G.; Sun, B.; Zheng, B.; Zhang, L.; et al. A perivascular origin for mesenchymal stem cells in multiple human organs. *Cell Stem Cell* **2008**, *3*, 301–313. [CrossRef] [PubMed]
37. Heathman, T.R.J.; Rafiq, Q.A.; Chan, A.K.C.; Coopman, K.; Nienow, A.W.; Kara, B.; Hewitt, C.J. Characterization of human mesenchymal stem cells from multiple donors and the implications for large scale bioprocess development. *Biochem. Eng. J.* **2016**, *108*, 14–23. [CrossRef]
38. Lo Surdo, J.; Bauer, S.R. Quantitative approaches to detect donor and passage differences in adipogenic potential and clonogenicity in human bone marrow-derived mesenchymal stem cells. *Tissue Eng. Part C Methods* **2012**, *18*, 877–889. [CrossRef] [PubMed]
39. Kobayashi, C.I.; Suda, T. Regulation of reactive oxygen species in stem cells and cancer stem cells. *J. Cell. Physiol.* **2012**, *227*, 421–430. [CrossRef] [PubMed]
40. Zhou, D.; Shao, L.; Spitz, D.R. Reactive oxygen species in normal and tumor stem cells. *Adv. Cancer Res.* **2014**, *122*, 1–67. [PubMed]
41. Zhang, Q.; Nguyen, A.L.; Shi, S.; Hill, C.; Wilder-Smith, P.; Krasieva, T.B.; Le, A.D. Three-dimensional spheroid culture of human gingiva-derived mesenchymal stem cells enhances mitigation of chemotherapy-induced oral mucositis. *Stem Cells Dev.* **2012**, *21*, 937–947. [CrossRef] [PubMed]
42. Friedenstein, A.J.; Piatetzky, S., II; Petrakova, K.V. Osteogenesis in transplants of bone marrow cells. *J. Embryol. Exp. Morphol.* **1966**, *16*, 381–390. [PubMed]
43. Woodbury, D.; Schwarz, E.J.; Prockop, D.J.; Black, I.B. Adult rat and human bone marrow stromal cells differentiate into neurons. *J. Neurosci. Res.* **2000**, *61*, 364–370. [CrossRef]
44. Toma, C.; Pittenger, M.F.; Cahill, K.S.; Byrne, B.J.; Kessler, P.D. Human mesenchymal stem cells differentiate to a cardiomyocyte phenotype in the adult murine heart. *Circulation* **2002**, *105*, 93–98. [CrossRef] [PubMed]
45. Sun, Y.; Li, W.; Lu, Z.; Chen, R.; Ling, J.; Ran, Q.; Jilka, R.L.; Chen, X.D. Rescuing replication and osteogenesis of aged mesenchymal stem cells by exposure to a young extracellular matrix. *FASEB J.* **2011**, *25*, 1474–1485. [CrossRef] [PubMed]
46. Tan, J.; Xu, X.; Tong, Z.; Lin, J.; Yu, Q.; Lin, Y.; Kuang, W. Decreased osteogenesis of adult mesenchymal stem cells by reactive oxygen species under cyclic stretch: A possible mechanism of age related osteoporosis. *Bone Res.* **2015**, *3*, 15003. [CrossRef] [PubMed]
47. Wilson, A.; Shehadeh, L.A.; Yu, H.; Webster, K.A. Age-related molecular genetic changes of murine bone marrow mesenchymal stem cells. *BMC Genom.* **2010**, *11*, 229. [CrossRef] [PubMed]
48. Siegel, G.; Kluba, T.; Hermanutz-Klein, U.; Bieback, K.; Northoff, H.; Schafer, R. Phenotype, donor age and gender affect function of human bone marrow-derived mesenchymal stromal cells. *BMC Med.* **2013**, *11*, 146. [CrossRef] [PubMed]
49. Hildebrandt, C.; Buth, H.; Thielecke, H. A scaffold-free in vitro model for osteogenesis of human mesenchymal stem cells. *Tissue Cell* **2011**, *43*, 91–100. [CrossRef] [PubMed]
50. Guo, L.; Zhou, Y.; Wang, S.; Wu, Y. Epigenetic changes of mesenchymal stem cells in three-dimensional (3D) spheroids. *J. Cell. Mol. Med.* **2014**, *18*, 2009–2019. [CrossRef] [PubMed]
51. Cheng, N.C.; Wang, S.; Young, T.H. The influence of spheroid formation of human adipose-derived stem cells on chitosan films on stemness and differentiation capabilities. *Biomaterials* **2012**, *33*, 1748–1758. [CrossRef] [PubMed]

52. Crowder, S.W.; Gupta, M.K.; Hofmeister, L.H.; Zachman, A.L.; Sung, H.J. Modular polymer design to regulate phenotype and oxidative response of human coronary artery cells for potential stent coating applications. *Acta Biomater.* **2012**, *8*, 559–569. [CrossRef] [PubMed]

53. Deskins, D.L.; Bastakoty, D.; Saraswati, S.; Shinar, A.; Holt, G.E.; Young, P.P. Human mesenchymal stromal cells: Identifying assays to predict potency for therapeutic selection. *Stem Cells Transl. Med.* **2013**, *2*, 151–158. [CrossRef] [PubMed]

54. Krause, U.; Seckinger, A.; Gregory, C.A. Assays of osteogenic differentiation by cultured human mesenchymal stem cells. *Methods Mol. Biol.* **2011**, *698*, 215–230. [PubMed]

55. Fink, T.; Zachar, V. Adipogenic differentiation of human mesenchymal stem cells. *Methods Mol. Biol.* **2011**, *698*, 243–251. [PubMed]

International Journal of
Molecular Sciences

MDPI

Article

In Situ Forming Gelatin Hydrogels-Directed Angiogenic Differentiation and Activity of Patient-Derived Human Mesenchymal Stem Cells

Yunki Lee [1,†], Daniel A. Balikov [1,†], Jung Bok Lee [1], Sue Hyun Lee [1], Seung Hwan Lee [2], Jong Hun Lee [3], Ki Dong Park [4] and Hak-Joon Sung [1,5,*]

[1] Department of Biomedical Engineering, Vanderbilt University, Nasville, TN 37235, USA;
    yunki.lee@vanderbilt.edu (Y.L.); daniel.a.balikov@vanderbilt.edu (D.A.B.);
    jung.bok.lee@vanderbilt.edu (J.B.L.); hj72sung@gmail.com (S.H.L.)
[2] Severance Biomedical Science Institute, College of Medicine, Yonsei University, Seoul 120-752, Korea;
    leeseh@yuhs.ac
[3] Department of Urology, College of Medicine, Yonsei University, Seoul 120-752, Korea;
    mushroom14@gmail.com
[4] Department of Food Science and Biotechnology, College of Life Science, CHA University, Gyeonggi 443-742,
    Korea; kdp@ajou.ac.kr
[5] Department of Molecular Science and Technology, Ajou University, Suwon 443-749, Korea
* Correspondence: hj72sung@yuhs.ac; Tel.: +1-615-322-6986
† These authors contributed equally to this work.

Received: 16 July 2017; Accepted: 1 August 2017; Published: 4 August 2017

Abstract: Directing angiogenic differentiation of mesenchymal stem cells (MSCs) still remains challenging for successful tissue engineering. Without blood vessel formation, stem cell-based approaches are unable to fully regenerate damaged tissues due to limited support for cell viability and desired tissue/organ functionality. Herein, we report in situ cross-linkable gelatin−hydroxyphenyl propionic acid (GH) hydrogels that can induce pro-angiogenic profiles of MSCs via purely material-driven effects. This hydrogel directed endothelial differentiation of mouse and human patient-derived MSCs through integrin-mediated interactions at the cell-material interface, thereby promoting perfusable blood vessel formation in vitro and in vivo. The causative roles of specific integrin types (α_1 and $\alpha_v\beta_3$) in directing endothelial differentiation were verified by blocking the integrin functions with chemical inhibitors. In addition, to verify the material-driven effect is not species-specific, we confirmed in vitro endothelial differentiation and in vivo blood vessel formation of patient-derived human MSCs by this hydrogel. These findings provide new insight into how purely material-driven effects can direct endothelial differentiation of MSCs, thereby promoting vascularization of scaffolds towards tissue engineering and regenerative medicine applications in humans.

Keywords: injectable gelatin hydrogels; patient-derived mesenchymal stem cells; integrin-mediated interactions; material-driven endothelial differentiation; angiogenesis

1. Introduction

Over the past few decades, stem cell therapies have demonstrated a certain degree of clinical success for thin tissues such as skin and cartilage [1–3]. These favorable outcomes have generated additional enthusiasm that these therapies could be translated to regenerate metabolically active tissues and organs with support from biomaterials. These therapeutic targets, however, require a high degree of vascularization to support influx and outflux of nutrients, oxygen, and waste products because if there are no blood vessels within any 200 µm (maximum) area, cells and tissue undergo necrosis [4]. Without blood vessel formation, stem cell-based approaches will not be fully successful in regenerating damaged tissues, particularly thick tissues, due to limited support for cell viability and desired tissue/organ functionality.

Continuous progress has been made to promote vascularization of scaffolds by incorporating or conjugating vascular growth/signaling factors (e.g., vascular endothelial growth factor; VEGF) [4–7]. While these approaches are effective in vitro, their in vivo effects are still questionable due to a short half-life (<30 min) of biological molecules under physiological conditions, and side effects associated with hypotension and edema [8–10]. Therefore, several methods have been considered in order to increase growth factor bioactivity over extended periods of time such as (i) incorporating components of the normal extracellular matrix (ECM) (e.g., heparin and fibronectin) that stabilize these factors and/or promote their activity; (ii) encapsulating growth factors in protease-resistant reservoirs that serve as a physical barrier against protease attack; and (iii) engineering proteolytic cleavage sites within growth factor proteins to inhibit their degradation [11,12]. However, there are still concerns about the expensive supply of growth factors and paracrine side effects associated with uncontrolled release into systemic circulation.

To address this unmet need, purely scaffold material-driven effects have been explored [13–16]. Controlling material properties at the cell–material interface (e.g., matrix stiffness and integrin–matrix interactions) can direct stem cell differentiation, which includes endothelial differentiation of mesenchymal stem cells (MSCs) [1,17]. For example, hydrogels have many advantages to mimic tissue properties due to tunable properties, including hydration and stiffness [18,19]. Moreover, their injectable format can provide a 3D microenvironment when cells are encapsulated and delivered to a target site. We previously developed an in situ forming gelatin–hydroxyphenyl propionic acid (GH) hydrogel. This hydrogel type was used as an injectable/sprayable platform upon a horseradish peroxidase (HRP)-mediated cross-linking reaction to deliver therapeutic cells and drugs [20–22]. In particular, this gelatin hydrogel, within a specific range of mechanical stiffness (0.5–3 kPa), was found to direct robust differentiation of mouse bone marrow-derived MSCs (mMSCs) into endothelial cells, and induce extensive vasculogenesis without any biological supplementation in vitro and in vivo [23].

In the present follow-up study, we accomplished the following three objectives. First, we identified a mechanism at the cell–matrix interface that is responsible for directed differentiation of mMSCs into endothelial cells. The causative roles of specific integrin types (α_1 and $\alpha_v\beta_3$) in directing endothelial differentiation were verified by blocking the integrin functions with chemical inhibitors. Second, to verify the material-driven effect is not species-specific, we confirmed in vitro endothelial differentiation of patient-derived human MSCs (hMSCs) by this hydrogel. Third, we confirmed consequent, improved blood vessel formation of these patient hMSCs when delivered through the gelatin hydrogel system in a subcutaneous implantation model of immune compromised mice.

While only about ~300 genes out of approximately 20,000 are unique to either humans or mice as revealed from whole genome sequencing, interspecies differences have been rather difficult to predict and understand at times, as evidenced by countless clinical trials that have ultimately failed after successful pre-clinical studies in mice [24]. Therefore, robust validation in human patient-derived MSCs is required for successful translation. To this end, we have isolated bone marrow-derived MSCs from three patients, particularly from old patients (>65 years old) as they stand to benefit the most from regenerative medicine approaches. This approach is necessary to provide new insight into how purely material-driven effects can direct endothelial differentiation of MSCs, thereby promoting vascularization of scaffolds towards tissue engineering and regenerative medicine applications in humans.

2. Results and Discussion

2.1. Characterization of In Situ Forming Gelatin–Hydroxyphenyl Propionic Acid (GH) Hydrogels

To prepare in situ cross-linkable gelatin hydrogels, we first synthesized HRP-reactive phenol-conjugated gelatin polymer (GH conjugate, phenolic content = 143 μmol/g of polymer).

This GH polymer solution (7 wt %) dissolved in PBS (pH 7.4) was subjected to HRP/H$_2$O$_2$-tiggered oxidative cross-linking (Figure 1A) [21]. Based upon our previous findings that GH hydrogels with <3 kPa mechanical stiffness led to robust vasculogenic induction of mMSCs in vitro and in vivo, we prepared two different GH hydrogels with distinct degrees of mechanical strength (GH-7-L; 1.4 kPa and GH-7-H; 3.0 kPa) for the present study [23]. The softer hydrogels (<1.0 kPa) were excluded because of fast degradation within a few days in in vitro cell experiments [23].

To characterize the mechanical stiffness of GH hydrogels formed with HRP (2.5 µg/mL) and H$_2$O$_2$ (0.005 wt % for GH-7-L and 0.006 wt % for GH-7-H), we investigated H$_2$O$_2$ concentration-dependent viscoelastic modulus (G' and G") as a function of time (Figure 1B,C). The time point at which G' and G" intersect each other (G' > G") is defined as a hydrogel formation [25]. The HRP/H$_2$O$_2$ interactions catalyzed in situ cross-linking among phenolic groups of GH conjugates, and thus resulted in a rapid gelation where an intersected point of G'/G" was observed within 15 s. In addition, higher H$_2$O$_2$ concentration led to a greater G' of hydrogels due to the increased cross-linking density where it was confirmed that G' of GH-7-L and GH-7-H were 1.4 and 3.0 kPa, respectively. We also characterized proteolytic degradation rates of GH-7-L and GH-7-H gels incubated in the media with or without collagenase solution (0.4 µg/mL) as a function of time. As shown in Figure 1D, GH-7-L and GH-7-H hydrogels completely degraded within 24 and 53 h, respectively, whereas the GH-7-L gels incubated in the media without collagenase were stable for 53 h. This result indicates that GH hydrogels can be degraded through the decomposition of gelatin backbone by matrix metalloproteases present in the body, and their degradation rate is dependent on the cross-linking density.

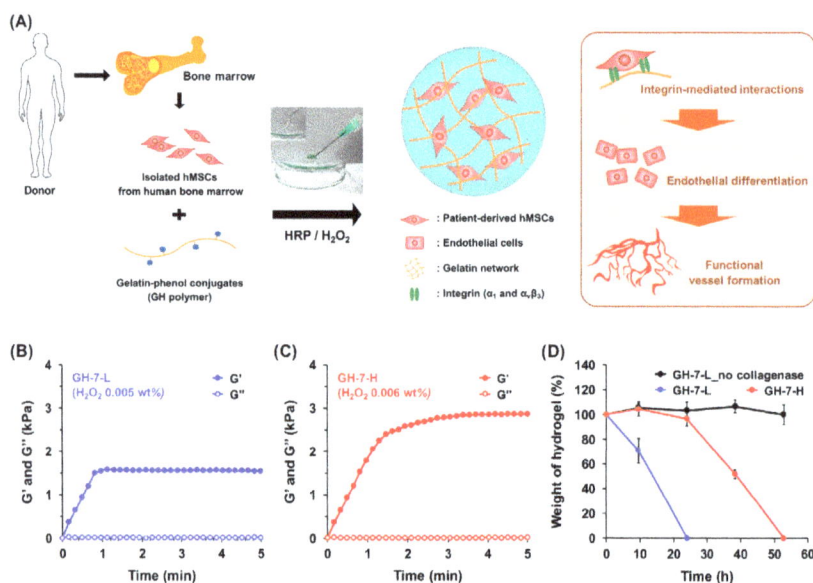

Figure 1. In situ forming gelatin−hydroxyphenyl propionic acid (GH) hydrogels as a translatable platform for mesenchymal stem cell (MSC) delivery. Schematic illustration of GH hydrogels that direct endothelial differentiation of MSCs and induce robust vascularization via integrin-mediated interactions. MSCs are collected from three patients' bone marrow (>65 years old) and loaded in GH hydrogel matrices during HRP/H$_2$O$_2$ cross-linking reaction (**A**); Time-sweep elastic modulus (G') and viscous modulus (G") of GH hydrogels with different concentration of H$_2$O$_2$ (0.005 wt % for GH-7-L and 0.006 wt % for GH-7-H) and HRP (2.5 µg/mL) measured by rheometer (**B,C**); In vitro degradation profiles of GH-7-L and GH-7-H hydrogels in the presence or absence of collagenase (0.4 µg/mL) treatment (*n* = 3) (**D**).

2.2. Gelatin Hydrogel Material-Driven Differentiation of MSCs (Mesenchymal Stem Cells) into Endothelial Cells

We next investigated the interfacial mechanism behind purely gelatin material-driven differentiation of MSCs into an endothelial lineage. While a number of previous studies have shown purely material-driven differentiation of MSCs into osteocytes, neural cells, and chondrocytes, endothelial differentiation most commonly required an extensive use of biochemical agents such as vascular endothelial growth factor (VEGF) and fibroblast growth factor (FGF) [26–28]. In our previous study, we also observed an increased expression of VEGF receptor 2 in mMSCs cultured in GH hydrogels without any addition of growth factors [23]. Therefore, we hypothesized that integrins at the cell-matrix interface might play a key role in inducing the endothelial differentiation of MSCs. Integrins are extracellular matrix receptors expressed on the cell membrane, and they serve as a physical anchoring point for adherent cells. Seminal studies have shown that many integrins are responsible for mediating cell–matrix interaction [29,30]. For example, integrins α_{1-3} and α_6 with their downstream signaling are crucial in driving MSC differentiation to an endothelial lineage by binding to a mixed laminin I and collagen IV substrate. Therefore, inquiring if integrin binding to the GH hydrogel caused endothelial differentiation of mMSCs, as seen in our previous work, would suggest that translation to hMSCs is possible as they express the same integrins.

In order to identify the types of integrin involved in the material-driven effect on mMSCs, we profiled expression of various integrins through qRT-PCR (Figure 2A). Of 26 integrins known to date, we focused on specific integrins that bind to collagen and other ECM components involved in angiogenesis. Integrin α_1 (collagen receptor) expression of mMSCs in GH-7-L and GH-7-H gels were significantly upregulated (3.8−4.9 times) compared to the TCPS condition. Additionally, integrin α_v and β_3, together forming a heterodimer that plays a significant role in angiogenesis, were shown to be upregulated. Lastly, ERK1, a key downstream signaling molecule in endothelial differentiation, was significantly upregulated as well.

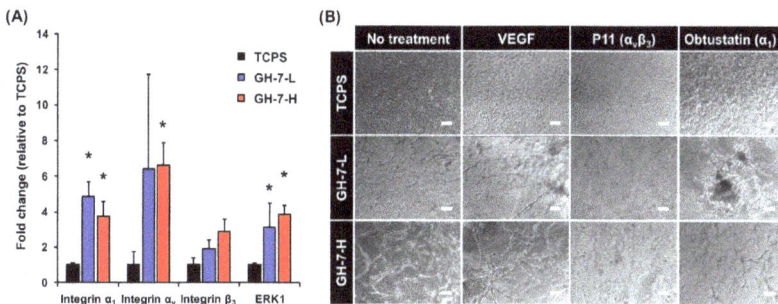

Figure 2. Integrin-mediated mechanisms at the cell–material interface. mRNA expression levels of mMSCs cultured for 15 days either on tissue culture polystyrene (TCPS) (control) or embedded in GH-7-L and GH-7-H hydrogels, * $p < 0.05$ vs. TCPS ($n = 3$) (**A**); Integrin inhibition effects on connectivity of blood vessels formed when mMSCs were cultured in GH-7-L and GH-7-H hydrogels. The experiment groups include mMSCs cultured on TCPS, GH-7-L, and GH-7-H gels for 15 days with no treatment, soluble vascular endothelial growth factor (VEGF), P11 (integrin $\alpha_v\beta_3$ inhibitor), and obtustatin (integrin α_1 inhibitor). Scale bars indicate 200 μm (**B**).

We then observed sprouting morphology of mMSCs cultured in GH hydrogels when integrin functions were selectively blocked with chemical inhibitors in order to verify their causative roles for directing endothelial differentiation (i.e., integrin $\alpha_v\beta_3$ inhibitor; P11 and integrin α_1 inhibitor; obtustatin). We also included a positive control where soluble VEGF was added in media. As shown in Figure 2B, addition of VEGF improved the degree and connectivity of vasculogenesis in GH hydrogel matrix. However, inhibitor treatment (P11 or obtustatin) attenuated the material-driven effect. P11-treated cells stayed rounded for 15 days, while obtustatin-treated cells tended to aggregate into large clumps with limited vessel sprouting. These results suggest that both integrins α_1 and $\alpha_v \beta_3$ are crucial in driving endothelial differentiation of mMSCs in gelatin hydrogels.

2.3. In Vitro Endothelial Differentiation of Patient-Derived MSCs in GH Hydrogels

Because we confirmed that material-mediated engagement of integrins triggered mMSC differentiation into endothelial cells, we set out to demonstrate translation to hMSCs from representative human donors. Patient-derived hMSCs (donor 1–3; >65 years old-patients), were harvested from bone marrow biopsies, and characterized for purity of isolation. For in vitro experiments, hMSCs were encapsulated in GH gel matrix, and their viability, spreading, and endothelial differentiation of hMSCs were investigated. As seen in Figure 3A,B, live/dead staining images demonstrated excellent survival (>80%) of hMSCs in both GH-7-L and GH-7-H gels throughout the culture period. hMSCs appeared rounded at day 1, but began spreading through the GH matrix over 14 days. Although there was a certain degree of donor-to-donor variation, GH-7-L with lower stiffness revealed better well-elongated and interconnected cell morphology than GH-7-H. The reason why hMSCs cultured in GH-7-L gels spread more is likely due to a faster degradation rate of cross-linked gelatin matrix, which leaves more room for cell spreading and nutrient/oxygen transport. In addition, other previous studies also demonstrated that lower cross-linking density resulted in an increase in mesh size, swelling degree, and degradation rate, thus enhanced cellular activities such as cell migration and growth in 3D cell culture system [31,32].

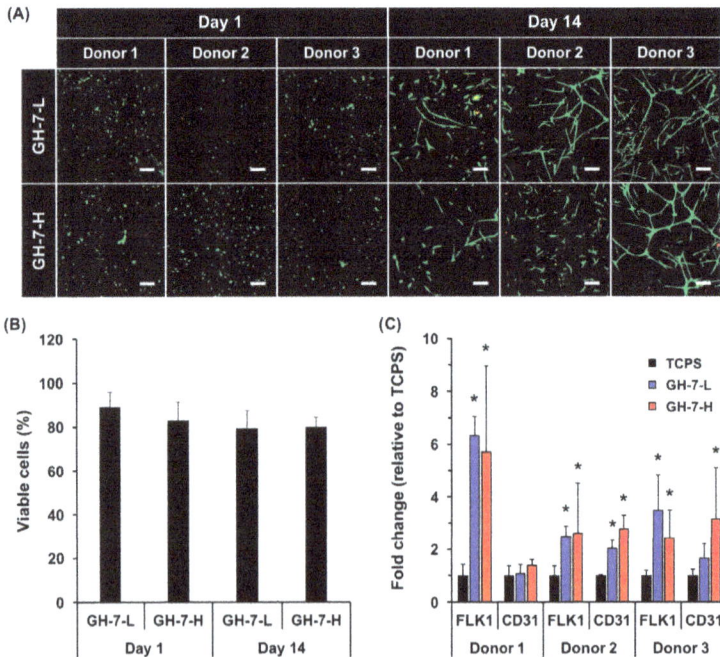

Figure 3. In vitro endothelial differentiation of patient-derived MSCs cultured in GH hydrogels. Live/dead staining images of hMSCs in GH-7-L and GH-7-H gels on days 1 and 14 post culture. Scale bars = 100 μm (**A**); Quantification of viable cells (%) at day 1 and 14 (**B**); mRNA expression levels of endothelial cell markers (FLK1 and CD31) in hMSCs determined by qRT-PCR after 21 days of culture in GH gels. As a control, the same number of cells was seed on TCPS. * $p < 0.05$ vs. TCPS ($n = 3$) (**C**).

To verify endothelial differentiation of hMSCs in GH hydrogels, we analyzed gene expression of vascular-endothelial lineage markers (i.e., FLK1 and CD31). For comparison, the same density of cells was also cultured on TCPS as a control. Overall, the expression of endothelial markers in hMSCs cultured in GH gel was significantly upregulated (2.4–6.3-fold for FLK1 and 1.1–3.2-fold CD31, donor 1−3) as compared to TCPS control (Figure 3C). However, no significant difference between GH-7-L and GH-7-H was observed. Taken together, our results demonstrate that the GH hydrogels can promote endothelial differentiation of patient-derived hMSCs in vitro without addition of pro-angiogenic supplements, which is comparable to what was shown with mMSCs [23].

2.4. In Vivo Vascularization of hMSCs Delivered with GH Hydrogel

Finally, to examine if hMSCs (donor 1−3) in GH hydrogels induced angiogenesis in vivo, we subcutaneously delivered cells in GH gel loaded on PVA scaffolds in immunodeficient mice [23]. The use of immunodeficient mice was necessary in order to avoid severe immune response and eventual rejection of hMSCs derived from unmatched donor to recipient species. Two weeks post-implantation, mice were perfused with a heparinized fluorescent microbead solution for imaging and quantification of perfusable vascular formation. Figure 4A shows fluorescence images of surface and cross-section of scaffolds by fluorescent angiography, and Figure 4B shows the quantification results on relative ratio of blood vessel area % to control. While the non-crosslinked control showed limited vascularization both on the surface and cross-section of the implants, hMSCs in cross-linked GH-7 gels substantially increased vessel formation. In particular, GH-7-H hydrogels most induced blood vessel formation (2.3-fold for surface and 1.9-fold for cross-section, * $p < 0.05$) compared to the control. When GH hydrogels were incubated in vitro with the collagenase as one of the matrix metalloproteinase (MMP) existing in the body, it was found that GH-7-H hydrogels was relatively stable than GH-7-L hydrogel. Accordingly, we speculate that the GH-7-H hydrogel provided more effective biomechanical structure than the non-crosslinked control and GH-7-L gel for cell survival/retention, and consequently supported functional vascularization of hMSCs.

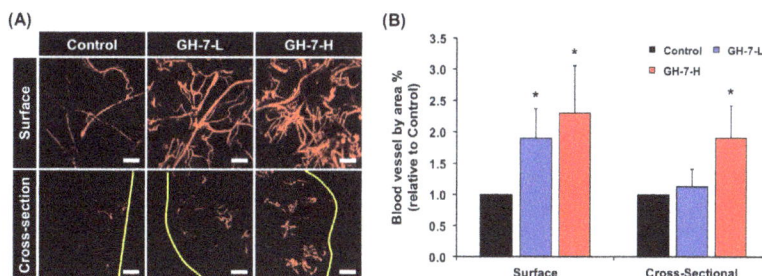

Figure 4. In vivo vascularization of hMSCs with GH hydrogels subcuntaneously delivered into nude mice. Representative images (surface and cross-section) of perfusable vasculature from delivered hMSCs in the GH gel loaded on polyvinyl alcohol (PVA) implants at two weeks post implantation. Yellow lines mark the surface boundaries of implants, and scale bars indicate 200 μm (**A**); Relative ratio of functional blood vessels by crosslinked GH-7-L and GH-7-H gels compared to non-crosslinked GH control (ratio = 1). * $p < 0.05$ vs. Control ($n = 3$, from mixed donor cell groups) (**B**).

3. Materials and Methods

3.1. Materials

Gelatin (from porcine skin, Type A, >300 bloom), 3-(4-hydroxyphenyl) propionic acid (HPA), 1-ethyl-3-(3-dimethylaminopropyl)-carbodiimide (EDC), N-hydroxysuccinimide (NHS), hydrogen peroxide (H_2O_2), and horseradish peroxidase (HRP, type VI, 250-330 units/mg solid) were purchased from Sigma-Aldrich (St. Louis, MO, USA). Dimethylformamide (DMF) was obtained from Junsei (Junsei, Tokyo, Japan).

HRP-reactive GH polymer was synthesized by conjugating HPA to the gelatin backbone as previously described [23]. In brief, HPA (20 mmol) was activated with EDC (20 mmol) and NHS (27.8 mmol) in 15 mL of co-solvent (volume ratio of deionized water and DMF = 3:2). After HPA activation for 1 h, the mixture was added to the pre-heated gelatin solution (5 g in 150 mL of deionized water). The reaction was carried out at 40 °C for 24 h. The resulting solution was purified using a dialysis membrane (MWCO = 3.5 kDa) against deionized water for 3 days, and lyophilized to obtain the GH polymer. The conjugated HPA amount of GH polymer was measured by UV–VIS spectrophotometer (V-750, Jasco, Japan) at 275 nm, and determined to be 143 µmol/g of polymer for this study.

3.2. Preparation and Characterization of GH Hydrogels with Different Mechanical Stiffness

GH hydrogels were prepared by simply mixing aqueous GH polymer solution in presence of HRP and H_2O_2. The pre-heated GH solution (7 wt %) at 40 °C was divided into two aliquots: (1) HRP (2.5 µg/mL) was added to one aliquot, and (2) H_2O_2 (0.005 wt % for GH-7-L and 0.006 wt % for GH-7-H) was added to another aliquot (volume ratio of GH:HRP and GH:H_2O_2 = 9:1, respectively). Lastly, two aliquots were mixed to generate in situ cross-linked GH hydrogels.

For characterization of gelation kinetics and mechanical stiffness of GH hydrogels formed with different H_2O_2 concentrations, the time-sweep elastic (G′) and viscous (G″) modulus of hydrogels (300 µL) were measured at 37 °C for 5 min using an Advanced Rheometer GEM-150-050 (Bohlin Instruments, East Brunswick, NJ, USA) in oscillation mode (strain = 0.01% and frequency = 0.1 Hz, gap = 0.5 mm).

For in vitro proteolytic degradation testing, GH-7-L and GH-7-H gels were formed in a microtube, and incubated in 1 mL PBS (pH 7.4) with or without 0.4 µg/mL of collagenase (type II, Sigma-Aldrich). At a predetermined time point, the weight of each hydrogel was recorded after removing media, and fresh media was then added for the next time point. The degradation degree of hydrogels was determined by measuring the weight change of initially formed gels with respect to the degraded gels at each time point [20].

3.3. In Vitro 3D Culture of hMSC in GH Hydrogels

hMSCs were isolated from the bone marrow of male patients (>65 in age) free from any blood disorders and cancer (hMSC harvest approved by Vanderbilt University Medical Center IRB#150133 February, 2015). FACS was used to isolate hMSCs that are CD14-/CD20-/CD34-/CD73+/ CD90+/CD105+. Cells (donor 1−3) were suspended in GH + HRP solution at the concentration of 2×10^6 cells/mL, and this suspension was then mixed with GH + H_2O_2 solution to make GH hydrogels incorporating cells. As a control, the same number of cells was also seeded on tissue culture plate (TCPS) without GH hydrogel. After GH hydrogels were formed on the well plate, DMEM supplemented with 10% FBS and 1% penicillin-streptomycin was added, and media was changed every 1–2 days. For the inhibition study, murine MSCs (mMSCs, GIBCO) were used with same conditions and procedures as described above. P11 (EMD Millipore, Billerica, MA, USA) was used at 10 µM, obtustatin (Tocris Biosciences, Bristol, UK) at 10 nM, and mVEGF (Sino Biological Inc., North Wales, PA, USA) at 50 ng/mL as supplements [33,34]. The cellular morphological changes of cells cultured in GH hydrogels for 15 days were observed by a Nikon Eclipse Ti microscope.

For live/dead staining assay, hMSCs (donor 1–3) cultured in GH-7-L and GH-7-H hydrogels for 1 and 14 days were stained with media containing 1 µg/mL propidium iodide (Sigma-Aldrich, St. Louis, MO, USA) and 1 µM calcein AM (Invitrogen, Carlsbad, CA, USA) for 15 min, and then the live/dead cells were identified using a Zeiss 710 confocal laser microscope.

For quantitative polymerase chain reaction (qRT-PCR) analysis, cells encapsulated within GH hydrogels were homogenized with the Trizol reagent (Life Technologies) mixed with chloroform (volume ratio of Trizol and chloroform = 1:5), and phase-separated by centrifugation (15 min, 4 °C). The aqueous phase containing RNA was isolated with RNeasy columns (Bio-Rad, Hercules, CA, USA) according to the manufacturer's instructions. cDNA was synthesized using a high-capacity cDNA reverse transcriptase kit (Applied Biosystems, Life Technologies, Foster City, CA, USA), and qRT-PCR was performed with a SYBR Green master mix (Bio-Rad) with 15–20 ng cDNA and 500 nM each of forward and reverse primers, using a CFX Real-Time PCR System (Bio-Rad). The qRT-PCR protocol included: 95 °C for 3 min, followed by 40 cycles of denaturation at 95 °C for 30 s, annealing at 58 °C for 30 s, and extension at 72 °C for 30 s. Expression of each gene of interest was normalized to expression of glyceraldehyde 3-phosphate dehydrogenase (GAPDH) as a housekeeping gene, generating the ΔC_t value, and expression of $2^{-\Delta\Delta Ct}$ relative to the TCPS control with $n \geq 3$ biological replicates for each experiment is reported. Primer sequences are listed in Table 1, and only those that showed single, specific amplicons were used for qRT-PCR experiments.

Table 1. Primer sequences used for quantitative real-time polymerase chain reaction (qRT-PCR).

Gene	Accession Number	Forward Primer (5′–3′)	Reverse Primer (3′–5′)	Species
Integrin α_1	NM_001033228.3	TCAGTGGAGAGCAGATCGGA	CCCACAGGGCTCATTCTTGT	Mouse
Integrin α_v	NM_008402.3	GTGCCAGCCCATTGAGTTTG	TGGAGCACAGGCCAAGATTT	Mouse
Integrin β_3	NM_016780.2	GCCTGGTGCTCAGATGAGACT	GATCTTCGAATCATCTGGCCG	Mouse
ERK1	NM_011952.2	CAACCCAAACAAGCGCATCA	AGGAGCAGGACCAGATCCAA	Mouse
GAPDH	NM_001289726	TGAAGCAGGCATCTGAGGG	CGAAGGTGGAAGAGTGGGAG	Mouse
FLK1	NM_002253.2	GAGGGGAACTGAAGACAGGC	GGCCAAGAGGCTTACCTAGC	Human
CD31	NM_000442.4	CCAAGCCCGAACTGGAATCT	CACTGTCCGACTTTGAGGCT	Human
GAPDH	NM_002046.4	GCACCGTCAAGGCTGAGAAC	TGGTGAAGACGCCAGTGGA	Human

3.4. hMSC Delivery in GH Hydrogels on Polyvinyl Alcohol (PVA) Scaffolds In Vivo

The animal experiment was approved by Vanderbilt Institutional Animal Care and Use Committee (IACUC) in accordance with the NIH Guide for the Care and Use of Laboratory Animals (IACUC number: M1600003 approved on 2 June 2015). GH polymer (7 wt %) and H_2O_2 (0.005 and 0.006 wt %) were dissolved in DMEM media as described above, while a constant HRP concentration (2.5 µg/mL) was used for all conditions. The procedures for in vivo animal study are comparable to that previously reported [23]. hMSCs (5×10^5 cells)-containing GH hydrogel solutions in a total volume of 60 µL were loaded on porous PVA scaffolds (6 mm in diameter, Medtronics, Dublin, Ireland). As a control, porous PVA scaffolds loaded with non-crosslinked GH solution containing hMSCs were implanted. The gel-scaffold complexes were then subcutaneously implanted on the ventral side of immunodeficient NU/J mice (male, five-months-old) for two weeks. A longitudinal incision (15 mm) was made on the ventral side of mice, and three different gel-scaffold complexes (donor 1−3) were inserted into individual subcutaneous pockets. The skin incision was closed with sutures.

At two-weeks post implantation, mice were perfused under heavy, near-lethal level of anesthesia with 4% isoflurane in 2 L/min oxygen. First, heparin sulfate (0.1 mg/mL) solution dissolved in PBS was injected into the left ventricle to exsanguinate through the cut inferior vena cava. Mice were then perfused with PBS containing fluorescent micro-beads (Invitrogen) for micro-angiography. Scaffolds were subsequently harvested, and analyzed for angiogenesis by micro-angiography using as described previously [35]. Fluorescence images were obtained using a Zeiss 710 confocal laser microscope. ImageJ software (National Institutes of Health; NIH, USA, Version 1.48) was used for all image preparation and analysis, including z-stacking fluorescence images and quantification.

3.5. Statistical Analysis

All results are expressed as mean $\pm$ standard deviation. Comparisons among samples in in vitro and in vivo quantitative analysis were performed by a Student's *t*-test, and $p < 0.05$ was considered statically significant.

4. Conclusions

Based on these results, we suggest in situ cross-linked GH hydrogels as a promising tissue engineering template to promote scaffold vascularization using hMSCs. Not only was it found that the material engaged with integrins to trigger endothelial differentiation, these results adds a substantial value to ongoing research in the field as perfusable vasculature can be generated with hMSCs sourced from older donors who would more likely take advantage of stem cell therapies. A future study will be designed to tune the GH hydrogel system to encapsulate more than one type of stem cell or co-culture MSCs with other somatic cell types for tissue type-specific regeneration with improved vascularization. As seen in Figure 4A, vascular networks were shown to branch out within the core of the GH hydrogel, which is a crucial requirement to maintain metabolically active tissues and organs that could be developed in more advanced co-culture experiments. Therefore, continuing and gradual improvements to the chemical design of the GH hydrogel system remain a long-term goal. In conclusion, the findings reported here illustrate an easy-to-use hydrogel system that can serve as a translatable platform technique for generating stem cell-derived endothelialization for future tissue engineering therapies.

Acknowledgments: This study was financially supported by the Basic Science Research Program through the National Research Foundation of Korea (NRF) funded by the Ministry of Science, ICT & Future Planning (NRF-2016M3A9E9941743 and NRF-2015R1A2A1A14027221).

Author Contributions: Yunki Lee, Ki Dong Park, and Hak-Joon Sung conceived and designed the study; Yunki Lee, Daniel A. Balikov, and Sue Hyun Lee performed the in vitro and in vivo experiments; Yunki Lee, Daniel A. Balikov, Jung Bok Lee, Seung Hwan Lee, and Jong Hun Lee performed the data analysis; Hak-Joon Sung provided guidance on the whole study. Yunki Lee wrote the paper. All authors read and approved the manuscript.

Conflicts of Interest: The authors declare no conflict of interest.

References

1. Han, H.; Ning, H.; Liu, S.; Lu, Q.; Fan, Z.; Lu, H.; Lu, G.; Kaplan, D.L. Silk Biomaterials with Vascularization Capacity. *Adv. Funct. Mater.* **2016**, *26*, 421–432. [CrossRef] [PubMed]
2. Chuang, C.H.; Lin, R.Z.; Tien, H.W.; Chu, Y.C.; Li, Y.C.; Melero-Martin, J.M.; Chen, Y.C. Enzymatic regulation of functional vascular networks using gelatin hydrogels. *Acta Biomater.* **2015**, *19*, 85–99. [CrossRef] [PubMed]
3. Levenberg, S.; Rouwkema, J.; Macdonald, M.; Garfein, E.S.; Kohane, D.S.; Darland, D.C.; Marini, R.; van Blitterswijk, C.A.; Mulligan, R.C.; D'Amore, P.A.; Langer, R. Engineering vascularized skeletal muscle tissue. *Nat. Biotechnol.* **2005**, *23*, 879–884. [CrossRef] [PubMed]
4. Griffith, L.G.; Naughton, G. Tissue engineering—current challenges and expanding opportunities. *Science* **2002**, *295*, 1009–1014. [CrossRef] [PubMed]
5. Lovett, M.; Lee, K.; Edwards, A.; Kaplan, D.L. Vascularization strategies for tissue engineering. *Tissue Eng. Part B* **2009**, *15*, 353–370. [CrossRef] [PubMed]
6. Linh, N.T.; Abueva, C.D.; Lee, B.T. Enzymatic in situ formed hydrogel from gelatin-tyramine and chitosan-4-hydroxylphenyl acetamide for the co-delivery of human adipose-derived stem cells and platelet-derived growth factor towards vascularization. *Biomed. Mater.* **2017**, *12*, 015026. [CrossRef] [PubMed]
7. Eke, G.; Mangir, N.; Hasirci, N.; MacNeil, S.; Hasirci, V. Development of a UV crosslinked biodegradable hydrogel containing adipose derived stem cells to promote vascularization for skin wounds and tissue engineering. *Biomaterials* **2017**, *129*, 188–198. [CrossRef] [PubMed]
8. Martino, M.M.; Briquez, P.S.; Guc, E.; Tortelli, F.; Kilarski, W.W.; Metzger, S.; Rice, J.J.; Kuhn, G.A.; Muller, R.; Swartz, M.A.; et al. Growth factors engineered for super-affinity to the extracellular matrix enhance tissue healing. *Science* **2014**, *343*, 885–888. [CrossRef] [PubMed]
9. Li, B.; Ogasawara, A.K.; Yang, R.; Wei, W.; He, G.W.; Zioncheck, T.F.; Bunting, S.; de Vos, A.M.; Jin, H. KDR (VEGF Receptor 2) Is the Major Mediator for the Hypotensive Effect of VEGF. *Hypertension* **2002**, *39*, 1095–1100. [CrossRef] [PubMed]
10. Martin, A.; Komada, M.R.; Sane, D.C. Abnormal angiogenesis in diabetes mellitus. *Med. Res. Rev.* **2003**, *23*, 117–145. [CrossRef] [PubMed]

11. Petreaca, M.; Martins-Green, M. The Dynamics of Cell-ECM Interactions, with Implications for Tissue Engineering. *Princ. of Tissue Eng.* **2014**, 161–187.
12. Eming, S.A.; Hubbell, J.A. Extracellular matrix in angiogenesis: Dynamic structures with translational potential. *Exp. Dermatol.* **2011**, *20*, 605–613. [CrossRef] [PubMed]
13. Rico, P.; Mnatsakanyan, H.; Dalby, M.J.; Salmerón-Sánchez, M. Material-Driven Fibronectin Assembly Promotes Maintenance of Mesenchymal Stem Cell Phenotypes. *Adv. Funct. Mater.* **2016**, *26*, 6563–6573. [CrossRef]
14. Sun, H.; Zhu, F.; Hu, Q.; Krebsbach, P.H. Controlling stem cell-mediated bone regeneration through tailored mechanical properties of collagen scaffolds. *Biomaterials* **2014**, *35*, 1176–1184. [CrossRef] [PubMed]
15. Park, K.M.; Gerecht, S. Hypoxia-inducible hydrogels. *Nat. Commun.* **2014**, *5*, 4075. [CrossRef] [PubMed]
16. Fu, J.; Wiraja, C.; Muhammad, H.B.; Xu, C.; Wang, D.A. Improvement of endothelial progenitor outgrowth cell (EPOC)-mediated vascularization in gelatin-based hydrogels through pore size manipulation. *Acta Biomater.* **2017**, *58*, 225–237. [CrossRef] [PubMed]
17. Dalby, M.J.; Gadegaard, N.; Oreffo, R.O.C. Harnessing nanotopography and integrin–matrix interactions to influence stem cell fate. *Nat. Mater.* **2014**, *13*, 558–569. [CrossRef] [PubMed]
18. Ko, D.Y.; Shinde, U.P.; Yeon, B.; Jeong, B. Recent progress of in situ formed gels for biomedical applications. *Prog. Polym. Sci.* **2013**, *38*, 672–701. [CrossRef]
19. Sivashanmugam, A.; Arun Kumar, R.; Vishnu Priya, M.; Nair, S.V.; Jayakumar, R. An overview of injectable polymeric hydrogels for tissue engineering. *Eur. Polym. J.* **2015**, *72*, 543–565. [CrossRef]
20. Lee, Y.; Bae, J.W.; Lee, J.W.; Suh, W.; Park, K.D. Enzyme-catalyzed in situ forming gelatin hydrogels as bioactive wound dressings: Effects of fibroblast delivery on wound healing efficacy. *J. Mater. Chem. B* **2014**, *2*, 7712–7718. [CrossRef]
21. Lee, Y.; Bae, J.W.; Oh, D.H.; Park, K.M.; Chun, Y.W.; Sung, H.-J.; Park, K.D. In situ forming gelatin-based tissue adhesives and their phenolic content-driven properties. *J. Mater. Chem. B* **2013**, *1*, 2407–2414. [CrossRef]
22. Knopf-Marques, H.; Barthes, J.; Wolfova, L.; Vidal, B.; Koenig, G.; Bacharouche, J.; Francius, G.; Sadam, H.; Liivas, U.; Lavalle, P.; Vrana, N.E. Auxiliary Biomembranes as a Directional Delivery System To Control Biological Events in Cell-Laden Tissue-Engineering Scaffolds. *ACS Omega* **2017**, *2*, 918–929. [CrossRef]
23. Lee, S.H.; Lee, Y.; Chun, Y.W.; Crowder, S.W.; Young, P.P.; Park, K.D.; Sung, H.J. In Situ Crosslinkable Gelatin Hydrogels for Vasculogenic Induction and Delivery of Mesenchymal Stem Cells. *Adv. Funct. Mater.* **2014**, *24*, 6771–6781. [CrossRef] [PubMed]
24. Waterston, R.H.; Lindblad-Toh, K.; Birney, E.; Rogers, J.; Abril, J.F.; Agarwal, P.; Agarwala, R.; Ainscough, R.; Alexandersson, M.; An, P.; et al. Initial sequencing and comparative analysis of the mouse genome. *Nature* **2002**, *420*, 520–562. [PubMed]
25. Choi, B.; Loh, X.J.; Tan, A.; Loh, C.K.; Ye, E.; Joo, M.K.; Jeong, B. Introduction to In Situ Forming Hydrogels for Biomedical Applications. In *In-Situ Gelling Polymers*; Springer: Berlin, Germany, 2015; pp. 5–35.
26. Trappmann, B.; Gautrot, J.E.; Connelly, J.T.; Strange, D.G.; Li, Y.; Oyen, M.L.; Cohen Stuart, M.A.; Boehm, H.; Li, B.; Vogel, V.; et al. Extracellular-matrix tethering regulates stem-cell fate. *Nat. Mater.* **2012**, *11*, 642–649. [CrossRef] [PubMed]
27. Engler, A.J.; Sen, S.; Sweeney, H.L.; Discher, D.E. Matrix elasticity directs stem cell lineage specification. *Cell* **2006**, *126*, 677–689. [CrossRef] [PubMed]
28. Oswald, J.; Boxberger, S.; Jørgensen, B.; Feldmann, S.; Ehninger, G.; Bornhäuser, M.; Werner, C. Mesenchymal Stem Cells Can Be Differentiated Into Endothelial Cells In Vitro. *Stem Cells* **2004**, *22*, 377–384. [CrossRef] [PubMed]
29. Kubota, Y.; Kleinman, H.K.; Martin, G.R.; Lawley, T.J. Role of laminin and basement membrane in the morphological differentiation of human endothelial cells into capillary-like structures. *J. Cell. Biol.* **1988**, *107*, 1589–1598. [CrossRef] [PubMed]
30. Wang, C.H.; Wang, T.M.; Young, T.H.; Lai, Y.K.; Yen, M.L. The critical role of ECM proteins within the human MSC niche in endothelial differentiation. *Biomaterials* **2013**, *34*, 4223–4234. [CrossRef] [PubMed]
31. Nichol, J.W.; Koshy, S.T.; Bae, H.; Hwang, C.M.; Yamanlar, S.; Khademhosseini, A. Cell-laden microengineered gelatin methacrylate hydrogels. *Biomaterials* **2010**, *31*, 5536–5544. [CrossRef] [PubMed]
32. Choi, M.-Y.; Kim, J.-T.; Lee, W.-J.; Lee, Y.; Park, K.M.; Yang, Y.-I.; Park, K.D. Engineered extracellular microenvironment with a tunable mechanical property for controlling cell behavior and cardiomyogenic fate of cardiac stem cells. *Acta Biomater.* **2017**, *50*, 234–248. [CrossRef] [PubMed]

Int. J. Mol. Sci. **2017**, *18*, 1705

33. Marcinkiewicz, C.; Weinreb, P.H.; Calvete, J.J.; Kisiel, D.G.; Mousa, S.A.; Tuszynski, G.P.; Lobb, R.R. Obtustatin: A potent selective inhibitor of alpha1beta1 integrin in vitro and angiogenesis in vivo. *Cancer Res.* **2003**, *63*, 2020–2023. [PubMed]

34. Choi, Y.; Kim, E.; Lee, Y.; Han, M.H.; Kang, I.C. Site-specific inhibition of integrin alpha v beta 3-vitronectin association by a ser-asp-val sequence through an Arg-Gly-Asp-binding site of the integrin. *Proteomics* **2010**, *10*, 72–80. [CrossRef] [PubMed]

35. Zachman, A.L.; Crowder, S.W.; Ortiz, O.; Zienkiewicz, K.J.; Bronikowski, C.M.; Yu, S.S.; Giorgio, T.D.; Guelcher, S.A.; Kohn, J.; Sung, H.J. Pro-angiogenic and anti-inflammatory regulation by functional peptides loaded in polymeric implants for soft tissue regeneration. *Tissue Eng. Part A* **2013**, *19*, 437–447. [CrossRef] [PubMed]

International Journal of
Molecular Sciences

MDPI

Article

Collagen-Based Medical Device as a Stem Cell Carrier for Regenerative Medicine

Léa Aubert [1,2], Marie Dubus [1,3], Hassan Rammal [1,3], Camille Bour [1,3], Céline Mongaret [1,2,4],
Camille Boulagnon-Rombi [5], Roselyne Garnotel [6], Céline Schneider [7], Rachid Rahouadj [8],
Cedric Laurent [8], Sophie C. Gangloff [1,2], Frédéric Velard [1,3], Cedric Mauprivez [1,3,†] and
Halima Kerdjoudj [1,3,*,†]

1   Equipe d'Accueil 4691, Biomatériaux et Inflammation en Site Osseux (BIOS), Pôle Santé, UFR d'Odontologie,
    SFR-CAP Santé (FED 4231), Université de Reims Champagne-Ardenne, 1 Avenue du Maréchal Juin,
    51100 Reims, France; l.aubert@outlook.fr (L.A.); marie.dubus@hotmail.com (M.D.);
    hkrammal@hotmail.com (H.R.); camille.bour@univ-reims.fr (C.B.); cmongaret@chu-reims.fr (C.M.);
    sophie.gangloff@univ-reims.fr (S.C.G.); frederic.velard@univ-reims.fr (F.V.);
    mauprivezcedric@gmail.com (C.M.)
2   UFR de Pharmacie, Université de Reims Champagne-Ardenne, 51100 Reims, France
3   UFR d'Odontologie, Université de Reims Champagne-Ardenne, 51100 Reims, France
4   Pole Pharmacie Pharmacovigilance CHU Reims, 51100 Reims, France
5   Laboratoire d'Anatomie et Cytologie Pathologiques, Centre Hospitalier-Universitaire, Hôpital Robert Debré,
    51100 Reims, France; cboulagnon-rombi@chu-reims.fr
6   CNRS, UMR 7369, Medyc, Université de Reims Champagne-Ardenne, 51100 Reims, France;
    roselyne.garnotel@univ-reims.fr
7   Equipe d'Accueil A 3795, Groupe d'Étude des Géomatériaux et Environnement Naturels,
    Anthropiques et Archéologiques (GEGENAA), Université de Reims Champagne Ardenne, 51100 Reims,
    France; celine.schneider@univ-reims.fr
8   CNRS, UMR 7563, LEMTA, Université de Lorraine, 54500 Vandœuvre-Lès-Nancy, France;
    rachid.rahouadj@univ-lorraine.fr (R.R.); cedric.laurent@univ-lorraine.fr (C.L.)
*   Correspondence: halima.kerdjoudj@univ-reims.fr
†   These authors contributed equally to this work.

Received: 26 September 2017; Accepted: 18 October 2017; Published: 21 October 2017

Abstract: Maintenance of mesenchymal stem cells (MSCs) requires a tissue-specific microenvironment
(i.e., niche), which is poorly represented by the typical plastic substrate used for two-dimensional
growth of MSCs in a tissue culture flask. The objective of this study was to address the potential
use of collagen-based medical devices (HEMOCOLLAGENE®, Saint-Maur-des-Fossés, France)
as mimetic niche for MSCs with the ability to preserve human MSC stemness in vitro. With a
chemical composition similar to type I collagen, HEMOCOLLAGENE® foam presented a porous
and interconnected structure (>90%) and a relative low elastic modulus of around 60 kPa. Biological
studies revealed an apparently inert microenvironment of HEMOCOLLAGENE® foam, where 80% of
cultured human MSCs remained viable, adopted a flattened morphology, and maintained their
undifferentiated state with basal secretory activity. Thus, three-dimensional HEMOCOLLAGENE®
foams present an in vitro model that mimics the MSC niche with the capacity to support viable and
quiescent MSCs within a low stiffness collagen I scaffold simulating Wharton's jelly. These results
suggest that haemostatic foam may be a useful and versatile carrier for MSC transplantation for
regenerative medicine applications.

Keywords: stem cell niche; medical device; regenerative medicine; biocompatibility; paracrine activities

1. Introduction

In the organism, adult stem cells guarantee the maintenance and repair of tissues and organs. Among them, mesenchymal stem cells (MSCs) are emerging as hopeful candidates for cell-based therapy of numerous diseases (i.e., myocardial infarction, Crohn's disease, Graft versus host disease, osteoarthritis, etc.) [1,2]. Indeed, along with their differentiation potential and the production of several humoral factors, MSCs are thought to exert regenerative effects by increasing healing rates, modulating inflammation and immune response, promoting angiogenesis, and enhancing tissue remodelling [3].

Adult tissues, such as bone marrow and adipose tissue, have proven to be sources for effective MSCs; however, several disadvantages exist, including availability and invasive and painful procedures required for their isolation. In addition, the health status and the age of the donor may affect MSC expansion capability, function, and/or survival after transplantation [4]. Consequently, scientists are looking for stable, safe, and highly accessible stem cell sources with great potential for regenerative medicine. Since 2004, human MSCs derived from umbilical cord Wharton's jelly (WJ-MSCs) have become popular in the regenerative medicine community, due to the ease of cell harvesting from scraps of perinatal tissues, which is neither painful nor invasive [5,6]. In addition, their primitive nature sheds light on their significant high proliferation capability and increased in vitro expansion ability when compared to adult MSCs [7].

In the natural environment, stem cells reside in a niche, a microenvironment regulating MSC quiescence, self-renewal, and differentiation. MSCs are isolated from their niche (typically composed of extracellular matrix, niche cells, and secreted stimulants such as growth factors, chemokines, and cytokines), and are traditionally expanded in culture plastic in a two-dimensional (2D) monolayer. However, one of the major limitations in using MSCs ex vivo is that they lose their fundamental properties, such as quiescence and multipotency, limiting their usefulness. The preservation of MSCs in the natural environment is unique, and researchers have been trying to understand how nature maintains the MSCs' fundamental features.

Advances in material sciences have garnered interest in providing a temporary three-dimensional environment (scaffold) for MSCs, allowing the diffusion of nutrients to ensure cell survival [8]. However, success of these structures is strongly dependent on their structural and mechanical features; ensuring stem cell interactions with the surrounding environment [9,10]. Naturally-derived polymers, such as collagen, are appealing for biological applications due to their enzymatic degradation and proven safety through long-term applications in clinical trials [11].

In light of these thoughts, we address the potential use of collagen-based medical devices (HEMOCOLLAGENE®) as WJ-MSCs in vitro niche model. Composition, structure, and mechanical properties of HEMOCOLLAGENE® foam were first investigated before assessment of WJ-MSCs behaviour and morphology once seeded in HEMOCOLLAGENE®. The potency of HEMOCOLLAGENE® as an inert WJ-MSCs niche was confirmed through MSC basal cytokines production and maintenance of their undifferentiated state, revealing that HEMOCOLLAGENE® presents suitable properties as a scaffold for WJ-MSCs culture.

2. Results and Discussion

Collagen-based foams are the most used materials in medical applications. In the present study, we selected bovine collagen foam, manufactured by Septodont (Saint-Maur-des-Fossés, France), as a carrier for stem cell-based therapy. The company´s main product, a foam-like collagen scaffold (HEMOCOLLAGENE®), has already been used in dental surgery as a haemostatic device since 1999. HEMOCOLLAGENE® was first characterized in terms of chemical composition. Fourier transform infrared (FTIR) spectroscopy analysis revealed three main peaks at 1237 cm^{-1}, 1544 cm^{-1}, and 1630 cm^{-1} (Figure 1, red line) characteristics of amide III, II, and I, respectively, matching with the spectra of type I collagen (Figure 1, black line). Additional FTIR spectra attributed to collagen were given in Table S1 in the Supporting Information. A further deep biochemical characterization, performed by ion exchange chromatography, showed that HEMOCOLLAGENE® foam had close similarities with type I collagen with a high amount of glycine, followed by proline, alanine, and hydroxyproline in each specimen (Table 1).

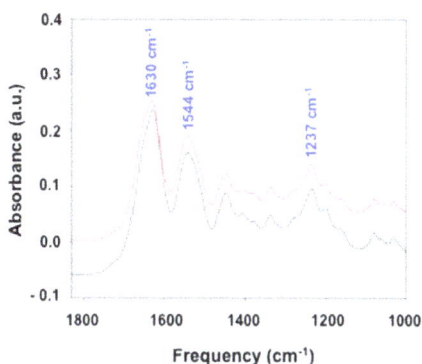

Figure 1. Composition of HEMOCOLLAGENE®. Spectra obtained by Fourier transform infrared (FTIR) spectroscopy showing similarities of HEMOCOLLAGENE® foam (red line) with type I collagen (black line).

Table 1. Amino acid content assessed by ion exchange chromatography (residue numbers/1000).

Amino-Acid	HEMOCOLLAGENE®	Type I Collagen (Reference)
Hydroxyproline	91.2	101.4
Aspartic acid	49.7	49.7
Threonine	19.4	18.9
Serine	33.7	28.8
Glutamic acid	75.2	72.1
Proline	126.8	117.3
Glycine	322.5	327.5
Alanine	106.5	114.8
Valine	23.8	20.9
Methionine	3.1	7.5
Isoleucine	13.5	11.9
Leucine	26.9	25.8
Tyrosine	2.0	2.0
Phenylalanine	14.7	13.9
Hydroxylysine	6.3	6.5
Lysine	24.6	26.3
Histidine	5.3	5
Arginine	54.6	49.7

From a structural viewpoint, HEMOCOLLAGENE® exhibited a high interconnectivity and micro/macroporous morphology as depicted in the scanning electron microscopy (SEM) and cross-histological section results (Figure 2A,B). Macro-pore diameters were estimated to be 150 to 500 μm. The total porosity assessed by the mercury intrusion porosimetry method [12] was high, about 95%. The pore access distribution from 165 μm to 3 nm showed a mean radius corresponding to the main peak on Figure 2C of around 9 μm. The pore threshold determined by the intersection of the two tangents of the cumulative intrusion curve was around 6.5 μm (Figure 2D). These values, close to each other, showed that the porous network of HEMOCOLLAGENE® was unimodal and that the pore threshold was significant. To our knowledge, in regenerative medicine applications, porous structures with radii around 150 μm and high porosity (>90%), along with well-interconnected open pores, are required for allowing cell infiltration and efficient nutrient and oxygen diffusion into the structure [8].

In addition to having suitable architecture, 3D cell carrier foam must exhibit appropriate physical properties to accommodate cell survival. Indentation experiments were used to explore the mechanical features of HEMOCOLLAGENE® foams (Figure 3). The values for the Young's modulus were 69.6 kPa and 62.3 kPa for foams at indentation rates of 1 mm/min and 10 mm/min, respectively. No significant difference was observed between the identified moduli at the different speeds, indicating no pronounced viscoelastic effect. The obtained values fall well within the values reported in the literature for 3D cell carrier foam [13,14].

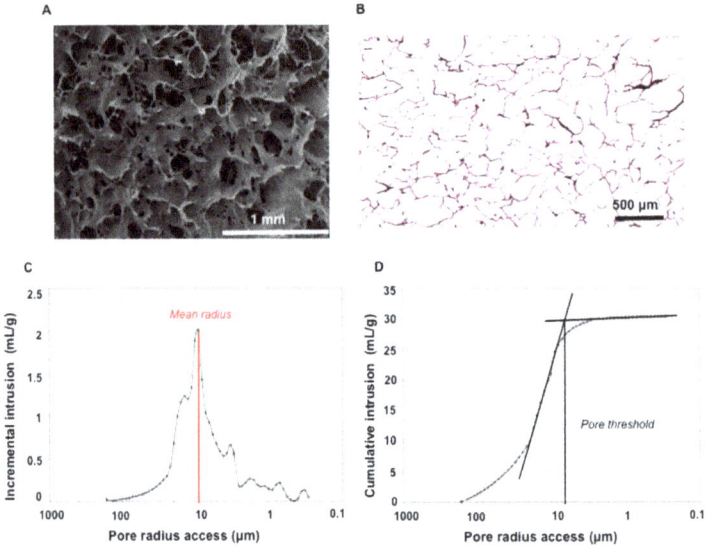

Figure 2. Structure of HEMOCOLLAGENE®. (**A**) Representative scanning electron microscopy image (SEM, scale bar indicates 1 mm); (**B**) Hematoxylin-Eosin-Saffron (HES) staining of paraffin-embedded HEMOCOLLAGENE® cross-section (scale bar indicates 500 μm); (**C,D**) Pore distribution obtained by mercury intrusion porosimetry showing the mean radius on the incremental curve and the pore threshold on the cumulative curve.

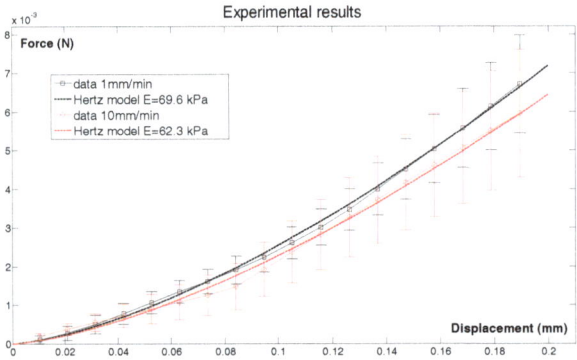

Figure 3. Mechanical features of HEMOCOLLAGENE®. Force-displacement responses obtained using indentation tests at 1 mm/min (*n* = 4) and 10 mm/min (*n* = 4). The Young's modulus has been identified using the Hertz model. No significant difference was observed between the two different testing speeds.

The above results highlighted that HEMOCOLLAGENE® foams possess the carrier-required criteria for stem cell based therapy [8]. In the following the results of assays involving the association of HEMOCOLLAGENE® foam with MSCs are presented. A sufficient number of WJ-MSCs from six human umbilical cords were expanded within approximately eight weeks. To reduce heterogeneity of the extracted cell populations [15], WJ-MSCs from three passages with fibroblast-like spindle shapes on culture plastic were used (Figure S1A,B in supporting information section). Moreover, these cells expressed the putative mesenchymal markers CD 44, CD 73, CD 90, and CD 105, but not the hematopoietic markers CD 34, CD 14, and HLA-DR (Figure S1C in the Supplementary Materials). These results, in agreement with previous studies, confirmed their MSC phenotype [16].

WJ-MSCs were injected in HEMOCOLLAGENE® foam at density of 8×10^4 cells/foam and placed in 24-well treated-chamber culture to avoid cell migration from foam to plastic. Cytocompatibility of HEMOCOLLAGENE® was firstly monitored by WST-1 (water-soluble tetrazolium salt-1) assay, DNA quantification and Zombie® labelling after 4, 7, and 10 days of culture, using independent foam for each test and time point. While WST-1 and DNA quantification did not show significant variation of measured values, Zombie® labelling revealed a cell survival rate of around 80% over the experimental time (Figure 4).

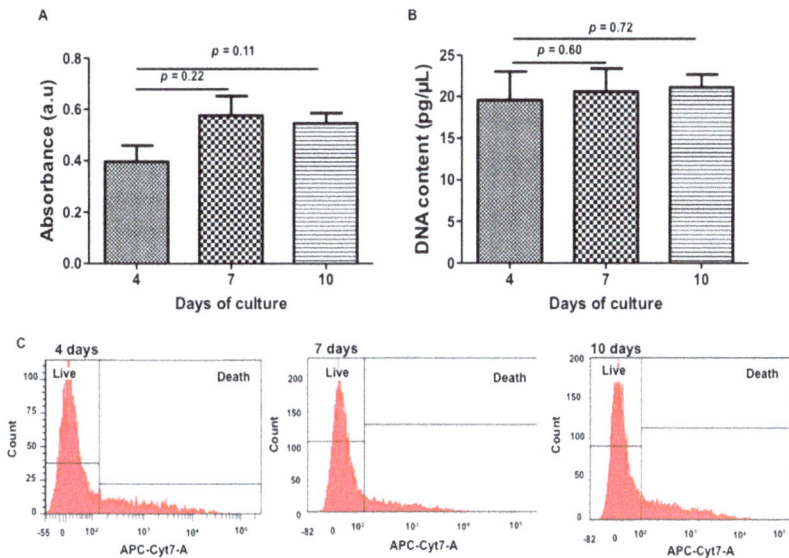

Figure 4. Cell viability over the time of the study. (**A,B**): Histogram reflecting WST-1 assay and DNA quantification, respectively; and (**C**): Flow cytometry results obtained after Zombie® labelling. Kinetic study performed after 4, 7, and 10 days of culture showing live and non-proliferating WJ-MSCs cultured within HEMOCOLLAGENE® foam (*n* = 6, Mann and Whitney test).

These results indicated that HEMOCOLLAGENE® foams are a suitable environment for stem cell culture. It is generally accepted that an increase in the WST-1 optical density at 450 nm and in DNA values reflects a proliferating state of cells; in our study, the number of WJ-MSCs in HEMOCOLLAGENE® foam remained stable over the culture time, a signature of the non-proliferating state. Moreover, the high survival rate reflected non-toxic culture conditions; meeting the basic requirements of tissue engineering scaffolds [11]. In addition to cell survival, a uniformity of cell distribution within scaffolds is required for 3D culture systems. A general WJ-MSCs distribution within HEMOCOLLAGENE® up to 10 days of culture was given by HES and Masson's trichrome staining (Figure 5). Histological cross-sections showed randomly-distributed WJ-MSCs embedded within the fibrous extracellular matrix (Figure 5A,B, yellow arrow). Few apoptotic WJ-MSCs revealed by cleaved caspase-3 immunohistochemistry were detected in the inner region of the foam (Figure 5C, red arrow), strengthening the Zombie® results. Finally, we noticed the absence of squamous cells within the HEMOCOLLAGENE® foam, the absence of glycosaminoglycane synthesis, and the absence of mineralization nodules, signature of the absence of spontaneous cell differentiation into adipose, cartilaginous and bone lineage, respectively.

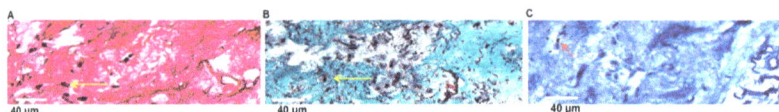

Figure 5. Cell distribution and apoptosis after 10 days of culture. (**A**,**B**) Haematoxylin-eosin-saffron (HES) and Masson's trichrome staining of paraffin-embedded cellularized HEMOCOLLAGENE® cross-sections, respectively. HES staining showing nuclei in blue (yellow arrow) and collagen in orange. Masson's trichrome staining, showing collagen in green and nuclei in brown (yellow arrow); and (**C**) cleaved caspase-3 immunohistochemistry showing few apoptotic cells (red arrow) within HEMOCOLLAGENE®. (Scale bars indicate 40 μm).

The low percentage of apoptotic/necrotic cells within foam might be attributed to the preservation of porosity and pore interconnectivity over the culture time, allowing mass transfer exchanges between foam and culture medium (i.e., nutriment diffusion and cell metabolite release). Furthermore, pore size and pore interconnectivity influence cell shape. Cells residing in open-pore scaffold exhibit an elongated shape in contrast to closed-pore scaffold where cells adopt a rounded morphology [8,17]. A deeper investigation of WJ-MSCs morphology within HEMOCOLLAGENE® foam was performed by SEM and CLSM experiments. SEM imaging showed that WJ-MSCs adopted flattened morphology within HEMOCOLLAGENE® pores (Figure 6). These observations were further supported by CLSM micrographs of Phalloidin® stained cytoskeleton, showing actin bundles of aligned long filaments and highly elongated morphology. These results clearly demonstrate that WJ-MSCs establish physical interactions with the HEMOCOLLAGENE® foam.

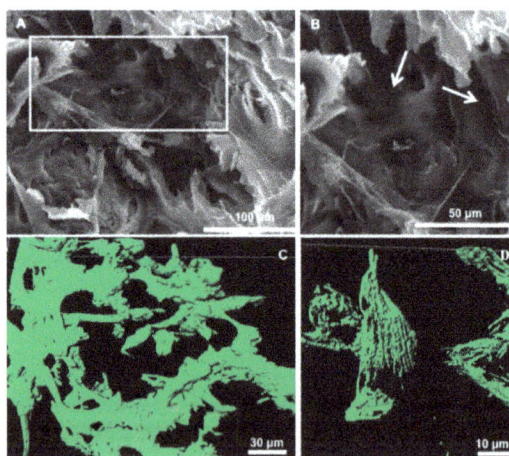

Figure 6. Cell morphology after 10 days of culture. (**A**,**B**) Over and closer (rectangle) scanning electron microscopy views (scale bars indicate 100 and 50 μm, respectively), highlighting cell distribution within pores (white arrows); (**C**,**D**) Over and closer Imaris 3D views of cytoskeleton-labelled cells (scale bars indicate 30 and 10 μm, respectively), showing actin bundles and highly-elongated cell morphology.

Cell spreading is a complicated process, which, besides biophysical cues (mechanical and structure), is also associated with the chemical composition of materials [8]. MSCs have been shown to possess a strong attachment to type I collagen through arginine-glycine-aspartate (RGD) binding domains with the cell membrane integrins [18,19]. Gathering HEMOCOLLAGENE® composition and porosity (Figures 1 and 2), the expected high cellular bond is confirmed. During experimental studies, we noticed that WJ-MSC-loaded HEMOCOLLAGENE® foams shrank drastically (Figure S2), which could be due to their enzymatic degradation. However, the lack of significant variation of the hydroxyproline rate in the supernatant (12, 16, 18, and 21 μM, after 1, 4, 7, and 10 days of culture, respectively, versus 12,680 μM for totally hydrolyzed foam), suggests a possible mechanical instability of HEMOCOLLAGENE® foams or active contractile forces generated by WJ-MSCs [20].

Mesenchymal stem cells within niches respond to biochemical, structural, and mechanical cues from their surrounding microenvironment, affecting their constitutive cytokines secretion/ self-renewal/differentiation, thus contributing to tissue homeostasis [21]. The present study was designed to determine if WJ-MSCs cultured with HEMOCOLLAGENE® foams maintain a constitutive secretory profile. Therefore, their ability to regulate and release cytokines, over the experimental time, such as IL-6, IL-8, IL-10, and vascular endothelial growth factor (VEGF), was followed at mRNA (i.e., *IL6*, *CXCL8*, *IL10*, and *VEGFA*) and protein levels by qRT-PCR and ELISA, respectively.

Among quantified cytokines, only IL-6 and IL-8 were detected in supernatants, whereas no regulation of *IL6* and *CXCL8* was noticed, suggesting a constitutive secretion of these cytokines over the kinetic study (Figure 7). While IL-6 and IL-8 releases were constant over the experimental time, we were able to distinguish a slight increase at day 7 (Figure 7A–D). Related to an anti-inflammatory profile, neither *IL10* regulation nor IL-10 release in supernatant were detected, probably due to the absence of a pro-inflammatory environment stimuli [22]. Despite a significant down-regulation of *VEGFA* at day 10, released VEGF was not detected in the culture supernatant over the time (Figure 7E), suggesting that VEGF production is under the detection threshold of the kit or VEGF is accumulated within the cell cytoplasm. Finally, as bone is composed of type I collagen, we looked for the BMP-2 release in the medium, which was not detected in the supernatant over the study time.

IL-6 and IL-8 are currently described as pro-inflammatory cytokines, but in our experiments, their basal level secretion suggests that HEMOCOLLAGENE® represents an "inert" environment for WJ-MSCs and cell supports secretory function. While the basic biological role of IL-8 is attracting and activating neutrophils, Boyden migration assays did not show significant increase in the recruitment of neutrophils by WJ-MSC loaded-HEMOCOLLAGENE® conditioned media compared to cell-free HEMOCOLLAGENE® conditioned media (Figure S3), confirming both the neutral role of HEMOCOLLAGENE® foam and constitutive IL-8 cytokine production of WJ-MSCs cultured in HEMOCOLLAGENE®. IL-6 cytokine is described to maintain the "stemness" of MSCs [23,24]. Therefore, the expression of NT5E, THY1 and ENG (genes corresponding to CD 73, CD 90, and CD 105, respectively) was followed by qRT-PCR after 4, 7, and 10 days of culture. With respect to stem cell associated markers, WJ-MSCs cultured in HEMOCOLLAGENE® did not show any gene variations over the time (Figure 8A–C), suggesting a role of HEMOCOLLAGENE® in maintaining WJ-MSCs under the undifferentiated state. Furthermore, once placed on plastic culture, WJ-MSCs were able to migrate and proliferate without any cell damage and spontaneous differentiation (Figure 8D–F).

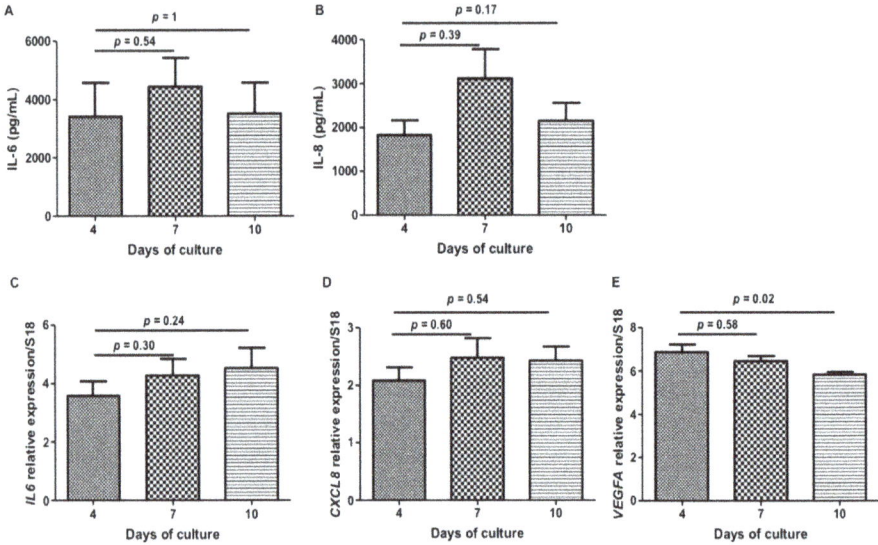

Figure 7. Cell paracrine activity over the time of the study. (**A,B**) Histograms of IL-6 and IL-8 ELISA quantification, respectively, showing non-significant production of related proteins; (**C–E**) Histograms of *IL6*, *CXCL8*, and *VEGFA* qRT-PCR analysis, showing non-significant regulation of *IL6* and *CXCL8*, and significant *VEGFA* down-regulation ($n = 6$, Mann and Whitney statistical test, $p < 0.05$ for four versus 10 days of culture).

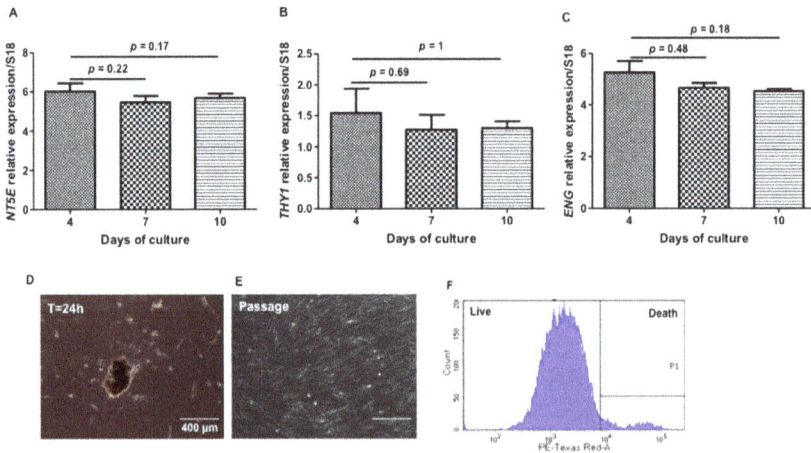

Figure 8. Cell phenotype over the time of the study. (**A–C**) Histograms of *NT5E* (CD 73), *THY1* (CD 90), and *ENG* (CD 105) qRT-PCR analysis, respectively, showing non-significant regulation of MSCs markers; (**D**) WJ-MSC migration on plastic; (**E**) WJ-MSC morphology after passage; and (**F**) flow cytometry results obtained after Zombie® labelling of amplified cells ($n = 6$, Mann and Whitney statistical test).

3. Materials and Methods

3.1. Materials

A haemostatic medical device named HEMOCOLLAGENE® foam was provided from Septodont, France. According to the manufacturer, HEMOCOLLAGENE® is made of bovine non-denatured collagen and obtained by a freeze-dried process.

3.1.1. Scanning Electron Microscopy (SEM)

Foam morphology was investigated by SEM with a LaB6 electron microscope (JEOL JSM-5400LV), on sputter-coated foam with thin gold–palladium film (JEOL ion sputter JFC 1100, Croissy Sur Seine, France). Images were acquired from secondary electrons at a primary beam energy of 10 kV.

3.1.2. Fourier Transform Infrared (FTIR)

FTIR spectra were obtained by a Fourier transform infrared-attenuated total reflection (FTIR-ATR, Vertex 70 spectrometer, Bruker, Ettlingen, Germany) using a DTGS detector. Type I collagen from bovine (medical grade; Symatese, Lyon, France) was used as control.

3.1.3. Ion Exchange Chromatography

Amino acid composition of HEMOCOLLAGENE® foam was determined by ion exchange chromatography (HITACHI 8800 analyzer, Science Tec, Tokyo, Japan). One milligram of foam was hydrolysed in HCl 6 M at 110 °C for 18 h. Dried hydrolysates obtained under nitrogen stream evaporation were then resuspended in 100 μL of a buffer composed of lithium 13.86 mM, lithium citrate 55 mM, citric acid 207 mM, ethanol 6% (*v/v*), and thiodiglycol 1% (*v/v*) pH 2.8. Type I collagen from bovine (medical grade; Symatese) was used as control. The chromatography was performed according to the manufacturer's instructions. Amino acid composition was expressed as residue numbers per 1000.

3.1.4. Mercury Intrusion Porosimetry

Pore access radii distribution and the total porosity of HEMOCOLLAGENE® foam were assessed by mercury intrusion porosimetry (Micromeretics AutoPore IV 9500, Hexton, UK). The measured pore access radius ranges from 165 μm (0.005 MPa) to 0.003 μm (274 MPa). Thus, pores larger or thinner than these sizes are not considered by this technique. The incremental curve gives the mean pore radius for which the intrusive volume is maximal. The cumulative curve allows plotting the pore threshold that corresponds to the pore access allowing filling of the main part of the porous network. When both radii are close, the pore distribution can be considered as unimodal.

3.1.5. Micro-Indentation

Mechanical properties were explored by a custom micro-indentation setup (LEMTA), as proposed in the literature [25]. HEMOCOLLAGENE® foam was placed on an electrical balance used as a force sensor, and compressed with a spherical indenter (radius $r = 0.75$ mm). Indentation speeds of 1 mm/min ($n = 4$) and 10 mm/min ($n = 4$) were used.

The force applied to the foam was then measured as a function of the applied displacement u of the spherical indenter (Figure 9). A simple Hertz contact model was used to fit with the experimental force-displacement curves using a least-square method, and the Young's modulus E of the foam was identified with this model:

$$F = \frac{16}{9} E \, r^{\frac{1}{2}} \, u^{\frac{3}{2}}$$

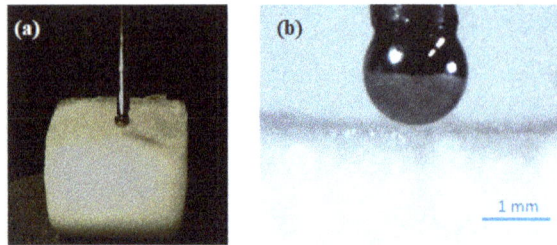

Figure 9. Micro-indentation tests performed on HEMOCOLLAGENE® foam. (**a**) Global view of the indentation test on a typical specimen (10 × 10 × 10 mm); (**b**) Zoom of the spherical indenter (radius = 0.75 mm).

3.2. Biological Experiments

3.2.1. Cell Culture

Mesenchymal stem cells were enzymatically isolated from fresh human umbilical cords obtained after full-term births as previously described [5] and amplified at a density of 3×10^3 cell/cm^2 in α-MEM culture medium (Lonza, Saint Quentin, France) supplemented with 10% decomplemented FBS, 1% Penicillin/Streptomycin/Amphotericin B, and 1% Glutamax (v/v, Gibco, Villebon-sur-Yvette, France), and maintained in a humidified atmosphere of 5% CO_2 at 37 °C with a medium change every three days. Human umbilical cord harvesting was approved ethically and methodologically by our local research institution and was conducted with informed patients (written consent, non-opposition) in accordance with the usual ethical legal regulations (Article R 1243-57). All procedures were done in accordance with our authorization and registration number DC-2014-2262 given by the National "Cellule de Bioéthique". At third passage, Wharton's jelly mesenchymal stem cells (WJ-MSCs) were characterized by flow cytometry (FACSCalibur; BD Bioscience, le Pont de Claix, France) through the expression of CD 73, CD 90, CD 44, CD 105, CD 34, CD 14, and HLA-DR (BD, le Pont de Claix, France), Table S2 in Supplementary Materials) and then used in our experimental procedure at the fourth passage.

WJ-MSCs suspension (8×10^4 cells) were injected with a 1 mL syringe and through a 21 G needle, in the middle of HEMOCOLLAGENE® foam. After 4 h of incubation at 37 °C, WJ-MSCs loaded HEMOCOLLAGENE® foams were transferred to 24-well coated plates preventing any cell migration and adhesion (Nunclon Sphera, Thermo scientific, Villebon-sur-Yvette, France). WJ-MSC-loaded HEMOCOLLAGENE® foam was cultured in 1 mL of α-MEM culture medium and changed after 1, 4, 7, and 10 days of culture. Supernatants after 4, 7, and 10 days were collected, centrifuged at $250 \times g$, and conserved at −20 °C (conditioned media).

3.2.2. WST-1

Mitochondrial activity, followed by WST-1 cell proliferation assay (Roche Diagnostics, Meylan , France), was performed after 4, 7, and 10 days of WJ-MSCs culture in HEMOCOLLAGENE® foam in accordance with the manufacturer protocol. Absorbance was measured at 440 nm using a FLUOstar Omega microplate reader (BMG Labtech, Champigny-sur-Marne, France) against a background control as blank. A wavelength of 750 nm was used as the reference wavelength.

3.2.3. DNA Quantification

DNA quantification was performed on extracted DNA after 4, 7, and 10 days of WJ-MSCs culture in HEMOCOLLAGENE® foam, using a MasterPure™ DNA Purification Kit (Epicentre, Biotechnologies, Strasbourg, France) in accordance with the manufacturer's protocol. The quantity of extracted DNA was assessed by measuring the absorbance at 260 nm (Nanodrop, Thermo Scientific,

Villebon-sur-Yvette, France) and the 260/280 nm absorbance ratio for all measured samples was between 1.8 and 2.

3.2.4. Zombie® Labeling

Cell membrane integrity was assessed by Zombie® labelling (Ozyme, Montigny-le-Bretonneux, France) in accordance with the manufacturer's protocol. Labelled cells within HEMOCOLLAGENE® foams were released using collagenase treatment for 5 min, fixed with 4% (*w/v*) paraformaldehyde (Sigma-Aldrich, Guyancourt, France), and analysed by an LSRFortessa flow cytometer (BD Bioscience, France).

3.2.5. SEM

SEM was performed after 10 days of culture on fixed WJ-MSCs loaded HEMOCOLLAGENE® foam with 2.5% (*w/v*) glutaraldehyde (Sigma-Aldrich) at room temperature for 1 h. Samples were dehydrated in graded ethanol solutions from 50% to 100% and in hexamethyldisilazane (Sigma, France) for 10 min. After air-drying at room temperature, samples were immersed in liquid nitrogen, fractured, then sputtered with a thin gold–palladium film under a JEOL JFC 1100 ion sputter and viewed using a LaB6 electron microscope (JEOL JSM-5400 LV, France). Images were acquired from secondary electrons at a primary beam energy of 20 kV.

3.2.6. Confocal Laser Scanning Microscopy (CLSM)

Confocal laser scanning microscopy (CLSM) was performed for visualization of Phalloidin®-labelled cytoskeletons after 10 days of culture of WJ-MSCs loaded HEMOCOLLAGENE® foam. Samples were fixed in 4% (*w/v*) paraformaldehyde (Sigma-Aldrich, France) at 37 °C for 10 min and permeabilized with 0.5% (*v/v*) Triton X-100 for 5 min. Then, Phalloidin® coupled to AlexaFluor® 488 (Invitrogen, 1:100 dilution in 0.1% Triton X-100) was incubated for 30 min at room temperature, rinsed twice, and labelled cytoskeletons were imaged by CLSM (Zeiss microscopy, Oberkochen, Germany, objectives × 20 and × 63).

3.2.7. Histology and Immunohistochemistry

Cell distribution and apoptotic cells were assessed by histology and immunohistochemistry, respectively. After 10 days of culture, WJ-MSCs loaded HEMOCOLLAGENE® foams were fixed in 4% (*w/v*) paraformaldehyde for 1h and paraffin embedded after ethanol dehydration using a Shandon Excelsior Tissue Processor (Thermo Fisher Scientific, Waltham, MA, USA). Five micrometer thick sections were performed on paraffin-embedded samples (rotation microtome AP280, Leica Microsystems). Hematoxylin-eosin-saffron (HES) and Masson's trichrome staining were performed separately on consecutive tissue sections and images were taken using a scanner iScan Coreo AU (Roche Ò Ventana). For immunohistochemistry, after deparaffinization, sections were incubated with the Cell Conditioner 1 (EDTA, pH 8.4) for 64 min, followed by preprimary peroxidase inhibition and incubation with the primary antibody anti-cleaved Caspase-3 (rabbit polyclonal, Cell Signaling Technology, Danvers, MA, USA) at a 1:600 dilution at 37 °C for 32 min on the automated staining instrument BenchMark XT (Ventana Medical System). Then, the staining reaction was performed using the UltraView Universal DAB v3 Kit (Ventana Medical System, Tucson, AZ, USA). The counterstain and post-counter-stain comprised haematoxylin and bluing reagent. Images were taken using an iScan Coreo AU scanner.

3.2.8. ELISA Quantification

Cytokine and growth factor releases were assessed by ELISA. The quantification of IL-6, IL-8, IL-10, and VEGF proteins at 4, 7, and 10 days in conditioned supernatants was assessed using respectively human IL-6, IL-8, IL-10 and human VEGF Duoset® (R&D systems, Lille, France). Absorbance was measured at 450 nm with correction of non-specific background at 570 nm according to the manufacturer's instructions.

3.2.9. Chemotaxis Assay

Neutrophil migration assay was followed by Boyden chamber chemotaxis assay. Neutrophils were collected as previously described [26]. Conditioned media from WJ-MSCs loaded Hemocollagene® foam cultured for 4, 7, and 10 days were deposited on the lower compartment, whereas 5×10^4 neutrophils were seeded on a polycarbonate membrane (5 µm pores, Nucleopore Track-etch membrane, Whatman, Maidstone, UK) in the upper compartment. After 45 min of incubation at 37 °C in 5% CO_2, non-migrating neutrophils were removed from the top of the membrane and migrated cells at the bottom were stained with May-Grünwald Giemsa (RAL555 kit) and imaged (Axiovert 200M microscope, Zeiss, Oberkochen, Germany, Objective × 40). Conditioned media from cell-free Hemocollagen® foam cultured for 4, 7, and 10 days were used as control.

3.2.10. Quantitative Real Time Polymerase Chain Reaction (qRT-PCR)

Mesenchymal markers, cytokines, and growth factor gene expressions were assessed by qRT-PCR. After 4, 7, and 10 days of culture in Hemocollagene® foams, total RNAs of WJ-MSCs were extracted using MasterPureTM RNA Purification Kit (Epicentre® Biotechnologies, Strasbourg, France) in accordance with the manufacturer protocol. RNA purity was assessed by measuring the absorbance ratio at 260/280 nm (Nanodrop 2000C, Thermo Scientific, France), which was comprised between 1.8 and 2. Total RNAs (500 ng) were reverse transcribed into cDNA using a high-capacity cDNA reverse transcription kit (Applied Biosystems, Villebon-sur-Yvette, France) following the manufacturer instructions. Ten nanograms of reverse transcription product were amplified by qRT-PCR on a StepOne Plus TM system (Applied Biosystems, Villebon-sur-Yvette, France). Using this approach, the transcriptional levels of *RPS18* (internal control), *NT5E*, *THY1*, and *ENG* (corresponding to CD73, CD90, CD105 MSC markers, respectively), *IL6*, *CXCL8*, *IL10* (corresponding to IL-6, IL-8, IL-10 cytokines, respectively), and *VEGFA* (vascular endothelial growth factor) were determined using the double strand-specific Power SYBR® Green dye system (Applied Biosystems, Villebon-sur-Yvette, France). After a first denaturation step at 95 °C for 10 min, qRT-PCR reactions were performed according to a thermal profile that corresponds to 40 cycles of denaturation at 95 °C for 15 s, annealing and extension at 60 °C for 1 min. Data collection was performed at the end of each annealing/extension step. The third step that consists in a dissociation process is performed to ensure the specificity of the amplicons by measuring their melting temperature (Tm). Data analysis was performed with the StepOneTM Software v2.3 (Applied Biosystems, Villebon-sur-Yvette, France).

3.3. Statistical Analysis

All the results were obtained with six independent umbilical cords. Results were represented on histograms as mean ± standard error of the mean using GraphPad Prism version 5.00 for Windows, (GraphPad Software, San Diego, CA USA, www.graphpad.com). All statistical analyses were performed using GraphPad Prism version 5.00 and, for Mann and Whitney tests, a value of $p < 0.05$ was accepted as statistically significant (rejection level of the null-hypothesis of equal means).

4. Conclusions

The mesenchymal stem cell niche represents a microenvironment regulating MSCs' quiescence, self-renewal, and differentiation. Thus, an ideal scaffold would keep MSCs under an undifferentiated state until host-derived paracrine signals induce their activation and commitment. In this work, the association of a dense and porous clinical haemostatic device with human MSCs was proposed as a versatile in vitro MSCs niche presenting suitable intrinsic properties (i.e., composition, porosity, and elastic modulus) for human MSC cultures. Biological investigations revealed an apparently

Int. J. Mol. Sci. **2017**, *18*, 2210

inert microenvironment of HEMOCOLLAGENE® foam, where cells are able to remain viable, in an undifferentiated state with basal cytokine secretion.

MSCs are used in regenerative medicine in two contexts: autologous and allogeneic [27,28]. Integration of autologous MSCs into the damaged tissues is sought, whereas a long-term integration of allogeneic MSCs is not expected, even if MSCs express only low levels of the histocompatibility markers. Their paracrine activity along with activation of resident stem cells and mobilization of circulating ones into the damaged tissues are, thus, relevant [29]. In view of application of HEMOCOLLAGENE® as a useful WJ-MSC carrier in the regenerative medicine field, additional studies under in vitro and in vivo mimicking pathological conditions (i.e., hypoxic and/or inflammatory stimuli) are required to identify whether host derived paracrine signals could stimulate regenerative properties of WJ-MSCs loaded in HEMOCOLLAGENE®.

Supplementary Materials: Supplementary materials can be found at www.mdpi.com/1422-0067/18/10/2210/s1.

Acknowledgments: The authors are very grateful to the staff of Reims Maternity Hospital headed by O. Graesslin for providing umbilical cords and the staff of the Core URCACyt (S. Audonnet) and PICT (C. Terryn). This work was partially supported by «L'agence Régionale de Santé de la Région Grand Est». We thank N. Bouland for technical help concerning histology and SEPTODONT for providing foams.

Author Contributions: Léa Aubert and Marie Dubus (Ph.D. students), Hassan Rammal (stem cell banking and characterization), Camille Bour (ELISA experiments), Céline Mongaret (pharmaceutical expertise), Camille Boulagnon-Rombi (Histology), Roselyne Garnotel (ion exchange chromatography), Céline Schneider (porosity experiments), Rachid Rahouadj and Cedric Laurent (indentation experiments and data treatment), Sophie C. Gangloff (financial contribution), Frédéric Velard (microscopy staff), and Cedric Mauprivez and Halima Kerdjoudj (scientific supervisors).

Conflicts of Interest: The authors declare that they have no competing interests.

Abbreviations

3D	Three-Dimensional
MSCs	Mesenchymal Stem Cells
WJ-MSCs	Wharton Jelly-Derived Mesenchymal Stem Cells
IL	Interleukin
SEM	Scanning Electron Microscopy
CLSM	Confocal Laser Scanning Microscopy
FTIR	Fourier Transform Infrared spectra
CD	Cluster of Differentiation
qRT-PCR	Quantitative Real Time Polymerase Chain Reaction
HES	Hematoxylin-Eosin-Saffron
WST-1	Water-Soluble Tetrazolium Salt-1
DNA	Deoxyribonucleic Acid
VEGF	Vascular Endothelial Growth Factor
ELISA	Enzyme-Linked Immunosorbent Assay

References

1. Wei, X.; Yang, X.; Han, Z.P.; Qu, F.F.; Shao, L.; Shi, Y.F. Mesenchymal stem cells: A new trend for cell therapy. *Acta Pharmacol. Sin.* **2013**, *34*, 747–754. [CrossRef] [PubMed]
2. Trounson, A.; McDonald, C. Stem cell therapies in clinical trials: Progress and challenges. *Cell. Stem Cell* **2015**, *17*, 11–22. [CrossRef] [PubMed]
3. Wang, Y.; Chen, X.; Cao, W.; Shi, Y. Plasticity of mesenchymal stem cells in immunomodulation: Pathological and therapeutic implications. *Nat. Immunol.* **2014**, *15*, 1009–1016. [CrossRef] [PubMed]
4. Siegel, G.; Kluba, T.; Hermanutz-Klein, U.; Bieback, K.; Northoff, H.; Schäfer, R. Phenotype, donor age and gender affect function of human bone marrow-derived mesenchymal stromal cells. *BMC Med.* **2013**, *11*, 146. [CrossRef] [PubMed]

5.	Smith, J.R.; Pfeifer, K.; Petry, F.; Powell, N.; Delzeit, J.; Weiss, M.L. Standardizing umbilical cord mesenchymal stromal cells for translation to clinical use: Selection of GMP-compliant medium and a simplified isolation method. *Stem Cells Int.* **2016**, *2016*, e6810980. [CrossRef] [PubMed]
6.	Rammal, H.; Harmouch, C.; Lataillade, J.J.; Laurent-Maquin, D.; Labrude, P.; Menu, P.; Kerdjoudj, H. Stem cells: A promising source for vascular regenerative medicine. *Stem Cells Dev.* **2014**, *23*, 2931–2949. [CrossRef] [PubMed]
7.	Hass, R.; Kasper, C.; Böhm, S.; Jacobs, R. Different populations and sources of human mesenchymal stem cells (MSC): A comparison of adult and neonatal tissue-derived MSC. *Cell Commun. Signal. CCS* **2011**, *9*, 12. [CrossRef] [PubMed]
8.	Li, Y.; Xiao, Y.; Liu, C. The horizon of materiobiology: A perspective on material-guided cell behaviors and tissue engineering. *Chem. Rev.* **2017**, *117*, 4376–4421. [CrossRef] [PubMed]
9.	Schmidt, T.; Stachon, S.; Mack, A.; Rohde, M.; Just, L. Evaluation of a thin and mechanically stable collagen cell carrier. *Tissue Eng. Part C Methods* **2011**, *17*, 1161–1170. [CrossRef] [PubMed]
10.	Murphy, W.L.; McDevitt, T.C.; Engler, A.J. Materials as stem cell regulators. *Nat. Mater.* **2014**, *13*, 547–557. [CrossRef] [PubMed]
11.	Strauss, K.; Chmielewski, J. Advances in the design and higher-order assembly of collagen mimetic peptides for regenerative medicine. *Curr. Opin. Biotechnol.* **2017**, *46*, 34–41. [CrossRef] [PubMed]
12.	Vennat, E.; Bogicevic, C.; Fleureau, J.M.; Degrange, M. Demineralized dentin 3D porosity and pore size distribution using mercury porosimetry. *Dent. Mater.* **2009**, *25*, 729–735. [CrossRef] [PubMed]
13.	Lee, J.; Choi, W.; Tae, G.; Kim, Y.H.; Kang, S.S.; Kim, S.E.; Jung, Y.; Kim, S.H. Enhanced regeneration of the ligament-bone interface using a poly(L-lactide-*co*-ε-caprolactone) scaffold with local delivery of cells/BMP-2 using a heparin-based hydrogel. *Acta Biomater.* **2011**, *7*, 244–257. [CrossRef] [PubMed]
14.	Gao, G.; Schilling, A.F.; Hubbell, K.; Yonezawa, T.; Truong, D.; Hong, Y.; Dai, G.; Cui, X. Improved properties of bone and cartilage tissue from 3D inkjet-bioprinted human mesenchymal stem cells by simultaneous deposition and photocrosslinking in PEG-GelMA. *Biotechnol. Lett.* **2015**, *37*, 2349–2355. [CrossRef] [PubMed]
15.	Mechiche Alami, S.; Velard, F.; Draux, F.; Siu Paredes, F.; Josse, J.; Lemaire, F.; Gangloff, S.C.; Graesslin, O.; Laurent-Maquin, D.; Kerdjoudj, H. Gene screening of Wharton's jelly derived stem cells. *Biomed. Mater. Eng.* **2014**, *24*, 53–61. [PubMed]
16.	Dominici, M.; Le Blanc, K.; Mueller, I.; Slaper-Cortenbach, I.; Marini, F.; Krause, D.; Deans, R.; Keating, A.; Prockop, D.J.; Horwitz, E. Minimal criteria for defining multipotent mesenchymal stromal cells. The International Society for Cellular Therapy position statement. *Cytotherapy* **2016**, *8*, 315–317. [CrossRef] [PubMed]
17.	Viswanathan, P.; Ondeck, M.G.; Chirasatitsin, S.; Ngamkham, K.; Reilly, G.C.; Engler, A.J.; Battaglia, G. 3D surface topology guides stem cell adhesion and differentiation. *Biomaterials* **2015**, *52*, 140–147. [CrossRef] [PubMed]
18.	Han, S.; Zhao, Y.; Xiao, Z.; Han, J.; Chen, B.; Chen, L.; Dai, J. The three-dimensional collagen scaffold improves the stemness of rat bone marrow mesenchymal stem cells. *J. Genet. Genom.* **2012**, *39*, 633–641. [CrossRef] [PubMed]
19.	Dhaliwal, A.; Brenner, M.; Wolujewicz, P.; Zhang, Z.; Mao, Y.; Batish, M.; Kohn, J.; Moghe, P.V. Profiling stem cell states in three-dimensional biomaterial niches using high content image informatics. *Acta Biomater.* **2016**, *45*, 98–109. [CrossRef] [PubMed]
20.	Heitz, J.; Plamadeala, C.; Wiesbauer, M.; Freudenthaler, P.; Wollhofen, R.; Jacak, J.; Klar, T.A.; Magnus, B.; Köstner, D.; Weth, A.; et al. Bone-forming cells with pronounced spread into the third dimension in polymer scaffolds fabricated by two-photon polymerization. *J. Biomed. Mater. Res. A* **2017**, *105*, 891–899. [CrossRef] [PubMed]
21.	Lo, Y.P.; Liu, Y.S.; Rimando, M.G.; Ho, J.H.; Lin, K.H.; Lee, O.K. Three-dimensional spherical spatial boundary conditions differentially regulate osteogenic differentiation of mesenchymal stromal cells. *Sci. Rep.* **2016**, *6*, 21253. [CrossRef] [PubMed]
22.	Yoo, K.H.; Jang, I.K.; Lee, M.W.; Kim, H.E.; Yang, M.S.; Eom, Y.; Lee, J.E.; Kim, Y.J.; Yang, S.K.; Jung, H.L.; et al. Comparison of immunomodulatory properties of mesenchymal stem cells derived from adult human tissues. *Cell Immunol.* **2009**, *259*, 150–156. [CrossRef] [PubMed]

23. Pricola, K.L.; Kuhn, N.Z.; Haleem-Smith, H.; Song, Y.; Tuan, R.S. Interleukin-6 maintains bone marrow-derived mesenchymal stem cell stemness by an ERK1/2-dependent mechanism. *J. Cell Biochem.* **2009**, *108*, 577–588. [CrossRef] [PubMed]

24. Li, P.; Li, S.H.; Wu, J.; Zang, W.F.; Dhingra, S.; Sun, L.; Weisel, R.D.; Li, R.K. Interleukin-6 downregulation with mesenchymal stem cell differentiation results in loss of immunoprivilege. *J. Cell. Mol. Med.* **2013**, *17*, 1136–1145. [CrossRef] [PubMed]

25. McKee, C.T.; Last, J.A.; Russell, P.; Murphy, C.J. Indentation versus tensile measurements of young's modulus for soft biological tissues. *Tissue Eng. Part B Rev.* **2011**, *17*, 155–164. [CrossRef] [PubMed]

26. Velard, F.; Laurent-Maquin, D.; Braux, J.; Guillaume, C.; Bouthors, S.; Jallot, E.; Nedelec, J.M.; Belaaouaj, A.; Laquerriere, P. The effect of zinc on hydroxyapatite-mediated activation of human polymorphonuclear neutrophils and bone implant-associated acute inflammation. *Biomaterials* **2010**, *31*, 2001–2009. [CrossRef] [PubMed]

27. Ezquer, F.E.; Ezquer, M.E.; Vicencio, J.M.; Calligaris, S.D. Two complementary strategies to improve cell engraftment in mesenchymal stem cell-based therapy: Increasing transplanted cell resistance and increasing tissue receptivity. *Cell Adhes. Migr.* **2017**, *11*, 110–119. [CrossRef] [PubMed]

28. Lee, D.E.; Ayoub, N.; Agrawal, D.K. Mesenchymal stem cells and cutaneous wound healing: Novel methods to increase cell delivery and therapeutic efficacy. *Stem Cell Res. Ther.* **2016**, *7*, 37. [CrossRef] [PubMed]

29. Caplan, A.I.; Correa, D. The MSC: An injury drugstore. *Cell Stem Cell* **2011**, *9*, 11–15. [CrossRef] [PubMed]

International Journal of
Molecular Sciences

MDPI

Article

Application of Millifluidics to Encapsulate and Support Viable Human Mesenchymal Stem Cells in a Polysaccharide Hydrogel

Fabien Nativel [1,2], Denis Renard [3,*], Fahd Hached [1,4], Pierre-Gabriel Pinta [1,2], Cyril D'Arros [1], Pierre Weiss [1,5,6], Catherine Le Visage [1,6], Jérôme Guicheux [1,5,6], Aurélie Billon-Chabaud [1,4] and Gael Grimandi [1,2,4]

[1] Inserm, UMR 1229, RMeS, Regenerative Medicine and Skeleton, Université de Nantes, ONIRIS, F-44042 Nantes, France; fabien.nativel90@gmail.com (F.N.); fahd.hached@univ-nantes.fr (F.H.); Pierre.Pinta@chu-nantes.fr (P.-G.P.); cyril.d-arros@inserm.fr (C.D.); Pierre.Weiss@univ-nantes.fr (P.W.); catherine.levisage@inserm.fr (C.L.V.); jerome.guicheux@univ-nantes.fr (J.G.); Aurelie.Billon@univ-nantes.fr (A.B.-C.); gael.grimandi@univ-nantes.fr (G.G.)
[2] CHU Nantes, Pharmacie Centrale, PHU 11, F-44093 Nantes, France
[3] INRA UR1268, Biopolymères Interactions Assemblages, F-44300 Nantes, France
[4] UFR Sciences Pharmaceutiques et Biologiques, Université de Nantes, F-44035 Nantes, France
[5] CHU Nantes, PHU4 OTONN, F-44093 Nantes, France
[6] UFR Odontologie, Université de Nantes, F-44042 Nantes, France
[*] Correspondence: denis.renard@inra.fr

Received: 14 June 2018; Accepted: 29 June 2018; Published: 3 July 2018

Abstract: Human adipose-derived stromal cells (hASCs) are widely known for their immunomodulatory and anti-inflammatory properties. This study proposes a method to protect cells during and after their injection by encapsulation in a hydrogel using a droplet millifluidics technique. a biocompatible, self-hardening biomaterial composed of silanized-hydroxypropylmethylcellulose (Si-HPMC) hydrogel was used and dispersed in an oil continuous phase. Spherical particles with a mean diameter of 200 μm could be obtained in a reproducible manner. The viability of the encapsulated hASCs in the Si-HPMC particles was 70% after 14 days in vitro, confirming that the Si-HPMC particles supported the diffusion of nutrients, vitamins, and glucose essential for survival of the encapsulated hASCs. The combination of droplet millifluidics and biomaterials is therefore a very promising method for the development of new cellular microenvironments, with the potential for applications in biomedical engineering.

Keywords: droplet millifluidics; encapsulation; human adipose-derived stromal cells; hydrogel; self-hardening; silanized-hydroxypropylmethylcellulose; biomedical; degenerative disease

1. Introduction

Mesenchymal stromal cells (MSCs) are of significant medical interest as they have the ability to differentiate into several cell types (including chondrocytes, osteocytes, and adipocytes). They have already been exploited to treat several pathologies, including osteo-articular diseases, diabetes, cancer, cardiovascular pathologies, angiogenic diseases, and skin injuries [1–6]. In recent years, they have become known for their potent immunomodulatory and anti-inflammatory activities, which stem from their ability to secrete bioactive trophic factors and to release extracellular vesicles [7–9]. MSCs can be isolated from a broad range of adult tissues, including bone marrow, adipose tissue, and articular synovial fluid [10].

The literature to date indicates that MSCs injection is subject to two main limitations: extensive cell death due to the mechanical forces during injection, thereby making it difficult to detect the injected cells over a sustained period of time [11,12], and the risk of cell leakage after their injection due to the propensity of MSCs to migrate [13,14]. The encapsulation of MSCs in biomaterials prior to their injection appears to be an alternative strategy for their administration that overcomes these limitations and that facilitates the delivery of therapeutic factors in several pathologies [15,16]. The optimal injection of encapsulated MSCs requires spherical devices with a size that is compatible with a standard needle characteristic.

Cell encapsulation consists in entrapping viable and functional cells within a matrix in order to enable immune isolation by the creation of a physical barrier between the host immune system and the transplanted tissue [17]. Moreover, cell encapsulation enhances the retention of the cells at the targeted tissue and protects them from the mechanical damage that occurs with injection. Indeed, the viability of the cells is compromised as they are subjected first to shear stress during syringe needle flow and secondly to stretching forces and deformations due to extensional flow during syringe needle ejection [12]. The matrix must be biocompatible and semi-permeable to obtain the intended bio-functionality. Indeed, it must not hinder the diffusion of essential factors required for cell survival (e.g., oxygen and nutrients derived from the blood and waste products of cellular metabolism). Most of the biomaterials used for cell encapsulation are hydrogels, in which the cells are embedded in a fully hydrated matrix. This allows for the survival of the encapsulated cells and their therapeutic actions. Cell encapsulation is carried out by mixing the cells with polymeric solutions that are then crosslinked under operating conditions that depend on the nature of the polymer. By protecting the transplanted tissue, encapsulation can improve the safety of cell therapies [17]. This approach was first proposed by Lim et al. In 1980, who encapsulated pancreatic cells in alginate prior to their injection into diabetic rats so as to be able to avoid immunosuppressive therapy [18].

Alginate is the most extensively investigated and characterized polymer for cell encapsulation due to its intrinsic properties, such as biocompatibility, and the requirements of calcium ions for crosslinking [19]. Other natural polymers such as agarose may support viable cells by constituting microenvironments that mimic natural tissues [20]. However, alginate hydrogels are sensitive to non-gelling agents (such as sodium ions) in physiological solutions. Indeed, a sodium-calcium exchange occurs under physiological conditions, leading to the disruption of the hydrogel [21]. To overcome these limitations, our laboratory favors the use of an injectable, biocompatible, self-hardening hydrogel composed of silanized-hydroxypropylmethylcellulose (Si-HPMC). This labeled hydrogel has been reported to support both cell viability and bio-functionality after encapsulation [22–24]. In 2008, in vivo evaluations after implantation of a ruthenium-labeled Si-HPMC hydrogel into rabbit bone defects showed that this polymer is biocompatible and not degraded until at least the eighth week [25].

Several techniques are currently used for cell encapsulation [26,27]. For example, emulsions can be used to encapsulate cells within spherical structures referred to as "particles" [28]. Recently, Hached et al. [29] encapsulated cells in Si-HPMC particles using a water-in-oil dispersion protocol. The MSCs were found to have long-term viability and the ability to secrete immunomodulatory and anti-inflammatory factors [29]. However, this method has not been able to reproducibly generate Si-HPMC particles with a monodispersed size that are optimal for in vivo studies. In contrast to dispersion/emulsification techniques, microfluidics technology allows the highly reproducible generation of uniform microparticles with a controlled size [30,31]. Microfluidics has generated significant interest in the research of cell encapsulation [32]. Another approach, called millifluidics, which bears similarities to microfluidics, has ample potential for use in cell encapsulation.

Millifluidics is characterized by the assembly of inexpensive and commercially available millimeter-sized tubing and chromatography connectors. Aside from the need for syringe pumps, this method is easy to implement and does not require specific equipment. The emergence of millifluidics is a relatively new phenomenon. The first millifluidic device, which was referred to as a "simplified microfluidic device", was developed by Quevedo et al. In 2005 [33]. It offers a high level of versatility and rapid creation of a modular setup. It is an efficient tool to investigate polymerization reactions or to create microparticles with controlled and fine-tuned sizes and shapes [34–37]. Moreover, Amine et al. recently showed that droplet-based millifluidics represents an efficient means to probe the liquid-liquid phase separation of various biopolymers mixtures [38]. Fluidic technologies consist of the introduction of the polymeric solution to be dispersed through a capillary or a needle into the co-flowing continuous phase to generate droplets [39]. The structure of the particles can be determined by the flow properties in the devices while the chemical composition is dictated by the selected fluids. With the use of a basic junction such as a co-axial or T-junction, millifluidics offers more advantages than soft lithography and microfluidic techniques. The millifluidic apparatus is an assembly of capillaries or flexible tubes (plastic or silica tubing with diameters ranging from 50 μm to a few millimeters) connected by elementary home-made or commercial (Upchurch®) modules. Commercially available tubes have a range of wettability and transparency properties and they generally exhibit good chemical resistance to pressure and temperature [40]. The elementary modules are able to achieve the basic functions used in microfluidic devices, such as the formation of periodic trains of monodisperse droplets with very good control over their size or the dilution-concentration of these trains while keeping the volume of the droplets unchanged [41]. Modular millifluidic setups can then be designed to produce newly controlled integrated configurations, limited only by the number of combinations possible and one's creativity [41]. The connecting capillary tubes and the various modules can readily be assembled and disassembled so that modular setups can be designed as needed in a short period of time. The great versatility of this method provided the millifluidics strategy several advantages over microfluidic synthesis, while retaining its suitability for the in-depth study of critical parameters involved in microparticle production. Millifluidics is therefore an inexpensive and versatile method that allows for the production of particles with optimal injection properties for in vivo studies: reproducibility and size monodispersity. The literature indicates that following the optimization of experimental conditions, the particles produced by millifluidics generally have a polydispersity below 2–3% [42].

In this context, millifluidics appears to be a promising approach to control the granulometry of encapsulated MSCs in polysaccharide hydrogel droplets and to prevent the polydispersity of the particles observed in previous studies [29]. This approach provides better control of the numbers of injectable cells and optimizes their viability. The objectives of this study were (i) to verify that the rheological properties of the Si-HPMC hydrogel are consistent with millifluidic techniques, (ii) to characterize the Si-HPMC particles generated, and (iii) to assess the in vitro viability and proliferation of MSCs after their encapsulation.

2. Results

2.1. Rheological Assessments of the Si-HPMC Solution and Gel

The flow behavior of the Si-HPMC solution showed that the steady shear viscosity decreased as the shear rate increased (Figure 1A). a Newtonian plateau observed at low shear rates was followed by a shear-thinning behavior. At a high shear rate, the viscosity exhibited power-law dependence with the shear rate. The flow curve could readily be fitted to the simplified Cross model using Equation (2) (correlation coefficient of 0.989). The limiting Newtonian viscosity (η_0) was 38.6 ± 1.2 Pa.s for a 4% (w/w) non-sterile Si-HPMC.

Instantaneously, after the initiation of Si-HPMC crosslinking upon pH neutralization by the mixing of one volume of HEPES buffer (pH 3.6) with one volume of the Si-HPMC polymer solution, the influence of temperature on the gelation time t_{gel} was monitored using dynamic frequency measurements (Figure 1B,C). According to the criterion of Winter and Chambon, the cross-over of tan δ with time at the five applied frequencies yielded a gelation time t_{gel} of 701 ± 29.7 s at 23 °C (Figure 1B) and of 123.3 ± 15.3 s at 37 °C (Figure 1C) for the 2% (w/w) Si-HPMC hydrogel [43,44].

To further characterize the Si-HPMC hydrogel, the average mesh size, corresponding to the average distance between entanglements in the hydrogel network, was evaluated using Equation (3). The equilibrium storage modulus G' had a value of 124.7 ± 17.9 Pa for 2% Si-HPMC gel at pH 7 (Figure 1D). From this value, an average mesh size (ξ) of 32.1 ± 1.6 nm was calculated.

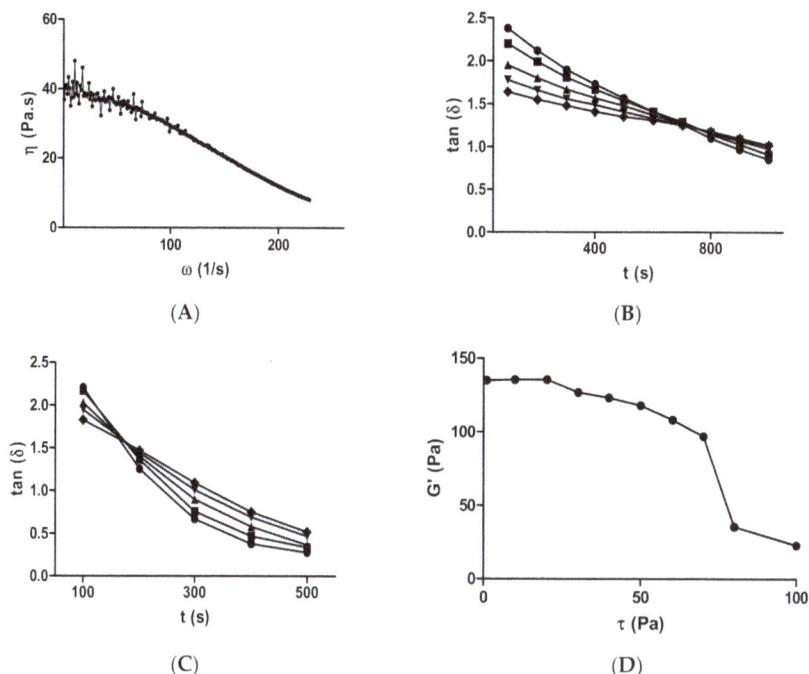

Figure 1. Rheological characterization of silanized-hydroxypropylmethylcellulose (Si-HPMC) solution and gel. (**A**) Flow curves (viscosity vs. shear rate) of a 4% Si-HPMC solution (pH 7.0). (**B**) Tan (δ) vs. time (in order to determine t_{gel}) of a 2% Si-HPMC hydrogel (pH 7.0) at 23 °C and (**C**) at 37 °C. Tan (δ) was determined at five oscillation frequencies: 0.30 Hz (●), 0.50 Hz (■), 1 Hz (▲), 1.80 Hz (▼), and 3.2 Hz (♦). (**D**) The equilibrium storage modulus (G') was determined for applied stress amplitudes (τ) ranging from 0.1 to 1000 Pa and a fixed frequency of 1 Hz after 24 h for a 2% Si-HPMC hydrogel (pH 7.0) at 37 °C. Each rheological test was repeated three times.

2.2. Characterization of the Shape and the Size of the Si-HPMC Particles

The droplet-based millifluidics process was optimized by the application of a dispersed flow rate of 16 µL/min and a continuous flow rate of 150 µL/min. Under these operating conditions, after the washing and collection of the Si-HPMC particles in phosphate buffered salt (PBS), light microscopy observations revealed that the Si-HPMC particles were spherical and uniform with smooth surfaces and a size of 200 µm (Figure 2A). Laser-based particle size analyses revealed a monomodal population of particles that were between 170 and 210 µm in size, with an average diameter of 192 ± 16 µm (Figure 2B).

2.3. Diffusion Properties of the Si-HPMC Particles

Following the incubation of the Si-HPMC particles (with an average diameter of 192 ± 16 µm) in 1 mg/mL solutions of fluorescein isothiocyanate (FITC)-dextran of different molecular weights for 18 h, the diffusion properties of Si-HPMC hydrogels were assessed using confocal laser scanning microscopy (CLSM).

At the beginning of the experiment, the Si-HPMC particles did not exhibit any fluorescence, which is in agreement with the absence of fluorescence signals emanating from the Si-HPMC polymer. an increase in fluorescence intensity was then noticed as a function of time for the Si-HPMC particles incubated with 20 kDa and 250 kDa FITC-dextrans (Figure 3). After 18 h, the internal to external ratio was 0.39 and 0.1 for the 20 kDa and the 250 kDa fluorescently labeled dextrans, respectively. In addition, no fluorescence was detected inside the Si-HPMC particles after 18 h for the fluorescently labeled 2000 kDa dextran.

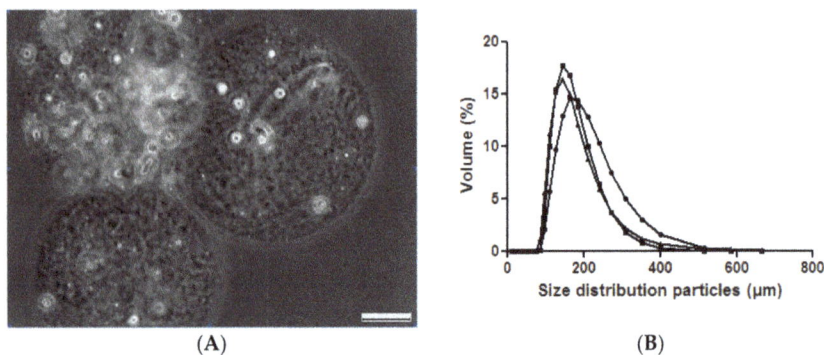

(A)   (B)

Figure 2. Characterization of the Si-HPMC particles. (**A**) Representative light microscopy image of the Si-HPMC particles produced using droplet-based millifluidics; (**B**) Size distribution of three batches of Si-HPMC particles produced using droplets-based millifluidics and as determined by laser diffraction. Scale bar: 50 μm.

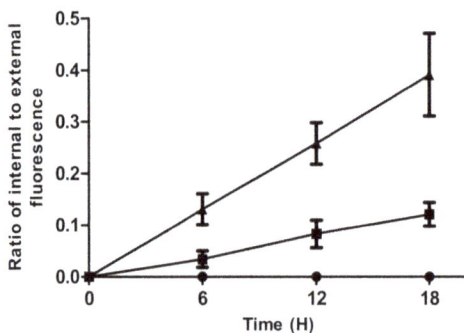

Figure 3. Diffusion properties of the Si-HPMC particles. Particles of Si-HPMC were incubated with FITC-dextran (M_W 20 kDa (▲), 250 kDa (■), and 2000 kDa (●)) solutions for 18 h. The ratio of the maximum fluorescence intensity inside and outside the particles was calculated, after assessment of the fluorescence intensities of the particles (inside) and the FITC-dextran solutions (outside) using confocal laser scanning microscopy (CLSM). Si-HPMC particles with 192 ± 16 μm diameters were selected for this study. Each test was performed for one particle at a time and repeated three times.

2.4. Evaluation of Encapsulated Human Adipose-Derived Stromal Cells (hASCs) Viability and Estimation of the Average Number of Encapsulated hASCs

The viability of the encapsulated hASCs in Si-HPMC particles was $71 \pm 2.9\%$ on average, irrespective of the time after encapsulation (Figure 4C), with the differences in percentage lacking statistical significance. The distribution of the cell viability along the radial axis of the Si-HPMC particles was uniform and no accumulation of dead cells in the center of the particles was detected (Figure 4A,B). In addition, the viable cells did not appear to interact with each other and there was no indication that the cells had clustered.

To determine the number of encapsulated hASCs in the Si-HPMC particles, several CSLM images with steps of 10 μm were analyzed using ImageJ® software. The fluorescent cells were manually scored and an average number of 69 ± 10, 66 ± 19, and 67 ± 8 hASCs per Si-HPMC particle was found after 1, 7, and 14 days, respectively, after the encapsulation (Figure 4D). These results were not statistically different.

A **B**

C **D**

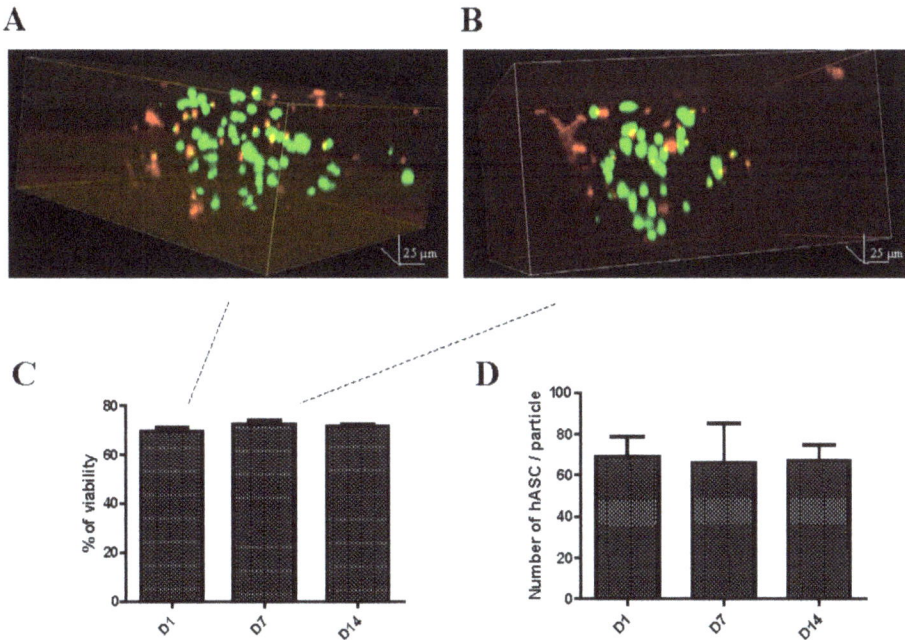

Figure 4. Human adipose-derived stromal cells (hASCs) viability after encapsulation in a Si-HPMC particle. Viable (green) and dead (red) cells were imaged using confocal microscopy and a Live/Dead assay kit at D1 (**A**) and D7 (**B**). hASCs viability in the Si-HPMC particles was monitored over 14 days of culture using a Live/Dead assay kit and manually determined using ImageJ® software. (**C**) Determination of the number of cells per particle was performed after using the Live/Dead assay kit and manually scored using ImageJ® software (**D**). Scale bar: 25 μm.

3. Discussion

Cell encapsulation in biomaterials facilitates the injection of MSCs and decreases both the extensive cell death that tends to occur upon injection and the capacity of the MSCs to migrate [15,45]. In this study, MSCs were isolated from human adipose tissue, which contains relatively large numbers (5%) of MSCs [10]. For potential applications in biomedical engineering, it is essential to generate particles with optimal injection properties for in vivo studies: sphericity, reproducibility, and size monodispersity.

Therefore, this study sought to generate reproducible Si-HPMC hydrogel particles by a droplet-based millifluidics method. The aims of this work were to (i) verify the compatibility of Si-HPMC hydrogel rheological properties with millifluidic techniques, (ii) determine the feasibility of generating monodispersed and reproducible particles from a Si-HPMC hydrogel using a novel and original droplet-based millifluidics method, and (iii) assess the ability of this technique to support the in vitro viability of encapsulated hASCs in Si-HPMC particles.

Si-HPMC is a semi-synthetic polymer that undergoes condensation and crosslinking when the pH decreases by the addition of an acidic N-(2-Hydroxyethyl)piperazine-N′-(2-ethanesulfonic acid)(HEPES) buffer [46]. This step needs to be undertaken with great care as numerous air bubbles can become embedded inside Si-HPMC hydrogels during the mixing step using Luer-Lock syringes containing Si-HPMC solution and HEPES buffer to initiate the decrease of pH and therefore the Si-HPMC crosslinking. Moreover, a slight chemical modification of the glycidoxypropyltrimethoxysilane (GPTMS) grafting rate on HPMC (0.6% w/w of silane) can lead to pronounced changes in the macroscopic behavior [24].

Our rheological studies confirmed previous results that demonstrate that this hydrogel has characteristics of a shear-thinning fluid, with a decrease in the viscosity observed as the shear rate increases [24]. Shear-thinning fluids are better integrated in capillary flow methods as no flow velocity fluctuations occur during the flow within the tubing. In 2005, Vinatier et al. showed that Si-HPMC hydrogel became a solid hydrogel 30 min after the initiation of crosslinking and that complete crosslinking occurred after 12 days with a maximum G' of 190 Pa [46], which is in keeping with the results of our study. Regarding Si-HPMC crosslinking, increasing temperature reduced the gelation time by a factor of ~4, which is qualitatively in keeping with the findings reported by Fatimi et al. [47]. These authors also found that there was a linear relationship between Ln (t_{gel}) and $1/T$ applied with an activation energy of the condensation reaction $E_a = 74.3$ kJ·mol^{-1}. The decrease in the gelation time with the increase in temperature at a fixed pH could be explained by the catalytic action of temperature on the silanol condensation [48]. Crosslinking of the Si-HPMC was therefore carried out at room temperature in order to reduce the rate of the chemical condensation reaction and to avoid curing in the millifluidic device. The particles were then collected in complete medium at 37 °C in order to ensure better cell survival and to complete the crosslinking of the Si-HPMC chains. The choice of the cell encapsulation method appears to be suitable for Si-HPMC as a result of its physicochemical properties and crosslinking mechanism.

The advantages of the droplet-based millifluidics method are that it requires a small amount of engaged volumes and that it generates particles with a uniform spherical shape and monodispersed size. Our results show that this encapsulation method supported hASCs survival and that it is suitable for hydrophilic biomaterials such as Si-HPMC. The difficulty in generating reproducible Si-HPMC particles using droplet-based millifluidics lies not only in finding the optimal dispersed and continuous flow rates, but also in the optimization of crosslinking off-line in the collection bath in order to avoid the coalescence of the particles. an appropriate stirring rate using a stirring paddle and a controlled temperature allowed for the production of uniform and spherical particles. In this study, the discrepancies observed in the sizes determined by light microscopy and laser diffraction could be due to the heterogeneous swelling of the Si-HPMC particles, depending on the solvent used (PBS vs. water). Tuning dispersed and continuous flow rates in conjunction with the variation of the internal diameter of the capillary tubing was successfully applied by Martins et al. To generate alginate capsules with diameters ranging from 140 µm to 1.4 mm, according to the variation of flow rates [37]. These results led to the conclusion that droplet-based millifluidics is a versatile and easy-to-use method for the production of a broad range of microparticles of different sizes for encapsulation purposes. In addition, the use of a T-junction configuration for the droplet production allowed for more than 500 particles to be generated per hour.

In order to facilitate the development of hydrogel-assisted hASCs therapies, the particles size can be modulated according to their application. The particles size is governed by a compromise between three criteria: (i) the number of cells that need to be injected in order to achieve the desired therapeutic effect, (ii) the site of the injection, and (iii) the quality of the exchange between the particles and their external environment. In this study, droplet-based millifluidics allowed the generation of 200-µm Si-HPMC monodispersed particles. This is compatible with potential human articular injection for inflammatory disease. The application of a Q_d/Q_c ratio of 0.1 in droplet-based millifluidics was hence a good compromise for the production of monodispersed Si-HPMC particles with this size, proving to be compatible with this objective. This size is controllable and can be modified by the variation of the flow rates of the different phases for numerous animal models [41]. Particle shape is also a parameter that has an impact on the injectability and the biocompatibility of hydrogels. Indeed, it has been reported that a non-spherical particle shape did not promote their injectability and, more importantly, induced in vivo inflammation [49,50]. The volume of each spherical particle was therefore of 4.2 nL, calculated based on a particle radius of 100 μm in PBS, allowing the in vivo injection of several thousand Si-HPMC particles loaded with hASCs.

The present study also relies on the diffusion of macromolecules of different sizes into the particles. It has been shown that diffusion is affected by the mechanical stress applied on the particles in vivo, which depends greatly on the elasticity, the degree of swelling, and the charge density of the hydrogel. These parameters and their interactions create a complex environment that determines the diffusion and duration kinetics [51]. In the present study, FITC-dextrans (M_w 20, 250, and 2000 kDa) were selected due to their extensive application and ease of use in diffusion studies [52]. These results were analyzed in terms of the hydrodynamic size (instead of the molecular weight) of the dextrans (i.e., branched polysaccharides). The relation between the molecular weight and the hydrodynamic radius of branched polysaccharides is represented by the following expression from Wyatt Technologies (data not shown):

$$M_w = [1.4782 \times R_h]^{(1.8136)} \tag{1}$$

where M_w is the molecular weight (kDa) and R_h is the hydrodynamic radius (nm). Using this equation, the 20, 250, and 2000 kDa fluorescently labeled dextrans were determined to have hydrodynamic diameters of 7, 28.4, and 89.4 nm, respectively.

Given that the average mesh size of the 2% Si-HPMC hydrogels was 32 nm, it would be reasonable to assume that the 20 and 250 kDa fluorescently labeled dextrans should be able to freely diffuse into the Si-HPMC particles while the 2000 kDa dextran should be maintained outside the particles due to steric hindrance. The results shown in Figure 4 indicate that this hypothesis is at least partially true, although the kinetics of the permeability of the 20 and 250 kDa FITC-dextrans appear to be very slow. This slow diffusion process can also arise from heterogeneity in the pore size at the surface of the Si-HPMC particles. The average mesh size compatible with the free diffusion of macromolecules with sizes less than 32.1 nm would, however, not reflect the discrepancies that could exist between the mesh size inside the hydrogel and the pore size in proximity to the surface of the hydrogel. Recently, the diffusion of FITC-dextrans in 1-mm Si-HPMC particles, obtained using a dispersion-emulsification process, revealed that the 20 and 250 kDa FITC-dextrans diffused faster and that, after 150 min of incubation with the Si-HPMC particles, the fluorescence intensity of the 20 kDa FITC-dextran reached equilibrium (i.e., ratio = 1), while the fluorescence intensity ratio was 0.7 for the 250 kDa FITC-dextran. This faster diffusion was therefore attributed to the larger size of the Si-HPMC particles. In addition, no fluorescence intensity was detected with the 2000 kDa FITC-dextran, which is in accordance with the present study [29]. The 20 kDa FITC-dextran was of particular interest, as the therapeutic factors secreted by stimulated hASCs have molecular weights that range from about 10 to about 45 kDa [53,54]. a thorough study using environmental microscopy could be very useful to probe the internal and external structures of Si-HPMC particles. These particles would, however, be adapted to allow the diffusion of essential nutrients for the viability of encapsulated hASCs, as well as the diffusion of therapeutic factors secreted in an inflammatory environment by encapsulated hASCs.

As hASCs are thought to mainly exert their therapeutic potential by the secretion of immunomodulatory, pro-angiogenic, anti-apoptotic, anti-fibrotic, and anti-inflammatory factors, encapsulated hASCs must remain viable and the average number of cells per particle has to be sufficient [55]. In the present study, the viability of encapsulated hASCs in Si-HPMC particles was estimated to be 70% after 14 days. As the encapsulation technique involves some degree of mechanical shear, it probably results in a slight reduction of cell viability. Unlike the direct injection of cells into the body with a conventional injection system, millifluidics can substantially reduce the extent of cell death. This result confirms that Si-HPMC particles obtained by millifluidics support the diffusion of nutrients, vitamins, and glucose essential for the survival of the encapsulated hASCs. In 2018, Figueiredo et al. showed that the diffusion of glucose through Si-HPMC hydrogels was correlated directly with the average distance between the polymer nodes in the hydrogel network, while the diffusion of oxygen was found to be the limiting factor for cell viability in Si-HPMC hydrogels [56]. In addition, the present study did not find that the dead cells were specifically localized at the center of the particles. Thus, it would appear that a sufficient level of nutrients, glucose, and oxygen reached the center of the particles by diffusion. This strongly suggests that, due to the suitable extent of diffusion obtained, the particle size did not appear to constitute a limiting factor for cell viability in this range of particle sizes. At 24 h after the encapsulation, the average number of live hASCs per Si-HPMC particle was estimated to be approximately 70. This average cell number remained constant for two weeks, suggesting that the cells in the Si-HPMC particles did not proliferate. This result is in line with the lack of hASC adhesion when encapsulated in a Si-HPMC hydrogel, as demonstrated by Moussa et al. [57]. With a negatively charged cytoplasm membrane, hASCs adhesion depends on the charge of the hydrogel matrix. The overall neutral behavior of the Si-HPMC polymer does not provide a favorable environment for the adhesion and proliferation of the encapsulated hASCs. For most cell therapy applications, the apparent inability of the encapsulated hASCs to proliferate is, however, not a drawback. Therefore, the findings of the present study are a further indication that cellular microenvironments can be developed that permit the release of soluble therapeutic factors after the injection of cells, while also preventing the triggering of inflammation in degenerative diseases such as osteoarthritis.

4. Materials and Methods

4.1. Materials

Hydroxypropylmethylcellulose (HPMC) (Methocel™E4M) was purchased from Colorcon-Dow chemical (Bougival, France). Glycidoxypropyltrimethoxysilane (GPTMS) was obtained from Acros (Geel, Belgium). Hank's Balanced Sodium Salt (HBSS), Dulbecco's Modified Eagle Medium high glucose (4.5 g/L) (DMEM), phosphate buffered salt (PBS) without calcium chloride and magnesium chloride, penicillin/streptomycin, and trypsin/EDTA (0.05%/0.53 mM) were obtained from Invitrogen (Paisley, UK). 4-(2-hydroxyethyl)-1-piperazineethanesulfonic acid (HEPES), olive oil, fluorescein isothiocyanate (FITC)-dextrans, collagenase crude type I A, and trypan blue were obtained from Sigma-Aldrich (St. Louis, MO, USA). Fetal calf serum (FCS) was purchased from Dominique Dutscher (Brumath, France). The Live/Dead assay kit was obtained from Molecular Probes (Leiden, The Netherlands). Twelve-well plates (ref. 3512) were purchased from Corning (Boulogne Billancourt, France).

PolyEtherEtherKetone (PEEK) tubing for millifluidics were purchased from CIL, Cluzeau Info Labo (Sainte Foy la Grande, France).

4.2. Synthesis of the Hydrogel

The synthesis of the silanized-HPMC (Si-HPMC) was performed by grafting 14.24% (w/w) GPTMS onto HPMC in a heterogeneous medium as previously described [23]. Lyophilized Si-HPMC powder was solubilized (4% w/v) in 0.1 M NaOH under constant stirring for 24 h. The solution was then sterilized by steam autoclave (121 °C, 20 min). a Si-HPMC solution can be made to undergo crosslinking by a decrease in the pH. The sterilized solution was therefore mixed with one volume of HEPES buffer (v/v) (pH 3.55), in order to initiate the formation of a crosslinking of Si-HPMC chains at a final concentration of 2%.

4.3. Rheological Characterization of the Si-HPMC Hydrogel

4.3.1. Characterization of the Si-HPMC Polymeric Solution

Steady shear measurements were carried out to determine the viscosity of the unsterilized 4% Si-HPMC solution according to the simplified Cross equation [58].

$$\eta = \frac{\eta_0}{1 + (\lambda \dot{\gamma}_c)^n} \tag{2}$$

where η_0 is the limiting Newtonian viscosity at a low shear rate (Pa.s), λ is the relaxation time (the inverse of a critical shear rate $\dot{\gamma}_c$) (s), and n is the exponent of the power law.

Flow measurements were performed at 23 °C, with a fixed shear stress of 1 Pa, using a Rheostress 300 rheometer (ThermoHaake®, Karlsruhe, Germany) equipped with a titanium cone-plate geometry (60 mm in diameter, 1° cone angle). The gap between the truncation and the plate was 0.052 mm.

4.3.2. Characterization of the Si-HPMC Hydrogel

To study the gel times of 2% Si-HPMC, dynamic frequency experiments were carried out using a Rheostress 300 rheometer (ThermoHaake®, Karlsruhe, Germany) equipped with a titanium cone-plate geometry (60 mm in diameter, 1° cone angle), immediately after the initiation of the Si-HPMC crosslinking. The storage (G') and loss (G'') moduli were determined as a function of time at five oscillation frequencies (0.30 Hz, 0.50 Hz, 1 Hz, 1.80 Hz, and 3.2 Hz). This assessment was operated under a stress amplitude of 1 Pa at 23 or 37 °C. The temperature was controlled by an external thermal bath. The gelation time (t_{gel}) was calculated as the time at which $tan\,\delta(=G''/G')$ becomes independent of the frequency, in accordance with the criterion defined and proposed by Winter and Chambon [43,44].

To estimate the Si-HPMC hydrogel mesh size ξ (i.e., the average distance between the polymer entanglements in the hydrogel network), the Si-HPMC solution was crosslinked in a 12-well plate for 24 h at 37 °C. Hydrogel with a height of 5 mm and a diameter of 22 mm was then removed from the 12-well plate. Dynamic shear stress sweep measurements, with stress amplitudes ranging from 0.1 to 1000 Pa and a fixed frequency of 1 Hz, were carried out at 23 °C using a Mars rheometer (ThermoHaake®, Karlsruhe, Germany) equipped with a plate-plate geometry (20 mm in diameter). The Si-HPMC hydrogel mesh size ξ (m) was determined according to the Flory equation:

$$\xi = [\frac{k_B T}{G'}]^{(\frac{1}{3})} \tag{3}$$

with k_B representing the Boltzmann constant (J/K) and T the temperature (K) [59]. The average storage modulus (G') (Pa) in the linear regime on triplicate samples was determined to calculate the mesh size of the hydrogel.

4.4. Preparation of Si-HPMC Particles Using Millifluidics

A millifluidics device with a T-junction configuration was used to produce Si-HPMC particles (Figure 5) [60]. The dispersed phase, the mixing of the Si-HPMC solution with freshly prepared HEPES buffer, was pumped (Pilote C, Fresenius Kabi®, France) through a fused silica capillary tube (interior diameter (ID) = 150 μm and outside diameter (OD) = 375 μm) at a rate that varied between 5 and 30 μL/min. The continuous phase, olive oil compatible with biomedical applications, was pumped through a Teflon tube (ID = 0.5 mm and OD = 1.571 mm) at a rate that varied between 50 and 400 μL/min. The Si-HPMC hydrogel and the oil co-flowed in the Teflon tube (ID = 0.5 mm, OD = 1.571 mm, and length = 10 cm). The Si-HPMC drops were formed and dispersed in the continuous oil phase. The crosslinking of Si-HPMC, starting in the drops due to the decrease of pH after the addition of HEPES buffer, continued in the collection bath. a controlled stirring rate of 100 rpm using a rotating paddle and a temperature of 37 °C in the collection bath was shown to be crucial to avoid the coalescence of the Si-HPMC particles and to maintain the viability of the cells. After stirring for 3 h, the particles were sieved using a 100-μm mesh filter unit and rinsed using complete medium (DMEM containing 1% penicillin/streptomycin and 10% FCS). The Si-HPMC particles were then incubated at 37 °C in PBS until use.

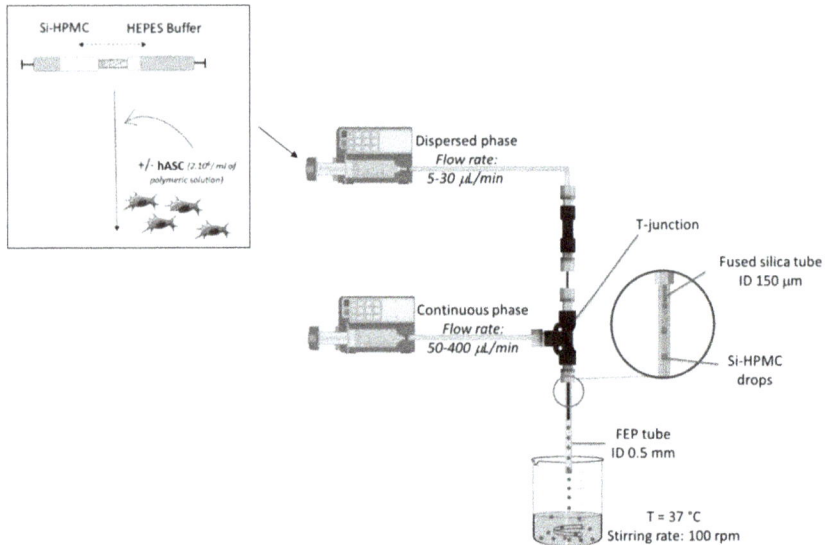

Figure 5. The droplet-based millifluidics device used to produce Si-HPMC particles. The control of temperature and stirring rate off-line in the collection bath were critical parameters that were optimized to avoid the coalescence of the particles and to maintain cell viability. The dispersed phase was comprised of Si-HPMC solution in the presence of freshly prepared HEPES buffer +/− loaded hASCs.

4.5. Characterization of the Si-HPMC Particles

4.5.1. Shape and Size

Particles were observed by light microscopy (Leica microsystems CMS GmbH, Type 11 090 137 002, Leica Biosystems, Nussloch, Germany) to investigate their shape, while the particle size was measured using a Mastersizer 3000 Laser (Malvern Instruments, Malvern, UK).

4.5.2. Diffusion Properties of the Si-HPMC Particles

The diffusion properties were studied by immobilizing Si-HPMC particles at the bottom of Lab-Tek chambers and followed by their incubation in 1 mg/mL solutions of fluorescently labeled dextran (molecular weights (M_w) of 20, 250, or 2000 kDa), for 18 h at room temperature. After incubation (6 h, 12 h, and 18 h), the particles were observed by confocal laser scanning microscopy (CLSM) (Nikon A1R Si, Champigny sur Marne, France; excitation wavelength of 488 nm, emission wavelength of 520 nm) in order to quantify the amount of fluorescently labeled dextran that had diffused into the Si-HPMC particles. The images were analyzed with NIS-Elements software to determine the fluorescence intensities inside the particles and outside (i.e., in the FITC-dextran solution). The ratio between the internal and the external fluorescence was then calculated. a ratio of 1 indicates that the fluorescence intensity was identical inside and outside the particles and that equilibrium was reached. The results were expressed as the internal/external fluorescence ratio over time.

4.6. Isolation and Culture of the hASCs

Human adipose-derived mesenchymal stem cells (hASCs) were isolated from subcutaneous adipose tissue of patients undergoing liposuction [61]. All of the protocols were approved by the biomedicine agency. ASCs were obtained from human patients undergoing liposuction and who had given written consent (Agence de BioMédecine n° PFS08-018, legislation codes L.1211-3 to L.1211-9, approval date: 9 September 2008). Briefly, lipoaspirate was washed five times in HBSS and then digested for one hour at 37 °C under constant stirring, in a solution of 0.025% collagenase in HBSS. The collagenase treatment was inactivated by the addition of an equal volume of complete medium. After 5 min of centrifugation ($260\times g$, 4 °C), the lower phase containing the stromal vascular fraction was collected, homogenized, filtered through a 70-μm cell strainer, and centrifuged for 8 min ($260\times g$, 4 °C). The cells were suspended in complete medium, seeded at 5000 cells/cm^2, and incubated at 37 °C in a humidified atmosphere containing 5% CO_2. After 2–3 days of incubation, the non-adherent cells were removed by successive washes.

4.7. hASCs Encapsulation

The cells were used at passage 5 (population doubling level (PDL) of 13.6). After being harvested using a trypsin/EDTA solution, the hASCs were counted and then loaded into Si-HPMC hydrogel. After the induction of crosslinking by the addition of HEPES buffer, 2.10^6 hASC (150 μL) was suspended in 1 mL of a sterile Si-HPMC hydrogel [46]. After droplets were generated in the millifluidics device, the hASCs loaded in the Si-HPMC particles were placed in complete medium (DMEM containing 1% penicillin/streptomycin and 10% FCS) for 3 h with stirring at 100 rpm and 37 °C. The Si-HPMC particles were then collected and incubated in complete medium, after the removal of the oily continuous phase.

4.8. hASCs Viability

The cells encapsulated in Si-HPMC were cultured in complete medium for up to 14 days at 37 °C in a humidified atmosphere containing 5% CO_2. The medium was changed every 2 days. After encapsulation, the hASCs viability in the Si-HPMC particles was followed from 24 h to 14 days of culture using a Live/Dead assay kit. The Si-HPMC particles were recovered, washed in PBS, and incubated for 45 min in the combined Live/Dead assay reagents. The labeled cells were imaged by confocal microscopy using an inverted fluorescence microscope (Nikon Eclipse TE 200 E, Badhoevedorp, The Netherlands). ImageJ® Software (version 1.8.0, NIH, Bethesda, MD, USA) was used to perform hASC viability calculations (the ratio of the number of live cells and the total number cells).

The average number of encapsulated hASCs per particle was manually assessed by confocal laser scanning microscopy (CLSM) (Nikon A1R Si, Champigny sur Marne, France) after analyses of approximately 500 images sections (section thickness of 10 μm) using ImageJ® Software.

4.9. Statistical Analysis

All of the experiments were performed with replicate samples from independent conditions ($n = 3$). Results of a representative experiment are presented as the mean of three independent replicates, and the error bars represent the standard error of the mean. The comparative studies of means were performed with GraphPad® software by using one-way ANOVA followed by a post hoc test with a statistical significance of $p < 0.05$.

5. Conclusions

In light of the considerable amount of encouraging data in the literature, ASCs encapsulation has a promising future in the treatment of inflammatory diseases. However, hASCs encapsulation remains rarely used for cell therapies. In this context, a novel and original approach based on droplet-based millifluidics was developed for the encapsulation of hASCs in injectable and biocompatible particles produced from Si-HPMC hydrogel. This versatile and effective tool allowed the generation of particles that fulfill three essential requirements for further in vivo study: sphericity, reproducibility, and size monodispersity. Spherical particles of 200 μm in diameter with an excellent reproducibility were obtained, which is a size that is suitable for injection in a large animal model. In addition, the successful encapsulation of hASCs in Si-HPMC particles was achieved and the cells remained viable for 14 days in vitro. The Si-HPMC particles' diffusion properties suggest that the diffusion of low-molecular-weight FITC-dextran was compatible with the diffusion of nutrients essential for the survival of the hASCs after their encapsulation. Droplet-based millifluidics therefore appears to be a promising non-cytotoxic method for the encapsulation of hASCs. Further developments are needed, however, in order to ensure better encapsulation performance, particularly by adding a third channel in the millifluidic setup, while ensuring cell sterility and the reduction of the "cellular stress" phenomenon.

Author Contributions: F.N., F.H., and P.-G.P. performed all of the experiments and analyzed the data; D.R. conceived and designed the droplet-based millifluidics experiments and conducted a critical read of the manuscript; C.D. conceived the rheological experiments; P.W. analyzed and interpreted the rheological data; C.L., J.G., and A.B.-C. thoroughly reviewed the manuscript; G.G. supervised F.N.'s work and conducted a critical read of the manuscript.

Funding: This work was supported by the INSERM and grants from the Research on OsteoArthristis Diseases network (ROAD network), the research programs "Longévité Mobilité Autonomie" and "Monomer" through the Bioregate Recherche Formation Innovation (RFI), both funded by the Region Pays de la Loire, and a pharmacy residency grant from the Nantes University Hospital.

Acknowledgments: We thank F. Lejeune (Clinique Brétéché, Nantes, France) for harvesting human lipoaspirates. We also would like to thank Joelle Davy (URBIA) for their helpful advice and technical assistance with the design of the millifluidics setup and the droplet-based millifluidics experiments. We acknowledge the help of the imaging core facilities of the University of Nantes (Micropicell platform).

Conflicts of Interest: The authors declare that there is no conflict of interest.

References

1. Perrot, P.; Heymann, D.; Charrier, C.; Couillaud, S.; Rédini, F.; Duteille, F. Extraosseous bone formation obtained by association of mesenchymal stem cells with a periosteal flap in the rat. *Ann. Plast. Surg.* **2007**, *59*, 201–206. [CrossRef] [PubMed]
2. Davis, N.E.; Beenken-Rothkopf, L.N.; Mirsoian, A.; Kojic, N.; Kaplan, D.L.; Barron, A.E.; Fontaine, M.J. Enhanced function of pancreatic islets co-encapsulated with ECM proteins and mesenchymal stromal cells in a silk hydrogel. *Biomaterials* **2012**, *33*, 6691–6697. [CrossRef] [PubMed]
3. Kauer, T.M.; Figueiredo, J.-L.; Hingtgen, S.; Shah, K. Encapsulated therapeutic stem cells implanted in the tumor resection cavity induce cell death in gliomas. *Nat. Neurosci.* **2011**, *15*, 197–204. [CrossRef] [PubMed]
4. Godier-Furnémont, A.F.G.; Tekabe, Y.; Kollaros, M.; Eng, G.; Morales, A.; Vunjak-Novakovic, G.; Johnson, L.L. Noninvasive imaging of myocyte apoptosis following application of a stem cell-engineered delivery platform to acutely infarcted myocardium. *J. Nucl. Med.* **2013**, *54*, 977–983. [CrossRef] [PubMed]
5. Paul, A.; Cantor, A.; Shum-Tim, D.; Prakash, S. Superior cell delivery features of genipin crosslinked polymeric microcapsules: Preparation, in vitro characterization and pro-angiogenic applications using human adipose stem cells. *Mol. Biotechnol.* **2011**, *48*, 116–127. [CrossRef] [PubMed]
6. Desando, G.; Cavallo, C.; Sartoni, F.; Martini, L.; Parrilli, A.; Veronesi, F.; Fini, M.; Giardino, R.; Facchini, A.; Grigolo, B. Intra-articular delivery of adipose derived stromal cells attenuates osteoarthritis progression in an experimental rabbit model. *Arthritis Res. Ther.* **2013**, *15*, R22. [CrossRef] [PubMed]
7. Pittenger, M.F.; Mackay, A.M.; Beck, S.C.; Jaiswal, R.K.; Douglas, R.; Mosca, J.D.; Moorman, M.A.; Simonetti, D.W.; Craig, S.; Marshak, D.R. Multilineage potential of adult human mesenchymal stem cells. *Science* **1999**, *284*, 143–147. [CrossRef] [PubMed]
8. Vizoso, F.J.; Eiro, N.; Cid, S.; Schneider, J.; Perez-Fernandez, R. Mesenchymal Stem Cell Secretome: Toward Cell-Free Therapeutic Strategies in Regenerative Medicine. *Int. J. Mol. Sci.* **2017**, *18*, 1852. [CrossRef] [PubMed]
9. Toh, W.S.; Lai, R.C.; Hui, J.H.P.; Lim, S.K. MSC exosome as a cell-free MSC therapy for cartilage regeneration: Implications for osteoarthritis treatment. *Semin. Cell Dev. Biol.* **2017**, *67*, 56–64. [CrossRef] [PubMed]
10. Vinatier, C.; Mrugala, D.; Jorgensen, C.; Guicheux, J.; Noël, D. Cartilage engineering: a crucial combination of cells, biomaterials and biofactors. *Trends Biotechnol.* **2009**, *27*, 307–314. [CrossRef] [PubMed]
11. Aguado, B.A.; Mulyasasmita, W.; Su, J.; Lampe, K.J.; Heilshorn, S.C. Improving viability of stem cells during syringe needle flow through the design of hydrogel cell carriers. *Tissue Eng. Part A* **2012**, *18*, 806–815. [CrossRef] [PubMed]

12. Amer, M.H.; Rose, F.R.A.J.; Shakesheff, K.M.; Modo, M.; White, L.J. Translational considerations in injectable cell-based therapeutics for neurological applications: Concepts, progress and challenges. *NPJ Regen. Med.* **2017**, *2*, 23. [CrossRef] [PubMed]

13. Detante, O.; Moisan, A.; Dimastromatteo, J.; Richard, M.-J.; Riou, L.; Grillon, E.; Barbier, E.; Desruet, M.-D.; De Fraipont, F.; Segebarth, C.; et al. Intravenous administration of 99mTc-HMPAO-labeled human mesenchymal stem cells after stroke: In vivo imaging and biodistribution. *Cell Transplant.* **2009**, *18*, 1369–1379. [CrossRef] [PubMed]

14. Toupet, K.; Maumus, M.; Peyrafitte, J.-A.; Bourin, P.; van Lent, P.L.E.M.; Ferreira, R.; Orsetti, B.; Pirot, N.; Casteilla, L.; Jorgensen, C.; et al. Long-term detection of human adipose-derived mesenchymal stem cells after intraarticular injection in SCID mice. *Arthritis Rheum.* **2013**, *65*, 1786–1794. [CrossRef] [PubMed]

15. Blocki, A.; Beyer, S.; Dewavrin, J.-Y.; Goralczyk, A.; Wang, Y.; Peh, P.; Ng, M.; Moonshi, S.S.; Vuddagiri, S.; Raghunath, M.; et al. Microcapsules engineered to support mesenchymal stem cell (MSC) survival and proliferation enable long-term retention of MSCs in infarcted myocardium. *Biomaterials* **2015**, *53*, 12–24. [CrossRef] [PubMed]

16. Leijs, M.J.; Villafuertes, E.; Haeck, J.C.; Koevoet, W.J.; Fernandez-Gutierrez, B.; Hoogduijn, M.J.; Verhaar, J.A.; Bernsen, M.R.; van Buul, G.M.; van Osch, G.J. Encapsulation of allogeneic mesenchymal stem cells in alginate extends local presence and therapeutic function. *Eur. Cell Mater.* **2017**, *33*, 43–58. [CrossRef] [PubMed]

17. Smith, K.E.; Johnson, R.C.; Papas, K.K. Update on cellular encapsulation. *Xenotransplantation* **2018**, e12399. [CrossRef] [PubMed]

18. Lim, F.; Sun, A.M. Microencapsulated islets as bioartificial endocrine pancreas. *Science* **1980**, *210*, 908–910. [CrossRef] [PubMed]

19. Orive, G.; Santos, E.; Poncelet, D.; Hernández, R.M.; Pedraz, J.L.; Wahlberg, L.U.; De Vos, P.; Emerich, D. Cell encapsulation: Technical and clinical advances. *Trends Pharmacol. Sci.* **2015**, *36*, 537–546. [CrossRef] [PubMed]

20. Forget, A.; Blaeser, A.; Miessmer, F.; Köpf, M.; Campos, D.F.D.; Voelcker, N.H.; Blencowe, A.; Fischer, H.; Shastri, V.P. Mechanically Tunable Bioink for 3D Bioprinting of Human Cells. *Adv. Healthc. Mater.* **2017**, *6*. [CrossRef] [PubMed]

21. LeRoux, M.A.; Guilak, F.; Setton, L.A. Compressive and shear properties of alginate gel: Effects of sodium ions and alginate concentration. *J. Biomed. Mater. Res.* **1999**, *47*, 46–53. [CrossRef]

22. Weiss, P.; Guicheux, J.; Daculsi, G.; Grimandi, G.; Vinatier, C. Use of a Hydrogel for the Culture of Chondrocytes. EP20030292759, 4 November 2013.

23. Bourges, X.; Weiss, P.; Daculsi, G.; Legeay, G. Synthesis and general properties of silated-hydroxypropyl methylcellulose in prospect of biomedical use. *Adv. Colloid Interface Sci.* **2002**, *99*, 215–228. [CrossRef]

24. Fatimi, A.; Tassin, J.F.; Quillard, S.; Axelos, M.A.V.; Weiss, P. The rheological properties of silated hydroxypropylmethylcellulose tissue engineering matrices. *Biomaterials* **2008**, *29*, 533–543. [CrossRef] [PubMed]

25. Laïb, S.; Fellah, B.H.; Fatimi, A.; Quillard, S.; Vinatier, C.; Gauthier, O.; Janvier, P.; Petit, M.; Bujoli, B.; Bohic, S.; et al. The in vivo degradation of a ruthenium labelled polysaccharide-based hydrogel for bone tissue engineering. *Biomaterials* **2009**, *30*, 1568–1577. [CrossRef] [PubMed]

26. Lee, B.-B.; Ravindra, P.; Chan, E.-S. Size and Shape of Calcium Alginate Beads Produced by Extrusion Dripping. *Chem. Eng. Technol.* **2013**, *36*, 1627–1642. [CrossRef]

27. Elbert, D.L. Liquid-liquid two-phase systems for the production of porous hydrogels and hydrogel microspheres for biomedical applications: a tutorial review. *Acta Biomater.* **2011**, *7*, 31–56. [CrossRef] [PubMed]

28. Schmit, A.; Courbin, L.; Marquis, M.; Renard, D.; Panizza, P. a pendant drop method for the production of calibrated double emulsions and emulsion gels. *RSC Adv.* **2014**, *4*, 28504–28510. [CrossRef]

29. Hached, F.; Vinatier, C.; Pinta, P.-G.; Hulin, P.; Le Visage, C.; Weiss, P.; Guicheux, J.; Billon-Chabaud, A.; Grimandi, G. Polysaccharide Hydrogels Support the Long-Term Viability of Encapsulated Human Mesenchymal Stem Cells and Their Ability to Secrete Immunomodulatory Factors. *Stem Cells Int.* **2017**, *2017*, 9303598. [CrossRef] [PubMed]

30. Duncanson, W.J.; Lin, T.; Abate, A.R.; Seiffert, S.; Shah, R.K.; Weitz, D.A. Microfluidic synthesis of advanced microparticles for encapsulation and controlled release. *Lab Chip* **2012**, *12*, 2135–2145. [CrossRef] [PubMed]

31. Theberge, A.B.; Courtois, F.; Schaerli, Y.; Fischlechner, M.; Abell, C.; Hollfelder, F.; Huck, W.T.S. Microdroplets in microfluidics: an evolving platform for discoveries in chemistry and biology. *Angew. Chem. Int. Ed. Engl.* **2010**, *49*, 5846–5868. [CrossRef] [PubMed]

32. Hidalgo San Jose, L.; Stephens, P.; Song, B.; Barrow, D. Microfluidic Encapsulation Supports Stem Cell Viability, Proliferation, and Neuronal Differentiation. *Tissue Eng. Part C Methods* **2018**. [CrossRef] [PubMed]

33. Quevedo, E.; Steinbacher, J.; McQuade, D.T. Interfacial polymerization within a simplified microfluidic device: Capturing capsules. *J. Am. Chem. Soc.* **2005**, *127*, 10498–10499. [CrossRef] [PubMed]

34. Engl, W.; Tachibana, M.; Panizza, P.; Backov, R. Millifluidic as a versatile reactor to tune size and aspect ratio of large polymerized objects. *Int. J. Multiph. Flow* **2007**, *33*, 897–903. [CrossRef]

35. Tadmouri, R.; Romano, M.; Guillemot, L.; Mondain-Monval, O.; Wunenburger, R.; Leng, J. Millifluidic production of metallic microparticles. *Soft Matter* **2012**, *8*, 10704–10711. [CrossRef]

36. Lukyanova, L.; Séon, L.; Aradian, A.; Mondain-Monval, O.; Leng, J.; Wunenburger, R. Millifluidic synthesis of polymer core-shell micromechanical particles: Toward micromechanical resonators for acoustic metamaterials. *J. Appl. Polym. Sci.* **2013**, *128*, 3512–3521. [CrossRef]

37. Martins, E.; Poncelet, D.; Marquis, M.; Davy, J.; Renard, D. Monodisperse core-shell alginate (micro)-capsules with oil core generated from droplets millifluidic. *Food Hydrocoll.* **2016**, *63*. [CrossRef]

38. Amine, C.; Boire, A.; Davy, J.; Marquis, M.; Renard, D. Droplets-based millifluidic for the rapid determination of biopolymers phase diagrams. *Food Hydrocoll.* **2017**, *70*, 134–142. [CrossRef]

39. Sun, X.-T.; Liu, M.; Xu, Z.-R. Microfluidic fabrication of multifunctional particles and their analytical applications. *Talanta* **2014**, *121*, 163–177. [CrossRef] [PubMed]

40. Lorber, N.; Sarrazin, F.; Guillot, P.; Panizza, P.; Colin, A.; Pavageau, B.; Hany, C.; Maestro, P.; Marre, S.; Delclos, T.; et al. Some recent advances in the design and the use of miniaturized droplet-based continuous process: Applications in chemistry and high-pressure microflows. *Lab Chip* **2011**, *11*, 779–787. [CrossRef] [PubMed]

41. Engl, W.; Backov, R.; Panizza, P. Controlled production of emulsions and particles by milli- and microfluidic techniques. *Curr. Opin. Colloid Interface Sci.* **2008**, *13*, 206–216. [CrossRef]

42. Tumarkin, E.; Kumacheva, E. Microfluidic generation of microgels from synthetic and natural polymers. *Chem. Soc. Rev.* **2009**, *38*, 2161–2168. [CrossRef] [PubMed]

43. Winter, H.H.; Chambon, F. Analysis of Linear Viscoelasticity of a Crosslinking Polymer at the Gel Point. *J. Rheol.* **1986**, *30*, 367–382. [CrossRef]

44. Chambon, F.; Winter, H.H. Linear Viscoelasticity at the Gel Point of a Crosslinking PDMS with Imbalanced Stoichiometry. *J. Rheol.* **1987**, *31*, 683–697. [CrossRef]

45. Bidarra, S.J.; Barrias, C.C.; Granja, P.L. Injectable alginate hydrogels for cell delivery in tissue engineering. *Acta Biomater.* **2014**, *10*, 1646–1662. [CrossRef] [PubMed]

46. Vinatier, C.; Magne, D.; Weiss, P.; Trojani, C.; Rochet, N.; Carle, G.F.; Vignes-Colombeix, C.; Chadjichristos, C.; Galera, P.; Daculsi, G.; et al. a silanized hydroxypropyl methylcellulose hydrogel for the three-dimensional culture of chondrocytes. *Biomaterials* **2005**, *26*, 6643–6651. [CrossRef] [PubMed]

47. Fatimi, A.; Tassin, J.-F.; Turczyn, R.; Axelos, M.A.V.; Weiss, P. Gelation studies of a cellulose-based biohydrogel: The influence of pH, temperature and sterilization. *Acta Biomater.* **2009**, *5*, 3423–3432. [CrossRef] [PubMed]

48. Bourges, X.; Weiss, P.; Coudreuse, A.; Daculsi, G.; Legeay, G. General properties of silated hydroxyethylcellulose for potential biomedical applications. *Biopolymers* **2002**, *63*, 232–238. [CrossRef] [PubMed]

49. Kondiah, P.J.; Choonara, Y.E.; Kondiah, P.P.D.; Marimuthu, T.; Kumar, P.; du Toit, L.C.; Pillay, V. a Review of Injectable Polymeric Hydrogel Systems for Application in Bone Tissue Engineering. *Molecules* **2016**, *21*. [CrossRef] [PubMed]

50. Matlaga, B.F.; Yasenchak, L.P.; Salthouse, T.N. Tissue response to implanted polymers: The significance of sample shape. *J. Biomed. Mater. Res.* **1976**, *10*, 391–397. [CrossRef] [PubMed]

51. Kim, S.W.; Bae, Y.H.; Okano, T. Hydrogels: Swelling, drug loading, and release. *Pharm. Res.* **1992**, *9*, 283–290. [CrossRef] [PubMed]

52. Lee, A.G.; Arena, C.P.; Beebe, D.J.; Palecek, S.P. Development of macroporous poly(ethylene glycol) hydrogel arrays within microfluidic channels. *Biomacromolecules* **2010**, *11*, 3316–3324. [CrossRef] [PubMed]

53. Opal, S.M.; DePalo, V.A. Anti-inflammatory cytokines. *Chest* **2000**, *117*, 1162–1172. [CrossRef] [PubMed]

54. Feghali, C.A.; Wright, T.M. Cytokines in acute and chronic inflammation. *Front. Biosci.* **1997**, *2*, d12–26. [PubMed]
55. Caplan, A.I.; Correa, D. The MSC: an injury drugstore. *Cell Stem Cell* **2011**, *9*, 11–15. [CrossRef] [PubMed]
56. Figueiredo, L.; Pace, R.; d'Arros, C.; Réthoré, G.; Guicheux, J.; Le Visage, C.; Weiss, P. Assessing glucose and oxygen diffusion in hydrogels for the rational design of 3D stem cell scaffolds in regenerative medicine. *J. Tissue Eng. Regen. Med.* **2018**. [CrossRef] [PubMed]
57. Moussa, L.; Pattappa, G.; Doix, B.; Benselama, S.-L.; Demarquay, C.; Benderitter, M.; Sémont, A.; Tamarat, R.; Guicheux, J.; Weiss, P.; et al. a biomaterial-assisted mesenchymal stromal cell therapy alleviates colonic radiation-induced damage. *Biomaterials* **2017**, *115*, 40–52. [CrossRef] [PubMed]
58. Cross, M. Rheology of Non-Newtonian Fluids: a New Flow Equation for Pseudoplastic Systems. *J. Colloid Sci.* **1965**, *20*, 417–437. [CrossRef]
59. Flory, F. *1953 Principles of Polymer Chemistry*; Cornell University Press: Ithaca, NY, USA, 2017.
60. Mazzitelli, S.; Capretto, L.; Quinci, F.; Piva, R.; Nastruzzi, C. Preparation of cell-encapsulation devices in confined microenvironment. *Adv. Drug Deliv. Rev.* **2013**, *65*, 1533–1555. [CrossRef] [PubMed]
61. Merceron, C.; Portron, S.; Vignes-Colombeix, C.; Rederstorff, E.; Masson, M.; Lesoeur, J.; Sourice, S.; Sinquin, C.; Colliec-Jouault, S.; Weiss, P.; et al. Pharmacological modulation of human mesenchymal stem cell chondrogenesis by a chemically oversulfated polysaccharide of marine origin: Potential application to cartilage regenerative medicine. *Stem Cells* **2012**, *30*, 471–480. [CrossRef] [PubMed]

International Journal of
Molecular Sciences

MDPI

Article

Preparation and Characterization of Resorbable Bacterial Cellulose Membranes Treated by Electron Beam Irradiation for Guided Bone Regeneration

Sung-Jun An [1,†], So-Hyoun Lee [2,†], Jung-Bo Huh [2,*], Sung In Jeong [1], Jong-Seok Park [1], Hui-Jeong Gwon [1], Eun-Sook Kang [3], Chang-Mo Jeong [2] and Youn-Mook Lim [1]

[1]   Advanced Radiation Technology Institute, Korea Atomic Energy Research Institute, 1266 Sinjeong-dong, Jeongeup-si, Jeollabuk-do 56212, Korea; asj@kaeri.re.kr (S.-J.A.); sijeong@kaeri.re.kr (S.I.J.); jspark75@kaeri.re.kr (J.-S.P.); hjgwon@kaeri.re.kr (H.-J.G.); ymlim71@gmail.com (Y.-M.L.)
[2]   Department of Prosthodontics, Dental Research Institute, Institute of Translational Dental Sciences, BK21 PLUS Project, School of Dentistry, Pusan National University, Yangsan 50612, Korea; romilove7@hanmail.net (S.-H.L.); cmjeong@pusan.ac.kr (C.-M.J.)
[3]   Department of Prosthodontics, In-Je University Haeundae Paik Hospital, Busan 48108, Korea; prosth-kang@hanmail.net
*   Correspondence: huhjb@pusan.ac.kr; Tel.: +82-55-360-5146
†   These authors contributed equally to this work.

Received: 26 September 2017; Accepted: 18 October 2017; Published: 25 October 2017

Abstract: Bacterial cellulose (BC) is an excellent biomaterial with many medical applications. In this study, resorbable BC membranes were prepared for guided bone regeneration (GBR) using an irradiation technique for applications in the dental field. Electron beam irradiation (EI) increases biodegradation by severing the glucose bonds of BC. BC membranes irradiated at 100 kGy or 300 kGy were used to determine optimal electron beam doses. Electron beam irradiated BC membranes (EI-BCMs) were evaluated by scanning electron microscopy (SEM), attenuated total reflectance-Fourier transform infrared (ATR-FTIR) spectroscopy, thermal gravimetric analysis (TGA), and using wet tensile strength measurements. In addition, in vitro cell studies were conducted in order to confirm the cytocompatibility of EI-BCMs. Cell viabilities of NIH3T3 cells on 100k and 300k EI-BCMs (100 kGy and 300 kGy irradiated BC membranes) were significantly greater than on NI-BCMs after 3 and 7 days ($p < 0.05$). Bone regeneration by EI-BCMs and their biodegradabilities were also evaluated using in vivo rat calvarial defect models for 4 and 8 weeks. Histometric results showed 100k EI-BCMs exhibited significantly larger new bone area (NBA; %) than 300k EI-BCMs at 8 weeks after implantation ($p < 0.05$). Mechanical, chemical, and biological analyses showed EI-BCMs effectively interacted with cells and promoted bone regeneration.

Keywords: bacterial cellulose membrane; guided bone regeneration; electron beam irradiation; resorbable barrier membrane; optimal radiation dose

1. Introduction

Various techniques have been used to increase the success rate of tissue regeneration in the dental field [1–7]. In particular, guided bone regeneration (GBR) is a well-known and widely used technique that uses barrier membranes to prevent the infiltration of soft tissue to bone augmented regions [8].The barrier membranes used for GBR should have the characteristics of cell occlusiveness, wound stabilization, space-making, and provide a stable environment for bone regeneration [9]. Barrier membranes are classified as resorbable or non-resorbable [10]. Non-resorbable membrane materials include polytetrafluoroethylene (PTFE) and titanium mesh, whereas, resorbable membrane materials include polyglycolic acid (PGA), alginate, polylactic acid (PLA), and collagen [11–14].

Non-resorbable membranes need an additional surgical removal procedure for their removal to prevent wound dehiscence [15]. However, this additional procedure can cause infection, undesirable bone resorption, and have other undesirable side effects [16,17]. Resorbable membranes offer many other advantages over non-resorbable membranes, for example, they provide better soft tissue healing, are cheaper to produce, and have lower complication risks [18].

Bacterial cellulose (BC) has been produced by in vitro synthesis using the Gram-negative bacterium Gluconacetobacter xylinum [19]. BC membranes (BCMs) are composed of 3-dimensional (3D) nano-fibrous networks of linear polysaccharide polymer linked by β-(1,4) glycosidic linkages [20]. Typically, BCMs have good mechanical properties, high levels of crystallinity, high water holding capacities, interconnected 3D porous nanostructures, and excellent biocompatibilities [21–24]. These characteristics can be advantageous for regeneration of body organs, such as, skin, bone, cartilage, nerves, heart, and blood vessels [25–27]. Recently, many studies have examined potential uses for BCM in the dental field [28,29]. The structure of BCMs is similar to that of collagen membrane, which is the material most widely used for resorbable barrier membranes for GBR [30].

However, there is an important limitation to the use of BCM to replace the collagen membrane. BCM is not biodegradable in the human body because of a lack of cellulose degrading enzymes (cellulases) [31]. To overcome this problem, various methods, such as, acid hydrolysis, alkaline hydrolysis, delignification by oxidation organosolv pretreatment and pretreatment with ionic liquids, have been proposed to accelerate the hydrolysis of cellulose [32–36]. However, these methods have disadvantages, such as, difficulty accurately controlling degradation and potential cytotoxicity due to residual chemicals in BCMs for clinical application [31,33,36].

Therefore, in the present study, an electron beam irradiation (EI) processes was used to control the biodegradability of BCMs. Several radiation-based techniques based on gamma ray, electron beam, or ion beam irradiation have been used to crosslink, graft, or degrade polymers and thus modify their properties [37]. In particular, it has been reported that high energy EI effectively reduces natural polymer molecular weight and mechanical properties without the need for additional chemicals [38–41]. The purpose of this study was to prepare BCMs irradiated with different electron beam doses and to determine their mechanical, chemical, and biological properties. In addition, an in vivo study was conducted using a rat calvarial defect model to optimize the electron beam irradiation process in terms of bone regeneration and biodegradability.

2. Results and Discussion

2.1. Characterization of EI-BCMs

2.1.1. Scanning Electron Microscopy

Figure 1 presents SEM images of NI-BCMs, and 100k and 300k EI-BCMs. All BCMs had a porous multilayered structure of entangled nanofibers and cross-linked by nanofibers between layers. Cleaved BC nanofibers were observed in 100k and 300k EI-BCMs, but 3D porous structures were not affected. BC nanofiber cleavages are considered to result from D-glucose chain and hydrogen bond cleavage [42].

2.1.2. Mechanical Properties Analyses

As depicted in Figure 2, mechanical properties, such as wet tensile stress, wet tensile strain, and Young's modulus of 100k and 300k EI-BCMs were significantly lower than those of NI-BCMs ($p < 0.05$), and those of 300k BCM were the lowest, which were attributed to cleavages observed by SEM [42]. It has been previously reported that chain cleavage of BC nanofibers resulted in significant reductions in mechanical properties [38–41], and that resorbable membranes used for GBR procedures should have sufficient mechanical strength to attach tightly to bone defects, to prevent sagging and avoid rupture during surgery [43,44]. Although the mechanical properties of EI-BCMs were reduced by irradiation, the mechanical properties of 100k EI-BCMs were similar to those of collagen membranes, as we previously reported [30].

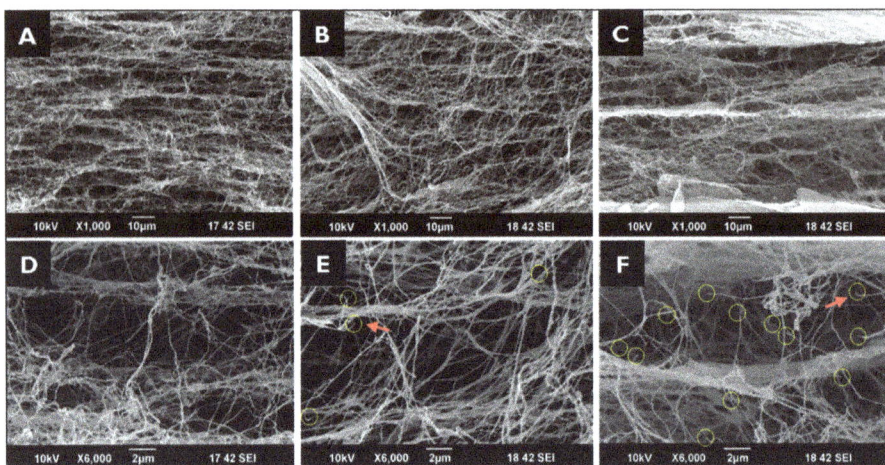

Figure 1. Cross-sectional SEM images: (**A,D**) NI-BCM; (**B,E**) 100k EI-BCM; (**C,F**) 300k EI-BCM. Yellow circles and red arrows indicate cleaved BC nanofibers in 100k (**E**) and 300k EI-BCMs (**F**).

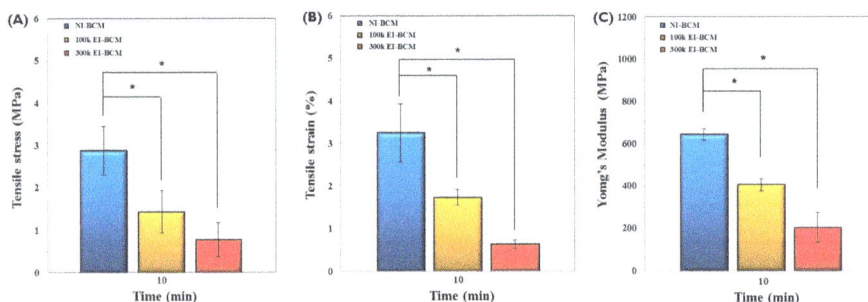

Figure 2. Mechanical properties of BCMs with respect to radiation dose rate after a 10-min soak in water. (**A**) Tensile stress (MPa), (**B**) Tensile strain (%), and (**C**) Young's modulus (MPa). The mechanical properties of 100k and 300k EI-BCMs were significantly lower than those of NI-BCMs (* $p < 0.05$).

2.1.3. Attenuated Total Reflection-Fourier Transform Infrared Spectroscopy (ATR-FTIR)

ATR-FTIR was performed in order to determine the effect of irradiation dose on molecular changes in BCMs. The IR spectra of all three BCMs had peaks at 3410, 2900, 1642, and 1060 cm^{-1}. Figure 3 illustrates that the intensity of the O–H group at 3410 cm^{-1}, of C–H stretch at 2900 cm^{-1}, of H–O–H bending of absorbed water at 1642 cm^{-1}, and of the C–O–C pyranose ring skeletal vibration at 1060 cm^{-1} decreased on increasing the irradiation dose. This indicates that the cleavage occurred in the main chain of the BC through change of chemical characteristics [45,46]. It has been previously reported that irradiation causes the degradations of polysaccharides and natural polymers [31,47].

2.1.4. Thermogravimetric Analyses (TGAs)

Thermal gravimetric analysis was used to determine the thermal properties of NI-BCM, 100k BCM and 300k BCM. Figure 4 shows the temperature decomposition profiles of these BCMs. NI-BCMs had a high decomposition temperature of near 300 °C. This was attributed to strong inter-chain hydrogen bonds in the crystalline regions of BCM [48]. As the radiation dose was increased, weight loss rates increased (NI-BCM, 7%; 100k EI-BCM, 15%; and 300k EI-BCM, 22%).

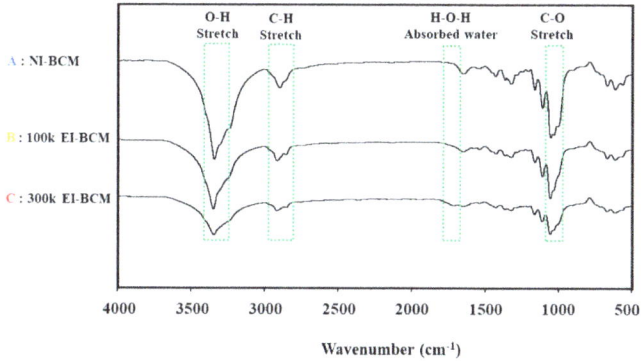

Figure 3. ATR-FTIR spectra of **A**: NI-BCM, **B**: 100k EI-BCM and **C**: 300k EI-BCM. The results obtained indicated that cleavage occurred in the main chain of the BC through change of chemical characteristics.

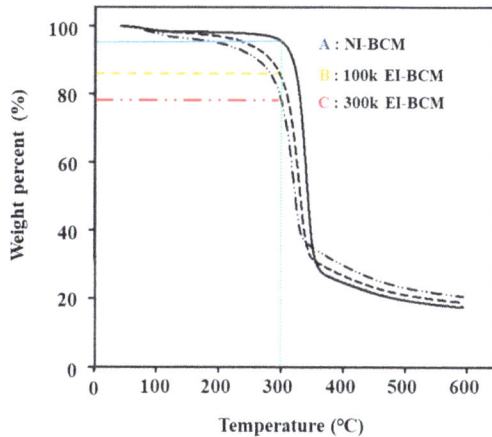

Figure 4. Thermal gravimetric analysis (TGA). NI-BCM had a high decomposition temperature near 300 °C. As radiation dose increased, weight loss rates increased (**A**: NI-BCM, 7%; **B**: 100k EI-BCM, 15%; and **C**: 300k EI-BCM, 22%).

2.1.5. In Vitro Degradation

Figure 5 shows the effect of electron beam dose on BCM degradation in PBS. The degradation rates of NI-BCMs and of 100k and 300k EI-BCMs were measured at 4, 8, and 16 weeks. The degradation rate of EI-BCMs was observed to increase with the radiation dose, whereas the degradation rate of NI-BCMs did not change. In particular, 300k EI-BCMs exhibited a weight reduction of ~70% after 16 weeks in PBS. It has been reported mechanical and chemical changes in biopolymers caused by radiation promote biodegradation [32,42]. This result confirmed that the hydrolysis of the BC polysaccharide chain by EI can be effective for reduction of the molecular weight of BC and accelerated the degradability of the BCMs [39,49,50].

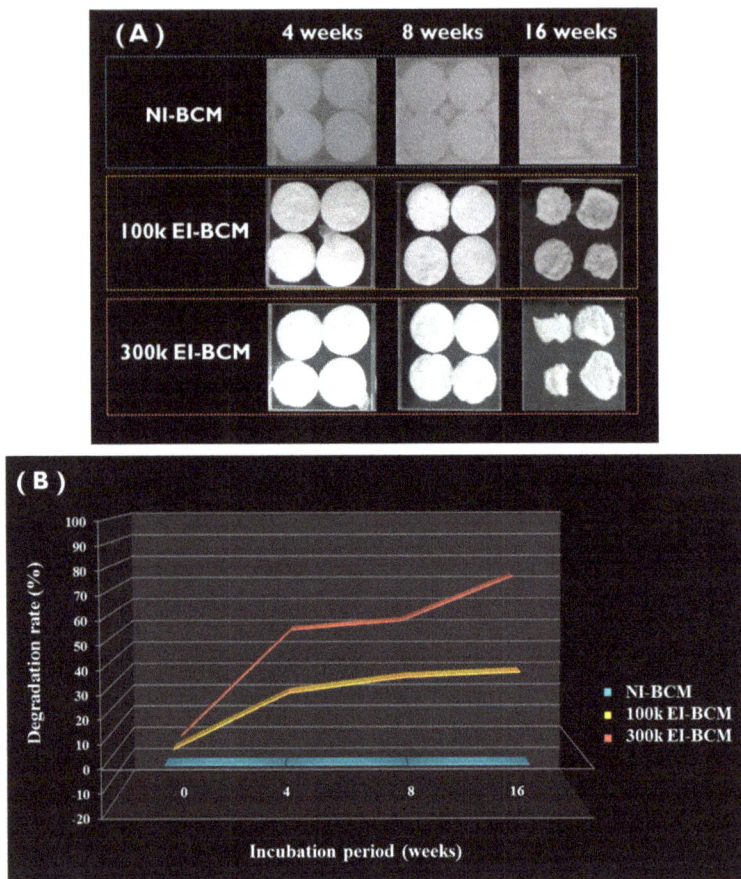

Figure 5. In vitro degradation of BCMs. (**A**) The degradation rates of NI-BCMs, and 100k and 300k EI-BCMs were measured after immersion in PBS for 4, 8, or 16 weeks. (**B**) Degradation rates of EI-BCMs increased with radiation dose, whereas the degradation rate of NI-BCM did not change. 300k EI-BCMs exhibited a weight loss of approximately 70% after 16 weeks.

2.2. In Vitro Cell Studies

2.2.1. Cell Proliferation Assay

CCk-8 assays of cells on the NI-BCMs and 100k and 300k EI-BCMs were conducted in order to determine cell viabilities, adhesions, and proliferations. As depicted in Figure 6, cell viabilities of NIH3T3 cells on 100k and 300k EI-BCMs were significantly greater than on NI-BCMs after 3 and 7 days ($p < 0.05$). After one day, although the initial cell proliferation of NIH3T3 cells on all samples were poorer, cells on 100k EI-BCM proliferated slightly better than cells on NI-BCM. Furthermore, these results demonstrate that the cell viabilities, adhesions, and proliferations are bioactive after the electron beam irradiation process. Because the surface biochemical characteristics of barrier membranes used for GBR influence cell adhesion, the effects of surface modification of BCMs using radiation [51], plasma [52], small signaling peptides [53], and of amino acid (e.g., Arg-Gly-Asp (RGD)) [54] have been investigated with the aim of improving interactions between cells and BC. These modifications of BCM surfaces can change the density in neutral polysaccharides of NI-BCM surface [55]. In the present study, the hydrophilic surfaces of EI-BCMs modified by electron beam irradiation were found to be more bioactive and to promote cell adhesion, viability and proliferation on BCMs.

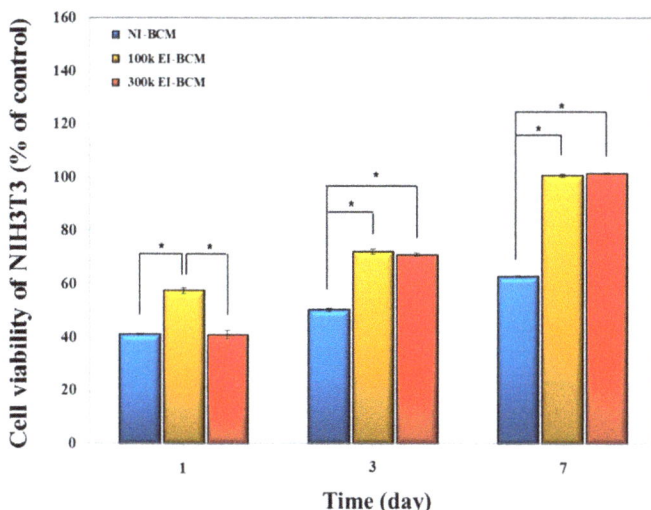

Figure 6. Cell viabilities of NIH3T3 cells cultured on BCMs. CCK-8 assays showed that the viabilities of cells on 100k EI-BCMs and 300k BCMs were significantly greater than those on NI-BCMs after 3 and 7 days (* $p < 0.05$).

2.2.2. Immunofluorescent Staining and FE-SEM Analyses of Cells on BCMs

The abilities of BCMs to support NIH3T3 cell adhesion were evaluated by F-actin staining (Figure 7) and FE-SEM (Figure 8). Cells adherent on EI-BCMs were more differentiated and had noticeable long, straight f-actin stress fibers than cells on NI-BCMs, which were circular. While the cells on NI-BCMs were mainly localized, the development of cell growth on EI-BCMs appeared to be guided by their nano-fibrous structures. In addition, cells on 300k EI-BCMs were more differentiated than cells on 100k EI-BCMs.

2.3. In Vivo Animal Studies

2.3.1. Histologic Findings

During the initial healing period, mild signs of inflammation, such as, exudate and edema, were observed around grafted BCMs in some rats, but no sign of foreign body or microscopic inflammation was observed. These tissue responses of grafted BC materials have been reported in previous studies [24,56–58]. Interestingly, the infection rate of BC in man is so low it is used to produce dressings for wounds and burns [26,55,59].

During the 8-week healing period, grafted EI-BCMs did not induce inflammatory responses and integrated with surrounding tissues (Figure 9). Both 100k and 300k EI-BCMs maintained adequate space for bone regeneration and these spaces under membranes were filled with fibrous connective tissue and bone-like materials (Figure 10). No EI-BCM was completely degraded after 8 weeks, but the degradation of 300k EI-BCMs was greater than that of 100k EI-BCMs at 4 and 8 weeks (Figure 11).

Although the EI process did not change the thickness or nanoporous structure of BCMs [41,42,49] and both EI-BCM specimens for present study were fabricated with similar thicknesses, the more hydrophilic surfaces of the 300k EI-BCM caused more active tissue reactions, leading to more degradation and the remaining membrane was thinner than that of 100k EI-BCM [39,42,47,49,50].

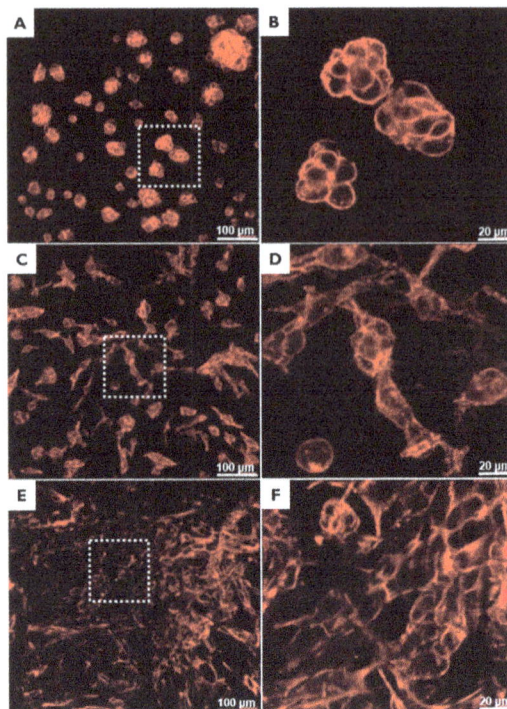

Figure 7. Immunofluorescent staining images obtained by confocal microscopy of adherent cells on NI-BCMs (**A**,**B**); 100k EI-BCMs (**C**,**D**); and 300k EI-BCMs (**E**,**F**). The adherent cells on EI-BCMs were more differentiated and possessed long, straight f-actin stress fibers whereas those on NI-BCMs were circular. Furthermore, degree of cell differentiation increased with irradiation dose.

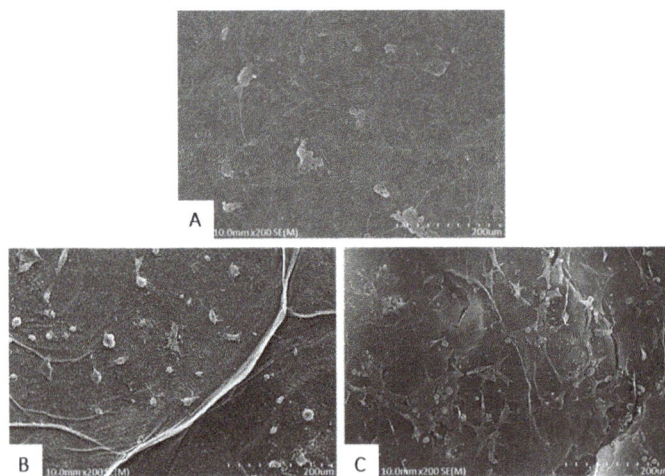

Figure 8. FE-SEM images of adherent cells on NI-BCMs (**A**), 100k EI-BCMs (**B**) and 300k EI-BCMs (**C**). Cells on NI-BCMs were localized, whereas the cell growth on EI-BCMs was induced by the nano-fibrous structure of BC and spread in random directions.

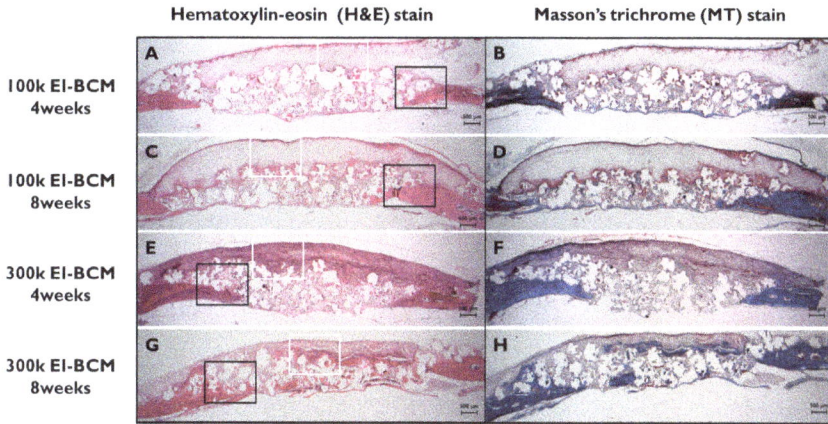

Figure 9. Histological views of defect sites in the 100k and 300k EI-BCM groups. New bone formation and fibrous connective tissue were observed at 4 weeks (**A**,**B**,**E**,**F**) and at 8 weeks after surgery (**C**,**D**,**G**,**H**), and were mainly observed around membranes and old bone. The black rectangles indicate the new bone area in Figure 10 and the white boxes represent the EI-BCMs in Figure 11 (original magnification: 12.5×; (**A**,**C**,**E**,**G**) H&E stained; (**B**,**D**,**F**,**H**) M&T stained).

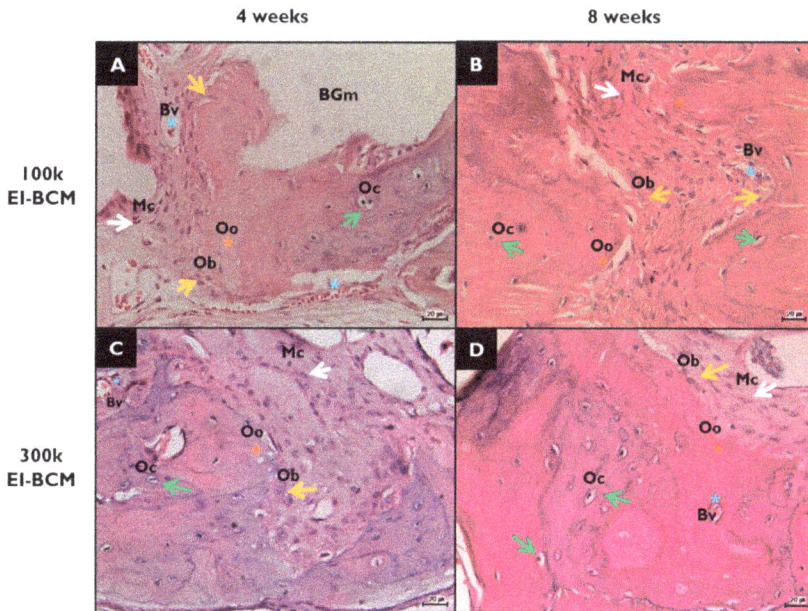

Figure 10. Histological view of new bone areas (**NBAs**) in H&E stained defect sites. (**A**) 100k EI-BCM group at 4 weeks after surgery; (**B**) 100k EI-BCM group at 8 weeks; (**C**) 300k EI-BCM group at 4 weeks; (**D**) 300k EI-BCM group at 8 weeks; **NB**, new bone; **BGm**, remaining graft materials; white arrow, mesenchymal cell (**Mc**); blue asterisk, blood vessel (**Bv**); yellow asterisk, osteoid (**Oo**); yellow arrow, osteoblast (**Ob**); green arrow, osteocyte (**Oc**) within lacuna (Original magnifications: 100×).

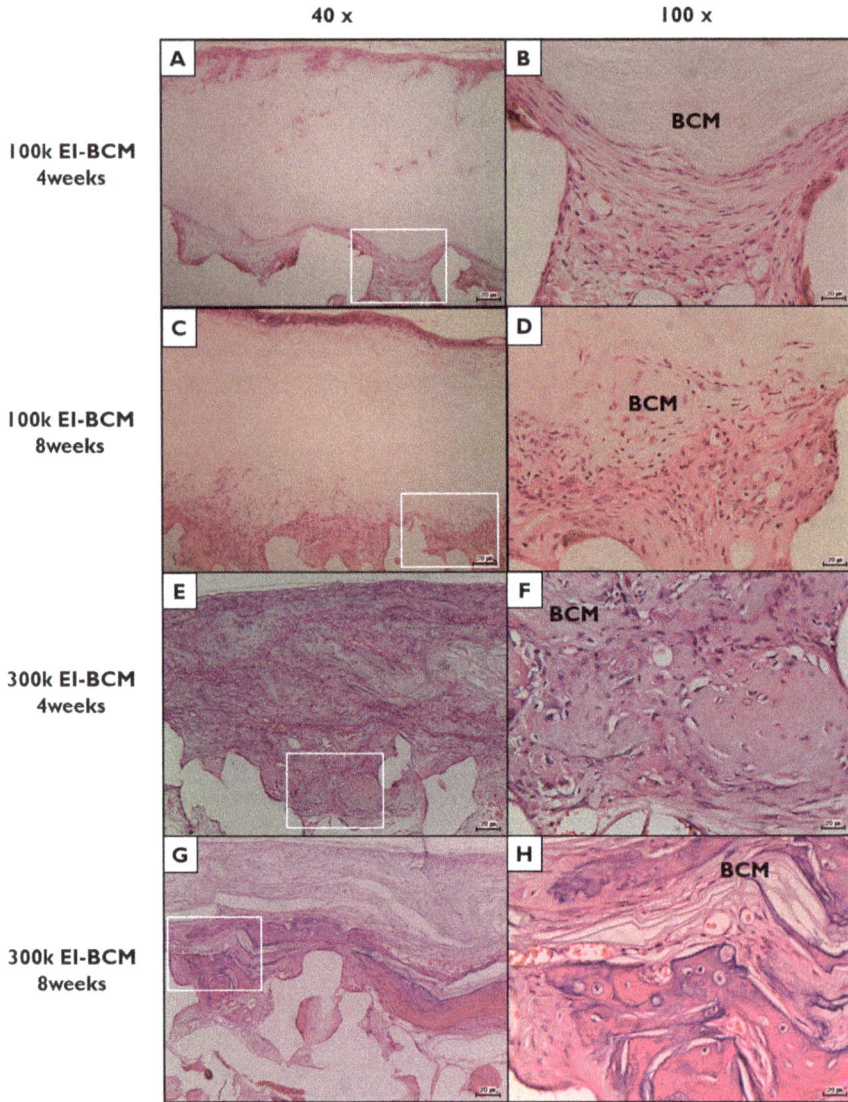

Figure 11. Histological view of electron beam irradiated bacterial cellulose membranes (EI-BCMs) in H&E stained defect sites. (**A,B**) 100k EI-BCM group at 4 weeks after surgery; (**C,D**) 100k EI-BCM group at 8 weeks; (**E,F**) 300k EI-BCM group at 4 weeks; (**G,H**) 300k EI-BCM group at 8 weeks; **BCM**, bacterial cellulose membrane. (Original magnifications: (**A,C,E,G**) 40×; (**B,D,F,H**) 100×).

2.3.2. Histometric Analyses

Results regarding comparisons of the collagen membrane (CM) and NI-BCM group include results of our previous study [30] (Table 1). The present study was carried out under identical conditions. At 4 weeks, the 100k and 300k EI-BCM groups showed significantly greater new bone areas (NBA; %) than the CM ($p < 0.001$) or NI-BCM groups ($p < 0.05$), but no significant difference was observed between the two EI-BCMs (Figure 12). At 8 weeks, the NBA (%) of 100k EI-BCM group was significantly greater than in the CM group ($p < 0.001$), the NI-BCM group ($p < 0.001$), and in the 300k EI-BCM group ($p < 0.05$) (Figure 13). Based on this analysis, EI-BCM was found to be more effective at promoting bone regeneration than CM or NI-BCM, and 100 kGy was better than 300 kGy for producing resorbable BCMs.

Table 1. New bone area percentages within areas of interest ($n = 6\%$).

Weeks	Group (Membrane)	Mean ± SD	Median	*** p
4	CM	17.13 ± 9.65	15.51	
	NI-BCM	15.82 ± 2.94	16.46	
	100k EI-BCM	26.48 ± 3.78	25.44	<0.001
	300k EI-BCM	26.55 ± 4.56	25.92	
8	CM	17.47 ± 5.09	17.07	
	NI-BCM	16.78 ± 5.27	16.88	
	100k EI-BCM	30.79 ± 3.86	30.70	<0.001
	300k EI-BCM	26.47 ± 6.77	23.45	

CM: collagen membrane; NI-BCM: unirradiated bacterial cellulose membrane; 100k EI-BCM: 100 kGy irradiated bacterial cellulose membrane; 300k EI-BCM: 300 kGy irradiated bacterial cellulose membrane. The symbols '***' indicate statistically significant at p values of <0.001).

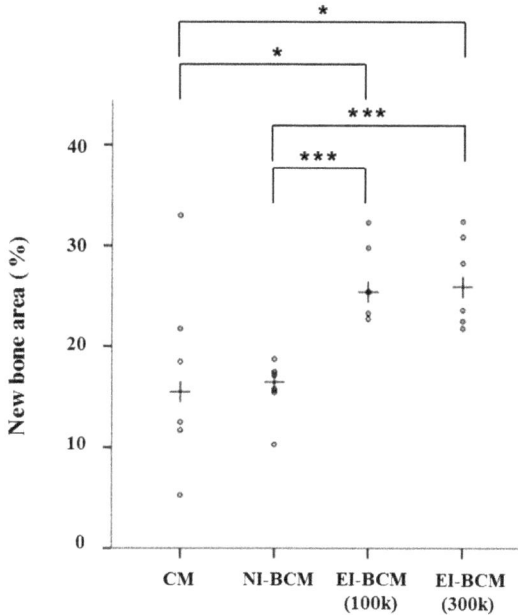

Figure 12. Scatter plots and median (crosses) new bone area percentages (%) at 4 weeks after surgery (* $p < 0.05$, *** $p < 0.001$).

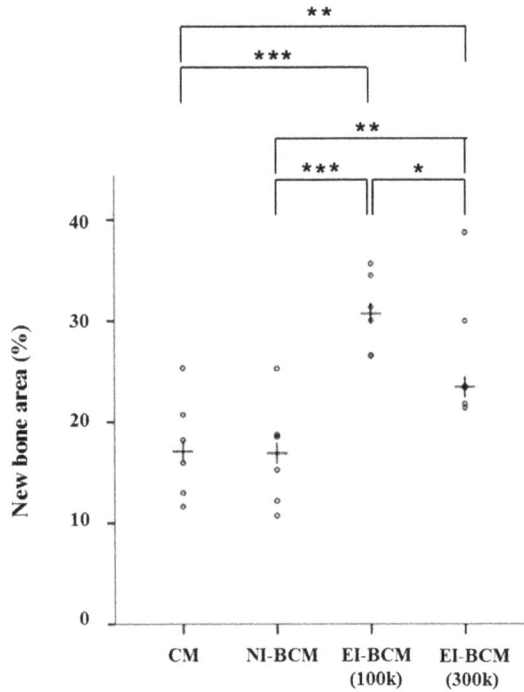

Figure 13. Scatter plots and median (crosses) new bone area percentages (%) at 8 weeks after surgery (* $p < 0.05$, ** $p < 0.01$, *** $p < 0.001$).

3. Materials and Methods

3.1. Preparation of Bacterial Cellulose Membranes (BCMs)

BCMs (Jadam Co., Jeju, Korea) were produced using the bacterial strain *Gluconacetobacter hansenii TL-2C*, which was incubated for 7 days in a static culture containing 0.3% (w/w) citrus fermented solution and 5% (w/w) sucrose at pH 4.5 (adjusted using acetic acid). The obtained gel-like pellicles of BC were purified by immersion in deionized water at 90 °C for 2 h, and then boiled in 0.5 M NaOH for 15 min in order to remove bacterial cell remnants. The BC obtained was washed several times with deionized water and soaked in 1% NaOH for 2 days. Finally, the alkali was removed from the pellicles by washing. All other reagents and solvents were of analytical grade and used without further purification.

3.2. Fabrication of Electron Beam Irradiated BCMs (EI-BCMs)

Initially, BC pellicles washed with distilled water were irradiated at room temperature using an electron beam linear accelerator (10 MeV, 0.5 mA) at the Korea Atomic Energy Research Institute (Jeongup, Korea) at a dose rate of 5 kGy/min to doses ranging from 100 kGy to 300 kGy. Pellicles were the washed with deionized water, fixed between stainless steel wire meshes to remove water and then compressed for 5 min into sheets using a press (Carver 3969, Wabash, IN, USA) and dried in a freeze dryer at −80 °C for 48 h. Finally, the 100k and 300k EI-BCMs (100 kGy and 300 kGy irradiated BC membrane) and non-irradiated BCMs (NI-BCMs; controls) were prepared (Figure 14).

Figure 14. Bacterial cellulose (BC) pellicles were irradiated in distilled water by Electron beam linear accelerator (**A**: NI-BCMs, **B**: 100k EI-BCMs, **C**: 300k EI-BCMs). After the radiation, the BC pellicles were lyophilized (**D**: NI-BCMs, **F**: 100k EI-BCMs, **H**: 300k EI-BCMs) and then compressed into sheet (**E**: NI-BCMs, **G**: 100k EI-BCMs, **I**: 300k EI-BCMs).

3.3. Characterization of EI-BCMs

3.3.1. Scanning Electron Microscope (SEM) Image Analysis of BCMs

SEM images of NI-BCMs, 100k and 300k EI-BCMs were obtained using a JSM-6390 unit (JEOL, Tokyo, Japan) at 10 kV and distance of 10–12 mm. Samples were placed on steel plates and coated with gold for 60 s.

3.3.2. Mechanical Properties

The mechanical properties of NI-BCMs and 100k and 300k EI-BCMs were determined using a Universal Testing Instrument (Instron 5569, Instron Corp., Canton, OH, USA) equipped with a 5 kN load cell at a crosshead speed of 10 mm/min. Samples were cut into 2 mm × 15 mm pieces. ASTM standard method D 882-88 was used to determine wet tensile strengths after soaking samples in water for 10 min.

3.3.3. Attenuated Total Reflection-Fourier Transform Infrared Spectroscopy (ATR-FTIR)

NI-BCM and 100k and 300k EI-BCMs also subjected to FTIR spectrophotometry using a Bruker Temsor 37 unit (Bruker AXS Inc., Ettlingen, Germany) over the range 500–4000 cm^{-1} at a resolution of 4 cm^{-1} using >32 scans. Specimens were examined in triplicate to ensure reproducibility.

3.3.4. Thermogravimetric Analysis (TGA)

Thermogravimetric analysis of NI-BCMs and EI-BCMs was performed using a thermal gravimeter (TA Q600, TA Instruments, New Castle, DE, USA). All specimens were dried at 45 °C for 12 h prior to conducting the tests. Specimens (15.9 mg) were placed in a platinum pan and heated at 10 °C/min from 40 °C to 800 °C under a nitrogen flow.

3.3.5. In Vitro Degradation of EI-BCMs

The in vitro degradations of NI-BCMs and 100k and 300k EI-BCMs was undertaken by immersing samples in phosphate buffered saline (PBS) solution at pH 7.4 and simulated body fluid (SBF) at 37 °C. These pre-wetted, irradiated BC membranes were then placed in a 20 mL weighing bottle containing 15 mL PBS and SBF solution. Samples were removed, rinsed with deionized water and freeze dried [60]. Samples were cut into 10 mm diameter circles and immersed in PBS and SBF at 37 °C for 4, 8, or

16 weeks, when they were rinsed, freeze dried, and weighed. The averages and standard deviations were recorded, and rates of weight loss were calculated. The compositions of $1\times$ PBS and $5\times$ SBF are provided in Table 2 [61].

Table 2. Amounts of reagents used to create $1\times$ PBS and $5\times$ SBF.

$1\times$ PBS (Phosphate Buffered Saline)			$5\times$ SBF (Simulated Body Fluid)		
137.0	mM	NaCl	710.0	mM	Na^+
2.70	mM	KCl	25.0	mM	K^+
10	mM	$Na_2HPO_4\cdot H_2O$	12.5	mM	Ca^{2+}
2.00	mM	KH_2PO_4	7.5	mM	Mg^{2+}
1.00	mM	$CaCl_2$	21.0	mM	HCO_3^-
0.50	mM	$MgCl_2$	740	mM	Cl^-
			5.0	mM	HPO_4^{2-}
			2.5	mM	SO_4^{2-}

3.4. In Vitro Cell Studies

NIH3T3 cells (ATCC® CRL-1658™) were cultured in Dulbecco's Modified Eagle Medium containing 4.5 g/L glucose (DMEM-HG, Gibco BRL, Grand Island, NY, USA) and supplemented with 10% fetal bovine serum and 1% penicillin/streptomycin in a 5% CO_2 incubator at 37 °C and RH 95%. The medium was changed every two days.

3.4.1. Cell Proliferation Assay

Cell proliferation was measured using a Cell Counting Kit-8 assay (CCK-8, Dojindo Laboratories, Kumamoto, Japan). Briefly, NIH3T3 cells were seeded at a density of 1×10^5 cells/well on NI-BCMs and 100k and 300k EI-BCMs, and cultured for 1, 3, and 7 days then normalized to the day 1 value to calculate the growth percentage of cells cultured in control media. After incubation, culture media were exchanged with culture medium containing 10% CCK-8 solution. Then, while maintaining the same conditions for 90 min, absorbances were measured at 450 nm using a UV-Vis spectrophotometer (MQX 200 model, Bio-Tek Instruments, Winooski, VT, USA). All experiments were performed in triplicate.

3.4.2. Immunofluorescent Staining

Cells were stained in order to evaluate their morphologies on BCMs. After 24 h of cell culture, BCM samples were fixed using 3.7% MeOH-free formaldehyde in PBS for 10 min at 37 °C, washed in PBS, and permeabilized in cytoskeleton (CSK) buffer (10.3 g sucrose, 0.292 g NaCl, 0.06 g $MgCl_2$, 0.476 g HEPES buffer, and 0.5 mL Triton X-100 in 100 mL water, pH 7.2) for 10 min at 4 °C. The cell was then blocked using blocking buffer (1% BSA in PBS) for 1 h at 37 °C, and samples were incubated with rhodamine phalloidin (1:100) and Hoechst 33258 (1:1000; a nuclear stain) (Molecular Probes, Eugene, OR, USA) for 1 h at 37 °C. After washing in PBS, samples were mounted on glass slides, and fluorescent images of stained cells on BCMs were acquired using a Laser Scanning Confocal Microscope (LSM 510, Zeiss, Jena, Germany). Cell areas were obtained from acquired images using Imagepro Plus 4.5 (Media Cybernetics, Silver Springs, MD, USA).

3.4.3. Field Emission-Scanning Electron Microscopy(FE-SEM) of Surface Cells

Samples of NI-BCMs and 100k and 300k EI-BCMs were punched out and sterilized with 70% EtOH. NIH3T3 cells were then seeded at a concentration of 1×10^5 cells/well on sample surfaces. After 24 h of cell culture, samples were washed three times with PBS, and cells were fixed in PBS containing 4% paraformaldehyde for 30 min at room temperature. Cells were then rinsed with PBS for 5 min and dehydrated using an ethanol gradient (50%, 70%, 80%, 95%, and 100% EtOH) for 10 min per step. Samples were dried using a hexamethyldisilazane (HMDS) chemical drying series (3:1, 1:1, and 1:3 EtOH:HMDS followed by 100% HMDS at 15 min each and allowed to air dry). Finally, samples were examined under a field emission-scanning electron microscope (S-4800, Hitachi, Tokyo, Japan).

3.5. In Vivo Animal Studies

3.5.1. Experimental Animals

Twenty-four Sprague-Dawley rats (males; weight 250–300 g) were chosen. Animals were housed individually in plastic cages under standard laboratory conditions and had ad libitum access to water and standard laboratory pellets. Animal selection, care, and management, and the surgical protocol and preparation for surgery were conducted in accordance with the guidelines issued by the Ethics Committee on Animal Experimentation at the Korea Atomic Energy Research Institute (KAERI-IACUC-2013-004).

3.5.2. Surgical Procedures

After intramuscularly injecting a mixture of xylazine (Rumpun, Bayer Korea, Seoul, Korea) and tiletamine-zolazepam (Zoletil, Vibac Laboratories, Carros, France), surgery was performed under general anesthesia. In each case, the shaved cranial surgical site was disinfected with betadine, and 2% lidocaine HCL (Yu-Han Co., Gunpo, South Korea) was administered for local anesthesia. After making a U-shaped incision, the full-thickness of flap of skin and periosteum was removed. In the middle of the cranium, a standardized 8 mm circular transosseous defect was created with a trephine bur (3i Implant Innovation, Palm Beach Garden, FL, USA). During drilling, the surgical site was washed with saline. After removing the trephinated bony disk, the experimental and control materials were applied. Twelve animals were allocated to each study group. After applying 0.12 mg hydroxyapatite (HA)/β-tricalcium phosphate (TCP) bone graft material (Bio-C, Cowellmedi Implant, Seoul, Korea), the defect site was covered with a 10×10 mm membrane of 100k or 300k EI-BCM. The surgical site was closed with 4-0 absorbable sutures (Vicryl®, Ethicon, Somerville, NJ, USA) (Figure 15).

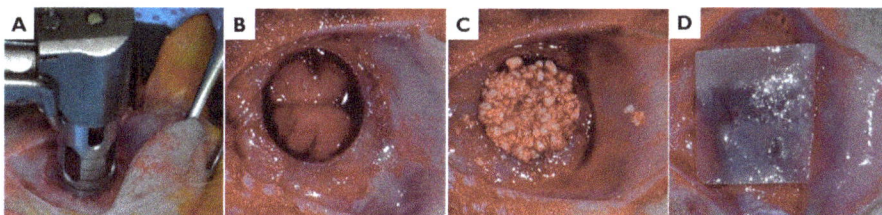

Figure 15. In vivo surgical procedure used to produce rat calvarial defects. (**A,B**) In the middle of the cranium, an 8 mm-diameter defect was created with a trephine bur. (**C**) The defect site was treated with HA/β-TCP bone graft material, and then (**D**) covered with 100 or 300 kGy irradiated BCM.

3.5.3. Post-Operative Care and Sacrifice

After surgery, animals received 1 mg/kg gentamicin (Kookje Co., Seoul, Korea) and 0.5 mL/kg pyrin (Green Cross Veterinary Products Co., Seoul, Korea) intramuscularly three times daily for 3 days. Animals were individually caged and received food and water ad libitum. Six animals in each group were allocated a healing period of 4 weeks, and the remaining six animals a healing period of 8 weeks. Animals were sacrificed by CO_2 inhalation. To collect specimens, defect sites were harvested along with surrounding bone and membranes. Harvested specimens were fixed in neutral buffered formalin (Sigma Aldrich Co., St. Louis, MO, USA) for 2 weeks.

3.5.4. Histometric Analysis

Calvarial specimens were decalcified using 14% ethylenediaminetetraacetic acid (EDTA) and rapid acid decalcification reagents, embedded in paraffin, and sectioned at a thickness of 5 μm in the centers of calvarial defects. The two centermost sections in each block were selected and stained with Hematoxylin-eosin and Masson's trichrome. Prepared histologic slides were observed under a light microscope (BX50, Olympus, Tokyo, Japan), and images were captured using a CCD camera (Spot Insight 2 Mp, Diagnostic Instruments, Inc., Sterling Heights, MI, USA) fitted with an adaptor (U-CMA3, Olympus, Tokyo, Japan). In order to calculate areas of new bone and residual biomaterials in images, computer-assisted histometric measurements were obtained and percentages of new-bone and residual biomaterials in defect areas were calculated using an image analysis program (Image-Pro Plus, Media Cybernetic, and Silver Spring, MD, USA).

3.5.5. Statistical Analyses

All quantitative results were obtained by analyzing samples in triplicate. In vitro study results are expressed as means ± SDs. Because data were not normally distributed, non-parametric tests were performed. The statistical analysis was performed using SPSS ver. 23.0 (SPSS, Chicago, IL, USA). Results obtained from the in vivo studies were expressed as means, standard deviations, and medians and statistical analysis was performed using R ver. 3.2.5 (The R Foundation, Vienna, Austria). To compare group histometric results, we used the non-parametric analysis devised by Brunner & Langer. The statistical significance was accepted for p values of <0.05.

4. Conclusions

The optimal radiation dose required to achieve a suitable level of biodegradation of bacterial cellulose membranes (BCMs) by electron beam irradiation (EI) process is important. In the present study, the mechanical, chemical, and biological characterizations of EI-BCMs prepared at different doses were investigated. High energy electron beams applied to BCMs reduced wet tensile strength, but increased in vitro cell responses and in vivo bone regeneration on calvarial defects. Within the limits of these experiments, it is suggested that BCMs irradiated at 100 kGy are more effective than BCMs irradiated at 300 kGy for clinical application as resorbable membrane for GBR. With regard to clinical utility, further studies are needed including a sufficient period of animal study, a more specific optimal radiation dose, and control of membrane thickness and porosity.

Acknowledgments: This study was supported by iPET (Korea Institute of Planning and Evaluation for Technology in Food, Agriculture, Forestry and Fisheries), Ministry of Agriculture, Food and Rural Affairs (No. 2013100659) and supported by the National Research Foundation of Korea (NRF) grant funded by the Korea government (MSIP) (No. 2017R1A2B4005820).

Author Contributions: Sung-Jun An, So-Hyoun Lee, Jung-Bo Huh and Youn-Mook Lim conceived and designed the experiments; So-Hyoun Lee, Jung-Bo Huh, Eun-Sook Kang and Chang-Mo Jeong performed experiments; Sung In Jeong, Hui-Jeong Gwon and Jong-Seok Park analyzed the data; Sung-Jun An, So-Hyoun Lee, Jung-Bo Huh and Youn-Mook Lim wrote the manuscript. All authors reviewed the final manuscript.

Conflicts of Interest: The authors declare no conflict of interest.

Abbreviations

BC	Bacterial cellulose
EI	Electron beam
EI-BCM	Electron beam irradiated BC membrane
NI-BCM	Unirradiated BC membrane
GBR	Guided bone regeneration
CM	Collagen membrane
SEM	Scanning electron microscopy

FE-SEM   Field emission-scanning electron microscope
ATR-FTIR   Attenuated total reflection-Fourier transform infrared spectroscopy analyses
TGA   Thermal gravimetric analyses
NBA   New bone area
PTFE   Polytetrafluoroethylene
PGA   Polyglycolic acid
PLA   Polylactic acid
3D   Three dimension
PBS   Phosphate buffered saline
SBF   Simulated body fluid
HA/β-TCP Hydroxyapatite/β-tricalcium phosphate
EDTA   Ethylenediaminetetraacetic acid

References

1. Gottlow, J.; Nyman, S.; Karring, T.; Lindhe, J. New attachment formation as the result of controlled tissue regeneration. *J. Clin. Periodontol.* **1984**, *11*, 494–503. [CrossRef] [PubMed]
2. Misch, C.M. Comparison of intraoral donor sites for onlay grafting prior to implant placement. *Int. J. Oral Maxillofac. Implant* **1997**, *12*, 767–776.
3. Hämmerle, C.H.; Karring, T. Guided bone regeneration at oral implant sites. *Periodontol 2000* **1998**, *17*, 151–175. [CrossRef] [PubMed]
4. Laino, L.; Iezzi, G.; Piattelli, A.; Muzio, L.L.; Cicciù, M. Vertical ridge augmentation of the atrophic posterior mandible with sandwich technique: Bone block from the chin area versus corticocancellous bone block allograft—Clinical and histological prospective randomized controlled study. *BioMed Res. Int.* **2014**, *2014*, 982104. [CrossRef] [PubMed]
5. Cicciù, M.; Herford, A.S.; Cicciù, D.; Tandon, R.; Maiorana, C. Recombinant human bone morphogenetic protein-2 promote and stabilize hard and soft tissue healing for large mandibular new bone reconstruction defects. *J. Craniofac. Surg.* **2014**, *3*, 860–862. [CrossRef] [PubMed]
6. Petrauskaite, O.; de Sousa Gomes, P.; Fernandes, M.H.; Juodzbalys, G.; Stumbras, A.; Maminskas, J.; Liesiene, J.; Cicciù, M. Biomimetic mineralization on a microporous cellulose-based matrix for bone regeneration. *BioMed Res. Int.* **2013**, *2013*, 452750. [CrossRef] [PubMed]
7. Herford, A.S.; Tandon, R.; Stevens, T.W.; Stoffella, E.; Cicciu, M. Immediate distraction osteogenesis: The sandwich technique in combination with rhBMP-2 for anterior maxillary and mandibular defects. *J. Craniofac. Surg.* **2013**, *4*, 1383–1387. [CrossRef] [PubMed]
8. Adell, R.; Lekholm, U.; Rockler, B.; Brånemark, P.I. A 15-year study of osseointegrated implants in the treatment of the edentulous jaw. *Int. J. Oral Surg.* **1981**, *10*, 387–416. [CrossRef]
9. Rispoli, L.; Fontana, F.; Beretta, M.; Poggio, C.E.; Maiorana, C. Surgery Guidelines for Barrier Membranes in Gudied Bone Regeneration (GBR). *J. Otolaryngol. Rhinol.* **2015**, *1*, 1–8. [CrossRef]
10. Kasaj, A.; Reichert, C.; Götz, H.; Röhrig, B.; Smeets, R.; Willershausen, B. In vitro evaluation of various bioabsorbable and nonresorbable barrier membranes for guided tissue regeneration. *Head Face Med.* **2008**, *4*, 22. [CrossRef] [PubMed]
11. Her, S.; Kang, T.; Fien, M.J. Titanium mesh as an alternative to a membrane for ridge augmentation. *J. Oral Maxillofac. Surg.* **2012**, *70*, 803–810. [CrossRef] [PubMed]
12. Gentile, P.; Chiono, V.; Tonda-Turo, C.; Ferreira, A.M.; Ciardelli, G. Polymeric membranes for guided bone regeneration. *Biotechnol. J.* **2011**, *6*, 1187–1197. [CrossRef] [PubMed]
13. Gentile, P.; Frongia, M.E.; Cardellach, M.; Miller, C.A.; Stafford, G.P.; Leggett, G.J.; Hatton, P.V. Functionalised nanoscale coatings using layer-by-layer assembly for imparting antibacterial properties to polylactide-co-glycolide surfaces. *Acta Biomater.* **2015**, *21*, 35–43. [CrossRef] [PubMed]
14. Von Arx, T.; Buser, D. Horizontal ridge augmentation using autogenous block grafts and the guided bone regeneration technique with collagen membranes: A clinical study with 42 patients. *Clin. Oral Implants Res.* **2006**, *17*, 359–366. [CrossRef] [PubMed]
15. Rakhmatia, U.D.; Ayukawa, Y.; Furuhashi, A.; Koyano, K. Current barrier membranes: Ti mesh and other membranes for guided bone regeneration in dental applications. *J. Prosthodont. Res.* **2013**, *57*, 3–14. [CrossRef] [PubMed]

16. Piattelli, A.; Scarano, A.; Corigliano, M.; Piattelli, M. Comparison of bone regeneration with the use of mineralized and demineralized freeze-dried bone allografts: A histological and histochemical study in man. *Biomaterials* **1996**, *17*, 1127–1131. [CrossRef]

17. Kellomäki, M.; Niiranen, H.; Puumanen, K.; Ashammakhi, N.; Waris, T.; Törmälä, P. Bioabsorbable scaffolds for guided bone regeneration and generation. *Biomaterials* **2000**, *21*, 2495–2505. [CrossRef]

18. Zitzmann, N.U.; Naef, R.; Schärer, P. Resorbable versus nonresorbable membranes in combination with bio-oss for guided bone regeneration. *Int. J. Oral Maxillofac. Implants.* **1997**, *12*, 844–852. [PubMed]

19. Embuscado, M.E.; Marks, J.S.; BeMiller, J.N. Bacterial cellulose. II. Optimization of cellulose production by Acetobacterxylinum through response surface methodology. *Food Hydrocoll.* **1994**, *8*, 419–430. [CrossRef]

20. Nishiyama, Y.; Sugiyama, J.; Chanzy, H.; Langan, P. Crystal structure and hydrogen bonding system in cellulose iα from synchrotron x-ray and neutron fiber diffraction. *J. Am. Chem. Soc.* **2003**, *125*, 14300–14306. [CrossRef] [PubMed]

21. Svensson, A.; Nicklasson, E.; Harrah, T.; Panilaitis, B.; Kaplan, D.; Brittberg, M.; Gatenholm, P. Bacterial cellulose as a potential scaffold for tissue engineering of cartilage. *Biomaterials* **2005**, *26*, 419–431. [CrossRef] [PubMed]

22. Gayathry, G.; Gopalaswamy, G. Production and characterisation of microbial cellulosic fibre from acetobacterxylinum. *Indian J. Fibre Text. Res.* **2014**, *39*, 93–96.

23. Wu, Z.Y.; Liang, H.W.; Chen, L.F.; Hu, B.C.; Yu, S.H. Bacterial cellulose: A robust platform for design of three dimensional carbon-based functional nanomaterials. *Acc. Chem. Res.* **2015**, *49*, 96–105. [CrossRef] [PubMed]

24. Helenius, G.; Bäckdahl, H.; Bodin, A.; Nannmark, U.; Gatenholm, P.; Risberg, B. In vivo biocompatibility of bacterial cellulose. *J. Biomed. Mater. Res. A* **2006**, *76*, 431–438. [CrossRef] [PubMed]

25. Wan, Y.; Gao, C.; Han, M.; Liang, H.; Ren, K.; Wang, Y.; Luo, H. Preparation and characterization of bacterial cellulose/heparin hybrid nanofiber for potential vascular tissue engineering scaffolds. *Polym. Adv. Technol.* **2011**, *22*, 2643–2648. [CrossRef]

26. Rajwade, J.M.; Paknikar, K.M.; Kumbhar, J.V. Applications of bacterial cellulose and its composites in biomedicine. *Appl. Microbiol. Biotechnol.* **2015**, *99*, 2491–2511. [CrossRef] [PubMed]

27. Müller, F.A.; Müller, L.; Hofmann, I.; Greil, P.; Wenzel, M.M.; Staudenmaier, R. Cellulose-based scaffold materials for cartilage tissue engineering. *Biomaterials* **2006**, *27*, 3955–3963. [CrossRef] [PubMed]

28. Dugan, J.M.; Gough, J.E.; Eichhorn, S.J. Bacterial cellulose scaffolds and cellulose nanowhiskers for tissue engineering. *Nanomedicine* **2013**, *8*, 287–298. [CrossRef] [PubMed]

29. Wan, Y.Z.; Huang, Y.; Yuan, C.D.; Raman, S.; Zhu, Y.; Jiang, H.J.; He, F.; Gao, C. Biomimetic synthesis of hydroxyapatite/bacterial cellulose nanocomposites for biomedical applications. *Mater. Sci. Eng. C* **2007**, *27*, 855–864. [CrossRef]

30. Lee, S.H.; Lim, Y.M.; Jeong, S.I.; An, S.J.; Kang, S.S.; Jeong, C.M.; Huh, J.B. The effect of bacterial cellulose membrane compared with CM on guided bone regeneration. *J. Adv. Prosthodont.* **2015**, *7*, 484–495. [CrossRef] [PubMed]

31. Zaborowska, M.; Bodin, A.; Bäckdahl, H.; Popp, J.; Goldstein, A.; Gatenholm, P. Microporous bacterial cellulose as a potential scaffold for bone regeneration. *Acta Biomater.* **2010**, *6*, 2540–2547. [CrossRef] [PubMed]

32. Chen, Y.M.; Xi, T.F.; Zheng, Y.F.; Zhou, L.; Wan, Y.Z. In vitro structural changes of nano-bacterial cellulose immersed in phosphate buffer solution. *J. Biomim. Biomater. Tissue Eng.* **2011**, *10*, 55–66. [CrossRef]

33. Li, J.; Wan, Y.; Li, L.; Liang, H.; Wang, J. Preparation and characterization of 2,3-dialdehyde bacterial cellulose for potential biodegradable tissue engineering scaffolds. *Mater. Sci. Eng. A* **2009**, *29*, 1635–1642. [CrossRef]

34. Czaja, W.K.; Kyryliouk, D.; DePaula, C.A.; Buechter, D.D. Oxidation of γ-irradiated microbial cellulose results in bioresorbable, highly conformable biomaterial. *J. Appl. Polym. Sci.* **2014**, *131*, 39995. [CrossRef]

35. Hu, Y.; Catchmark, J.M. In vitro biodegradability and mechanical properties of bioabsorbable bacterial cellulose incorporating cellulases. *Acta Biomater.* **2011**, *7*, 2835–2845. [CrossRef] [PubMed]

36. Lee, H.; Hamid, S.; Zain, S. Conversion of lignocellulosic biomass to nanocellulose: Structure and chemical process. *Sci. World J.* **2014**, *2014*, 631013. [CrossRef] [PubMed]

37. Darwis, D.; Khusniya, T.; Hardiningsih, L.; Nurlidar, F.; Winarno, H. In Vitro degradation behaviour of irradiated bacterial cellulose membrane. *Atom Indones.* **2013**, *38*, 78–82. [CrossRef]

38. Chmielewski, A. Worldwide developments in the field of radiation processing of materials in the down of 21st century. *Nukleonika* **2006**, *51*, 3–9.

39. Földváry, C.M.; Takács, E.; Wojnarovits, L. Effect of high-energy radiation and alkali treatment on the properties of cellulose. *Radiat. Phys. Chem.* **2003**, *67*, 505–508. [CrossRef]

40. Petryaev, Y.P.; Boltromeyuk, V.; Kovalenko, N.; Shadyro, O. Mechanism of radiation-initiated degradation of cellulose and derivatives. *Polym. Sci. USSR* **1988**, *30*, 2208–2214. [CrossRef]

41. Lee, S.H.; An, S.J.; Lim, Y.M.; Huh, J.B. The Efficacy of electron beam irradiated bacterial cellulose membranes as compared with collagen membranes on guided bone regeneration in peri-implant bone defects. *Materials* **2017**, *10*, 1018. [CrossRef] [PubMed]

42. Khan, R.A.; Beck, S.; Dussault, D.; Salmieri, S.; Bouchard, J.; Lacroix, M. Mechanical and barrier properties of nanocrystalline cellulose reinforced poly (caprolectone) composites: Effect of gamma radiation. *J. Appl. Polym. Sci.* **2013**, *129*, 3038–3046. [CrossRef]

43. Bartee, B.K.; Carr, J. Evaluation of a high-density polytetrafluoroethylene (n-PTFE) membrane as a barrier material to facilitate guided bone regeneration in the rat mandible. *J. Oral Implantol.* **1995**, *21*, 88–95. [PubMed]

44. Fujihara, K.; Kotaki, M.; Ramakrishna, S. Guided bone regeneration membrane made of polycaprolactone/calcium carbonate composite nano-fibers. *Biomaterials* **2005**, *26*, 4139–4147. [CrossRef] [PubMed]

45. Hegazy, E.; Abdel-Rehim, H.; Diaa, D.; El-Barbary, A. Controlling of degradation effects in radiation processing of polymers. In *Controlling of Degradation Effects in Radiation Processing of Polymers*; IAEA: Vienna, Austria, May 2009; pp. 67–84.

46. Ershov, B.G. Radiation-chemical degradation of cellulose and other polysaccharides. *Russ. Chem. Rev.* **1998**, *67*, 315–334. [CrossRef]

47. Von Sonntag, C. Free-radical-induced chain scission and cross-linking of polymers in aqueous solution—An overview. *Radiat. Phys. Chem.* **2003**, *67*, 353–359. [CrossRef]

48. Webb, P.A. *An Introduction to the Physical Characterization of Materials by Mercury Intrusion Porosimetry with Emphasis on Reduction and Presentation of Experimental Data*; Micromeritics Instrument Corp.: Norcross, GA, USA, 2001.

49. Eo, M.Y.; Fan, H.; Cho, Y.J.; Kim, S.M.; Lee, S.K. Cellulose membrane as a biomaterial: From hydrolysis to depolymerization with eletron beam. *Biomater. Res.* **2016**, *20*, 16. [CrossRef] [PubMed]

50. Matsuhashi, S.; Kume, T.; Hashimoto, S.; Awang, M.R. Effect of gamma irradiation on enzymatic digestion of oil palm empty fruit bunch. *J. Sci. Food Agric.* **1995**, *69*, 265–267. [CrossRef]

51. Ahn, S.J.; Shin, Y.M.; Kim, S.E.; Jeong, S.I.; Jeong, J.O.; Park, J.S.; Kim, C.Y. Characterization of hydroxyapatite-coated bacterial cellulose scaffold for bone tissue engineering. *Biotechnol. Bioprocess Eng.* **2015**, *20*, 948–955. [CrossRef]

52. Chu, P.K.; Chen, J.Y.; Wang, L.P.; Huang, N. Plasma-surface modification of biomaterials. *Mater. Sci. Eng. R Rep.* **2002**, *36*, 143–206. [CrossRef]

53. Pertile, R.; Moreira, S.; Andrade, F.; Domingues, L.; Gama, M. Bacterial cellulose modified using recombinant proteins to improve neuronal and mesenchymal cell adhesion. *Biotechnol. Prog.* **2012**, *28*, 526–532. [CrossRef] [PubMed]

54. Andrade, F.K.; Moreira, S.M.; Domingues, L.; Gama, F.M. Improving the affinity of fibroblasts for bacterial cellulose using carbohydrate—Binding modules fused to RGD. *J. Biomed. Mater. Res. A* **2010**, *28*, 526–532. [CrossRef] [PubMed]

55. De Oliveira Barud, H.G.; da Silva, R.R.; da Silva Barud, H.; Tercjak, A.; Gutierrez, J.; Lustri, W.R.; Wilton, R.L.; Ribeiro, S.J. A multipurpose natural and renewable polymer in medical applications: Bacterial cellulose. *Carbohydr. Polym.* **2016**, *153*, 406–420. [CrossRef] [PubMed]

56. Mendes, P.N.; Rahal, S.C.; Pereira-Junior, O.C.M.; Fabris, V.E.; Lenharo, S.L.R.; de Lima-Neto, J.F.; da Cruz Landim-Alvarenga, F. In vivo and in vitro evaluation of an *Acetobacter xylinum* synthesized microbial cellulose membrane intended for guided tissue repair. *Acta Vet. Scand.* **2009**, *51*, 12. [CrossRef] [PubMed]

57. Mello, L.R.; Feltrin, L.T.; Neto, P.T.F.; Ferraz, F.A.P. Duraplasty with biosynthetic cellulose: An experimental study. *J. Neurosurg.* **1997**, *86*, 143–150. [CrossRef] [PubMed]

58. Herford, A.S.; Cicciù, M.; Eftimie, L.F.; Miller, M.; Signorino, F.; Famà, F.; Cervino, G.; Giudice, G.L.; Bramanti, E.; Lauritano, F.; et al. rhBMP-2 applied as support of distraction osteogenesis: A split-mouth histological study over nonhuman primates mandibles. *Int. J. Clin. Exp. Med.* **2016**, *9*, 17187–17194.

59. Czaja, W.K.; Young, D.J.; Kawecki, M.; Brown, R.M. The Future prospects of microbial cellulose in biomedical applications. *Biomacromolecules* **2007**, *8*, 1–12. [CrossRef] [PubMed]
60. Oyane, A.; Kim, H.M.; Furuya, T.; Kokubo, T.; Miyazaki, T.; Nakamura, T. Preparation and assessment of revised simulated body fluids. *J. Biomed. Mater. Res. Part A* **2003**, *65*, 188–195. [CrossRef] [PubMed]
61. Wang, Y.Y.; Lü, L.X.; Shi, J.C.; Wang, H.F.; Xiao, Z.D.; Huang, N.P. Introducing RGD peptides on PHBV films through peg-containing cross-linkers to improve the biocompatibility. *Biomacromolecules* **2011**, *12*, 551–559. [CrossRef] [PubMed]

International Journal of
Molecular Sciences

MDPI

Article

Collagen as Coating Material for 45S5 Bioactive Glass-Based Scaffolds for Bone Tissue Engineering

Jasmin Hum * and Aldo R. Boccaccini *

Institute of Biomaterials, Department of Materials Science and Engineering, University of Erlangen-Nuremberg, Cauerstrasse 6, 91058 Erlangen, Germany
* Correspondence: jasmin.hum@fau.de (J.H.); aldo.boccaccini@ww.uni-erlangen.de (A.R.B.);
  Tel.: +49-9131-85-28601 (A.R.B.)

Received: 15 April 2018; Accepted: 4 June 2018; Published: 19 June 2018

Abstract: Highly porous 45S5 bioactive glass-based scaffolds were fabricated by the foam replica technique and coated with collagen by a novel method. After an initial cleaning step of the bioactive glass surface to expose reactive –OH groups, samples were surface functionalized by (3-aminopropyl)triethoxysilane (APTS). Functionalized scaffolds were immersed in a collagen solution, left for gelling at 37 °C, and dried at room temperature. The collagen coating was further stabilized by crosslinking with 1-ethyl-3-(3-dimethylaminopropyl)carbodiimide (EDC) and *N*-hydroxysuccinimide (NHS). Applying this coating method, a layer thickness of a few micrometers was obtained without affecting the overall scaffold macroporosity. In addition, values of compressive strength were enhanced by a factor of five, increasing from 0.04 ± 0.02 MPa for uncoated scaffolds to 0.18 ± 0.03 MPa for crosslinked collagen-coated scaffolds. The composite material developed in this study exhibited positive cell (MG-63) viability as well as suitable cell attachment and proliferation on the surface. The combination of bioactivity, mechanical competence, and cellular response makes this novel scaffold system attractive for bone tissue engineering.

Keywords: bioactive glass; collagen; scaffolds; bone tissue engineering; surface functionalization

1. Introduction

The interdisciplinary field of tissue engineering combines approaches from biology, engineering, and material science to develop alloplastic grafts or scaffolds for substitution or repair of tissue damaged by disease or trauma [1,2]. Especially in bone tissue engineering, there is an important demand for scaffolds as the availability of autografts, the current gold standard for bone transplants [3,4], is strongly limited. For the fabrication of a nonbiological matrix, which acts as a temporary scaffold structure to support cell attachment, growth, and proliferation [5,6], different biomaterials are available. The scaffold structure should offer suitable mechanical properties, adequate pore size, and a high degree of porosity to enable tissue ingrowth and vascularization, whereas the applied biomaterial should also possess osteoconductivity, osteoinductivity, and should be biodegradable [7]. In this context, bioactive glasses (BGs) are attracting increasing attention based on their high bioactivity, osteogenic potential, biodegradability, and angiogenic effects [8,9]. However, due to the brittle nature of glasses, scaffolds produced from these materials are usually not suitable for load-bearing applications. A common strategy to improve the mechanical performance of such scaffolds is the development of composite materials. By reinforcement with a polymer phase, the mechanical competence can be improved [10]. Collagen is the most abundant protein in the extracellular matrix, being also a very popular biomaterial [11]. Due to its natural origin, collagen offers suitable binding sites for cellular attachment, being thus a very interesting bone substitute material, usually in combination with bioactive, inorganic phases [12].

Collagen is a very versatile biomaterial and can be processed in different ways to form sheets, sponges, or injectable scaffolds, among others [13,14]. There is a broad range of applications of collagen in the biomedical field. Collagen films are used, for example, as graft material for corneal replacement or for treatment of infections when loaded with anti-inflammatory drugs [14,15]. Furthermore, collagen can be combined with bone morphogenetic proteins (BMPs), which can be used to stimulate osteoinduction [16]. For wound dressing, three-dimensional collagen sponges are often applied as they can absorb exudates and offer a natural barrier for bacterial infections [11]. Collagen is also applied as a coating material [17,18]. For example, Douglas and Haugen [19] described the coating of polyurethane scaffolds with collagen to improve the cell response. In most cases, collagen applied as a coating material aims to increase the biological activity. However, the mechanical improvement that can be achieved by collagen coatings, based on the high tensile strength of collagen fibers, is not usually investigated.

In this study, bioactive glass-based scaffolds were produced by the foam replica technique [20] and coated by collagen. The approach of using collagen as a coating material for BG-based scaffolds has been put forward before [21]. However, the previously reported method [21] usually leads to extremely thin coatings as only a collagen monolayer can attach on the scaffold surface. Therefore, a new coating method is introduced in this work, which enables application of a homogenous collagen layer of a few microns thickness on the surface of BG-based scaffolds. It is expected that such novel collagen coating will enhance the biologically activity of the scaffolds, maintaining a suitable mechanical competence for bone tissue engineering applications.

Samples were analyzed in terms of bioactivity, release behavior, and mechanical performance. In addition, biocompatibility was studied in vitro using a human osteosarcoma cell line (MG-63).

2. Results

2.1. Morphological and Microstructural Characterization

The micro- and macrostructure of as-fabricated BG-based scaffolds was investigated by SEM and can be seen in Figure 1. The images show the typical structure of scaffolds fabricated by the foam replica technique. Samples exhibit a highly interconnected porous structure similar to cancellous bone (Figure 1A). As the sacrificial polyurethane (PU) foam is burned out during sintering, the struts exhibit a hollow nature (Figure 1C). Pore size ranges between 250 and 500 μm, whereas struts show diameters from 50 to 100 μm (Figure 1B).

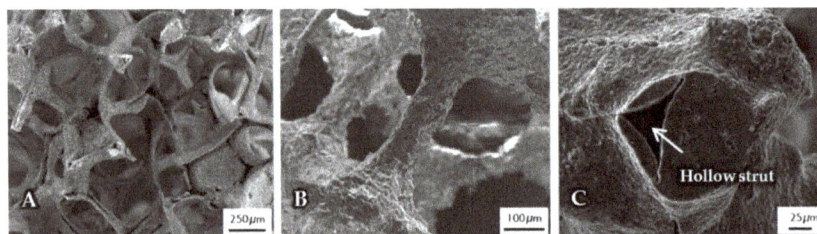

Figure 1. SEM images of as-fabricated bioactive glass-based scaffolds after sintering, at lower (**A**) and higher (**B**) magnifications. The hollow nature of the struts can be clearly seen (**C**) and can be attributed to the burn-out of the sacrificial polyurethane (PU) foam.

After the surface functionalization of the BG-based scaffolds, a collagen coating was applied. The structure, observed by SEM, is shown in Figure 2. Due to the high shrinkage rate during the drying process of the collagen gel, collagen fibers are contracted and orientated along the structure of the scaffold, which results in a dense and homogenous layer wrapped around the struts (Figure 2B), while the macroporosity of the structure is not affected (Figure 2A). Cross-section images show a layer thickness of a few micrometers (Figure 2C). At the interface (Figure 2D–F), the difference between the fibrous collagen structure and the rough bioactive glass surface can be clearly observed. In addition, by the presence of the coating, the handling of the scaffolds was improved, which is also demonstrated by their higher compressive strength in comparison to uncoated scaffolds (see Section 2.6).

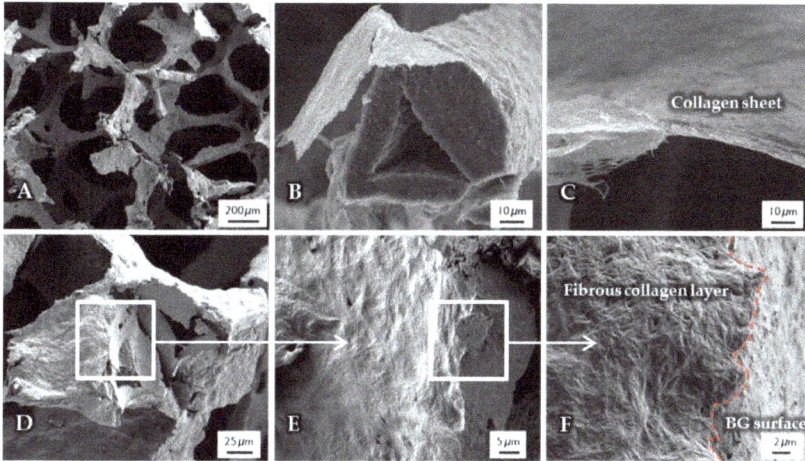

Figure 2. SEM images of uncrosslinked, collagen-coated bioactive glass-based scaffolds. The fibrous collagen layer can be clearly seen (**B**). After the coating process, the overall macroporosity of the scaffold is not affected (**A**). The collagen layer exhibits a thickness of a few micrometers (**C**). At the interface ((**D–F**), different magnifications), the rough bioactive glass surface can be clearly distinguished from the fibrous collagen layer.

2.2. Surface Analysis

After the silanization process of the bioactive glass surface, the presence of APTS molecules was detected by X-ray Photoelectron Spectroscopy (XPS) measurements. Different markers can be applied. As carbon can represent a source of contamination, nitrogen (N) was chosen as representative marker which is not part of the 45S5 bioactive glass substrate [22]. As a reference, as-received samples were chosen. The recorded spectra are presented in Figure 3. The associated atomic concentration of each element is shown in Table 1. A noticeable change in the spectra can be seen after functionalization as the intensity of the peaks is reduced after 1 h of immersion, indicating the fast dissolution process of the bioactive glass substrate. This fact is supported by the decreased atomic concentration of bioactive glass-related elements such as calcium, sodium, and phosphorous (Table 1). Due to the reaction of APTS molecules with free –OH groups on the surface, the peak at 533 eV (O1s) decreases after functionalization. In addition, the peak at 400 eV (N1s), which appears in return, confirms the presence of N-containing groups, attributed to the successful coupling of APTS molecules to the surface. Also, an increased atomic concentration of nitrogen and silicon can be found.

Table 1. X-ray Photoelectron Spectroscopy (XPS) analysis showing the atomic concentration for each element present on the surface before and after the functionalization process.

	Atomic Concentration				
	N1s	Si2p	Ca2p	Na1s	P2p
Before	0.01	6.58	5.83	24.68	4.25
After	3.29	11.36	3.07	9.42	2.54

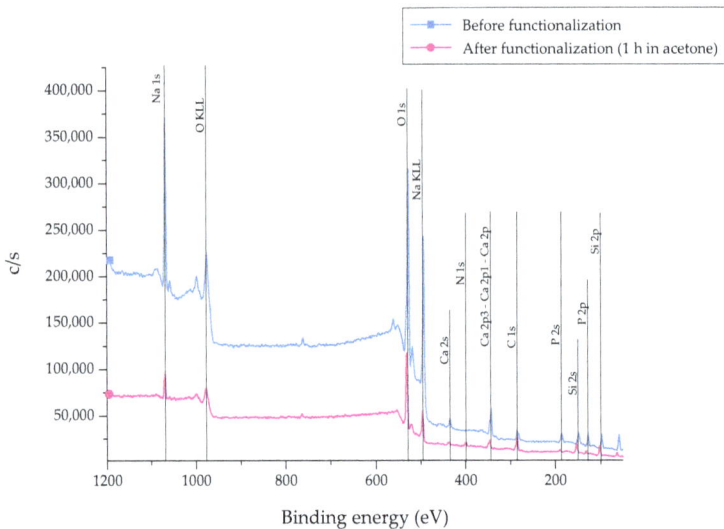

Figure 3. XPS spectra of 45S5 bioactive glass (BG) surfaces before and after the silanization process in acetone +2 vol % APTS for 1 h.

2.3. Crosslinking Process

For further stabilization of the collagen coating, chemical crosslinks were introduced by treatment with EDC ((*N*-3-Dimethylaminopropyl)-*N*′ethylcarbodiimide) and NHS (*N*-Hydroxysuccinimide). The success of crosslinking was evaluated by FTIR. Figure 4 shows the FTIR spectra of collagen-coated bioactive glass-based scaffolds with and without crosslinking. For comparison, results of as-fabricated scaffolds are also shown and the most relevant peaks are labeled.

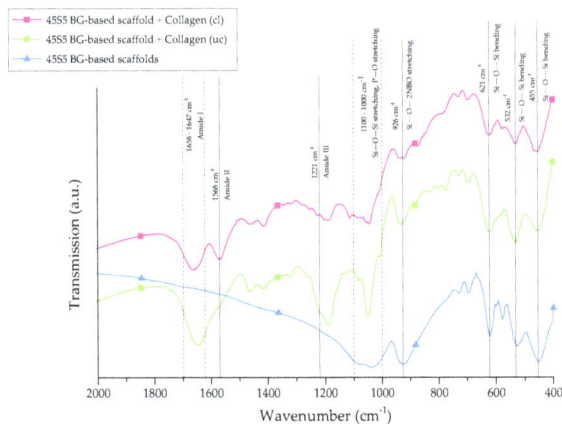

Figure 4. FTIR spectra of 45S5 BG-based scaffolds and collagen-coated 45S5 BG-based scaffolds before and after crosslinking. Relevant peaks are discussed in the text.

Uncoated 45S5 bioactive glass-based scaffolds show typical peaks at 455 cm^{-1} (Si–O–Si bending) and in the region between 1100 and 1000 cm^{-1} (Si–O–Si stretching), which can be attributed to the Si–O–Si vibrational modes in the glass network [23,24]. Due to the overlap of Si–O–Si stretching and

P–O stretching in the area between 1100 and 1000 cm^{-1}, these bands are difficult to distinguish [25]. Sharp peaks at 455, 532, and 621 cm^{-1} indicate a high crystallinity of the bioactive glass-based samples attributed to the Si–O–Si vibrational mode of the developed crystal phase [26]. The vibrational mode of the Si–O–2NBO (nonbridging oxygen) bond is visible by the peak at 926 cm^{-1}, resulting from the formation of SiO$^-$ groups in the glass network [23,25,27]. After the collagen coating, the formation of new peaks in the area between 1700 and 1200 cm^{-1} confirms the presence of the protein on the surface. These peaks can be attributed to amide I [25,28,29] (C=O stretching), typical for the triple helical structure in nondenatured collagen, and amide III [30] (N–H deformation). The additional amide-type bond (amide II) at 1568 cm^{-1}, which is present in the spectrum of crosslinked samples, confirms the success of the crosslinking process [28,30].

In order to quantify the amount of collagen before and after crosslinking, thermogravimetric measurements were carried out. Temperature and weight loss were continuously recorded and are presented in Figure 5.

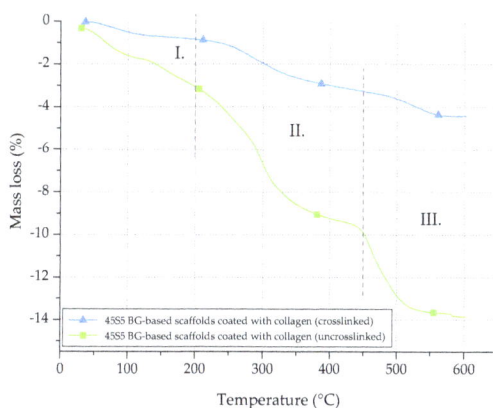

Figure 5. Mass loss of uncrosslinked and crosslinked collagen on 45S5 BG-based scaffolds during TGA.

The mass loss of collagen can be divided into three different stages [31]. Between room temperature (RT) and 200 °C (stage I), the mass loss can be attributed to the evaporation of water. In stage II (200–450 °C), collagen molecules are decomposed. In the last step (stage III), between 450 and 600 °C, residual organic components of the collagen coating are pyrolysed. Considering the mass loss in stage II, the collagen amount of uncrosslinked samples can be determined as 6.5 wt %, whereas only 2 wt % of collagen is left after crosslinking. The difference can be explained by the release of uncrosslinked collagen during the process of crosslinking, whereas only collagen fibrils, fixed in the crosslinked network, remain. Indeed, during the crosslinking process, the collagen fibrils are stabilized. However, the crosslinking process itself takes a few hours. During this time, some collagen fibrils may be released and are not fixed in the network. Thus, a given amount of collagen is lost during the crosslinking process, resulting in a lower amount of collagen after crosslinking.

2.4. Evaluation of Bioactivity

For evaluation of the bone-bonding ability of uncoated and collagen-coated scaffolds, the bioactive behavior was investigated by immersion in simulated body fluid (SBF). The formation of hydroxyapatite on the surface of different samples was observed by SEM and confirmed by FTIR measurements. Figure 6 shows SEM images of as-fabricated bioactive glass-based scaffolds after immersion in SBF for 1, 3, and 7 days. After one day, calcium phosphate precipitates are already visible. The typical morphology of hydroxyapatite (HA) can be clearly recognized after three days of immersion in SBF, qualitatively assessed by the cauliflower-like shape [32]. After seven days, the entire surface is covered by a dense layer of HA. Also, the FTIR spectra, which can be seen in Figure 7, confirm the formation of hydroxyapatite on the surface of bioactive glass-based scaffolds during immersion in SBF. The growing double peak at 600 and 570 cm^{-1} can be attributed to the P–O bending vibrations related to the PO_4^{3-} group in the crystalline layer of HA [33]. Also, the growing peak in the area between 1100 and 1000 cm^{-1} and the small shoulder at 960 cm^{-1} can be related to the phosphate group (P–O stretching) [33,34]. Developed peaks at 875 cm^{-1} and in the region between 1500 and 1400 cm^{-1} can be attributed to the C–O bending and stretching, respectively, and belong to the CO_3^{2-} groups in the carbonated HA layer [33,35]. Due to the formation of a silica-rich layer, a peak at 800 cm^{-1} can be observed [27].

Figure 6. SEM images of as-fabricated 45S5 BG-based samples after immersion in simulated body fluid (SBF) for 1, 3, and 7 days (at different magnifications).

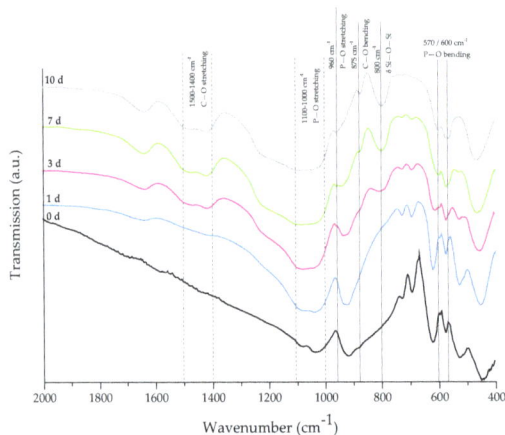

Figure 7. FTIR spectra of as-fabricated 45S5 BG-based samples after 0, 1, 3, 7, and 10 days of immersion in SBF. Relevant peaks are discussed in the text.

Figure 8 shows SEM images demonstrating the bioactive behavior of 45S5 bioactive glass-based scaffolds coated by collagen (crosslinked). The formation of small hydroxyapatite nodules along the collagen fibers can be observed after one day. With longer immersion time in SBF, HA microcrystals are directly deposited on the fibrous structure and remain well-embedded in the collagen matrix. After 10 days, the layer of collagen is mostly mineralized. The literature describes the bioactive behavior of collagen as being due to the existence of nucleation sites in the collagen network. Nucleation sites such as –COOH groups interact with the Ca^{2+} ions in the SBF solution and support the formation of hydroxyapatite [36–38]. This interaction can be also seen in the FTIR spectra (Figure 9), due to the shifted peak of amide I and the disappearance of amide II and III peaks after immersion in SBF. Characteristic peaks, indicating the formation of HA on the surface, are developed in the wavenumber region between 1100 and 1000 cm^{-1} and the small shoulder at 960 cm^{-1}, and at 875, 800, 600, and 570 cm^{-1}, as explained above. It should be mentioned that the amount of HA formed on the different scaffold types (shown in Figures 6 and 8) was not quantified in this study, so the results remain qualitative. Nevertheless, summarizing the results, it can be stated that the bioactive behavior of 45S5 bioactive glass-based scaffolds is not compromised by the presence of collagen, but it is even enhanced due to the presence of nucleation sites in the collagen structure.

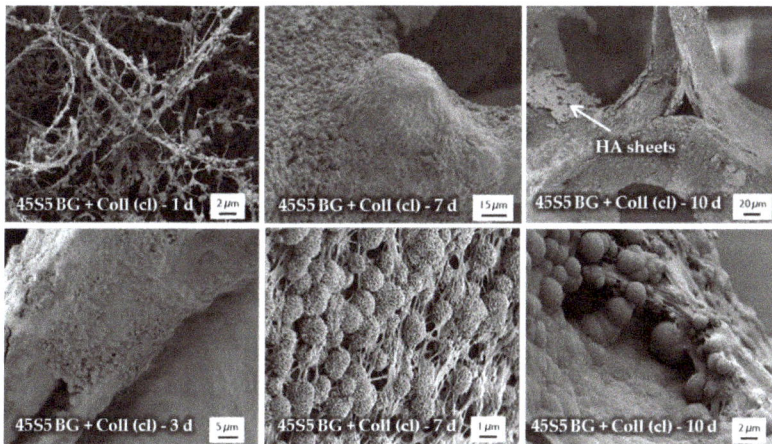

Figure 8. SEM images of collagen (Coll)-coated 45S5 BG-based scaffolds (cl) after immersion in SBF for 1, 3, 7, and 10 days (at different magnifications). The formation of a mineralized collagen layer can be observed.

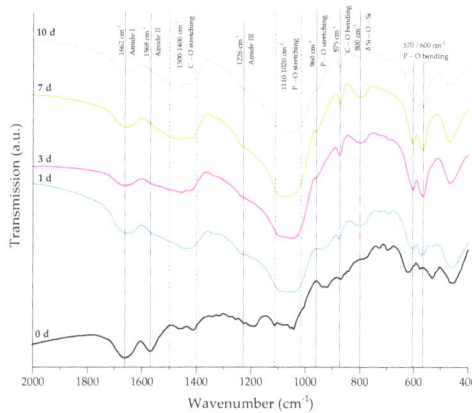

Figure 9. FTIR spectra of collagen-coated 45S5 BG-based scaffolds (crosslinked) after 0, 1, 3, 7, and 10 days of immersion in SBF. Relevant peaks are discussed in the text.

2.5. Release Behavior

To investigate the stability of the collagen coating on the surface of the bioactive glass-based scaffolds and to evaluate the delivery capability, release studies were carried out in different media. An overview of the investigated samples is given in Table 2. Collagen release was determined depending on the cleaning and functionalization process.

Table 2. Selected scaffolds. Collagen release was determined depending on the cleaning and functionalization process (- not applied, + applied).

Sample	Cleaning	Functionalization	Collagen	Crosslinking
● (A)	-	-	+	-
◆ (B)	+	-	+	-
■ (C)	-	+	+	-
▲ (D)	+	+	+	-

In relation to the scaffolds listed in Table 2, Figure 10 shows the cumulative amount of released collagen in phosphate buffered saline (PBS) up to 28 days. All curves show an initial burst release followed by a plateau. During the process of collagen coating, a collagen mesh of single collagen fibers is wrapped around the struts of the 45S5 bioactive glass-based scaffold (Figure 2). In the case of nonfunctionalized samples (A and B), the collagen is only physically adsorbed on the surface due to missing binding sites. Because of missing crosslinks, water molecules penetrate the collagen mesh very fast, leading to swelling of the fibrous structure. Macromolecules are released over time and diffuse, which is shown by a high release rate of around 90% in the first 24 h. After 28 days, less than 10% of collagen is left on the surface of the scaffolds. By introducing a silanized surface (C and D), collagen can bind covalently to the bioactive glass surface, which results in a reduced release rate. Similar results were also reported before [21]. In addition, results show that the amount of covalently bonded collagen could be increased by introducing an initial cleaning step, due to more reactive –OH groups on the surface. However, initial burst release up to 50% can be observed as the functionalization process only affects collagen close to the surface. Collagen molecules, which are not covalently bonded to the surface, still diffuse and are released very fast.

Figure 11 shows the cumulative collagen release in SBF for up to 28 days. The collagen release kinetics is similar to that in PBS, although the overall amount is remarkably lower. Even without surface functionalization, max 60% of the initial collagen is released. As the collagen is mineralized due to the immersion in SBF (Figure 8), nodules of hydroxyapatite are embedded in the collagen matrix, which could explain the reduced release rate of the protein.

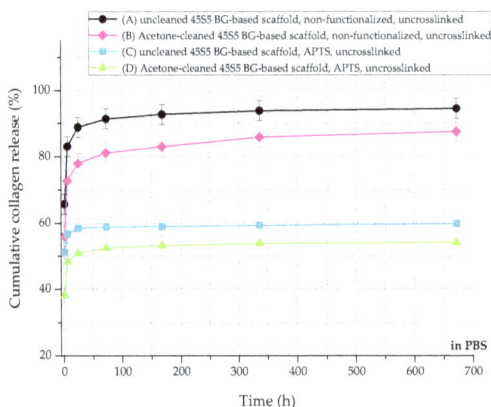

Figure 10. Cumulative collagen release from different types of 45S5 BG-based scaffolds in PBS.

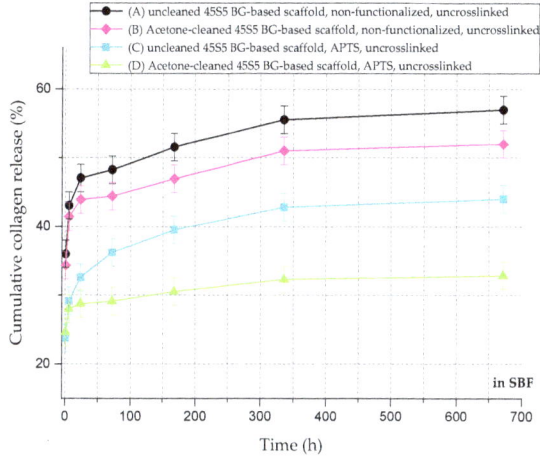

Figure 11. Cumulative collagen release from different types of 45S5 BG-based scaffolds in SBF.

By the crosslinking process, the cumulative release rate of collagen can be further decreased, as visible in Figure 12. The amount of released collagen could be further reduced from 54% to 36% in PBS and from 32% to 26% in SBF, respectively. In addition to covalently bonded collagen molecules on the bioactive glass-based surface, the crosslinked network of collagen exhibits a lower release rate during the 28 days of immersion. In conclusion, the collagen release from 45S5 bioactive glass-based scaffolds can be adapted introducing a silanized surface and by varying the degree of crosslinks, depending on the application, thus providing a very versatile system.

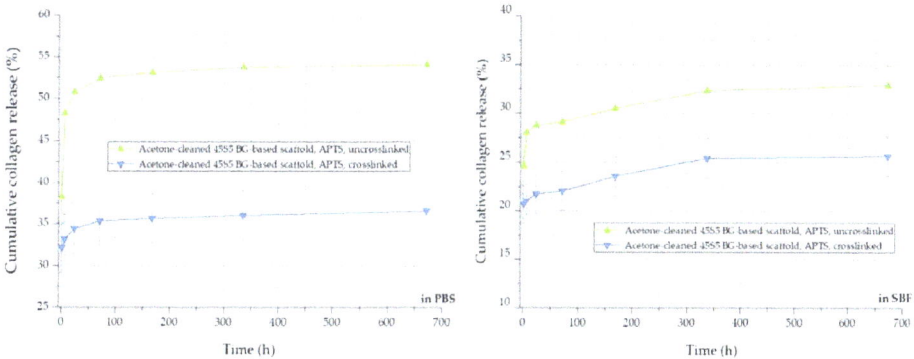

Figure 12. Collagen release kinetics of collagen-coated 45S5 BG-based scaffolds in PBS (**left**) and SBF (**right**) before and after crosslinking.

2.6. Mechanical Characterization

In order to investigate the potential of collagen-coated bioactive glass-based scaffolds for application in bone tissue engineering, the compressive strength (σ) of uncoated and collagen-coated samples was determined. Figure 13 shows exemplary stress-displacement curves for uncoated and collagen-coated samples, which can be divided into three different regions [20]. The compressive stress increases constantly in region I until a maximum stress is reached and the scaffolds' struts fail. As a result, the curve shows a negative slope in region II. Due to progressive compression in region III, the broken struts are densified and the compressive stress increases. As the struts exhibit a hollow nature, uncoated bioactive glass-based scaffolds show low compressive strength values (0.04 ± 0.02 MPa). The shape of the stress-displacement curve and such jagged characteristics are common in this type of scaffolds tested in compression [10,20–22]. Moreover, the compressive strength values for uncoated bioactive glass-based scaffolds are comparable with results found in the literature [39–41] and are typical for scaffolds produced by the foam replica technique. However, Bellucci et al. [42] describe compressive strength values of up to 0.8 MPa for similar bioactive scaffolds (based on a modified replication method), which can be attributed to the reduced porosity of the samples.

Due to the collagen coating, microcracks present on the surface of the scaffolds are filled by collagen, resulting in enhanced mechanical strength. The effect of polymer infiltration and the associated reinforcement is well-known and has been described elsewhere [10]. Similar values of compressive strength of 0.21 ± 0.03 MPa and 0.18 ± 0.03 MPa were achieved for uncrosslinked and crosslinked collagen-coated scaffolds, respectively. In order to simulate the environment in vitro, samples were additionally soaked in PBS. After 1 h, stress-displacement curves were recorded and compared with those of dry samples. It was observed that due to the rehydration of the coating, collagen transforms into a hydrogel, which results in a drop of the compressive strength (to values of <0.1 MPa). However, even if the compressive strength is reduced after the scaffold comes into contact with an aqueous medium, the biomedical application of collagen-coated BG-based scaffolds in vivo is not compromised, since due to the collagen coating, the scaffolds can be safely handled and processed.

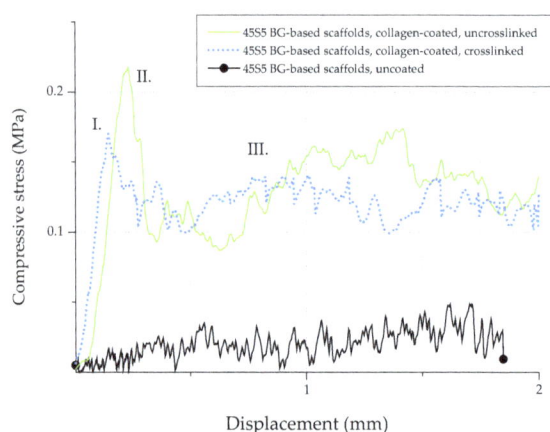

Figure 13. Exemplary stress-displacement curves for 45S5 BG-based scaffolds with and without collagen coating.

2.7. Cell Studies

2.7.1. Cell Viability and Relative Proliferation

The biocompatibility of scaffolds was evaluated by investigating their interaction with osteoblast-like cells (MG-63). Uncoated and collagen-coated samples (with and without crosslinks) were seeded with MG-63 cells and cultivated in static conditions for 7, 14, and 21 days. Figure 14 shows the viability of seeded MG-63 cells. After 21 days, uncrosslinked collagen-coated samples exhibit significant lower cell viability (105%) compared to as-fabricated scaffolds without coating (130%). Because uncrosslinked collagen is released very fast (see Section 2.5), a high cell number is lost at the beginning of cultivation as cells directly attach to the collagen matrix. This is already visible after seven days, where a noticeable difference in the cell viability of samples coated with uncrosslinked collagen can be seen compared to the uncoated reference. Therefore, higher cell viability can be detected for crosslinked samples.

Because a relatively high cell number is lost in uncrosslinked collagen-coated samples, a reduced proliferation rate can also be detected in this case (Figure 15). After seven days, samples showed a relative proliferation below 75%. A significant difference can be seen after 14 days comparing uncrosslinked and crosslinked collagen-coated scaffolds. However, remaining cells were found to be able to recover, which results in a similar proliferation rate compared to pure 45S5 bioactive glass-based scaffolds after 21 days of incubation.

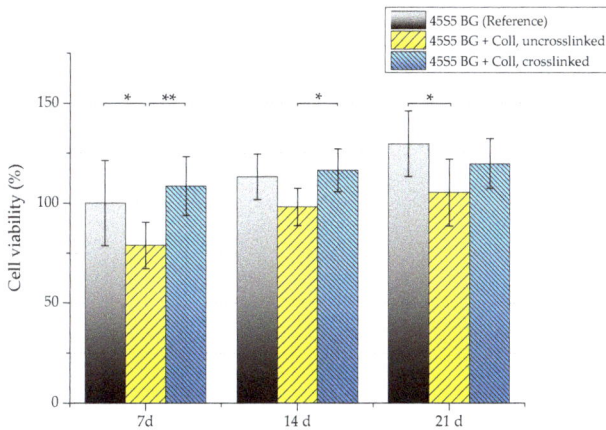

Figure 14. Cell viability of MG-63 cells of different scaffold types (absorbance at 450 nm) after 7, 14, and 21 days. Significance levels: * $p < 0.05$, ** $p < 0.01$ (Bonferroni's post-hoc test was used).

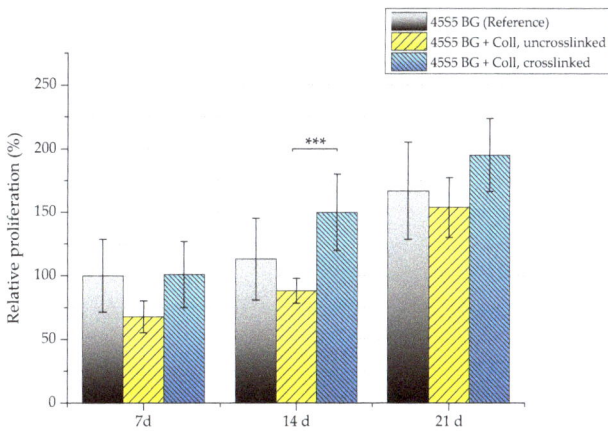

Figure 15. Relative proliferation of MG-63 cells on different scaffold types (absorbance at 450 nm) after 7, 14, and 21 days. Significance level: *** $p < 0.001$ (Bonferroni's post-hoc test was used).

2.7.2. Cell Morphology

After the cultivation of scaffolds with MG-63 cells, the morphology of cells was observed by SEM, as shown in Figure 16. SEM images show the cell morphology after different time points and confirm the results which were achieved by the evaluation of cell viability and relative proliferation. Cells with their typical spindle-shaped morphology can be found on the surface of every sample, indicating the high biocompatibility of the applied collagen coating. After 21 days of incubation, even multilayer growth of MG-63 cells can be observed. Due to the homogenous distribution of cells, it is assumed that scaffolds exhibit suitable pore size and porosity, which enables the infiltration of cells. However, a denser layer of cells can be found on the outer layer of the scaffolds, attributed to the gradient of nutrient and oxygen supply related to the static cultivation system.

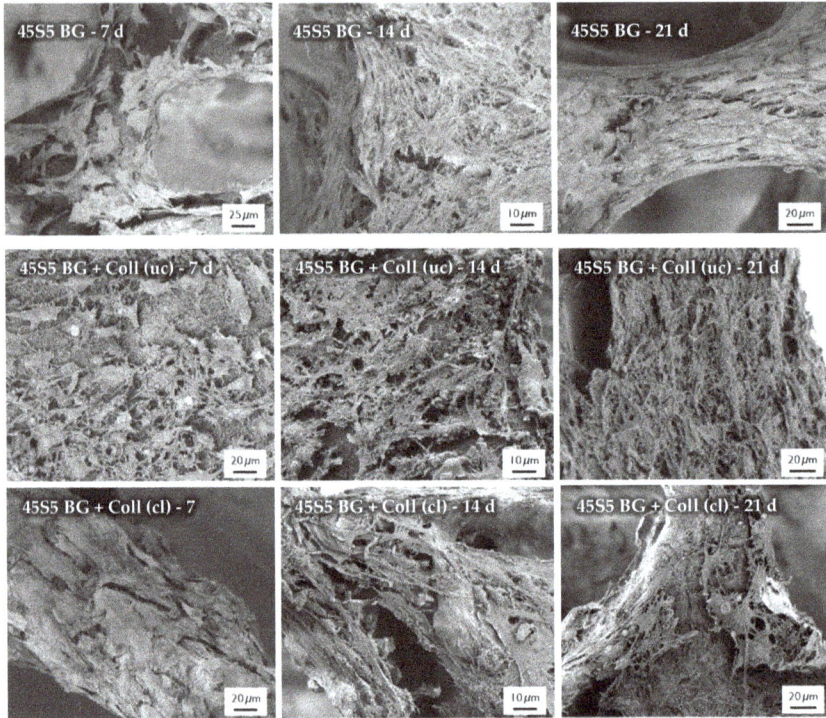

Figure 16. SEM images of seeded MG-63 cells on different types of scaffolds for 7, 14, and 21 days (at different magnifications).

3. Discussion

Due to its highly bioactive character, 45S5 bioactive glass shows great potential to be used as bone-substitute material. However, scaffolds produced from this material are highly brittle and the relative low fracture strength of 45S5 bioactive glass-based scaffolds is a significant drawback for their application in bone tissue engineering, particularly in load-bearing sites. Even the handling of such highly brittle scaffolds is very challenging. On the other hand, polymers are in general flexible and provide chemical versatility, thus the formation of bioactive glass–polymer composites is being explored to provide bioactive and mechanically sound scaffolds [2].

For the reinforcement of highly porous glass or ceramic structures and to overcome their lack of mechanical stability, scaffolds can be toughened by a polymer coating, as reviewed by Philippart et al. [10] and Yunos et al. [2]. The related reinforcement can be attributed to the filling and bridging of microcracks present on the scaffold surface by a polymeric phase. The literature mainly describes synthetic polymers, for example, polycaprolactone (PCL) [43,44] or poly (D,L-lactic acid) (PDLLA) [45,46], as the coating material. For example, to improve the compressive strength of bioactive glass-based scaffolds fabricated by the foam replica technique, Chen and Boccaccini [45] used

PDLLA as the coating material. By dip-coating, the mechanical performance of the brittle samples could be considerably increased.

Natural polymers are an attractive alternative to synthetic polymers and are drawing attention as coating materials for scaffolds, taking into account that synthetic polymers can comprise negative byproducts when degrading. Natural polymers such as gelatin, silk, or collagen are often used as biomaterials and offer a wide range of properties, for example, biocompatibility, degradability, and nontoxicity. In addition, special sequences of amino acids (RGD) present in such biopolymers can improve cell attachment. For example, Metze et al. [41] used gelatin, a close relative of collagen, as a coating for 45S5 bioactive glass-based scaffolds to improve the compressive strength. Li et al. [47] showed increased mechanical performance as well as enhanced osteogenic differentiation behavior when coating ceramic scaffolds with silk.

In the present study, collagen was the polymer of choice due to its high biocompatibility and bioactivity. The collagen coating was applied by a combined process of surface functionalization and collagen immersion, leading to homogenous and thick collagen layers on the surface of BG-based scaffolds, while the overall porosity was not significantly hampered. By filling and bridging microcracks on the surface of the struts, which are caused by the sintering process during scaffold fabrication, the compressive strength of the scaffolds increased from 0.04 ± 0.02 MPa to 0.18 ± 0.02 MPa for uncoated and collagen-coated scaffolds, respectively. These values are at the lower boundary of the compressive strength of natural bone, which is reported to be between 0.2 and 4 MPa [20], and thus the collagen-coated scaffolds become attractive for application in bone tissue engineering. In addition, it is assumed that ingrowth of tissue in vivo will provide additional mechanical support, as reported in the literature [48].

Even though collagen transforms into a hydrogel in wet conditions, which could imply a loss of mechanical strength in vivo, the biomedical application of collagen-coated BG-based scaffolds is not negatively affected, as the in situ degradation of the collagen coating will occur in conjunction with the formation of hydroxyapatite on the surface of the scaffolds, due to the high bioactivity of the bioactive glass, providing local support to the structure. Indeed, when using scaffolds as bone-substituting materials, one important requirement is to ensure a safe handling of the alloplastic graft before implantation, for which the presence of the collagen coating is required. Due to the collagen coating, the otherwise brittle bioactive glass-based scaffolds can be safely handled and processed, and such a tough and stable matrix is suitable to fill bone defects during operation. The degradation of the collagen coating and the bioactive glass matrix starts immediately after implantation. Release products of collagen should enhance tissue regeneration and angiogenesis [49–51], while the ionic dissolution products of bioactive glass will further stimulate osteogenesis and angiogenesis by the enhanced expression of genes in cells, leading to bone formation and vascularization [52]. In addition, collagen offers many nucleation sites for the formation of hydroxyapatite, which is an important parameter for the interaction of a material with bone tissue, enhancing the intrinsic bioactivity of the bioactive glass scaffold. Thus, composite scaffolds that result from combining 45S5 bioactive glass and collagen offer high biocompatibility and bioactivity coupled with adequate structural integrity for safe handling of the scaffolds. In addition, collagen as a coating material not only influences cell attachment and proliferation, but it also offers a platform for the incorporation of biomolecules and drugs. By introducing a functionalized surface, combined with a crosslinked network of collagen molecules, the release rate of collagen (and incorporated drugs) can be easily controlled and applied as a drug delivery vehicle.

Overall, this study has introduced an alternative approach for the future development of composite scaffolds based on collagen and bioactive glasses. Even if significant challenges remain to enable the translation of the present scaffolds to clinical practice, the present results provide a well-founded basis for considering the newly developed scaffolds as a valid alternative for further investigations in this field.

4. Materials and Methods

4.1. Scaffolds

4.1.1. Scaffold Production

Cylindrical and highly porous scaffolds (diameter = 0.75 mm, height = 0.4 mm) were produced by the foam replica process, originally described by Chen et al. [20]. Briefly summarized, sacrificial polyurethane (PU) foams (45 pores per inch (ppi), PL Bulpren S28133, Eurofoam Deutschland GmbH, Germany) were dip-coated in a BG-based slurry. For the preparation of the slurry, 0.3 wt % of binder (polyvinyl alcohol (PVA), fully hydrolyzed, M_w ~30,000, Merck Millipore, Darmstadt, Germany) was dissolved at 80 °C in ultrapure water (UPW) under stirring conditions. After cooling down to room temperature (RT), 2 wt % of disperser (KV 9062, Zschimmer & Schwarz GmbH & Co. KG, Lahnstein, Germany) and 50 wt % of BG powder were added. For the production of scaffolds, a commercially available melt-derived bioactive glass powder of the 45S5 composition (45S5 BG, in wt %: 45 SiO_2, 24.5 CaO, 24.5 Na_2O, and 6 P_2O_5) with particle size of ~2 μm was used (Vitryxx®, Schott, Mainz, Germany). After mixing the slurry homogenously, precut PU foams were immersed into the slurry, retrieved after 5 min, and squeezed to remove excess slurry. After drying at 60 °C for 2 h, the coating procedure was repeated. The samples were dried again at 60 °C and the received green bodies were finally sintered at 1050 °C to densify the structure.

4.1.2. Surface Functionalization

As already mentioned, cells attach to material surfaces in contact with proteins. Proteins reveal chemical complexity and a fragile nature and can change their native conformation considerably [21,53] due to electrostatic and hydrophobic interactions. To prevent the alteration of the biological functionality, proteins (in this case, collagen) can be grafted to the scaffold surface by chemical bonding. In this study, the surface of the bioactive glass-based scaffolds was functionalized using (3-aminopropyl)triethoxysilane (APTS), which is one of the most studied coupling agents and has been used in several studies before [21,54,55]. The following scheme (Figure 17) describes the surface functionalization of 45S5 BG-based scaffolds using APTS. The silanization process can be divided into four stages, namely: (I) hydrolysis, (II) condensation reaction, (III) hydrogen bonding, and (IV) bond formation.

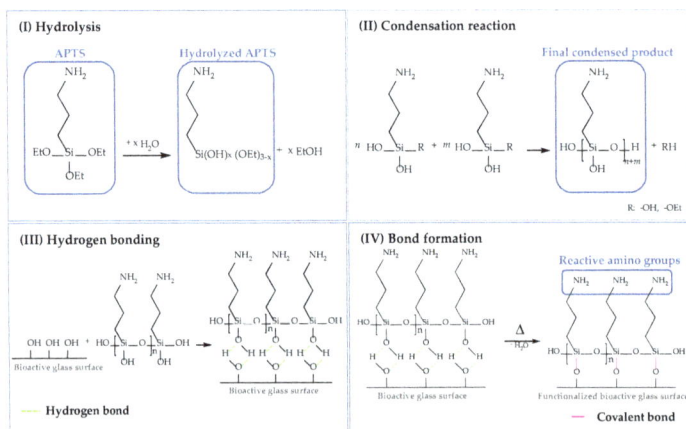

Figure 17. Surface functionalization of 45S5 bioactive glass-based scaffolds with APTS divided into (**I**) hydrolysis, (**II**) condensation reaction, (**III**) hydrogen bonding, and (**IV**) bond formation [50,54–56].

To start the process of hydrolysis, adsorbed moisture on the surface is sufficient. Water molecules attack the ester bonds in the ethoxy groups (Si–OEt), which are hydrolyzed and replaced by hydroxyl groups (–OH) to form reactive silanol groups (Si–OH). Through a self-condensation process, silanol groups react with each other in step (II) to form stable siloxane bonds (Si–O–Si). The final condensed product can now react with –OH groups located on the BG surface. By hydrogen bonding, a network of siloxanes assembles on the surface of 45S5 BG-based scaffolds. As these bonds are not very stable, water is removed by a heat treatment and hydrogen bonds are replaced by covalent bonds in step (IV). To ensure a uniform deposition of reactive amino groups on the surface of bioactive glass-based scaffolds, an initial cleaning step was performed to remove contaminations and to activate reactive hydroxyl groups on the surface [56]. Based on previous work [56], an initial cleaning step was introduced before the surface of the samples was functionalized. Therefore, bioactive glass-based scaffolds were gently swiveled in acetone for 5 min, followed by rinsing in UPW. This process was repeated three times. Afterwards, samples were dried at 60 °C.

For surface functionalization, cleaned samples were immersed in a solution with 2 vol % APTS (99%, Sigma-Aldrich, Schnelldorf, Germany) in acetone (Acetone Technical, VWR Chemicals, Ismaning, Germany). After 1 h, scaffolds were removed, rinsed two times in UPW, and heat-treated at 100 °C for 2 h.

4.1.3. Collagen Coating

Surface-functionalized bioactive glass-based scaffolds were coated with collagen using the following protocol. Solution A was prepared by mixing 1 M NaOH (sodium hydroxide, Merck Millipore, Darmstadt, Germany), 1 M HEPES buffer (4-(2-Hydroxyethyl)piperazine-1-ethanesulfonic acid dissolved in UPW, Sigma-Aldrich, Schnelldorf, Germany), and 10× DMEM (10× concentrated Dulbecco's Modified Eagle's Medium, Biochrom, Berlin, Germany) at a ratio of 1:1:2. Solution A was mixed with four parts of 0.5% collagen solution (Collagen G1 from bovine calf skin, Matrix BioScience, Mörlenbach, Germany) to obtain solution B. To avoid early gelation of the neutralized collagen solution, all chemicals were refrigerated (4–8 °C). The surface-functionalized scaffolds were placed in a 48-well plate, dripped by the collagen solution, and incubated over night at 37 °C to initiate and complete the collagen fibrillogenesis. After gelation, samples were completely dried under the hood.

4.1.4. Crosslinking

For further stabilization of the collagen coating, samples were additionally crosslinked (cl). According to Powell et al. [57], a 50 mM MES (4-Morpholineethanesulfonic) acid solution in 40 vol % ethanol (Ethanol EMSURE®, Merck Millipore, Darmstadt, Germany) was prepared using MES hydrate (Sigma-Aldrich, Schnelldorf, Germany). 60 mM EDC (N-3-Dimethylaminopropyl)-N'ethylcarbodiimide, Sigma-Aldrich, Schnelldorf, Germany) and 60 mM NHS (N-Hydroxysuccinimide, Sigma-Aldrich, Schnelldorf, Germany) were dissolved in the MES buffer solution. Collagen-coated bioactive glass-based scaffolds were immersed in the crosslinking solution for 4 h at RT. Afterwards, samples were removed and washed in 0.1 M Na_2HPO_4 (disodium hydrogen phosphate, Sigma-Aldrich, Schnelldorf, Germany) for 2 h to hydrolyze remaining O-acylisourea of the carbodiimide [58,59]. Subsequently, scaffolds were washed in UPW and left to dry in air. Uncrosslinked samples are labeled as "uc".

4.2. Evaluation of Bioactivity

For the assessment of acellular bioactivity in vitro, uncoated and collagen-coated samples were immersed in simulated body fluid (SBF), applying the well-known protocol introduced by Kokubo and Takadama [60]. Samples were put in closable containers of polypropylene, immersed in 50 mL of SBF, and placed in an orbital shaker (90 rpm, 37 °C). After different time points, samples were removed, washed in UPW, and left for drying. The formation of hydroxyapatite on the surface of the scaffolds, as a marker of bioactivity, was evaluated by scanning electron microscopy (SEM) and Fourier transform infrared spectroscopy (FTIR).

4.3. Release Behavior

The release behavior of collagen-coated BG-based scaffolds with and without crosslinks was evaluated in different media. Samples were immersed in 5 mL of PBS or SBF, respectively, and placed in an orbital shaker at 37 °C and 90 rpm. After different time points, the solution was completely removed and refilled by fresh solution. The released collagen was determined colorimetrically based on a modification of Lowry's method [61,62]. For this analysis, 250 μL of protein solution was incubated with 250 μL of Lowry reagent solution (Sigma-Aldrich, Schnelldorf, Germany) for 20 min at RT. Afterwards, 125 μL of 0.3 N Folin & Ciocalteu's phenol reagent (Sigma-Aldrich, Schnelldorf, Germany) was added and incubated for an additional 30 min to allow the color to develop. The solution was irradiated by visible light at 595 nm using a single beam spectrometer (Specord 40, Analytik Jena, Jena, Germany) and absorbance values were recorded. The collagen concentration was determined from a standard calibration curve.

4.4. Cell Studies

4.4.1. Cell Culture and Seeding

For static cell culture experiments, a human osteosarcoma cell line (MG-63, Sigma-Aldrich, Schnelldorf, Germany) was used. MG-63 are adherent cells received from a malignant bone tumor and are commonly used as osteoblastic model cells [63]. Cells were cultured in low glucose Dulbecco's Modified Eagle Medium (DMEM, Thermo Fisher Scientific, Schwerte, Germany) supplemented with 10 vol % of fetal bovine serum (FBS, Sigma-Aldrich, Schnelldorf, Germany) and 1 vol % of penicillin–streptomycin (PenStrep, Thermo Fisher Scientific, Schwerte, Germany) in an incubator at 37 °C with 5% of CO_2 and 95% humidity. Cells were collected by trypsinization at confluency between 80% and 100%. Scaffolds were seeded with 600,000 cells/mL. The medium was changed twice a week.

4.4.2. Cell Viability and Relative Proliferation

Cell viability was measured after 7, 14, and 21 days as an indirect measurement of the viable cell number using a cell counting kit (Cell Counting Kit-8, Sigma-Aldrich). Samples were washed with PBS and incubated with 1 vol % WST-8 in DMEM. After 4 h of incubation, absorbance was measured at 450 nm (PHOmo, Autobio Labtec Instruments, Zhengzhou, China).

The relative proliferation was determined by an enzyme-linked immunosorbent assay (ELISA) using a cell proliferation kit (Cell Proliferation ELISA, Bromodeoxyuridine (BrdU), Roche, Mannheim, Germany). Shortly summarized, samples were incubated with a BrdU labeling solution for 2 h at 37 °C. During incubation, BrdU is incorporated into the DNA of proliferating cells. Afterwards, cells were fixed and incubated with an antibody. As the antibody was detected immunohistochemically, the reaction was stopped by addition of 1 M H_2SO_4 (Sulfuric acid, EMSURE® ISO, Merck Millipore, Darmstadt, Germany) and absorbance was measured at 450 nm.

4.4.3. Cell Morphology

The morphology of seeded MG-63 cells on top of uncoated and collagen-coated bioactive glass-based scaffolds was investigated by SEM. After incubation, samples were washed in PBS and fixed with fixative I and II (Table 3) for 1 h, respectively. After fixation, samples were dehydrated in a graded ethanol series (30, 50, 70, 80, 90, 95, and 99.8 vol %) and dried in a critical point dryer (EM EPD300, Leica, Germany).

Table 3. Composition of fixative I and II.

Reactant	Fixative I	Fixative II
Sodium cacodylate trihydrate (Sigma-Aldrich, Germany)	0.2 M	0.2 M
Glutaraldehyde (AppliChem, Germany)	0.1 wt %	0.3 wt %
Paraformaldehyde (Sigma-Aldrich, Germany)	2 wt %	3 wt %
Sucrose (Sigma-Aldrich, Germany)	5 wt %	-

4.5. Methods

4.5.1. Scanning Electron Microscopy (SEM)

For the observation by SEM, an Ultra plus scanning electron microscope (Auriga, Zeiss, Jena, Germany) was used. Samples were fixed with conductive silver, sputtered with gold (Quorum Q150T S, Quorum Technology, Darmstadt, Germany), and examined with 2 kV at a working distance of ~5 mm.

4.5.2. X-ray Photoelectron Spectroscopy (XPS)

To confirm the presence of APTS molecules on the bioactive glass surface, X-ray photoelectron spectroscopy (XPS) was applied. The measurement was carried out by a multi-technique XPS (Phi-5600, Al-Kα radiation).

4.5.3. Fourier Transform Infrared Spectroscopy (FTIR)

For chemical analysis, FTIR was used. 1 wt % of the sample to be analyzed was mixed with KBr (potassium bromide for IR spectroscopy Uvasol®, Merck Millipore, Darmstadt, Germany) and pressed into a pellet. The spectra were recorded in absorbance mode (Nicolet 6700 FTIR spectrometer, Thermo Scientific, Waltham, MA, USA) and collected between 4000 and 400 cm^{-1}. Pure KBr was used to correct the background noise.

4.5.4. Thermogravimetric Analysis (TGA)

In order to quantify the amount of collagen on the surface of coated scaffolds, thermogravimetric analysis was carried out. Heating was performed in air with a heating rate of 5 K/min (TGA, STA 449 F3 Jupiter, Netzsch, Selb, Germany).

4.5.5. Mechanical Characterization

The compressive strength of uncoated and collagen-coated scaffolds (with and without crosslinks) was evaluated by uniaxial compression tests. Samples were compressed with a crosshead velocity of 1 mm/min, a preload of 0.1 N, and a maximum applied force of 50 N (Z050, Zwick, Ulm, Germany).

4.6. Statistical Analysis

Statistic evaluation of the results of cell studies was performed by one-way ANOVA (Analysis of Variance, Origin 8.6) with significance levels at $p < 0.05$, ** $p < 0.01$, and *** $p < 0.001$, using Bonferroni's post-hoc test.

5. Conclusions

It can be concluded that collagen is an advantageous coating material for bioactive glass-based scaffolds. Relatively thick collagen layers (of a few micrometers) were applied on scaffolds without affecting the scaffold macroporosity. The compressive strength of BG-based scaffolds was increased by the presence of collagen, in particular by a factor of five for crosslinked collagen-coated scaffolds. Cell culture studies (MG-63 cells) demonstrated cell viability and suitable cell attachment and proliferation. Comparative studies considering alternative natural biopolymers as coatings of BG scaffolds are required to fully assess the relative advantages of collagen in this application. The results of this study provide a well-founded basis for future investigations in this field, in particular for further in vivo characterization of the produced composite scaffolds.

Author Contributions: J.H. and A.R.B. conceived the idea of the project; J.H. designed and performed the experiments; J.H. and A.R.B. analyzed the data; A.R.B. contributed reagents, materials, and analysis tools; J.H. wrote the first draft of the paper. A.R.B. corrected the paper. Both authors approved the paper.

Acknowledgments: The authors thank Judith Roether, Yaping Ding, and Dirk Dippold (Institute of Polymer Materials, University Erlangen-Nuremberg) as well as Sabine Fiedler (Institute of Glass and Ceramics, University Erlangen-Nuremberg), Helga Hildebrand (Institute for Surface Science and Corrosion, University Erlangen-Nuremberg), and Giulia Rella (Institute of Biomaterials, University Erlangen-Nuremberg) for their support during relevant measurements. The results presented in this paper are part of the doctoral thesis of Jasmin Hum presented at the University of Erlangen-Nuremberg.

Conflicts of Interest: The authors declare no conflict of interest.

References

1. Salgado, A.J.; Coutinho, O.P.; Reis, R.L. Bone tissue engineering: State of the art and future trends. *Macromol. Biosci.* **2004**, *4*, 743–765. [CrossRef] [PubMed]
2. Yunos, D.M.; Bretcanu, O.; Boccaccini, A.R. Polymer-bioceramic composites for tissue engineering scaffolds. *J. Mater. Sci.* **2008**, *43*, 4433–4442. [CrossRef]
3. García-Gareta, E.; Coathup, M.J.; Blunn, G.W. Osteoinduction of bone grafting materials for bone repair and regeneration. *Bone* **2015**, *81*, 112–121. [CrossRef] [PubMed]
4. Lareau, C.R.; Deren, M.E.; Fantry, A.; Donahue, R.M.J.; DiGiovanni, C.W. Does autogenous bone graft work? A logistic regression analysis of data from 159 papers in the foot and ankle literature. *Foot Ankle Surg.* **2015**, *21*, 150–159. [CrossRef] [PubMed]
5. Bose, S.; Roy, M.; Bandyopadhyay, A. Recent advances in bone tissue engineering scaffolds. *Trends Biotechnol.* **2012**, *30*, 546–554. [CrossRef] [PubMed]
6. Bruder, S.P.; Fox, B.S. Tissue engineering of bone. Cell based strategies. *Clin. Orthop. Relat. Res.* **1999**, *367*, S68–S83. [CrossRef]
7. Hutmacher, D.W. Scaffolds in tissue engineering bone and cartilage. *Biomaterials* **2000**, *21*, 2529–2543. [CrossRef]
8. Jones, J.R. Review of bioactive glass: From Hench to hybrids. *Acta Biomater.* **2013**, *9*, 4457–4486. [CrossRef] [PubMed]
9. El-Rashidy, A.A.; Roether, J.A.; Harhaus, L.; Kneser, U.; Boccaccini, A.R. Regenerating bone with bioactive glass scaffolds: A review of in vivo studies in bone defect models. *Acta Biomater.* **2017**, *62*, 1–28. [CrossRef] [PubMed]
10. Philippart, A.; Boccaccini, A.R.; Fleck, C.; Schubert, D.W.; Roether, J.A. Toughening and functionalization of bioactive ceramic and glass bone scaffolds by biopolymer coatings and infiltration: A review of the last 5 years. *Expert Rev. Med. Devices* **2015**, *12*, 93–111. [CrossRef] [PubMed]
11. Lee, C.H.; Singla, A.; Lee, Y. Biomedical applications of collagen. *Int. J. Pharm.* **2001**, *221*, 1–22. [CrossRef]
12. Sarker, B.; Hum, J.; Nazhat, S.N.; Boccaccini, A.R. Combining collagen and bioactive glasses for bone tissue engineering: A review. *Adv. Healthc. Mater.* **2015**, *4*, 176–194. [CrossRef] [PubMed]
13. Friess, W. Collagen—Biomaterial for drug delivery. *Eur. J. Pharm. Biopharm.* **1998**, *45*, 113–136. [CrossRef]
14. Khan, R.; Khan, M.H. Use of collagen as a biomaterial: An update. *J. Indian Soc. Periodontol.* **2013**, *17*, 539–542. [CrossRef] [PubMed]
15. Minabe, M.; Uematsu, A.; Nishijima, K.; Tomomatsu, E.; Tamura, T.; Hori, T.; Umemoto, T.; Hino, T. Application of a local drug delivery system to periodontal therapy: I. Development of collagen preparations with immobilized tetracycline. *J. Periodontol.* **1989**, *60*, 113–117. [CrossRef] [PubMed]
16. Nakagawa, T.; Tagawa, T. Ultrastructural study of direct bone formation induced by BMPs-collagen complex implanted into an ectopic site. *Oral Dis.* **2000**, *6*, 172–179. [CrossRef] [PubMed]
17. Peterson, D.R.; Ohashi, K.L.; Aberman, H.M.; Piza, P.A.; Crockett, H.C.; Fernandez, J.I.; Lund, P.J.; Funk, K.A.; Hawes, M.L.; Parks, B.G.; et al. Evaluation of a collagen-coated, resorbable fiber scaffold loaded with a peptide basic fibroblast growth factor mimetic in a sheep model of rotator cuff repair. *J. Should. Elb. Surg.* **2015**, *24*, 1764–1773. [CrossRef] [PubMed]

18. Lee, H.; Kim, G. Three-dimensional plotted PCL/β-TCP scaffolds coated with a collagen layer: Preparation, physical properties and in vitro evaluation for bone tissue regeneration. *J. Mater. Chem.* **2011**, *21*, 6305. [CrossRef]

19. Douglas, T.; Haugen, H.J. Coating of polyurethane scaffolds with collagen: Comparison of coating and cross-linking techniques. *J. Mater. Sci. Mater. Med.* **2008**, *19*, 2713–2719. [CrossRef] [PubMed]

20. Chen, Q.Z.; Thompson, I.D.; Boccaccini, A.R. 45S5 Bioglass®-derived glass-ceramic scaffolds for bone tissue engineering. *Biomaterials* **2006**, *27*, 2414–2425. [CrossRef] [PubMed]

21. Chen, Q.Z.; Ahmed, I.; Knowles, J.C.; Nazhat, S.N.; Boccaccini, A.R.; Rezwan, K. Collagen release kinetics of surface functionalized 45S5 Bioglass-based porous scaffolds. *J. Biomed. Mater. Res. A* **2008**, *86*, 987–995. [CrossRef] [PubMed]

22. Chen, Q.-Z.; Rezwan, K.; Françon, V.; Armitage, D.; Nazhat, S.N.; Jones, F.H.; Boccaccini, A.R. Surface functionalization of Bioglass-derived porous scaffolds. *Acta Biomater.* **2007**, *3*, 551–562. [CrossRef] [PubMed]

23. Filgueiras, M.R.; La Torre, G.; Hench, L.L. Solution effects on the surface reactions of a bioactive glass. *J. Biomed. Mater. Res.* **1993**, *27*, 445–453. [CrossRef] [PubMed]

24. Mukundan, L.M.; Nirmal, R.; Vaikkath, D.; Nair, P.D. A new synthesis route to high surface area sol gel bioactive glass through alcohol washing: A preliminary study. *Biomatter* **2013**, *3*, e24288. [CrossRef] [PubMed]

25. Marelli, B.; Ghezzi, C.E.; Barralet, J.E.; Boccaccini, A.R.; Nazhat, S.N. Three-dimensional mineralization of dense nanofibrillar collagen-bioglass hybrid scaffolds. *Biomacromolecules* **2010**, *11*, 1470–1479. [CrossRef] [PubMed]

26. Peitl Filho, O.; LaTorre, G.P.; Hench, L.L. Effect of crystallization on apatite-layer formation of bioactive glass 45S5. *J. Biomed. Mater. Res.* **1996**, *30*, 509–514. [CrossRef]

27. Cerruti, M.; Greenspan, D.; Powers, K. Effect of pH and ionic strength on the reactivity of Bioglass 45S5. *Biomaterials* **2005**, *26*, 1665–1674. [CrossRef] [PubMed]

28. Doyle, B.B.; Bendit, E.G.; Blout, E.R. Infrared spectroscopy of collagen and collagen-like polypeptides. *Biopolymers* **1975**, *14*, 937–957. [CrossRef] [PubMed]

29. Lazarev, Y.A.; Grishkovsky, B.A.; Khromova, T.B. Amide I band of IR spectrum and structure of collagen and related polypeptides. *Biopolymers* **1985**, *24*, 1449–1478. [CrossRef] [PubMed]

30. Chang, M.C.; Tanaka, J. FT-IR study for hydroxyapatite/collagen nanocomposite cross-linked by glutaraldehyde. *Biomaterials* **2002**, *23*, 4811–4818. [CrossRef]

31. Iafisco, M.; Foltran, I.; Sabbatini, S.; Tosi, G.; Roveri, N. Electrospun Nanostructured Fibers of Collagen-Biomimetic Apatite on Titanium Alloy. *Bioinorg. Chem. Appl.* **2012**, *2012*, 8. [CrossRef] [PubMed]

32. Oliveira, J.M.; Silva, S.S.; Malafaya, P.B.; Rodrigues, M.T.; Kotobuki, N.; Hirose, M.; Gomes, M.E.; Mano, J.F.; Ohgushi, H.; Reis, R.L. Macroporous hydroxyapatite scaffolds for bone tissue engineering applications: Physicochemical characterization and assessment of rat bone marrow stromal cell viability. *J. Biomed. Mater. Res. A* **2009**, *91*, 175–186. [CrossRef] [PubMed]

33. Koutsopoulos, S. Synthesis and characterization of hydroxyapatite crystals: A review study on the analytical methods. *J. Biomed. Mater. Res.* **2002**, *62*, 600–612. [CrossRef] [PubMed]

34. Rehman, I.; Knowles, J.C.; Bonfield, W. Analysis of in vitro reaction layers formed on Bioglass using thin-film X-ray diffraction and ATR-FTIR microspectroscopy. *J. Biomed. Mater. Res.* **1998**, *41*, 162–166. [CrossRef]

35. Stoch, A.; Jastrzebski, W.; Brożek, A.; Trybalska, B.; Cichocińska, M.; Szarawara, E. FTIR monitoring of the growth of the carbonate containing apatite layers from simulated and natural body fluids. *J. Mol. Struct.* **1999**, *511–512*, 287–294. [CrossRef]

36. Scharnweber, D.; Born, R.; Flade, K.; Roessler, S.; Stoelzel, M.; Worch, H. Mineralization behaviour of collagen type I immobilized on different substrates. *Biomaterials* **2004**, *25*, 2371–2380. [CrossRef] [PubMed]

37. Hunter, G.K.; Hauschka, P.V.; Poole, A.R.; Rosenberg, L.C.; Goldberg, H.A. Nucleation and inhibition of hydroxyapatite formation by mineralized tissue proteins. *Biochem. J.* **1996**, *317*, 59–64. [CrossRef] [PubMed]

38. Zhang, W.; Huang, Z.L.; Liao, S.S.; Cui, F.Z. Nucleation sites of calcium phosphate crystals during collagen mineralization. *J. Am. Chem. Soc.* **2003**, *86*, 1052–1054. [CrossRef]

39. Li, W.; Wang, H.; Ding, Y.; Scheithauer, E.C.; Goudouri, O.M.; Grünewald, A.; Detsch, R.; Agarwal, S.; Boccaccini, A.R. Antibacterial 45S5 Bioglass®-based scaffolds reinforced with genipin cross-linked gelatin for bone tissue engineering. *J. Mater. Chem. B* **2015**, *3*, 3367–3378. [CrossRef]

40. Li, W.; Garmendia, N.; de Larraya, U.P.; Ding, Y.; Detsch, R.; Grünewald, A.; Roether, J.A.; Schubert, D.W.; Boccaccini, A.R. 45S5 bioactive glass-based scaffolds coated with cellulose nanowhiskers for bone tissue engineering. *RSC Adv.* **2014**, *4*, 56156–56164. [CrossRef]

41. Metze, A.L.; Grimm, A.; Nooeaid, P.; Roether, J.A.; Hum, J.; Newby, P.J.; Schubert, D.W.; Boccaccini, A.R. Gelatin Coated 45S5 Bioglass®-Derived Scaffolds for Bone Tissue Engineering. *Key Eng. Mater.* **2013**, *541*, 31–39. [CrossRef]

42. Bellucci, D.; Chiellini, F.; Ciardelli, G.; Gazzarri, M.; Gentile, P.; Sola, A.; Cannillo, V. Processing and characterization of innovative scaffolds for bone tissue engineering. *J. Mater. Sci. Mater. Med.* **2012**, *23*, 1397–1409. [CrossRef] [PubMed]

43. Dorozhkin, S.; Ajaal, T. Toughening of porous bioceramic scaffolds by bioresorbable polymeric coatings. *Proc. Inst. Mech. Eng. H* **2009**, *223*, 459–470. [CrossRef] [PubMed]

44. Kim, H.-W.; Knowles, J.C.; Kim, H.-E. Hydroxyapatite/poly(epsilon-caprolactone) composite coatings on hydroxyapatite porous bone scaffold for drug delivery. *Biomaterials* **2004**, *25*, 1279–1287. [CrossRef] [PubMed]

45. Chen, Q.Z.; Boccaccini, A.R. Poly(D,L-lactic acid) coated 45S5 Bioglass-based scaffolds: Processing and characterization. *J. Biomed. Mater. Res. A* **2006**, *77*, 445–457. [CrossRef] [PubMed]

46. Yunos, D.M.; Ahmad, Z.; Salih, V.; Boccaccini, A.R. Stratified scaffolds for osteochondral tissue engineering applications: Electrospun PDLLA nanofibre coated Bioglass®-derived foams. *J. Biomater. Appl.* **2013**, *27*, 537–551. [CrossRef] [PubMed]

47. Li, J.J.; Roohani-Esfahani, S.-I.; Kim, K.; Kaplan, D.L.; Zreiqat, H. Silk coating on a bioactive ceramic scaffold for bone regeneration: Effective enhancement of mechanical and in vitro osteogenic properties towards load-bearing applications. *J. Tissue Eng. Regen. Med.* **2017**, *11*, 1741–1753. [CrossRef] [PubMed]

48. Tamai, N.; Myoui, A.; Tomita, T.; Nakase, T.; Tanaka, J.; Ochi, T.; Yoshikawa, H. Novel hydroxyapatite ceramics with an interconnective porous structure exhibit superior osteoconduction in vivo. *J. Biomed. Mater. Res.* **2002**, *59*, 110–117. [CrossRef] [PubMed]

49. Yang, Y.I.; Seol, D.L.; Kim, H.I.; Cho, M.H.; Lee, S.J. Composite fibrin and collagen scaffold to enhance tissue regeneration and angiogenesis. *Curr. Appl. Phys.* **2007**, *7*, e103–e107. [CrossRef]

50. Quinlan, E.; Partap, S.; Azevedo, M.M.; Jell, G.; Stevens, M.M.; O'Brien, F.J. Hypoxia-mimicking bioactive glass/collagen glycosaminoglycan composite scaffolds to enhance angiogenesis and bone repair. *Biomaterials* **2015**, *52*, 358–366. [CrossRef] [PubMed]

51. He, Q.; Zhao, Y.; Chen, B.; Xiao, Z.; Zhang, J.; Chen, L.; Chen, W.; Deng, F.; Dai, J. Improved cellularization and angiogenesis using collagen scaffolds chemically conjugated with vascular endothelial growth factor. *Acta Biomater.* **2011**, *7*, 1084–1093. [CrossRef] [PubMed]

52. Hoppe, A.; Güldal, N.S.; Boccaccini, A.R. A review of the biological response to ionic dissolution products from bioactive glasses and glass-ceramics. *Biomaterials* **2011**, *32*, 2757–2774. [CrossRef] [PubMed]

53. Lundqvist, M.; Sethson, I.; Jonsson, B.-H. Protein adsorption onto silica nanoparticles: Conformational changes depend on the particles' curvature and the protein stability. *Langmuir* **2004**, *20*, 10639–10647. [CrossRef] [PubMed]

54. Lenza, R.F.S.; Jones, J.R.; Vasconcelos, W.L.; Hench, L.L. In vitro release kinetics of proteins from bioactive foams. *J. Biomed. Mater. Res. A* **2003**, *67*, 121–129. [CrossRef] [PubMed]

55. Verné, E.; Vitale-Brovarone, C.; Bui, E.; Bianchi, C.L.; Boccaccini, A.R. Surface functionalization of bioactive glasses. *J. Biomed. Mater. Res. A* **2009**, *90*, 981–992. [CrossRef] [PubMed]

56. Cras, J.; Rowe-Taitt, C.; Nivens, D.; Ligler, F. Comparison of chemical cleaning methods of glass in preparation for silanization. *Biosens. Bioelectron.* **1999**, *14*, 683–688. [CrossRef]

57. Powell, H.M.; Boyce, S.T. EDC cross-linking improves skin substitute strength and stability. *Biomaterials* **2006**, *27*, 5821–5827. [CrossRef] [PubMed]

58. Olde Damink, L.H.; Dijkstra, P.J.; Van Luyn, M.J.A.; Van Wachem, P.B.; Nieuwenhuis, P.; Feijen, J. Cross-linking of dermal sheep collagen using a water-soluble carbodiimide. *Biomaterials* **1996**, *17*, 765–773. [CrossRef]

59. Zeeman, R.; Dijkstra, P.J.; van Wachem, P.B.; van Luyn, M.J.; Hendriks, M.; Cahalan, P.T.; Feijen, J. Successive epoxy and carbodiimide cross-linking of dermal sheep collagen. *Biomaterials* **1999**, *20*, 921–931. [CrossRef]

60. Kokubo, T.; Takadama, H. How useful is SBF in predicting in vivo bone bioactivity? *Biomaterials* **2006**, *27*, 2907–2915. [CrossRef] [PubMed]

61. Lowry, O.H.; Rosebrough, N.J.; Farr, A.L.; Randall, R.J. Protein measurement with the Folin phenol reagent. *J. Biol. Chem.* **1951**, *193*, 265–275. [PubMed]

62. Peterson, G.L. A simplification of the protein assay method of Lowry et al. which is more generally applicable. *Anal. Biochem.* **1977**, *83*, 346–356. [CrossRef]

63. Pautke, C.; Schieker, M.; Tischer, T.; Kolk, A.; Neth, P.; Mutschler, W.; Milz, S. Characterization of osteosarcoma cell lines MG-63, Saos-2 and U-2 OS in comparison to human osteoblasts. *Anticancer Res.* **2004**, *24*, 3743–3748. [PubMed]

International Journal of
Molecular Sciences

MDPI

Article

Healing of Osteochondral Defects Implanted with Biomimetic Scaffolds of Poly(ε-Caprolactone)/Hydroxyapatite and Glycidyl-Methacrylate-Modified Hyaluronic Acid in a Minipig

Yi-Ho Hsieh [1,2], Bo-Yuan Shen [3], Yao-Horng Wang [4], Bojain Lin [1,5], Hung-Maan Lee [1,6] and Ming-Fa Hsieh [1,*]

[1]   Department of Biomedical Engineering, Chung Yuan Christian University, 200 Chung Pei Road, Chung-Li District, Taoyuan City 320, Taiwan; dilantin11@gmail.com (Y.-H.H.); linbojain@gmail.com (B.L.); hml@hinet.net (H.-M.L.)
[2]   Department of Orthopedics, Min-Sheng General Hospital, 168, Ching Kuo Road, Taoyuan 330, Taiwan
[3]   Mater Program for Nanotechnology, Chung Yuan Christian University, 200 Chung Pei Road, Chung-Li District, Taoyuan City 320, Taiwan; dunshore1@gmail.com
[4]   Department of Nursing, Yuanpei University of Medical Technology, 306, Yuanpei Street, Hsinchu 300, Taiwan; pigmodel@gmail.com
[5]   Department of Orthopedics, Taoyuan Armed Forces General Hospital, No. 168, Zhongxing Road, Longtan District, Taoyuan City 325, Taiwan
[6]   Department of Orthopedics, Hualien Tzu Chi General Hospital, No. 707, Sec. 3, Chung Yang Road, Hualien 970, Taiwan
*   Correspondence: mfhsieh@cycu.edu.tw; Tel.: +886-3265-4550

Received: 4 March 2018; Accepted: 4 April 2018; Published: 9 April 2018

Abstract: Articular cartilage is a structure lack of vascular distribution. Once the cartilage is injured or diseased, it is unable to regenerate by itself. Surgical treatments do not effectively heal defects in articular cartilage. Tissue engineering is the most potential solution to this problem. In this study, methoxy poly(ethylene glycol)-block-poly(ε-caprolactone) (mPEG-PCL) and hydroxyapatite at a weight ratio of 2:1 were mixed via fused deposition modeling (FDM) layer by layer to form a solid scaffold. The scaffolds were further infiltrated with glycidyl methacrylate hyaluronic acid loading with 10 ng/mL of Transforming Growth Factor-β1 and photo cross-linked on top of the scaffolds. An in vivo test was performed on the knees of Lanyu miniature pigs for a period of 12 months. The healing process of the osteochondral defects was followed by computer tomography (CT). The defect was fully covered with regenerated tissues in the control pig, while different tissues were grown in the defect of knee of the experimental pig. In the gross anatomy of the cross section, the scaffold remained in the subchondral location, while surface cartilage was regenerated. The cross section of the knees of both the control and experimental pigs were subjected to hematoxylin and eosin staining. The cartilage of the knee in the experimental pig was partially matured, e.g., few chondrocyte cells were enclosed in the lacunae. In the knee of the control pig, the defect was fully grown with fibrocartilage. In another in vivo experiment in a rabbit and a pig, the composite of the TGF-β1-loaded hydrogel and scaffolds was found to regenerate hyaline cartilage. However, scaffolds that remain in the subchondral lesion potentially delay the healing process. Therefore, the structural design of the scaffold should be reconsidered to match the regeneration process of both cartilage and subchondral bone.

Keywords: cartilage; biomimetic; fused deposition modeling; poly(ε-caprolactone); hyaluronic acid; large animal models

1. Introduction

Articular cartilage is structurally deficient in vascular distribution and cannot regenerate once the cartilage is injured [1]. The uneven articular surface will gradually develop into arthritis over time, and this process is irreversible. Thus, treating blemishes of joint surfaces has been a major challenge for orthopedic surgeons. Current surgical approaches have been developed and range from arthroscopic ablation, osteotomy, and arthroplasty to cartilage restoration such as microfracture surgery, osteochondral autografts (mosaicplasty), autologous chondrocyte implantation (ACI), and fresh osteochondral allograft. However, these treatments do not fully heal the bone and cartilage tissue, and the regenerated cartilage is of inferior quality [2].

Microfracture surgery is a marrow stimulation technique that involves the migration of mesenchymal stem cells (MSCs) from the bone marrow to the defective site. This technique results in fibrocartilage regeneration and the mechanical property is inferior to native hyaline cartilage [3]. Mosaicplasty is a surgical procedure involving transplantation of cylindrical osteochondral grafts from a non-weight bearing region of the knee to the defects. The limitation of this technique is considered the scarce donor tissue source and donor site morbidity [4]. ACI is a cell-based approach for treating cartilage lesions, where autologous chondrocytes are implanted into the cartilage defect region. However, it has been reported that full restoration of the functional knee cannot be achieved [5]. In cartilage injury or osteoarthritis of joints, subchondral bone changes are a distinctive feature and can consist in, for example, sclerosis, cyst formation, bone attrition, bone marrow lesions, osteophytes, and subchondral bone remodeling [6]. Therefore, subchondral bone should be taken into consideration in the treatment of cartilage defects.

The application of tissue engineering to the repair of defects in cartilage and subchondral bone may serve as a solution to the above-mentioned inadequacies in treatment [7]. Tissue engineering to promote cartilage regeneration must take on three elements: scaffolds, cells, and growth agents [8–10]. Scaffolds can provide cell adhesion and proliferation. The scaffold material selection, mechanical properties, and pore architecture of the scaffold can affect the cell growth following implantation. Rapid prototyping (RP) can be used for scaffold fabrication, with various geometrical structures, controllable pore size and porosity, connectivity, and other advantages [11,12]. With RP, scaffolds that closely match defect shapes can be constructed. Poly(ε-caprolactone) (PCL) has good biocompatibility and has been employed in the fused deposition modeling (a type of RP) of porous scaffolds [13,14]. However, because of the synthetic nature of PCL, the cells are not prone to adhere to and grow on the surface of PCL scaffolds. It has been found that Arginylglycylaspartic acid (RGD) peptide can enhance cell adhesion by binding the integrin $\alpha v \beta 3$ in cell membranes. The use of RGD peptide grafted to the surface of PCL material promotes cells adhesion [15,16]. To enhance osteointegration and mimic the bony environment, hydroxyapatite (HAp) is typically added to the scaffold [17]. To mimic the cartilaginous environment, hyaluronic acid (HA) combined with transforming growth factor $\beta 1$ (TGF-$\beta 1$) can induce the differentiation of bone marrow mesenchymal stem cells to hyaline cartilage [18,19]. In consideration of both scaffold design and anatomical structure, glycidyl methacrylate was grafted to hyaluronic acid (GMHA), making a photo-curable hydrogel to carry growth factors [20]. In the present study, a solution containing TGF-$\beta 1$ was photo-crosslinked to form a hydrogel layer fixed on the upper side of an RP scaffold. With the biphasic-composite-designed scaffold, both the injured articular cartilage and subchondral bone were repaired and regrown.

Before bringing out such treatments into clinical practice, in vivo animal studies are required to close the gap between in vitro experiments and human clinical studies. In animal studies, the size of the joint and cartilage thickness are essential for simulating human disease. The animal size roughly corresponds to the size of the joint and cartilage thickness. In general, human cartilage lesions requiring treatment are at least 10 mm in diameter. The knee joints of New Zealand White rabbits are large enough to create defects of only about 3–4 mm, and the cartilage is relatively thin. Minipigs have larger knee joints such that 10 mm defects can be induced, and they have limited capability for the endogenous repair of chondral and osteochondral defects, which is similar to humans [21–23].

2. Results

The minipigs were sacrificed one year after scaffold implantation. Histological sections of the experimental group showed good regeneration of cartilage and subchondral bone in the defect area. The sections of the newly formed cartilage stained with hematoxylin and eosin presented a loose arrangement chondrocytes located in the lacunae surrounded by an extracellular matrix, which is a feature of hyaline cartilage. Beneath the cartilage layer, a good regeneration of the subchondral bone was noted. Undegraded scaffold in the deep part of the femur condyle as well as bone tissue growth in the pores of the scaffold was observed. In the control group, the defect region was observed to be filled with hypertrophic cartilage with invasion to the subchondral bone area, and the tissue's histology and morphology showed that the newly formed cartilage was fibrocartilage instead of hyaline cartilage.

2.1. Live Magnetic Resoance Imaging (MRI) Monitoring of Articular Cartilage Repair in Minipigs

Under general anesthesia, the magnetic resonance imaging (MRI) of the minipigs was performed six months after scaffold implantation. In the experimental knee joint, the bone defect was fully filled without free fluid. The scaffold could still be identified and maintained its structure. In the control, the bone defect region was filled with joint fluid, and partial bony ingrowth was noted (Figure 1).

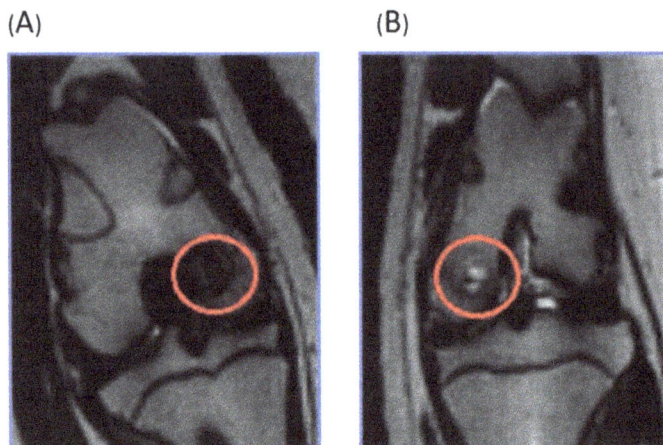

(A)   **(B)**

Figure 1. T2 weighted coronal plane MRI. The red circle was the operation area. (**A**) In the scaffold implantation group, the bone defect was fully filled by the scaffolds and tissue ingrowth. (**B**) In the control group, the bone defect region was filled with joint fluid, and partial bony ingrowth was noted.

2.2. Macroscopic Observation

Macroscopic observation provided a direct assessment of the cartilage defect repair. One year after engraftment, the minipigs were sacrificed, and their knees were harvested and analyzed. In the experimental group, near full-thickness defects were repaired with glossy white tissue (Figure 2A). The defects in the control group remained largely unfilled (Figure 2B). The International Cartilage Repair Society (ICRS) macroscopic score for the scaffold implantation group in terms of defect filling (3), integration to surrounding host cartilage (3), and macroscopic appearance (3) were all higher than those of the control group (3,1,1). The overall repair assessment of the experimental group was near normal (9, grade II), and the control group was abnormal (5, grade III).

(A)   (B)

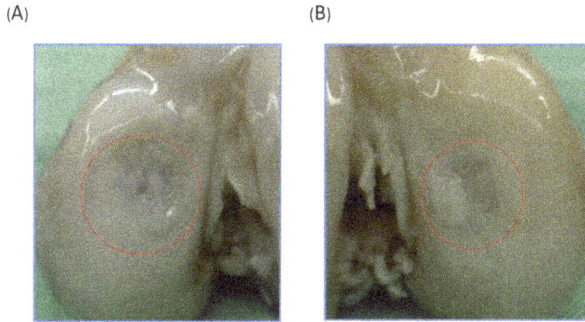

Figure 2. Gross observation of the repair of swine knees implanted with biphasic scaffolds of Poly(ε-caprolactone)/Hydroxyapatite and glycidyl-methacrylate-modified hyaluronic acid or without implants as the control group one year after surgery. The red circle was the operation area. (**A**) The scaffold implantation group showed successful regeneration of the previously removed cartilage. (**B**) The control group without implants showed soft tissue hypertrophy and cartilage defect.

To observe the repair tissue at the defect region beneath the surface of the regenerated cartilage, each condyle was sectioned along the sagittal and frontal plane. In the experimental group, the cartilage layer and subchondral bone were successfully regenerated, and undegraded scaffold in the deep part of the femur condyle was noted (Figure 3A). In the control group, the defect site was observed to be filled with hypertrophic cartilage-like tissue with invasion to the subchondral area (Figure 3B).

(A)   (B)

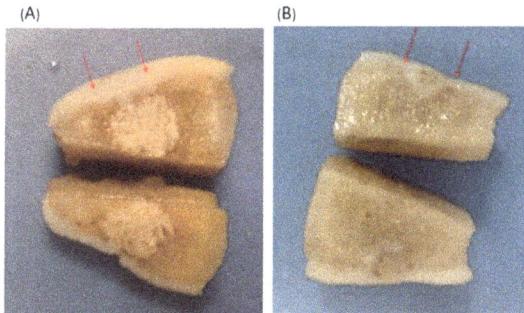

Figure 3. Cross-sectional view of the knee joint of the experimental group with scaffold implantation and the control group without implants. The operation area was between the red arrows. (**A**) With scaffold implantation, the defect of the surface was fully filled with hyaline cartilage, and subchondral bone regeneration was noted. Undegraded scaffold was noted in the deep layer. (**B**) Without implants, the defect region was filled with hypertrophic cartilage-like soft tissue with invasion to the subchondral bone area.

2.3. Histological Analysis

To identify the quality of the regenerated tissue, the experimental group and the control group were histologically evaluated. One year after implantation, the cartilage of the scaffold implantation group was successfully regeneration. Alcian blue staining was used to detect the presence of glycosaminoglycans, which parallels the degree of cartilage formation. The Alcian blue staining of the experimental group displayed an even concentrated blue color, the thickness was close to that of the surrounding normal cartilage, and the surface was smooth and uniform. The control group presented discontinuous and hypertrophic cartilage surface, with invasion to the subchondral bone area (Figure 4).

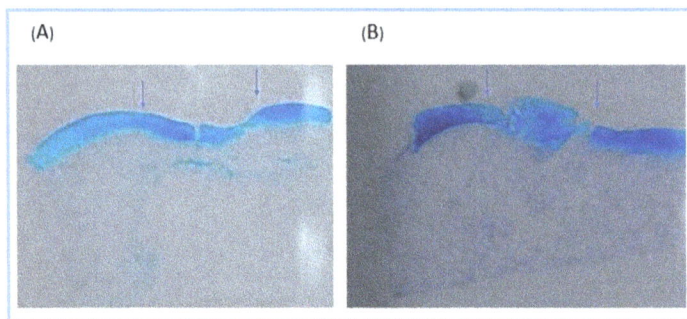

Figure 4. Alcian blue staining of the minipig knee joints one year after operation. The operation area was between the arrows. (**A**) The experimental group showed a smooth surface and an even thickness of the regenerated cartilage. (**B**) The control group presented a discontinuous and hypertrophic cartilage surface, and the subchondral bone area was replaced by cartilage-like soft tissue.

The second staining method, hematoxylin and eosin (H & E) staining, was used to show the distribution of the nucleus and cytoplasmic inclusions. The experimental group was shown in Figure 5A. Along the upper panel, cartilage successfully regenerated, and the displayed tissue morphology was resembled normal hyaline cartilage. The regenerated cartilage, compared to that of the control group, showed better columnar organization and integrated well with the surrounding cartilage. Good growth of the subchondral bone was also observed. However, the lower panel of the defect was filled with undegraded scaffold, with a partial ingrowth of bone tissue into the scaffold porous. In contrast, untreated defects in the control group (Figure 5B) were filled mostly with disorganized fibrocartilage that did not restore a continuous articular surface with adjacent host cartilage. The subchondral bone region was filled with fibrocartilage. These results suggest that an mPEG-PCL porous scaffold combined with GMHA hydrogel loading with TGF-β1 can help cartilage and subchondral bone formation and prevent fibrous tissue invagination.

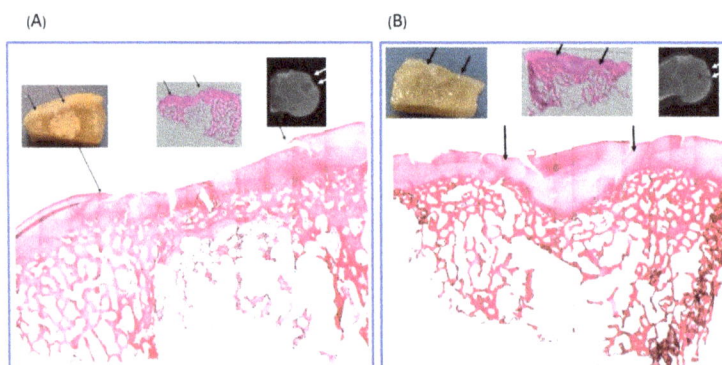

Figure 5. (**A**) Hematoxylin & Eosin staining of the biphasic scaffold implantation group (reorganized picture, original magnification ×5). The histology showed successfully cartilage regeneration over defect creating region (between two arrows). The repaired cartilage integrated well with the surrounding cartilage. Good growth of the subchondral bone was observed. The defect in the lower panel was filled with undegraded scaffold, and partial bony ingrowth in the scaffold porous was noted. (**B**) H & E staining of the untreated control group. Untreated defects in the control group were filled mostly with disorganized fibrocartilage that did not restore a smooth articular surface with adjacent host cartilage. The subchondral bone region was replaced by fibrocartilage.

To identify the regenerated tissue type, the cell morphology and matrix structure were analyzed. H & E staining of the normal minipig knee cartilage, the experimental group cartilage, and the control group cartilage were analyzed. The normal minipig knee cartilage showed native hyaline cartilage cell morphology with round chondrocytes in the lacunae surrounded by the matrix. A tidemark between cartilage and subchondral bone was observed (Figure 6A). Chondrocytes in the tissue regenerated from the biphasic scaffold implantation group were round, clustered, and surrounded by a matrix that is similar to those found in the surrounding hyaline cartilage. However, there were more chondrocytes in the lacunae than that of native hyaline cartilage, suggesting that the regenerated cartilage was hyaline cartilage albeit not completely differentiated (Figure 6B). In the defect-only control group, an abundance of spindle-shaped, fibroblast-like cells within the defect site was observed, suggesting that only fibrous or fibrocartilaginous tissues were formed (Figure 6C). Moreover, the tidemark between calcified and uncalcified cartilage was clear in the experimental group, but that in the control group could not be well distinguished.

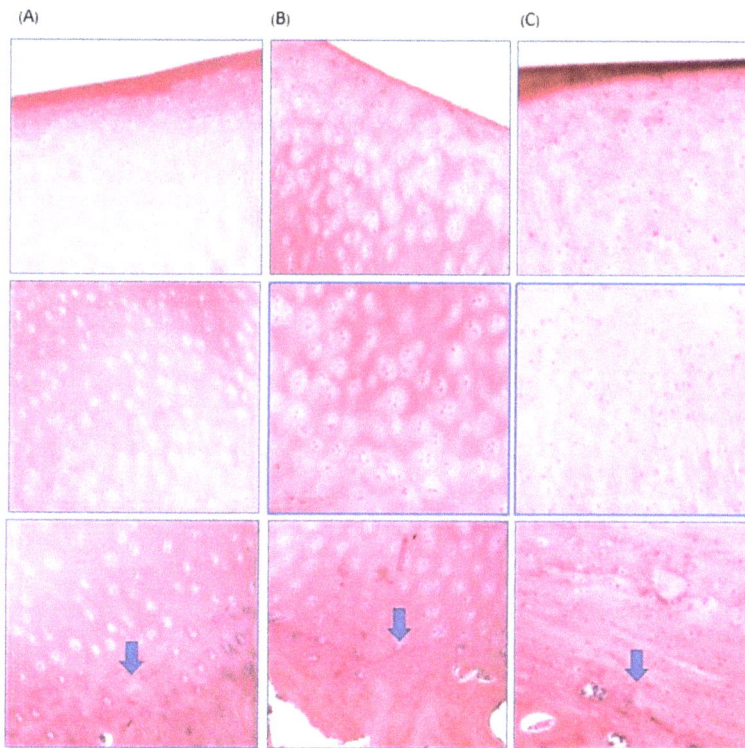

Figure 6. H & E staining (original magnification ×100) of (**A**) The normal minipig knee cartilage showed native hyaline cartilage cell morphology with round chondrocytes in the lacunae surrounding by the matrix. The tidemark is indicated by blue arrows. (**B**) The cell morphology of the regenerated cartilage from the biphasic scaffold implantation group was similar to the native hyaline cartilage with round chondrocytes in lacunae. The tidemark presented well. (**C**) The control group showed only spindle-shaped, fibroblast-like cells within the regenerated cartilage. The tidemark between calcified and uncalcified cartilage could not be well distinguished.

The regenerated bone tissue was also analyzed. With H & E staining, the normal minipig knee cancellous bone tissue showing the characteristic structure of cancellous bone with trabecular structure and numerous spaces containing bone marrow (Figure 7A). In the biphasic scaffold implantation group, the lower panel of the defect was filled with undegraded scaffolds. The scaffold porous was filled with mixed bone tissue and fibrotic tissue (Figure 7B). In the control group, the regenerated bone tissue showed a loose trabecular structure (Figure 7C).

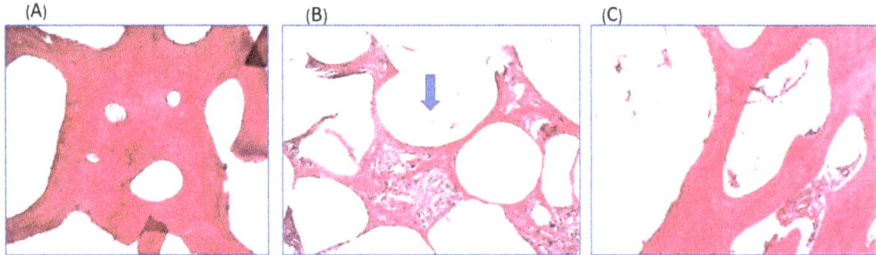

Figure 7. H & E staining (original magnification ×40) of (**A**) The normal minipig knee cancellous bone tissue showing the characteristic structure of cancellous bone with trabecular structure and numerous spaces containing bone marrow. (**B**) The experimental group showed undegraded scaffold (blue arrow) one year after implantation. The scaffold porous was filled with mixed bone tissue and fibrotic tissue. (**C**) The regenerated bone tissue of control group showed loose trabecular structure.

An ICRS Visual Histological Assessment was used to quantitatively compare the regenerative cartilage from the experimental group with that from the control group. For surface smoothness, the experimental group scored 2.5 and the control group scored 0.5, suggesting that the experimental group had a smoother and integrated cartilage surface. For matrix types, the experimental group scored 2 and the control group scored 0.5, indicating greater hyaline cartilage formation in the experimental group. In terms of cell distribution, the experimental group presented in a columnar/clustery arrangement with a score of 2, and the control group was more disorganized and presented with the score of 0.5. For cell viability, all two groups scored 2. For subchondral bone evaluation, the repair in the experimental group scored 2.5 and the no-implant group scored 1 due to fibrotic tissue invagination. Finally, for cartilage mineralization, no pathological mineralization was observed in either group, both had a score of 3 (Table 1). Based on the ICRS Visual Histological Assessment, we found that the experimental group can obtain better cartilage and subchondral bone regeneration compared with the control group.

Table 1. The result of ICRS Visual Histological Assessment Scale.

Feature	Score of Experimental Group	Score of Control Group
Surface	2.5	0.5
Matrix	2	0.5
Cell distribution	2	0.5
Cell viability	2	2
Subchondral bone	2.5	1
Cartilage mineralization (calcification)	3	3

2.4. Computed Tomography (CT) Scan Evaluation

The CT scan of the experimental group illustrated that the articular side of the scaffold was successfully degraded, subchondral bone regenerated, and the deep area of the created defect was filled with undegraded materials (Figure 8A). In the control group, there was more new bone formation in the defect site, but still some defect without bony growth was noted (Figure 8B).

(A)   (B)

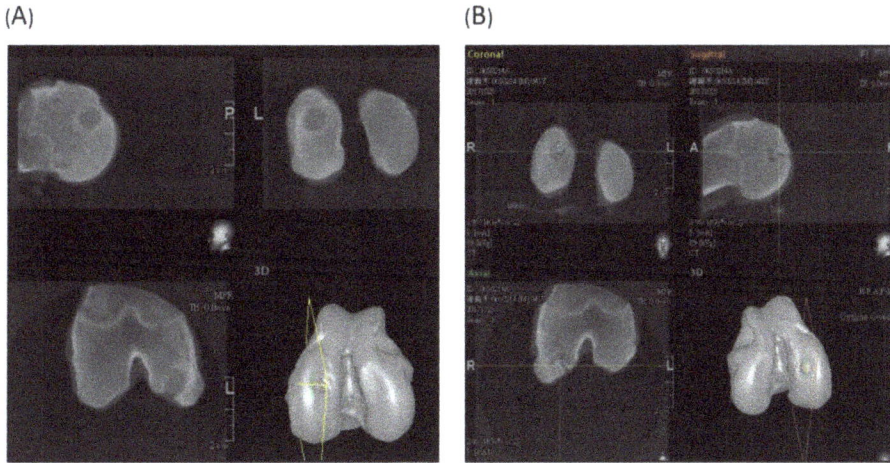

Figure 8. Sagittal, horizontal, and coronal view of the CT scan. (**A**) Experimental group showed an undegraded scaffold in the femur condyle. The articular part of the scaffold degraded well and was replaced by subchondral bone tissue. (**B**) The control group showed partial bony growth in the created defect region.

3. Discussion

Impact and torsional joint loading can injure articular cartilage. If joint injuries are localized to the upper layer of the avascular articular cartilage, no inflammation or effective healing can occur. When the lesions are too deep, and the subchondral bone vascular region is injured, the granulation tissue formed to fill the defect then changes to fibrocartilage [24]. Current treatments cannot replace damaged cartilage with new tissue with the same biomechanical properties as normal hyaline cartilage. Subchondral bone changes and remodeling have been observed in cartilage injury or in the osteoarthritis of joints and should be considered in treatment strategies [25]. Several chondrocyte implantation techniques have been used to treat cartilage injury, but mechanical stability cannot be provided by these grafts, and various side effects lead to treatment failure [26,27]. Subchondral bone remodeling and bone replaced by soft tissue and cartilage is another unresolved issue in chondrocyte implantation [28]. The tidemark separating cartilage and subchondral bone cannot be identified well in newly regenerated tissue. Wan-Ju Li et al. [29] noted that biodegradable poly(ε-caprolactone) (PCL) nanofibrous scaffolds seeded with allogeneic chondrocytes or xenogeneic human mesenchymal stem cells (MSCs), implanted to treat chondral defects, repaired the chondral defects, but the subchondral bone was replaced by soft tissue and cartilage. Pulliainen et al. [28] used chondrocyte-seeded poly-L,D-lactic acid scaffolds and found similar results. The phenomenon was similar to that observed in our control group. Chiang et al. [30] reported that fibrous tissue replaced the subchondral bone in a control, no-treatment, defect-created knees, suggesting that fibrous tissue formation is a natural wound-healing process. The recent treatment trend of osteochondral injury is to replace the injured osteochondral lesion by biomaterials, leading to in situ regeneration of not only cartilage but also subchondral bone. Various innovative scaffolds have been designed and reported [31]. Biphasic scaffolds with mimic structures of osteochondral tissues present an opportunity to close this chasm [32]. Frenkel et al. [33] developed biphasic scaffolds using poly(D,L-lactide) (PDLLA) mixed with HAp as an osteogenic layer and polyelectrolyte complex (PEC) hydrogel of HA and chitosan or a collagen type I scaffold as a cartilage layer. However, no growth factors consisted in the devices or the scaffold architecture, and pore size was difficult to control. To solve the problem of scaffold architecture and pore size design, a 3D printing technique was introduced. Sherwood et al. [34] designed 3D-printed scaffolds with an upper cartilage layer consisting of poly(D,L-lactide-*co*-glycolide) and poly(L-lactide) with a porosity of 90% and a lower bone layer composed of poly(L-lactide-*co*-glycolide)/tricalcium phosphate (TCP) with a porosity of 55%. Zhang et al. [35] also used a 3D printing technique to fabricate a biphasic poly(ethylene glycol) (PEG)/β-TCP scaffold. The cartilage layer of a PEG hydrogel was cured on a bone layer of β-TCP as a biphasic scaffold. With 3D printing techniques, architecture and pore size have been designed and controlled, but these studies have not tested for growth factors.

In our study, we demonstrated the successful regeneration of not only articular cartilage but also subchondral bone in an osteochondral defect of a minipig knee joint using biphasic mPEG-PCL porous scaffolds combined with GMHA hydrogel loading with TGF-β1. A tidemark was present, and the subchondral bone was not replaced by soft tissue macroscopically and histologically.

The velocity of degradation depends on the polymers and their molecular weights, as well as other properties. There are four major degradation mechanisms for the biodegradable polymers: hydrolysis, oxidation, enzymatic degradation, and physical degradation [36]. Degradation studies of three-dimensional PCL and PCL-based composite scaffolds have been conducted in vitro (in phosphate buffered saline), and in vivo (rabbit model) studies have revealed only about 7% scaffolds degraded in vivo six months after implantation [37]. In our study, the joint surface side of the mPEG-PCL porous scaffold was degraded and replaced by regenerated cartilage and subchondral bone successfully. The degradation mechanism may be due to hydrolysis by the joint fluid. Histology analysis showed that the scaffold in the deep bone layer was surrounded by regenerated fibrotic tissue and bone tissue. The regenerated tissue may shield the hydrolysis and lead to delayed degradation. The undegraded scaffolds that remain in bone potentially delay the healing process. Therefore, the structural design of the scaffold should be reconsidered to match the regeneration process of both the cartilage and subchondral bone [38].

4. Materials and Methods

4.1. Polymer Synthesis and Scaffold Fabrication

Raw materials used in the present study include biodegradable methoxy poly(ethylene glycol)-poly(ε-caprolactone) (mPEG-PCL) and hydroxyapatite. The synthesis of mPEG-PCL has been reported in our previous studies [13,39]. The physical and chemical properties and biological toxicity have been analyzed and described in our previous studies and in Supplementary Materials. The theoretical molecular weight of mPEG-PCL in those papers was 9450 Da. The diblock copolymer of mPEG-PCL was synthesized via ring opening polymerization of ε-caprolactone (CL) in the presence of mPEG as a macro-initiator and Sn(Oct)$_2$ as a catalyst.

Hyaluronic acid (HA) was employed to carry growth factor (TGF-β1) for the induction of hyaline cartilage in this study. To adhere TGF-β1-loaded HA to the top of the mPEG-PCL scaffold, photo-curable methacrylate groups were attached to HA by a method described in our previous studies. Briefly, 1 g of hyaluronic acid (molecular weight = 4.4×10^5) was dissolved in 100 mL phosphate buffered solution (PBS), and this mixture was stirred overnight. One hundred milliliters of *N*,*N*-dimethylformamide (DMF), 18.04 mL of trimethylamine (TEA), and 35.11 mL of glycidylmethacrylate (GM) were added to the HA solution and stirred for 10 days at room temperature. The resultant solution was then precipitated in acetone at a 10-fold volume with respect to the HA solution, then centrifuged to remove the acetone at 5000 rpm for 10 min. The precipitate was then dissolved in deionized water and subjected to dialysis (molecular weight cut off = 8 kDa) for 24 h and lyophilized to obtain photo-curable GMHA.

The scaffolds were fabricated with fused deposition modeling (FDM) by a custom-made air-pressure-aided deposition system. To improve the cellular attachment properties, mPEG-PCL was reacted with succinic anhydride, whereas the terminal group of the carboxylic acid group was attached to the end group of PCL. It was further grafted with the RGD peptide by condensation.

The carboxylic groups were introduced into the polymers. The scaffold surfaces were then modified via RGD peptide grafting. The powders of the synthesized mPEG-PCL diblock copolymers and HAp were mixed at a weight ratio of 60:40. FDM rapid prototyping technology was used to fabricate the scaffold by a home-made air pressure-aided deposition system. The home-made deposition system is composed of a stainless-steel sample container, a heating system, an air compressor, motors for *X*, *Y* and *Z* axes, and self-developed software (NI LabView 8.5, National Instruments Corporation, Austin, Texas, U.S). The mixed materials were melted by heating up to 60 °C. Subsequently, a pressure of 15 psi was applied to extrude the molten sample from a 0.4 mm nozzle onto a data processor-controlled *X*–*Y*–*Z* table. The construction of the scaffolds was computer-designed, and the extruded filament was laid down layer by layer in orientations of $0°/90°$. The printing speed was hand-optimized to a feed rate of 4 mm/s. The average pore size measurements of the scaffolds were 100~150 μm as our previous study. The RP fabricated scaffolds were immersed in a solution of dimethylaminopropyl-3-ethylcarbodiimide hydrochloride (EDC, 0.2 M) + *N*-hydroxysuccinimide (NHS, 0.1 M) in (2-(*N*-morpholino)-ethanesulfonic acid (MES buffer, 0.1 M in dd H_2O) for 1 h to activate the hydroxyl groups. Pure water was used to flush the scaffolds for 30 min. RGD peptides grafting was achieved via a condensation reaction by immersing the mPEG-PCL-COOH scaffolds in a solution of RGD peptides (10^{-3} M) and PBS for 24 h at 4 °C. Afterward, the scaffolds were rinsed with dd H_2O (100 mL) for 10 min to remove non-grafted peptides and freeze-dried to remove water.

The physical and chemical properties and biological toxicity have been analyzed and the results was put in the Supplementary file. The analysis including Differential scanning calorimetry (DSC), Fourier transform infrared spectroscopy (FTIR) (Supplementary Figure S1), Gel Permeation Chromatography(GPC), nuclear magnetic resonance spectrophotometry (^{1}H NMR) (Supplementary Figures S2 and S3), Thermogravimetric Analysis (TGA), In Vitro degradation (Supplementary Figure S4), and agar diffusion test for cytotoxicity analysis (Supplementary Figure S5).

4.2. Scaffold Implantation

The animal experimental protocol was approved by the Institutional Animal Experiment Committee of Chung Yuan Christian University and the Pigmodel® Animal Technology Co., Ltd. (Miaoli, Taiwan) (project identification code: PIG-105007, 15 April 2016). All animal surgeries were performed in a certified operating room at the Pigmodel® Animal Technology Co., Ltd., under general anesthesia using sterile techniques. Two male 2-year-old Lanyu miniature pigs weighing 25–28 kg were used for the evaluation of cartilage and bone regeneration. The pigs were put under general anesthesia, and their knees were shaved and disinfected with 2% chlorhexidine gluconate in 70% isopropyl alcohol. Operations performed were identical. Arthrotomy was made through a longitudinal medial parapatellar incision, and the femur condyle was exposed. A cylindrical osteochondral defect was created in the center of each medial femur condyle using an osteochondral coring device (OATS, Arthrex, Naples, FL, USA). A core 10 mm in diameter and 10 mm in depth was removed. Then, the scaffold was implanted into the defect and infiltrated with light-curing gel containing TGF-β1, cross-linking with 365 nm ultraviolet light for 10 min. In the control group, the defects were created without scaffold implantation.

4.3. Magnetic Resonance Imaging (MRI) and Computed Tomography (CT)

To monitor the progress of regeneration in a non-invasive method, magnetic resonance imaging (MRI) was performed 6 months after scaffold implantation under general anesthesia without any contrast agent. The minipigs were sacrificed, and joints were harvested one year after implantation. Computed tomography (CT) scans of the harvested knee joints were performed, and osteochondral specimens were evaluated pathologically.

4.4. Macroscopic Examination

One year after surgery, the animals were sacrificed for evaluation of the regeneration condition of the femoral condyles. Macroscopically, cartilage surface condition of the repair sites was observed and recorded. The repaired cartilage was scored using the ICRS scoring system for cartilage repair [40] (Table 2).

Table 2. International Cartilage Repair Society (ICRS) macroscopic evaluation of cartilage repair.

ICRS-Cartilage Repair Assessment System	Points
Degree of defect repair (Mossaicplasty; OAT; osteochondral allografts; others)	
100% survival of initially grafted surface	4
75% survival of initially grafted surface	3
50% survival of initially grafted surface	2
25% survival of initially grafted surface	1
0% (plugs are lost or broken)	0
Integration to border zone	
Complete integration with surrounding cartilage	4
Demarcating border < 1 mm	3
3/4th of graft integrated, 1/4th with a notable border >1 mm width	2
1/2 of graft integrated with surrounding cartilage, 1/2 with a notable border >1 mm	1
From no contact to 1/4th of graft integrated with surrounding cartilage	0
Macroscopic appearance	
Intact smooth surface	4
Fibrillated surface	3
Small, scattered fissures or cracs	2
Several, small or few but large fissures	1
Total degeneration of grafted area	0
Overall repair assessment	
Grade I: normal	12
Grade II: nearly normal	11–8
Grade III: abnormal	7–4
Grade IV: severely abnormal	3–1

4.5. Histological Evaluation

Each condyle was sectioned along the sagittal and frontal plane to evaluate the regeneration condition of the cartilage and subchondral bone. The specimens were fixed in 4% paraformaldehyde, decalcified in 10% Ethylenediaminetetraacetic acid, dehydrated, embedded in paraffin, sectioned to a 5 μm thickness, and stained with hematoxylin and eosin (H & E; Sigma, St. Louis, MO, USA), and Alcian blue. The regenerated cartilage quality was evaluated according to the Visual Histological Assessment Scale published by the ICRS [41]. The ICRS scale is based on the parameters of surface, matrix, cellular distribution, cell population viability, subchondral bone, and cartilage mineralization. High scores of the parameters indicate good-quality cartilage regeneration (Table 3).

Table 3. ICRS Visual Histological Assessment Scale, modified from the scale described by Mainil-Varlet et al. [41].

Feature	Score
Surface	
Smooth/continuous	3
Discontinuous/irregular	0
Matrix	
Hyaline cartilage	3
Hyaline cartilage/fibrocartilage	2
Fibrocartilage	1
Fibrous tissue	0
Cell distribution	
Columnar	3
Columnar/clustery	2
Clustery	1
Individual cells/disorganized	0
Cell viability	
Predominantly viable	3
Partially viable	1
Less than 10% viable	0
Subchondral bone	
Normal	3
Increased remodeling	2
Bone necrosis/granulation tissue	1
Detached/fracture/callus at base	0
Cartilage mineralization (calcification)	
Normal	3
Abnormal/inappropriate location	0

5. Conclusions

In our study, we obtained demonstrated in vivo healing of an osteochondral defect implanted with biphasic scaffolds of poly(ε-caprolactone)/hydroxyapatite and glycidyl methacrylate-modified hyaluronic acid in a minipig. One year after the scaffold was implanted into the Lanyu miniature pig's knee, the cartilage was regenerated successfully on the articular side of the scaffold and the morphology was significantly closer to hyaline cartilage compared to the control group. Although the scaffold was still not fully absorbed, bone tissue ingrowth to the pores of the scaffold was observed. Our study thus proposes a new clinical option to be considered alongside current treatments of cartilage injury.

Supplementary Materials: Supplementary materials can be found at http://www.mdpi.com/1422-0067/19/4/1125/s1.

Author Contributions: Ming-Fa Hsieh and Hung-Maan Lee conceived and designed the experiments; Yi-Ho Hsieh, Bo-Yuan Shen, Yao-Horng Wang, and Bojain Lin performed the experiments; Yi-Ho Hsieh and Bo-Yuan Shen analyzed the data; Yi-Ho Hsieh and Ming-Fa Hsieh wrote the paper.

Conflicts of Interest: The authors declare no conflict of interest.

References

1. Buckwalter, J.A.; Mankin, H.J. Articular cartilage: Tissue design and chondrocyte-matrix interactions. *Instr. Course Lect.* **1998**, *47*, 477–486. [PubMed]
2. Ozmeric, A.; Alemdaroglu, K.B.; Aydogan, N.H. Treatment for cartilage injuries of the knee with a new treatment algorithm. *World J. Orthop.* **2014**, *5*, 677–684. [CrossRef] [PubMed]
3. Kreuz, P.C.; Steinwachs, M.R.; Erggelet, C.; Krause, S.J.; Konrad, G.; Uhl, M.; Südkamp, N. Results after microfracture of full-thickness chondral defects in different compartments in the knee. *Osteoarthr. Cartil.* **2006**, *14*, 1119–1125. [CrossRef] [PubMed]

4. Robert, H. Chondral repair of the knee joint using mosaicplasty. *Orthop. Traumatol. Surg. Res.* **2011**, *97*, 418–429. [CrossRef] [PubMed]

5. Niemeyer, P.; Porichis, S.; Steinwachs, M.; Erggelet, C.; Kreuz, P.C.; Schmal, H.; Uhl, M.; Ghanem, N.; Südkamp, N.P.; Salzmann, G. Long-term outcomes after first-generation autologous chondrocyte implantation for cartilage defects of the knee. *Am. J. Sports Med.* **2014**, *42*, 150–157. [CrossRef] [PubMed]

6. Burr, D.B. The importance of subchondral bone in osteoarthrosis. *Curr. Opin. Rheumatol.* **1998**, *10*, 256–262. [CrossRef] [PubMed]

7. Makris, E.A.; Gomoll, A.H.; Malizos, K.N.; Hu, J.C.; Athanasiou, K.A. Repair and tissue engineering techniques for articular cartilage. *Nat. Rev. Rheumatol.* **2015**, *11*, 21–34. [CrossRef] [PubMed]

8. Wang, W.; Li, B.; Yang, J.; Xin, L.; Li, Y.; Yin, H.; Qi, Y.; Jiang, Y.; Ouyang, H.; Gao, C. The restoration of full-thickness cartilage defects with BMSCs and TGF-β1 loaded PLGA/fibrin gel constructs. *Biomaterials* **2010**, *31*, 8964–8973. [CrossRef] [PubMed]

9. Bhumiratana, S.; Eton, R.E.; Oungoulian, S.R.; Wan, L.Q.; Ateshian, G.A.; Vunjak-Novakovic, G. Large, stratified, and mechanically functional human cartilage grown in vitro by mesenchymal condensation. *Proc. Natl. Acad. Sci. USA* **2014**, *111*, 6940–6945. [CrossRef] [PubMed]

10. McCormick, F.; Cole, B.J.; Nwachukwu, B.; Harris, J.D.; Adkisson, H.D.; Farr, J. Treatment of Focal Cartilage Defects with a Juvenile Allogeneic 3-Dimensional Articular Cartilage Graft. *Oper. Tech. Sports Med.* **2013**, *21*, 95–99. [CrossRef]

11. Wang, X.; Yan, Y.; Zhang, R. Recent trends and challenges in complex organ manufacturing. *Tissue Eng. B Rev.* **2010**, *16*, 189–197. [CrossRef] [PubMed]

12. Hu, C.Z.; Tercero, C.; Ikeda, S.; Nakajima, M.; Tajima, H.; Shen, Y.J.; Fukuda, T.; Arai, F. Biodegradable porous sheet-like scaffolds for soft-tissue engineering using a combined particulate leaching of salt particles and magnetic sugar particles. *J. Biosci. Bioeng.* **2013**, *116*, 126–131. [CrossRef] [PubMed]

13. Jiang, C.P.; Chen, Y.Y.; Hsieh, M.F. Biofabrication and in vitro study of hydroxyapatite/mPEG-PCL-mPEG scaffolds for bone tissue engineering using air pressure-aided deposition technology. *Mater. Sci. Eng. C Mater. Boil. Appl.* **2013**, *33*, 680–690. [CrossRef] [PubMed]

14. Liao, H.T.; Chen, Y.Y.; Lai, Y.T.; Hsieh, M.F.; Jiang, C.P. The osteogenesis of bone marrow stem cells on mPEG-PCL-mPEG/hydroxyapatite composite scaffold via solid freeform fabrication. *BioMed Res. Int.* **2014**, *2014*, 321549. [CrossRef] [PubMed]

15. Gentile, P.; Ferreira, A.M.; Callaghan, J.T.; Miller, C.A.; Atkinson, J.; Freeman, C.; Hatton, P.V. Multilayer Nanoscale Encapsulation of Biofunctional Peptides to Enhance Bone Tissue Regeneration In Vivo. *Adv. Healthc. Mater.* **2017**, *6*. [CrossRef] [PubMed]

16. Wang, C.; Liu, Y.; Fan, Y.; Li, X. The use of bioactive peptides to modify materials for bone tissue repair. *Regen. Biomater.* **2017**, *4*, 191–206. [CrossRef] [PubMed]

17. Kutikov, A.B.; Song, J. An amphiphilic degradable polymer/hydroxyapatite composite with enhanced handling characteristics promotes osteogenic gene expression in bone marrow stromal cells. *Acta Biomater.* **2013**, *9*, 8354–8364. [CrossRef] [PubMed]

18. Fortier, L.A.; Barker, J.U.; Strauss, E.J.; McCarrel, T.M.; Cole, B.J. The role of growth factors in cartilage repair. *Clin. Orthop. Relat. Res.* **2011**, *469*, 2706–2715. [CrossRef] [PubMed]

19. Amann, E.; Wolff, P.; Breel, E.; van Griensven, M.; Balmayor, E.R. Hyaluronic acid facilitates chondrogenesis and matrix deposition of human adipose derived mesenchymal stem cells and human chondrocytes co-cultures. *Acta Biomater.* **2017**, *52*, 130–144. [CrossRef] [PubMed]

20. Baier Leach, J.; Bivens, K.A.; Patrick, C.W., Jr.; Schmidt, C.E. Photocrosslinked hyaluronic acid hydrogels: Natural, biodegradable tissue engineering scaffolds. *Biotechnol. Bioeng.* **2003**, *82*, 578–589. [CrossRef] [PubMed]

21. Chu, C.R.; Szczodry, M.; Bruno, S. Animal Models for Cartilage Regeneration and Repair. *Tissue Eng. Part B Rev.* **2010**, *16*, 105–115. [CrossRef] [PubMed]

22. Ahern, B.J.; Parvizi, J.; Boston, R.; Schaer, T.P. Preclinical animal models in single site cartilage defect testing: A systematic review. *Osteoarthr. Cartil.* **2009**, *17*, 705–713. [CrossRef] [PubMed]

23. Hunziker, E.B. Articular cartilage repair: Basic science and clinical progress. A review of the current status and prospects. *Osteoarthr. Cartil.* **2002**, *10*, 432–463. [CrossRef] [PubMed]

24. Hayes, D.W., Jr.; Brower, R.L.; John, K.J. Articular cartilage. Anatomy, injury, and repair. *Clin. Podiatr. Med. Surg.* **2001**, *18*, 35–53. [PubMed]

25. Intema, F.; Thomas, T.P.; Anderson, D.D.; Elkins, J.M.; Brown, T.D.; Amendola, A.; Lafeber, F.P.; Saltzman, C.L. Subchondral bone remodeling is related to clinical improvement after joint distraction in the treatment of ankle osteoarthritis. *Osteoarthr. Cartil.* **2011**, *19*, 668–675. [CrossRef] [PubMed]

26. Matricali, G.A.; Dereymaeker, G.P.; Luyten, F.P. Donor site morbidity after articular cartilage repair procedures: A review. *Acta Orthop. Belg.* **2010**, *76*, 669–674. [PubMed]

27. Darling, E.M.; Athanasiou, K.A. Rapid phenotypic changes in passaged articular chondrocyte subpopulations. *J. Orthop. Res.* **2005**, *23*, 425–432. [CrossRef] [PubMed]

28. Pulliainen, O.; Vasara, A.I.; Hyttinen, M.M.; Tiitu, V.; Valonen, P.; Kellomäki, M.; Jurvelin, J.S.; Peterson, L.; Lindahl, A.; Kiviranta, I.; et al. Poly-l,d-lactic acid scaffold in the repair of porcine knee cartilage lesions. *Tissue Eng.* **2007**, *13*, 1347–1355. [CrossRef] [PubMed]

29. Li, W.-J.; Chiang, H.; Kuo, T.-F.; Lee, H.-S.; Jiang, C.-C.; Tuan, R.S. Evaluation of articular cartilage repair using biodegradable nanofibrous scaffolds in a swine model: A pilot study. *J. Tissue Eng. Regen. Med.* **2009**, *3*, 1–10. [CrossRef] [PubMed]

30. Chiang, H.; Kuo, T.F.; Tsai, C.C.; Lin, M.C.; She, B.R.; Huang, Y.Y.; Lee, H.S.; Shieh, C.S.; Chen, M.H.; Ramshaw, J.A.; et al. Repair of porcine articular cartilage defect with autologous chondrocyte transplantation. *J. Orthop. Res.* **2005**, *23*, 584–593. [CrossRef] [PubMed]

31. Kon, E.; Filardo, G.; Roffi, A.; Andriolo, L.; Marcacci, M. New trends for knee cartilage regeneration: From cell-free scaffolds to mesenchymal stem cells. *Curr. Rev. Musculoskelet. Med.* **2012**, *5*, 236–243. [CrossRef] [PubMed]

32. Li, X.; Ding, J.; Wang, J.; Zhuang, X.; Chen, X. Biomimetic biphasic scaffolds for osteochondral defect repair. *Regen. Biomater.* **2015**, *2*, 221–228. [CrossRef] [PubMed]

33. Frenkel, S.R.; Bradica, G.; Brekke, J.H.; Goldman, S.M.; Ieska, K.; Issack, P.; Bong, M.R.; Tian, H.; Gokhale, J.; Coutts, R.D.; et al. Regeneration of articular cartilage—Evaluation of osteochondral defect repair in the rabbit using multiphasic implants. *Osteoarthr. Cartil.* **2005**, *13*, 798–807. [CrossRef] [PubMed]

34. Sherwood, J.K.; Riley, S.L.; Palazzolo, R.; Brown, S.C.; Monkhouse, D.C.; Coates, M.; Griffith, L.G.; Landeen, L.K.; Ratcliffe, A. A three-dimensional osteochondral composite scaffold for articular cartilage repair. *Biomaterials* **2002**, *23*, 4739–4751. [CrossRef]

35. Zhang, W.; Lian, Q.; Li, D.; Wang, K.; Hao, D.; Bian, W.; Jin, Z. The effect of interface microstructure on interfacial shear strength for osteochondral scaffolds based on biomimetic design and 3D printing. *Mater. Sci. Eng. C Mater. Biol. Appl.* **2015**, *46*, 10–15. [CrossRef] [PubMed]

36. Lyu, S.; Untereker, D. Degradability of Polymers for Implantable Biomedical Devices. *Int. J. Mol. Sci.* **2009**, *10*, 4033–4065. [CrossRef] [PubMed]

37. Lam, C.X.; Hutmacher, D.W.; Schantz, J.-T.; Woodruff, M.A.; Teoh, S.H. Evaluation of polycaprolactone scaffold degradation for 6 months in vitro and in vivo. *J. Biomed. Mater. Res. A* **2009**, *90*, 906–919. [CrossRef] [PubMed]

38. Leja, K.; Lewandowicz, G. Polymer Biodegradation and Biodegradable Polymers—A Review. *Pol. J. Environ. Stud.* **2010**, *19*, 255–266.

39. Hsieh, Y.H.; Hsieh, M.F.; Fang, C.H.; Jiang, C.P.; Lin, B.J.; Lee, H.M. Osteochondral Regeneration Induced by TGF-β Loaded Photo Cross-Linked Hyaluronic Acid Hydrogel Infiltrated in Fused Deposition-Manufactured Composite Scaffold of Hydroxyapatite and Poly(Ethylene Glycol)-Block-Poly(ε-Caprolactone). *Polymers* **2017**, *9*, 182. [CrossRef]

40. Van den Borne, M.P.; Raijmakers, N.J.; Vanlauwe, J.; Victor, J.; de Jong, S.N.; Bellemans, J.; Saris, D.B. International cartilage repair society (ICRS) and Oswestry macroscopic cartilage evaluation scores validated for use in autologous chondrocyte implantation (ACI) and microfracture. *Osteoarthr. Cartil.* **2007**, *15*, 1397–1402. [CrossRef] [PubMed]

41. Mainil-Varlet, P.; Aigner, T.; Brittberg, M.; Bullough, P.; Hollander, A.; Hunziker, E.; Kandel, R.; Nehrer, S.; Pritzker, K.; Roberts, S.; et al. Histological assessment of cartilage repair: A report by the Histology Endpoint Committee of the International Cartilage Repair Society (ICRS). *J. Bone Joint Surg. Am.* **2003**, *85*, 45–57. [CrossRef] [PubMed]

International Journal of
Molecular Sciences

MDPI

Article

The Osteogenic Differentiation Effect of the FN Type 10-Peptide Amphiphile on PCL Fiber

Ye-Rang Yun [1], Hae-Won Kim [2,3,4,*] and Jun-Hyeog Jang [5,*]

[1] Industrial Technology Research Group, Research and Development Division, World Institute of Kimchi, Nam-Gu, Gwangju 61755, Korea; yunyerang@wikim.re.kr
[2] Institute of Tissue Regeneration Engineering (ITREN), Dankook University, Cheonan 330-714, Korea
[3] Department of Nanobiomedical Science & BK21 PLUS NBM Global Research Center for Regenerative Medicine, Dankook University, Cheonan 330-714, Korea
[4] Department of Biomaterials Science, School of Dentistry, Dankook University, Cheonan 330-714, Korea
[5] Department of Biochemistry, Inha University School of Medicine, Incheon 22212, Korea
*   Correspondence: kimhw@dku.edu (H.-W.K.); juhjang@inha.ac.kr (J.-H.J.);
    Tel.: +82-41-550-3081 (H.-W.K.); +82-32-890-0930 (J.-H.J.);
    Fax: +82-41-550-3085 (H.-W.K.); +82-32-882-1877 (J.-H.J.)

Received: 1 November 2017; Accepted: 3 January 2018; Published: 4 January 2018

Abstract: The fibronectin type 10-peptide amphiphile (FNIII10-PA) was previously genetically engineered and showed osteogenic differentiation activity on rat bone marrow stem cells (rBMSCs). In this study, we investigated whether FNIII10-PA demonstrated cellular activity on polycaprolactone (PCL) fibers. FNIII10-PA significantly increased protein production and cell adhesion activity on PCL fibers in a dose-dependent manner. In cell proliferation results, there was no effect on cell proliferation activity by FNIII10-PA; however, FNIII10-PA induced the osteogenic differentiation of MC3T3-E1 cells via upregulation of bone sialoprotein (*BSP*), collagen type I (*Col I*), osteocalcin (*OC*), osteopontin (*OPN*), and runt-related transcription factor 2 (*Runx2*) mitochondrial RNA (mRNA) levels; it did not increase the alkaline phosphatase (*ALP*) mRNA level. These results indicate that FNIII10-PA has potential as a new biomaterial for bone tissue engineering applications.

Keywords: FNIII10; peptide amphiphile; PCL fiber; osteogenic differentiation activity

1. Introduction

Recently, various attempts have been made to find new biomaterial for tissue engineering. Among the attempts, research on the peptide amphiphile (PA) has been widely reported. PA has hydrophilic and hydrophobic components and a self-assembly ability as a peptide-based molecule. There is increased interest in the extracellular matrix (ECM) molecule mimetic PA because it can promote certain cellular activities including adhesion, proliferation, and differentiation [1–5]. Based on these characteristics, biomimetic PA is extensively utilized for tissue engineering as a new biomaterial. For example, the Arg-Gly-Asp (RGD) peptide, composed of L-arginine, glycine, and L-aspartic acid, participates in cellular attachment via integrin [6]. For this reason, numerous studies on PA-RGD have been reported in the pharmaceutical field [7–9] as well as in the tissue engineering field [10–12]. For instance, the combination of hydroxyapatite (HA) and PA-RGD enhances the osteoinductive and osteoconductive activities of the scaffold [13].

Moreover, studies on amphiphile containing fibronectin (FN) are commonly demonstrated and reported. We also previously demonstrated that the engineered fibronectin type 10-peptide amphiphile sequence (FNIII10-PA) potentiates utilization for bone tissue engineering by enhancing the adhesion, proliferation, and differentiation of rat bone marrow stem cells (rBMSCs) [14]. In another study, the tenascin-C mimetic PA (TN-C PA) showed self-assembly activity. Nanofiber gels containing TN-C

PA were shown to promote neurite outgrowth, which could potentially be used for artificial matrix therapy in neuronal regeneration [15].

Polycaprolactone (PCL) is a biomaterial that is used mostly for bone tissue engineering [16,17]. PCL is a biodegradable polyester with a low melting point (60 °C) and a low glass transition temperature (−60 °C). Because of its biodegradation properties, PCL is regarded as a suitable biomaterial for long-term implantable devices [18]. Above all, PCL has advantages in tissue engineering applications for drug delivery devices, sutures, or adhesion barriers without safety problems for the Food and Drug Administration (FDA) approval [19]. However, it is difficult for PCL to influence cellular activity because it does not provide an ECM type structure for the cells [20]. Recently, nanofibrous PCL has been the preferred choice for bone tissue engineering applications by improving cellular activities and mechanical properties. PCL is indisputably the most used biomaterial in bone tissue engineering research.

Previously, we revealed that genetically engineered FNIII10-PA promoted cell adhesion, proliferation, and differentiation of rBMSCs. In the present study, we investigated whether or not FNIII10-PA with PCL fiber promoted the MC3T3-E1's cellular activities. FNIII10 was used as the negative control in all experiments. Initially, the protein adhesion activity of FNIII10-PA was measured. Then, the effects of FNIII10-PA with PCL fiber on MC3T3-E1 cells adhesion and proliferation were investigated. Furthermore, the osteogenic differentiation activity of FNIII10-PA with PCL fibers on MC3T3-E1 cells was investigated by measuring the osteogenic differentiation marker mRNA level.

2. Results and Discussion

2.1. Morphology of PCL Fibers

PCL has been commonly used as a biomaterial in the tissue engineering field because of its non-toxic and biodegradable characteristics. On the other hand, PCL demonstrated a slow degradation rate, poor mechanical properties, and low cell adhesion. Particularly, PCL shows low affinity and stiffness in cells due to poor hydrophilicity, leading to the reduction in cell proliferation, migration, differentiation, and regeneration [20]. However, nanofibrous PCL was reported to improve the cellular activities in recent studies [21,22]. FNIII10 was connected to the PA sequence LLLLCCCGGDSDS (L, leucine; C, cysteine; G, glycine; D, aspartic acid; S, serine) to create FNIII10-A sequence as illustrated in Figure 1A. PCL fiber was fabricated by the electrospinning technique. Figure 1B shows that the electrospun PCL fiber was uniform with straight fibers. Average diameter of PCL fibers was approximately 1 µM (Figure 1C). Hence, PCL fiber was suitable to investigate the cell adhesion activity of FNIII10 in combination with biomaterial in the present study.

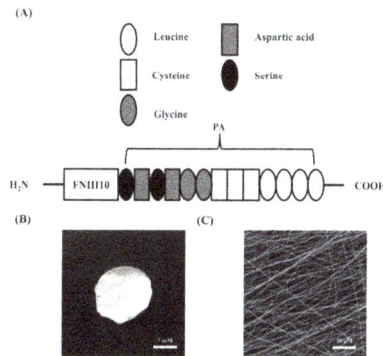

Figure 1. Structure of fibronectin type 10-peptide amphiphile (FNIII10-PA) (**A**) and morphology of polycaprolactone(PCL) fiber (**B**,**C**). Peptide amphiphile (PA) sequence, LLLLCCCGGDS (L, leucine; C, cysteine; G, glycine; D, aspartic acid; S, serine) was connected with fibronectin type III10. Scale is 7 mM in panel B, 50 µM in panel C (resolution 500×).

2.2. Protein Adhesion Activity of FNIII10-PA on PCL Fibers

Previously, FNIII10-PA was shown to increase protein adhesion activity on tissue culture plates as compared with FNIII10. As shown in Figure 2, the protein adhesion activity of FNIII10-PA was significantly increased in a dose-dependent manner (*** $p < 0.001$). Particularly, 5 μg·mL^{-1} of FNIII10-PA increased protein adhesion 3-fold compared with that of FNIII10. In many studies, PCL scaffold coated with FN remarkably enhanced cell adhesion, proliferation, and differentiation [23–25]. Drevelle et al. found that the PCL film functionalized with RGD peptide (Ac-CGGNGEPRGDTYRAY-NH$_2$ derived from the bone sialoprotein) and easily adsorbed the serum adhesive proteins FN and vitronectin (VN), leading to increased cell adhesion and spreading by activating focal adhesion kinase (FAK) signaling [26]. In another study, FN-immobilized nanobioactiveglass (nBG)/PCL (FN-nBG/PCL) scaffolds also increased cell differentiation as well as cell proliferation [27]. In the present study, FNIII10-PA showed higher protein adhesion activity on PCL fibers than that of FNIII10 alone.

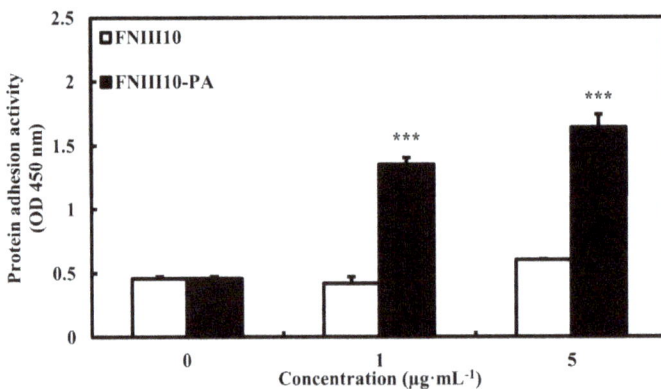

Figure 2. Protein adhesion activity of FNIII10-PA on PCL fiber. PCL fibers were placed in 24-well plates and coated with 0, 1, or 5 μg·mL^{-1} of FNIII10 or FNIII10-PA overnight at 4 °C. Protein adhesion activities are expressed as means $\pm$ SD ($n = 3$). *** $p < 0.001$. OD = optical density.

2.3. Cell Adhesion Activity of FNIII10-PA on PCL Fibers

As mentioned above, cells are unable to perform many cellular activities on PCL fibers, because PCL fibers do not provide a structure similar to the natural ECM. To evaluate whether or not FNIII10-PA enhances cellular activity on PCL fibers, cell adhesion activity of FNIII10-PA was performed on PCL fibers. In addition, MC3T3-E1 cell spreading on PCL fibers by FNIII10-PA was observed by scanning electron microscopy (SEM). Figure 3A shows that 5 μg·mL^{-1} of FNIII10-PA significantly enhanced cell adhesion activity on PCL fibers (Figure 3A, * $p < 0.05$). Hence, SEM analysis was performed in 5 μg·mL^{-1} of FNIII10 or FNIII10-PA without a treatment group for comparison. In the SEM image, 5 μg·mL^{-1} of FNIII10-PA induced more cell spreading and attachment on PCL fiber compared to that of FNIII10 (Figure 3B). Generally, PCL in the nanofiber form is preferred in bone tissue engineering studies, because PCL shows low cellular activities such as cell adhesion, proliferation, and differentiation. For instance, Yun et al. reported that PCL nanofibers enhanced cellular activities by providing a three-dimensional structure similar to the ECM [28]. In this study, FNIII10-PA enhanced cell adhesion activity on PCL fibers, even though PCL was in the microfiber form. Based on these results, FNIII10-PA may improve the cell interaction with PCL fiber.

(A)

(B)

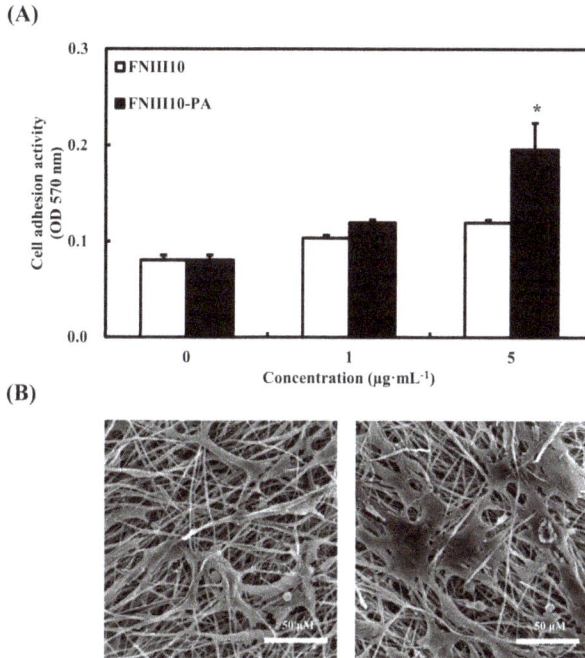

Figure 3. Cell adhesion activity (**A**) and cell spreading (**B**) of MC3T3-E1 on PCL fibers. PCL fibers were placed in 24-well plates and coated with 0, 1, or 5 $\mu g \cdot mL^{-1}$ of FNIII10 or FNIII10-PA overnight at 4 °C. MC3T3-E1 cells were cultured at a density of 1×10^5 cells and incubated for 30 min at 37 °C. Cell adhesion activities are expressed as means ± SD ($n = 3$). * $p < 0.05$. For SEM analysis, 5 $\mu g \cdot mL^{-1}$ of FNIII10 and FNIII10-PA were used. Scale is 50 μM (resolution 500×).

2.4. Cell Proliferation Activity of FNIII10-PA on PCL Fiber

We previously revealed that FNIII10-PA was shown to promote rBMSCs proliferation activity. However, MC3T3-El cell proliferation activity of FNIII10-PA on PCL fibers was not observed in the present study. As shown in Figure 4, there were no differences in cell proliferative activity between FNIII10 and FNIII10-PA. Cell proliferative activity of FNIII10-PA could be weak in MC3T3-E1 cells compared to that in rBMSC cells. In the next study, application of FNIII10-PA needs to confirm the rBMSC proliferative activity on PCL fiber. FN was shown to have a powerful effect on both cell adhesion and differentiation of stem cells. For instance, Mohamadyar-Toupkanlou et al. reported that nanofibrous PCL/HA scaffold coated with FN provided a suitable environment for proliferation as well as attachment and differentiation in mouse MSCs (mMSCs) [23]. Shekaran et al. found that ECM-coated PCL microcarriers promoted human early mesenchymal stem cell growth, leading to cell expansion [24]. Similarly, Mousavi et al. revealed that three dimensional nanoscaffold coated with FN induced the human cord blood hematopoietic stem cell expansion [25].

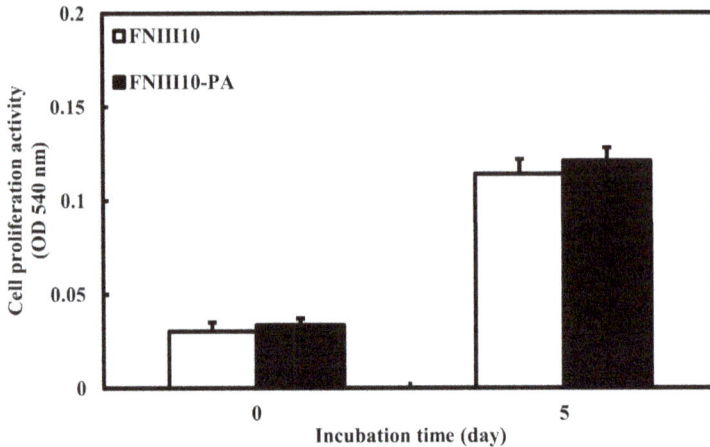

Figure 4. Cell proliferation activity of FNIII10-PA on PCL fibers at 0 and 5 days. PCL fibers were placed in 24-well plates and coated with 5 μg·mL^{-1} of FNIII10 or FNIII10-PA overnight at 4 °C. MC3T3-E1 cells were cultured at a density of 1×10^5 cells and incubated for 5 days at 37 °C. Cell proliferation activity is expressed as mean ± SD ($n = 3$).

2.5. The Osteogenic Differentiation Activity of FNIII10-PA on PCL Fibers

To investigate the osteogenic differentiation activity of FNIII10-PA on PCL fibers, we measured the mitochondrial RNA (mRNA) levels of alkaline phosphatase (*ALP*), bone sialoprotein (*BSP*), collagen type I (*Col I*), osteocalcin (*OC*), osteopontin (*OPN*), and runt-related transcription factor 2 (*Runx2*) genes at 5 and 10 days. At both time points, the mRNA levels of *BSP*, *Col I*, *OC*, *OPN*, and *Runx2* were significantly upregulated by FNIII10-PA treatment (Figure 5, * $p < 0.05$, ** $p < 0.005$, and *** $p < 0.001$). However, there was no alteration of *ALP* mRNA level at either 5 or 10 days. Interestingly, relative mRNA level of *Col I* was slightly decreased at 10 days compared with that at 5 days. These results were possibly caused by an increase of detached cells. Generally, these markers are expressed with early- or late-stage osteogenic differentiation as follows. *ALP* is a key marker that is responsible for the mineralization of the ECM [29]. *Col I* is the dominant collagen and exists in many tissues such as tendons and ligaments. *ALP* and *Col I* are commonly expressed in the early stage of osteogenic differentiation [30] and are considered early differentiation markers. *OPN* is also an early marker of osteogenic differentiation and is part of non-collagenous bone ECM proteins [31]. *OC* is associated with the late stage of osteogenic differentiation and is known as bone Gla protein (BGP), although *OC* is also a non-collagenous bone ECM protein [32]. As another late differentiation marker, *Runx2* is a master switch for the osteogenic differentiation [33]. In our results, FNIII10-PA significantly induced the osteogenic differentiation of MC3T3-E1 cells on PCL fibers by upregulation of most marker's mRNA levels at both 5 and 10 days. These results are similar to our previous results [14]. Taken together, we reconfirmed the osteogenic differentiation activity of FNIII10-PA on PCL fibers.

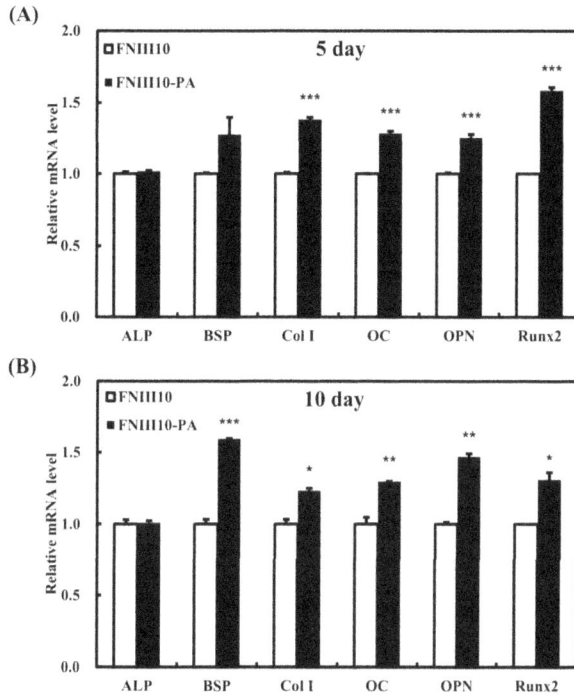

Figure 5. The osteogenic differentiation activity of FNIII10-PA on PCL fibers at 5 days (**A**) and 10 days (**B**). PCL fibers were placed in 24-well plates and coated with 5 μg·mL^{-1} of FNIII10 or FNIII10-PA overnight at 4 °C. MC3T3-E1 cells were cultured at a density of 5 × 10^4 cells and incubated for 5 and 10 days at 37 °C. Quantitative real-time PCR results were analyzed. * $p < 0.05$, ** $p < 0.005$, and *** $p < 0.001$. Genes: alkaline phosphatase (*ALP*), bone sialoprotein (*BSP*), collagen type I (*Col I*), osteocalcin (*OC*), osteopontin (*OPN*), and runt-related transcription factor 2 (*Runx2*).

3. Materials and Methods

3.1. Protein Expression and Purification

After transformation into TOP10 *Escherichia coli*, cells were grown overnight at 37 °C in Luria-Bertani (LB) medium containing ampicillin. Induction was initiated using 0.1% (*w/v*) L-arabinose at A600 = 0.6 followed by incubation at 20 °C for 6 h. Bacteria were pelleted by centrifugation at 6000× *g* for 10 min and then lysed and sonicated. A soluble extract was prepared by centrifugation for 30 min at 14,000× *g* in a refrigerated centrifuge, and the supernatant obtained was purified using a nickel-nitrilotriacetic acid resin (Invitrogen, Carlsbad, CA, USA).

3.2. Electrospinning of PCL Fibers

A PCL (MW = 80,000; Sigma-Aldrich, St. Louis, MO, USA) solution was prepared by dissolving 10% *w/v* PCL in dichloromethane (DCM) and ethanol (4:1 ratio). PCL solutions were ultrasonicated and loaded into a 10 mL plastic syringe equipped with a 21-gauge needle made of stainless steel. The needle was connected to a high-voltage power supply. The tip-to-collector distance was kept at 10 cm. The voltage and injection rate were 15 kV and 0.5 mL·h^{-1}, respectively. PCL solution was eletrospun to foil. After drying, PCL fiber (15 mm diameter) was cut from the foil using a punch. All experiments were performed at room temperature (RT). In each experiment, PCL fibers coated with FNIII10 or FNIII10-PA were used.

3.3. Morphology of PCL Fibers and Cells on PCL Fibers

The structure of the PCL fibers was observed by SEM at an accelerating voltage of 15 kV after fixing with glutaraldehyde (2.5%), dehydrating with a graded series of ethanol (75%, 90%, 95%, and 100% (v/v) for 10 min each), treating with hexamethyldisilazane, and coating with platinum.

To observe the cell spreading and expansion on PCL fibers, PCL fibers were placed in 24-well plates and were coated only once with FNIII10 or FNIII10-PA (5 $\mu g \cdot mL^{-1}$) overnight at 4 °C. Cells were seeded with 1×10^5 cells, incubated for 30 min, washed with Dulbecco's phosphate-buffered saline (DPBS), the cells on the PCL fibers were fixed with a 3.7% (w/v) formalin solution for 15 min at RT, dehydrated with a graded series of ethanol (75%, 90%, 95%, and 100% (v/v) for 10 min each), treated with hexamethyldisilazane, and coated with platinum.

3.4. Protein Adhesion Assay

To evaluate the protein adhesion activity, PCL fibers were placed in 24-well plates and were coated only once with FNIII10 or FNIII10-PA (0, 1, or 5 $\mu g \cdot mL^{-1}$) overnight at 4 °C. Each well was washed with phosphate-buffered saline (PBS) and blocked with 1% (w/v) bovine serum albumin (BSA) solution for 1 h at RT. After washing with PBS, a peroxidase conjugate of a monoclonal anti-polyhistidine antibody was added and incubated for 1 h at RT. After washing with tris-buffered saline tween (TBS-T), 200 µL Turbo TMB-enzyme-linked immunoserological assay (ELISA) was added and incubated for 30 min at RT. H_2SO_4 (100 µL, 2 M) was added to stop the reaction, and the absorbance was read at 450 nm. A protein adhesion assay was performed for the negative control group (un-treated), the positive control group (FNIII10), and the experimental group (FNIII10-PA). The value of the negative control was approximately equal to the value seen with FNIII10 and FNIII10-PA. Hence, results were expressed as the comparison between FNIII10 and FNIII10-PA without negative control.

3.5. Cell Culture

MC3T3-E1 is a pre-osteoblastic cell line established from newborn mouse calvarias. We purchased from the American Type Culture Collection (ATCC). Cells were cultured in α-minimum essential media (α-MEM) containing 10% (v/v) heat-inactivated fetal bovine serum (FBS, WELGENE, Daegu, Korea), 100 units·mL^{-1} penicillin G sodium, 10 $\mu g \cdot mL^{-1}$ streptomycin, and 25 $\mu g \cdot mL^{-1}$ amphotericin B (WELGENE, Daegu, Korea) at 37 °C in a humidified atmosphere with 5% CO_2.

3.6. Cell Adhesion Assay

Cell adhesion activity on PCL fibers was measured using the crystal violet assay. PCL fibers were placed in 24-well plates and were coated with FNIII10 or FNIII10-PA (0, 1, or 5 $\mu g \cdot mL^{-1}$) overnight at 4 °C. Each well was washed with DPBS and blocked with 1% (w/v) BSA solution for 30 min. Cells were cultured with 1×10^5 cells, incubated for 30 min, washed twice with DPBS, and fixed with 3.7% (w/v) formalin solution for 15 min at RT. Cells were then stained with 0.25% (w/v) crystal violet (Sigma, St. Louis, MO, USA) in 2% (v/v) ethanol/water for 1 h at 37 °C and gently washed three times with DPBS. Cells were then lysed with 2% sodium dodecyl sulfate (SDS) solution and transferred to 96-well plates. Absorbance was read at 570 nm. Results were expressed as the comparison between FNIII10 and FNIII10-PA.

3.7. Cell Proliferation Assay

Cell proliferation activity was evaluated using the 3-(4,5-dimethylthiazol-2-yl)-2,5-diphenyltetrazolium bromide (MTT) assay, according to the manufacturer's instructions (Promega, Madison, WI, USA). FNIII10 and FNIII10-PA protein solutions were coated at 5 $\mu g \cdot mL^{-1}$ on 24-well plates; cells were cultured with 1×10^5 cells and incubated for 0 and 5 days at 37 °C. Cells were then washed three times with DPBS, and 500 µL of MTT (5 $mg \cdot mL^{-1}$ in PBS) was added to each well. After incubation for 4 h, media were removed, and formazan crystals were dissolved in

200 μL dimethyl sulfoxide (DMSO). Absorbance was read at 540 nm. The cell proliferation assay was performed in the negative control group (un-treated), the positive control group (FNIII10), and the experimental group (FNIII10-PA). After identifying the effect of FNIII10 and FNIII10-PA as compared with that of the negative control, the results were expressed as the comparison between FNIII10 and FNIII10-PA.

3.8. Quantitative Real-Time PCR Analysis

Initially, PCL fibers were placed in 24-well plates and were coated with FNIII10 or FNIII10-PA (5 μg·mL^{-1}) overnight at 4 °C. Cells were cultured at a density of 5×10^4 cells and incubated for 5 and 10 days at 37 °C. Total RNA was extracted using the Easy-spin RNA Extraction kit (iNtRON, Seoul, Korea), and circular (cDNA) was synthesized. The expression levels of the genes *ALP*, *BSP*, *Col I*, *OC*, *OPN*, *Runx2* were confirmed by quantitative real-time PCR. All real-time PCR analyses were performed using the ABI Step One real-time PCR system. Each reaction was performed in a 20-μL reaction mixture containing 0.1 μM of each primer, 10 μL of 2× SYBR Green PCR master mix (Applied Biosystems, including AmpliTaq Gold DNA polymerase in buffer, a dNTP mix, SYBR Green I dye, Rox dye, and 10 mM MgCl$_2$), and 1 μL of template cDNA. The *Ct* (cycle threshold) value for each gene was determined using the automated threshold analysis function in the ABI instrument and was normalized with respect to *Ct*(GAPDH) to obtain d*Ct* (d*Ct* = *Ct*$_{(GAPDH)}$ − *Ct*$_{(specific gene)}$). Finally, the *Ct* value of FNIII10-PA was normalized by the *Ct* value of FNIII10. The primers used for quantitative real-time PCR are shown in Table 1. Quantitative real-time PCR results were also expressed as a comparison between FNIII10 and FNIII10-PA.

Table 1. Sequences of primers used for quantitative real-time PCR.

Genes	Forward Primer	Reverse Primer
GAPDH	TCCACTCACGGCAAATTCAAC	AGCCCAAGATGCCCTTCAGT
ALP	GGGCAATGAGGTCACATCC	GTCACAATGCCCACGGACTT
BSP	CAGAGGAGGCAAGCGTCACT	CTGTCTGGGTGCCAACACTG
Col I	GAGGCATAAAGGGTATCGTGG	CATTAGGCGCAGGAAGGTCAGC
OC	CATCACTGCCACCCAGAAGAC	CAGTGGATGCAGGGATGATGT
OPN	CCAATGAAAGCCATGACCAC	CGACTGTAGGGACGATTGGA
Runx2	GGCCGGGAATGATGAGAACTA	GGCCCACAAATCTCAGATCGT

3.9. Statistical Analysis

All experiments were conducted in triplicate. Experimental data are expressed as mean ± standard deviation (SD). The Student's *t*-test with paired data sets was used to determine the significances of differences between FNIII10 and FNIII10-PA. Statistical significance was accepted for *p* values <0.05.

4. Conclusions

Previously, we revealed that FNIII10-PA has potential in the osteogenic differentiation activity of rBMSCs. In present study, we also found FNIII10-PA enhanced cell adhesion and osteogenic differentiation activities of MC3T3-E1 cells on PCL fibers. Although the PCL microfiber form had low affinity to cells, the combination with FNIII10-PA significantly enhanced cellular activities including adhesion and differentiation. Taken together, FNIII10-PA potentiates the appropriate utilization for bone tissue engineering with biomaterials as well as working FNIII10-PA alone.

Acknowledgments: This work was supported by the Priority Research Centers Program (grant No. 2009-0093829), by the Korean Healthcare Technology R&D Project, Ministry of Health & Welfare, Republic of Korea (HI14C0522), and by a National Research Foundation of Korea (NRF) grant funded by the Korean government (MSIP) (NRF-2016R1A2B4008811).

Author Contributions: Ye-Rang Yun performed the experiments and wrote the manuscript; Hae-Won Kim and Jun-Hyeog Jang organized this work and helped write the manuscript.

Conflicts of Interest: The authors have no conflicts of interest to declare.

References

1. Shin, H.; Jo, S.; Mikos, A.G. Biomimetic materials for tissue engineering. *Biomaterials* **2003**, *24*, 4353–4364. [CrossRef]
2. Goktas, M.; Cinar, G.; Orujalipoor, I.; Ide, S.; Tekinay, A.B.; Guler, M.O. Self-assembled peptide amphiphile nanofibers and peg composite hydrogels as tunable ECM mimetic microenvironment. *Biomacromolecules* **2015**, *16*, 1247–1258. [CrossRef] [PubMed]
3. Anderson, J.M.; Patterson, J.L.; Vines, J.B.; Javed, A.; Gilbert, S.R.; Jun, H.W. Biphasic peptide amphiphile nanomatrix embedded with hydroxyapatite nanoparticles for stimulated osteoinductive response. *ACS Nano* **2011**, *5*, 9463–9479. [CrossRef] [PubMed]
4. Shroff, K.; Pearce, T.R.; Kokkoli, E. Enhanced integrin mediated signaling and cell cycle progression on fibronectin mimetic peptide amphiphile monolayers. *Langmuir* **2012**, *28*, 1858–1865. [CrossRef] [PubMed]
5. Kushwaha, M.; Anderson, J.M.; Bosworth, C.A.; Andukuri, A.; Minor, W.P.; Lancaster, J.R., Jr.; Anderson, P.G.; Brott, B.C.; Jun, H.W. A nitric oxide releasing, self-assembled peptide amphiphile matrix that mimics native endothelium for coating implantable cardiovascular devices. *Biomaterials* **2010**, *31*, 1502–1508. [CrossRef] [PubMed]
6. Jeschke, B.; Meyer, J.; Jonczyk, A.; Kessler, H.; Adamietz, P.; Meenen, N.M.; Kantlehner, M.; Goepfert, C.; Nies, B. RGD-peptides for tissue engineering of articular cartilage. *Biomaterials* **2002**, *23*, 3455–3463. [CrossRef]
7. Liu, Z.; Yu, L.; Wang, X.; Zhang, X.; Liu, M.; Zeng, W. Integrin ($\alpha_v\beta_3$) tageted RGD peptide based probe for cancer optical imaging. *Curr. Protein Pept. Sci.* **2016**, *17*, 570–581. [CrossRef] [PubMed]
8. Hou, J.; Diao, Y.; Li, W.; Yang, Z.; Zhang, L.; Chen, Z.; Wu, Y. RGD peptide conjugation results in enhanced antitumor activity of PD0325901 against glioblastoma byboth tumor-targeting delivery and combination therapy. *Int. J. Pharm.* **2016**, *505*, 329–340. [CrossRef] [PubMed]
9. Song, Z.; Lin, Y.; Zhang, X.; Feng, C.; Lu, Y.; Gao, Y.; Dong, C. Cyclic RGD peptide-modified liposomal drug delivery system for targeted oral apatinib administration: Enhanced cellular uptake and improved therapeutic effects. *Int. J. Nanomed.* **2017**, *12*, 1941–1958. [CrossRef] [PubMed]
10. Chen, W.; Zhou, H.; Weir, M.D.; Tang, M.; Bao, C.; Xu, H.H. Human embryonic stem cell-derived mesenchymal stem cell seeding on calcium phosphate cement-chitosan-RGD scaffold for bone repair. *Tissue Eng. Part A* **2013**, *19*, 915–927. [CrossRef] [PubMed]
11. Chen, L.; Li, B.; Xiao, X.; Meng, Q.; Li, W.; Yu, Q.; Bi, J.; Cheng, Y.; Qu, Z. Preparation and evaluation of an Arg-Gly-Asp-modified chitosan/hydroxyapatite scaffold for application in bone tissue engineering. *Mol. Med. Rep.* **2015**, *12*, 7263–7270. [CrossRef] [PubMed]
12. Kim, H.D.; Heo, J.; Hwang, Y.; Kwak, S.Y.; Park, O.K.; Kim, H.; Varghese, S.; Hwang, N.S. Extracellular-matrix-based and Arg-Gly-Asp-modified photopolymerizing hydrogels for cartilage tissue engineering. *Tissue Eng. Part A* **2015**, *21*, 757–766. [CrossRef] [PubMed]
13. Çakmak, S.; Çakmak, A.S.; Gümüşderelioğlu, M. RGD-bearing peptide-amphiphile-hydroxyapatite nanocomposite bone scaffold: An in vitro study. *Biomed. Mater.* **2013**, *8*. [CrossRef] [PubMed]
14. Yun, Y.R.; Pham, L.B.H.; Yoo, Y.R.; Lee, S.; Kim, H.W.; Jang, J.H. Engineering of self-assembled fibronectin matrix protein and its effects on mesenchymal stem cells. *Int. J. Mol. Sci.* **2015**, *16*, 19645–19656. [CrossRef] [PubMed]
15. Berns, E.J.; Álvarez, Z.; Goldberger, J.E.; Boekhoven, J.; Kessler, J.A.; Kuhn, H.G.; Stupp, S.I. A tenascin-C mimetic peptide amphiphile nanofiber gel promotes neurite outgrowth and cell migration of neurosphere-derived cells. *Acta Biomater.* **2016**, *37*, 50–58. [CrossRef] [PubMed]
16. Gómez-Lizárraga, K.K.; Flores-Morales, C.; Del Prado-Audelo, M.L.; Álvarez-Pérez, M.A.; Piña-Barba, M.C.; Escobedo, C. Polycaprolactone- and polycaprolactone/ceramic-based 3D-bioplotted porous scaffolds for bone regeneration: A comparative study. *Mater. Sci. Eng. C Mater. Biol. Appl.* **2017**, *79*, 326–335. [CrossRef] [PubMed]
17. Xu, T.; Miszuk, J.M.; Zhao, Y.; Sun, H.; Fong, H. Electrospun polycaprolactone 3D nanofibrous scaffold with interconnected and hierarchically structured pores for bone tissue engineering. *Adv. Healthc. Mater.* **2015**, *4*, 2238–2246. [CrossRef] [PubMed]

18. Dash, T.K.; Konkimalla, V.B. Poly-ε-caprolactone based formulations for drug delivery and tissue engineering: A review. *J. Control. Release* **2012**, *158*, 15–33. [CrossRef] [PubMed]

19. Li, Z.; Tan, B.H. Towards the development of polycaprolactone based amphiphilic block copolymers: Molecular design, self-assembly and biomedical applications. *Mater. Sci. Eng. C Mater. Biol. Appl.* **2014**, *45*, 620–634. [CrossRef] [PubMed]

20. Ghasemi-Mobarakeh, L.; Prabhakaran, M.P.; Morshed, M.; Nasr-Esfahani, M.H.; Ramakrishna, S. Electrospun poly(ε-caprolactone)/gelatin nanofibrous scaffolds for nerve tissue engineering. *Biomaterials* **2008**, *29*, 4532–4539. [CrossRef] [PubMed]

21. Kim, M.; Kim, G. Electrospun PCL/phlorotannin nanofibres for tissue engineering: Physical properties and cellular activities. *Carbohydr. Polym.* **2012**, *90*, 592–601. [CrossRef] [PubMed]

22. Chen, J.P.; Chang, Y.S. Preparation and characterization of composite nanofibers of polycaprolactone and nanohydroxyapatite for osteogenic differentiation of mesenchymal stem cells. *Colloids Surf. B Biointerfaces* **2011**, *86*, 169–175. [CrossRef] [PubMed]

23. Mohamadyar-Toupkanlou, F.; Vasheghani-Farahani, E.; Hanaee-Ahvaz, H.; Soleimani, M.; Dodel, M.; Havasi, P.; Ardeshirlajimi, A.; Taherzadeh, E.S. Osteogenic differentiation of MSCs on fibronectin-coated and nHA-modified scaffolds. *ASAIO J.* **2017**, *63*, 684–691. [CrossRef] [PubMed]

24. Shekaran, A.; Lam, A.; Sim, E.; Jialing, L.; Jian, L.; Wen, J.T.; Chan, J.K.; Choolani, M.; Reuveny, S.; Birch, W.; et al. Biodegradable ECM-coated PCL microcarriers support scalable human early MSC expansion and in vivo bone formation. *Cytotherapy* **2016**, *18*, 1332–1344. [CrossRef] [PubMed]

25. Mousavi, S.H.; Abroun, S.; Soleimani, M.; Mowla, S.J. Expansion of human cord blood hematopoietic stem/progenitor cells in three-dimensional Nanoscaffold coated with Fibronectin. *Int. J. Hematol. Oncol. Stem Cell Res.* **2015**, *9*, 72–79. [PubMed]

26. Drevelle, O.; Bergeron, E.; Senta, H.; Lauzon, M.A.; Roux, S.; Grenier, G.; Faucheux, N. Effect of functionalized polycaprolactone on the behaviour of murine preosteoblasts. *Biomaterials* **2010**, *31*, 6468–6476. [CrossRef] [PubMed]

27. Won, J.E.; Mateos-Timoneda, M.A.; Castano, O.; Planell, J.A.; Seo, S.J.; Lee, E.J.; Han, C.M.; Kim, H.W. Fibronectin immobilization on to robotic-dispensed nanobioactive glass/polycaprolactone scaffolds for bone tissue engineering. *Biotechnol. Lett.* **2015**, *37*, 935–942. [CrossRef] [PubMed]

28. Yun, Y.P.; Kim, S.J.; Lim, Y.M.; Park, K.; Kim, H.J.; Jeong, S.I.; Kim, S.E.; Song, H.R. The effect of alendronate-loaded polycarprolactone nanofibrous scaffolds on osteogenic differentiation of adipose-derived stem cells in bone tissue regeneration. *J. Biomed. Nanotechnol.* **2014**, *10*, 1080–1090. [CrossRef] [PubMed]

29. Marom, R.; Shur, I.; Solomon, R.; Benayahu, D. Characterization of adhesion and differentiation markers of osteogenic marrow stromal cells. *J. Cell. Physiol.* **2005**, *202*, 41–48. [CrossRef] [PubMed]

30. Tsai, M.T.; Li, W.J.; Tuan, R.S.; Chang, W.H. Modulation of osteogenesis in human mesenchymal stem cells by specific pulsed electromagnetic field stimulation. *J. Orthop. Res.* **2009**, *27*, 1169–1174. [CrossRef] [PubMed]

31. Zohar, R.; Cheifetz, S.; McCulloch, C.A.; Sodek, J. Analysis of intracellular osteopontin as a marker of osteoblastic cell differentiation and mesenchymal cell migration. *Eur. J. Oral Sci.* **1998**, *106* (Suppl. S1), 401–407. [CrossRef] [PubMed]

32. Candeliere, G.A.; Liu, F.; Aubin, J.E. Individual osteoblasts in the developing calvaria express different gene repertoires. *Bone* **2001**, *28*, 351–361. [CrossRef]

33. Li, S.; Kong, H.; Yao, N.; Yu, Q.; Wang, P.; Lin, Y.; Wang, J.; Kuang, R.; Zhao, X.; Xu, J.; et al. The role of runt-related transcription factor 2 (Runx2) in the late stage of odontoblast differentiation and dentin formation. *Biochem. Biophys. Res. Commun.* **2011**, *410*, 698–704. [CrossRef] [PubMed]

International Journal of
Molecular Sciences

MDPI

Article

Human Mesenchymal Stem Cells Growth and Osteogenic Differentiation on Piezoelectric Poly(vinylidene fluoride) Microsphere Substrates

R. Sobreiro-Almeida [1,†], M. N. Tamaño-Machiavello [2,†], E. O. Carvalho [1], L. Cordón [3,4], S. Doria [2], L. Senent [3,4,5], D. M. Correia [1,6], C. Ribeiro [1,7,*], S. Lanceros-Méndez [8,9], R. Sabater i Serra [2,10], J. L. Gomez Ribelles [2,10] and A. Sempere [3,4,5]

[1] Centro/Departamento de Física, Universidade do Minho, 4710-057 Braga, Portugal; rita.isaias.almeida@hotmail.com (R.S.-A.); estela.carvalhoo@hotmail.com (E.O.C.); d.correia@fisica.uminho.pt (D.M.C.)
[2] Centre for Biomaterials and Tissue Engineering, CBIT, Universitat Politècnica de València, 46022 Valencia, Spain; noeltm@gmail.com (M.N.T.-M.); sergidobe@msn.com (S.D.); rsabater@die.upv.es (R.S.iS.); jlgomez@ter.upv.es (J.L.G.R.)
[3] Hematology Research Group, Instituto de Investigación Sanitaria La Fe, 46026 Valencia, Spain; lou.cordon@gmail.com (L.C.); senent_leo@gva.es (L.S.); sempere_amp@gva.es (A.S.)
[4] Centro de Investigación Biomédica en Red de Cáncer (CIBERONC), Instituto Carlos III, 28029 Madrid, Spain
[5] Hematology Department, Hospital Universitario y Politécnico La Fe, 46026 Valencia, Spain
[6] Centro/Departamento de Química, Universidade do Minho, Campus de Gualtar, 4710-057 Braga, Portugal
[7] CEB—Centre of Biological Engineering, University of Minho, Campus de Gualtar, 4710-057 Braga, Portugal
[8] BCMaterials, Parque Científico y Tecnológico de Bizkaia, 48160 Derio, Spain; lanceros@fisica.uminho.pt
[9] IKERBASQUE, Basque Foundation for Science, 48013 Bilbao, Spain
[10] Biomedical Research Networking Center on Bioengineering, Biomaterials and Nanomedicine (CIBER-BBN), 46022 Valencia, Spain
* Correspondence: cribeiro@fisica.uminho.pt; Tel.: +351-253-604-061
† These authors contributed equally to this work.

Received: 29 September 2017; Accepted: 9 November 2017; Published: 11 November 2017

Abstract: The aim of this work was to determine the influence of the biomaterial environment on human mesenchymal stem cell (hMSC) fate when cultured in supports with varying topography. Poly(vinylidene fluoride) (PVDF) culture supports were prepared with structures ranging between 2D and 3D, based on PVDF films on which PVDF microspheres were deposited with varying surface density. Maintenance of multipotentiality when cultured in expansion medium was studied by flow cytometry monitoring the expression of characteristic hMSCs markers, and revealed that cells were losing their characteristic surface markers on these supports. Cell morphology was assessed by scanning electron microscopy (SEM). Alkaline phosphatase activity was also assessed after seven days of culture on expansion medium. On the other hand, osteoblastic differentiation was monitored while culturing in osteogenic medium after cells reached confluence. Osteocalcin immunocytochemistry and alizarin red assays were performed. We show that flow cytometry is a suitable technique for the study of the differentiation of hMSC seeded onto biomaterials, giving a quantitative reliable analysis of hMSC-associated markers. We also show that electrosprayed piezoelectric poly(vinylidene fluoride) is a suitable support for tissue engineering purposes, as hMSCs can proliferate, be viable and undergo osteogenic differentiation when chemically stimulated.

Keywords: tissue engineering; bone differentiation; poly(vinylidene fluoride); microspheres

1. Introduction

Tissue engineering combines specially designed biomaterials with cells to augment, replace or reconstruct damaged or diseased tissues. Human Mesenchymal Stem Cells (hMSCs) have attracted strong interest in the scientific community in the past few years due to their differentiation potential towards cells belonging to musculoskeletal lineages, such as osteoblasts, adipocytes and chondrocytes [1,2]. In order to induce hMSC differentiation to the desired lineage, extracellular stimuli are required, including chemical and physical cues. In addition, biomaterial properties have been demonstrated to modulate short-term cell functions, and in the last years they have evolved from basic supporting materials to biofunctional materials [3]. It has been extensively proven that hMSCs can be expanded in adherent culture flasks to high cell numbers while maintaining their multipotentiality, i.e., their ability to differentiate to osteoblasts, adipocytes and chondrocytes in specific differentiation media [4]. Nevertheless, the range of biomaterials used as supports in cell culture and as a vehicle for cell transplant to the organism in tissue engineering therapies are very broad, including many different chemical structures, surface topographies and 3D arrangements such as gels or scaffolds. During culture, the influence of cell–material interaction on the fate of hMSCs is not well understood. In particular, if cell culture into 3D supports compromises the cells towards one of the particular lineages [5].

Synthetic polymers have been increasingly used as biomaterials to serve as a supports for cell growth and differentiation due to their intrinsic attractive properties compared to inorganic materials; they can be manufactured into complex shapes and their physicochemical properties can easily be tailored [6]; as well as for their surface properties. It has been demonstrated that biomaterial design and surface characteristics (including roughness, surface charge or chemistry and wettability) can modulate the cell function [7,8]. In particular, biomaterial surface properties are linked to cell adhesion, morphology, proliferation and differentiation [9].

Furthermore, new structures are being produced in order to offer adequate cell supports which could mimic their natural niche in a more realistic way. In particular, polymer microspheres have been increasingly used as supports for cell expansion and differentiation, holding and populating a higher number of cells than the traditional 3D scaffolds [10,11].

Furthermore, the intrinsic properties of specific polymers (such as smart polymers) can be taken to advantage of in order to influence the behavior and differentiation of cells. In particular, it has been demonstrated that the use of piezoelectric polymers as substrates/scaffolds enhances cell response since they have the ability to vary their surface charge when a mechanical load is applied, without the need of a power source or wires [12,13]. This electrical response can also be induced in related magnetic composites through the application of a magnetic field [14], allowing the tuning of cell response. Among these polymers, poly(vinylidene fluoride) (PVDF) is the one with the largest piezo-, pyro- and ferroelectric responses [15]. Also, this biocompatible polymer shows good osteoinductive results, not only on dynamic, but also on static cultures, proving that it is suitable for stem cell culture and differentiation into the osteogenic lineage [16–20].

However, it is important to have a precise and defined method to assess hMSC differentiation when cultured on diverse biomaterials. There are various methods for the identification, verification and characterization of hMSCs, such as immunofluorescence/immunocytochemistry, Western blot, protein arrays and real-time polymerase chain reaction (RT-PCR) [21–23]. Cytometry has emerged as a simple and more precise method to identify and evaluate hMSC differentiation as they change the expression of their classical surface antigens—CD105, CD73 and CD90—after culture on the biomaterial [24,25].

In this work, PVDF substrates have been produced that are somewhere in between 2D and 3D configurations; the cells can grow on flat surface and 3D spheres. In this way, the cells do not find just a flat surface to attach to, but do not find a typical 3D scaffold with attachments points all around the cells, either. They consist of a flat PVDF film on which PVDF microspheres have been randomly distributed with varying surface density. It is important to mention that with lower quantity of microspheres deposited on the film, the structures are closer to 2D, while structures with a higher quantity of microspheres deposited on the films are closer to 3D. Furthermore, by keeping the processing parameters constant, including deposition time, structures with the same characteristics area always obtained. The objective is to show if an increasing three-dimensional interactions between cells and biomaterial surfaces influence cell fate either in expansion or in an osteoblastic culture medium.

2. Results

2.1. PVDF Samples Characterization

PVDF microspheres have been produced by electrospray from a 7% w/v solution in N,N-dimethylformamide (DMF) and obtained according to [26] (Figure S1), with diameters between 0.81 ± 0.34 and 5.55 ± 2.34 μm, and an average diameter of 3.04 ± 1.70 μm.

The β-PVDF electroactive phase has shown superior results in MSC and MC3T3-E1 preosteoblasts cell culturing [12,18]. Thus, Fourier transformed infrared (FTIR) measurements were carried out in order to confirm the presence of this phase in the microspheres. The characteristic FTIR bands of PVDF provide an accurate confirmation that the β-phase was obtained [15]. The relative fraction of the β-phase of the microspheres was determined as explained in [26], showing an electroactive phase content around 70%.

Complementary to FTIR measurements, differential scanning calorimetry (DSC) was performed in order to identify and quantify the crystalline phase of PVDF microspheres following a procedure similar to the one in [26] (data not shown). The obtained degree of crystallinity for the electrosprayed microspheres is around 52%.

2.2. Characterization of the Mesenchymal Cells

The mesenchymal origin of the hMSCs was confirmed by the adherence of the cells to tissue culture polystyrene plates (TCPS), and their surface marker expression was analyzed by multiparametric flow cytometry (MFC).

The hMSCs are characterized by the expression of the monoclonal antibodies CD90, CD105 and CD73, and the lack of expression of CD14, CD19, CD34, CD45 and HLADR as proposed by The International Society of Cellular Therapy (ISCT) [24]. We show in Figure 1 MFC analysis of the hMSCs, performed after culture in TCPS at passage 4 when merging the data from the unstained and stained cells with the negative or the positive markers employing FACSDiva version 8 software (Becton Dickinson, San Jose, CA, USA). The histograms show the percentage of each antigen fluorochrome conjugated expressed by the hMSCs. The higher the value the higher the expression of cell antigens.

The percentage of expression of cell surface proteins (TCPS, passage 4) is shown in Table 1. Unstained hMSCs were employed as negative controls, showing no expression for the indicated markers. Stained hMSCs were positive for CD90, CD105 and CD73 whereas they were negative for hematopoietic markers. Notably, CD90 expression was lower than expected.

Table 1. Percentage of expression of unstained and stained hMSCs surface proteins and number of events analyzed.

hMSCs	CD90 (%)	CD105 (%)	CD73 (%)	HLADR (%)	CD19 (%)	CD34 (%)	CD45 (%)	CD14 (%)	Events
Unstained	0.1	0.1	0.2	0.1	0.1	0.1	0.2	0.1	11,637
Stained	77.7	98.6	99.7	2.4	0.2	0.2	0.1	0.1	12,217

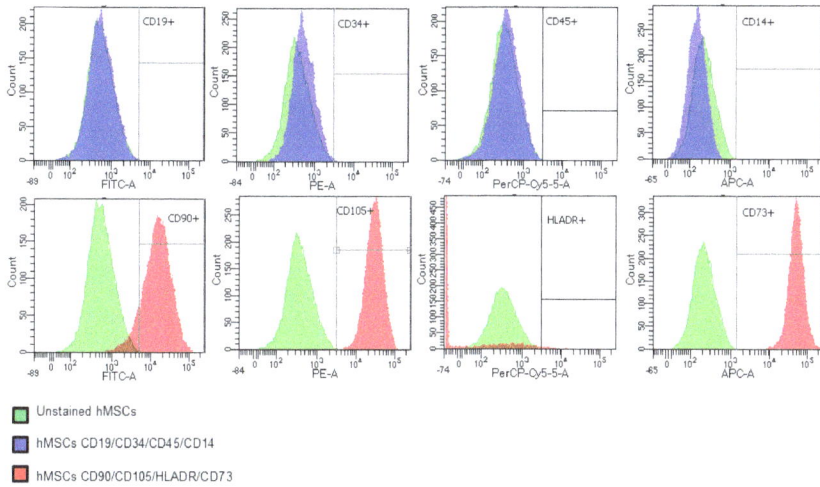

Unstained hMSCs

hMSCs CD19/CD34/CD45/CD14

hMSCs CD90/CD105/HLADR/CD73

Figure 1. Flow cytometry histograms of merged samples of the hMSCs (passage 4) at day 0 of cell culture. Unstained hMSCs (**green**); hMSCs stained with CD19/CD34/CD45/CD14 (**blue**); hMSCs stained with CD90/CD105/HLADR/CD73 (**red**). Fluorescein isothiocyanate (FITC): CD19 and CD90; phycoerythrin (PE): CD34 and CD105; peridinin chlorophyll protein-cyanine5.5 (PerCP-Cy5.5): CD45 and HLA-DR; allophycocyanin (APC): CD14 and CD73.

2.3. hMSCs Culture in Expansion Medium

2.3.1. Cell Morphology

Overall cell morphology of the hMSCs seeded on the PVDF samples was visualized after four days of cell culture by scanning electron microscopy (SEM) (Figure 2). SEM images revealed that cells cultured on the microsphere films seem to elongate their adhesion points in order to find a suitable place to hold on to. In the film with high density of microspheres (HD-M), the cells became thinner and their bodies became less flattened and more elongated, compared to the film with low density of microspheres (LD-M). Figure 2c also shows that the cells were able to attach with their elongated filopodia while showing adhesion within the cell body to the film and microspheres. As there is no visible film in Figure 2d, the film with HD-M resembles a 3D environment, where the cells can only attach to the agglomerates of microspheres, and because of that, hMSCs adopt a particular shape. Contrarily, it is possible to verify that cells cultured in flat substrates (glass and β-PVDF film) showed a more spread and flattened morphology (Figure 2a,b respectively).

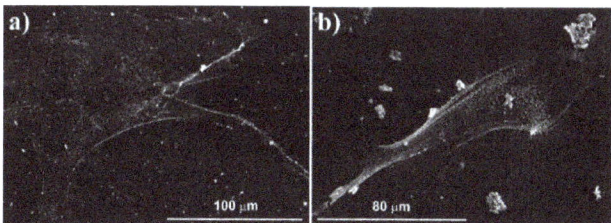

Figure 2. *Cont.*

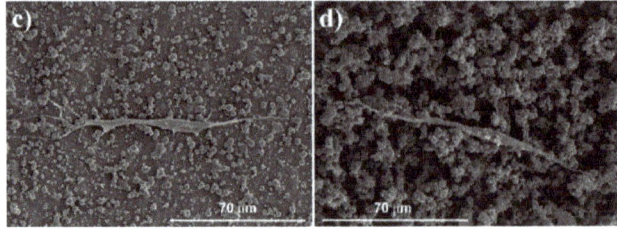

Figure 2. Overall cell morphology of hMSCs analyzed by SEM after four days of cell culture on: (**a**) glass; (**b**) β-PVDF film; (**c**) film with low density of PVDF microspheres; (**d**) film with high density of PVDF microspheres.

2.3.2. Viability

The viability of the attached cells on PVDF samples after four days of culture was also studied by 3-(4,5-dimethylthiazol-2-yl)-5-(3-carboxymethoxyphenyl)-2-(4-sulfophenyl)-2H-tetrazolium (MTS) assays (Figure 3). The results showed that PVDF is a suitable biomaterial for hMSC growth and survival, since the PVDF samples have an increase of ≅400% in the measured absorbance compared to cells seeded on glass covers. Furthermore, there are no significant statistical differences among the PVDF samples. This result corroborates other cell studies performed with flat PVDF substrates [12,26,27].

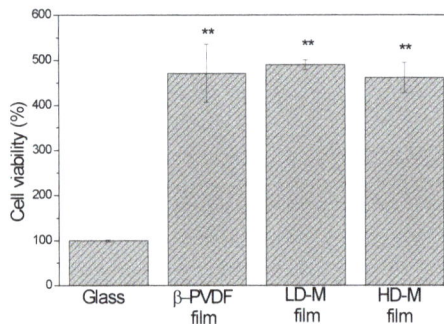

Figure 3. Cell viability for cells seeded on the PVDF samples and on glass covers (positive control) after four days of cell culture. Results are expressed as mean ± standard deviation with $n = 3$. ** $p \leq 0.01$ vs. glass.

2.3.3. MFC Analysis

In order to evaluate the loss or maintenance of the hMSC antigens after four days of cell culture on TCPS and PVDF samples, cells were submitted to MFC analysis (Figure 4).

To assess whether films with adsorbed microparticles induce changes in cell marker expression, hMSCs were cultured for four days on different biomaterials: TCPS, low density, high density and film-beta were compared to stained cell control at day 0 (Figure 4).

The percentage of expression of CD90, CD105 and CD73 antigens in hMSCs is shown in Table 2. We observed that CD90 expression decreased significantly in hMSCs cultured on all samples, being more significantly diminished in HD-M, LD-M and film-beta with respect to the control at day 0. CD105 expression was lower in HD-M samples compared to the control at day 0. In the rest of the samples the decrease of this expression was inferior. HD-M sample showed less expression of CD73 compared with the other samples, which maintained similar levels to control at day 0.

Table 2. Percentage of expression of the hMSC specific markers cultured on different biomaterials (TCPS, HD-M, LD-M and film-beta) compared with control at day 0 and number of events analyzed.

hMSCs	CD90 (%)		CD105 (%)		CD73 (%)		Events
	Sample	Control	Sample	Control	Sample	Control	
TCPS	35.4		87.3		99.4		46,873
High Density	0.8	77.7	51.2	98.6	79.0	99.7	16,893
Low Density	3.9		81.1		97.9		7,156
Film beta	2.7		82.0		99.1		27,355

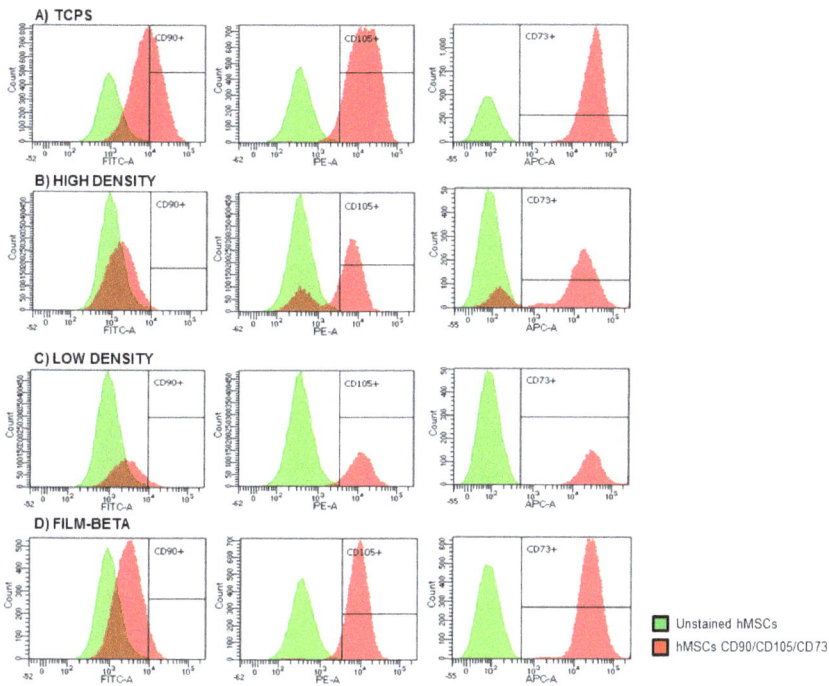

Figure 4. Flow cytometry histograms of merged samples of the unstained and stained hMSCs cultured at day 4 on (**a**) tissue culture polystyrene (TCPS); (**b**) low density of microparticles (LD-M) film; (**c**) high density of microparticles (HD-M) film; and (**d**) film-beta. Fluorescein isothiocyanate (FITC): CD90; phycoerythrin (PE): CD105; allophycocyanin (APC): CD73.

2.3.4. Alkaline Phosphatase Activity

Alkaline phosphatase (ALP) is an enzyme that increases the local concentration of inorganic phosphatase and has been shown to be important in hard tissue formation and mineralization [28]. Furthermore, ALP activity is higher in mature osteoblastic cells than in preosteoblastic and mesenchymal cells [29]. Figure 5 shows ALP activity in hMSCs. The ALP activity on β-PVDF films was significantly higher than that on HD-M film and on glass ($p < 0.05$), but there was no significant difference among the other samples.

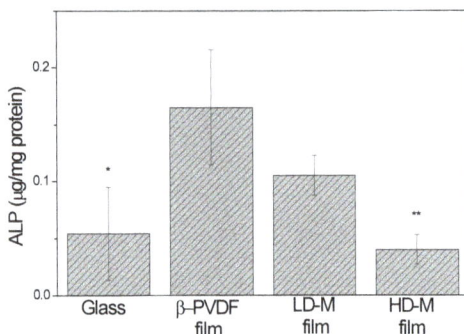

Figure 5. Alkaline phosphatase (ALP) activity in hMSCs cultured on different substrates for seven days. * The ALP activity of cells on β-PVDF film was significantly higher than that on high density (HD-M) film and glass ($p < 0.05$). Data are expressed as the mean standard deviation with $n = 3$. * $p \leq 0.05$ vs. β-PVDF film and ** $p \leq 0.01$ vs. β-PVDF film.

2.4. hMSCs Culture in Differentiation Medium

Expansion medium was changed by osteogenic medium when cells become confluent. Cell culture that continued with expansion medium for the same number of days was used as control.

2.4.1. Osteocalcin Expression

Osteocalcin (OC) is a major bone protein and has an important function in the metabolism of mineralized tissues [30]. Therefore, to corroborate the results obtained by MFC analysis, after 14 days of osteogenic medium addition, an immunocytochemistry localization of osteocalcin was performed. The staining in β-phase PVDF film or in the two PVDF microsphere samples are similar to each other (Figure 6b shows the case of the (HD-M support) which is clearer than in the glass control (Figure 6a). The elongated morphology after osteogenic induction can also be observed through actin green staining. It is noteworthy that cells seeded on glass showed a much more organized morphology, when compared to those cultured on PVDF samples, which seem to swirl. Also, in the HD-M film this effect seemed to be enhanced, probably due to the fact that the cells are forced to elongate and grow depending on the relative placement of the microspheres. Given that these films have higher amounts of microspheres, cells do not have a flat surface to hold on and spread freely.

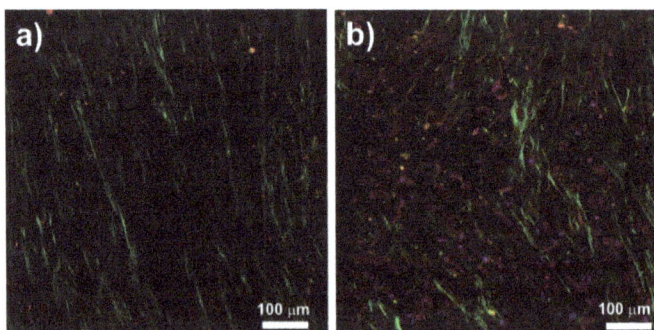

Figure 6. Confocal fluorescence microscopy images of cells after 14 days of culture (with differentiation medium) in: (**a**) glass; (**b**) high density of microparticles (HD-M) film. The picture shows bone-specific osteocalcin (red), the cells actin cytoskeleton (green) and the nucleus (blue). The scale bar (100 μm) is valid for all images.

2.4.2. Quantitative Analysis of Alizarin Red Staining

Figure 7 shows the alizarin red staining (ARS) acid extraction in hMSCs cultured in different substrates with osteogenic supplements (differentiation medium) and without them (expansion medium) after 14 days. The mineralization was detected in all materials and in both media. Comparing the PVDF samples, higher ARS concentration was found in the HD-M samples in differentiation medium.

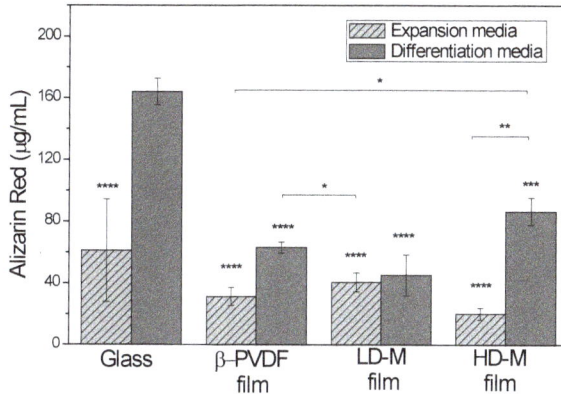

Figure 7. Alizarin red staining (ARS) acid extraction in hMSCs cultured in different substrates with osteogenic supplements (differentiation media) and without them (expansion media) after 14 days. Data are expressed as the mean ± standard deviation with $n = 3$. * $p \leq 0.05$, ** $p < 0.01$, *** $p \leq 0.001$ vs. Glass with differentiation media and **** $p \leq 0.0001$ vs. glass with differentiation media.

3. Discussion

In this work, the effect of PVDF's surface topography for tissue engineering purposes has been studied. New quasi-3D substrates of microspheres were produced and hMSC osteogenic differentiation was induced in order to observe the potential of each biomaterial.

The morphology adopted by the cells on different materials has already proven to be fundamental for cell proliferation and fate [31]. The images have shown that different PVDF topographies were successfully obtained and that cells adopt different morphologies regarding the substrate they are cultured on (Figure 2). It has been demonstrated previously that disordered structures promote hMSCs to undergo osteogenic differentiation and that mechanotransductive events between the cell and the biomaterial are a key factor influencing cell fate [32]. In this context, it has been reported that increased contractility of hMSCs leads preferentially to osteogenesis, while low contractility leads to adipogenesis [33]. Throughout the years, a series of studies have been performed demonstrating that cell culture conditions that increased cytoskeletal tension promote osteogenesis; other studies have linked cytoskeletal tension to cell spreading. For instance, previous studies related matrix elasticity to cell differentiation showing that stiffer matrices increase cytoskeleton tension, thereby promoting osteogenesis, while softer matrices lead hMSCs to differentiate towards alternative lineages [34]. Furthermore, it has been proven that changes in cell shape can alone influence hMSC commitment between osteogenic and adipogenic differentiation [35]. Although the assessment of cell morphology in this study was done with few days of cell culture, it has been shown that cells cultured on flat supports already have a more spread and flattened morphology than cells seeded on microsphere substrates. Although the disordered nature of the microspheres do not induce any preferential orientation of the cells, those that are grown on these substrates in a similar way find a place to elongate their filopodia and start to spread between the microspheres. The differences between these two substrates rely on the fact that on the LD-M film, cells still have film to adhere to and seem to be flattened when compared to the HD-M film, where cells find themselves on a niche full of microspheres. The tension provided by the LD-M and the flat film on the cells can be larger and cells are able to spread more on these substrates, making them suitable for osteogenic differentiation.

Complementary to these results, the MFC technique was applied to observe how the surface markers evolved over time and on different substrates. It was previously shown that MFC can identify cell differentiation into distinct cell lineages [25]. All cells cultured on the produced biomaterials were analyzed compared to cells of the same passage and at the same day of culture.

The percentages of the expression of CD90, CD105 and CD73 antigens in hMSCs are shown in Table 2. We observed that CD90 expression decreased significantly in hMSCs cultured on all samples, while being more significantly diminished in HD-M, LD-M and film-beta with respect to the control at day 0. The CD105 expression was lower in HD-M samples compared to the control at day 0. In the rest of samples, the decrease of this expression was inferior. HD-M samples showed less expression of CD73 compared with the other samples, which maintained similar levels to control at day 0.

The histograms of hMSCs cultured on TCPS, HD-M, LD-M and β-PVDF films (shown in Figure 4) revealed a large loss of the CD90 surface marker expression when compared to cells cultured at day 0. Furthermore, the HD-M films even presented a subpopulation of cells that lost all of the markers (CD105 and CD73) and had a larger loss of CD90 expression (Figure 4b). The loss of these three hMSC markers together was already reported to be related with their differentiation [22,25]. This leads to the conclusion that both substrate and irregular topography induce the cells to differentiate more than when cultured on flatter surfaces. In Figure 4, it can be observed that there is a larger loss of expression of CD90 and CD105 when compared to CD73 in all the samples, and these differences might be relevant. The loss of CD105 expression is related to multi-lineage differentiation of stem cells [25]. On the other hand, it has been reported that hMSCs lose CD90 expression as cells mature towards osteoblastic-like cells [36]. Low down-regulation of CD73 could be explained by its adenosine production that promotes osteoblast differentiation and by being expressed in mature osteoblasts [37,38]. Thus, its expression may vary, as seen in [25], but not as much as the other positive markers, possibly because more time is still needed for these cells to become totally differentiated, or even because cells could be differentiating into another lineage. Overall, we can conclude that microsphere films, and particularly the HD-M films represent appropriate topography to induce cell differentiation. The diverse topography of biomaterials induces different cell shapes, and these shapes have been shown to indirectly regulate differentiation into the osteoblast phenotype [35]. Therefore, these substrates can be providing the cells with a specific tension that directly stimulates their differentiation, even without addition of supplements or even without reaching confluence, only in the first four days of culture.

To investigate if the osteogenic loss observed with the MFC analysis indicates differentiation towards the osteogenic lineage, the activity of ALP enzyme was examined. This evaluation was performed at day 7, when cells were not yet supplemented with differentiation medium. Furthermore, the results obtained concerning the HD-M films suggest that the adopted cell morphology on these substrates does not favor osteogenic differentiation, and that the loss of mesenchymal stem cell markers observed on the MFC analysis should indicate cell differentiation into another lineage. Also, the LD-M films corroborate this theory, mainly because (as seen in Figure 2) cells still have some gaps of flat film where they can spread further, which is a substrate stiffer than the microspheres (on HD-M films). Stiffer substrates have already been proven to favor osteogenic differentiation [34], which is in agreement with the results for the flat β-phase films. Therefore, the results shown for ALP activity are in agreement with the morphology adopted by the cells and microspheres seem to be inductive of other differentiation pathways.

Finally, after 14 days of osteogenic induction with supplemented medium, osteocalcin localization by immunocytochemistry methods and the quantification of mineralization by alizarin red assay were performed. Regarding the osteocalcin staining (Figure 6), osteogenic differentiation can be confirmed. Also, the microsphere substrates are expressing this major bone protein even more than the glass control. This proves that the support itself has the appropriate features to induce this differentiation state. However, without chemical stimulus, these supports do not directly induce this kind of differentiation. These results are confirmed by the alizarin red results, in which we can confirm that without addition of osteogenic medium, the stiffer or flattened substrates have more calcific deposition. However, it can be seen that, when induced, all of the supports were able to demonstrate calcific deposition, characteristic of osteogenic lineage cells. These results show the capacity of β-PVDF films to enhance hMSC differentiation in early phase (Figure 5) without osteogenic medium. However, based on the alizarin red results, the HD-M film shows higher mineralization (Figure 7). In this way, it is possible to deduce that the β-PVDF films have higher effect at the beginning of differentiation whereas as differentiation progresses, cells grown on HD-M films have an enhanced higher differentiation.

In conclusion, it is possible to say that MFC is a valuable technique for the study of the differentiation of hMSCs. Also, when the topography of the substrate approaches three-dimensionality, the expression of characteristic osteogenic markers of hMSCs decrease. In this way, this study shows the viability of the use of electrosprayed piezoelectric poly(vinylidene fluoride) for tissue engineering applications, which can be used in future in other systems, such as injectable hydrogels.

4. Materials and Methods

4.1. Processing of Films Adsorbed with Microspheres

PVDF, Solef 1010, was acquired from Solvay and DMF was purchased from Merk. PVDF was dissolved in DMF with a concentration of 7% (w/v) under magnetic stirring at 60 °C, according the previous work of Correia et al. [26]. An α-phase PVDF film from Measurement Specialties was then subjected to PVDF electrospray, with the previously prepared solution with the same conditions indicated in a former study [26]. Briefly, the polymer solution was placed in a plastic syringe fitted with a steel needle with an inner diameter of 0.25 mm. A syringe pump (NE-1000, Syringepump) fed the polymer solution into the tip at a rate of 2 mL·h^{-1}. The distance between the tip of the needle and the collector was 20 cm, the needle being in horizontal position and the collector in vertical position. The experiment was conducted by applying a voltage of 20 kV with a high-voltage power supply (Glassman FC Series 120 W). Two different densities of microparticles adsorbed on film were produced: low (LD-M) and high (HD-M) density. The low density was obtained after 15 min of electrodeposition and the high density after 45 min. "Poled –" β-phase PVDF films (cells cultured on the negatively charged side of the material) from Measurement Specialties were used as control.

The films were cut in 8 mm diameter circles and placed on 48-well non-treated tissue culture polystyrene plates (TCPS, VRW).

4.2. Samples Characterization

Electrosprayed samples were coated with a gold layer using a sputter coating (EM MED020, Leica, Wetzlar, Germany) and their morphology was observed by SEM (JSM6300, JEOL, Peabody, MA, USA), with an accelerating voltage of 10 kV. Then, the average diameter of approximately 550 microspheres was measured with the ImageJ Software using the SEM images (https://imagej.nih.gov/ij/).

FTIR was performed at room temperature in a Thermo Nicolet Nexus apparatus in Attenuated Total Reflectance (ATR) mode (GMI, Ramsey, MN, USA). The spectrum was obtained from 4000 to 400 cm^{-1}, using 128 scans at a resolution of 8 cm^{-1}.

DSC measurements were performed in a PerkinElmer DSC 8000 (PerkinElmer, Villepinte, France) apparatus using a heating rate of 20 °C/min under nitrogen purge.

4.3. Materials Sterilization

For sterilization purposes, all the samples were subjected to ultra violet (UV) light overnight and then washed three times for 10 min with Dulbecco's Phosphate-Buffered Saline (DPBS) (Thermo Fisher, Waltham, MA, USA).

4.4. Fibronectin Adsorption

Fibronectin (FN) from human plasma (Sigma-Aldrich, St. Louis, MO, USA) was adsorbed onto the PVDF samples. The biomaterials were immersed in a FN solution of 20 µg/mL for 1 h under constant shaking. After protein adsorption, the samples were rinsed in saline solution to eliminate the non-adsorbed protein.

4.5. Extraction of Human Mesenchymal Stem Cells

Bone marrow (BM) from patients without hematological malignancies and normal cytomorphological study was collected at the Hematology Department of the Hospital Universitario y Politécnico La Fe of Valencia. This procedure was performed according to established protocols after informed approval of the Local Ethics Committee of the Hospital. The isolation of mononuclear cells (MNC's) of BM samples was performed by ficoll density gradient centrifugation. Briefly, the BM samples were diluted in Dulbecco's Modified Eagle's Medium (DMEM, Thermo Fisher, Waltham, MA, USA) in a proportion 1:2. After, 3 mL of Histopaque®-1077 (Sigma-Aldrich) were added to the sample and the mixture was centrifuged at $1000\times g$ for 25 min at RT. MNCs at the interphase were collected and washed twice in DMEM at $400\times g$ for 10 min. Finally, the MNCs were diluted on DMEM with 10% fetal bovine serum (FBS, Biowest, Labclinic, Nuaillé, France) and counted. Cells were then seeded on T25 cm^2 flasks (Becton Dickinson, San Jose, CA, USA) with DMEM culture medium composed with 10% FBS, 100 U/mL penicillin-streptomycin (P/S, Invitrogen) and 2.5 mg/L amphotericin B (Sigma Aldrich, St. Louis, MO, USA) at 37 °C in a 95% humidified air containing 5% CO_2.

After 48 h, the medium was changed and non-adherent cells were discarded. The isolation of hMSCs from the MNCs relies on their ability to adhere on plastic between 24–48 h [39]. The medium changes were performed every 4 days.

4.6. Primary Cell Culture and Cell Characterization by Flow Cytometry

When 90% of confluence was reached, cells were trypsinized (trypsin-EDTA 0.25%, Thermo Fisher, Waltham, MA, USA). Then it was centrifuged at $400\times g$ for 10 min. After discarding the supernatant, the cells were diluted in low-glucose DMEM supplemented with 10% of FBS. Cells were counted and cultured at a density of 100,000 cells/flask.

The mesenchymal origin of the cells was confirmed by their adherence to tissue culture polystyrene (TCPS) and by surface markers expression using MFC. Freshly obtained and expanded hMSCs (day 0) were characterized by MFC in a FACSCanto-II (Becton Dickinson, San Jose, CA, USA). For this purpose, FcR Blocking Reagent (Miltenyi Biotec, Auburn, CA, USA) was added to the cells in order to block unspecific binding and hMSCs were stained using monoclonal antibodies (MoAbs) fluorochrome conjugated (30 min, 2–8 °C in dark) (Table 1).

A minimum of 50,000 cells were acquired and data were subsequently analyzed using FACSDiva software (Becton Dickinson, San Jose, CA, USA).

4.7. Cell Culture

The isolated hMSCs were maintained and expanded in maintenance medium (DMEM containing 1 g/L glucose supplemented with 0.5% amphotericin B, 1% P/S and 10% FBS) at 37 °C in a 95% humidified air containing 5% CO_2. The medium was changed every 3 days.

The experiments were performed at passages 4–6. A density of 1×10^4 cells/cm^2 was seeded onto each one of the films ("poled –" β-PVDF films and PVDF films with high and low density of microspheres). Cells were kept under expansion medium until confluence was reached.

The cells that were not used immediately in experiments were placed in cryovials (Thermo Scientific, Waltham, MA, USA) and frozen in liquid nitrogen with FBS supplemented with 10% dimethyl sulfoxide (DMSO) after trypsinization and centrifugation at $400\times g$ for 5 min.

Additionally, a differentiation culture medium (osteogenic medium) was added after the hMSCs reached 100% confluence on the biomaterials. The osteogenic medium was composed of DMEM medium containing 1 g/L glucose supplemented with 0.5% amphotericin B, 1% P/S, 10% FBS, 8 mM of β-Glycerophosphate disodium salt hydrate (Sigma Aldrich), 10 nM of dexamethasone-water soluble (Sigma Aldrich, St. Louis, MO, USA) and 50 μg/L of L-Ascorbic acid 2-phosphate sesquimagnesium salt hydrate (Sigma Aldrich, St. Louis, MO, USA). The cell culture medium was replaced every 3 days during 14 days.

4.8. Cell Viability

For quantification of viable cells in proliferation, after 4 days of cell seeding on the supports, MTS assay was carried out. To do so, cells were incubated with a 5:1 proportion of MTS (Promega, Madison, WI, USA) to DMEM without phenol red (Thermo Fisher, Waltham, MA, USA) for 3 h at 37 °C in the dark. Then, the supernatant was used to determine the absorbance at 490 nm. For this study, the solution of MTS + DMEM without phenol red was used as reference (blank) and the supernatant of cells cultured in 12 mm glass coverslips were considered to be the positive control.

All the quantitative results will be presented as mean ± standard deviation (SD) of triplicate samples. Statistical differences were determined by ANOVA using Tukey test for the evaluation of different groups (Graphpad Prism 5.0, GraphPad Software). p values < 0.05 were considered to be statistically significant.

4.9. Cell Morphology

At the fourth day of culture, the morphology of the cells on the different produced PVDF supports was analyzed. First, the samples were fixed with formalin (Sigma Aldrich) at 4 °C for 1 h. The samples were then washed in phosphate buffer (PB) (ThermoFisher, Waltham, MA, USA) before incubation with 1% osmium tetraoxide (Aname) in PB for 45 min in the dark. Then, the biomaterials were again washed to assure total removal of osmium tetraoxide, before being dehydrated through a graded series of alcohol (50%, 60%, 70%, 80%, 96% and 100%) and submitted to critical-point drying (E3000, Polaron, Quorum Technologies, East Sussex, UK). The dried samples were coated with a gold layer using a sputter coating (EM MED020, Leica, Wetzlar, Germany) and their morphology was observed by SEM (JSM6300, JEOL) with an accelerating voltage of 10 kV.

4.10. Assessment of Osteogenic Differentiation

4.10.1. Flow Cytometry Study

To analyze the hMSC differentiation, aliquots of cells cultured onto the biomaterials were studied by MFC at day 4 (when approximately 90% of confluence was reached). At this time point, cells do not have osteogenic supplements nor mechanical stress or cell-cell interactions. Therefore, the differentiation produced on cells was due to the PVDF nature and topography.

In order to have a suitable number of cells to perform MFC analysis, biomaterials were cut to a diameter of 3.4 cm and placed in the 6-well non-treated TCPS. Cells were cultured with the same density mentioned previously and, at the day of analysis, they were treated with 1 mL of trypsin for 5 min. Each one of the biomaterial-cultured cells were divided in three aliquots and stained with the antibodies combinations according to Table 3. The following protocol was the same as described above for cells before seeding (day 0).

Table 3. MoAbs fluorochrome conjugated employed for surface staining to characterize hMSCs: fluorescein isothiocyanate (FITC); phycoerythrin (PE); peridinin chlorophyll protein-cyanine5.5 (PerCP-Cy5.5); allophycocyanin (APC).

Tube	FITC	PE	PerCP-Cy5.5	APC
1	CD90 [1]	CD105 [1]	HLA-DR [2]	CD73 [1]
2	CD19 [2]	CD34 [2]	CD45 [2]	CD14 [2]
Unstained	NA	NA	NA	NA

[1] Miltenyi Biotec; [2] Becton Dickinson; NA: no antibody.

4.10.2. Study of Alkaline Phosphatase Activity

After 7 days of cell culture (in expansion medium), the cells were lysed and collected. The level of ALP present in the cells was analyzed using a SensoLyte® pNPP Alkaline Phosphatase Assay Kit (ANASPEC) and the total protein of the lysate on each sample was determined using a Micro BCA™ Protein Assay Kit (Thermo Scientific, Waltham, MA, USA). The alkaline phosphatase was expressed as per microgram of total protein for each sample.

4.10.3. Osteocalcin Immunocytochemistry

After 14 days of culture in osteogenic differentiation medium, the content of bone-specific OC was measured by immunocytochemistry methods. First, the cells were washed in DPBS and fixed with formalin as described before. After, hMSCs were washed 3 times with DPBS++ (+calcium, +magnesium, Sigma Aldrich, St. Louis, MO, USA) and permeabilized with 0.5% Triton X-100 (Sigma Aldrich, St. Louis, MO, USA) in DPBS during 5 min at RT. After washing the samples with DBPS++, a protein solution with 5% of bovine serum albumin (BSA, Sigma Aldrich, St. Louis, MO, USA) and 0.1% Triton X-100 in DPBS was added. After 30 min at 37 °C, the solution was removed and the samples were incubated with anti-osteocalcin antibody (Abcam, Cambridge, UK) at a 1:200 dilution in a solution of 5% BSA and 0.1% Triton X-100 in DPBS for 1 h at 37 °C. Then, the primary antibody was removed and the samples were washed with 0.1% Triton X-100 in DPBS++. At this point, the samples were incubated with the secondary goat antibody anti-rabbit Alexa 488® (Invitrogen, Waltham, MA, USA), at a 1:200 dilution in the previously termed BSA solution for 1 h at 37 °C. Finally, the solution was removed and the samples were once again washed in 0.1% Triton X-100 in DPBS++ before being mounted in a microscope slide with aqueous mounting medium containing DAPI (Vector Laboratories, Peterborough, UK). For this study, cells cultured in 12 mm glass coverslips were used as a reference. Cells relative content of OC was studied using a confocal microscope (DMi8, Leica, Wetzlar, Germany) and ImageJ, Photoshop and Leica Application Suite X software were used for treatment and analysis of the obtained images.

4.10.4. Quantitative Analysis of Alizarin Red Staining

A quantitative analysis of ARS (Sigma-Aldrich, St. Louis, MO, USA) was performed after 14 days of the addition of differentiation medium (expansion, differentiation). For quantification of mineralization, the protocol described in Gregory et al. was followed [40]. This assay is based on alizarin red S staining of the mineral followed by the extraction with 10% acetic acid (Panreac). The acidified ARS is then neutralized by the addition of ammonium hydroxide (Sigma Aldrich, St. Louis, MO, USA). Also, alizarin red standards were prepared with known solution concentrations for the standard curve elaboration. Finally, 50 µL of the sample/standard was added to an opaque-walled, transparent bottom 96-well plate and the absorbance was spectrophotometrically measured at 405 nm. Quantitative analysis was calculated from the standard curve.

4.11. Statistical Analysis

All quantitative data were analyzed using GraphPad Prism (v6.00, La Jolla, CA, USA). The results were analyzed statistically using the one way ANOVA test for the cell viability and ALP assays and two-way ANOVA test for the alizarin red assays followed by Tukey's test. Differences were considered to be significant when $p < 0.05$.

5. Conclusions

MFC has proven to be a powerful and valuable technique for the study of hMSC differentiation in cells cultured on biomaterials, giving a quantitative reliable analysis of hMSC-associated markers and the loss, gain or maintenance of marker expression. It has been shown that the decrease of the expression of characteristic markers of multipotentiality in hMSCs is more apparent when the topography of the substrate approaches three-dimensionality. Since these cells lose their specific markers, we can conclude that the supports are naturally inducing cells to differentiate into other lineages. It has been shown that electrosprayed piezoelectric poly(vinylidene fluoride) are valuable for tissue engineering purposes, as hMSCs can proliferate, be viable and undergo osteogenic differentiation when chemically stimulated.

Supplementary Materials: Supplementary materials can be found at www.mdpi.com/1422-0067/18/11/2391/s1.

Acknowledgments: The authors thank the Portuguese Foundation for Science and Technology (FCT) for financial support under project PTDC/EEI-SII/5582/2014, Strategic Funding UID/FIS/04650/2013 and grants SFRH/BPD/90870/2012 (C.R.) and SFRH/BPD/121526/2016 (D.M.C). The authors acknowledge funding by the Spanish Ministry of Economy and Competitiveness (MINECO) through the project MAT2016-76039-C4-3-R (AEI/FEDER, UE) and from the Basque Government Industry Department under the ELKARTEK program. JLGR, LC, RSS and AS acknowledge funding by the Conselleria de Educación, Investigación, Cultura y Deporte of the Generalitat Valenciana through PROMETEO/2016/063 project. CIBER-BBN is an initiative funded by the VI National R&D&i Plan 2008–2011, Iniciativa Ingenio 2010, Consolider Program, CIBER Actions and financed by the Instituto de Salud Carlos III with assistance from the European Regional Development. This work was partially financed with FEDER funds (CIBERONC (CB16/12/00284)). The authors acknowledge the assistance and advice of Electron Microscopy Service of the UPV.

Author Contributions: C. Ribeiro, S. Lanceros-Méndez and J. L. Gomez Ribelles conceived and designed the experiments; R. Sobreiro Almeida, M. N. Tamaño-Machiavello, E. O. Carvalho, L. Cordón, S. Doria, L. Senent, D. M. Correia performed the experiments; R. Sobreiro Almeida, L. Cordón, R. Sabater i Serra and A. Sempere analyzed the data; All authors have read and approved the final submitted manuscript.

Conflicts of Interest: The authors declare no conflict of interest.

Abbreviations

ALP	Alkaline phosphatase
APC	Allophycocyanin
ARS	Alizarin red staining
ATR	Attenuated Total Reflectance
BM	Bone marrow
DMEM	Dulbecco's Modified Eagle's Medium
DMF	*N,N*-dimethylformamide
DPBS	Dulbecco's Phosphate-Buffered Saline
DSC	Differential scanning calorimetry
FBS	Fetal bovine serum
FITC	Fluorescein isothiocyanate
FN	Fibronectin
FTIR	Fourier transformed infrared
HD-M	High density of microparticles
HMSCs	Human mesenchymal stem cells
LD-M	Low density of microparticles
MNC's	Mononuclear cells

MTS	3-(4,5-dimethylthiazol-2-yl)-5-(3-carboxymethoxyphenyl)-2-(4-sulfophenyl)-2H-tetrazolium
P/S	Penicillin-streptomycin
PE	Phycoerythrin
PerCP-cY5.5	Peridinin chlorophyll protein-cyanine5.5
PVDF	Poly(vinylidene fluoride)
RT-PCR	Real-time polymerase chain reaction
SEM	Scanning electron microscope
TCPS	Tissue culture polystyrene plates

References

1. Godara, P.; Nordon, R.E.; McFarland, C.D. Mesenchymal stem cells in tissue engineering. *J. Chem. Technol. Biotechnol.* **2008**, *83*, 397–407. [CrossRef]
2. Nombela-Arrieta, C.; Ritz, J.; Silberstein, L.E. The elusive nature and function of mesenchymal stem cells. *Nat. Rev. Mol. Cell Biol.* **2011**, *12*, 126–131. [CrossRef] [PubMed]
3. Fu, R.H.; Wang, Y.C.; Liu, S.P.; Huang, C.M.; Kang, Y.H.; Tsai, C.H.; Shyu, W.C.; Lin, S.Z. Differentiation of stem cells: Strategies for modifying surface biomaterials. *Cell Transp.* **2011**, *20*, 37–47. [CrossRef] [PubMed]
4. Ullah, I.; Subbarao, R.B.; Rho, G.J. Human mesenchymal stem cells—Current trends and future prospective. *Biosci. Rep.* **2015**, *35*, e00191. [CrossRef] [PubMed]
5. Leferink, A.M.; Santos, D.; Karperien, M.; Truckenmüller, R.K.; Van Blitterswijk, C.A.; Moroni, L. Differentiation capacity and maintenance of differentiated phenotypes of human mesenchymal stromal cells cultured on two distinct types of 3D polymeric scaffolds. *Integr. Biol.* **2015**, *7*, 1574–1586. [CrossRef] [PubMed]
6. Duncan, R. The dawning era of polymer therapeutics. *Nat. Rev. Drug Discov.* **2003**, *2*, 347–360. [CrossRef] [PubMed]
7. Phillips, J.E.; Petrie, T.A.; Creighton, F.P.; García, A.J. Human mesenchymal stem cell differentiation on self-assembled monolayers presenting different surface chemistries. *Acta Biomater.* **2010**, *6*, 12–20. [CrossRef] [PubMed]
8. Alves, N.M.; Pashkuleva, I.; Reis, R.L.; Mano, J.F. Controlling cell behavior through the design of polymer surfaces. *Small* **2010**, *6*, 2208–2220. [CrossRef] [PubMed]
9. Chen, Y.; Cho, M.R.; Mak, A.F.T.; Li, J.S.; Wang, M.; Sun, S. Morphology and adhesion of mesenchymal stem cells on PLLA, apatite and apatite/collagen surfaces. *J. Mater. Sci. Mater. Med.* **2008**, *19*, 2563–2567. [CrossRef] [PubMed]
10. Hong, S.J.; Yu, H.S.; Kim, H.W. Preparation of porous bioactive ceramic microspheres and in vitro osteoblastic culturing for tissue engineering application. *Acta Biomater.* **2009**, *5*, 1725–1731. [CrossRef] [PubMed]
11. Wang, H.; Leeuwenburgh, S.C.G.; Li, Y.; Jansen, J.A. The use of micro-and nanospheres as functional components for bone tissue regeneration. *Tissue Eng. Part B Rev.* **2012**, *18*, 24–39. [CrossRef] [PubMed]
12. Ribeiro, C.; Moreira, S.; Correia, V.; Sencadas, V.; Rocha, J.G.; Gama, F.M.; Gómez Ribelles, J.L.; Lanceros-Méndez, S. Enhanced proliferation of pre-osteoblastic cells by dynamic piezoelectric stimulation. *RSC Adv.* **2012**, *2*, 11504–11509. [CrossRef]
13. Ribeiro, C.; Sencadas, V.; Correia, D.M.; Lanceros-Méndez, S. Piezoelectric polymers as biomaterials for tissue engineering applications. *Colloids Surf. B Biointerfaces* **2015**, *136*, 46–55. [CrossRef] [PubMed]
14. Ribeiro, C.; Correia, V.; Martins, P.; Gama, F.M.; Lanceros-Mendez, S. Proving the suitability of magnetoelectric stimuli for tissue engineering applications. *Colloids Surf. B Biointerfaces* **2016**, *140*, 430–436. [CrossRef] [PubMed]
15. Martins, P.; Lopes, A.C.; Lanceros-Mendez, S. Electroactive phases of poly(vinylidene fluoride): Determination, processing and applications. *Prog. Polym. Sci.* **2014**, *39*, 683–706. [CrossRef]
16. Ribeiro, C.; Panadero, J.A.; Sencadas, V.; Lanceros-Méndez, S.; Tamaño, M.N.; Moratal, D.; Salmerón-Sánchez, M.; Gómez Ribelles, J.L. Fibronectin adsorption and cell response on electroactive poly(vinylidene fluoride) films. *Biomed. Mater.* **2012**, *7*, 035004. [CrossRef] [PubMed]
17. Damaraju, S.M.; Wu, S.; Jaffe, M.; Arinzeh, T.L. Structural changes in PVDF fibers due to electrospinning and its effect on biological function. *Biomed. Mater.* **2013**, *8*, 045007. [CrossRef] [PubMed]

18. Ribeiro, C.; Parssinen, J.; Sencadas, V.; Correia, V.; Miettinen, S.; Hytonen, V.P.; Lanceros-Mendez, S. Dynamic piezoelectric stimulation enhances osteogenic differentiation of human adipose stem cells. *J. Biomed. Mater. Res. Part A* **2015**, *103*, 2172–2175. [CrossRef] [PubMed]

19. Rodrigues, M.T.; Gomes, M.E.; Mano, J.F.; Reis, R.L. β-PVDF membranes induce cellular proliferation and differentiation in static and dynamic conditions. *Mater. Sci. Forum* **2008**, *587–588*, 72–76. [CrossRef]

20. Pärssinen, J.; Hammarén, H.; Rahikainen, R.; Sencadas, V.; Ribeiro, C.; Vanhatupa, S.; Miettinen, S.; Lanceros-Méndez, S.; Hytönen, V.P. Enhancement of adhesion and promotion of osteogenic differentiation of human adipose stem cells by poled electroactive poly(vinylidene fluoride). *J. Biomed. Mater. Res. Part A* **2015**, *103*, 919–928. [CrossRef] [PubMed]

21. Martinez, C.; Hofmann, T.J.; Marino, R.; Dominici, M.; Horwitz, E.M. Human bone marrow mesenchymal stromal cells express the neural ganglioside GD2: A novel surface marker for the identification of MSCs. *Blood* **2007**, *109*, 4245–4248. [CrossRef] [PubMed]

22. Kern, S.; Eichler, H.; Stoeve, J.; Klüter, H.; Bieback, K. Comparative analysis of mesenchymal stem cells from bone marrow, umbilical cord blood, or adipose tissue. *Stem Cells* **2006**, *24*, 1294–1301. [CrossRef] [PubMed]

23. Delorme, B.; Ringe, J.; Gallay, N.; Vern, Y.L.; Kerboeuf, D.; Jorgensen, C.; Rosset, P.; Sensebé, L.; Layrolle, P.; Häpupl, T.; et al. Specific plasma membrane protein phenotype of culture-amplified and native human bone marrow mesenchymal stem cells. *Blood* **2008**, *111*, 2631–2635. [CrossRef] [PubMed]

24. Dominici, M.; Le Blanc, K.; Mueller, I.; Slaper-Cortenbach, I.; Marini, F.C.; Krause, D.S.; Deans, R.J.; Keating, A.; Prockop, D.J.; Horwitz, E.M. Minimal criteria for defining multipotent mesenchymal stromal cells. The International Society for Cellular Therapy position statement. *Cytotherapy* **2006**, *8*, 315–317. [CrossRef] [PubMed]

25. Jin, H.J.; Park, S.K.; Oh, W.; Yang, Y.S.; Kim, S.W.; Choi, S.J. Down-regulation of CD105 is associated with multi-lineage differentiation in human umbilical cord blood-derived mesenchymal stem cells. *Biochem. Biophys. Res. Commun.* **2009**, *381*, 676–681. [CrossRef] [PubMed]

26. Correia, D.M.; Goncalves, R.; Ribeiro, C.; Sencadas, V.; Botelho, G.; Ribelles, J.L.G.; Lanceros-Mendez, S. Electrosprayed poly(vinylidene fluoride) microparticles for tissue engineering applications. *RSC Adv.* **2014**, *4*, 33013–33021. [CrossRef]

27. Costa, R.; Ribeiro, C.; Lopes, A.C.; Martins, P.; Sencadas, V.; Soares, R.; Lanceros-Mendez, S. Osteoblast, fibroblast and in vivo biological response to poly(vinylidene fluoride) based composite materials. *J. Sci. Mater. Med.* **2013**, *24*, 395–403. [CrossRef] [PubMed]

28. Golub, E.E.; Boesze-Battaglia, K. The role of alkaline phosphatase in mineralization. *Curr. Opin. Orthop.* **2007**, *18*, 444–448. [CrossRef]

29. El-Amin, S.F.; Botchwey, E.; Tuli, R.; Kofron, M.D.; Mesfin, A.; Sethuraman, S.; Tuan, R.S.; Laurencin, C.T. Human osteoblast cells: Isolation, characterization, and growth on polymers for musculoskeletal tissue engineering. *J. Biomed. Mater. Res. Part A* **2006**, *76*, 439–449. [CrossRef]

30. Roach, H.I. Why does bone matrix contain non-collagenous proteins? The possible roles of osteocalcin, osteonectin, osteopontin and bone sialoprotein in bone mineralisation and resorption. *Cell Biol. Int.* **1994**, *18*, 617–628. [CrossRef] [PubMed]

31. Cha, K.J.; Hong, J.M.; Cho, D.W.; Kim, D.S. Enhanced osteogenic fate and function of MC3T3-E1 cells on nanoengineered polystyrene surfaces with nanopillar and nanopore arrays. *Biofabrication* **2013**, *5*, 025007. [CrossRef] [PubMed]

32. Dalby, M.J.; Gadegaard, N.; Tare, R.; Andar, A.; Riehle, M.O.; Herzyk, P.; Wilkinson, C.D.W.; Oreffo, R.O.C. The control of human mesenchymal cell differentiation using nanoscale symmetry and disorder. *Nat. Mater.* **2007**, *6*, 997–1003. [CrossRef] [PubMed]

33. Kilian, K.A.; Bugarija, B.; Lahn, B.T.; Mrksich, M. Geometric cues for directing the differentiation of mesenchymal stem cells. *Proc. Natl. Acad. Sci. USA* **2010**, *107*, 4872–4877. [CrossRef] [PubMed]

34. Engler, A.J.; Sen, S.; Sweeney, H.L.; Discher, D.E. Matrix Elasticity Directs Stem Cell Lineage Specification. *Cell* **2006**, *126*, 677–689. [CrossRef] [PubMed]

35. McBeath, R.; Pirone, D.M.; Nelson, C.M.; Bhadriraju, K.; Chen, C.S. Cell shape, cytoskeletal tension, and RhoA regulate stem cell lineage commitment. *Dev. Cell* **2004**, *6*, 483–495. [CrossRef]

36. Wiesmann, A.; Bühring, H.-J.; Mentrup, C.; Wiesmann, H.-P. Decreased CD90 expression in human mesenchymal stem cells by applying mechanical stimulation. *Head Face Med.* **2006**, *2*, 8. [CrossRef] [PubMed]

37. Takedachi, M.; Oohara, H.; Smith, B.J.; Iyama, M.; Kobashi, M.; Maeda, K.; Long, C.L.; Humphrey, M.B.; Stoecker, B.J.; Toyosawa, S.; et al. CD73-generated adenosine promotes osteoblast differentiation. *J. Cell. Physiol.* **2012**, *227*, 2622–2631. [CrossRef] [PubMed]

38. Ode, A.; Schoon, J.; Kurtz, A.; Gaetjen, M.; Ode, J.E.; Geissler, S.; Duda, G.N. CD73/5′-ecto-nucleotidase acts as a regulatory factor in osteo-/chondrogenic differentiation of mechanically stimulated mesenchymal stromal cells. *Eur. Cells Mater.* **2012**, *25*, 37–47. [CrossRef]

39. Williams, A.R.; Hare, J.M. Mesenchymal stem cells: Biology, pathophysiology, translational findings, and therapeutic implications for cardiac disease. *Circ. Res.* **2011**, *109*, 923–940. [CrossRef] [PubMed]

40. Gregory, C.A.; Grady Gunn, W.; Peister, A.; Prockop, D.J. An Alizarin red-based assay of mineralization by adherent cells in culture: Comparison with cetylpyridinium chloride extraction. *Anal. Biochem.* **2004**, *329*, 77–84. [CrossRef] [PubMed]

International Journal of
Molecular Sciences

MDPI

Article

3D Biomimetic Magnetic Structures for Static Magnetic Field Stimulation of Osteogenesis

Irina Alexandra Paun [1,2], **Roxana Cristina Popescu** [3,4], **Bogdan Stefanita Calin** [1,2],
Cosmin Catalin Mustaciosu [3,4], **Maria Dinescu** [5] and **Catalin Romeo Luculescu** [1,*]

[1] Center for Advanced Laser Technologies (CETAL), National Institute for Laser,
    Plasma and Radiation Physics, Magurele, RO-077125 Bucharest, Romania; irina.paun@inflpr.ro (I.A.P.);
    bogdan.calin@inflpr.ro (B.S.C.)
[2] Faculty of Applied Sciences, University Politehnica of Bucharest, RO-060042 Bucharest, Romania
[3] Horia Hulubei National Institute for Physics and Nuclear Engineering IFIN-HH, Magurele,
    RO-077125 Bucharest, Romania; roxpopescu@yahoo.co.uk (R.C.P.); cosmin@nipne.ro (C.C.M.)
[4] Faculty of Applied Chemistry and Materials Science, University Politehnica of Bucharest,
    RO-011061 Bucharest, Romania
[5] National Institute for Laser, Plasma and Radiation Physics, Magurele, RO-077125 Bucharest, Romania;
    dinescum@nipne.ro
[*] Correspondence author: catalin.luculescu@inflpr.ro; Tel.: +40-21-457-4550

Received: 27 December 2017; Accepted: 29 January 2018; Published: 7 February 2018

Abstract: We designed, fabricated and optimized 3D biomimetic magnetic structures that stimulate the osteogenesis in static magnetic fields. The structures were fabricated by direct laser writing via two-photon polymerization of IP-L780 photopolymer and were based on ellipsoidal, hexagonal units organized in a multilayered architecture. The magnetic activity of the structures was assured by coating with a thin layer of collagen-chitosan-hydroxyapatite-magnetic nanoparticles composite. In vitro experiments using MG-63 osteoblast-like cells for 3D structures with gradients of pore size helped us to find an optimum pore size between 20–40 μm. Starting from optimized 3D structures, we evaluated both qualitatively and quantitatively the effects of static magnetic fields of up to 250 mT on cell proliferation and differentiation, by ALP (alkaline phosphatase) production, Alizarin Red and osteocalcin secretion measurements. We demonstrated that the synergic effect of 3D structure optimization and static magnetic stimulation enhances the bone regeneration by a factor greater than 2 as compared with the same structure in the absence of a magnetic field.

Keywords: static magnetic field stimulation; 3D biomimetic structures; bone cell growth and differentiation

---

1. Introduction

Nowadays, the term tissue engineering is frequently used in order to address the solutions for tissue replacement following an accident, surgical excision or organ loss of function. The concept refers to the "development of biological substitutes that restore, maintain or improve the tissue function" [1]. Starting from this concept, the strategy of developing 3D biomimetic structures that simulate the architecture of the natural tissue has evolved into implementing specific properties to the structures for stimulating the cells growth, development and differentiation into functional tissue.

Bone tissue engineering has been one of the hot topics in biomaterials science, due to the increased need of tissue replacement in traumas, tumor excision, skeletal abnormalities or resection. A successful regeneration of this tissue depends on the interface interactions that take place between the osteoblasts and the 3D structure, on the ability of cells penetration, growth and development inside the construct, as well as on their access to growth factors and nutrients.

Engineered materials for bone implants make use of different physical external stimuli, such as magnetic, electric or mechanic, in order to accelerate the repair and regeneration in the affected tissue [2]. In particular, magnetic field stimulation has been proved to promote the integration of the implant, to determine an increased bone density of the newly developed tissue, by increasing the calcium content, thus promoting a more rapid and better healing of the affected bone [2,3]. Static magnetic fields were found to accelerate the proliferation, migration, orientation or differentiation of osteoblast-like cells [4–9], and to induce the osteogenic differentiation in bone marrow-derived mesenchymal stem cells [10–12]. These effects can be correlated to the fact that the cell membrane has diamagnetic properties and the membrane flux can be modified by exposure to static magnetic field [13–15]. In addition, the extracellular matrix proteins have diamagnetic properties, their structure and orientation being also affected by static magnetic fields [16]. Weak static magnetic or pulsed electromagnetic fields are also effective stimuli for bone fracture healing, spinal fusion, bone ingrowths into ceramics in animal models. Strong static magnetic fields of 5–10 T were also found to regulate the orientation of matrix proteins and cells in vitro and in vivo [12,13,16].

Researchers have previously studied the effects of static magnetic fields on cells cultured on different non-magnetic substrates [5], but implantable structures incorporating superparamagnetic nanoparticles have gained significantly more interest [2,17–19]. Owing to their intrinsic magnetic properties, they have the ability to improve the adhesion and growth of cells, even in the absence of an external magnetic field [20–24]. In particular, composites containing magnetic nanoparticles (MNPs) integrated in various matrices showed significant potentials as bone substitutes. Ceramic composites containing super-paramagnetic nanoparticles, hydroxyapatite and tricalcium phosphate had good biocompatibility with the bone cells, and the presence of the nanoparticles did not affect the function of the bone morphological protein binding to the composites [25]. Magnetic, biodegradable Fe_3O_4/chitosan/poly (vinyl alcohol) nanofibrous films fabricated by electrospinning, with average fiber diameters ranging from 230 to 380 nm and porosity of 83.9–85.1%, facilitated the osteogenesis in MG-63 human osteoblast-like cells [26]. Magnetic hydroxyapatite coatings with oriented nanorod arrays using magnetic bioglass coatings as sacrificial templates were also fabricated and used as substrates for bone growth. To date, the magnetic implantable structures have been fabricated either by dip-coating conventional structures in aqueous ferrofluids containing iron oxide nanoparticles coated with various biopolymers or by direct nucleation of biomimetic phase and super-paramagnetic nanoparticles on self-assembling collagen fibers [27].

The major goal of this work is to accelerate the osteogenesis via the synergic effect of magnetic 3D structures in response to weak static magnetic fields. For this, we designed and fabricated novel complex 3D structures with unitary elements that mimic the shape of native osteoblast-like cells, using laser direct writing via two photons polymerization (LDW via TPP) method. For large-scale use of these structures, as, for example, future in vivo studies with envisaged outcome to be translated into clinics, an important point is that the material designs need to be reproduced with exactly the same dimensions for every application. LDW via TPP is a method with great advantages in manufacturing arbitrary three dimensional (3D) micro/nanostructures of polymers, hybrid materials, organically modified ceramics (Ormocer), and metals with high reproducibility and sub diffraction-limit resolution down to 100 nm [28]. The high reproducibility of the 3D structures is mostly provided by the fact that LDW via TPP uses three-dimensional computer aided design. LDW via TPP is a layer-by-layer method where every layer is stacked up by voxels. The two-photon solidified small volume elements called voxels are quite reproducible, their size fluctuating within less than 8% (from the actual sizes down to 1 μm depending on the multicomponent optical objective used for laser focusing). The micro/nanostructures are formed by the stack of the voxels, so the resolution and spatial arrangement of the voxels play an important role in fabricating high precise structures. The high accuracy and reproducibility of the 3D structures implies the use of small voxels and tight arrangement. Owing to these characteristics, LDW via TPP is generally recognized as having high reproducibility and fidelity in obtaining 3D structures with complex architectures [28]. After fabrication, we identified the

3D structures that provided the best micro-environmental conditions for cell attachment and growth and favored the cells interconnections in complex 3D architectures, similar with those encountered in vivo. Then, we provided the structures with the function of responding to applied static magnetic fields, by coating them with collagen (Col)-chitosan (Chi)-hydroxyapatite (HA)-magnetic nanoparticles (MNPs) composite. Thus, we obtained new biomimetic magnetically responsive structures aiming to accelerate the osteogenic response in vitro in osteoblast-like cells, when exposed to static magnetic fields. We stimulated the cell-seeded structures using a range of weak magnetic fields intensities (between 100–250 mT) that was not investigated in previous studies and we provide evidence of their osteogenic effect.

2. Results and Discussion

2.1. Structures Optimization

The fabricated structures closely followed the design, yet not perfectly, as there are some geometry variations determined by intrinsic material properties and development methodology (Figure 1). Voxel height accounts for stronger overlap on the Z axis. This results in better structural integrity, albeit along with lowering porosity and potentially hindering cell migration due to smaller transfer windows throughout the structure.

Variations at the edge of the structure were determined by both material properties and development methodology. During irradiation, a series of chemical reactions result in the formation of polymeric chains. The density of the resulting polymer is slightly higher compared to non-irradiated material. As such, there is mechanical tension of various strengths throughout the irradiated volume. Moreover, until the sample is developed and dried, the polymer possesses higher flexibility, adherence and surface charges. This results in the welding of neighboring structures which, in combination with other effects of the irradiation (mechanical tension and surface charges), induces small variations of geometry at every contact point. After development, during the drying phase of the sample, surface tension of the evaporating developer can also induce deformation of the still-flexible polymer. This can be observed in Figure 1a. Apart from edge effects, the structure presents high stability and integrity due to the high number of contact points. Negligible differences from the design can still be observed at contact points, yet these are not considered variations as they are well reproduced throughout the whole structure.

The exponential overlap is designed for the Y-axis. The structure is designed to have 4 rows with respect to this axis. As such, on the Y-axis, there are 3 contact points with different overlaps. This results in rows with different spatial parameters (such as porosity) in the same structure. Its purpose is to determine the optimal overlap for cell growth throughout the entire volume of the structure.

Figure 2 shows the morphology of the cells during the first days of culture on the ellipsoidal and hexagonal multilayered 3D structures. The aspect is heterogeneous, depending on the movement of the cells and their affinity for the area in which they have settled (morphology is given by the number and position of the attachment points on the structure). In the detailed pictures (Figure 2b,c) it is evidenced the tendency of the cells to migrate into the interior of the structure, to climb onto the lateral walls and to travel through the inner part of the structure, where the specific surface is higher and thus the higher the number of attachment points. We can also observe a tendency of the osteoblast-like cells surface density to increase with higher overlap. For medium and low overlaps of the ellipsoidal and hexagonal elements, the cells seem to have better migrated through whole volume of the scaffold. While this distribution is found for both the elliptic and the hexagonal structures, we observe a higher density for the elliptic version. This is determined by the shape of the edges, as the cells seem to attach better on rounded edges rather than straight ones.

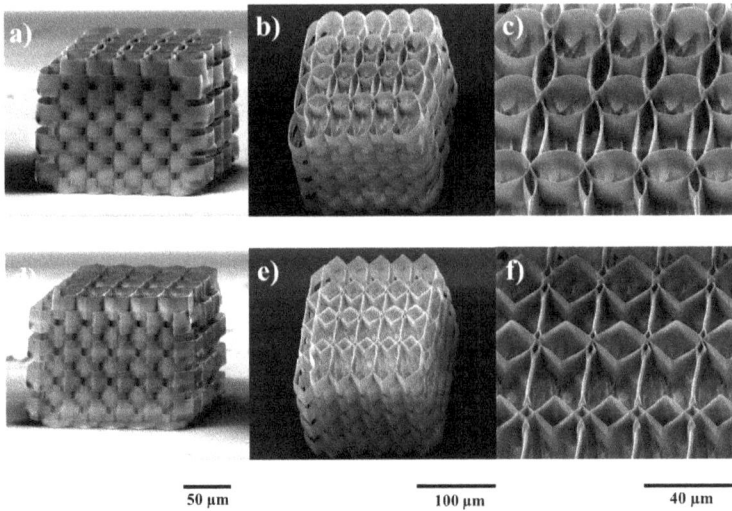

Figure 1. SEM micrographs of ellipsoidal (upper panel) and hexagonal (lower panel) multilayered 3D structures produced by LDW (laser direct writing) via TPP (two photon polymerization) of IP-L780 photopolymer. (**a,d**) Side overviews; (**b,e**) Tiled overviews; (**c,f**) Closer, tilted views of the structures.

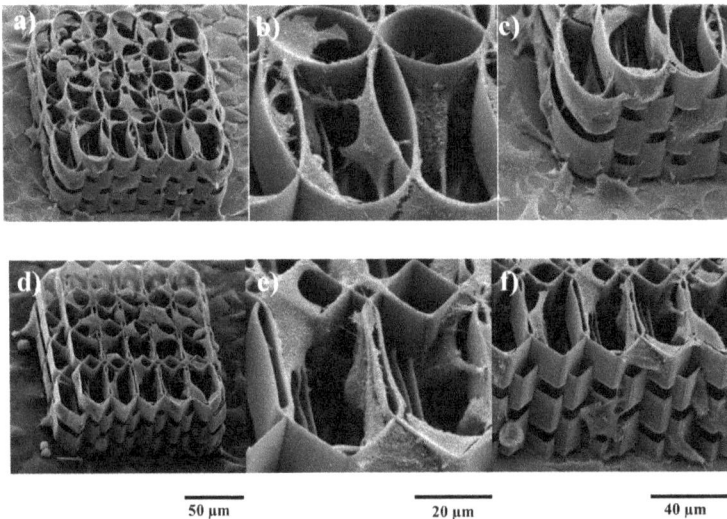

Figure 2. SEM (scanning electron microscopy) micrographs of MG-63 osteoblast like cells seeded for two days on ellipsoidal (upper panel) and hexagonal (lower panel) multilayered 3D structures. (**a,d**) Overviews of cells attachment on the structures; (**b,e**) Cells penetrating inside the structures; (**c,f**) Cells growing on the lateral walls of the structures.

SEM (scanning electron microscopy) images for MG-63 cells cultured during 7 days showed the high potential of the structures to support the cells growth. It can be clearly seen that the cells invaded both the ellipsoidal and hexagonal structures, on the outside as well as on the lateral walls (Figure 3b). In case of the ellipsoidal structure, the cells have a circular shape, given by the growth support, forming a continuous layer on the top surface of the scaffold (Figure 3a). The lateral walls show few cells attached to the exterior of the ellipsoidal elements, the attachment points for the osteoblast-like cells being mostly at their intersection. It seems as though the top cell layer continues until the bottom of the glass support, covering the 3D structure. In case of the hexagonal multilayered scaffold, the cells have a more fragmented morphology, star-shaped, guided by the morphology of the unit structures. Here, the lateral walls are fully invaded (Figure 3b), the cells display a 3D arrangement, but not necessary a methodical one. The fracture into the thick layer of cells covering the scaffold shows a 3D biomimetic displacement of the cells.

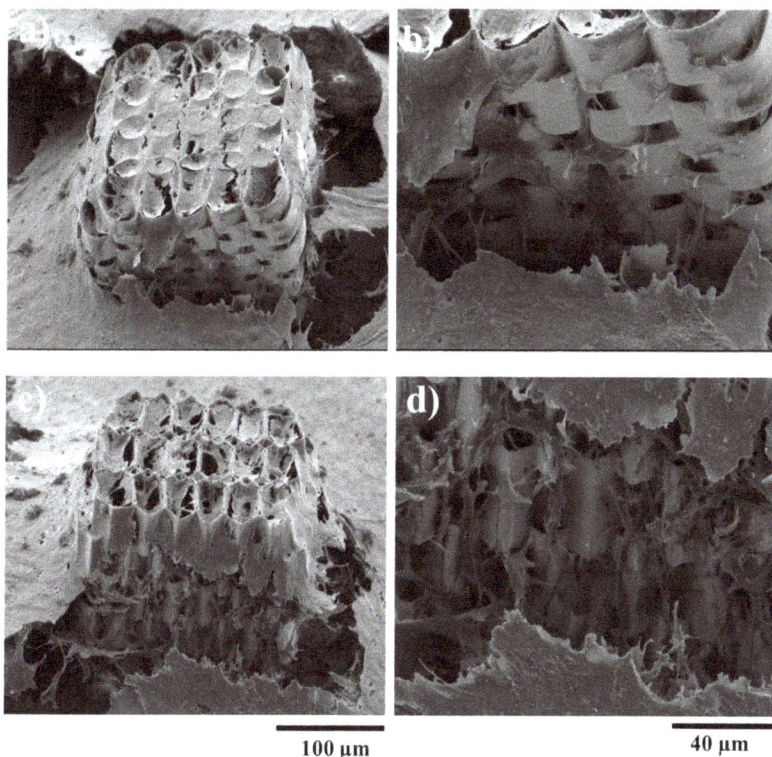

100 µm   40 µm

Figure 3. SEM (scanning electron microscopy) micrographs of MG-63 osteoblast like cells growing on ellipsoidal (upper panel) and hexagonal (lower panel) multilayered 3D structures with optimum horizontal arrangement, after 7 days in culture. (**a,c**) Overviews; (**b,d**) Cells growing on the lateral walls of the structures.

The above experimental studies were used to determine the optimal geometry in the XY plane for the cell growth. It is clear that the spacing between neighboring layers should be increased with respect to Z-axis, in order to enhance the cell migration throughout the volume of the structure. This was achieved by separating consecutive layers using appropriately placed cylindrical pillars. Each layer was designed to be 10 µm tall (~14 µm when accounting for voxel height), while pillars were designed to be 20 µm tall. The optimum overlap of consecutive rows in the XY plane was determined from previous parametrization. Pillars were placed in overlap regions for enhanced stability. Resulting structures are presented in Figure 4, for both ellipsoidal and hexagonal variants.

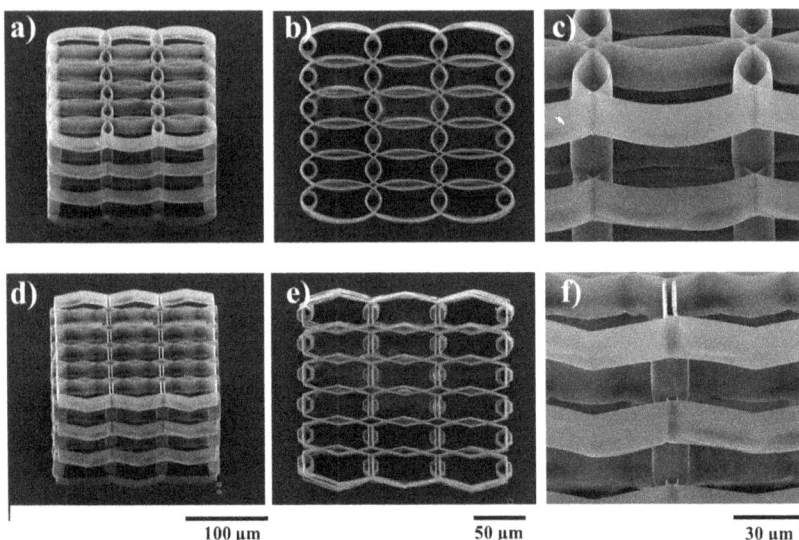

Figure 4. SEM (scanning electron microscopy) micrographs of ellipsoidal (upper panel) and hexagonal (lower panel) multilayered 3D structures having the layers spatially separated by cylindrical pillars. (**a,d**) Tilted overviews; (**b,e**) Top views; (**c,f**) Closer, tilted side views.

To test the efficiency of this design, the structures were seeded with osteoblast-like cells and observed by SEM. It can be seen that the larger space created by the pillars between the stories of the structures allowed the cells to penetrate the interior of it. Figure 5b,d shows the 3D displacement of the cells, inside both ellipsoidal and hexagonal structures, forming a tissue-like morphology. However, the density of cells inside the structure was higher in case of the hexagonal one, as it provides more attachment points for the cells. In case of the ellipsoidal structures, the cells have a fragmented appearance, with a star-shaped morphology, for both the inside and outside of the structure. On the top wall of the structure, the cells form a fragmented layer, because the unit cylinders have a larger inner diameter, thus the cells need to stretch to have enough attachment points. In the case of the hexagonal structure, the cells on the lateral walls and inside the structure have a star-shaped morphology, but the ones on the top layer are hexagonally shaped, forming a more compact layer covering the structure. These findings are rather intriguing, considering the previous statements related to better adhesion to the elliptical shaped walls. It is highly likely that a larger top surface area in the case of elliptical shaped structure induced the collapse of the top cellular layer.

In order to find a suitable architecture for bone tissue engineering, one must take into consideration a series of factors, starting from the biocompatibility of the materials, porosity, mechanical properties and osteointegration. Considering the porosity of the structure, this not only refers to the density of pores and their dimension, which are important to allow the penetration of cells into the scaffold, as well as nutrient perfusion inside, but it can also refer to the general displacement of these pores and their ability to guide the attachment and growth of the bone cells, so that the resulting new tissue can mimic the architecture of the natural bone.

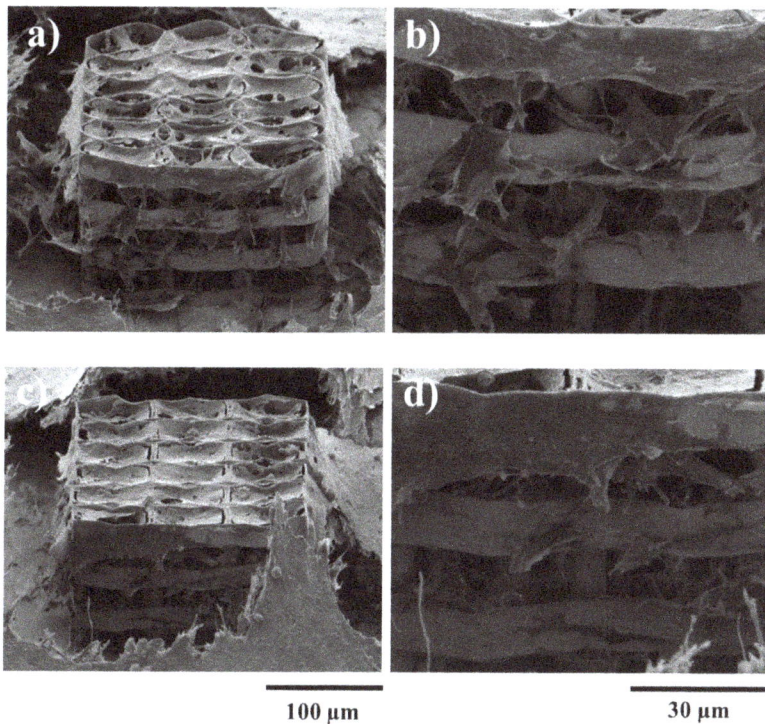

Figure 5. SEM (scanning electron microscopy) micrographs of MG-63 osteoblast-like cells growing on ellipsoidal (upper panel) and hexagonal (lower panel) multilayered 3D structures having the layers spatially separated by cylindrical pillars, after 7 days in cell culture. (**a**,**c**) Overviews; (**b**,**d**) Closer, tilted side view, showing cells penetration inside the structure.

Loh et al. [29] discussed how the displacement and morphology of the pores can affect the properties and architecture of the extracellular matrix in the resulting tissue. Thus, low porosity materials initially can exhibit high proliferation rates, but compared to high porosity structures they do not allow cell differentiation [30]. However, Mandal et al. [31] proved that the proliferation of fibroblasts on porous scaffolds is facilitated not only when the dimension of the pores is higher (around 200–250 µm), but also when the dimension of pores is lower (100–150 µm) accompanied by a higher pore density. Loh et al. [29] reported data on scaffolds with various pore sizes, while the optimal dimension of pore size and density is far from being established. 3D columnar layered structures were obtained by Mata et al. [32] using microfabrication and soft lithography approaches. The structures were seeded with adult human stem cells and connective tissue progenitor cells, which showed an osteoblastic phenotype at 9 days of culturing. The cells were able to invade the interior of the structure and form colonies. Mohanti et al. [33] used a 3D printing technique to obtain layered woodpile-like structures, made of spaced polymer filaments. The porosity was varied from 20–80% and channel distances from 78–1482 µm. Also, they designed the scaffolds to exhibit both elliptical and hexagonal architecture of the pore structures. High porosity enabled a high specific surface area and thus improved the ligand density for cell attachment and spreading, as showed by the group. In our case, both elliptical and hexagonal 3D structures with optimum pore size allowed good cell attachment and proliferation with a small difference in cell density related to the top layer. Mohanti et al. [33] stated that the improved cell viability and proliferation are linked to the layered architecture of the scaffold, i.e., the network of periodic channels allowing the perfusion and mass transport inside the scaffold.

2.2. Structure Functionalization

The structures with optimized architectures were coated with Col-HA-MNPs:Chit-HA-MNPs (Collagen-Hydroxyapatite-Magnetic nanoparticles: Chitosan-Hydroxyapatite-Magnetic nanoparticles) (Figure 6). A conformal coating was relatively well achieved. Morphological investigations reveal that the coated/functionalized structure allowed the cells to attach to the inner and outer parts of the structures, in a similar fashion as for the non-coated samples.

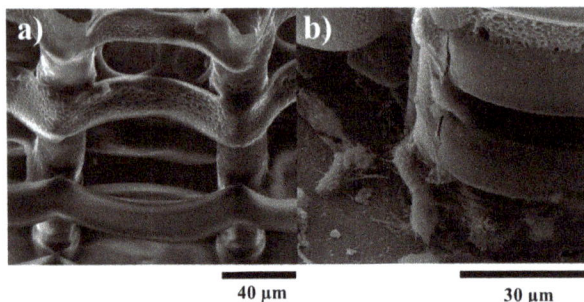

Figure 6. SEM (scanning electron microscopy) micrographs of: (**a**) 3D structures coated with Col-Chit-HA-MNPs (Collagen-Chitosan-Hydroxyapatite-Magnetic nanoparticles); (**b**) MG-63 cells growing on the magnetic structures.

2.3. Static Magnetic Field Stimulation

Previous studies on SMF simulation of the osteogenesis have mainly focused on magnetic fields above 250 M or below 50 mT. For example, Yamamoto et al. [4] reported that weak SMF between 280 and 340 mT applied to osteoblast cultures stimulated bone formation by promoting osteoblastic differentiation and/or activation. On the other hand, Cunha et al. [8] showed that increasing the magnetic field intensity up to 320 mT resulted in detrimental effects on cell proliferation and osteocalcin secretion. On the opposite, Feng et al. [6] reported that MG63 cells seeded on a PLLA discs and exposed to SMF of 400 mT showed a more differentiated phenotype. Much lower SMFs, from 50 mT down to even 3 mT, were used as biophysical stimulators of proliferation and osteoblastic differentiation of human bone marrow-derived mesenchymal stem cells [10]. Within this framework, in the present study, we cover a range of magnetic fields (100–250 mT) in between of those previously reported. In this way, we aim to explore new possibilities to optimize the cell osteogenic differentiation and to gain more insight into the roles of SMF for the stimulation of the osteogenesis.

2.4. Biological Assessments

Based on the above findings, the optimum design for the 3D structures was the one with ellipsoidal elements and having the layers spatially separated by cylindrical pillars (Figure 4a–c, upper panel). These structures were seeded with MG-63 osteoblast-like cells. External static magnetic fields were applied by positioning permanent magnets in the vicinity of the samples. The osteogenic effect of the magnetic structures synergizing with the static magnetic field was investigated within 30 days, by morphological evaluation, viability tests, differentiation and mineralization assays. We found evidence that the super-paramagnetic structures accelerate bone tissue formation under the external magnetic field in reference to the ones without external magnetic field.

Viability assay showed a reduced proliferation rate for the stimulated samples compared to non-stimulated ones (Figure 7). The proliferation rate was reduced with increasing magnetic field, in relation to the control i.e., unstimulated samples.

A question to be raised is why some previous studies showed stimulation of proliferation yet ours did not. Cooper [34] stated that there are three types of differentiated cells: the terminally differentiated cells that do not have any precursor left (e.g., heart cells), the cells arrested in G0, that replace death cells when needed (e.g., skin fibroblasts, smooth muscle cells, endothelial cells in blood vessels, epithelial cells in organs) and the rest of differentiated cells in organs that exhibit their function, which are not differentiating, but are replaced by stem cells undergoing differentiation (if needed). Noda [35] stated that, during the first steps of bone cell differentiation, the proliferation gene expression is supported, then the down-regulation of proliferation happens. Zhang et al. [36] used hyperoside, a flavonoid compound to study its effects on U2OS and MG63 cell lines. The group proved that the compound induces differentiation of the cells which is accompanied by cell cycle arrest in G_0/G_1. Whang et al. [37] showed similar results for cinnamic acid, after 7 days of culture.

In our experiments, we evaluated the proliferative activity of the MG63 cells at 4 weeks of culture, the inhibition of proliferation being associated with an advanced stage of cell differentiation. Considering the papers that we have cited, Panseri et al. [38] has evaluated the proliferation and differentiation of human osteoblast-like cells on magnetic hydroxyapatite-based scaffolds at 7, 14, and 21 days of culturing and magnetic stimulation. However, by comparing the graphs for cell proliferation measurements and ALP (Alkaline Phosphatase) measurements (differentiation), we can see that cells exhibiting higher ALP content were not undergoing proliferation anymore (this can be especially observed at day 10 and day 20). Li et al. [39] evaluated the proliferation of the cells in magnetic scaffolds just until 7 days of culturing, so these are quite early time points associated with the first steps in the differentiation process. Similar results were reported by Zheng et al. [40].

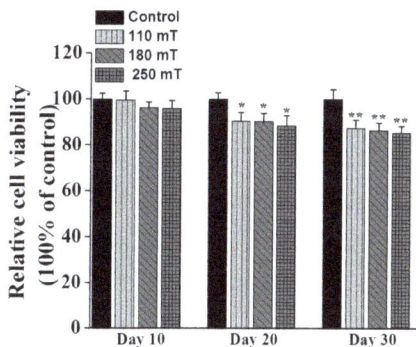

Figure 7. MTS viability of MG-63 osteoblast-like cells growing on ellipsoidal multilayered 3D structures having the layers spatially separated by cylindrical pillars, as a function of the applied static magnetic fields. Results for unstimulated samples (controls) are shown for comparison. Each bar represents the mean ± STD. Statistical significance was determined by Student's *t*-test (* $p \leq 0.05$, ** $p \leq 0.001$).

ALP (Alkaline Phosphatase) is one of the substances in the ECM (extracellular matrix) that indicates if the osteoblast cells have entered the period of ECM development and maturation. Over the whole investigation period, the cells growing on magnetically stimulated structures produced significantly more ALP than those growing on the non-stimulated samples (Figure 8). Moreover, the ALP activity increased with increasing strength of the applied magnetic field. At 10 days of culture, the difference in ALP production for the different groups of samples did not exceed 0.3 fold, regardless the intensity of the applied static magnetic field. The difference between each group of samples became significant starting from 20 days of cells stimulation. The stimulated cells showed more than a twofold increase of ALP production compared to the control (unstimulated samples). Moreover, the ALP production increased with the intensity of the static magnetic field i.e., up to almost 3 fold for the group of samples stimulated at 250 mT. The proportion between the ALP productions for each group of samples was maintained after 30 days of magnetic stimulation, the values increasing as compared to 20 days.

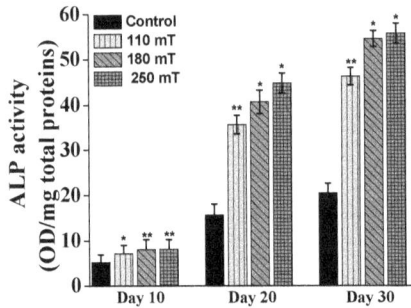

Figure 8. ALP (Alkaline Phosphatase) activity normalized to protein content for MG-63 osteoblast-like cells growing on ellipsoidal multilayered 3D structures having the layers spatially separated by cylindrical pillars, as a function of the applied static magnetic fields. Results for unstimulated samples (controls) are shown for comparison. Each bar represents the mean ± STD. Statistical significance was determined by Student's *t*-test (* $p \leq 0.05$, ** $p \leq 0.001$).

Alizarin Red staining was used to examine mineral deposition in the newly developed extracellular matrix. The samples exposed to magnetic fields exhibited more mineral content that the unstimulated ones (Figure 9). After 10 days of static magnetic stimulation, for the highest intensity of the magnetic field the cells exhibited an increased Alizarin Red coloring up to 0.7 fold. Similar to ALP activity measurements, the Alizarin Red dye attached more to the stimulated samples beginning with 20 days groups (almost a 2.7-fold increase in the case of the 250 mT group). The increase of Alizarin Red absorbance with stimulation time and intensity of magnetic field was more evident for the 30-day groups. These results indicate a higher number of osteoblast cells differentiating when exposed to static magnetic fields, leading to more new bone tissue formation on day 30.

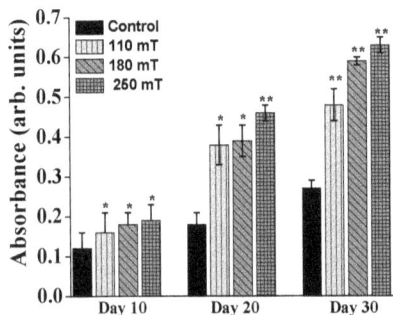

Figure 9. Absorbance measurements for Alizarin Red marking of the mineral deposits in MG-63 osteoblast-like cells growing on ellipsoidal multilayered 3D structures having the layers spatially separated by cylindrical pillars, as a function of the applied static magnetic fields. Results for unstimulated samples (controls) are shown for comparison. Each bar represents the mean ± STD. Statistical significance was determined by Student's *t*-test (* $p \leq 0.05$, ** $p \leq 0.001$).

For further confirmation, the level of osteocalcin was measured using immunohistochemically staining. Osteocalcin is a bone-specific extracellular matrix protein produced by the osteoblast cells during the process of the new bone formation. At each testing time point, the samples exposed to magnetic fields exhibited higher osteocalcin formation than the unstimulated samples (Figure 10). The samples showed the highest level of osteocalcin production on the day 30 (almost a 0.7-fold increase compared to controls). The level of osteocalcin secretion increased with increasing intensity of the applied magnetic field.

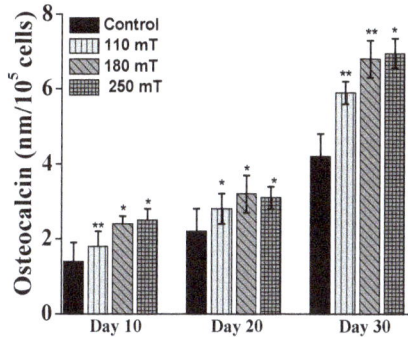

Figure 10. Absorbance measurements for the Osteocalcin secretion from MG-63 osteoblast-like cells growing on ellipsoidal multilayered 3D structures having the layers spatially separated by cylindrical pillars, as a function of the applied static magnetic fields. Results for unstimulated samples (controls) are shown for comparison. Each bar represents the mean ± STD. Statistical significance was determined by Student's *t*-test (* $p \leq 0.05$, ** $p \leq 0.001$).

All evaluated responses (cell viability, ALP, Alizarin Red staining and Osteocalcin secretion) with respect to magnetic field stimulation were statistically significant as compared with the controls (unstimulated) samples.

Our study followed the effects of static magnetic stimulation of osteoblast-like cells cultured on 3D biomimetic structures during 30 days of stimulation and incubation under standard conditions of humidity and temperature. The results showed a decreased cell proliferation in stimulated samples compared to non-stimulated samples, as measured by MTT tetrazolium salt viability assay at 30 days. This can be explained by the fact that the cells undergoing differentiation do not proliferate anymore and thus have a reduced metabolic activity [41,42]. The other results showed a more than 2-fold increase of ALP production and Alizarin Red coloring of the mineral depositions in the 250 mT sample group. These results were supported by the Osteocalcin level measurements, suggesting that the cells underwent differentiation, the degree and advancement being directly influenced by the intensity of the static magnetic field. All these results indicate that the innovative magnetic structure accelerated new bone tissue formation via the synergic action of the magnetic 3D biomimetic architecture and the applied static magnetic field, emerging as a promising approach for guiding and enhancing the process of bone growth and regeneration.

Our results are consistent with previously reported studies on static magnetic field stimulated bone tissue regeneration. Previous in vitro studies on magnetic responsive scaffolds showed the stimulating effect of the static magnetic field to the proliferation and differentiation of cells. Tampieri et al. reported on porous ceramic composite made of HA and magnetite, with enhanced in vitro cell proliferation at early stage under the external magnetic field [38]. Zhou's group fabricated a nanofibrous scaffold composed of PLA and iron oxide nanoparticles, with good biocompatibility and guided cells orientation along the fibers under the external magnetic field [39]. A weak magnetic force with intensity of 10–50 mT was reported to accelerate osteoblast differentiation, the effect being assigned to the increased phosphorylation. Porous hydroxyapatite scaffolds containing magnetic nanoparticles enhanced the in vitro osteoblast cells growth when a magnetic field was applied [39,40]. A composite of a polyester matrix magnetically functionalized with iron oxide nanoparticles showed good ability to support and enhance the osteogenic differentiation of mesenchymal stem cells [23].

Despite these positive results, the mechanism of bone cell stimulation in magnetically responsive structures is not yet understood. It was hypothesized that the MNPs generate the microdeformation of the structure under the magnetic field, providing strain stimulation to the seeded cells. The strain stimulation would activate the cells to proliferate and differentiate and form new bone tissue. The synergy effect of magnetically responsive biomimetic structures in response to the external applied magnetic field to fasten the osteogenesis may be further amplified by combining it with chemical signaling provided by growth factors and osteogenic drugs. Moreover, the magnetism of the structures can be tailored by controlling the MNPs content in the composite.

3. Materials and Methods

3.1. Structures Design

The 3D structures were aimed to mimic the shape of a typical osteoblast cell. To this end, we designed repetitive ellipsoidal and hexagonal units, organized in a multilayered architecture. We used Python (SciPy pack, Python Software Foundation, Beaverton, OR, USA) for structure generation. After all parameters were calculated, the Python script wrote .gwl files, which are specific to the equipment used for fabrication. From a computing point of view, both geometries represent ellipses. Hexagonal structures were derived from elliptic structures, as we used only 6 points on each elliptic cell to define the hexagonal element. We employed different geometries, for different purposes, in an iterative fashion. The complete structure is composed of elements of specific geometry and position, which are repeated on each axis independently. Elements on the X-axis are positioned with a constant distance between the centers of neighboring elements, 30 μm. In order to determine the optimal geometry, the distanced between neighboring elements on the Y axis is varied according to the following equation:

$$C_{Y-axis} = i \cdot inc_x \cdot 2^{\left(\frac{i}{10}\right)} + const \tag{1}$$

where C_{Y-axis} represents the center of the specified element, i is the row number, inc_x is the distance between the centers of the first two rows of elements, and *const* is an arbitrary constant used for positioning the whole structure with respect to the 0 position. This results in an exponential variation of the distance between the centers of neighboring elements. In other words, the overlap of elements varies exponentially with respect to the Y-axis. The $2^{(i/10)}$ term is used to adjust the exponential overlap to the overall size of the structure. Element diameters were 40 μm on X-axis and 80 μm on the Y-axis, respectively. The height of each layer was designed to be 20 μm (~24 μm when accounting for the voxel height). Neighboring layers intertwine for higher structural resistance. The overlap is ~8 μm depth-wise for two consecutive layers. Moreover, the layers are dislocated to the left and right, consecutively, by half the diameter of a cell on the X-axis. In order to compensate for this dislocation, even-numbered layers are comprised of 6 cells on the X-axis, while odd-numbered layers are comprised of 5 cells on the X-axis. After determining the optimal distance between centers of neighboring cells on each axis, we fabricated structures with fixed overlap in the X and Y directions. Height optimization is done after experimental results for optimized XY geometry.

3.2. Structure Fabrication and Characterization

Structure fabrication was achieved using the Photonic Professional 3D Lithography system from Nanoscribe GmbH (Eggenstein-Leopoldshafen, Germany). This installation relies on two-photon polymerization to create 3D structures by laser direct writing (LDW by TPP). We used IP-780 as the material of choice for the structures. This material is a liquid photoresist that results in a biocompatible polymer after TPP-induced chain reaction and development. This formulation is optimized for high sensitivity in the case of rapid 3D structuring. Laser irradiation was achieved with 150 fs pulses with an 80 MHz repetition rate and centered on $\lambda = 780$ nm. The light was focused with a $63\times$ microscope objective. The positioning was made using a hybrid system comprised of motorized and piezoelectric stages. Coarse positioning was achieved with the motorized stages, while the laser writing was done using the piezoelectric stages. Each sample was prepared by drop-casting the photoresist on a 170 μm thick glass substrate, previously cleaned by ultrasonication for 30 min in ethanol. It was then inserted into the positioning system and the laser was focused on the surface. The structure was written in the polymer drop while maintaining contact points to the substrate for adherence. The polymer was formed rapidly after irradiation without the need for additional processing. After the laser writing, the sample was taken out and immersed in Propylene Glycol Mono-methyl Ether Acetate (PGMEA) solvent for up to 15 min in order to remove the non-photopolymerized material. After removing the samples from the solvent, they were allowed to dry, in air, at room temperature.

Bare iron oxide nanoparticles (MNPs) with physical dimension of 4–20 nm have been obtained using a modified chemical co-precipitation as described in [43]. It has been proved that iron oxide nanoparticles express super-paramagnetic behavior that is preserved when incorporated in nanocomposite materials [19]. Collagen (Col), chitosan (Chit) and hydroxyapatite (HA) were acquired from Sigma Aldrich, St. Louis, MO, USA. The 3D structures were coated by spin coating at 6000 rpm with solutions containing 2 wt % chitosan, 2 wt % collagen, 2 wt % HA, and 4 wt % MNPs. Preliminary studies highlighted the differences in cell viability, density and morphology as a function of the Col:Chit ratio in the mixture. Specifically, the cell viability and density were higher for higher Col concentration, decreasing progressively with increasing Chit content. Also, in the case of compounds with a higher concentration of Col, the cells retained their native morphology, while on structures with higher Chit content, cell morphology was altered, showing specific signs of apoptosis. Based on these preliminary biological assessments, the optimal composition of the Col:Chit nanocomposite was established to be 80:20.

SEM

The structures were investigated by Scanning Electron Microscopy (SEM, FEI InspectS model, Hillsboro, OR, USA). Prior to SEM examination, the samples were coated with ~10 nm gold. After the cell-seeding, the samples were fixed and dehydrated using the protocol described in the next section.

3.3. Biological Assessments

3.3.1. Cell Cultures

MG-63 osteoblast-like cells, were purchased from the European Collection of Cell Cultures (ECACC, Salisbury, United Kingdom). The cells were cultured in MEM growth medium (Biochrom, Berlin, Germany), supplemented with 10% fetal bovine serum (FBS, Biochrom), 2 mML-glutamine (Biochrom), 1% (*v*/*v*) non-essential amino-acids and 100 IU/mL of penicillin/streptomycin (Biochrom) under standard conditions of temperature and humidity (37 °C, 5% CO_2). When confluent, the cells were detached with 1% Trypsin and seeded onto the UV-sterilized structures (5000 cells/structure) and cultured under standard conditions for 4 weeks. Before preparation for MTS assay, ALP production measurement, Alizarin Red staining and Human Osteocalcin Immunoassay, the samples were checked under an Axio Imager 2, Zeiss microscope with AxioCam MRm camera and the cells surrounding the structures were removed using a cell scraper (TPP, Trasadingen, Switzerland).

3.3.2. Cells Morphological Investigations by SEM

After being cultured for 7 days under standard conditions on the 3D structures and controls, the cells were gently washed with PBS and fixed with 2.5% glutaraldehyde in PBS, during 1 h, at room temperature. After this, the cells were washed again and proceeded to the dehydration procedure. First, the samples were dehydrated in ethanol (EtOH) solutions with the indicated concentrations (2 times, during 15 min wash with Et OH 70%, 90%, respectively 100%). Then, the samples were immersed in EtOH-HMDS solutions (50%:50%; 25%:75%, respectively 0%:100% ratios, 2 times, during 3 min). Finally, the samples were let to dry prior to SEM analysis.

3.3.3. MTS

5000 cells/ sample were cultured in complete MEM for 4 weeks under standard conditions of temperature and humidity. After this time, the culture medium was replaced with 16.67% MTS (Cell Titer 96® Aqueous One Solution Cell Proliferation Assay, Promega, Madison, WI, USA) and 83.33% MEM (5% FBS). After 2–3 h of incubation, the supernatant was collected and 100 μL from each sample was distributed in a 96-well plate. The absorbance was measured at 490 nm, using the Mitras LB 940 (Berthold Technologies, Bad Wildbad, Germany) spectrophotometer. The viability was calculated as percent from controls (non-stimulated samples i.e., 0 mT).

3.3.4. ALP

Alkaline Phosphatase production was spectrophotometrically measured at 405 nm, using the Alkaline Phosphatase Assay Kit (Colorimetric) (ab83369) (Abcam, Cambridge, UK), which uses p-Nitrophenyl Phosphate Liquid Substrate (pNPP) for cell lysate. For this, the cells were cultured similarly as for MTS. The Assay standards were prepared as following: 40 μL pNPP 5mM liquid standard solution were mixed with 160 μL Assay Buffer and serial dilution were further prepared (0, 4, 8, 12, 16, 20 nmol/well of pNPP). For the sample preparation, the cells were harvested by trypsinisation, gently washed for several times with cold PBS and resuspended in 100 μL Assay Buffer and were then centrifuged at 7000 rpm, for 15 min, to remove the insoluble components. The supernatant was transferred into 96 well-plates (100 μL/well) and completed with 50 μL of 5mM pNPP solution; 10 μL of ALP enzyme solution was only added into the standard wells. All standards and samples were incubated in the dark, at room temperature, for 60 min. Next, 20 μL of Stop Solution was added into each well, standards and samples and the absorbance was measured at 405 nm using the Mithras (Berthold Technologies) spectrophotometer. The results were expressed as units per milligram of protein in cell lysate, where the protein was assayed by the Bradford (B6916, Sigma Aldrich) method, using serum bovine albumin as standard.

3.3.5. Osteocalcin

In order to measure the Human Osteocalcin protein, the cells were seeded similarly as for SE; the samples were prepared using Quantikine®ELISA Human Osteocalcin Immunoassay (Catalog Number DSTCN0 (R&D SYSTEMS, Minneapolis, MN, USA)), according to the producer's specifications. The standard Osteocalcin solution in the kit was used in order to obtain a standard curve for Osteocalcin calibration. A total of 50 μL from the supernatant of each sample was added to a 96-well plate, together with 100 μL of Assay Diluent. The samples were incubated while shaking during 2 h; after this time, they were washed 3 times using the washing buffer; 200 μL from the conjugate were added in each well. After another 2 h of shaking at room temperature, the samples were washed 4 times using the washing buffer; 200 μL from the Substrate solution was added to each well and then allowed to incubate for 30 min, in the dark. At the end of this period, the reaction was finished using 50 μL of the Stop solution. The osteocalcin protein secretion was measured spectrophotometrically at 450 nm with a correction at 570 nm, using the Mitras LB 940 (Berthold Technologies).

3.3.6. Alizarin Red (ARS) Assay

For mineral distribution evaluation, the cells were seeded similarly as for SEM. After 4 weeks of incubation, the samples were washed twice with double-distilled water; after this 1 mL of 40 mM ARS (pH 4.1) was added to each well. Following this, the samples were incubated for 20 min, at room temperature and then washed several times with double-distilled water while shaking for

Int. J. Mol. Sci. **2018**, *19*, 495

5 min. The quantification of mineralization was done by extracting the calcified mineral at low pH, followed by neutralization with ammonium hydroxide and absorbance measurement at 405 nm.

3.3.7. Statistical Analysis

The values were presented as mean ± STD (standard deviation) of 3 measurements. The data were analyzed statistically using a two-tailed Student's test, where p values ≤ 0.05 were accepted as statistically significant. Each data point in the relative cell viability, ALP, Alizarin and Osteocalcin estimations was calculated as the mean of 3 different measurements performed in 3 different experiments. The standard deviation was shown as an error bar. The calculated probability that resulted in significant differences from the control samples was calculated based on the t-statistic of the variance of differences between individual observations as related to control. We defined * $p < 0.05$ and ** $p < 0.001$.

3.3.8. Static Magnetic Fields Stimulation

The samples were positioned in the vicinity of gold-plated cubic Neodymium magnets (5 mm). The strengths of the magnetic fields were measured using a Phywe digital teslameter with tangential and axial Hall probes. For obtaining strengths of the magnetic field between 100 and 250 mT, we employed 1 to 3 magnets positioned in particular configurations. The Petri dishes containing the cell-seeded structures were placed on top of the magnets.

4. Conclusions

We demonstrated the synergistic effect of 3D magnetic structures on enhancing cell differentiation in response to static magnetic fields. We fabricated innovative, complex 3D structures of ellipsoidal and hexagonal repetitive units that mimic the native shape of osteoblast-like cells, by LDW via TPP of IP-L780 photoresist. This is the first report in the literature of the use of this technique in obtaining biomimetic 3D structures for bone tissue engineering, with this specific architectural design. The structures were coated with a magnetic composite made of collagen, chitosan, HA and MNPs. The nanoparticles provided the structures with magnetization ability suitable for cell guiding in the vicinity and inside the structure, as well as for cells differentiation and mineralization. The static magnetic field applied to the 3D structures accelerated the cell differentiation in vitro, in relation to the structures without stimulation using a magnetic field. Moreover, the cells exposed to the most intense magnetic field reached the end of proliferation period and started their differentiation faster than those in the other samples. Thus, we have succeeded in obtaining novel 3D biomimetic structures with potential for bone tissue engineering that are specifically designed to offer the best architectural features that enable the support and growth of osteoblast cells. Moreover, adding magnetic properties to the structures via the nanocomposite biocompatible coating made of collagen, chitosan, hydroxiapatite and MNPs, accelerated cell differentiation, with the potential to promote earlier development of new bone. These results provide encouraging perspectives for pre-clinical applications and set the basis for further research in developing clinically available smart engineered materials for the management of bone injures.

Acknowledgments: This work was supported by a grant of the Romanian National Authority for Scientific Research and Innovation, CNCS/CCCDI-UEFISCDI, project number PN-III-P2-2.1-PED-2016-1787, within PNCD III. A part of this work was performed in CETAL facility, supported by the National Program National Program PN 16 47-LAPLAS IV.

Author Contributions: Irina Alexandra Paun contributed to the concept of the experiment, to the fabrication of the structures, to data interpretation and to the correlation of the results in the context of the in vitro studies; Roxana Cristina Popescu contributed to in vitro analysis of the structures and to data interpretation; Bogdan Stefanita Calin worked on the structures design and fabrication; Cosmin Catalin Mustaciosu participated to the in vitro biological assays and to the analysis of the in vitro results; Maria Dinescu contributed to data interpretation; Catalin Romeo Luculescu contributed to the concept of the experiments, to the morphological investigations of the samples by Scanning Electron Microscopy and to data analysis.

Int. J. Mol. Sci. **2018**, *19*, 495

Conflicts of Interest: The authors declare no conflict of interest.

References

1. Langer, R.; Vacanti, J.P. Tissue engineering. *Science* **1993**, *260*, 920–926. [CrossRef] [PubMed]
2. Yun, H.M.; Ahn, S.J.; Park, K.R.; Kim, M.J.; Kim, J.J.; Jin, G.Z.; Kim, H.W.; Kim, E.C. Magnetic nanocomposite scaffolds combined with static magnetic field in the simulation of osteoblastic differentiation and bone formation. *Biomaterials* **2016**, *85*, 88–98. [CrossRef] [PubMed]
3. Fini, M.; Cadossi, R.; Canè, V.; Cavani, F.; Giavaresi, G.; Krajewski, A.; Martini, L.; Aldini, N.N.; Ravaglioli, A.; Rimondini, L.; et al. The effect of pulsed electromagnetic fields on the osteointegration of hydroxyapatite implants in cancellous bone: A morphologic and microstructural in vivo study. *J. Orthop. Res.* **2002**, *20*, 756–763. [CrossRef]
4. Yamamoto, Y.; Ohsaki, Y.; Goto, T.; Nakashima, A.; Iijima, T. Effects of static magnetic fields on bone formation in rat osteoblast cultures. *J. Dent. Res.* **2003**, *82*, 926–966. [CrossRef] [PubMed]
5. Ba, X.; Hadjiargyrou, M.; DiMasi, E.; Meng, Y.; Simon, M.; Tan, Z.; Rafailovich, M.H. The role of moderate static magnetic fields on biomineralization of osteoblasts on suflonated polystyrene films. *Biomaterials* **2011**, *32*, 7831–7838. [CrossRef] [PubMed]
6. Feng, S.W.; Lo, Y.J.; Chang, W.J.; Lin, C.T.; Lee, S.Y.; Abiko, Y.; Huang, H.M. Static magnetic field exposure promotes differentiation of osteoblastic cells grown on the surface of a poly-L-lactide substrate. *Med. Biol. Eng. Comput.* **2010**, *48*, 793–798. [CrossRef] [PubMed]
7. Chiu, K.H.; Ou, K.L.; Lee, S.Y.; Lin, C.T.; Chang, W.J.; Chen, C.C.; Huang, H.M. Static magnetic fields promote osteoblast-like cells differentiation via increasing the membrane rigidity. *Ann. Biomed. Eng.* **2007**, *35*, 1932–1939. [CrossRef] [PubMed]
8. Cunha, C.; Panseri, S.; Marcacci, M.; Tampieri, A. Evaluation of the Effects of a Moderate Intensity Static Magnetic Field Application on Human Osteoblast-like Cells. *Am. J. Biomed. Eng.* **2012**, *2*, 263–268. [CrossRef]
9. Lin, S.L.; Chang, W.J.; Hsieh, S.C.; Lin, C.T.; Chen, C.C.; Huang, H.M. Mechanobiology of MG63 osteoblast-like cells adaptation to static magnetic forces. *Electromagn. Biol. Med.* **2008**, *27*, 55–64. [CrossRef] [PubMed]
10. Kim, E.C.; Leesunbok, R.; Lee, S.W.; Park, S.H.; Mah, S.J.; Ahn, S.J. Effects of moderate intensity static magnetic fields on human bone marrow-derived mesenchymal stem cells. *Bioelectromagnetics* **2015**, *36*, 267–276. [CrossRef] [PubMed]
11. Schäfer, R.; Bantleon, R.; Kehlbach, R.; Siegel, G.; Wiskirchen, J.; Wolburg, H.; Kluba, T.; Eibofner, F.; Northoff, H.; Claussen, C.D.; et al. Functional investigations on human mesenchymal stem cells exposed to magnetic fields and labeled with clinically approved iron nanoparticles. *BMC Cell Biol.* **2010**, *11*–22. [CrossRef] [PubMed]
12. Huang, J.; Wang, D.; Chen, J.; Liu, W.; Duan, L.; You, W.; Xiong, J.; Wang, D. Osteogenic differentiation of bone marrow mesenchymal stem cells by magnetic nanoparticle composite scaffolds under a pulsed electromagnetic field. *Saudi Pharm. J.* **2017**, *25*, 575–579. [CrossRef] [PubMed]
13. Rosen, A.D. Mechanism of action of moderate-intensity static magnetic fields on biological systems. *Cell Biochem. Biophys.* **2003**, *39*, 163–173. [CrossRef]
14. Rosen, A.D. Membrane response to static magnetic fields: Effect of exposure duration. *Biochim. Biophys. Acta* **1993**, *1148*, 317–320. [CrossRef]
15. Aoki, H.; Yamazaki, H.; Yoshino, T.; Akagi, T. Effects of static magnetic fields on membrane permeability of a cultured cell line. *Res. Commun. Chem. Pathol. Pharmacol.* **1990**, *69*, 103–106. [PubMed]
16. Kotani, H.; Iwasaka, M.; Ueno, S. Magnetic orientation of collagen and bone mixture. *J. Appl. Phys.* **2000**, *87*, 6191–6193. [CrossRef]
17. Bañobre-López, M.; Piñeiro-Redondo, Y.; De Santis, R.; Gloria, A.; Ambrosio, L.; Tampieri, A.; Dediu, V.; Rivas, J. Poly(caprolactone) based magnetic scaffolds for bone tissue engineering. *J. Appl. Phys.* **2011**, *109*, 07B313. [CrossRef]
18. Bock, N.; Riminucci, A.; Dionigi, C.; Russo, A.; Tampieri, A.; Landi, E.; Goranov, V.A.; Marcacci, M.; Dediu, V. A novel route in bone tissue engineering: Magnetic biomimetic scaffolds. *Acta Biomater.* **2010**, *6*, 786–796. [CrossRef] [PubMed]

19. Meng, J.; Xiao, B.; Zhang, Y.; Liu, J.; Xue, H.; Lei, J.; Kong, H.; Huang, Y.; Jin, Z.; Gu, N.; et al. Super-paramagnetic responsive nanofibrous scaffolds under static magnetic field enhance osteogenesis for bone repair in vivo. *Sci. Rep.* **2013**, *3*, 2655. [CrossRef] [PubMed]

20. De Santis, R.; Gloria, A.; Russo, T.; D'Amora, U.; Zeppetelli, S.; Dionigi, C.; Sytcheva, A.; Herrmannsdörfer, T.; Dediu, V.; Ambrosio, L. A basic approach toward the development of nanoscomposite magnetic scaffolds for advanced bone tissue engineering. *J. Appl. Polym. Sci.* **2011**, *122*, 3599–3605. [CrossRef]

21. Tampieri, A.; D'Alessandro, T.; Sandri, M.; Sprio, S.; Landi, E.; Bertinetti, L.; Panseri, S.; Pepponi, G.; Goettlicher, J.; Bañobre-López, M.; et al. Intrinsic magnetism and hyperthermia in bioactive Fe-doped hydroxyapatite. *Acta Biomater.* **2012**, *8*, 843–851. [CrossRef] [PubMed]

22. Tampieri, A.; Landi, E.; Valentini, F.; Sandri, M.; D'Alessandro, T.; Dediu, V.; Marcacci, M. A conceptually new type of bio-hybrid scaffold for bone regeneration. *Nanotechnology* **2011**, *22*, 015104. [CrossRef] [PubMed]

23. Gloria, A.; Russo, T.; D'Amora, U.; Zeppetelli, S.; D'Alessandro, T.; Sandri, M.; Bañobre-López, M.; Piñeiro-Redondo, Y.; Uhlarz, M.; Tampieri, A.; et al. Magnetic poly(ε-caprolactone)/iron-doped hydroxyapatite nanocomposite substrates for advanced bone tissue engineering. *J. R. Soc. Interface* **2013**, *10*, 20120833. [CrossRef] [PubMed]

24. Singh, R.K.; Patel, K.D.; Lee, J.H.; Lee, E.J.; Kim, T.H.; Kim, H.W. Potential of magnetic nanofiber scaffolds with mechanical and biological properties applicable for bone regeneration. *PLoS ONE* **2014**, *9*, e91584. [CrossRef] [PubMed]

25. Wu, Y.; Jiang, W.; Wen, X.; He, B.; Zeng, X.; Wang, G.; Gu, Z. A novel calcium phosphate ceramic-magnetic nanoparticle composite as a potential bone substitute. *Biomed. Mater.* **2010**, *5*, 015001. [CrossRef] [PubMed]

26. Wei, Y.; Zhang, X.; Song, Y.; Han, B.; Hu, X.; Wang, X.; Lin, Y.; Deng, X. Magnetic biodegradable Fe$_3$O$_4$/CS/PVA nanofibrous membranes for bone regeneration. *Biomed. Mater.* **2011**, *6*, 055008. [CrossRef] [PubMed]

27. Panseri, S.; Russo, A.; Giavaresi, G.; Sartori, M.; Veronesi, F.; Fini, M.; Salter, D.M.; Ortolani, A.; Strazzari, A.; Visani, A.; et al. Inovative magnetic scaffolds for orthopedic tissue engineering. *J. Biomed. Mater. Res.* **2012**, *100*, 2278–2286. [CrossRef]

28. Zhou, X.; Hou, Y.; Lin, J. A review on the processing accuracy of two-photon polymerization. *AIP Adv.* **2015**, *5*, 030701. [CrossRef]

29. Loh, Q.L.; Choong, C. Three-Dimensional scaffolds for Tissue Engineering Applications: Role of Porosity and Pore Size. *Tissue Eng. Part B Rev.* **2013**, *19*, 485–502. [CrossRef] [PubMed]

30. Ma, T.; Li, Y.; Yang, S.T.; Kniss, D.A. Effects of pore size in 3-D fibrous matrix on human trophoblast tissue development. *Biotechnol. Bioeng.* **2000**, *70*, 606–618. [CrossRef]

31. Mandal, B.B.; Kundu, S.C. Cell proliferation and migration in silk fibroin 3D scaffolds. *Biomaterials* **2009**, *30*, 2956–2965. [CrossRef] [PubMed]

32. Mata, A.; Kim, E.J.; Boehm, C.A.; Fleischman, A.J.; Muschler, G.F.; Roy, S. A three-dimensional scaffold with precise micro-architecture and surface micro-textures. *Biomaterials* **2009**, *30*, 4610–4617. [CrossRef] [PubMed]

33. Mohanty, S.; Larsen, L.B.; Trifol, J.; Szabo, P.; Burri, H.V.; Canali, C.; Dufva, M.; Emnéus, J.; Wolff, A. Fabrication of scalable and structured tissue engineering scaffolds using water dissolvable sacrificial 3D printed moulds. *Mater. Sci. Eng. C Mater. Biol. Appl.* **2015**, *55*, 569–578. [CrossRef] [PubMed]

34. Cooper, G.M.; Hausman, R.E. *The Cell: A Molecular Approach*, 4th ed.; Sinauer Associates: Sunderland, MA, USA, 2007; pp. 599–648, ISBN 0-87893-219-4.

35. Noda, M. *Cellular and Molecular Biology of the Bone*; Academic Press: London, UK, 1993; pp. 49–63, ISBN 0-12-520225-3.

36. Zhang, N.; Ying, M.-D.; Wu, Y.-P.; Zhou, Z.-H.; Ye, Z.-M.; Li, H.; Lin, D.-S. Hyperoside, a flavonoid compound, inhibits proliferation and stimulates osteogenic differentiation of human osteosarcoma cells. *PLoS ONE* **2014**, *9*, e98973. [CrossRef] [PubMed]

37. Wang, G.-H.; Guo, Z.-Y.; Shi, S.-L.; Li, Q.-F. Effect of cinnamic acid on proliferation and differentiation of human osteosarcoma MG-63 cells. *Chin. Pharm. Bull.* **2012**, *9*, 1262–1266. [CrossRef]

38. Panseri, S.; Cunha, C.; D'Alessandro, T.; Sandri, M.; Russo, A.; Giavaresi, G.; Marcacci, M.; Hung, C.T.; Tampieri, A. Magnetic hydroxyapatite bone substitutes to enhance tissue regeneration: Evaluation in vitro using osteoblast-like cells and in vivo in a bone defect. *PLoS ONE* **2012**, *7*, e38710. [CrossRef] [PubMed]

39. Li, L.; Yang, G.; Li, J.; Ding, S.; Zhou, S. Cell behaviors on magnetic electrospun poly-D, L-lactide nanofibers. *Mater. Sci. Eng. C Mater. Biol. Appl.* **2014**, *34*, 252–261. [CrossRef] [PubMed]

40. Zeng, X.B.; Hu, H.; Xie, L.Q.; Lan, F.; Jiang, W.; Wu, Y.; Gu, Z.W. Magnetic responsive hydroxyapatite composite scaffolds construction for bone defect reparation. *Int. J. Nanomed.* **2012**, *7*, 3365–3378. [CrossRef] [PubMed]

41. Pautke, C.; Schieker, M.; Tischer, T.; Kolk, A.; Neth, P.; Mutschler, W.; Milz, S. Characterization of osteosarcoma cell lines MG-63, Saos-2 and U-2 OS in comparison to human osteoblasts. *Anticancer Res.* **2004**, *24*, 3743–3748. [PubMed]

42. Tsai, S.W.; Liou, H.M.; Lin, C.J.; Kuo, K.L.; Hung, Y.S.; Weng, R.C.; Hsu, F.Y. MG63 osteoblast-like cells exhibit different behaviour when grown on electrospun collagen matrix versus electrospun gelatin matrix. *PLoS ONE* **2012**, *7*, e31200. [CrossRef]

43. Popescu, R.C.; Andronescu, E.; Vasile, B.Ş.; Truşcă, R.; Boldeiu, A.; Mogoantă, L.; Mogoşanu, G.D.; Temelie, M.; Radu, M.; Grumezescu, A.M.; Savu, D. Fabrication and Cytotoxicity of Gemcitabine-Functionalized Magnetite Nanoparticles. *Molecules* **2017**, *22*, 1080. [CrossRef] [PubMed]

International Journal of
Molecular Sciences

MDPI

Article

Effects and Mechanisms of Total Flavonoids from *Blumea balsamifera* (L.) DC. on Skin Wound in Rats

Yuxin Pang [1,2,3,4,†], **Yan Zhang** [1,2,†], **Luqi Huang** [1,2,*], **Luofeng Xu** [3,4], **Kai Wang** [3,4], **Dan Wang** [3,4], **Lingliang Guan** [3,4], **Yingbo Zhang** [3,4], **Fulai Yu** [3,4], **Zhenxia Chen** [3,4] and **Xiaoli Xie** [3,4]

[1]   National Resource Center for Chinese Materia Medica, China Academy of Chinese Medical Sciences, Beijing 100700, China; yxpang@catas.cn (Y.P.); zhangyan8669@126.com (Y.Z.)
[2]   Center for Post-Doctoral Research, China Academy of Chinese Medical Sciences, Beijing 100700, China
[3]   Tropical Crops Genetic Resources Institute, Chinese Academy of Tropical Agricultural Sciences, Danzhou 571737, China; x1531298865@126.com (L.X.); jimojijie29@163.com (K.W.); wang_dan1414@163.com (D.W.); gllgirl123@163.com (L.G.); zhangyingbo1984@catas.cn (Y.Z.); flyu@catas.cn (F.Y.); hnchenzhenxia@126.com (Z.C.); xiexiaoli198883@163.com (X.X.)
[4]   Hainan Provincial Engineering Research Center for Blumea Balsamifera, Danzhou 571737, China
*   Correspondence: huangluqi01@126.com; Tel.: +86-10-8404-4340
†   These authors contributed equally to this work.

Received: 2 December 2017; Accepted: 16 December 2017; Published: 19 December 2017

Abstract: Chinese herbal medicine (CHM) evolved through thousands of years of practice and was popular not only among the Chinese population, but also most countries in the world. *Blumea balsamifera* (L.) DC. as a traditional treatment for wound healing in Li Nationality Medicine has a long history of nearly 2000 years. This study was to evaluate the effects of total flavonoids from *Blumea balsamifera* (L.) DC. on skin excisional wound on the back of Sprague-Dawley rats, reveal its chemical constitution, and postulate its action mechanism. The rats were divided into five groups and the model groups were treated with 30% glycerol, the positive control groups with Jing Wan Hong (JWH) ointment, and three treatment groups with high dose (2.52 g·kg^{-1}), medium dose (1.26 g·kg^{-1}), and low dose (0.63 g·kg^{-1}) of total flavonoids from *B. balsamifera*. During 10 consecutive days of treatment, the therapeutic effects of rates were evaluated. On day 1, day 3, day 5, day 7, and day 10 after treatment, skin samples were taken from all the rats for further study. Significant increases of granulation tissue, fibroblast, and capillary vessel proliferation were observed at day 7 in the high dose and positive control groups, compared with the model group, with the method of 4% paraformaldehyde for histopathological examination and immunofluorescence staining. To reveal the action mechanisms of total flavonoids on wound healing, the levels of CD68, vascular endothelial growth factor (VEGF), transforming growth factor-β_1 (TGF-β_1), and hydroxyproline were measured at different days. Results showed that total flavonoids had significant effects on rat skin excisional wound healing compared with controls, especially high dose ones ($p < 0.05$). Furthermore, the total flavonoid extract was investigated phytochemically, and twenty-seven compounds were identified from the total flavonoid sample by ultra-high-performance liquid chromatography coupled with quadrupole time-of-flight mass spectrometry/diode array detector (UPLC-Q-TOF-MS/DAD), including 16 flavonoid aglucons, five flavonoid glycosides (main peaks in chromatogram), five chlorogenic acid analogs, and 1 coumarin. Reports show that flavonoid glycoside possesses therapeutic effects of curing wounds by inducing neovascularization, and chlorogenic acid also has anti-inflammatory and wound healing activities; we postulated that all the ingredients in total flavonoids sample maybe exert a synergetic effect on wound curing. Accompanied with detection of four growth factors, the upregulation of these key growth factors may be the mechanism of therapeutic activities of total flavonoids. The present study confirmed undoubtedly that flavonoids were the main active constituents that contribute to excisional wound healing, and suggested its action mechanism of improving expression levels of growth factors at different healing phases.

Keywords: *Blumea balsamifera* (L.) DC.; total flavonoids; skin wound; VEGF; TGF-β_1

1. Introduction

Acute wound healing proceeds through four stages: inflammation response, migration, proliferation, and tissue remodeling [1]. When the skin is injured, the normal healing response begins. During inflammation and initial stages of wound healing, the process of repair is largely mediated by cytokines or growth factors such as tumor necrosis factor alpha (TNF-α), transforming growth factor-beta (TGF-β) [2,3], platelet derived growth factor (PDGF), and vascular endothelial growth factor (VEGF) that orchestrate the manifold cellular activities [4–6]. Simultaneously, specialized cells move to the wound site. In the inflammatory phase, polymorphonuclear neutrophilic leukocytes and macrophages appear around the wounds. Macrophages, as the principal phagocytic cells in wound repair, provide an effective local barrier against bacterial invasion and wound debridement. Macrophages can also induce production of TNF-α, TGF-β, CD68, and other factors [7–9]. The proliferative phase is characterized by angiogenesis, collagen deposition, epithelialization, and wound contraction [10]. In this phase, fibroblasts act as the principal cells responsible for collagen deposition to form a new, provisional extracellular matrix [11]. Collagen is the most abundant protein, accounting for 30% of the total protein in the human body [12]. Collagen contains substantial amounts of hydroxyproline, which is used as a biochemical marker for tissue collagen [13]. The aforementioned contents resulted in our selection of four growth factors of TGF-β, VEGF, CD68, and hydroxyproline at different stages as biomarkers to elucidate the wound healing mechanism.

Blumea balsamifera (L.) DC., belongs to *Blumea*, Compositae, widely distributed in Hainan, Guizhou, Yunnan, Guangdong, and Taiwan provinces in China [14,15]. Its leaves, twigs, and roots were widely used for many diseases, for instance, rheumatism, dermatitis, beriberi, and lumbago, and especially for treatment of snake bites and bruises in some ancient Chinese minorities such as Li, Miao, and Zhuang [16]. Recent pharmacological studies have showed that *B. balsamifera* possesses a broad spectrum of pharmacological activities such as coagulation [17], antibacterial, free radical scavenging [18], antioxidant [19], and anticancer [20,21], while few studies have investigated and reported on the injure curing effects of this plant. Phytochemical analysis revealed that *B. balsamifera* leaves contained considerable amounts of flavonoids [22], and related literature studies also show that many flavonoids possess direct or indirect wound healing effects, such as soy isoflavone on scalded mice [23], and corylin on human fibroblast cells in an in vitro model of wound healing [24].

However, flavonoids, once discarded in residue after L-borneol was distilled, were proved to be the main ingredients in *B. balsamifera*, and may contribute to many traditional usages of this herb. As part of our ongoing search for skin recovery candidates, we investigated the chemical constituents of total flavonoids from *B. balsamifera* using UPLC-Q-TOF-MS/DAD, evaluated the its wound curing effects, and elucidated the underlying mechanisms of total flavonoids in the process of wound healing.

2. Results

2.1. Content and Identification of Total Flavonoids

The typical calibration plot conducted with rutin as standard resulted in the regression equation: $y = 5.1907x - 0.0098$, and accordingly, the content of total flavonoids was deduced to be 81.1% (Figures S1 and S2). From literature reported [22], *B. balsamifera* is rich in flavonoids, and nearly forty flavonoid analogs were isolated and identified from this plant, including structure types of flavonoid glycoside and aglucon.

Investigation of the chemical constitution of total flavonoids recorded with UPLC-Q-TOF-MS/DAD led to the identification of 27 compounds (two unidentified of 29 peaks) (Figure 1). The base peak intensity (BPI) and ultra-high-performance liquid chromatography (UPLC) chromatogram at 254 nm of total flavonoids of *Blumea balsamifera* (L.) DC. with positive ion electrospray ionization (ESI) and negative ESI were recorded (as shown in Figure S3). The measured MS data exhibited high consistency with theoretical values, with deviation limited within 5 ppm, which provide valuable information for the determination of constituents. Twenty-seven compounds, including 21 flavonoids, were tentatively confirmed on the basis of their retention behaviors (Figure S4), accurate molecular weight, and MS^E fragment data, and comparison with closely related substances reported in literature (chemical structures are shown in Figure S5, corresponding quasi-molecular ions are listed in Table S1).

The peaks of 3, 4, 5, 6, and 7, ascribed to the chromatogram, accounted for a large proportion and were identified to be the same structure type of flavonoid glycoside (Figure S5), which indicated flavonoid glycosides may be the main effective ingredients of total flavonoids based on related literature [25].

2.2. Effects of Different Doses of Total Flavonoids on Wound Healing Rates in Rats

Treatments were carried out after surgery, and photos of wounds were taken on indicated days. All rats lost weight on the first day after excision, while two days later, their weights steadily increased, and the wound tissues were ruddy and edema disappeared (Figure 2). There was no significant difference between all groups. On day 4, there were no wound infections except the model control, and all rats in the different groups developed granulation tissue growth and significantly reduced wound area. However, the groups with high and medium doses of total flavonoids demonstrated accelerated re-epithelization compared with the model groups. The black crusts were residue resulting from the absorption of total flavonoids. On day 6, the black crusts exfoliated in some rats of the high dose and Jing Wan Hong (JWH) groups. On day 8, the high and medium dose groups showed significantly accelerated wound contraction and closures compared with the model groups. On day 10, the percentage of wound contraction was nearly 100% in the high dose groups, and wound contraction in the other groups were also fully completed, but surprisingly, rats with total flavonoids showed better effects on the wound area and epithelization than the other groups.

As shown in Figure 3 (Table S2), on day 4 and 6, the high dose and JWH groups showed significantly better effects of wound healing than the model groups ($p < 0.01$). On day 8, wound healing was obviously better in high dose groups ($p = 0.011$) and JWH groups ($p = 0.032$) than the model groups. Until day 10, the wound healing rates of high dose and JWH groups were approximately 95.0%, while the rate of model groups was lower than 85.0%, which indicated the potent efficiency of total flavonoids, especially at high doses. The rates in the medium dose and low dose groups were also higher than the model group, but not statistically significant.

2.3. Effects of Different Doses of Total Flavonoids on CD68 Levels on Rats

CD68 is a special antigen expressed by macrophages and cells of myeloid/mononuclear lineage [9]. As shown in Figure 4 (Table S3), in each of the total flavonoids groups, the increased CD68 expression in skin wounds indicated that total flavonoids contributed much to the increasing numbers of macrophages. On day 3, the average integral optical density (IOD) of CD68 in high dose groups were higher than that of control groups ($p < 0.05$). On day 5, CD68 levels of all total flavonoids groups reached a peak ($p = 0.005$, 0.009, and 0.036, respectively). On day 7, the average IOD values of CD68 in high and medium dose groups were still higher than control ($p = 0.003$ and 0.015, respectively). However, on day 10, there were no significant differences between all groups. The changing tendency of CD68 levels indicated that macrophages were very active in the inflammatory stage of high dose treatments before day 5, and downregulated in the following days.

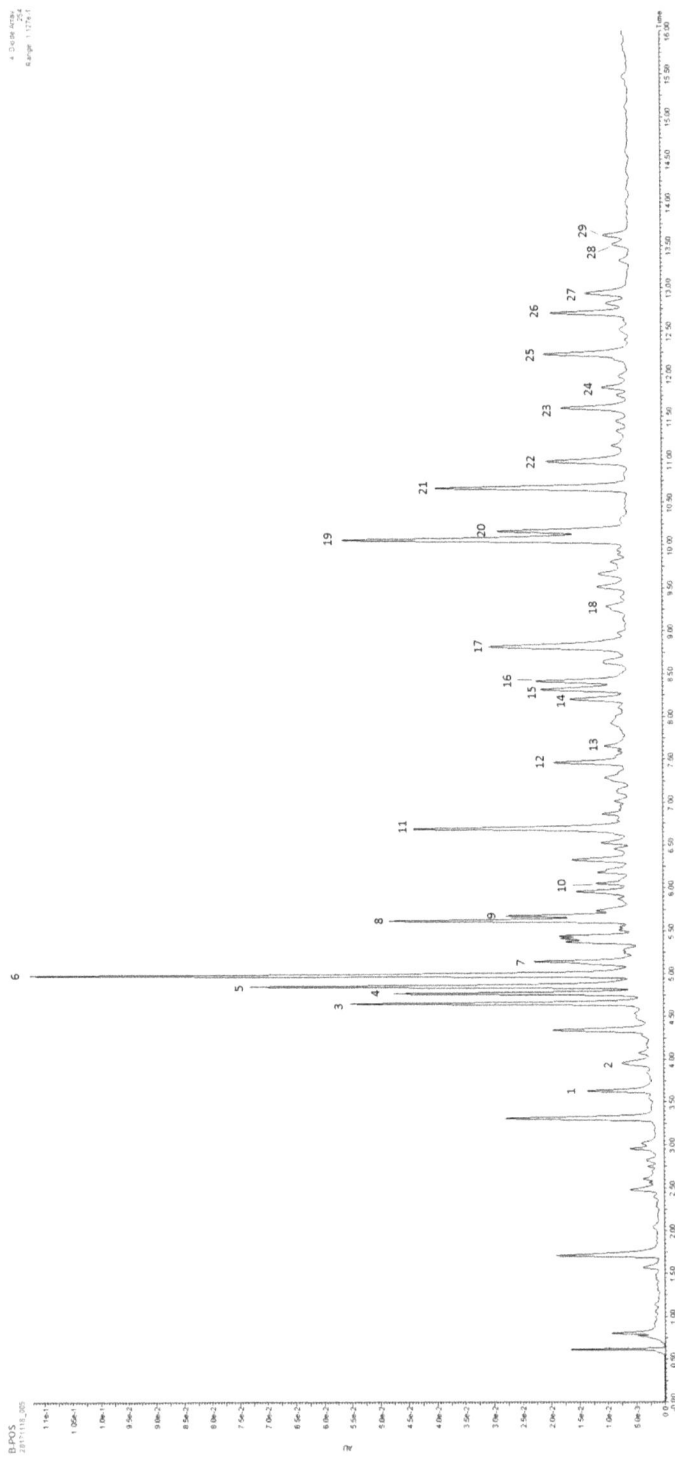

Figure 1. UPLC chromatograms at 254 nm of total flavonoids sample in positive ion modes analyzed by UPLC-Q-TOF/MS/DAD.

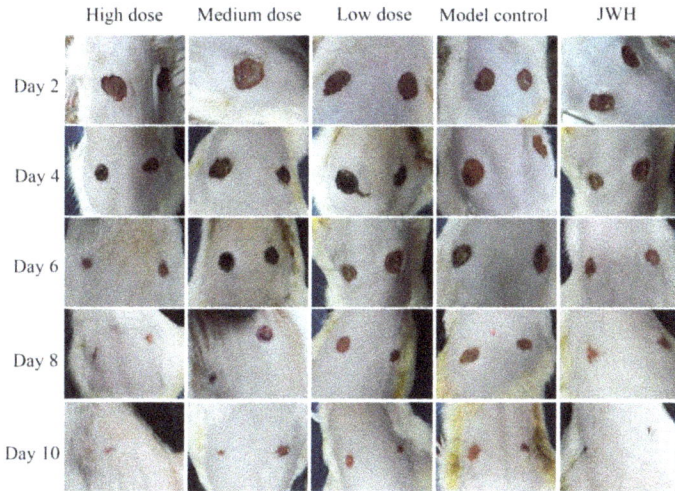

Figure 2. Recovery of wounds at different times. JWH, Jing Wan Hong.

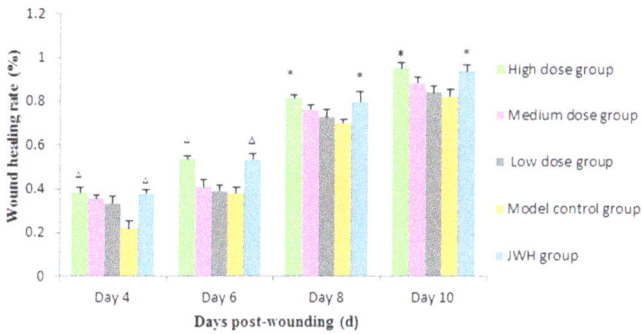

Figure 3. Effects of total flavonoids from *B. balsamifera* on wound healing rate of rats. Values are expressed as mean $\pm$ SD ($n = 6$) as compared to the control group, $* p < 0.05$, $^\Delta p < 0.01$.

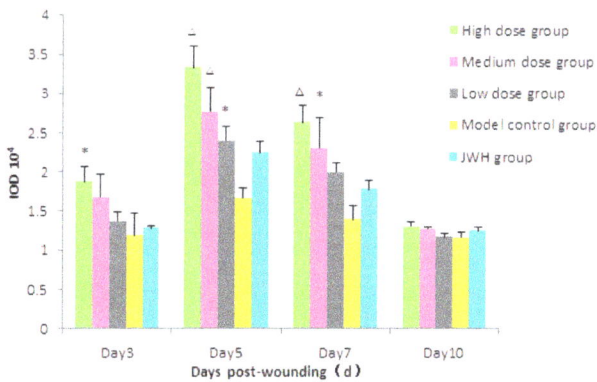

Figure 4. Effects of total flavonoids from *B. balsamifera* on CD68 levels in wound tissues of rats. Values are expressed as mean $\pm$ SD ($n = 6$) as compared to control group, $* p < 0.05$, $^\Delta p < 0.01$.

2.4. Effects of Total Flavonoids on VEGF and TGF-β₁ of the Wound Tissues on Rats

The expression of VEGF was able to drive the proliferation of endothelial cells and formation of new vessels. As shown in Figure 5 (Table S4), VEGF expression of total flavonoid groups increased in a dose-dependent manner ($p < 0.05$) compared to controls. Their tendencies were in an obviously descending order: high dose > medium dose > low dose. VEGF expression reached a peak in high dose, medium dose, and JWH groups on day 5. However, the peaks converted to low dose and control groups after day 7. These results also revealed that VEGF is very active in the crucial curing related stage of day 3 to 7. The peak appearance of high and medium doses ahead of time indicated their acceleration of wound curing.

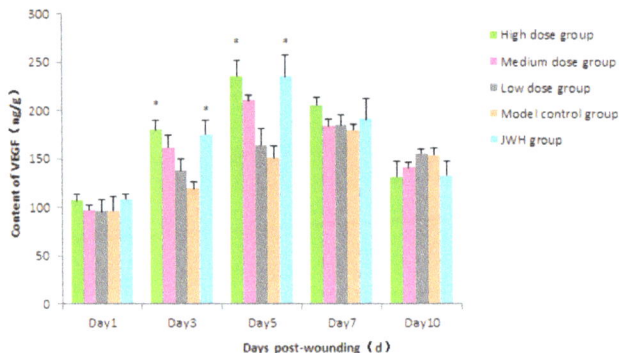

Figure 5. Effects of total flavonoids from *B. balsamifera* on vascular endothelial growth factor (VEGF) levels in wound tissues of rats. Values are expressed as mean $\pm$ SD ($n = 6$) as compared to the control group, * $p < 0.05$.

TGF-β_1 could induce gathering of inflammatory cells to cuts, which is important to the curing process [26]. On day 1, there was no significant difference in TGF-β_1 levels in wound tissues between each group. On day 3 and 5, total flavonoids of low dose were able to mildly stimulate the expression of TGF-β_1, but its effect was obvious as compared to the model control ($p > 0.05$) (Figure 6 and Table S5). The results indicated the enrichment and proliferation of macrophages and fibroblasts in this stage. On day 7, total flavonoids still were able to promote TGF-β_1 levels at all concentrations tested as compared to model control groups ($p < 0.05$ or $p < 0.01$), but the TGF-β_1 levels of all groups began to drop. On day 10, TGF-β_1 expression induced by wounding was markedly decreased to normal level.

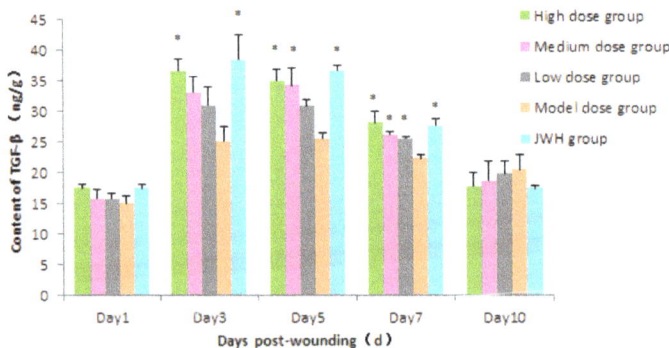

Figure 6. Effects of total flavonoids from *B. balsamifera* on TGF-β_1 levels in wound tissues of rats. Values are expressed as mean $\pm$ SD ($n = 6$) as compared to the control group, * $p < 0.05$.

2.5. Effects of Total Flavonoids on Hydroxyproline Level in Wound Tissues of Rats

Hydroxyproline content is a stable parameter of collagen, which correlated with the growth of granulation tissue. On day 3 after the wound was treated, the high dose and JWH groups began to show statistically significant differences in hydroxyproline contents compared to the model control ($p < 0.05$) (Figure 7 and Table S6). Hydroxyproline expression peaked on day 10. On day 5 and 7, hydroxyproline content maintained an ascending tendency. Leading up to day 10, hydroxyproline contents in the granulation tissue of high dose and JWH groups reached a peak and were approximately 0.80 mg/g, while the contents in model groups were less than 0.65 mg/g. The contents in the medium dose and low dose groups were also higher than the model groups, but there were no significant differences in wound tissues between groups. These results revealed that the total flavonoids of *B. balsamifera* were able to promote synthesis of collagen and accelerate formation of granulation tissue in the middle and late period of wound recovery.

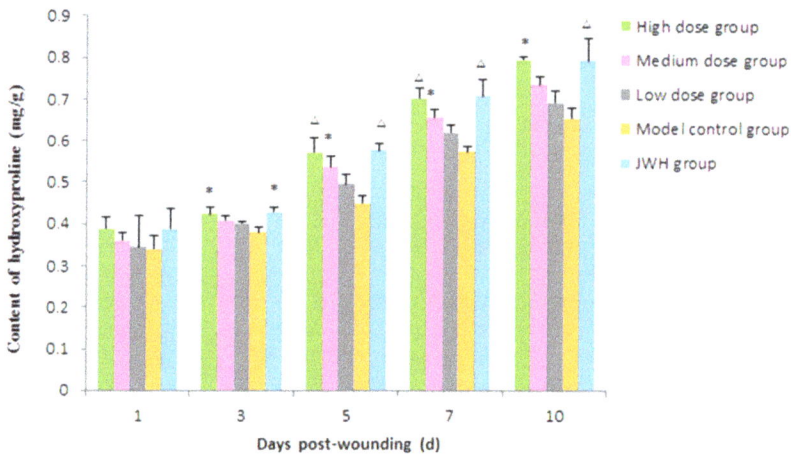

Figure 7. Effects of total flavonoids from *B. balsamifera* on hydroxyproline level in wound tissues of rats. Values are expressed as mean ± SD ($n = 6$) as compared to the control group, * $p < 0.05$, $^\Delta$ $p < 0.01$.

3. Discussion

Skin trauma is very prevalent in the world, research studies showed that more than one million people each year in America may acquire skin injuries, so a search for effective drugs with low side effects is desperately needed. Nowadays, Traditional Chinese Medicine (TCM) is widely used in treatments of many diseases due to its safety and minimum side effects. Within considering both the traditional uses of *B. balsamifera* for treatments of traumatic injury [16] as well as the main ingredients of flavonoids it contains, these aforementioned factors indicated the potential therapeutic effects of total flavonoids in wound healing, which led to our present ongoing study. Chemical investigation, wound healing effects, and the pharmacological mechanism of the total flavonoids were studied. In the present study, our data revealed that total flavonoids could significantly promote wound healing, improve wound contraction, and accelerate epithelialization.

Analysis of the total flavonoids extract sample with UPLC-Q-TOF-MS/DAD resulted in the identification of 27 compounds (2 unidentified of 29 peaks), including 16 flavonoid aglucons, five flavonoid glycosides, five chlorogenic acid (CQA) derivatives, and one coumarin, which indicated flavonoid derivatives as the main effective constituents. Considering the reported activity of flavonoid glycoside on wound healing [27] and the efficiency of chlorogenic acid on anti-inflammatory and wound healing [28,29], it is suggested that 16 flavonoid aglucons, five flavonoid glycosides, and five chlorogenic acid analogs may possess some joint synergetic effects in the wound curing process. This also provides a novel insight into the therapeutic effects of total flavonoids.

Wound healing was a coordinated effort of several growth factors, cytokines, and chemokines. The TGF-β family and VEGF family played significant roles in this process, and proper wound healing was guided by stringent regulation of these agents as well as a wound environment that favors their activity [26]. Reports show that vaccarin, a flavonoid glycoside, can induce neovascularization and accelerate wound healing by promoting the expression of CD31 levels and enhanced protein expression of p-Akt and p-Erk [27], which indicates that the flavonoid glycosides (accounting for a large proportion of chromatogram in this study) may be the main active ingredients for wound healing; together with other flavonoids and chlorogenic acids, they may contribute to joint therapeutic effects by prompting growth factors, accelerating neovascularization, and inducing tissue formation in different stages. Our data demonstrated that total flavonoids of *B. balsamifera* can promote the expression of TGF-β_1 compared to control groups, while TGF-β_1 plays a key role in the recruitment of additional inflammatory cells, enhancing tissue debridement of macrophages, and promoting collagen and granulation tissue formation [30–32]. Increased expression of TGF-β_1 in *B. balsamifera* groups may be due to stimulation of the macrophages induced by total flavonoids in wounds, because earlier studies suggested that macrophages can enhance the contents of TGF-β_1 [33]. CD68 antigen, as a pan-macrophage marker, can be used as an essential indicator to evaluate quantities of macrophages, and macrophages contribute greatly to inflammatory reaction of wound. Our data showed an increased expression of CD68, which suggested an increasing number of macrophages and was in accordance with the increased expression of TGF-β_1.

The granulation tissue of wounds is primarily composed of fibroblasts, collagen, and new small blood vessels. Hydroxyproline, the major component of collagen, has been used as a biochemical marker for tissue collagen [34]. Our results showed that hydroxyproline content was significantly different to control. TGF-β_1 is known for promoting fibroblast proliferation [35,36], therefore, it is likely that high expression of TGF-β_1 resulted in an increase in hydroxyproline. VEGF can promote angiogenesis and increase vascular permeability. It gathers macrophages, fibroblasts, and some other cells around the wound site [37,38]. VEGF expression increased in total flavonoids groups compared to controls, which indicated that the total flavonoids could significantly promote wound healing not only due to increased collagen synthesis, but also due to its inducing expression of cytokines. Flavonoids in *B. balsamifera* are well known for their astringent, free radical-scavenging activity and antimicrobial properties [21,39,40]. Our present investigations further enlarged total flavonoids' activities, and elucidated the pharmacological mechanisms in wound contraction and promotion of epithelialization.

Collectively, these results suggested that total flavonoids of *B. balsamifera* possess significant therapeutic effects on skin injuries. A phytochemical investigation uncovered the composition of total flavonoids, including twenty-one flavonoids, five chlorogenic acids, and one coumarin. We postulated their joint synergistic therapeutic mechanism as a whole agent based on measurement and analysis of several representative biomarkers, accompanied with related literature studies, which will be helpful for understanding action mechanisms of TCM and selecting candidates for clinical therapy for skin injuries in the future.

4. Materials and Methods

4.1. Plant Collection and Total Flavonoids Preparation

Leaves of *Blumea balsamifera* (L.) DC. were collected from Danzhou, Hainan, China and authenticated by Dr. Yuxin Pang, professor of the Tropical Crops Genetic Resources Institute, Chinese Academy of Tropical Agricultural Sciences. Jing Wan Hong cream was produced by Tianjin Darentang Jingwanhong Pharmaceutical Co., Ltd. (Tianjin, China). The air dried *Blumea balsamifera* (L.) DC. leaves (400 g) were extracted with 95% methanol reflux (2X3L, each for two hours). The organic solvent was evaporated in vacuo to afford a dark residue, the residue was then suspended in water and fractionated with petroleum ether three times, the water part was evaporated until there were no organic solvent odors, and then subjected to polyamide macroporous resin column chromatography, eluting with a MeOH/H$_2$O gradient (0/100, 80/20, 100/0) to afford three fractions (F1–F3). The F2, eluted with 80% methanol, was evaluated for total flavonoids content of 81.1% with the method of UV-Vis spectrophotometry [41,42].

4.2. Establishment of Rutin Standard Curve

The total flavonoids sample was prepared (100 mg) by dissolving with 2 mL 75% methanol and 100 μL was transferred from pipette to 25 mL volumetric flask, to which a series of solvents, including 75% methanol 10 mL, 5% NaNO$_2$ 1 mL, 10% Al(NO$_3$)$_3$ 1 mL, 4% NaOH 10 mL, and then 75% methanol, were added into the measuring flask accurately in turns.

The standard rutin (18 mg) was dissolved with 75% methanol and transferred accurately to a 10-mL measuring flask, with different volumes of rutin solution: 0.25 mL, 0.5 mL, 1 mL, 2 mL, 2.5 mL, and 3.35 mL were transferred to 10-mL measuring flasks respectively with solvents added the same as the total flavonoids to form different concentrations of standard rutin samples. All the different solutions of rutin had their absorbance recorded via UV-Vis spectrophotometer at the maximum absorbance wavelength of 500 nm (Figure S1), and a standard curve was formed with concentrations listed on the *X*-axis, and absorbance listed on the *Y*-axis (Figure S2).

4.3. Qualitative Characteristics of Chemical Constituents of Total Flavonoids Extract

Identification of chemical constituents in the total flavonoids extract was performed by UPLC-Q-TOF/MS/DAD analysis and the UPLC-MS spectra of samples were acquired in positive and negative modes. The optimized UPLC-MS condition is shown in the Supplementary Materials.

4.4. Animals

A total of 150 healthy Sprague-Dawley (SD) rats of specific pathogen-free (SPF) grade, weighing 200–240 g, were supplied by Changsha Tianqin Biotech Ltd. (Certificate of quality No. SCXK (xiang) 2014-0011, date: 4 September 2014 to 4 September 2019), Changsha, China. All rats were maintained under 24 °C, with 55–65% humidity and a 12-h light/dark cycle in the Laboratory of Tropical Medicinal Plants Resources, Tropical Crops Genetic Resources Institute, Chinese Academy of Tropical Agricultural Sciences, Danzhou, China before use. The handling and care of the rats abided by the National Institutes of Health (NIH) guidelines for animal research, and all experimental protocols were approved by the National Research Institute for Child Health and Development Animal Care and Use Committee (Permit Number: S24018). All animal experiments were performed according to these guidelines. Many efforts were made to minimize the suffering of the rats.

4.5. Animal Modeling and Drug Treatments

Full-thickness skin excision wounds of 1 cm diameter were created by removing the whole dorsal skin layer on both sides of the backbone of rats. All rats were randomly divided into five groups, with 30 rats per group, and treated with the random number table method: the model group treated with 30% glycerol solution, the positive control treated with JWH cream, and three total flavonoids treatments, including high dose (2.52 g·kg^{-1}), medium dose (1.26 g·kg^{-1}), and low dose (0.63 g·kg^{-1}). All the flavonoids extracts were dissolved in 30% glycerol. These treatments were sustained for 10 consecutive days.

4.6. Measurements of Wound Healing

To measure (horizontal) wound progression, the skin wound healing rates (WHRs) of each group were measured on day 2, 4, 6, 8, and 10. The wound was covered with transparent film and labeled along the wound edge. Then, the required area was excised and weighed.

$$WHR = [(W_O - W_u)/W_O] \times 100$$

W_O: on day 2, wound area weight; W_u: Unhealed wound area weight.

4.7. Immunohistochemistry

On day 1, 3, 5, 7, and 10, the full-thickness of wound skin and surrounding normal skin was removed from different treated rats. One part of the tissue samples was fixed in 4% paraformaldehyde and placed in paraffin blocks for sectioning and 4-μm sections were sliced to evaluate macrophage contents. The wound tissue sections were stained with anti-mouse CD68 antibody (Boster Biological Technology Co., Ltd., Wuhan, China), followed with biotinylated anti-rabbit IgG-HRP antibody (Boster Biological Technology Co., Ltd., Wuhan, China). Four random views of each slice were observed under a microscope (×40). Then, the integral optical density (IOD) of each view was assessed using the Image-Pro plus 6 software to determinate the CD68 content.

4.8. Clinical Chemistry

The frozen full-thickness samples were subsequently homogenized, centrifuged, and the supernatant was isolated for analysis of VEGF, TGF-β_1, and hydroxyproline levels using the VEGF ELISA kit (Nanjing Jiancheng Bioengineering Institute, Nanjing, China), the TGF-β_1 ELISA kit (Nanjing Jiancheng Bioengineering Institute, Nanjing, China), and the hydroxyproline acid hydrolysis kit (Suzhou Comin Biotechnology Co., Ltd., Suzhou, China), respectively.

4.9. Statistical Analysis

Results were expressed as means ± standard deviation (SD). Comparisons between the groups were performed using one-way ANOVA followed by least significant difference (LSD) post hoc test with SPSS 22.0; $p < 0.05$ was considered to be statistically significant.

5. Conclusions

The present work elucidated that the total flavonoids from *B. balsamifera* could promote wound healing on rats significantly. Twenty-seven compounds were identified from the twenty-nine peaks of total flavonoids of *Blumea balsamifera* (L.) DC., including twenty-one flavonoid analogs, five CQA derivatives and one coumarin. The mechanisms of therapeutic effects were attributed to wound contraction, capillary regeneration, collagen deposition, and re-epithelialization. On day 10, the healing rate of the high dose group was a bit better than the JWH group, and both of them reached nearly 95%. The CD68 levels of all total flavonoids groups reached a peak at day 5 after treatment, suggesting that macrophages were active in the inflammatory stage. This study suggested that total flavonoids of *Blumea balsamifera* (L.) DC. represented an appropriate candidate for skin injuries, and this study also indicated that the flavonoid analogs and CQA derivatives may exert joint synergistic therapeutic effects on skin wound healing rates. Nowadays, with increasing global demands for medicines of botanical origin medicine, this work has opened a window for further exploration of ethnomedicine.

Supplementary Materials: Supplementary materials can be found at www.mdpi.com/1422-0067/18/12/2766/s1.

Acknowledgments: This work has been financially supported by the National Natural Science Foundation of China (81374065) and Basic Special Fund of Science & Technology (2015FY111500-060).

Author Contributions: Yuxin Pang: literature search, data collection, and interpretation; Yan Zhang: literature search, design of the work, figures, data analysis, and writing; Luqi Huang: editing and final approval of the version to be published; Luofeng Xu: literature search, study design, and data collection; Kai Wang: data collection and analysis; Dan Wang: data collection; Lingliang Guan: study design; Yingbo Zhang: study design; Fulai Yu: literature search, study design; Zhenxia Chen: data collection and analysis; Xiaoli Xie: data interpretation and revision of the draft.

Conflicts of Interest: The authors declare no conflict of interest.

References

1. Li, Y.; Jalili, R.B.; Ghahary, A. Accelerating skin wound healing by M-CSF through generating SSEA-1 and -3 stem cells in the injured sites. *Sci. Rep.* **2016**, *6*, 28979. [CrossRef] [PubMed]
2. Wahl, S.M.; Sporn, M.B. Transforming growth factor type beta induces monocyte chemotaxis and growth factor production. *Proc. Natl. Acad. Sci. USA* **1987**, *84*, 5788–5792. [CrossRef] [PubMed]
3. Russell, P.K.; Zhang, H.P.; Breit, S.N. *TGF-β and Related Cytokines in Inflammation*; Birkhäuser: Basel, Switzerland, 2001.
4. Esser, S.; Wolburg, K.; Wolburg, H.; Breier, G.; Kurzchalia, T.; Risau, W. Vascular endothelial growth factor induces endothelial fenestrations in vitro. *Brain Inj.* **2014**, *140*, 947–959. [CrossRef]
5. Ferrara, N. Role of vascular endothelial growth factor in regulation of physiological angiogenesis. *Am. J. Physiol. Cell Physiol.* **2001**, *280*, C1358–C1366. [PubMed]
6. Ashcrof, G.S.; Mills, S.J.; Ashworth, J.J. Ageing and wound healing. *Biogerontology* **2002**, *3*, 337–345. [CrossRef]
7. Leibovich, S.J.; Ross, R. The role of the macrophage in wound repair. A study with hydrocortisone and antimacrophage serum. *Am. J. Pathol.* **1975**, *78*, 71–100. [PubMed]
8. Mantovani, A.; Sica, A.; Sozzani, S.; Allavena, P.; Vecchi, A.; Locati, M. The chemokine system in diverse forms of macrophage activation and polarization. *Trends Immunol.* **2004**, *25*, 677–686. [CrossRef] [PubMed]
9. Hameed, A.; Hruban, R.H.; Gage, W.; Pettis, G.; Fox, W.M. Immunohistochemical expression of CD68 antigen in human peripheral blood T cells. *Hum. Pathl.* **1994**, *25*, 872–876. [CrossRef]
10. Nayak, B.S.; Pereira, L.M.P. Catharanthus roseus flower extract has wound-healing activity in Sprague Dawley rats. *BMC Complement. Altern. Med.* **2006**, *6*, 41. [CrossRef] [PubMed]
11. Prockop, D.J.; Kivirikko, K.I. Collagens: Molecular biology, diseases, and potentials for therapy. *Biochemistry* **1995**, *64*, 403–434. [CrossRef] [PubMed]
12. Diegelmann, R.F.; Evans, M.C. Wound healing: An overview of acute, fibrotic and delayed healing. *Front. Biosci.* **2004**, *9*, 283–289. [CrossRef] [PubMed]
13. Nayak, B.S.; Sandiford, S.; Maxwell, A. Evaluation of the Wound-healing Activity of Ethanolic Extract of *Morinda citrifolia* L. Leaf. *Evid.-Based Complement. Altern. Med.* **2007**, *6*, 351–356. [CrossRef] [PubMed]
14. Yuan, Y.; Pang, Y.X.; Wang, W.Q.; Zhang, Y.B.; Yu, J.B. Investigation on the Plants Resources of *Blumea Balsamifera* (L.) DC. in China. *J. Trop. Org.* **2011**, *2*, 78–82.
15. Guan, L.L.; Pang, Y.X.; Wang, D.; Zhang, Y.B.; Wu, K.Y. Research progress on Chinese Minority Medicine of *Blumea balsamifera* (L.) DC. *J. Plant Genet. Resour.* **2012**, *13*, 695–698.
16. Nanjing University of Traditional Chinese Medicine. *Dictionary of Chinese Medicine*; Shanghai Science and Technology Press: Shanghai, China, 2006.
17. De Boer, H.J.; Cotingting, C. Medicinal plants for women's healthcare in southeast Asia: A meta-analysis of their traditional use, chemical constituents, and pharmacology. *J. Ethnopharmacol.* **2014**, *151*, 747–767. [CrossRef] [PubMed]
18. Li, J.; Zhao, G.Z.; Chen, H.H.; Wang, H.B.; Qin, S.; Zhu, W.Y.; Xu, L.H.; Jiang, C.L.; Li, W.J. Antitumour and antimicrobial activities of endophytic streptomycetes from pharmaceutical plants in rainforest. *Lett. Appl. Microbiol.* **2008**, *47*, 574–580. [CrossRef] [PubMed]
19. Fazilatun, N.; Nornisah, M.; Zhari, I. Superoxide radical scavenging properties of extracts and flavonoids isolated from the leaves of *Blumea balsamifera*. *Pharm. Biol.* **2008**, *42*, 404–408. [CrossRef]
20. Norikura, T.; Kojima-Yuasa, A.; Shimizu, M.; Huang, X.; Xu, S.; Kametani, S.; Rho, S.N.; Kennedy, D.O.; Matsui-Yuasa, I. Anticancer activities and mechanisms of *Blumea balsamifera* extract in hepatocellular carcinoma Cells. *Am. J. Chin. Med.* **2012**, *36*, 411–424. [CrossRef] [PubMed]
21. Hasegawa, H.; Yamada, Y.; Komiyama, K.; Hayashi, M.; Ishibashi, M.; Yoshida, T.; Sakai, T.; Koyano, T.; Kam, T.S.; Murata, K.; et al. Dihydroflavonol BB-1, an extract of natural plant *Blumea balsamifera*, abrogates TRAIL resistance in leukemia cells. *Blood* **2006**, *107*, 679–688. [CrossRef] [PubMed]
22. Pang, Y.; Wang, D.; Fan, Z.; Chen, X.; Yu, F.; Hu, X.; Wang, K.; Yuan, L. *Blumea balsamifera*—A Phytochemical and Pharmacological Review. *Molecules* **2014**, *19*, 9453–9477. [CrossRef] [PubMed]
23. Zhang, L.; Chen, J.; Su, W.S.; Huang, J.J. Influence of soy isoflavone tincture on wound healing of deep partial-thickness scald in mice. *Chin. J. Tissue Eng. Res.* **2013**, *17*, 264–269.
24. Liu, G.L.; Li, J.M.; Yao, Y.; Zhang, N.; Jiang, Y.; Zhang, M.L.; Niu, C.Y.; Yu, Y.J. Effect of corylin on human fibroblast cells in an in vitro model of wound healing. *Acta Chin. Med. Pharmacol.* **2016**, *44*, 37–40.

25. Qiu, Y.Y.; Cai, W.W.; Qiu, L.Y.; Wang, Q.Q.; Wei, Q.F. Preparation and study of vaccarin-loaded nanofibers used as wound healing material. *J. Biomed. Eng.* **2017**, *34*, 394–400.

26. Barrientos, S.; Stojadinovic, O.; Golinko, M.S.; Brem, H.; Tomic-Canic, M. Growth factors and cytokines in wound healing. *Wound Repair Regener.* **2008**, *16*, 585–601. [CrossRef] [PubMed]

27. Xie, F.; Feng, L.; Cai, W.; Qiu, Y.; Liu, Y.; Li, Y.; Du, B.; Qiu, L. Vaccarin promotes endothelial cell proliferation in association with neovascularization in vitro and in vivo. *Mol. Med. Rep.* **2015**, *12*, 1131–1136. [CrossRef] [PubMed]

28. Song, Y.L.; Wang, H.M.; Ni, F.Y.; Wang, X.J.; Zhao, Y.W.; Huang, W.Z.; Wang, Z.Z.; Xiao, W. Study on anti-inflammatory activities of phenolic acids from *Lonicerae japonicae* Flos. *Chin. Tradit. Herb. Drugs* **2015**, *46*, 490–495.

29. Zhang, J.; Huang, W.; Zhang, L. The Application of Chlorogenic Acid in Drug Preparation of Promoting Fibroblast Proliferation. Chinese Patent: CN104825436A, 12 August 2015.

30. Lim, J.S.; Yoo, G. Effects of adipose-derived stromal cells and of their extract on wound healing in a mouse model. *J. Korean Med. Sci.* **2010**, *25*, 746–751. [CrossRef] [PubMed]

31. Greenwel, P.; Inagaki, Y.; Hu, W.; Walsh, M.; Ramirez, F. Sp1 is required for the early response of α2 (I) collagen to transforming growth factor-β1. *J. Biol. Chem.* **1997**, *272*, 19738–19745. [CrossRef] [PubMed]

32. Mauviel, A.; Chung, K.Y.; Agarwal, A.; Tamai, K.; Uitto, J. Cell-specific Induction of Distinct Oncogenes of the Jun Family Is Responsible for Differential Regulation of Collagenase Gene Expression by Transforming Growth Factor-β in Fibroblasts and Keratinocytes. *J. Biol. Chem.* **1996**, *271*, 10917–10923. [CrossRef] [PubMed]

33. Li, X.F.; Wang, H.J.; Luo, H. Tenporal relation of transforming growth factor-β mRNA expression with injury in the healing process of mouse skin wounds. *J. Nanjing Univ.* **2006**, *26*, 60–61.

34. James, O.; Victoria, I.A. Excision and incision wound healing potential of *Saba florida* (Benth) leaf extract in rattus novergicus. *Int. J. Pharm. Biomed. Res.* **2010**, *1*, 101–107.

35. Sapudom, J.; Rubner, S.; Martin, S.; Thoenes, S.; Anderegg, U.; Pompe, T. The interplay of fibronectin functionalization and TGF-β1 presence on fibroblast proliferation, differentiation and migration in 3D matrices. *Biomater. Sci.* **2015**, *3*, 1291–1301. [CrossRef] [PubMed]

36. Meran, S.; Thomas, D.W.; Stephens, P.; Enoch, S.; Martin, J.; Steadman, R.; Phillips, A.O. Hyaluronan facilitates transforming growth factor-beta1-mediated fibroblast proliferation. *J. Biol. Chem.* **2008**, *283*, 6530–6545. [CrossRef] [PubMed]

37. Morbidelli, L.; Chang, C.H.; Douglas, J.G.; Granger, H.J.; Ledda, F.A.; Ziche, M.A. Nitric oxide mediates mitogenic effect of VEGF on coronary venular endothelium. *Am. J. Physiol.* **1996**, *270*, 411–415.

38. Corral, C.J.; Siddiqui, A.; Wu, L.; Farrell, C.L.; Lyons, D.; Mustoe, T.A. Vascular endothelial growth factor is more important than basic fibroblastic growth factor during ischemic wound healing. *Arch. Surg.* **1999**, *134*, 200–205. [CrossRef] [PubMed]

39. Nessa, F.; Ismail, Z.; Mohamed, N.; Haris, M.R. Free radical-scavenging activity of organic extracts and of pure flavonoids of *Blumea balsamifera* DC leaves. *Food Chem.* **2004**, *88*, 243–252. [CrossRef]

40. Nessa, F.; Ismail, Z.; Mohamed, N. Xanthine oxidases inhibitory activities of extracts and flavonoids of the leaves of *Blumea balsamifera*. *Pharm. Biol.* **2010**, *48*, 1405–1412. [CrossRef] [PubMed]

41. Jingjing, W.; Ming, J.; Zhengjun, C.; Hui, C. Technical study on separation and purification of total flavonoid in *Gentianopsis paludosa* (Mum.) Ma by polyamide resin. *J. Gansu Coll. Tradit. Chin. Med.* **2013**, *30*, 35–38.

42. Cai, W.R.; Gu, X.H.; Tang, J. Extraction, Purification, and Characterisationof the Flavonoids from Opuntia milpa alta skin. *Czech J. Food Sci.* **2010**, *28*, 108–116.

International Journal of
Molecular Sciences

MDPI

Review

Tissue Engineering to Improve Immature Testicular Tissue and Cell Transplantation Outcomes: One Step Closer to Fertility Restoration for Prepubertal Boys Exposed to Gonadotoxic Treatments

Federico Del Vento [1], Maxime Vermeulen [1], Francesca de Michele [1,2], Maria Grazia Giudice [1,2], Jonathan Poels [2], Anne des Rieux [3] and Christine Wyns [1,2,*

[1] Gynecology-Andrology Unit, Medical School, Institut de Recherche Expérimentale et Clinique, Université Catholique de Louvain, 1200 Brussels, Belgium; federico.delvento@uclouvain.be (F.D.V.); vermeulen.maxime@live.be (M.V.); francesca.demichele@uclouvain.be (F.d.M.); giudicemariagrazia@gmail.com (M.G.G.)

[2] Department of Gynecology-Andrology, Cliniques Universitaires Saint-Luc, 1200 Brussels, Belgium; jonathan.poels@uclouvain.be

[3] Advanced Drug Delivery and Biomaterials Unit, Louvain Drug Research Institute, Université Catholique de Louvain, 1200 Brussels, Belgium; anne.desrieux@uclouvain.be

*   Correspondence: christine.wyns@uclouvain.be; Tel.: +32-2-764-95-01

Received: 30 December 2017; Accepted: 16 January 2018; Published: 18 January 2018

Abstract: Despite their important contribution to the cure of both oncological and benign diseases, gonadotoxic therapies present the risk of a severe impairment of fertility. Sperm cryopreservation is not an option to preserve prepubertal boys' reproductive potential, as their seminiferous tubules only contain spermatogonial stem cells (as diploid precursors of spermatozoa). Cryobanking of human immature testicular tissue (ITT) prior to gonadotoxic therapies is an accepted practice. Evaluation of cryopreserved ITT using xenotransplantation in nude mice showed the survival of a limited proportion of spermatogonia and their ability to proliferate and initiate differentiation. However, complete spermatogenesis could not be achieved in the mouse model. Loss of germ cells after ITT grafting points to the need to optimize the transplantation technique. Tissue engineering, a new branch of science that aims at improving cellular environment using scaffolds and molecules administration, might be an approach for further progress. In this review, after summarizing the lessons learned from human prepubertal testicular germ cells or tissue xenotransplantation experiments, we will focus on the benefits that might be gathered using bioengineering techniques to enhance transplantation outcomes by optimizing early tissue graft revascularization, protecting cells from toxic insults linked to ischemic injury and exploring strategies to promote cellular differentiation.

Keywords: prepubertal; male fertility; fertility preservation; fertility after cancer; spermatogenesis; testicular tissue; spermatogonial stem cells; transplantation; tissue engineering; nanoparticles

1. Introduction

While oncological treatments can cure more than 80% of pediatric cancers in Europe [1], chemotherapeutic agents and radiotherapy have deleterious effects on the gonads of prepubertal boys [2,3]. Moreover, the risk of permanent infertility is also high when preconditioning therapies are applied before bone marrow transplantation to cure benign conditions such as hemoglobinopathies [4].

Loss of fertility significantly compromises quality of life [5], and surveys performed in cancer survivors have shown that 80% of these patients are considering parenthood, although only 10% of them would use donor sperm or choose adoption [6,7]. Between 2005 and 2013 in our institution, the acceptance rates of the fertility preservation procedure justified by an oncological diagnosis were 74% for boys under 12 years and 78.6% for boys aged 12 to 18 years [7]. Similar rates were reported in USA centers [8] and are expected to be alike in Europe based on a questionnaire offered to child cancer survivors' parents [9]. This illustrates the significant importance of fertility for minor patients and their parents and fully justifies the development of fertility preservation strategies in pediatric populations.

If, for post-pubertal patients, sperm cryopreservation is the gold standard for fertility preservation, for young patients who do not yet produce mature gametes, cryopreservation of immature testicular tissue (ITT) (containing spermatogonial stem cells (SSCs), which are spermatozoa precursors [10]) or germ cells suspension can be offered before gonadotoxic therapies to preserve their fertility [3].

A controlled slow freezing procedure using dimethyl sulfoxide as the main cryoprotectant is commonly applied to cryopreserve small ITT fragments (2–4 mm^3) taken from one testis, with sampling limited to less than 5% of the testicular volume [11–14]. Using a xenotransplantation assay in nude mice to evaluate the cryopreservation procedure, the integrity of the SSC niche that regulates stem cell self-renewal and differentiation [15] appeared to be well preserved [16].

Restoring patient's compromised fertility by obtaining mature gametes production from stored cells or tissue might be achieved through auto-transplantation or in vitro maturation once the cytotoxic treatment is completed (Figure 1).

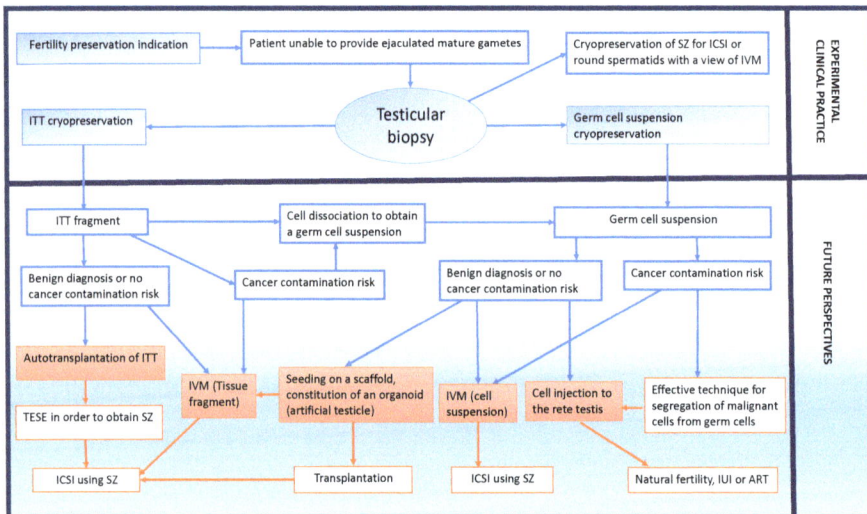

Figure 1. Fertility preservation for peri- and pre-pubertal male patients: experimental clinical practice and future perspectives. SZ: Spermatozoa; ICSI: Intra-Cytoplasmic Sperm Injection; ART: Assisted Reproductive Technology; IVM: In Vitro Maturation; TESE: TEsticular Sperm Extraction; IUI: Intra-Uterine Insemination.

The choice between different options must be made considering the disease of the patient. Since intratesticular contamination by malignant cells was described in approximately 21% of boys with leukemia [17] and xenotransplantation in nude mice of rat testicular cells contaminated by leukemia cells brought tumor transmission in the host [18], auto-transplantation of frozen-thawed gonadal tissue fragments may only be considered for fertility restoration purposes when there is no risk of retransmission of cancer cells to the cured patient. Such risk must also be evaluated for other blood malignancies and for metastasizing cancers. Selection of SSCs before transplantation may therefore be an alternative for these patients if proof can be made that the cell sorting technique using post-purification examination of the prepared samples is safe [19].

Otherwise, in vitro maturation (IVM) of human prepubertal spermatogonia [20], which is currently under investigation, will be the only option; however, it will not be addressed in this review.

2. Results and Discussion

2.1. Lessons Learned from Transplantation of Immature Testicular Tissue Fragments

While fertility restoration with transplantation of cryopreserved human ITT is still at a preclinical stage, autotransplantation of frozen-thawed ovarian cortex [21] in the context of fertility preservation has already proven clinically efficient [22]. Over 100 live births have been reported worldwide whether using adult [23] or prepubertal tissue [24]. However, further progress is still awaited considering the important loss of ovarian follicles following transplantation assays in mice due to hypoxia/reoxygenation injuries [25] and to a massive primordial follicle activation [26]. Different strategies, based on tissue engineering approaches, have been tested to limit these phenomena, like follicle encapsulation in fibrin hydrogel and vascular growth factor administration [23].

With regard to transplantation of cryopreserved testicular tissue, grafting of human ITT has been so far exclusively performed in host nude mice [12,16,27–31]. A summary of human prepubertal and postpubertal testicular tissue grafting experiments is provided in Table 1.

Poor graft survival [32] and massive germ cells loss [27] were observed especially after mature testicular tissue ectopic xenotransplantation. After orthotopic ITT xenotransplantation using the same nude mouse model, there seemed to be a worsening of spermatogonial survival rates over time (61% at 5 days, 14.5% at 3 weeks, and 3.7% at 6 months [12,16,31]). Such reduction in germ cell numbers might likely be the result of tissue degeneration caused by the hypoxia that occurs before the generation of an early blood supply, as was observed in ovarian transplantation experiments [25]. Moreover, further germ cell loss continues with the highest proportion of the loss happening within the first three weeks, suggesting the importance of a stable vascularization for an efficient grafting technique [33]. Interestingly, as spermatogonial recovery rates decrease, seminiferous tubules' structures partially recover over time (18–21% at 5 days, 82.2 $\pm$ 16.5% at three weeks, 89.7 $\pm$ 17.9% at 6 months), although presenting more niches that do not contain spermatogonia [12,16,30,31]. As qualitative analysis of Leydig cells showed a good preservation at the protein and ultrastructural levels with maintenance of their steroidogenic potential [16], and as Sertoli cells proliferation was found to be superior to spermatogonial proliferative activity three weeks after xenografting [12,31], it seems that the germ cell population is mainly affected. It has thus been postulated that the transplantation procedure and the recipient environment could be responsible for an increased recruitment of SSCs entering the differentiation process, thus reducing the number of SSCs capable of auto-renewal [31].

Table 1. Reports on xenotransplantation of post- and pre-pubertal human testicular tissue in nude mice.

Reference	Donor	Grafting Site	Graft Size	Cryopreserved or Fresh	Castrated Host	Outcome
Wyns et al., 2007; [12]	Prepubertal	Peritoneum scrotal bursa	2–9 mm^3	F/T	yes	SG survival after 3 weeks: 14 %
Wyns et al., 2008; [16]	Prepubertal	Peritoneum scrotal bursa	2–8 mm^3	F/T	yes	SG survival 3.7%, numerous premeiotic spermatocytes, a few spermatocytes at the pachytene stage and spermatid and spermatozoa-like cells, without expression of the meiotic and post-meiotic markers
Goossens et al., 2008; [27]	Prepubertal 10 and 11 y.o.	Back skin	2–10 mm^3	fresh	no	Some rare SG survival after 4–9 months
Sato et al., 2010; [28]	3 y.o. testicular hemangioma	Back skin	0.5–1 mm^3	fresh	yes	Pachytene spermatocytes after one year
Van Saen et al., 2013; [29]	Prepubertal	Intra-testicular	1.5–3 mm^3	Fresh and F/T	No	No effect of FSH administration or slow-freezing primary pachytene spermatocytes 9 and 12 months after grafting
Poels et al., 2013; [30]	Prepubertal	Peritoneum scrotal bursea	1 mm^3	Fresh, slow-frozen and vitrified	yes	SG survival after 6 months: 3.4%, 4.1%, and 7.3%, respectively, for fresh, slow-frozen-thawed and vitrified-warmed tissue. No statistical significant difference between three groups.
Poels et al 2014; [31]	Prepubertal	Peritoneum scrotal bursa	1 mm^3	F/T	yes	SG survival after 5 days, 67%, 63%, and 53%, respectively, for slow-frozen tissue, slow-frozen tissue supplemented with NAC, and slow-frozen tissue supplemented with FSH and testosterone. No impact of NAC or FSH/Testosterone supplementation on SG survival
Schlatt et al., 2006; [32]	Adult4 Azoospermic patients, 1 Cancer survivor, 1 Testicular cancer patient, 3 transsexual patients.	Back skin	0.5–1 mm^3	fresh	yes	Occasional Type A SG survival No correction of spermatogenesis disruption after xenografting
Geens et al., 2006; [34]	Adult	Back skin	4 mm^3	fresh	yes	SG survival 33% < 120 d 14% > 120 d
Van Saen et al., 2011; [35]	Prepubertal POST-CHEMOTHERAPY Postpubertal (12 and 13 y.o.)	Intra-testicular	6 mm^3	F/T	no	4 months: SG survival 0.2/ST. 9 months: SG survival 0.6/ST. Higher SG survival in xenografts from the postpubertal donors. No SG differentiation in younger patients' tissue, two older donors' tissue with differentiation up to primary spermatocyte and secondary spermatocytes in the oldest donor after 9 months.

SG: = spermatogonia; y.o.: years old; ST: seminiferous tubule; F/T: Frozen/Thawed; hCG: human Chorionic Gonadotropin; FSH: Follicle Stimulating Hormone; NAC: N-Acetyl-Cysteine.

Furthermore, while the survival of proliferating spermatogonia was observed, their differentiation potential was limited to the primary pachytene spermatocyte stage, although spermatid-like and spermatozoon-like structures were described using standard and transmission electron microscopy [16]. The phylogenetic distance between species was suggested to be responsible for the abnormal differentiation, but this should be further confirmed after the development of a validated model able to assess tissue and cell functionality including its maturation (see Section 2.11).

Moreover, the pachytene spermatocyte stage was reached within 6 months after orthotopic xenotransplantation in castrated nude mice of tissue obtained from boys under the age of 10 years [16,30]. However, the endocrine exposure needed for the physiological pubertal transition period and onset of spermatogenesis in boys may last between two and four years [36], which may also be a factor that could influence the maturation stage achieved after 6 months xenotransplantation.

By contrast, in other species, complete spermatogenesis was reported after both autologous transplantation [37–39] and xenotransplantation [37,39–43] and even led to offspring in some species (including non-human primates) after prepubertal testicular tissue autotransplantation [39,44] and xenotransplantation [39,45–47]; for a complete review, see [48].

Besides the phylogenetic distance between species and the length of the pubertal transition period, a number of other parameters such as the avascular grafting procedure, the grafting site [12], the maturational stage of donor's tissue [16,49], and other donors' characteristics [50] may have an impact on graft development and functionality.

2.1.1. Transplantation Technique

In 2002, Honaramooz et al. grafted on the back skin of nude mice fresh small ITT biopsies retrieved form prepubertal mice, pigs, and goats obtaining a complete spermatogenesis with all three species' tissue. According to the report, ITT fragments were simply fixated subcutaneously without the constitution of an artificial vascular anastomosis [37]. A vascular connection between the grafted tissue and the host was then spontaneously established, with vessels outgrowing from the transplanted fragment and then connecting to the host's vascular system [51]. As the cell-to-capillary distance is limited to 200 µm, survival of any tissue exceeding this dimension requires a mature and stable vascularization [52]. Therefore, even very small testicular tissue biopsies ($\pm$1 mm^3) are exposed to a period without blood supply characterized by hypoxia, followed by a phenomenon of reoxygenation as already demonstrated for ovarian tissue grafting [25].

2.1.2. Transplantation Site

Transplantation of rodent testicular tissue has been performed in several sites, such as the anterior chamber of the eye [53], the back skin [37], the testis [39], and the ear tip [54], with results in terms of germ cells differentiation that suggested a negative impact of higher temperatures on grafts, hence pointing out orthotopic grafting as the best option [12].

Marmoset ITT autologous ectopic transplantation encountered a differentiation blockade at the spermatocyte stage [55], although interestingly, complete spermatogenesis was detected after autologous orthotopic transplantation in prepubertal castrated host [56].

Intra-testicular grafting has also been performed, and according to the authors of these reports, removing testicular parenchyma could create rupture of seminiferous tubules continuity, facilitating recolonization by SSCs from the donor. Therefore, intra-testicular graft (though technically more complicated) has been proposed as a useful alternative option for intra-scrotal ITT grafting [57].

2.1.3. Maturational Stage of Donor Tissue

The possible impact of the age of the donor on testicular tissue graft outcome has been postulated in many reports on animal experiments [49,58,59]. In agreement with these observations, a higher degree of maturity of human prepubertal tissue also appeared to play a role, as reported by Wyns et al. in 2008. The authors observed a very low recovery rate in the two 14-year-old patients, where focal spermatogenesis was already present in fresh tissue, compared to the better results obtained with younger (12 and 7 years old) donor tissue after xenografting [16].

2.1.4. The Host Environment

In addition to all these non-modifiable characteristics, the nude mouse model itself might be responsible for the germ cell maturation arrest encountered after ITT xenotransplantation. Experiments with non-human primate tissue provided useful and challenging information. Rhesus monkeys' ITT led to complete spermatogenesis after autologous transplantation [38] but not after xenotransplantation in nude mice [60], proving host environment to be critical for spermatogenic differentiation.

These results suggest that differences in gonadotrophins structure and the endocrine environment could be involved in graft's germ cell differentiation.

Several strategies have been proposed to reach further progress using available study models, such as acting on the endocrine environment or sheltering spermatogonia from hypoxic insults either by reducing the ischemic period before revascularization or by using protective molecules. These reports not only pioneered in ways to improve the transplantation techniques, but they also provided insight and information on the physiology of (xeno)-transplanted testicular tissue.

2.2. Acting on the Endocrine Environment

The hypothalamic–pituitary–testis axis manages the induction and maintenance of spermatogenesis [36], as well as the endocrine function of the testicle.

Spermatogonia lie in the seminiferous tubule in strict contact with surrounding cells, in a functional unit called the stem cell niche, and it is the relationship between soluble molecules, cells cohesion within the niche, endocrine feedback, and paracrine environment that modulates spermatogenesis and spermatogonial self-renewal (for review see Potter et al., 2017) [61]. Better results for mouse ITT intra-testicular grafting compared to SSCs rete testis injection were ascribed to the preservation of the spermatogonial paracrine microenvironment, which allows spermatogenesis to be supported by the original Sertoli cells [12].

During xenografting experiments castrated hosts have frequently been used. After bilateral orchiectomy, the absence of testosterone production causes an interruption of the relationship between the testicle and the pituitary gland, leading to increased gonadotrophin levels above those present at physiological pubertal onset [62]. This affects testicular maturation and might contribute to the acceleration of the differentiation process of the immature tissue previously reported in non-human primates and human tissue [28,63].

The restoration of a functional endocrine feedback between the grafted tissue and the host hypothalamic-pituitary-gonadal axis has been proven. Indeed, 4 weeks after allotransplantation of prepubertal mouse tissue in castrated mice, a decrease of FSH levels to precastration values was observed [62], and grafted ITT supplied normal to elevated levels of androgens to the castrated donors [37]. This was further corroborated by increased weights of seminal vesicles in grafted hosts [37]. However, mouse hormonal hyper-gonadotropic environment after bilateral orchiectomy has been exonerated from the responsibility of the spermatogenic blockade, as similar results were obtained after human ITT xenografting in non-castrated mice (when intra-testicular transplantation was performed) [29].

Questions have been raised about the actual benefit that external intervention on the endocrine environment might add to the grafting procedure. Indeed, hamster ITT xenografts were proven to be mainly regulated by intrinsic mechanisms, as they grew to an analogous size when grafted to nude mice with different degrees of castration (hemi or complete) [64].

In addition, co-grafting in nude mice of ITT from two species, Hamster and Marmoset, where opposite results after xenografting were obtained (the prior going through complete spermatogenesis and the latter who did not and did not even achieve androgens release) [49] showed that the increase in testosterone supplied by the Hamster tissue was not able to overcome the limitations of xenografting of

Marmoset ITT [65]. Furthermore, exogenous hCG also remained inefficacious to achieve spermatogenic differentiation [65].

On the contrary, administration of exogenous gonadotropins to recipient castrated nude mice led to complete spermatogenesis using infant rhesus monkeys ITT xenografts [66,67], and it improved tissue maturation and differentiation after xenografting of ITT from prebubertal horses [68].

Thyroid hormones play a well-known role in spermatogenesis and hemato-testicular-barrier function [69], and interaction between grafted testicular tissue and the recipients' thyroid was established. Indeed, administration of propylthiouracile, a disruptor of thyroid axis function, reduced graft efficiency and germ cells differentiation in bovine ITT xenografts in nude mice [70].

As far as it concerns human ITT, two reports attempted to artificially modify the recipient mouse endocrine environment to assess its influence on the xenografts. In a first experiment, subcutaneous administration of exogenous recombinant human follicle stimulating hormone (FSH) was performed after human ITT intratesticular xenografting in nude mice [29], and in a second experiment, ITT xenografting in the scrotal bursa of nude mice was followed by intraperitoneal testosterone supplementation with a fixed dose of intramuscular human FSH in order to counterbalance the negative feedback on gonadotropins secretion [31]. None of the graft hormone supplementations promoted spermatogonial differentiation nor restored spermatogenesis, although combined administration of FSH and testosterone improved intra-tubular cellular proliferation of both spermatogonia and Sertoli cells.

The way hormones were administered might have been unsuitable to elicit any desired effect but other molecular mechanisms might be implicated, such as a disrupted cellular response caused by the grafting procedure, or an interaction with the host's own gonadotropins or with its thyroid-gonadal axis [70].

2.3. Reducing Ischemia Due to the Avascular Transplantation Procedure

As germ cells, and more specifically differentiating spermatogonia, are highly vulnerable to the hypoxia insult [71], reducing the ischemic period through enhanced vascularization using vascular growth factors might have beneficial effects on SSC survival.

Vessels formation occurs by a complex dynamic process governed by several pro and anti-angiogenic molecules [72], and several growth factors such as vascular endothelial growth factor (VEGF), platelet derived growth factor (PDGF), and fibroblast derived growth factor (FGF) have been tested for distinct therapeutic applications including tissue regeneration and neovascularization of ischemic tissues, although clinical trials results were mainly disappointing [73,74].

Beyond its role in angiogenesis, VEGF is also involved in the support of germ cell survival and self-renewal [75] through receptors whose expression has been demonstrated on spermatogonia, and Sertoli and Leydig cells [76]. This makes VEGF an excellent candidate for supplementation during ITT grafting for both its role as a vascularization enhancer and spermatogenesis regulator [77]. VEGF administration, either with subcutaneous injection directly to bovine testicular tissue grafting site [78] or during in vitro culture prior to germ cells transplantation [79], increased the number of seminiferous tubules containing elongating spermatids and germ cells survival, respectively. Moreover, in vitro culture of mouse SSCs with VEGF improved seminiferous tubules and vascularization reconstitution after ectopic allotransplantation of SSCs on the back skin of nude mice [77].

These encouraging results endorsed the use of nanoparticles containing VEGF as described below (see Section 2.10).

2.4. Using Protective Molecules to Reduce Ischemic Injury

Another strategy to limit damages to the testicular cells consists in actively protecting them from external insults during experimental manipulation. Hypoxic injury is associated with generation of reactive oxygen species, and different treatments have been tested trying to reduce oxidative stress.

N-acetylcysteine acts as an antioxidant regenerating the pool of intracellular glutathione, it protects the cell from oxidative stress caused by reactive oxygen species and reduces cell membrane lipid peroxidation [80]. Systemic administration of *N*-acetylcysteine has proven useful for germ cells protection during testicular torsion [81,82], and media supplementation with acetylcysteine during in vitro culture reduced germ cells apoptosis [83]. However, immersion for 5 minutes of a 1 mm^3 fragment of frozen-thawed human ITT in a solution containing *N*-acetylcysteine before xenografting in nude mice associated with intraperitoneal administration of the same drug during the 5 days had no impact on spermatogonial survival, nor on seminiferous tubules cellular proliferation or apoptosis [31].

The impact of melatonin supplementation on the improvement of murine spermatogonial survival resulted in different outcomes. This drug has antioxidant and antiapoptotic properties that rely on the radical scavenger effect delivered by its indole derivate [84]. While melatonin appeared to be beneficial in SSC transplantation experiments [85,86], addition of melatonin to the vitrification-warming medium did not reduce expression of apoptotic genes in frozen/thawed murine ITT [87].

Further studies are therefore needed to investigate the impact of antioxidants molecules delivery on ITT, focusing on new molecules or on the way these drugs are administered.

Sustained and localized delivery systems might be useful to further improve the effects of already tested molecules (see Section 2.9).

2.5. Challenges to Achieve a Successful Transplantation of Human Immature Testicular Tissue

Because of an existing risk of zoonosis and epigenetic modification of genetic heritage, under no circumstances is xenotransplantation supposed to be considered as an option to produce sperm for future clinical use [88].

With regards to autotransplantation of cryoconserved ITT and before translation to clinical practice, the grafting technique needs to be improved with the objectives of increasing spermatogonial survival and achieving germ cells differentiation up to the haploid stage.

The main challenge lies in the absence of a useful model to study tissue transplantation, since the host environment of the mouse precludes proper interpretation of the outcomes.

However, efforts to mimic the physiologic endocrine environment of the peripubertal transition period and development of methods to accelerate angiogenesis and stabilization of the vascular connection between host and graft will probably represent the main challenges to explore.

The use of vascular growth factors in vivo is confronted with major hurdles, like their short half-life, the effects of distant distribution such as increased vascular permeability with risk of edema and hypotension [89,90], and the enhancement of tumor neo-angiogenesis [91,92]. These issues could be overcome by controlled release of biologically active substances confined to the site of interest [93] using encapsulation matrices and tailored nanoparticles. The latter approach also offers further options for testing local growth factors that act on the regulation of the transplanted spermatogonial niche.

2.6. Lessons Learned from Transplantation of Spermatogonial Stem Cells (SSCs)

An alternative option for fertility preservation and restoration is to produce and transplant functional germ stem cells into the patient once his cancer has been cured or gonadotoxic treatment has been completed.

Suspensions of isolated SSCs have been obtained from both animal [94] and human [95] tissue, offering the possibility of germ cell cryopreservation [96], as well as the perspective of fertility restoration through cell propagation and transplantation. According to the technique described by Brinster and Zimmerman, mouse ITT samples were incubated with collagenase and trypsin to digest the extracellular matrix, and the cell suspension subsequently isolated (at a cellular concentration ranging from 10^6 to 10^7 SSCs/mL) was micro-injected directly in the recipient mice seminiferous tubules [94]. The injection site has been the object of further investigations that pointed to ultrasound-guided injection in the rete testis as the best technique for SSCs transplantation to testicles of a size bigger than rodents', like humans' [97,98].

Mouse SSCs proved to be able to colonize the testicle stem cell niche after infusion into the seminiferous tubules of allogenic mice sterilized by means of busulfan exposure, further allowing spermatogenesis and progeny conception through natural mating [94]. Interestingly, xenotransplantation in nude mice testicles of rat [99] and hamster [100] testicular suspensions produced mature spermatozoa, although with structural abnormalities [101]. On the contrary, human adult [102] and prepubertal [103,104] testicular suspensions could only colonize the seminiferous tubules.

The overall encouraging results of SSC transplantation in animals endorsed an experimental setting involving human adult testicular tissue. Testicular cell suspensions of patients affected by non-Hodgkin lymphoma were obtained before chemotherapy, cryopreserved and retransplanted after cancer treatment, but unfortunately no follow-up information on the outcomes of this clinical trial has been reported [105,106].

2.7. Challenges to Achieve a Successful Transplantation of Human Immature SSCs

Several issues such as the isolation of a sufficient SSC number in the suspension, the lack of a complete knowledge of gonadotoxic treatment effects on the endocrine and paracrine environment of the patient's SSC niche, and the possible cancer cell contamination of the cell suspension are still under investigation.

For the procedure to be efficient in colonizing the recipient seminiferous tubules, cell suspension should have a sufficient concentration of SSC. While SSCs represent only a small proportion of testicular cells [107], less than 10% of mice transplanted spermatogonial stem cells can form colonies in the recipient seminiferous tubules [108], and this percentage is presumed to also be low in human testis [109].

Using mouse testicular-derived cell suspension, increased cell concentrations from 10^6 up to 10^7 and 10^8 cells/mL led to an increase in the colonization efficiency after transplantation to sterilized mice [108], while studies using human adult testicular derived cell suspensions met slightly different results. Suspensions obtained from human adult testicular tissue with different cells concentrations were compared between each other in two experiments. Cells concentrations of 49.7×10^6 cells/mL and 51.5×10^6 cells/mL, when compared respectively to 23.7×10^6 cells/mL and 27.4×10^6 cells/mL, did not improve the outcome of SSCs xenotransplantation to nude mice sterilized by Busulfan exposure [102,110]. Interestingly, in the second experiment, increasing the cell concentration from 10.3×10^6 to 27.4×10^6 cells/mL improved colonization of seminiferous tubules [110]. Therefore, SSC expansion would probably be necessary for clinical application, as the cellular concentration of the transplanted suspension influences the colonization efficiency.

Indeed, in vitro culture of SSC retrieved from cell suspensions could be used to obtain a sufficient spermatogonial number for a clinical application of SSC transplantation [104].

Enrichment of cell suspensions in SSCs is also possible using several methods, such as cell sorting followed by in vitro expansion [111] or differential plating [112]. However, these techniques still need to be refined because of partial results in terms of cell purity [113] and viability.

Besides germ cells depletion, testicular somatic cells may also be affected by gonadotoxic therapies. Indeed, direct damage of Sertoli cells has been demonstrated [114], and post-chemotherapy increased LH levels with reduced testosterone blood concentrations observed after cytotoxic therapies have been ascribed to Leydig cells' impairments [115].

The possibility of cancer cells contamination of the cellular suspension would forbid any attempt of transplantation. To overcome this issue, both cell sorting and in vitro culture techniques have been developed. While encouraging results were obtained using cell sorting for animal testicular tissue [116], previous attempts using multi-parameter cell sorting strategies for human SSCs were only partially successful [117–119].

An overview of previous studies attempting to separate cancer cells from SSC suspension is provided in Table 2.

Table 2. Segregation of cancerous cells from human and animal testicular tissue.

Reference	Species	Technique	Outcome (Residual Contamination/ Contamination of Samples or Contamination of Mice after Transplantation)
Fujita et al., 2005; [116]	Mouse	FACS	No contamination of recipient mice
Fujita et al., 2006; [117]	Human	FACS	Malignant cells in 1/8 in vitro cultures
Geens et al., 2007; [118]	Mouse	MACS + FACS	Malignant cells in 1/32 in vitro cultures 43% of mice contaminated after transplantation
	Human	FACS	10/11 contaminated cultures
Dovey et al., 2013; [119]	Human	FACS	Post FACS Purity check was only 98.8–99.9% No tumour formation after xenotransplantation of sorted cell suspension to 55 nude mice (but tumour formation after contaminated cell transplantation was only 23–55%)
Hou et al., 2007; [120]	Rat	FACS	Germ cells selection or leukaemia cells isolation: contamination of 2/3 and 2/2 recipient rats Germ cell selection and leukaemia cells isolation: survival of all recipient rats
Hermann et al., 2011; [121]	Non-human primates	FACS	No tumour after nude mouse transplantation in 3 of 4 cell colonies
Sadri-Aderkani et al., 2014; [122]	Human	In vitro culture	Acute lymphoblastic leukaemia cells undetectable after 26 d

FACS: fluorescence activated cell sorting; MACS: magnetic activated cell sorting; d: days.

Hence these techniques need to be improved, although there exist some limiting factors. Indeed, on one hand, a potential similarity between antigens expressed on the membrane of human SSCs and leukemia cells might interfere with the sorting techniques, and on the other hand, no phenotypic marker for SSCs can differentiate them from other spermatogonial cell populations [3]. Laborious immune-phenotyping analysis of the malignant cells would therefore be necessary to define individual surface markers suitable for negative selection [117].

Another strategy proposed by Sadri-Ardekani et al. involved the use of a specific culture system that was able to successfully eliminate contaminating leukemic cells when co-cultured together with male germ cells. However, cells from different types of tumours could behave in different manners once exposed to specific culture conditions; hence, the same technique might not be applicable in every circumstance [122]. Furthermore, dissimilar cellular behaviors between the co-culture model and the scenario where a tissue invaded by malignant cells is dissociated might be expected. Another important concern is the potential epigenetic modification of cultured germ cells. While long term cell culture did not appear to affect the genetic heritage of mouse SSCs [123], human SSCs presented modifications in DNA-methylation after 50 days of in vitro culture [124]. Further studies are thus needed to explore the procedure and its consequences.

2.8. Lessons Learned from Reports on Cellular and Tissue Encapsulation and Perspectives Using Scaffolds

The limitations and flaws of ITT and SSC transplantation techniques that we itemized in the first part of this paper could be overcome by interventions on cellular support and on integration of donor grafts to the host. Such objectives have been explored in several domains besides the field of fertility preservation and belong to bioengineering.

Tissue engineering is an interdisciplinary field that combines principles from chemistry, materials engineering, and life sciences [125]. It is based on the association of a scaffold, cells, and bioactive molecules [126]. It aims to support cell viability and functionality, and at the same time it provides controlled and sustained delivery of single or multiple biological active molecules, such as growth factors and therapeutic drugs [127]. Bioactive cues can be either incorporated in the construct as free molecules, as nanomedicines, or both, depending on the desired release profiles and, ultimately, on the expected effects. Such composites can reestablish, maintain, or improve the condition of tissues or cells by reproducing the architecture and biochemical characteristics of the original organ or tissue [128]. Tissue engineering could lead to multiple strategies to improve ITT and SSC transplantation, to develop a proper study model or to elaborate a transplantable artificial testis.

2.8.1. Cells or Tissues Encapsulation

Cells and tissue encapsulation in a three-dimensional environment using synthetic or biologically-derived matrices mainly aims to provide an environment that mimics the extracellular matrix (ECM) [129,130]. ECM is the non-cellular component that participates in the constitution of tissues and organs. It provides essential physical support for cells [131], allows cells communication and migration [132], and facilitates diffusion of cell nutrients and released products [133], and is thus required for tissue homeostasis. Testicular ECM plays a pivotal role in spermatogenesis through proteins like laminin and collagens that rule cellular interactions and thus differentiation of germ cells [134]. Modifications of this peculiar structure have indeed been observed when the normal function of the testis is compromised as in several pathologies associated with infertility [135].

Choosing the most appropriate material as an encapsulation matrix is essential for the outcome of tissue engineering constructs, and the designated components must satisfy many requirements, i.e., biocompatibility, biodegradability, and mechanical properties similar to those of the native tissue [131]. To be biocompatible, biomaterials must coexist and interact with the biological environment of the recipient without eliciting an excessive immune reaction [136] and eventually allowing a successful host integration [137]. Scaffold degradation products should also be non-toxic and should be eliminated from the body without interference with other organs [126,138].

Characteristics like stiffness and elasticity are influenced by the procedures used for matrix production, like the concentration of a given biomaterial, the degree of humidity, and temperature during preparation [139], or by the conditions to which the scaffold is subjected once implanted. For instance, an increase of temperature and hydration can reduce compressive modulus and compressive strength of poly(lactic-co-glycolic acid) (PLGA) scaffolds [140]. Furthermore, the material is supposed to allow artificial modifications of its physico-chemical characteristics in order to enhance specific cell behaviors and favor easy surgical manipulation of the graft [141]. An overview of characteristics to consider when a material is chosen and processed to be used for encapsulation is provided in Table 3.

Table 3. Matrix mechanical characteristics that impact cell or tissue function.

Matrix Mechanical Characteristics that Impact Cell or Tissue Function
• Pore size and morphology (Chan et al., 2008); [142]
• Elasticity (Janson et al., 2015); [131]
• Stiffness (Xia et al., 2017); [143]
• Hydration degree (Wu et al., 2006); [140]

2.8.2. Use of Scaffolds

Tissue engineering provides an option for the reproduction of the structure of a tissue or organ. It consists in using a synthetic composite that could eventually be seeded and repopulated with isolated cells, providing an effective scaffold [48].

An alternative option to provide a scaffold for cellular support relies on the decellularization of tissues in order to obtain acellular matrices. It is a strategy that could be summarized in removing cells while preserving biological activity, biochemical composition, and three-dimensional structure of the ECM [144].

2.9. Bioactive Molecules Supplementation Using Nanoparticles

Entrapment of active molecules in nanoparticles is an approach used to circumvent unsuitable bioavailability, inadequate stability, and secondary effects at distant sites that may be observed with conventional systemic administration systems like tablets, capsules, and solutions. Nanomedicines incorporation in scaffolds would provide sustained drug delivery directly to the target site, and the molecule bio-distribution would no longer be related only to the drug itself, but also to carrier physicochemical properties [145].

Nanoparticles are solid colloidal particles with a diameter below 1 μm, in which the drug is either confined within a cavity enveloped by a membrane composed of a polymer, or it is dissolved within the polymer matrix [146].

The strategies that are employed to enhance the binding between a drug and the polymer composing the nanoparticle are both physical (charge interaction) and chemical (covalent binding) and depend on both the molecule to deliver and the carrier [147]. For instance, Huang et al. developed an encapsulation method for VEGF that relies on its ability to bind heparin. VEGF interacts with dextran sulfate via its heparin binding site and the polyelectrolyte complexes are stabilized by coacervation with chitosan [148].

Different molecules can be encapsulated or co-encapsulated in nanoparticles that can be combined in an implant, providing simultaneous [149] or sequential [150] drug administration according to the pharmacokinetic characteristics required for the specific therapeutic purpose.

Enhancing angiogenesis through vascular growth factor delivery is one the most frequent application of nanomedicine, and the possibility of obtaining a sequential delivery of multiple growth factors might help the reproduction of the process of angiogenesis.

Simultaneous release of VEGF and PDGF incorporated into fibrin scaffolds boosted angiogenesis in pancreatic islets grafts [151]. Nanoparticles of PLGA and poly(L-lactide) (PLLA) allowed simultaneous release of VEGF and FGF and sequential release of PDGF. This system has effectively enhanced vascularization in the rat aortic ring assay [150].

Drug release rates can be orchestrated, as material degradation can be programmed using either different molecular weights of the same polymer, different polymer formulation, or multiple drugs encapsulation within the same matrix but with different interaction mechanisms between the drug and the material [152].

2.10. Tissue-Engineering Applications for Testicular Tissue Transplantation

Encapsulation and local drug delivery systems have been the object of several reports concerning the testicle [129,141,153–155].

In a study addressing support matrices to optimize orthotopic avascular ITT auto-grafting in mice, Poels et al. evaluated two different hydrogels, one made of 1% alginate and another made of fibrin (30 mg/mL fibrinogen/30 IU/mL thrombin). Results showed a two-fold improvement in the survival of the subpopulation of spermatogonia not committed to differentiation (including the SSCs) with the use of the alginate matrix compared to fibrin gel ($p < 0.05$) [141]. This could be explained by differences in the structure of the materials that may influence the diffusion of nutrients and the invasion of vascular cells [156]. Indeed, the alginate hydrogel used presented a honeycomb structure with pores of 200 μm diameter, while the fibrin hydrogel had a nano-fibrous network with 1 μm pores [141]. Another reason for the increase in spermatogonial cell survival could be the intrinsic antioxidant properties of the oligo- and polysaccharides originating from algae such as alginate [157]. In the only previous experiment concerning human testicular tissue, encapsulation with alginate of testicular cells dissociated from seminiferous tubules of adult azoospermic patients with maturation arrest led to maturation of differentiated haploid germ cells during in vitro culture [153].

Alginate hydrogel displayed low cytotoxicity in 3D culture of mice prepubertal male germ cells [155]. Moreover, when used for encapsulation of bull germ cells during in vitro culture, it allowed differentiation up to the stage of haploid cells [129].

Such results suggest that alginate is an ideal candidate for tissue engineering of the testicle.

The effects of VEGF-loaded nanoparticles have been explored in an experiment involving orthotopic auto-graft of fresh mouse ITT. Use of dextran/chitosan nanoparticles delivering VEGF led to an increased graft vascular density at 5 days. However, this result was not maintained at 21 days post-implantation, suggesting a lack of stabilization of the neovascularization [141].

Any action aimed at increasing, accelerating the formation, and stabilizing newly formed vessels might promote graft survival and function. It is thus an important target to further improve ITT transplantation technique using tissue engineering approaches.

2.11. Future Directions for Fertility Restoration in Boys Using Transplantation of Prepubertal Cells or Tissues

The many differences in the previous experimental settings, such as different xenografting sites, hormone environment of the host mice, and donors' characteristics like age, preexisting medical condition, concomitant gonadotoxic treatments, and donors' unknown fertility potential, make the results of these reports somehow difficult to compare. In addition, a major limitation of studies on fertility preservation in prepubertal patients is the limited availability of human ITT.

However, the development of models, relying on the use of nanotechnology, on bioengineering, and on organoïds, provides further perspectives to the field.

New drug delivery strategies also open a vast window of opportunities, like the evaluation of new molecules for vascularization enhancement, the prevention of oxidative stress, and hormonal environment modulation, which would directly improve ITT and SSCs transplantation outcome.

Other ways to support gonadal cells or tissue in vivo might also be taken into consideration, like cell therapy. For example, locally injected allogenic mesenchymal stem cells were shown to improve spermatogonial survival after testicular torsion-induced hypoxia-reoxygenation in the rat [158].

The heterogeneous behavior of the different testicular cells populations when exposed to stress in vivo and in experimental conditions produces different responses and is yet to be fully investigated. In rats injected with ethanol in order to reproduce a model of stress, germ cells apoptosis was found to be enhanced, while Sertoli cells could activate pathways such as autophagy and mitophagy [159,160]. These pro-survival mechanisms might have implications that should be considered in situations when germ cells are exposed to important stress (such as during transplantation).

Germ cells could be seeded after in vitro and in vivo maturation [20] in testicular tissue decellularized matrix [161,162] to produce organoids, relying on the ability of mammalian cells to reproduce multi-cellular structures outside the body [163]. The knowledge resulting from these studies will certainly play a major role in the studies focalizing on the physiology of the testicle and could offer alternative models to the cell xenografting assay.

The creation of an artificial testicle would involve isolating human male germ cells from cryostored ITT or testicular cell suspensions, manipulating them safely and effectively in vitro and incorporating them in scaffolds possibly loaded with specific bioactive molecules. An eventual transplantation of this construct directly to the cancer cured infertile patient would be an ultimate goal of tissue engineering application to prepubertal male fertility preservation. A key issue in this process would be the repopulation of the scaffold, a process that could be enhanced by use of encapsulation with materials like alginate or collagen. Collagen already proved itself useful, as testicular cells isolated from juvenile rats and cultured in vitro in collagen sponges could form clusters composed of Sertoli cells, peritubular cells, and undifferentiated spermatogonia [154]. Scaffold obtained from cadaveric adult human testicles showed an effective decellularization and a preserved 3-D structure [161]. However, such tissue availability is limited, and its supply is rather complicated. Therefore, using animal-derived tissue would probably be a better option. With a view to restore the fertility of young boys exposed to cytotoxic treatments, the differences in physical and 3D properties between prepubertal and mature testicular tissue should be kept in mind. The perspective of using decellularized scaffold obtained from multi-transgenic swine-derived testicle might be an option that would help circumvent complications linked to the transplantation of scaffolds from animal origin. Tissues derived from such animals have already been authorized for clinical trials (e.g., for swine-derived pancreatic islets transplantation to diabetic human patients) [164], or have even entered clinical practice (e.g., for cardiac valve replacement) [165], and both the safety and feasibility of this procedures have been proven.

Further perspectives using the cell transplantation assay may involve the use of suspensions containing germ cells obtained from pluripotent stem cells, a cellular population that can be induced through a dedifferentiation process of somatic adult cells [166]. Mouse pluripotent stem cells could successfully differentiate up to the stage of spermatogonial-like cells and colonize seminiferous tubules once injected into adult testicles. Albeit with alterations of DNA methylation, this cellular population led to spermatogenesis and generation of fertile progeny [167]. Similar encouraging results have been obtained when mice primordial germ cells (retrieved from mice embryos) [168], and primordial germ cell-like cells (obtained from in vitro culture of induced pluripotent stem cells) [169] were transplanted to neonatal testicles. Human pluripotent stem cells successfully differentiated in vitro to primordial germ cell-like cells [170], although the potential of these cells for generation of more mature germ line cells is yet to be explored. So far, the most relevant benefit of this approach lies in its use as a disease model to study fertility restoration possibilities in cases of genetic abnormalites responsible for infertility [171].

3. Materials and Methods

We conducted a search on PubMed database (PubMED, National Center for Biotechnology Information, US National Institutes of Health, Bethesda, MD, USA) for the terms "Testicular Tissue transplantation" and "Spermatogonial stem cells transplantation", representing the primary topic.

The studies linked to the principal subject of interest, published in English or French between 1984 and 15 December 2017 were referenced in this review. The bibliography of the cited study has been likewise reviewed and used as source for reports cited in the discussion.

In Figure 2, a flow chart describes the article selection process.

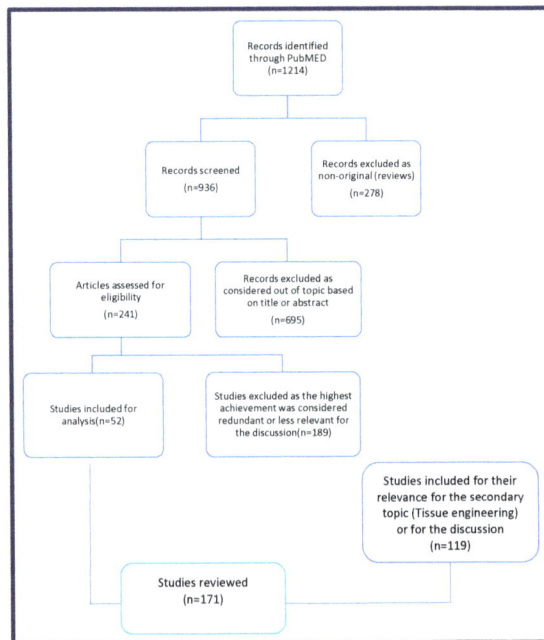

Figure 2. Flow chart that describes the articles selection process for inclusion in the review.

Int. J. Mol. Sci. **2018**, *19*, 286

4. Conclusions

The many perspectives we itemized could offer, alone or combined with each other, several hints for male prepubertal fertility restoration. Efforts have been made in the last decade for cryobanking of ITT, but a shift from an experimental to a clinical practice of these procedures needs further reassurance that one technique can become effective to restore a patient's fertility. Insights in the processes that rule spermatogenesis would help to clarify the reproductive potential of spermatogonia that survive after prepubertal tissue cryopreservation. Distinction between different populations of germ cells is mandatory to understand whether the germ cells precursors present in cryostored tissues can lead to fertility restoration. Indeed, in experimental conditions, a differentiation blockade at the diploid cell stage has always been encountered so far, while for the procedure to be considered effective, a differentiation up to the stage of haploid cells is necessary. Further studies are thus needed to explore the biology of spermatogenesis and to find out if the techniques that have already been effective with animal tissue could also be successful with human tissue. Cells sustainment and molecule supplementation could be options to explore.

In conclusion, tissue engineering might represent not only a progress in fertility restoration, but it could also offer a model to study prepubertal testicle physiology.

Acknowledgments: Studies conducted at the Université Catholique de Louvain (UCL)were supported by grants from the Fonds National de la Recherche Scientifique de Belgique (Télévie Grants 7.6511.16 and 7.4510.16 F), the Fondation Salus Sanguinis. Anne des Rieux is a research associate from the FRS-FNRS (Fonds de la Recherche Scientifique, Belgique) and is a recipient of subsidies from the Fonds Spéciaux de Recherche Scientifique (FSR, UCL).

Author Contributions: Federico Del Vento performed the literature review and wrote the manuscript. Maxime Vermeulen, Francesca de Michele, Maria Grazia Giudice, and Jonathan Poels participated in the review of the literature. Anne des Rieux was responsible for critical review of the manuscript. Christine Wyns was responsible for critical review of the manuscript and the discussion.

Conflicts of Interest: The authors declare no conflict of interest.

Abbreviations

SSC	Spermatogonial stem cells
ITT	Immature testicular tissue
SG	Spermatogonia
ST	Seminiferous tubules
F/T	Frozen/Thawed
FSH	Follicle stimulating Hormone
LH	Luteinizing Hormone
HcG	Human chorionic Gonadotropin
VEGF	Vascular Endothelial Growth Factor
PDGF	Platelet Derived Growth Factor
FGF	Fibroblast Growth Factor
MACS	Magnetic Activated Cell Sorting
FACS	Fluorescence Activated Cell Sorting

References

1. Gatta, G.; Zigon, G.; Capocaccia, R.; Coebergh, J.W.; Desandes, E.; Kaatsch, P.; Pastore, G.; Peris-Bonet, R.; Stiller, C.A.; Group, E.W. Survival of european children and young adults with cancer diagnosed 1995–2002. *Eur. J. Cancer* **2009**, *45*, 992–1005. [CrossRef] [PubMed]
2. Jahnukainen, K.; Mitchell, R.T.; Stukenborg, J.B. Testicular function and fertility preservation after treatment for haematological cancer. *Curr. Opin. Endocrinol. Diabetes Obes.* **2015**, *22*, 217–223. [CrossRef] [PubMed]
3. Wyns, C.; Curaba, M.; Vanabelle, B.; Van Langendonckt, A.; Donnez, J. Options for fertility preservation in prepubertal boys. *Hum. Reprod. Update* **2010**, *16*, 312–328. [CrossRef] [PubMed]

4. Lukusa, A.K.; Vermylen, C.; Vanabelle, B.; Curaba, M.; Brichard, B.; Chantrain, C.; Dupont, S.; Ferrant, A.; Wyns, C. Bone marrow transplantation or hydroxyurea for sickle cell anemia: Long-term effects on semen variables and hormone profiles. *Pediatr. Hematol. Oncol.* **2009**, *26*, 186–194. [CrossRef] [PubMed]

5. Lee, S.J.; Schover, L.R.; Partridge, A.H.; Patrizio, P.; Wallace, W.H.; Hagerty, K.; Beck, L.N.; Brennan, L.V.; Oktay, K. American society of clinical oncology recommendations on fertility preservation in cancer patients. *J. Clin. Oncol.* **2006**, *24*, 2917–2931. [CrossRef] [PubMed]

6. Schover, L.R.; Rybicki, L.A.; Martin, B.A.; Bringelsen, K.A. Having children after cancer. A pilot survey of survivors' attitudes and experiences. *Cancer* **1999**, *86*, 697–709. [CrossRef]

7. Wyns, C.; Collienne, C.; Shenfield, F.; Robert, A.; Laurent, P.; Roegiers, L.; Brichard, B. Fertility preservation in the male pediatric population: Factors influencing the decision of parents and children. *Hum. Reprod.* **2015**, *30*, 2022–2030. [CrossRef] [PubMed]

8. Ginsberg, J.P.; Li, Y.; Carlson, C.A.; Gracia, C.R.; Hobbie, W.L.; Miller, V.A.; Mulhall, J.; Shnorhavorian, M.; Brinster, R.L.; Kolon, T.F. Testicular tissue cryopreservation in prepubertal male children: An analysis of parental decision-making. *Pediatr. Blood Cancer* **2014**, *61*, 1673–1678. [CrossRef] [PubMed]

9. Van den Berg, H.; Repping, S.; van der Veen, F. Parental desire and acceptability of spermatogonial stem cell cryopreservation in boys with cancer. *Hum. Reprod.* **2007**, *22*, 594–597. [CrossRef] [PubMed]

10. Clermont, Y. Kinetics of spermatogenesis in mammals. *Arch. Anat. Microsc. Morphol. Exp.* **1967**, *56*, 7–60. [PubMed]

11. Keros, V.; Hultenby, K.; Borgstrom, B.; Fridstrom, M.; Jahnukainen, K.; Hovatta, O. Methods of cryopreservation of testicular tissue with viable spermatogonia in pre-pubertal boys undergoing gonadotoxic cancer treatment. *Hum. Reprod.* **2007**, *22*, 1384–1395. [CrossRef] [PubMed]

12. Wyns, C.; Curaba, M.; Martinez-Madrid, B.; Van Langendonckt, A.; Francois-Xavier, W.; Donnez, J. Spermatogonial survival after cryopreservation and short-term orthotopic immature human cryptorchid testicular tissue grafting to immunodeficient mice. *Hum. Reprod.* **2007**, *22*, 1603–1611. [CrossRef] [PubMed]

13. Wyns, C.; Curaba, M.; Petit, S.; Vanabelle, B.; Laurent, P.; Wese, J.F.; Donnez, J. Management of fertility preservation in prepubertal patients: 5 years' experience at the catholic university of louvain. *Hum. Reprod.* **2011**, *26*, 737–747. [CrossRef] [PubMed]

14. Picton, H.M.; Wyns, C.; Anderson, R.A.; Goossens, E.; Jahnukainen, K.; Kliesch, S.; Mitchell, R.T.; Pennings, G.; Rives, N.; Tournaye, H.; et al. A european perspective on testicular tissue cryopreservation for fertility preservation in prepubertal and adolescent boys. *Hum. Reprod.* **2015**, *30*, 2463–2475. [CrossRef] [PubMed]

15. Ogawa, T.; Ohmura, M.; Ohbo, K. The niche for spermatogonial stem cells in the mammalian testis. *Int. J. Hematol.* **2005**, *82*, 381–388. [CrossRef] [PubMed]

16. Wyns, C.; Van Langendonckt, A.; Wese, F.X.; Donnez, J.; Curaba, M. Long-term spermatogonial survival in cryopreserved and xenografted immature human testicular tissue. *Hum. Reprod.* **2008**, *23*, 2402–2414. [CrossRef] [PubMed]

17. Kim, T.H.; Hargreaves, H.K.; Chan, W.C.; Brynes, R.K.; Alvarado, C.; Woodard, J.; Ragab, A.H. Sequential testicular biopsies in childhood acute lymphocytic leukemia. *Cancer* **1986**, *57*, 1038–1041. [CrossRef]

18. Hou, M.; Andersson, M.; Eksborg, S.; Soder, O.; Jahnukainen, K. Xenotransplantation of testicular tissue into nude mice can be used for detecting leukemic cell contamination. *Hum. Reprod.* **2007**, *22*, 1899–1906. [CrossRef] [PubMed]

19. De Michele, F.; Vermeulen, M.; Wyns, C. Fertility restoration with spermatogonial stem cells. *Curr. Opin. Endocrinol. Diabetes Obes.* **2017**. [CrossRef] [PubMed]

20. De Michele, F.; Poels, J.; Weerens, L.; Petit, C.; Evrard, Z.; Ambroise, J.; Gruson, D.; Wyns, C. Preserved seminiferous tubule integrity with spermatogonial survival and induction of sertoli and leydig cell maturation after long-term organotypic culture of prepubertal human testicular tissue. *Hum. Reprod.* **2017**, *32*, 32–45. [CrossRef] [PubMed]

21. Donnez, J.; Dolmans, M.M.; Demylle, D.; Jadoul, P.; Pirard, C.; Squifflet, J.; Martinez-Madrid, B.; Van Langendonckt, A. Livebirth after orthotopic transplantation of cryopreserved ovarian tissue. *Lancet* **2004**, *364*, 1405–1410. [CrossRef]

22. Jadoul, P.; Guilmain, A.; Squifflet, J.; Luyckx, M.; Votino, R.; Wyns, C.; Dolmans, M.M. Efficacy of ovarian tissue cryopreservation for fertility preservation: Lessons learned from 545 cases. *Hum. Reprod.* **2017**, *32*, 1046–1054. [CrossRef] [PubMed]

23. Chiti, M.C.; Dolmans, M.M.; Donnez, J.; Amorim, C.A. Fibrin in reproductive tissue engineering: A review on its application as a biomaterial for fertility preservation. *Ann. Biomed. Eng.* **2017**, *45*, 1650–1663. [CrossRef] [PubMed]

24. Demeestere, I.; Simon, P.; Dedeken, L.; Moffa, F.; Tsepelidis, S.; Brachet, C.; Delbaere, A.; Devreker, F.; Ferster, A. Live birth after autograft of ovarian tissue cryopreserved during childhood. *Hum. Reprod.* **2015**, *30*, 2107–2109. [CrossRef] [PubMed]

25. Van Eyck, A.S.; Jordan, B.F.; Gallez, B.; Heilier, J.F.; Van Langendonckt, A.; Donnez, J. Electron paramagnetic resonance as a tool to evaluate human ovarian tissue reoxygenation after xenografting. *Fertil. Steril.* **2009**, *92*, 374–381. [CrossRef] [PubMed]

26. Dolmans, M.M.; Martinez-Madrid, B.; Gadisseux, E.; Guiot, Y.; Yuan, W.Y.; Torre, A.; Camboni, A.; Van Langendonckt, A.; Donnez, J. Short-term transplantation of isolated human ovarian follicles and cortical tissue into nude mice. *Reproduction* **2007**, *134*, 253–262. [CrossRef] [PubMed]

27. Goossens, E.; Geens, M.; De Block, G.; Tournaye, H. Spermatogonial survival in long-term human prepubertal xenografts. *Fertil. Steril.* **2008**, *90*, 2019–2022. [CrossRef] [PubMed]

28. Sato, Y.; Nozawa, S.; Yoshiike, M.; Arai, M.; Sasaki, C.; Iwamoto, T. Xenografting of testicular tissue from an infant human donor results in accelerated testicular maturation. *Hum. Reprod.* **2010**, *25*, 1113–1122. [CrossRef] [PubMed]

29. Van Saen, D.; Goossens, E.; Haentjens, P.; Baert, Y.; Tournaye, H. Exogenous administration of recombinant human fsh does not improve germ cell survival in human prepubertal xenografts. *Reprod. Biomed. Online* **2013**, *26*, 286–298. [CrossRef] [PubMed]

30. Poels, J.; Van Langendonckt, A.; Many, M.C.; Wese, F.X.; Wyns, C. Vitrification preserves proliferation capacity in human spermatogonia. *Hum. Reprod.* **2013**, *28*, 578–589. [CrossRef] [PubMed]

31. Poels, J.; Abou-Ghannam, G.; Herman, S.; Van Langendonckt, A.; Wese, F.X.; Wyns, C. In search of better spermatogonial preservation by supplementation of cryopreserved human immature testicular tissue xenografts with n-acetylcysteine and testosterone. *Front. Surg.* **2014**, *1*, 47. [CrossRef] [PubMed]

32. Schlatt, S.; Honaramooz, A.; Ehmcke, J.; Goebell, P.J.; Rubben, H.; Dhir, R.; Dobrinski, I.; Patrizio, P. Limited survival of adult human testicular tissue as ectopic xenograft. *Hum. Reprod.* **2006**, *21*, 384–389. [CrossRef] [PubMed]

33. Benjamin, L.E.; Hemo, I.; Keshet, E. A plasticity window for blood vessel remodelling is defined by pericyte coverage of the preformed endothelial network and is regulated by pdgf-b and vegf. *Development* **1998**, *125*, 1591–1598. [PubMed]

34. Geens, M.; De Block, G.; Goossens, E.; Frederickx, V.; Van Steirteghem, A.; Tournaye, H. Spermatogonial survival after grafting human testicular tissue to immunodeficient mice. *Hum. Reprod.* **2006**, *21*, 390–396. [CrossRef] [PubMed]

35. Van Saen, D.; Goossens, E.; Bourgain, C.; Ferster, A.; Tournaye, H. Meiotic activity in orthotopic xenografts derived from human postpubertal testicular tissue. *Hum. Reprod.* **2011**, *26*, 282–293. [CrossRef] [PubMed]

36. Kulin, H.E.; Frontera, M.A.; Demers, L.M.; Bartholomew, M.J.; Lloyd, T.A. The onset of sperm production in pubertal boys. Relationship to gonadotropin excretion. *Am. J. Dis. Child.* **1989**, *143*, 190–193. [PubMed]

37. Honaramooz, A.; Snedaker, A.; Boiani, M.; Scholer, H.; Dobrinski, I.; Schlatt, S. Sperm from neonatal mammalian testes grafted in mice. *Nature* **2002**, *418*, 778–781. [CrossRef] [PubMed]

38. Jahnukainen, K.; Ehmcke, J.; Nurmio, M.; Schlatt, S. Autologous ectopic grafting of cryopreserved testicular tissue preserves the fertility of prepubescent monkeys that receive sterilizing cytotoxic therapy. *Cancer Res.* **2012**, *72*, 5174–5178. [CrossRef] [PubMed]

39. Shinohara, T.; Inoue, K.; Ogonuki, N.; Kanatsu-Shinohara, M.; Miki, H.; Nakata, K.; Kurome, M.; Nagashima, H.; Toyokuni, S.; Kogishi, K.; et al. Birth of offspring following transplantation of cryopreserved immature testicular pieces and in vitro microinsemination. *Hum. Reprod.* **2002**, *17*, 3039–3045. [CrossRef] [PubMed]

40. Snedaker, A.K.; Honaramooz, A.; Dobrinski, I. A game of cat and mouse: Xenografting of testis tissue from domestic kittens results in complete cat spermatogenesis in a mouse host. *J. Androl.* **2004**, *25*, 926–930. [CrossRef] [PubMed]

41. Abrishami, M.; Anzar, M.; Yang, Y.; Honaramooz, A. Cryopreservation of immature porcine testis tissue to maintain its developmental potential after xenografting into recipient mice. *Theriogenology* **2010**, *73*, 86–96. [CrossRef] [PubMed]

42. Yildiz, C.; Mullen, B.; Jarvi, K.; McKerlie, C.; Lo, K.C. Effect of different cryoprotectant agents on spermatogenesis efficiency in cryopreserved and grafted neonatal mouse testicular tissue. *Cryobiology* **2013**, *67*, 70–75. [CrossRef] [PubMed]

43. Pukazhenthi, B.S.; Nagashima, J.; Travis, A.J.; Costa, G.M.; Escobar, E.N.; Franca, L.R.; Wildt, D.E. Slow freezing, but not vitrification supports complete spermatogenesis in cryopreserved, neonatal sheep testicular xenografts. *PLoS ONE* **2015**, *10*, e0123957. [CrossRef] [PubMed]

44. Liu, J.; Cheng, K.M.; Silversides, F.G. Production of live offspring from testicular tissue cryopreserved by vitrification procedures in japanese quail (coturnix japonica). *Biol. Reprod.* **2013**, *88*, 124. [CrossRef] [PubMed]

45. Ohta, H.; Wakayama, T. Generation of normal progeny by intracytoplasmic sperm injection following grafting of testicular tissue from cloned mice that died postnatally. *Biol. Reprod.* **2005**, *73*, 390–395. [CrossRef] [PubMed]

46. Kaneko, H.; Kikuchi, K.; Nakai, M.; Somfai, T.; Noguchi, J.; Tanihara, F.; Ito, J.; Kashiwazaki, N. Generation of live piglets for the first time using sperm retrieved from immature testicular tissue cryopreserved and grafted into nude mice. *PLoS ONE* **2013**, *8*, e70989. [CrossRef] [PubMed]

47. Liu, Z.; Nie, Y.H.; Zhang, C.C.; Cai, Y.J.; Wang, Y.; Lu, H.P.; Li, Y.Z.; Cheng, C.; Qiu, Z.L.; Sun, Q. Generation of macaques with sperm derived from juvenile monkey testicular xenografts. *Cell Res.* **2016**, *26*, 139–142. [CrossRef] [PubMed]

48. Vermeulen, M.; Poels, J.; de Michele, F.; des Rieux, A.; Wyns, C. Restoring fertility with cryopreserved prepubertal testicular tissue: Perspectives with hydrogel encapsulation, nanotechnology, and bioengineered scaffolds. *Ann. Biomed. Eng.* **2017**. [CrossRef]

49. Schlatt, S.; Kim, S.S.; Gosden, R. Spermatogenesis and steroidogenesis in mouse, hamster and monkey testicular tissue after cryopreservation and heterotopic grafting to castrated hosts. *Reproduction* **2002**, *124*, 339–346. [CrossRef] [PubMed]

50. Arregui, L.; Dobrinski, I. Xenografting of testicular tissue pieces: 12 years of an in vivo spermatogenesis system. *Reproduction* **2014**, *148*, R71–R84. [CrossRef] [PubMed]

51. Schlatt, S.; Westernstroer, B.; Gassei, K.; Ehmcke, J. Donor-host involvement in immature rat testis xenografting into nude mouse hosts. *Biol. Reprod.* **2010**, *82*, 888–895. [CrossRef] [PubMed]

52. Andrejecsk, J.W.; Cui, J.; Chang, W.G.; Devalliere, J.; Pober, J.S.; Saltzman, W.M. Paracrine exchanges of molecular signals between alginate-encapsulated pericytes and freely suspended endothelial cells within a 3d protein gel. *Biomaterials* **2013**, *34*, 8899–8908. [CrossRef] [PubMed]

53. Chan, F.; Allison, J.E.; Stanley, A.J.; Gumbreck, L.G. Reciprocal transplantation of testes between normal and pseudohermaphroditic male rats. *Fertil. Steril.* **1969**, *20*, 482–494. [CrossRef]

54. Johnson, L.; Suggs, L.C.; Norton, Y.M.; Zeh, W.C. Effect of developmental age or time after transplantation on sertoli cell number and testicular size in inbred fischer rats. *Biol. Reprod.* **1996**, *54*, 948–959. [CrossRef] [PubMed]

55. Wistuba, J.; Luetjens, C.M.; Wesselmann, R.; Nieschlag, E.; Simoni, M.; Schlatt, S. Meiosis in autologous ectopic transplants of immature testicular tissue grafted to callithrix jacchus. *Biol. Reprod.* **2006**, *74*, 706–713. [CrossRef] [PubMed]

56. Luetjens, C.M.; Stukenborg, J.B.; Nieschlag, E.; Simoni, M.; Wistuba, J. Complete spermatogenesis in orthotopic but not in ectopic transplants of autologously grafted marmoset testicular tissue. *Endocrinology* **2008**, *149*, 1736–1747. [CrossRef] [PubMed]

57. Van Saen, D.; Goossens, E.; De Block, G.; Tournaye, H. Regeneration of spermatogenesis by grafting testicular tissue or injecting testicular cells into the testes of sterile mice: A comparative study. *Fertil. Steril.* **2009**, *91*, 2264–2272. [CrossRef] [PubMed]

58. Oatley, J.M.; Reeves, J.J.; McLean, D.J. Establishment of spermatogenesis in neonatal bovine testicular tissue following ectopic xenografting varies with donor age. *Biol. Reprod.* **2005**, *72*, 358–364. [CrossRef] [PubMed]

59. Schmidt, J.A.; de Avila, J.M.; McLean, D.J. Grafting period and donor age affect the potential for spermatogenesis in bovine ectopic testis xenografts. *Biol. Reprod.* **2006**, *75*, 160–166. [CrossRef] [PubMed]

60. Jahnukainen, K.; Ehmcke, J.; Hergenrother, S.D.; Schlatt, S. Effect of cold storage and cryopreservation of immature non-human primate testicular tissue on spermatogonial stem cell potential in xenografts. *Hum. Reprod.* **2007**, *22*, 1060–1067. [CrossRef] [PubMed]

61. Potter, S.J.; DeFalco, T. Role of the testis interstitial compartment in spermatogonial stem cell function. *Reproduction* **2017**, *153*, R151–R162. [CrossRef] [PubMed]

62. Schlatt, S.; Honaramooz, A.; Boiani, M.; Scholer, H.R.; Dobrinski, I. Progeny from sperm obtained after ectopic grafting of neonatal mouse testes. *Biol. Reprod.* **2003**, *68*, 2331–2335. [CrossRef] [PubMed]
63. Honaramooz, A.; Li, M.W.; Penedo, M.C.; Meyers, S.; Dobrinski, I. Accelerated maturation of primate testis by xenografting into mice. *Biol. Reprod.* **2004**, *70*, 1500–1503. [CrossRef] [PubMed]
64. Schlatt, S.; Gassei, K.; Westernstroer, B.; Ehmcke, J. Modulating testicular mass in xenografting: A model to explore testis development and endocrine function. *Endocrinology* **2010**, *151*, 4018–4023. [CrossRef] [PubMed]
65. Wistuba, J.; Mundry, M.; Luetjens, C.M.; Schlatt, S. Cografting of hamster (phodopus sungorus) and marmoset (callithrix jacchus) testicular tissues into nude mice does not overcome blockade of early spermatogenic differentiation in primate grafts. *Biol. Reprod.* **2004**, *71*, 2087–2091. [CrossRef] [PubMed]
66. Rathi, R.; Zeng, W.; Megee, S.; Conley, A.; Meyers, S.; Dobrinski, I. Maturation of testicular tissue from infant monkeys after xenografting into mice. *Endocrinology* **2008**, *149*, 5288–5296. [CrossRef] [PubMed]
67. Ehmcke, J.; Gassei, K.; Westernstroer, B.; Schlatt, S. Immature rhesus monkey (macaca mulatta) testis xenografts show increased growth, but not enhanced seminiferous differentiation, under human chorionic gonadotropin treatment of nude mouse recipients. *Int. J. Androl.* **2011**, *34*, e459–e467. [CrossRef] [PubMed]
68. Rathi, R.; Honaramooz, A.; Zeng, W.; Turner, R.; Dobrinski, I. Germ cell development in equine testis tissue xenografted into mice. *Reproduction* **2006**, *131*, 1091–1098. [CrossRef] [PubMed]
69. Marchlewska, K.; Kula, K.; Walczak-Jedrzejowska, R.; Kula, W.; Oszukowska, E.; Filipiak, E.; Moszura, T.; Slowikowska-Hilczer, J. Maturational changes in connexin 43 expression in the seminiferous tubules may depend on thyroid hormone action. *Arch. Med. Sci.* **2013**, *9*, 139–145. [CrossRef] [PubMed]
70. Rodriguez-Sosa, J.R.; Costa, G.M.; Rathi, R.; Franca, L.R.; Dobrinski, I. Endocrine modulation of the recipient environment affects development of bovine testis tissue ectopically grafted in mice. *Reproduction* **2012**, *144*, 37–51. [CrossRef] [PubMed]
71. Turner, T.T.; Tung, K.S.; Tomomasa, H.; Wilson, L.W. Acute testicular ischemia results in germ cell-specific apoptosis in the rat. *Biol. Reprod.* **1997**, *57*, 1267–1274. [CrossRef] [PubMed]
72. Yancopoulos, G.D.; Davis, S.; Gale, N.W.; Rudge, J.S.; Wiegand, S.J.; Holash, J. Vascular-specific growth factors and blood vessel formation. *Nature* **2000**, *407*, 242–248. [CrossRef] [PubMed]
73. Lee, K.; Silva, E.A.; Mooney, D.J. Growth factor delivery-based tissue engineering: General approaches and a review of recent developments. *J. R. Soc. Interface* **2011**, *8*, 153–170. [CrossRef] [PubMed]
74. Mitchell, A.C.; Briquez, P.S.; Hubbell, J.A.; Cochran, J.R. Engineering growth factors for regenerative medicine applications. *Acta Biomater.* **2016**, *30*, 1–12. [CrossRef] [PubMed]
75. Caires, K.C.; de Avila, J.M.; Cupp, A.S.; McLean, D.J. Vegfa family isoforms regulate spermatogonial stem cell homeostasis in vivo. *Endocrinology* **2012**, *153*, 887–900. [CrossRef] [PubMed]
76. Ergun, S.; Kilic, N.; Fiedler, W.; Mukhopadhyay, A.K. Vascular endothelial growth factor and its receptors in normal human testicular tissue. *Mol. Cell. Endocrinol.* **1997**, *131*, 9–20. [CrossRef]
77. Tian, R.; Yang, S.; Zhu, Y.; Zou, S.; Li, P.; Wang, J.; Zhu, Z.; Huang, Y.; He, Z.; Li, Z. Vegf/vegfr2 signaling regulates germ cell proliferation in vitro and promotes mouse testicular regeneration in vivo. *Cells Tissues Organs* **2016**, *201*, 1–13. [CrossRef] [PubMed]
78. Schmidt, J.A.; de Avila, J.M.; McLean, D.J. Effect of vascular endothelial growth factor and testis tissue culture on spermatogenesis in bovine ectopic testis tissue xenografts. *Biol. Reprod.* **2006**, *75*, 167–175. [CrossRef] [PubMed]
79. Caires, K.C.; de Avila, J.; McLean, D.J. Vascular endothelial growth factor regulates germ cell survival during establishment of spermatogenesis in the bovine testis. *Reproduction* **2009**, *138*, 667–677. [CrossRef] [PubMed]
80. Zafarullah, M.; Li, W.Q.; Sylvester, J.; Ahmad, M. Molecular mechanisms of n-acetylcysteine actions. *Cell. Mol. Life Sci.* **2003**, *60*, 6–20. [CrossRef] [PubMed]
81. Cay, A.; Alver, A.; Kucuk, M.; Isik, O.; Eminagaoglu, M.S.; Karahan, S.C.; Deger, O. The effects of *N*-acetylcysteine on antioxidant enzyme activities in experimental testicular torsion. *J. Surg. Res.* **2006**, *131*, 199–203. [CrossRef] [PubMed]
82. Turkmen, S.; Mentese, A.; Karaguzel, E.; Karaca, Y.; Kucuk, A.; Uzun, A.; Yulug, E.; Turedi, S. A comparison of the effects of n-acetylcysteine and ethyl pyruvate on experimental testicular ischemia-reperfusion injury. *Fertil. Steril.* **2012**, *98*, 626–631. [CrossRef] [PubMed]
83. Erkkila, K.; Hirvonen, V.; Wuokko, E.; Parvinen, M.; Dunkel, L. *N*-acetyl-L-cysteine inhibits apoptosis in human male germ cells in vitro. *J. Clin. Endocrinol. Metab.* **1998**, *83*, 2523–2531. [CrossRef] [PubMed]

84. Poeggeler, B.; Reiter, R.J.; Hardeland, R.; Tan, D.X.; Barlow-Walden, L.R. Melatonin and structurally-related, endogenous indoles act as potent electron donors and radical scavengers in vitro. *Redox Rep.* **1996**, *2*, 179–184. [CrossRef] [PubMed]

85. Gholami, M.; Saki, G.; Hemadi, M.; Khodadadi, A.; Mohammadi-Asl, J. Melatonin improves spermatogonial stem cells transplantation efficiency in azoospermic mice. *Iran. J. Basic Med. Sci.* **2014**, *17*, 93–99. [PubMed]

86. Navid, S.; Rastegar, T.; Baazm, M.; Alizadeh, R.; Talebi, A.; Gholami, K.; Khosravi-Farsani, S.; Koruji, M.; Abbasi, M. In vitro effects of melatonin on colonization of neonate mouse spermatogonial stem cells. *Syst. Biol. Reprod. Med.* **2017**, 1–12. [CrossRef] [PubMed]

87. Gholami, M.; Saki, G.; Hemadi, M.; Khodadadi, A.; Mohamma-di-Asl, J. Effect of melatonin on the expression of apoptotic genes in vitrified-thawed spermatogonia stem cells type a of 6-day-old mice. *Iran. J. Basic Med. Sci.* **2013**, *16*, 906–909. [PubMed]

88. Kimsa, M.C.; Strzalka-Mrozik, B.; Kimsa, M.W.; Gola, J.; Nicholson, P.; Lopata, K.; Mazurek, U. Porcine endogenous retroviruses in xenotransplantation—Molecular aspects. *Viruses* **2014**, *6*, 2062–2083. [CrossRef] [PubMed]

89. Nomi, M.A.A.; Coppi, P.D.; Soker, S. Principals of neovascularization for tissue engineering. *Mol. Asp. Med.* **2002**, *23*, 463–483. [CrossRef]

90. Vajanto, I.; Rissanen, T.T.; Rutanen, J.; Hiltunen, M.O.; Tuomisto, T.T.; Arve, K.; Narvanen, O.; Manninen, H.; Rasanen, H.; Hippelainen, M.; et al. Evaluation of angiogenesis and side effects in ischemic rabbit hindlimbs after intramuscular injection of adenoviral vectors encoding vegf and lacz. *J. Gene Med.* **2002**, *4*, 371–380. [CrossRef] [PubMed]

91. Ferrara, N.; Davis-Smyth, T. The biology of vascular endothelial growth factor. *Endocr. Rev.* **1997**, *18*, 4–25. [CrossRef] [PubMed]

92. Folkman, J. Role of angiogenesis in tumor growth and metastasis. *Semin. Oncol.* **2002**, *29*, 15–18. [CrossRef] [PubMed]

93. Moulisova, V.; Gonzalez-Garcia, C.; Cantini, M.; Rodrigo-Navarro, A.; Weaver, J.; Costell, M.; Sabater, I.S.R.; Dalby, M.J.; Garcia, A.J.; Salmeron-Sanchez, M. Engineered microenvironments for synergistic vegf—Integrin signalling during vascularization. *Biomaterials* **2017**, *126*, 61–74. [CrossRef] [PubMed]

94. Brinster, R.L.; Zimmermann, J.W. Spermatogenesis following male germ-cell transplantation. *Proc. Natl. Acad. Sci. USA* **1994**, *91*, 11298–11302. [CrossRef] [PubMed]

95. Brook, P.F.; Radford, J.A.; Shalet, S.M.; Joyce, A.D.; Gosden, R.G. Isolation of germ cells from human testicular tissue for low temperature storage and autotransplantation. *Fertil. Steril.* **2001**, *75*, 269–274. [CrossRef]

96. Kanatsu-Shinohara, M.; Ogonuki, N.; Inoue, K.; Ogura, A.; Toyokuni, S.; Shinohara, T. Restoration of fertility in infertile mice by transplantation of cryopreserved male germline stem cells. *Hum. Reprod.* **2003**, *18*, 2660–2667. [CrossRef] [PubMed]

97. Schlatt, S.; Rosiepen, G.; Weinbauer, G.F.; Rolf, C.; Brook, P.F.; Nieschlag, E. Germ cell transfer into rat, bovine, monkey and human testes. *Hum. Reprod.* **1999**, *14*, 144–150. [CrossRef] [PubMed]

98. Ning, L.; Meng, J.; Goossens, E.; Lahoutte, T.; Marichal, M.; Tournaye, H. In search of an efficient injection technique for future clinical application of spermatogonial stem cell transplantation: Infusion of contrast dyes in isolated cadaveric human testes. *Fertil. Steril.* **2012**, *98*, 1443–1448. [CrossRef] [PubMed]

99. Clouthier, D.E.; Avarbock, M.R.; Maika, S.D.; Hammer, R.E.; Brinster, R.L. Rat spermatogenesis in mouse testis. *Nature* **1996**, *381*, 418–421. [CrossRef] [PubMed]

100. Ogawa, T.; Dobrinski, I.; Avarbock, M.R.; Brinster, R.L. Xenogeneic spermatogenesis following transplantation of hamster germ cells to mouse testes. *Biol. Reprod.* **1999**, *60*, 515–521. [CrossRef] [PubMed]

101. Russell, L.D.; Brinster, R.L. Ultrastructural observations of spermatogenesis following transplantation of rat testis cells into mouse seminiferous tubules. *J. Androl.* **1996**, *17*, 615–627. [PubMed]

102. Nagano, M.; Patrizio, P.; Brinster, R.L. Long-term survival of human spermatogonial stem cells in mouse testes. *Fertil. Steril.* **2002**, *78*, 1225–1233. [CrossRef]

103. Wu, X.; Schmidt, J.A.; Avarbock, M.R.; Tobias, J.W.; Carlson, C.A.; Kolon, T.F.; Ginsberg, J.P.; Brinster, R.L. Prepubertal human spermatogonia and mouse gonocytes share conserved gene expression of germline stem cell regulatory molecules. *Proc. Natl. Acad. Sci. USA* **2009**, *106*, 21672–21677. [CrossRef] [PubMed]

104. Sadri-Ardekani, H.; Akhondi, M.A.; van der Veen, F.; Repping, S.; van Pelt, A.M. In vitro propagation of human prepubertal spermatogonial stem cells. *JAMA* **2011**, *305*, 2416–2418. [CrossRef] [PubMed]

105. Radford, J.; Shalet, S.; Lieberman, B. Fertility after treatment for cancer. Questions remain over ways of preserving ovarian and testicular tissue. *BMJ* **1999**, *319*, 935–936. [CrossRef] [PubMed]

106. Radford, J. Restoration of fertility after treatment for cancer. *Horm. Res.* **2003**, *59* (Suppl. S1), 21–23. [CrossRef] [PubMed]

107. Muller, J.; Skakkebaek, N.E. Quantification of germ cells and seminiferous tubules by stereological examination of testicles from 50 boys who suffered from sudden death. *Int. J. Androl.* **1983**, *6*, 143–156. [CrossRef] [PubMed]

108. Dobrinski, I.; Ogawa, T.; Avarbock, M.R.; Brinster, R.L. Computer assisted image analysis to assess colonization of recipient seminiferous tubules by spermatogonial stem cells from transgenic donor mice. *Mol. Reprod. Dev.* **1999**, *53*, 142–148. [CrossRef]

109. Sadri-Ardekani, H.; Mizrak, S.C.; van Daalen, S.K.; Korver, C.M.; Roepers-Gajadien, H.L.; Koruji, M.; Hovingh, S.; de Reijke, T.M.; de la Rosette, J.J.; van der Veen, F.; et al. Propagation of human spermatogonial stem cells in vitro. *JAMA* **2009**, *302*, 2127–2134. [CrossRef] [PubMed]

110. Mirzapour, T.; Movahedin, M.; Koruji, M.; Nowroozi, M.R. Xenotransplantation assessment: Morphometric study of human spermatogonial stem cells in recipient mouse testes. *Andrologia* **2015**, *47*, 626–633. [CrossRef] [PubMed]

111. Conrad, S.; Azizi, H.; Hatami, M.; Kubista, M.; Bonin, M.; Hennenlotter, J.; Renninger, M.; Skutella, T. Differential gene expression profiling of enriched human spermatogonia after short- and long-term culture. *Biomed. Res. Int.* **2014**, *2014*, 138350. [CrossRef] [PubMed]

112. Gat, I.; Maghen, L.; Filice, M.; Kenigsberg, S.; Wyse, B.; Zohni, K.; Saraz, P.; Fisher, A.G.; Librach, C. Initial germ cell to somatic cell ratio impacts the efficiency of ssc expansion in vitro. *Syst. Biol. Reprod. Med.* **2017**, 1–12. [CrossRef] [PubMed]

113. Ko, K.; Arauzo-Bravo, M.J.; Tapia, N.; Kim, J.; Lin, Q.; Bernemann, C.; Han, D.W.; Gentile, L.; Reinhardt, P.; Greber, B.; et al. Human adult germline stem cells in question. *Nature* **2010**, *465*, 344–349. [CrossRef] [PubMed]

114. Bar-Shira Maymon, B.; Yogev, L.; Marks, A.; Hauser, R.; Botchan, A.; Yavetz, H. Sertoli cell inactivation by cytotoxic damage to the human testis after cancer chemotherapy. *Fertil. Steril.* **2004**, *81*, 1391–1394. [CrossRef] [PubMed]

115. Howell, S.J.; Radford, J.A.; Ryder, W.D.; Shalet, S.M. Testicular function after cytotoxic chemotherapy: Evidence of leydig cell insufficiency. *J. Clin. Oncol.* **1999**, *17*, 1493–1498. [CrossRef] [PubMed]

116. Fujita, K.; Ohta, H.; Tsujimura, A.; Takao, T.; Miyagawa, Y.; Takada, S.; Matsumiya, K.; Wakayama, T.; Okuyama, A. Transplantation of spermatogonial stem cells isolated from leukemic mice restores fertility without inducing leukemia. *J. Clin. Investig.* **2005**, *115*, 1855–1861. [CrossRef] [PubMed]

117. Fujita, K.; Tsujimura, A.; Miyagawa, Y.; Kiuchi, H.; Matsuoka, Y.; Takao, T.; Takada, S.; Nonomura, N.; Okuyama, A. Isolation of germ cells from leukemia and lymphoma cells in a human in vitro model: Potential clinical application for restoring human fertility after anticancer therapy. *Cancer Res.* **2006**, *66*, 11166–11171. [CrossRef] [PubMed]

118. Geens, M.; Van de Velde, H.; De Block, G.; Goossens, E.; Van Steirteghem, A.; Tournaye, H. The efficiency of magnetic-activated cell sorting and fluorescence-activated cell sorting in the decontamination of testicular cell suspensions in cancer patients. *Hum. Reprod.* **2007**, *22*, 733–742. [CrossRef] [PubMed]

119. Dovey, S.L.; Valli, H.; Hermann, B.P.; Sukhwani, M.; Donohue, J.; Castro, C.A.; Chu, T.; Sanfilippo, J.S.; Orwig, K.E. Eliminating malignant contamination from therapeutic human spermatogonial stem cells. *J. Clin. Investig.* **2013**, *123*, 1833–1843. [CrossRef] [PubMed]

120. Hou, M.; Andersson, M.; Zheng, C.; Sundblad, A.; Soder, O.; Jahnukainen, K. Decontamination of leukemic cells and enrichment of germ cells from testicular samples from rats with roser's t-cell leukemia by flow cytometric sorting. *Reproduction* **2007**, *134*, 767–779. [CrossRef] [PubMed]

121. Hermann, B.P.; Sukhwani, M.; Salati, J.; Sheng, Y.; Chu, T.; Orwig, K.E. Separating spermatogonia from cancer cells in contaminated prepubertal primate testis cell suspensions. *Hum. Reprod.* **2011**, *26*, 3222–3231. [CrossRef] [PubMed]

122. Sadri-Ardekani, H.; Homburg, C.H.; van Capel, T.M.; van den Berg, H.; van der Veen, F.; van der Schoot, C.E.; van Pelt, A.M.; Repping, S. Eliminating acute lymphoblastic leukemia cells from human testicular cell cultures: A pilot study. *Fertil. Steril.* **2014**, *101*, 1072–1078. [CrossRef] [PubMed]

123. Kanatsu-Shinohara, M.; Ogonuki, N.; Iwano, T.; Lee, J.; Kazuki, Y.; Inoue, K.; Miki, H.; Takehashi, M.; Toyokuni, S.; Shinkai, Y.; et al. Genetic and epigenetic properties of mouse male germline stem cells during long-term culture. *Development* **2005**, *132*, 4155–4163. [CrossRef] [PubMed]

124. Nickkholgh, B.; Mizrak, S.C.; van Daalen, S.K.; Korver, C.M.; Sadri-Ardekani, H.; Repping, S.; van Pelt, A.M. Genetic and epigenetic stability of human spermatogonial stem cells during long-term culture. *Fertil. Steril.* **2014**, *102*, 1700–1707. [CrossRef] [PubMed]

125. Fleischer, S.; Feiner, R.; Dvir, T. Cardiac tissue engineering: From matrix design to the engineering of bionic hearts. *Regen. Med.* **2017**, *12*, 275–284. [CrossRef] [PubMed]

126. O'Brien, F.J. Biomaterials & scaffolds for tissue engineering. *Mater. Today* **2011**, *14*, 88–96.

127. Richardson, T.P.; Peters, M.C.; Ennett, A.B.; Mooney, D.J. Polymeric system for dual growth factor delivery. *Nat. Biotechnol.* **2001**, *19*, 1029–1034. [CrossRef] [PubMed]

128. Gorain, B.; Tekade, M.; Kesharwani, P.; Iyer, A.K.; Kalia, K.; Tekade, R.K. The use of nanoscaffolds and dendrimers in tissue engineering. *Drug Discov. Today* **2017**, *22*, 652–664. [CrossRef] [PubMed]

129. Lee, D.R.; Kaproth, M.T.; Parks, J.E. In vitro production of haploid germ cells from fresh or frozen-thawed testicular cells of neonatal bulls. *Biol. Reprod.* **2001**, *65*, 873–878. [CrossRef] [PubMed]

130. Badylak, S.F. The extracellular matrix as a biologic scaffold material. *Biomaterials* **2007**, *28*, 3587–3593. [CrossRef] [PubMed]

131. Janson, I.A.; Putnam, A.J. Extracellular matrix elasticity and topography: Material-based cues that affect cell function via conserved mechanisms. *J. Biomed. Mater. Res. A* **2015**, *103*, 1246–1258. [CrossRef] [PubMed]

132. Qazi, T.H.; Mooney, D.J.; Duda, G.N.; Geissler, S. Biomaterials that promote cell-cell interactions enhance the paracrine function of mscs. *Biomaterials* **2017**, *140*, 103–114. [CrossRef] [PubMed]

133. Li, E.; Chang, C.C.; Zhang, Z.; Li, Q. Characterization of tissue scaffolds for time-dependent biotransport criteria—A novel computational procedure. *Comput. Methods Biomech. Biomed. Eng.* **2016**, *19*, 1210–1224. [CrossRef] [PubMed]

134. Cheng, C.Y.; Wong, E.W.; Yan, H.H.; Mruk, D.D. Regulation of spermatogenesis in the microenvironment of the seminiferous epithelium: New insights and advances. *Mol. Cell. Endocrinol.* **2010**, *315*, 49–56. [CrossRef] [PubMed]

135. Pollanen, P.P.; Kallajoki, M.; Risteli, L.; Risteli, J.; Suominen, J.J. Laminin and type iv collagen in the human testis. *Int. J. Androl.* **1985**, *8*, 337–347. [CrossRef] [PubMed]

136. Anderson, J.M.; Rodriguez, A.; Chang, D.T. Foreign body reaction to biomaterials. *Semin. Immunol.* **2008**, *20*, 86–100. [CrossRef] [PubMed]

137. Morris, A.H.; Stamer, D.K.; Kyriakides, T.R. The host response to naturally-derived extracellular matrix biomaterials. *Semin. Immunol.* **2017**. [CrossRef] [PubMed]

138. Chen, Y.; Zhou, S.; Li, Q. Microstructure design of biodegradable scaffold and its effect on tissue regeneration. *Biomaterials* **2011**, *32*, 5003–5014. [CrossRef] [PubMed]

139. Pan, Z.; Ding, J. Poly(lactide-co-glycolide) porous scaffolds for tissue engineering and regenerative medicine. *Interface Focus* **2012**, *2*, 366–377. [CrossRef] [PubMed]

140. Wu, L.; Zhang, J.; Jing, D.; Ding, J. "Wet-state" mechanical properties of three-dimensional polyester porous scaffolds. *J. Biomed. Mater. Res. A* **2006**, *76*, 264–271. [CrossRef] [PubMed]

141. Poels, J.; Abou-Ghannam, G.; Decamps, A.; Leyman, M.; Rieux, A.D.; Wyns, C. Transplantation of testicular tissue in alginate hydrogel loaded with vegf nanoparticles improves spermatogonial recovery. *J. Control. Release* **2016**, *234*, 79–89. [CrossRef] [PubMed]

142. Chan, B.P.; Leong, K.W. Scaffolding in tissue engineering: General approaches and tissue-specific considerations. *Eur. Spine J.* **2008**, *17* (Suppl. S4), 467–479. [CrossRef] [PubMed]

143. Xia, T.; Liu, W.; Yang, L. A review of gradient stiffness hydrogels used in tissue engineering and regenerative medicine. *J. Biomed. Mater. Res. A* **2017**, *105*, 1799–1812. [CrossRef] [PubMed]

144. Crapo, P.M.; Gilbert, T.W.; Badylak, S.F. An overview of tissue and whole organ decellularization processes. *Biomaterials* **2011**, *32*, 3233–3243. [CrossRef] [PubMed]

145. Bayat, A.; Dorkoosh, F.A.; Dehpour, A.R.; Moezi, L.; Larijani, B.; Junginger, H.E.; Rafiee-Tehrani, M. Nanoparticles of quaternized chitosan derivatives as a carrier for colon delivery of insulin: Ex vivo and in vivo studies. *Int. J. Pharm.* **2008**, *356*, 259–266. [CrossRef] [PubMed]

146. Letchford, K.; Burt, H. A review of the formation and classification of amphiphilic block copolymer nanoparticulate structures: Micelles, nanospheres, nanocapsules and polymersomes. *Eur. J. Pharm. Biopharm.* **2007**, *65*, 259–269. [CrossRef] [PubMed]

147. Hamidi, M.; Azadi, A.; Rafiei, P. Hydrogel nanoparticles in drug delivery. *Adv. Drug Deliv. Rev.* **2008**, *60*, 1638–1649. [CrossRef] [PubMed]

148. Huang, M.; Vitharana, S.N.; Peek, L.J.; Coop, T.; Berkland, C. Polyelectrolyte complexes stabilize and controllably release vascular endothelial growth factor. *Biomacromolecules* **2007**, *8*, 1607–1614. [CrossRef] [PubMed]

149. Baek, J.S.; Cho, C.W. A multifunctional lipid nanoparticle for co-delivery of paclitaxel and curcumin for targeted delivery and enhanced cytotoxicity in multidrug resistant breast cancer cells. *Oncotarget* **2017**, *8*, 30369–30382. [CrossRef] [PubMed]

150. Izadifar, M.; Kelly, M.E.; Chen, X. Regulation of sequential release of growth factors using bilayer polymeric nanoparticles for cardiac tissue engineering. *Nanomedicine* **2016**, *11*, 3237–3259. [CrossRef] [PubMed]

151. Najjar, M.; Manzoli, V.; Abreu, M.; Villa, C.; Martino, M.M.; Molano, R.D.; Torrente, Y.; Pileggi, A.; Inverardi, L.; Ricordi, C.; et al. Fibrin gels engineered with pro-angiogenic growth factors promote engraftment of pancreatic islets in extrahepatic sites in mice. *Biotechnol. Bioeng.* **2015**, *112*, 1916–1926. [CrossRef] [PubMed]

152. Felice, B.; Prabhakaran, M.P.; Rodriguez, A.P.; Ramakrishna, S. Drug delivery vehicles on a nano-engineering perspective. *Mater. Sci. Eng. C Mater. Biol. Appl.* **2014**, *41*, 178–195. [CrossRef] [PubMed]

153. Lee, J.H.; Gye, M.C.; Choi, K.W.; Hong, J.Y.; Lee, Y.B.; Park, D.W.; Lee, S.J.; Min, C.K. In vitro differentiation of germ cells from nonobstructive azoospermic patients using three-dimensional culture in a collagen gel matrix. *Fertil. Steril.* **2007**, *87*, 824–833. [CrossRef] [PubMed]

154. Reuter, K.; Ehmcke, J.; Stukenborg, J.B.; Simoni, M.; Damm, O.S.; Redmann, K.; Schlatt, S.; Wistuba, J. Reassembly of somatic cells and testicular organogenesis in vitro. *Tissue Cell* **2014**, *46*, 86–96. [CrossRef] [PubMed]

155. Jalayeri, M.; Pirnia, A.; Najafabad, E.P.; Varzi, A.M.; Gholami, M. Evaluation of alginate hydrogel cytotoxicity on three-dimensional culture of type a spermatogonial stem cells. *Int. J. Biol. Macromol.* **2017**, *95*, 888–894. [CrossRef] [PubMed]

156. Artel, A.; Mehdizadeh, H.; Chiu, Y.C.; Brey, E.M.; Cinar, A. An agent-based model for the investigation of neovascularization within porous scaffolds. *Tissue Eng. Part A* **2011**, *17*, 2133–2141. [CrossRef] [PubMed]

157. Wang, P.; Jiang, X.; Jiang, Y.; Hu, X.; Mou, H.; Li, M.; Guan, H. In vitro antioxidative activities of three marine oligosaccharides. *Nat. Prod. Res.* **2007**, *21*, 646–654. [CrossRef] [PubMed]

158. Hsiao, C.H.; Ji, A.T.; Chang, C.C.; Cheng, C.J.; Lee, L.M.; Ho, J.H. Local injection of mesenchymal stem cells protects testicular torsion-induced germ cell injury. *Stem Cell Res. Ther.* **2015**, *6*, 113. [CrossRef] [PubMed]

159. Horibe, A.; Eid, N.; Ito, Y.; Hamaoka, H.; Tanaka, Y.; Kondo, Y. Upregulated autophagy in sertoli cells of ethanol-treated rats is associated with induction of inducible nitric oxide synthase (inos), androgen receptor suppression and germ cell apoptosis. *Int. J. Mol. Sci.* **2017**, *18*. [CrossRef]

160. Eid, N.; Kondo, Y. Ethanol-induced mitophagy in rat sertoli cells: Implications for male fertility. *Andrologia* **2017**. [CrossRef] [PubMed]

161. Baert, Y.; Stukenborg, J.B.; Landreh, M.; De Kock, J.; Jornvall, H.; Soder, O.; Goossens, E. Derivation and characterization of a cytocompatible scaffold from human testis. *Hum. Reprod.* **2015**, *30*, 256–267. [CrossRef] [PubMed]

162. Vermeulen, M.; Del Vento, F.; de Michele, F.; Poels, J.; Wyns, C. Development of a cytocompatible scaffold from pig immature testicular tissue allowing human sertoli cell attachment, proliferation and functionality. *Int. J. Mol. Sci.* **2018**, *19*. [CrossRef] [PubMed]

163. Baert, Y.; Rombaut, C.; Goossens, E. Scaffold-based and scaffold-free testicular organoids from primary human testicular cells. *Methods Mol. Biol.* **2017**. [CrossRef]

164. Wang, W.; Mo, Z.; Ye, B.; Hu, P.; Liu, S.; Yi, S. A clinical trial of xenotransplantation of neonatal pig islets for diabetic patients. *Zhong Nan Da Xue Xue Bao Yi Xue Ban* **2011**, *36*, 1134–1140. [CrossRef] [PubMed]

165. Ciubotaru, A.; Cebotari, S.; Tudorache, I.; Beckmann, E.; Hilfiker, A.; Haverich, A. Biological heart valves. *Biomed. Tech.* **2013**, *58*, 389–397. [CrossRef] [PubMed]

166. Takahashi, K.; Yamanaka, S. Induction of pluripotent stem cells from mouse embryonic and adult fibroblast cultures by defined factors. *Cell* **2006**, *126*, 663–676. [CrossRef] [PubMed]

167. Ishikura, Y.; Yabuta, Y.; Ohta, H.; Hayashi, K.; Nakamura, T.; Okamoto, I.; Yamamoto, T.; Kurimoto, K.; Shirane, K.; Sasaki, H.; et al. In vitro derivation and propagation of spermatogonial stem cell activity from mouse pluripotent stem cells. *Cell Rep.* **2016**, *17*, 2789–2804. [CrossRef] [PubMed]

168. Chuma, S.; Kanatsu-Shinohara, M.; Inoue, K.; Ogonuki, N.; Miki, H.; Toyokuni, S.; Hosokawa, M.; Nakatsuji, N.; Ogura, A.; Shinohara, T. Spermatogenesis from epiblast and primordial germ cells following transplantation into postnatal mouse testis. *Development* **2005**, *132*, 117–122. [CrossRef] [PubMed]

169. Hayashi, K.; Ohta, H.; Kurimoto, K.; Aramaki, S.; Saitou, M. Reconstitution of the mouse germ cell specification pathway in culture by pluripotent stem cells. *Cell* **2011**, *146*, 519–532. [CrossRef] [PubMed]

170. Sasaki, K.; Yokobayashi, S.; Nakamura, T.; Okamoto, I.; Yabuta, Y.; Kurimoto, K.; Ohta, H.; Moritoki, Y.; Iwatani, C.; Tsuchiya, H.; et al. Robust in vitro induction of human germ cell fate from pluripotent stem cells. *Cell Stem Cell* **2015**, *17*, 178–194. [CrossRef] [PubMed]

171. Botman, O.; Wyns, C. Induced pluripotent stem cell potential in medicine, specifically focused on reproductive medicine. *Front. Surg.* **2014**, *1*, 5. [CrossRef] [PubMed]

International Journal of
Molecular Sciences

MDPI

Review

Biomimetic Layer-by-Layer Self-Assembly of Nanofilms, Nanocoatings, and 3D Scaffolds for Tissue Engineering

Shichao Zhang [1], Malcolm Xing [2,3] and Bingyun Li [1,4,*]

[1] Department of Orthopaedics, School of Medicine, West Virginia University, Morgantown, WV 26506, USA;
     Shichao.zhang@hsc.wvu.edu
[2] Department of Mechanical Engineering, University of Manitoba, Winnipeg, MB R3T 2N2, Canada;
     Malcolm.Xing@umanitoba.ca
[3] The Children's Hospital Research Institute of Manitoba, Winnipeg, MB R3E 3P4, Canada
[4] West Virginia University Cancer Institute, Morgantown, WV 26506, USA
*   Correspondence: bili@hsc.wvu.edu; Tel.: +1-304-293-1075; Fax: +1-304-293-7070

Received: 25 April 2018; Accepted: 30 May 2018; Published: 1 June 2018

Abstract: Achieving surface design and control of biomaterial scaffolds with nanometer- or micrometer-scaled functional films is critical to mimic the unique features of native extracellular matrices, which has significant technological implications for tissue engineering including cell-seeded scaffolds, microbioreactors, cell assembly, tissue regeneration, etc. Compared with other techniques available for surface design, layer-by-layer (LbL) self-assembly technology has attracted extensive attention because of its integrated features of simplicity, versatility, and nanoscale control. Here we present a brief overview of current state-of-the-art research related to the LbL self-assembly technique and its assembled biomaterials as scaffolds for tissue engineering. An overview of the LbL self-assembly technique, with a focus on issues associated with distinct routes and driving forces of self-assembly, is described briefly. Then, we highlight the controllable fabrication, properties, and applications of LbL self-assembly biomaterials in the forms of multilayer nanofilms, scaffold nanocoatings, and three-dimensional scaffolds to systematically demonstrate advances in LbL self-assembly in the field of tissue engineering. LbL self-assembly not only provides advances for molecular deposition but also opens avenues for the design and development of innovative biomaterials for tissue engineering.

Keywords: layer-by-layer; self-assembly; polyelectrolyte; multilayer; nanofilm; nanocoating; biomaterial; scaffold; tissue engineering

---

1. Introduction

Tissue engineering, an interdisciplinary area that evolved from biomaterials and engineering development, aims to assemble scaffolds, cells, and functional molecules into tissues to create biological alternatives for damaged tissues or organs [1–5]. To obtain the desired outcomes, biomaterial scaffolds that act as templates for tissue regeneration should induce appropriate cellular responses, guide the growth of new functional tissues, and support organ systems. To mimic the extracellular matrix (ECM) of native tissues, the ideal scaffolds must meet some specific requirements, involving structural, physical, chemical, and biological properties and functions [6–8]. Among these requirements, the surface properties of the scaffolds have long been recognized as being of utmost importance due to the direct interface between materials and cells as well as tissues [9–11]. Therefore, one major challenge in tissue engineering is to control the surface properties of biomaterials (especially at the molecular level), to modify the behavior of cells, and to tune the formation of new tissues.

Up to now, considerable efforts have been devoted to functionalizing biomaterial surfaces for tissue engineering. Due to the ability to regulate the assembly of coating at the nanometer- or micrometer-scale, self-assembled monolayer assembly and Langmuir-Blodgett deposition have shown remarkable capability in modifying material surfaces for tissue engineering applications [12–16]. However, several intrinsic limitations of these two techniques, including a long fabrication period, limited raw material types, low formation efficiency, limited stability, and expensive instrumentation, have restricted their practical applications. In contrast, layer-by-layer (LbL) assembly is a highly versatile and simple multilayer self-assembly technique; it has the capability to fabricate multilayer coatings with controlled architectures and compositions from extensive choices of usable materials for various biomedical applications (as shown in Figure 1a) [17–23].

Benefiting from different driving forces and assembly technologies of LbL self-assembly, various LbL assembly biomaterials have been prepared from different material species, including polyelectrolytes, biomolecules, colloids, particles, etc., and have shown remarkable physical, chemical, and biological properties/functions in the field of tissue engineering [20,24,25]. In this review, we attempt to present a brief overview of recent advances in designing and fabricating LbL self-assembly biomaterial scaffolds for tissue engineering applications. The advanced LbL assembly technique is introduced, ranging from origin, technology, and mechanisms to biomedical applications. Furthermore, we highlight recent advances in controllable fabrication, properties, and performance of LbL assembly in the forms of multilayer nanofilms, nanocoatings, and three-dimensional (3D) scaffolds for tissue engineering. Finally, we discuss the perspectives of further research directions in the development of LbL assembly for tissue engineering.

2. LbL Self-Assembly Technology

2.1. Origin and Definition

LbL assembly is an alternative to self-assembled monolayer assembly and Langmuir-Blodgett deposition, which are two dominant techniques for obtaining solid films at the molecular level. LbL assembly was first proposed by Iler in 1966 and achieved substantial development after the pioneering work of Decher et al. in the 1990s [18,20,26]. Since then, LbL assembly technique has become an efficient, facile, flexible, and versatile strategy to coat substrates with multilayers of controlled structures, properties, and functions for various applications [26]. The process of LbL assembly is simple and can be accurately controlled to create finely tailored structures [27,28]. Typically, the LbL assembly process includes the sequential adsorption of complementary molecules on a substrate surface, driven by multiple interactions involving electrostatic and/or nonelectrostatic interactions. Between the adsorption steps for each layer deposition, steps of washing and drying are usually introduced to avoid contamination of the next solution due to liquid adhering on substrates from the former solution, and to elute the loose molecules and stabilize them in the formed layers. These deposition and wash steps can be repeated to achieve the desired number of deposition layers. Moreover, fine control of composition, thickness, and topography can be achieved by adjusting the assembly parameters involving solution properties, like concentration, ionic strength, and pH, and process parameters, such as temperature, time, and drying conditions [18,29–34]. Various building blocks used for LbL assembly include, but are not limited to, natural polymers, synthetic polymers, peptides, clays, metal oxides, polymer gels, and complexes of such materials [20,29,35–37]. Compared to other methods for fabricating nanofilms, there are three prominent advantages of the LbL assembly technique, these include precise control of the composition and structure of nanofilms, large-scale fabrication capacity on various types of substrates regardless of size and shape, and mild and confined formation environments.

2.2. Technology Categories and Mechanisms

The widespread use of LbL assembly for various applications with different processing requirements and tools has led to the design and development of a variety of deposition technologies, for instance, dipping, centrifugation, roll-to-roll, calculated saturation, creaming, immobilization, atomization, spinning, spraying, magnetic assembly, high gravity, electrodeposition, electrocoupling, filtration, fluidics, and fluidized beds [17,38–43]. These technologies can be divided into five main categories, including (i) immersion; (ii) spin; (iii) spray; (iv) electromagnetic driven; and (v) fluidic assembly, as shown in Figure 1b. Taking into account the diversity of process properties and resulted nanofilm properties due to the assembly technologies, the proper choice of assembly technology is crucial for controllable fabrication and successful application of the assembled nanostructured materials. Different assembly methods usually result in different structures and properties of the assembled materials.

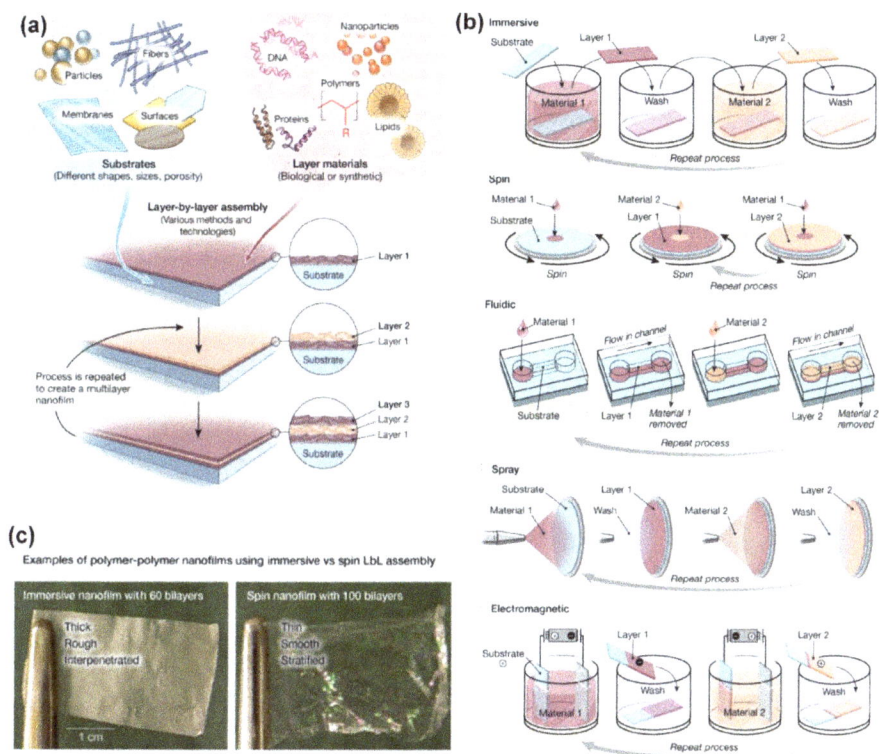

Figure 1. (**a**) Schematic overview of LbL assembly technique with fabrication capacity on any type of substrates and from an extensive choice of materials; (**b**) schematics showing the five main technology categories for LbL assembly; (**c**) a typical comparison of different films using immersive and spin assembly. Reprinted from [17] with permission from American Association for the Advancement of Science.

Immersive LbL assembly, also called dip assembly, including different techniques like dewetting, roll-to-roll, centrifugation, creaming, and so on, is the most widely used method and can form interpenetrated layered structures [17,44]. Immersive assembly allows for nanofilms on substrates of almost any shape or size. The challenge of this technique is to achieve reduced assembly times and to create automated systems with less manual intervention, which has attracted much attention and research. LbL spin assembly employs the common coating technology of spinning surface to construct nanostructured materials [17]. It usually results in humongous nanofilms with a short assembly process; however, it is limited to coating small planar substrates, as shown in Figure 1c. LbL assembly using spray coating can form films based on aerosolizing solutions and facilely spraying aerosols onto planar materials and particulates [45]. There are two main forces governing spray assembly: Bulk movement in the spray and random movement in the liquid film. Due to the unique manner, spray assembly can fabricate films on industrial-scale substrates with any surface topography and the resultant films usually show distinct layered structures. Electromagnetic assembly is a relatively new method based on applied electric or magnetic field to form layered structures, showing a substantially different driving force, like current-induced pH change and redox-reactions [43,46]. Although this technology usually requires special equipment and expertise, it does provide a new strategy for multilayer film assembly. By coating channel walls or substrates placed in channels, fluidic assembly technique is proposed to perform complicated assembly on designed 3D structures and surfaces that are not easily accessible to other technologies, providing a novel strategy for region-specific patterning and low reagent consumption [5,47]. The general method involves using vacuum or pressure to move polymer and washing solutions through the channels, like capillaries, tubing, microfluidic networks, etc. Especially, besides planar substrates, 3D aerogels and small particles (<5 µm) can also be coated using fluidic assembly, which can function as a valuable tool to coat-sensitive particulate substrates. In despite of the required special equipment and operation, fluidic assembly greatly increases the industrial capacity of multilayer assemblies by using various substrates along with reduced reagent consumption.

Conventionally, electrostatic interaction is the primary and most widely used driving force for the formation of nanostructured multilayer films, as presented in Figure 2a. Besides electrostatic interaction, LbL assembly can be driven by multiple other interactions, including hydrogen bonds, halogen bonds, charge-transfer interactions, covalent bonds, host-guest interactions, coordination chemistry interactions, biologically specific interactions, stereo complexation, surface sol-gel process, etc. [18]. Each of these intermolecular interactions has advantages and disadvantages. The driving forces and combinations among them can be utilized to significantly enrich the fabrication and application of LbL assembly materials for tissue engineering and biomedicine, bioelectronics, drug delivery, environment, energy, and information storage [48–50].

2.3. Biomedical Applications

The LbL self-assembly technique has been widely used for various biomedical applications including tissue engineering, medical implants, regenerative medicines, drug delivery, biosensors, bioreactors, and so on (Figure 2b) [35,51–55]. The capacity to manipulate the chemical, physical, and topographical properties of nanostructured architectures by facilely adjusting solution properties and assembly parameters, not only allows LbL self-assembly to be suitable to investigate the effects of external stimuli on cellular responsiveness but also provides a new strategy for creating two/three-dimensional (2/3D) nanostructured architectures and coatings for individual cells or scaffolds for tissue engineering (Figure 2c) [56–59]. The uses of LbL self-assembly biomaterials include nanostructured coating or materials to either promote or prevent cell adhesion, to maintain and direct cellular phenotypes, and to provide 3D scaffolds for cell culture or co-culture. Since the other biomedical applications of LbL assembled materials have been extensively discussed and reviewed, this review will focus on the modulation of structures and functions, and applications of LbL assembly in the field of tissue engineering.

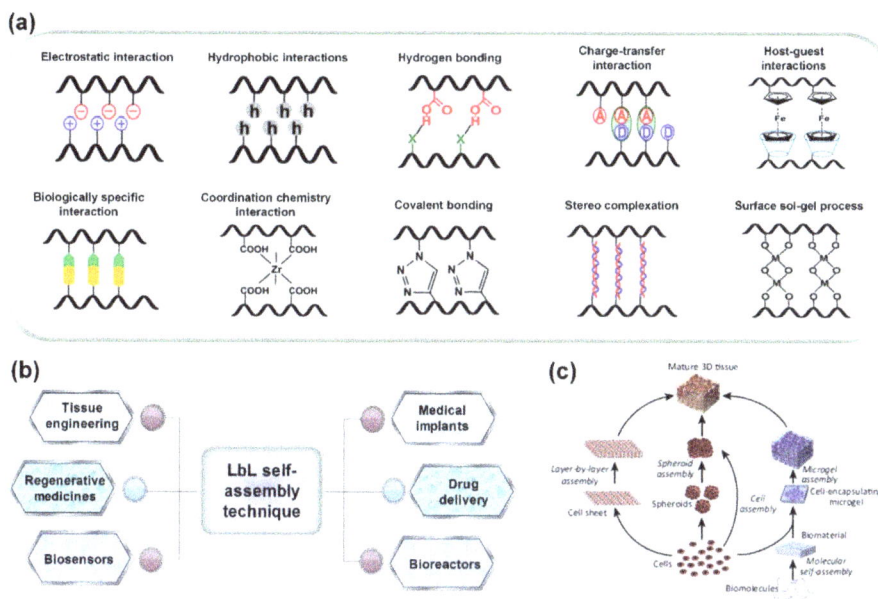

Figure 2. (a) Molecular interactions driving the LbL self-assembly of materials; (b) schematic showing the main biomedical application fields for LbL self-assembly technique; (c) schematic of multiscale assembly strategies for engineering tissue constructs. Reprinted from [18,56] with permissions from American Chemical Society and Elsevier.

3. LbL Self-Assembly of 2D Multilayer Nanofilms for Tissue Engineering

3.1. LbL Multilayer Nanofilms Directing Cellular Phenotypes

In biology, ECM is a collection of extracellular molecules secreted by cells that provide structural and biochemical support to the surrounding cells [60]. Taking into account the key role of ECM, various ECM molecules like collagen, polysaccharide, etc. have been used to fabricate multilayer nanofilms for tissue engineering using the LbL self-assembly technique [61–65]. Among these ECM molecules, collagen is most widely used in combination with other natural polyelectrolytes. The structures and properties involving surface morphology, thickness, zeta potential, surface roughness, and cellular phenotype like cell adhesion, cell growth, and cell differentiation have been investigated to demonstrate the feasibility of designing functional biomaterials by means of LbL assembly. For instance, Wittmer et al. fabricated various multilayer nanofilms composed of polysaccharides, polypeptides, and synthetic polymers, and investigated the adhesion and function of various hepatic cells in terms of terminal layer, film composition, charge, rigidity, and presence of biofunctional species [65]. This study offered the key variables in promoting attachment and function of hepatic cells and provided a promising candidate for in vivo human liver tissue engineering applications.

Single-walled carbon nanotubes (SWNTs) can not only exhibit a series of unique mechanical and electrical properties, but also demonstrate fascinating cell effect due to their electrical stimulation for neurological- or brain-related tissue engineering. Gheith et al. prepared SWNT multilayer films using charged nanotubes coated with designed copolymers, which exhibited high electrical conductivity to electrically stimulate excitable neuronal cells [66]. The resultant assembly films showed a clear demonstration of electrical excitation of neurons when a current was passed through the LbL films, revealing the capacity to adjust the biological activity of films and their interaction with cells. SWNT/polyelectrolyte multilayer films were also fabricated by Jan et al., on which environment-sensitive embryonic neural stem cells from the cortex were successfully differentiated into neurons, astrocytes, and oligodendrocytes, as shown in Figure 3a,b [24].

Up to now, considerable efforts have been devoted to optimizing the surface properties of LbL assembly films, such as rigidity, roughness, hydrophilicity, etc. [67–70]. Among them, cross-linking of layer components is the most widely used strategy. Hillberg et al. studied the effect of genipin cross-linking on the cellular adhesion properties of LbL polyelectrolyte films, and found that the cross-linking resulted in an increased cell adhesion and spreading on polymeric films [71]. They mainly focused on the investigation of film rigidity and cell adhesion of nanofilms with and without crossing-linking, providing an effective strategy for improving cell adhesion on nanofilms. However, no further detailed studies on surface chemistry of films and its effect on cell proliferation were carried out. By employing chitosan with/without cross-linking and/or coating with alginate, Silva et al. proposed a strategy to adjust the cell adhesion of LbL assembly films, as demonstrated in Figure 3c. The significant changes observed in cell adhesion, spreading, and proliferation can be attributed to the change of surface chemistry and mechanical properties due to cross-linking of multilayer films [72].

To further enhance the function of LbL assembly nanofilms, various functional fillers including nanoparticles, growth factors, and antibacterial agents, have been introduced to construct multilayer films as layer component and loading drugs [73–79]. For example, Hu and co-authors reported a strategy to construct hybrid chitosan/gelatin multilayers embedded with mesoporous silica nanoparticles on a titanium implant to regulate biological behavior of osteoblasts/osteoclasts in vitro, as presented in Figure 3d [77]. In addition, the free-standing multilayer nanofilms composed of chitosan and dopamine-modified hyaluronic acid were prepared by Sousa et al., and exhibited enhanced cell adhesion, viability, and proliferation for bone tissue engineering by optimizing their morphology, chemistry, and mechanical properties (Figure 3e,f) [80].

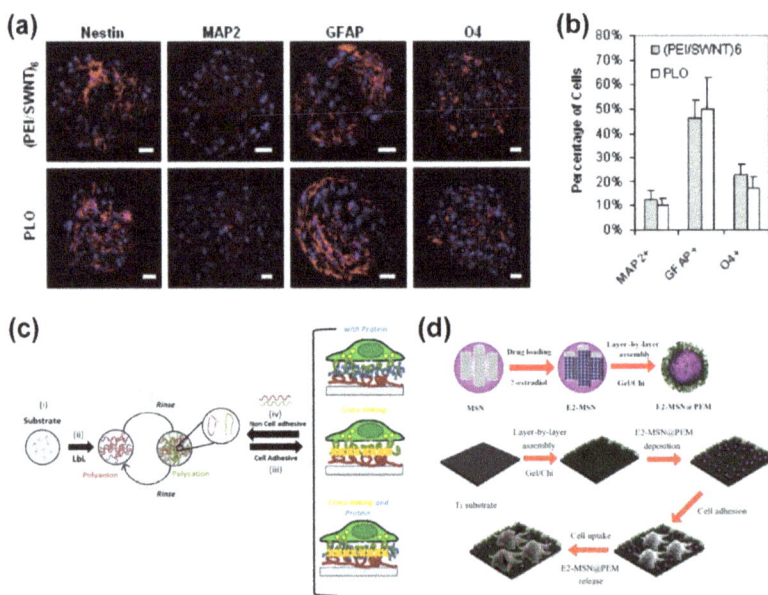

Figure 3. *Cont.*

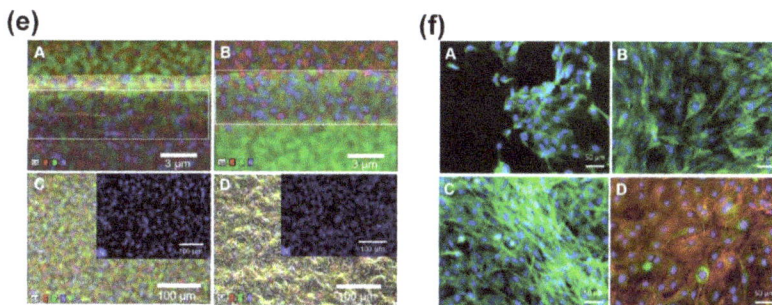

Figure 3. (**a**) Confocal microscopy images of differentiated neurospheres on day 7; (**b**) average percentages of differentiated cell phenotypes after 7 days in culture; (**c**) schematic showing LbL assembly of polyelectrolytes based on electrostatic interactions for tuning cell adhesive properties using cross-linking; (**d**) schematic illustration of the fabrication and cell uptake of LbL assembly multilayers embedded with β-estradiol-silica nanoparticles onto substrates; (**e**) mixed element map of the cross-section and upper surface of different catechol-based freestanding membranes; (**f**) Osteopontin immunofluorescence images of cells after 14 days cultured on different catechol-based freestanding membranes. Reprinted from [24,72,77] with permissions from American Chemical Society, Elsevier, and Wiley-VCH.

3.2. LbL Multilayer Nanofilms Encapsulating Cells and Tissues

Encapsulation of live cells and tissue offers an effective strategy to modulate cells or tune the response to their environment, especially for attenuating deleterious host responses toward transplanted cells [70,81]. Because of its ability to generate films of nanometer thickness on chemically and geometrically diverse substrates under mild formation environments, LbL assembly has emerged as an ideal technique compared to other methods, and can greatly minimize transplant volume for cell/tissue encapsulation in the field of tissue engineering. Wilson et al. reported the successful intraportal islet transplantation by LbL self-assembly of poly(L-lysine)-*g*-poly(ethylene glycol) (biotin) and streptavidin, and the resulted nanothin; PEG-rich conformal coatings can be tuned to coat inlets without loss of their viability and function, revealing a unique approach to modify the biochemical surfaces of living cells and tissues for various tissue engineering applications, as shown in Figure 4a,b [82]. Single living cell encapsulation in nano-organized polyelectrolyte shells were reported and the resultant shells could effectively protect cell integrity and preserve cell metabolic activities (Figure 4c,d) [83,84]. Recently, Mansouri et al. prepared nonimmunogenic polyelectrolyte multilayer films composed with alginate, chitosan-graft-phosphorylcholine and poly-L-lysine-graft-polyethylene glycol using LbL assembly; they investigated these nanofilms on fully functional human red blood cells in suspension for attenuated immune response, as exhibited in Figure 4e [85]. The resultant nonimmunogenic films not only create an advanced nanomaterial for production of universal red blood cells, but also provide an effective strategy for the design and development of functional multilayers for cell and tissue encapsulation.

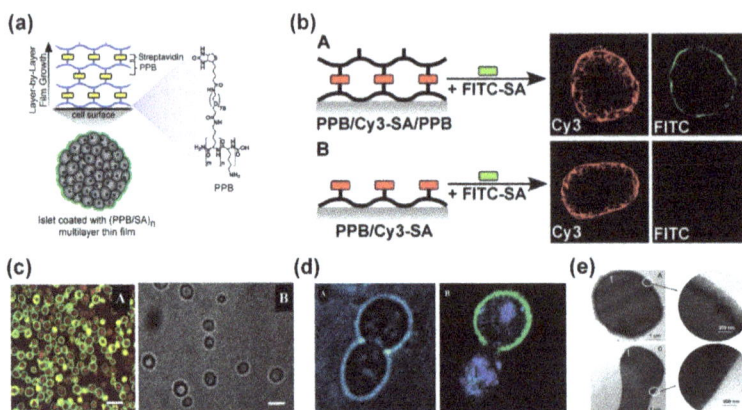

Figure 4. (a) Schematic illustration of the fabrication of PEG-rich, nanothin conformal islet nanofilms via LbL assembly; **(b)** Poly(L-lysine)-*g*-poly(ethylene glycol)(biotin)/streptavidin multilayer films assembled on individual pancreatic islets; **(c)** confocal images of freshly coated living cells and **(d)** transmission image of cells during their duplicating process; **(e)** transmission electron microscope images of coated and uncoated red blood cells with LbL assembly films. Reprinted from [82,84,85] with permissions from American Chemical Society.

4. LbL Self-Assembly of Scaffold Nanocoatings for Tissue Engineering

4.1. LbL Scaffold Coating Directing Cellular Phenotypes

To engineer tissues in vitro, cells are usually cultured on 3D scaffolds that provide the bioactive cues to guide their growth and differentiation into tissues. Designing the physical and chemical properties of scaffold surfaces has become a crucial aspect for tissue engineering as it can directly tune the cell responses to 3D scaffolds. Among various methods for modifying scaffold surfaces, LbL assembly is highly attractive due to its capability to conformally coat complicated geometries and its tunability of incorporation. Due to their good biocompatibility, ECM components including collagen, chitosan, and gelatin have been widely used to construct scaffold nanocoating using LbL assembly [86–91]. For example, the poly(styrene sulfonate)/chitosan multilayer films were prepared to modify poly(L-lactic acid) scaffold surfaces toward improving the scaffold cytocompatibility to human endothelial cells [86]. Various nanocoatings composed with positively charged macromolecules and negatively charged gelatin have been prepared to improve the surface biocompatibility of different tissue scaffolds, like titanium films, poly(D,L-lactide) films, poly(L-lactic acid) fibrous scaffolds, etc., and to promote cell adhesion, growth, and differentiation by modifying the structural and chemical properties of scaffolds [87–89]. For cartilage tissue engineering applications, collagen and chondroitin sulfate were deposited alternately on 3D scaffolds to optimize the cell-material interaction for enhancing chondrogenesis [90,91].

Besides the ECM components, biocompatible inorganic materials, such as clay, calcium phosphate, and hydroxyapatite have been used as building blocks to modify 3D scaffold surfaces using LbL assembly. Lee and co-workers prepared new LbL assembly of clay/poly(diallyldimethylammonium chloride) multilayers on inverted colloidal crystal scaffolds, and examined their effect on cell adhesion and differentiation from both experimental and modeling aspects (Figure 5a) [92]. This study not only created new nanofilm-coated materials as 3D microenvironments for cellular co-cultures, but also provided a strategy to efficiently simulate differentiation niches for the different components of hematopoietic systems. An electrospun fiber scaffold coated with LbL assembly of gelatin and calcium phosphate films was prepared by Li et al. for bone tissue engineering [93]. This modified scaffold could effectively mimic the structure, composition, and biological function of the bone extracellular matrix, resulting in a significantly higher cell proliferation rate due to the mineralized nanocoating. Hydroxyapatite-based LbL assembly nanocoatings were also developed and used to enhance the osteogenic capacity of human mesenchymal stem cells (hMSCs) by providing additional bioactive agents, as shown in Figure 5b,c [94].

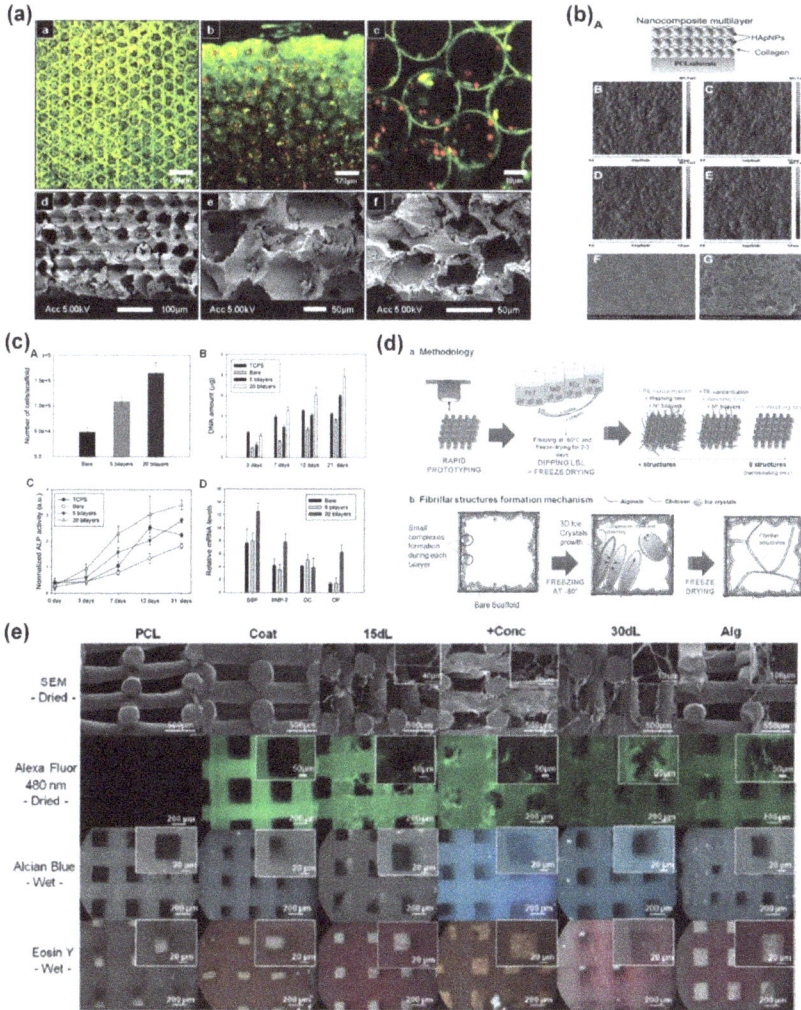

Figure 5. (**a**) Confocal images of inverted colloidal crystal scaffolds cultured with thymic epithelial cells and monocyte cells; (**b**) illustration of LbL multilayer nanocomposite coating with hydroxyapatite and collagen on substrates, and the AFM images of multilayers with different numbers of bilayers; (**c**) hMSCs adhesion and their quantification of DNA amounts to bare and coated scaffolds, and alkaline phosphatase activity and relative mRNA expression during the culture of hMSCs on various substrates; (**d**) steps and mechanism for developing the hierarchical and hybrid 3D scaffolds and (**e**) representative images of the structures of these scaffolds. Reprinted from [49,92,94] with permissions from Wiley-VCH and Royal Society of Chemistry.

Recently, many non-conventional LbL self-assembly techniques and hierarchical structures incorporated with multi-functions have been proposed to design and develop novel scaffold nanocoatings for tissue engineering. Oliveira et al. reported a new approach to develop inner structures inside 3D scaffolds for tissue engineering [49]. By combining the non-conventional LbL assembly having incomplete washing steps with ice crystal growth, hierarchical and fibrillar structures could be created in the interior of 3D scaffolds, which could effectively enhance the surface area available for cell growth and mimic the natural environment of fibrillar extracellular matrix, as shown in Figure 5d,e. Spray-assisted LbL assembly was also proposed to fabricate hyaluronic acid and poly-L-lysine multilayers on porous hyaluronic acid scaffold for the development of a single epidermal-dermal scaffold to treat full-thickness skin defects [95]. In addition, various growth factors and antibacterial agents were also immobilized on 3D scaffolds to form mirco/nano-hierarchical structures using the LbL assembly technique [23,29,79,96]. The effective combination of hierarchical structures and incorporated fillers may result in multi-functions for various tissue engineering, especially for bone regeneration and anti-infection.

4.2. LbL Scaffold Coating for Cell Co-Culture

The cell co-culture technique is of great importance in proof-of-principle studies, cell-cell interaction evaluation, and clinical transformation for various tissue engineering applications [97,98]. To achieve the complexity and organization of the in vivo cellular microenvironment, cell co-culture scaffolds with patterned coating and designed anisotropic properties are greatly needed, which can be effectively created using the LbL assembly technique due to its capacity of nanometer-scale engineering of hierarchical surfaces. Khademhosseini et al. reported an effective approach to prepare LbL assembly multilayers of hyaluronic acid and poly-L-lysine for patterned cell co-cultures, as presented in Figure 6a–c [99]. Based on the ionic adsorption of poly-L-lysine to hyaluronic patterns, the scaffold surface could be switched from cell repulsive to adherent, facilitating the adhesion of other type of cells. Furthermore, they examined the utility of this approach for co-culture patterns using different cell systems of hepatocytes or embryonic stem cells with fibroblasts. Micropatterned cell co-cultures using LbL deposition of ECM components were proposed by Fukuda et al. for creating effective tools for cell-cell interaction studies and tissue-engineering applications (Figure 6d,e) [100]. The co-culture scaffolds were coated with LbL assembly of ECM components (i.e., hyaluronic acid, fibronectin, and collagen) in which fibronectin was used to create cell-adhesive islands and the collagen was used to change the non-adherent hyaluronic acid pattern to cell adherent for the other cell type, respectively.

To achieve selective cell targeting, Zhou et al. prepared chitosan/alginate multilayer coatings on poly(lactide-co-glycolide) nanoparticles for antifouling protection and folic acid binding by LbL self-assembly (Figure 6f) [101]. They carried out cellular uptake measurements by co-culturing different cells, revealing a facile way to sequentially tailor nanoparticle surfaces to reduce unspecific interactions and to attach other molecules for successive cell selective adhesion. In this study, the nanoparticles were employed as substrates and endowed with selective targeting properties, fully indicating a promising design capability of nanodrugs using this strategy for targeted cell therapy. In addition, Kidambi et al. reported the fabrication of patterned cell co-cultures using LbL assembly of synthetic polymers without the aid of adhesive proteins/ligands [102]. As an alternative approach for co-culture scaffold fabrication, this strategy provides flexibility in design and development of cell-specific surfaces for tissue engineering.

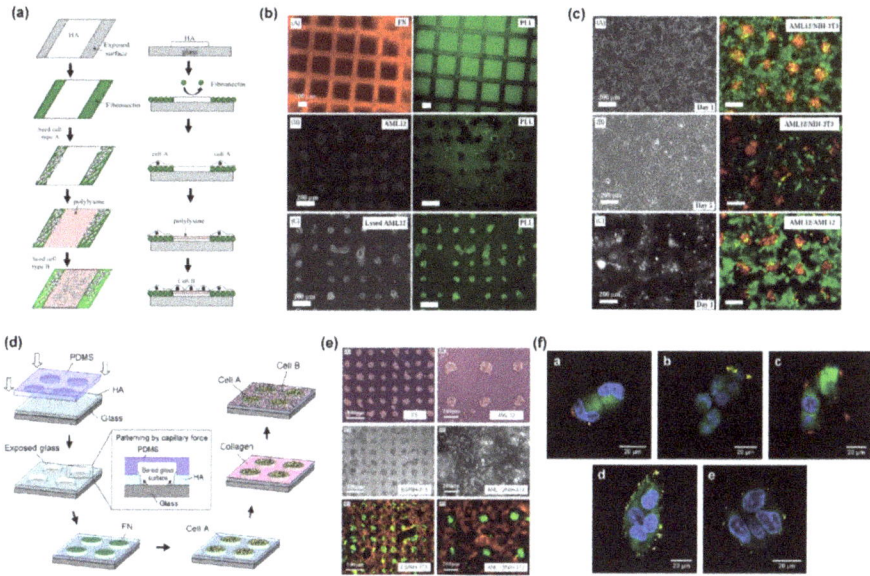

Figure 6. (**a**) Schematic diagram showing the layering approach to pattern cell co-cultures; (**b**) patterned hyaluronic acid surfaces attached with poly-L-lysine and immobilized cells; (**c**) patterned co-cultures of hepatocytes with fibroblasts; (**d**) schematic diagram of the fabrication of the co-culture system using LbL assembly technique; (**e**) patterned cell culture and co-culture on hyaluronic acid/collagen surface; (**f**) confocal laser scanning microscopy images of hepatocytes after co-culture on different chitosan/alginate LbL assembly films with nanoparticles. Reprinted from [99–101] with permissions from Elsevier.

5. LbL Self-Assembly of 3D Scaffolds for Tissue Engineering

5.1. 3D Scaffolds of ECM Films

The design of ECM multilayered scaffolds that resemble the hierarchical, lattice-like structure of tissues poses a big challenge for tissue engineering. Such ECM multilayer scaffolds can effectively regulate the interactions between different cells and function as natural tissues. Rajagopalan et al. described an effective approach to fabricate 3D scaffolds that mimic the micro-environment surrounding cells in vivo [103]. By using this strategy, different constructs composed of alternating layers of cells and biocompatible ECM scaffolds were fabricated, involving hepatocyte-scaffold-hepatocyte, hepatocyte-scaffold-endothelial cell, and hepatocyte-scaffold-fibroblast constructs, fully indicating the potential to generate constructs of various tissue types. Kim et al. developed the design of in vitro liver sinusoid mimics using LbL assembly scaffolds of chitosan and hyaluronic acid [104]. Silva et al. prepared nanostructured 3D constructs based on chitosan and chondroitin sulphate multilayers by combining LbL technology and template leaching for cartilage tissue engineering [50]. The obtained 3D scaffolds retrieved after paraffin leaching showed a high porosity and water uptake capacity of ~300%, and could maintain the chondrogenic phenotype and chondrogenic differentiation of multipotent bone marrow derived stromal cells, revealing the potential for clinical application in the field of cartilage tissue engineering. Although many studies have been carried out on 3D scaffolds of ECM films for tissue engineering, these studies were mainly focused on the material property investigation and in vitro studies; more detailed in vivo studies on 3D scaffolds of ECM films are still a big challenge and need to be carried out in the future.

5.2. 3D Scaffolds with Cell Composition Layers

To achieve the ideal spatial distribution of cultured cells with the ECM micro-environment, various cells have been used as building blocks to construct 3D scaffolds for tissue engineering using LbL assembly. Compared to other LbL assembly methods, microfluidic LbL patterning is widely used due to its region-specific capacity. By using microfluidic LbL patterning of 3D biopolymer matrices, Tan et al. created biologically relevant cellular arrangements on scaffold surfaces, revealing a robust strategy to enhance cellular pattern integrity and effectively control cellular microenvironment, as shown in Figure 7a,b [7,47]. They investigated the effects of channel size, cell type, and matrix composition on pattern integrity. During their subsequent studies, the hierarchical biomimetic multilayer constructs with layers of cells and biopolymers in micro-channels were created for blood vessel engineering, indicating the controllable patterning of cells and ECM environments in 3D [105]. Besides biopolymers, carbon-based materials like graphene oxide were also used to combine with cells to form 3D tissue scaffolds. Shin et al. proposed an effective strategy to mimic the structure and function of native ECM materials using LbL assembly of cells separated with graphene oxide [106]. The resultant graphene oxide-based structures were used as adhesive sheets for cells and allowed the formation of multilayer cell constructs, showing great potential in engineering 3D tissues with enhanced organization and mechanical integrity, as displayed in Figure 7c.

Furthermore, certain structural designs and new LbL assembly techniques have been proposed to mimic the native and complex 3D cellular architecture for tissue engineering. Feng et al. described LbL seeding of smooth muscle cells in deep micro-channels to create aligned multilayers for vascular engineering [107]. Different from most other methods, Choi and co-workers proposed a new strategy to fabricate microfluidic structures within 3D cell-seeded scaffolds [5]. This approach could control the chemical environment on a micrometer scale within a macroscopic scaffold, as shown in Figure 7d. The resultant microfluidic channels allowed efficient distribution control and exchange of soluble chemicals, indicating a new format for the design and development for tissue engineering.

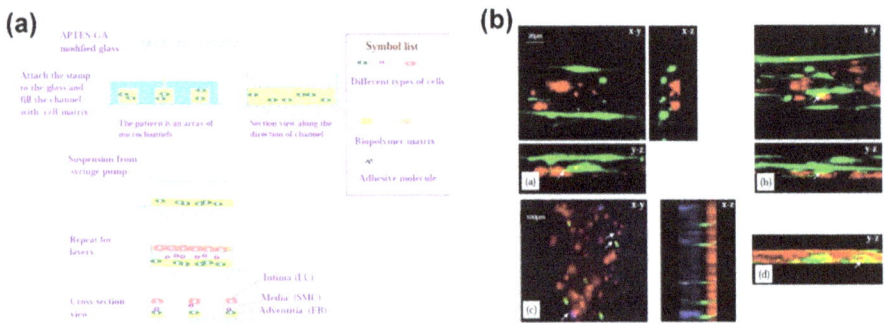

Figure 7. *Cont.*

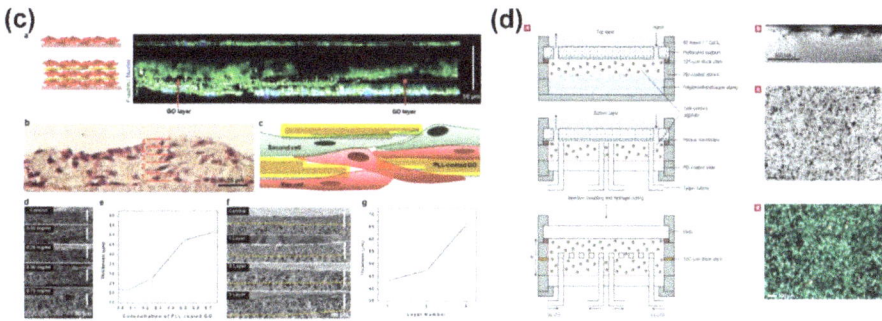

Figure 7. (**a**) Schematic illustration of the microfluidic LbL approach used to create 3D hierarchical systems with matrices and cells; (**b**) 3D images reconstituted from stacks demonstrate a two- or three-layer structure; (**c**) confocal cross-sectional images of the control group (top) and the 3 layer tissue constructs (bottom) after 2 days of culture. Hematoxylin and eosin stain images of 3 layer fibroblasts. Schematic illustration of the cross-section of the 2 layer construct. SEM images showing the cross-section and the thickness of 1, 2 and 3 layer constructs fabricated with various concentrations of poly-L-lysine-coated graphene oxide as interlayer films; (**d**) fabrication of cellular microfluidic scaffolds including the fabrication process and its resultant microstructures. Reprinted from [5,7,106] with permissions from Nature Publishing Group, Elsevier, and Wiley-VCH.

5.3. 3D Scaffolds with ECM Film Encapsulated Cells

To create ideal 3D cell-polymer material composites for tissue engineering, the methodology to construct 3D multilayers composed of various cells with a nanometer-sized ECM layer is highly desired [108–110]. The principle of this design is that the formation of a nanometer-sized ECM layer on the cell surface first provides a cell-adhesive surface for the second type of cells. Due to the unique biocompatibility, ECM components of fibronectin and gelatin have been widely selected as building blocks to create nanometer-sized ECM films on different cell surfaces. Matsusaki et al. described the fabrication of well-organized cellular multilayers by constructing fibronectin-gelatin nanocoatings with designed thickness on cell surfaces using LbL assembly [111]. Furthermore, xenogeneic cellular multilayers such as blood vessels were constructed. By constructing fibronectin-gelatin coating nanofilms on a single cell, Nishiguchi et al. developed a cell-accumulation-based LbL assembly technique to rapidly construct 3D multilayered tissues with endothelial tube networks, as shown in Figure 8a,b [112]. Especially, the layer number, cell type, and location could be controlled by adjusting the seeding cell number and order using this technique. Recently, Sasaki et al. successfully constructed homogenous, dense, and vascularized liver tissue thereby presenting the functional ability of using the LbL cell coating technique (Figure 8c) [113]. This approach allowed loading of cells sterically onto other cells coated with LbL assembly of fibronectin and gelatin, and led to improved cellular function in terms of human albumin production and cytochrome P450 activity in vitro. Besides the conventional methods, a new filtration LbL assembly technique was introduced by Amano et al. to develop vascularized pluripotent stem cell-derived 3D-cardiomyocyte tissues for pharmaceutical assays, as shown in Figure 8d [114]. Benefiting from the integrated capacity of high yield and low damage, the constructed vascularized tissues could be a promising tool for tissue regeneration and drug development.

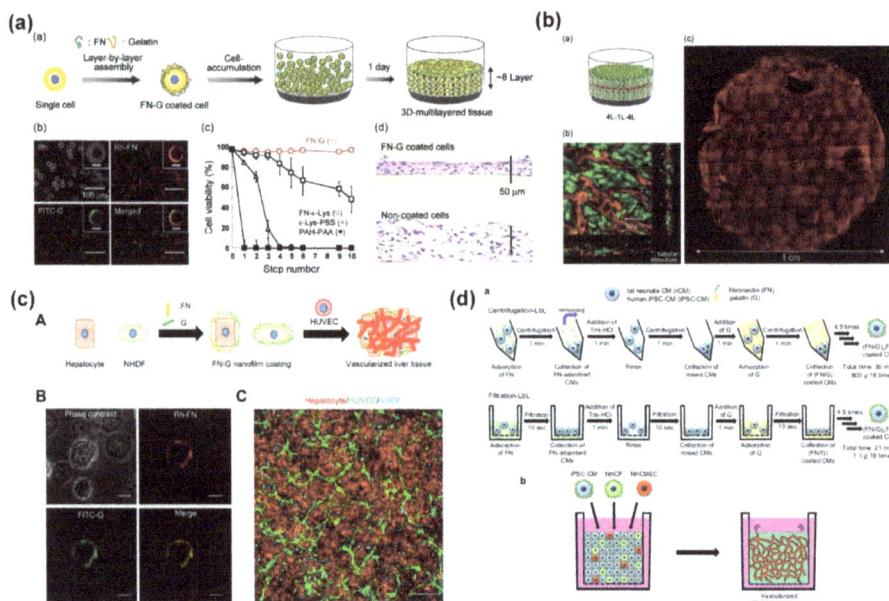

Figure 8. (**a**) Rapid construction of 3D multilayered tissues with endothelial tube networks by the cell-accumulation technique and the phase/fluorescent microscopic images, cell viability, and hematoxylin and eosin staining images of the resultant tissues; (**b**) schematic illustration, cross-section image, and reconstructed fluorescent image of 3D multilayered tissues; (**c**) the construction of vascularized liver tissue using LbL cell coating technique; (**d**) schematic illustrations of centrifugation-LbL and filtration-LbL for nanofilm coating on cell surfaces, and construction of vascularized 3D tissues by the cell accumulation technique. Reprinted from [112–114] with permissions from Wiley-VCH and Elsevier.

6. Concluding Remarks and Perspectives

LBL assembly, as a molecular-assembly technique, has been extensively used for the design and fabrication of biomaterial scaffolds for tissue engineering. The LbL assembly technique based on various driving forces and in the form of different technical strategies has been developed for a variety of biomedical applications, in particular, tissue engineering. To impart cell adhesive properties and biocompatibility to an existing substrate, to reassemble cells into specific tissues, or to function alone as tissue scaffolds, LbL assembly in the forms of multilayer nanofilms, scaffold nanocoatings, and 3D scaffolds have been created and served as new biomimetic matrices for tissue engineering. Yet important issues regarding LbL technologies, formation mechanisms, and applications, such as driving forces, regulation strategies, and biological functions still need further investigation. Here, we have reviewed the utilization of LbL assembly to design and create nanofilms, nanocoatings, and 3D scaffolds for tissue engineering.

In spite of remarkable progress in the development of LbL assembly biomaterials and related tissue engineering applications, some challenges still remain which may restrict their practical use. The challenges in the field of LbL assembly technique include the design and optimization of the technology parameters to obtain fast and stable coatings to enable long-term storage of multilayer systems. Moreover, the "black box" of LbL assembly techniques, focused on what materials are used (the input) for assembling the desired functional films (the output), has not yet been unpacked for specific tissue engineering applications. The current applications of LbL assembly in the field of tissue engineering are mainly focused on cell cultures in vitro, while the construction of complicated and functional tissues or organs and their practical applications in vivo still remain a great challenge. In addition, combinations between the LbL self-assembly technique and other multidisciplinary approaches should be explored to develop new LbL assembly biomaterials with multi-functions for tissue engineering. The potential for manufacture and application of LbL assembly biomaterials in the tissue engineering field is apparently unlimited, and we expect that continuous efforts in developing functional biomaterials using LbL assembly will address the current challenges and contribute to further advances in tissue engineering.

Author Contributions: S.Z. performed literature review and drafted the manuscript; S.Z., M.X. and B.L. discussed the ideas and outlines, and critically reviewed and revised the manuscript; All authors have approved the final version of the manuscript.

Acknowledgments: This work was supported by the Office of the Assistant Secretary of Defense for Health Affairs, through the Peer Reviewed Medical Research Program, Discovery Awards under Award Nos. W81XWH-17-1-0603 and W81XWH1810203. We also acknowledge financial support from AO Foundation, Osteosynthesis & Trauma Care Foundation, the West Virginia National Aeronautics and Space Administration Experimental Program to Stimulate Competitive Research (WV NASA EPSCoR), WVU PSCoR, and WVCTSI. In addition, we acknowledge the use of the WVU Shared Research Facilities that are supported by NIH grants 2U54GM104942-02, 5P20RR016477, U57GM104942, P30GM103488, P20GM109098, and P20GM103434. Opinions, interpretations, conclusions, and recommendations are those of the authors and are not necessarily endorsed by the funding agencies. We also acknowledge Suzanne Danley for proofreading.

Conflicts of Interest: The authors declare no conflict of interest.

References

1. Lutolf, M.; Hubbell, J. Synthetic biomaterials as instructive extracellular microenvironments for morphogenesis in tissue engineering. *Nat. Biotechnol.* **2005**, *23*, 47–55. [CrossRef] [PubMed]
2. Stevens, M.M.; George, J.H. Exploring and Engineering the Cell Surface Interface. *Science* **2005**, *310*, 1135–1138. [CrossRef] [PubMed]
3. Khademhosseini, A.; Langer, R. A decade of progress in tissue engineering. *Nat. Protoc.* **2016**, *11*, 1775–1781. [CrossRef] [PubMed]
4. Parpura, V. Nanoelectronics for the heart. *Nat. Nanotechnol.* **2016**, *11*, 738–739. [CrossRef] [PubMed]
5. Choi, N.W.; Cabodi, M.; Held, B.; Gleghorn, J.P.; Bonassar, L.J.; Stroock, A.D. Microfluidic scaffolds for tissue engineering. *Nat. Mater.* **2007**, *6*, 908–915. [CrossRef] [PubMed]
6. Green, J.J.; Elisseeff, J.H. Mimicking biological functionality with polymers for biomedical applications. *Nature* **2016**, *540*, 386–394. [CrossRef] [PubMed]
7. Tan, W.; Desai, T.A. Layer-by-layer microfluidics for biomimetic three-dimensional structures. *Biomaterials* **2004**, *25*, 1355–1364. [CrossRef] [PubMed]
8. Dvir, T.; Timko, B.P.; Kohane, D.S.; Langer, R. Nanotechnological strategies for engineering complex tissues. *Nat. Nanotechnol.* **2011**, *6*, 13–22. [CrossRef] [PubMed]
9. Matsusaki, M.; Ajiro, H.; Kida, T.; Serizawa, T.; Akashi, M. Layer-by-Layer Assembly through Weak Interactions and Their Biomedical Applications. *Adv. Mater.* **2012**, *24*, 454–474. [CrossRef] [PubMed]
10. Shen, M.; Zhu, S.; Wang, F. A general strategy for the ultrafast surface modification of metals. *Nat. Commun.* **2016**, *7*, 13797. [CrossRef] [PubMed]
11. Discher, D.E.; Janmey, P.; Wang, Y.L. Tissue cells feel and respond to the stiffness of their substrate. *Science* **2005**, *310*, 1139–1143. [CrossRef] [PubMed]
12. Langmuir, I.; Schaefer, V.J. Monolayers and Multilayers of Chlorophyll. *J. Am. Chem. Soc.* **1937**, *59*, 2075–2076. [CrossRef]
13. Nuzzo, R.G.; Allara, D.L. Adsorption of bifunctional organic disulfides on gold surfaces. *J. Am. Chem. Soc.* **1983**, *105*, 4481–4483. [CrossRef]
14. Bain, C.D.; Whitesides, G.M. Formation of monolayers by the coadsorption of thiols on gold: Variation in the length of the alkyl chain. *J. Am. Chem. Soc.* **1989**, *111*, 7164–7175. [CrossRef]
15. Mrksich, M. A surface chemistry approach to studying cell adhesion. *Chem. Soc. Rev.* **2000**, *29*, 267–273. [CrossRef]
16. Tsai, P.-S.; Yang, Y.-M.; Lee, Y.-L. Fabrication of Hydrophobic Surfaces by Coupling of Langmuir-Blodgett Deposition and a Self-Assembled Monolayer. *Langmuir* **2006**, *22*, 5660–5665. [CrossRef] [PubMed]

17. Richardson, J.J.; Bjornmalm, M.; Caruso, F. Multilayer assembly. Technology-driven layer-by-layer assembly of nanofilms. *Science* **2015**, *348*, aaa2491. [CrossRef] [PubMed]

18. Borges, J.; Mano, J.F. Molecular interactions driving the layer-by-layer assembly of multilayers. *Chem. Rev.* **2014**, *114*, 8883–8942. [CrossRef] [PubMed]

19. Ariga, K.; Yamauchi, Y.; Rydzek, G.; Ji, Q.; Yonamine, Y.; Wu, K.C.W.; Hill, J.P. Layer-by-layer nanoarchitectonics: Invention, innovation, and evolution. *Chem. Lett.* **2014**, *43*, 36–68. [CrossRef]

20. Tang, Z.; Wang, Y.; Podsiadlo, P.; Kotov, N.A. Biomedical applications of layer-by-layer assembly: From biomimetics to tissue engineering. *Adv. Mater.* **2006**, *18*, 3203–3224. [CrossRef]

21. Kim, B.S.; Park, S.W.; Hammond, P.T. Hydrogen-bonding layer-by-layer-assembled biodegradable polymeric micelles as drug delivery vehicles from surfaces. *ACS Nano* **2008**, *2*, 386–392. [CrossRef] [PubMed]

22. Krogman, K.C.; Lowery, J.L.; Zacharia, N.S.; Rutledge, G.C.; Hammond, P.T. Spraying asymmetry into functional membranes layer-by-layer. *Nat. Mater.* **2009**, *8*, 512–518. [CrossRef] [PubMed]

23. Likibi, F.; Jiang, B.; Li, B. Biomimetic nanocoating promotes osteoblast cell adhesion on biomedical implants. *J. Mater. Res.* **2011**, *23*, 3222–3228. [CrossRef]

24. Jan, E.; Kotov, N.A. Successful differentiation of mouse neural stem cells on layer-by-layer assembled single-walled carbon nanotube composite. *Nano Lett.* **2007**, *7*, 1123–1128. [CrossRef] [PubMed]

25. Miller, J.S.; Stevens, K.R.; Yang, M.T.; Baker, B.M.; Nguyen, D.H.; Cohen, D.M.; Toro, E.; Chen, A.A.; Galie, P.A.; Yu, X.; et al. Rapid casting of patterned vascular networks for perfusable engineered three-dimensional tissues. *Nat. Mater.* **2012**, *11*, 768–774. [CrossRef] [PubMed]

26. Decher, G. Fuzzy nanoassemblies: Toward layered polymeric multicomposites. *Science* **1997**, *277*, 1232–1237. [CrossRef]

27. Zhong, Y.; Li, B.; Haynie, D.T. Fine tuning of physical properties of designed polypeptide multilayer films by control of pH. *Biotechnol. Prog.* **2006**, *22*, 126–132. [CrossRef] [PubMed]

28. Li, B.; Haynie, D.T.; Palath, N.; Janisch, D. Nanoscale biomimetics: Fabrication and optimization of stability of peptide-based thin films. *J. Nanosci. Nanotechnol.* **2005**, *5*, 2042–2049. [CrossRef] [PubMed]

29. Jiang, B.; Li, B. Tunable drug loading and release from polypeptide multilayer nanofilms. *Int. J. Nanomed.* **2009**, *4*, 37–53.

30. Szabó, T.; Péter, Z.; Illés, E.; Janovák, L.; Talyzin, A. Stability and dye inclusion of graphene oxide/polyelectrolyte layer-by-layer self-assembled films in saline, acidic and basic aqueous solutions. *Carbon* **2017**, *111*, 350–357. [CrossRef]

31. Tang, K.; Besseling, N.A. Formation of polyelectrolyte multilayers: Ionic strengths and growth regimes. *Soft Matter* **2016**, *12*, 1032–1040. [CrossRef] [PubMed]

32. Sung, C.; Lutkenhaus, J.L. Effect of assembly condition on the morphologies and temperature-triggered transformation of layer-by-layer microtubes. *Korean J. Chem. Eng.* **2018**, *35*, 263–271. [CrossRef]

33. Chang, L.; Kong, X.; Wang, F.; Wang, L.; Shen, J. Layer-by-layer assembly of poly (*N*-acryloyl-*N*′-propylpiperazine) and poly (acrylic acid): Effect of pH and temperature. *Thin Solid Films* **2008**, *516*, 2125–2129. [CrossRef]

34. Zou, J.; Kim, F. Diffusion driven layer-by-layer assembly of graphene oxide nanosheets into porous three-dimensional macrostructures. *Nat. Commun.* **2014**, *5*, 5254. [CrossRef] [PubMed]

35. Gentile, P.; Carmagnola, I.; Nardo, T.; Chiono, V. Layer-by-layer assembly for biomedical applications in the last decade. *Nanotechnology* **2015**, *26*, 422001. [CrossRef] [PubMed]

36. Li, B.; Haynie, D.T. Multilayer biomimetics: Reversible covalent stabilization of a nanostructured biofilm. *Biomacromolecules* **2004**, *5*, 1667–1670. [CrossRef] [PubMed]

37. Li, B.; Rozas, J.; Haynie, D.T. Structural stability of polypeptide nanofilms under extreme conditions. *Biotechnol. Prog.* **2006**, *22*, 111–117. [CrossRef] [PubMed]

38. Shim, B.S.; Podsiadlo, P.; Lilly, D.G.; Agarwal, A.; Lee, J.; Tang, Z.; Ho, S.; Ingle, P.; Paterson, D.; Lu, W. Nanostructured thin films made by dewetting method of layer-by-layer assembly. *Nano Lett.* **2007**, *7*, 3266–3273. [CrossRef] [PubMed]

39. Grigoriev, D.; Bukreeva, T.; Möhwald, H.; Shchukin, D. New method for fabrication of loaded micro-and nanocontainers: Emulsion encapsulation by polyelectrolyte layer-by-layer deposition on the liquid core. *Langmuir* **2008**, *24*, 999–1004. [CrossRef] [PubMed]

40. Richardson, J.J.; Ejima, H.; Lörcher, S.L.; Liang, K.; Senn, P.; Cui, J.; Caruso, F. Preparation of nano-and microcapsules by electrophoretic polymer assembly. *Angew. Chem. Int. Ed.* **2013**, *52*, 6455–6458. [CrossRef] [PubMed]

41. Ma, L.; Cheng, M.; Jia, G.; Wang, Y.; An, Q.; Zeng, X.; Shen, Z.; Zhang, Y.; Shi, F. Layer-by-layer self-assembly under high gravity field. *Langmuir* **2012**, *28*, 9849–9856. [CrossRef] [PubMed]

42. Qi, A.; Chan, P.; Ho, J.; Rajapaksa, A.; Friend, J.; Yeo, L. Template-free synthesis and encapsulation technique for layer-by-layer polymer nanocarrier fabrication. *ACS Nano* **2011**, *5*, 9583–9591. [CrossRef] [PubMed]

43. Hong, X.; Li, J.; Wang, M.; Xu, J.; Guo, W.; Li, J.; Bai, Y.; Li, T. Fabrication of magnetic luminescent nanocomposites by a layer-by-layer self-assembly approach. *Chem. Mater.* **2004**, *16*, 4022–4027. [CrossRef]

44. Lee, D.; Rubner, M.F.; Cohen, R.E. All-nanoparticle thin-film coatings. *Nano Lett.* **2006**, *6*, 2305–2312. [CrossRef] [PubMed]

45. Schlenoff, J.B.; Dubas, S.T.; Farhat, T. Sprayed polyelectrolyte multilayers. *Langmuir* **2000**, *16*, 9968–9969. [CrossRef]

46. Sun, J.; Gao, M.; Feldmann, J. Electric field directed layer-by-layer assembly of highly fluorescent CdTe nanoparticles. *J. Nanosci. Nanotechnol.* **2001**, *1*, 133–136. [CrossRef] [PubMed]

47. Tan, W.; Desai, T.A. Microfluidic patterning of cells in extracellular matrix biopolymers: Effects of channel size, cell type, and matrix composition on pattern integrity. *Tissue Eng.* **2003**, *9*, 255–267. [CrossRef] [PubMed]

48. Khademhosseini, A.; Langer, R. Microengineered hydrogels for tissue engineering. *Biomaterials* **2007**, *28*, 5087–5092. [CrossRef] [PubMed]

49. Oliveira, S.M.; Silva, T.H.; Reis, R.L.; Mano, J.F. Hierarchical fibrillar scaffolds obtained by non-conventional layer-by-layer electrostatic self-assembly. *Adv. Health. Mater.* **2013**, *2*, 422–427. [CrossRef] [PubMed]

50. Silva, J.M.; Georgi, N.; Costa, R.; Sher, P.; Reis, R.L.; Van Blitterswijk, C.A.; Karperien, M.; Mano, J.F. Nanostructured 3D constructs based on chitosan and chondroitin sulphate multilayers for cartilage tissue engineering. *PLoS ONE* **2013**, *8*, e55451. [CrossRef] [PubMed]

51. Gentile, P.; Frongia, M.E.; Cardellach, M.; Miller, C.A.; Stafford, G.P.; Leggett, G.J.; Hatton, P.V. Functionalised nanoscale coatings using layer-by-layer assembly for imparting antibacterial properties to polylactide-co-glycolide surfaces. *Acta Biomater.* **2015**, *21*, 35–43. [CrossRef] [PubMed]

52. Li, W.; Guan, T.; Zhang, X.; Wang, Z.; Wang, M.; Zhong, W.; Feng, H.; Xing, M.; Kong, J. The effect of layer-by-layer assembly coating on the proliferation and differentiation of neural stem cells. *ACS Appl. Mater. Interfaces* **2015**, *7*, 3018–3029. [CrossRef] [PubMed]

53. Lai, Y.-T.; Reading, E.; Hura, G.L.; Tsai, K.-L.; Laganowsky, A.; Asturias, F.J.; Tainer, J.A.; Robinson, C.V.; Yeates, T.O. Structure of a designed protein cage that self-assembles into a highly porous cube. *Nat. Chem.* **2014**, *6*, 1065–1071. [CrossRef] [PubMed]

54. Zhang, Y.; Arugula, M.A.; Wales, M.; Wild, J.; Simonian, A.L. A novel layer-by-layer assembled multi-enzyme/CNT biosensor for discriminative detection between organophosphorus and non-organophosphrus pesticides. *Biosens. Bioelectron.* **2015**, *67*, 287–295. [CrossRef] [PubMed]

55. Hong, C.-Y.; Wu, S.-X.; Li, S.-H.; Liang, H.; Chen, S.; Li, J.; Yang, H.-H.; Tan, W. Semipermeable Functional DNA-Encapsulated Nanocapsules as Protective Bioreactors for Biosensing in Living Cells. *Anal. Chem.* **2017**, *89*, 5389–5394. [CrossRef] [PubMed]

56. Guven, S.; Chen, P.; Inci, F.; Tasoglu, S.; Erkmen, B.; Demirci, U. Multiscale assembly for tissue engineering and regenerative medicine. *Trends Biotechnol.* **2015**, *33*, 269–279. [CrossRef] [PubMed]

57. Han, L.; Wang, M.; Sun, H.; Li, P.; Wang, K.; Ren, F.; Lu, X. Porous titanium scaffolds with self-assembled micro/nano hierarchical structure for dual functions of bone regeneration and anti-infection. *J. Biomed. Mater. Res. Part A* **2017**, *105*, 3482–3492. [CrossRef] [PubMed]

58. Akiba, U.; Minaki, D.; Anzai, J.-I. Photosensitive layer-by-layer assemblies containing azobenzene groups: Synthesis and biomedical applications. *Polymers* **2017**, *9*, 553. [CrossRef]

59. Park, K.; Choi, D.; Hong, J. Nanostructured polymer thin films fabricated with brush-based layer-by-layer self-assembly for site-selective construction and drug release. *Sci. Rep.* **2018**, *8*, 3365. [CrossRef] [PubMed]

60. Wei, J.; Wang, G.; Li, X.; Ren, P.; Yu, H.; Dong, B. Architectural delineation and molecular identification of extracellular matrix in ascidian embryos and larvae. *Biol. Open* **2017**, *6*, 1383–1390. [CrossRef] [PubMed]

61. Grant, G.G.S.; Koktysh, D.S.; Yun, B.; Matts, R.L.; Kotov, N.A. Layer-by-layer assembly of collagen thin films controlled thickness and biocompatibility. *Biomed. Microdevices* **2001**, *3*, 301–306. [CrossRef]

62. Richert, L.; Lavalle, P.; Payan, E.; Shu, X.Z.; Prestwich, G.D.; Stoltz, J.-F.; Schaaf, P.; Voegel, J.-C.; Picart, C. Layer by layer buildup of polysaccharide films: Physical chemistry and cellular adhesion aspects. *Langmuir* **2004**, *20*, 448–458. [CrossRef] [PubMed]

63. Zhang, J.; Senger, B.; Vautier, D.; Picart, C.; Schaaf, P.; Voegel, J.C.; Lavalle, P. Natural polyelectrolyte films based on layer-by layer deposition of collagen and hyaluronic acid. *Biomaterials* **2005**, *26*, 3353–3361. [CrossRef] [PubMed]

64. Wu, Z.R.; Ma, J.; Liu, B.F.; Xu, Q.Y.; Cui, F.Z. Layer-by-layer assembly of polyelectrolyte films improving cytocompatibility to neural cells. *J. Biomed. Mater. Res. Part A* **2007**, *81*, 355–362. [CrossRef] [PubMed]

65. Wittmer, C.R.; Phelps, J.A.; Lepus, C.M.; Saltzman, W.M.; Harding, M.J.; Van Tassel, P.R. Multilayer nanofilms as substrates for hepatocellular applications. *Biomaterials* **2008**, *29*, 4082–4090. [CrossRef] [PubMed]

66. Gheith, M.K.; Pappas, T.C.; Liopo, A.V.; Sinani, V.A.; Shim, B.S.; Motamedi, M.; Wicksted, J.P.; Kotov, N.A. Stimulation of neural cells by lateral currents in conductive layer-by-layer films of single-walled carbon nanotubes. *Adv. Mater.* **2006**, *18*, 2975–2979. [CrossRef]

67. Shen, L.; Cui, X.; Yu, G.; Li, F.; Li, L.; Feng, S.; Lin, H.; Chen, J. Thermodynamic assessment of adsorptive fouling with the membranes modified via layer-by-layer self-assembly technique. *J. Colloid Interface Sci.* **2017**, *494*, 194–203. [CrossRef] [PubMed]

68. Manabe, K.; Matsuda, M.; Nakamura, C.; Takahashi, K.; Kyung, K.-H.; Shiratori, S. Antifibrinogen, antireflective, antifogging surfaces with biocompatible nano-ordered hierarchical texture fabricated by layer-by-layer self-assembly. *Chem. Mater.* **2017**, *29*, 4745–4753. [CrossRef]

69. Wang, L.; Wang, N.; Yang, H.; An, Q.; Zeng, T.; Ji, S. Enhanced pH and oxidant resistance of polyelectrolyte multilayers via the confinement effect of lamellar graphene oxide nanosheets. *Sep. Purif. Technol.* **2018**, *193*, 274–282. [CrossRef]

70. Hsu, S.-W.; Long, Y.; Subramanian, A.G.; Tao, A.R. Directed assembly of metal nanoparticles in polymer bilayers. *Mol. Syst. Des. Eng.* **2018**, *3*, 390–396. [CrossRef]

71. Hillberg, A.L.; Holmes, C.A.; Tabrizian, M. Effect of genipin cross-linking on the cellular adhesion properties of layer-by-layer assembled polyelectrolyte films. *Biomaterials* **2009**, *30*, 4463–4470. [CrossRef] [PubMed]

72. Silva, J.M.; Garcia, J.R.; Reis, R.L.; Garcia, A.J.; Mano, J.F. Tuning cell adhesive properties via layer-by-layer assembly of chitosan and alginate. *Acta Biomater.* **2017**, *51*, 279–293. [CrossRef] [PubMed]

73. Jiang, B.; Li, B. Polypeptide nanocoatings for preventing dental and orthopaedic device-associated infection: PH-induced antibiotic capture, release, and antibiotic efficacy. *J. Biomed. Mater. Res. Part B* **2009**, *88*, 332–338. [CrossRef] [PubMed]

74. Li, B.; Jiang, B.; Dietz, M.J.; Smith, E.S.; Clovis, N.B.; Rao, K.M. Evaluation of local MCP-1 and IL-12 nanocoatings for infection prevention in open fractures. *J. Orthop. Res.* **2010**, *28*, 48–54. [CrossRef] [PubMed]

75. Li, H.; Ogle, H.; Jiang, B.; Hagar, M.; Li, B. Cefazolin embedded biodegradable polypeptide nanofilms promising for infection prevention: A preliminary study on cell responses. *J. Orthop. Res.* **2010**, *28*, 992–999. [CrossRef] [PubMed]

76. Macdonald, M.L.; Samuel, R.E.; Shah, N.J.; Padera, R.F.; Beben, Y.M.; Hammond, P.T. Tissue integration of growth factor-eluting layer-by-layer polyelectrolyte multilayer coated implants. *Biomaterials* **2011**, *32*, 1446–1453. [CrossRef] [PubMed]

77. Hu, Y.; Cai, K.; Luo, Z.; Jandt, K.D. Layer-by-layer assembly of beta-estradiol loaded mesoporous silica nanoparticles on titanium substrates and its implication for bone homeostasis. *Adv. Mater.* **2010**, *22*, 4146–4150. [CrossRef] [PubMed]

78. Li, B.; Jiang, B.; Boyce, B.M.; Lindsey, B.A. Multilayer polypeptide nanoscale coatings incorporating IL-12 for the prevention of biomedical device-associated infections. *Biomaterials* **2009**, *30*, 2552–2558. [CrossRef] [PubMed]

79. Jiang, B.; Defusco, E.; Li, B. Polypeptide multilayer film co-delivers oppositely-charged drug molecules in sustained manners. *Biomacromolecules* **2010**, *11*, 3630–3637. [CrossRef] [PubMed]

80. Sousa, M.; Mano, J. Cell-Adhesive bioinspired and catechol-based multilayer freestanding membranes for bone tissue engineering. *Biomimetics* **2017**, *2*, 19. [CrossRef]

81. Hwang, J.; Choi, D.; Choi, M.; Seo, Y.; Son, J.; Hong, J.; Choi, J. Synthesis and characterization of functional nanofilm coated live immune cells. *ACS Appl. Mater. Interfaces* **2018**, *10*, 17685–17692. [CrossRef] [PubMed]

82. Wilson, J.T.; Cui, W.; Chaikof, E.L. Layer-by-layer assembly of a conformal nanothin PEG coating for intraportal islet transplantation. *Nano Lett.* **2008**, *8*, 1940–1948. [CrossRef] [PubMed]

83. Zhao, Q.; Li, H.; Li, B. Nanoencapsulating living biological cells using electrostatic layer-by-layer self-assembly: Platelets as a model. *J. Mater. Res.* **2011**, *26*, 347–351. [CrossRef] [PubMed]

84. Diaspro, A.; Silvano, D.; Krol, S.; Cavalleri, O.; Gliozzi, A. Single Living Cell Encapsulation in Nano-organized Polyelectrolyte Shells. *Langmuir* **2002**, *18*, 5047–5050. [CrossRef]

85. Mansouri, S.; Merhi, Y.; Winnik, F.M.; Tabrizian, M. Investigation of layer-by-layer assembly of polyelectrolytes on fully functional human red blood cells in suspension for attenuated immune response. *Biomacromolecules* **2011**, *12*, 585–592. [CrossRef] [PubMed]

86. Zhu, Y.; Gao, C.; He, T.; Liu, X.; Shen, J. Layer-by-layer assembly to modify poly(L-lactic acid) surface toward improving its cytocompatibility to human endothelial cells. *Biomacromolecules* **2003**, *4*, 446–452. [CrossRef] [PubMed]

87. He, X.; Wang, Y.; Wu, G. Layer-by-layer assembly of type I collagen and chondroitin sulfate on aminolyzed PU for potential cartilage tissue engineering application. *Appl. Surf. Sci.* **2012**, *258*, 9918–9925. [CrossRef]

88. Zhu, H.; Ji, J.; Shen, J. Biomacromolecules electrostatic self-assembly on 3-dimensional tissue engineering scaffold. *Biomacromolecules* **2004**, *5*, 1933–1939. [CrossRef] [PubMed]

89. Cai, K.; Rechtenbach, A.; Hao, J.; Bossert, J.; Jandt, K.D. Polysaccharide-protein surface modification of titanium via a layer-by-layer technique: Characterization and cell behaviour aspects. *Biomaterials* **2005**, *26*, 5960–5971. [CrossRef] [PubMed]

90. Liu, X.; Smith, L.; Wei, G.; Won, Y.; Ma, P.X. Surface engineering of nano-fibrous poly(L-lactic acid) scaffolds via self-assembly technique for bone tissue engineering. *J. Biomed. Nanotechnol.* **2005**, *1*, 54–60. [CrossRef]

91. Gong, Y.; Zhu, Y.; Liu, Y.; Ma, Z.; Gao, C.; Shen, J. Layer-by-layer assembly of chondroitin sulfate and collagen on aminolyzed poly(L-lactic acid) porous scaffolds to enhance their chondrogenesis. *Acta Biomater.* **2007**, *3*, 677–685. [CrossRef] [PubMed]

92. Lee, J.; Shanbhag, S.; Kotov, N.A. Inverted colloidal crystals as three-dimensional microenvironments for cellular co-cultures. *J. Mater. Chem.* **2006**, *16*, 3558–3564. [CrossRef]

93. Li, X.; Xie, J.; Yuan, X.; Xia, Y. Coating electrospun poly(epsilon-caprolactone) fibers with gelatin and calcium phosphate and their use as biomimetic scaffolds for bone tissue engineering. *Langmuir* **2008**, *24*, 14145–14150. [CrossRef] [PubMed]

94. Kim, T.G.; Park, S.-H.; Chung, H.J.; Yang, D.-Y.; Park, T.G. Microstructured scaffold coated with hydroxyapatite/collagen nanocomposite multilayer for enhanced osteogenic induction of human mesenchymal stem cells. *J. Mater. Chem.* **2010**, *20*, 8927–8933. [CrossRef]

95. Monteiro, I.P.; Shukla, A.; Marques, A.P.; Reis, R.L.; Hammond, P.T. Spray-assisted layer-by-layer assembly on hyaluronic acid scaffolds for skin tissue engineering. *J. Biomed. Mater. Res. Part A* **2015**, *103*, 330–340. [CrossRef] [PubMed]

96. Gao, P.; Nie, X.; Zou, M.; Shi, Y.; Cheng, G. Recent advances in materials for extended-release antibiotic delivery system. *J. Antibiot.* **2011**, *64*, 625–634. [CrossRef] [PubMed]

97. Ryu, S.; Yoo, J.; Han, J.; Kang, S.; Jang, Y.; Han, H.J.; Char, K.; Kim, B.-S. Cellular layer-by-layer coculture platform using biodegradable, nanoarchitectured membranes for stem cell therapy. *Chem. Mater.* **2017**, *29*, 5134–5147. [CrossRef]

98. Kook, Y.-M.; Jeong, Y.; Lee, K.; Koh, W.-G. Design of biomimetic cellular scaffolds for co-culture system and their application. *J. Tissue Eng.* **2017**, *8*, 1–17. [CrossRef] [PubMed]

99. Khademhosseini, A.; Suh, K.Y.; Yang, J.M.; Eng, G.; Yeh, J.; Levenberg, S.; Langer, R. Layer-by-layer deposition of hyaluronic acid and poly-L-lysine for patterned cell co-cultures. *Biomaterials* **2004**, *25*, 3583–3592. [CrossRef] [PubMed]

100. Fukuda, J.; Khademhosseini, A.; Yeh, J.; Eng, G.; Cheng, J.; Farokhzad, O.C.; Langer, R. Micropatterned cell co-cultures using layer-by-layer deposition of extracellular matrix components. *Biomaterials* **2006**, *27*, 1479–1486. [CrossRef] [PubMed]

101. Zhou, J.; Romero, G.; Rojas, E.; Ma, L.; Moya, S.; Gao, C. Layer by layer chitosan/alginate coatings on poly(lactide-co-glycolide) nanoparticles for antifouling protection and Folic acid binding to achieve selective cell targeting. *J. Colloid Interface Sci.* **2010**, *345*, 241–247. [CrossRef] [PubMed]

102. Kidambi, S.; Sheng, L.; Yarmush, M.L.; Toner, M.; Lee, I.; Chan, C. Patterned co-culture of primary hepatocytes and fibroblasts using polyelectrolyte multilayer templates. *Macromol. Biosci.* **2007**, *7*, 344–353. [CrossRef] [PubMed]

103. Rajagopalan, P.; Shen, C.J.; Berthiaume, F.; Tilles, A.W.; Toner, M.; Yarmush, M.L. Polyelectrolyte nano-scaffolds for the design of layered cellular architectures. *Tissue Eng.* **2006**, *12*, 1553–1563. [CrossRef] [PubMed]

104. Kim, Y.; Larkin, A.L.; Davis, R.M.; Rajagopalan, P. The design of in vitro liver sinusoid mimics using chitosan-hyaluronic acid polyelectrolyte multilayers. *Tissue Eng. Part A* **2010**, *16*, 2731–2741. [CrossRef] [PubMed]

105. Tan, W.; Desai, T.A. Microscale multilayer cocultures for biomimetic blood vessels. *J. Biomed. Mater. Res. Part A* **2005**, *72*, 146–160. [CrossRef] [PubMed]

106. Shin, S.R.; Aghaei-Ghareh-Bolagh, B.; Gao, X.; Nikkhah, M.; Jung, S.M.; Dolatshahi-Pirouz, A.; Kim, S.B.; Kim, S.M.; Dokmeci, M.R.; Tang, X.S.; et al. Layer-by-layer assembly of 3D tissue constructs with functionalized graphene. *Adv. Funct. Mater.* **2014**, *24*, 6136–6144. [CrossRef] [PubMed]

107. Feng, J.; Chan-Park, M.B.; Shen, J.; Chan, V. Quick layer-by-layer assembly of aligned multilayers of vascular smooth muscle cells in deep microchannels. *Tissue Eng.* **2007**, *13*, 1003–1012. [CrossRef] [PubMed]

108. Fukuda, Y.; Akagi, T.; Asaoka, T.; Eguchi, H.; Sasaki, K.; Iwagami, Y.; Yamada, D.; Noda, T.; Kawamoto, K.; Gotoh, K. Layer by layer cell coating technique using extracellular matrix facilitates rapid fabrication and function of pancreatic β-cell spheroids. *Biomaterials* **2018**, *160*, 82–91. [CrossRef] [PubMed]

109. Yang, J.; Li, J.; Li, X.; Wang, X.; Yang, Y.; Kawazoe, N.; Chen, G. Nanoencapsulation of individual mammalian cells with cytoprotective polymer shell. *Biomaterials* **2017**, *133*, 253–262. [CrossRef] [PubMed]

110. Liu, C.Y.; Matsusaki, M.; Akashi, M. Three-Dimensional Tissue Models Constructed by Cells with Nanometer-or Micrometer-Sized Films on the Surfaces. *Chem. Rec.* **2016**, *16*, 783–796. [CrossRef] [PubMed]

111. Matsusaki, M.; Kadowaki, K.; Nakahara, Y.; Akashi, M. Fabrication of cellular multilayers with nanometer-sized extracellular matrix films. *Angew. Chem. Int. Ed.* **2007**, *119*, 4773–4776. [CrossRef]

112. Nishiguchi, A.; Yoshida, H.; Matsusaki, M.; Akashi, M. Rapid construction of three-dimensional multilayered tissues with endothelial tube networks by the cell-accumulation technique. *Adv. Mater.* **2011**, *23*, 3506–3510. [CrossRef] [PubMed]

113. Sasaki, K.; Akagi, T.; Asaoka, T.; Eguchi, H.; Fukuda, Y.; Iwagami, Y.; Yamada, D.; Noda, T.; Wada, H.; Gotoh, K.; et al. Construction of three-dimensional vascularized functional human liver tissue using a layer-by-layer cell coating technique. *Biomaterials* **2017**, *133*, 263–274. [CrossRef] [PubMed]

114. Amano, Y.; Nishiguchi, A.; Matsusaki, M.; Iseoka, H.; Miyagawa, S.; Sawa, Y.; Seo, M.; Yamaguchi, T.; Akashi, M. Development of vascularized iPSC derived 3D-cardiomyocyte tissues by filtration layer-by-layer technique and their application for pharmaceutical assays. *Acta Biomater.* **2016**, *33*, 110–121. [CrossRef] [PubMed]

International Journal of
Molecular Sciences

MDPI

Article

3D Bioprinted Artificial Trachea with Epithelial Cells and Chondrogenic-Differentiated Bone Marrow-Derived Mesenchymal Stem Cells

Sang-Woo Bae [1], Kang-Woog Lee [2], Jae-Hyun Park [1], JunHee Lee [3], Cho-Rok Jung [4], JunJie Yu [3,5], Hwi-Yool Kim [1] and Dae-Hyun Kim [2,*]

[1] Department of Veterinary Surgery, College of Veterinary Medicine, Konkuk University, 120 Neungdong-ro, Gwangjin-gu, Seoul 05029, Korea; raphael0826@gmail.com (S.-W.B.); planetes1202@naver.com (J.-H.P.); hykim@konkuk.ac.kr (H.-Y.K.)

[2] Division of Cardiovascular Surgery, Severance Cardiovascular Hospital, Yonsei University College of Medicine, 50-1 Yonsei-ro, Seodaemun-gu, Seoul 03722, Korea; sysebg@naver.com

[3] Department of Nature-Inspired Nanoconvergence System, Korea Institute of Machinery and Materials (KIMM), 156 Gajeongbuk-Ro, Yuseong-Gu, Daejeon 34103, Korea; meek@kimm.re.kr (J.L.); junjie0801@kimm.re.kr (J.Y.)

[4] Gene Therapy Research Unit, Korea Research Institute of Bioscience and Biotechnology, 125 Gwahak-ro, Yuseong-gu, Daejeon 34141, Korea; crjung@kribb.re.kr

[5] Department of Biomedical Engineering, School of Integrative Engineering, Chung-Ang University, 84 Heukseok-Ro, Dongjak-Gu, Seoul 06974, Korea

* Correspondence: vet1982@hanmail.net

Received: 10 May 2018; Accepted: 29 May 2018; Published: 31 May 2018

Abstract: Tracheal resection has limited applicability. Although various tracheal replacement strategies were performed using artificial prosthesis, synthetic stents and tissue transplantation, the best method in tracheal reconstruction remains to be identified. Recent advances in tissue engineering enabled 3D bioprinting using various biocompatible materials including living cells, thereby making the product clinically applicable. Moreover, clinical interest in mesenchymal stem cell has dramatically increased. Here, rabbit bone marrow-derived mesenchymal stem cells (bMSC) and rabbit respiratory epithelial cells were cultured. The chondrogenic differentiation level of bMSC cultured in regular media (MSC) and that in chondrogenic media (d-MSC) were compared. Dual cell-containing artificial trachea were manufactured using a 3D bioprinting method with epithelial cells and undifferentiated bMSC (MSC group, $n = 6$) or with epithelial cells and chondrogenic-differentiated bMSC (d-MSC group, $n = 6$). d-MSC showed a relatively higher level of glycosaminoglycan (GAG) accumulation and chondrogenic marker gene expression than MSC in vitro. Neo-epithelialization and neo-vascularization were observed in all groups in vivo but neo-cartilage formation was only noted in d-MSC. The epithelial cells in the 3D bioprinted artificial trachea were effective in respiratory epithelium regeneration. Chondrogenic-differentiated bMSC had more neo-cartilage formation potential in a short period. Nevertheless, the cartilage formation was observed only in a localized area.

Keywords: bone marrow-derived mesenchymal stem cell; chondrogenic differentiation; three-dimensional bioprinting; artificial trachea; tissue engineering

1. Introduction

The trachea is a hollow cylindrical organ composed of 15–20 C-shaped cartilages with fibroelastic ligaments [1]. Permanent damage, stenosis and tumor in the trachea require surgical intervention with tracheal resection [2] and in such cases, circumferential resection and end-to-end anastomosis are generally considered the optimal surgical methods. However, these procedures are applicable only in particular conditions. For instance, when a lesion exceeds half of the trachea in adults, or one third in children, tracheal replacement is recommended rather than tracheal resection. Nevertheless, with the developments in tissue engineering, tracheal replacement strategies were diversified from the use of prostheses and synthetic stents to tissue transplantation [3].

The first case of allogenic tracheal transplantation in humans was published by Rose et al. in 1979 [4]. A cadaveric tracheal graft was heterotopically transplanted for 3 weeks to provide revascularization and subsequently orthotopically repositioned. Wurtz et al. reported on tracheal replacement with an aortic homograft combined with an intraluminal stent to support the structural integrity [5]. Macroscopically, the aortic graft was surrounded by a thick wall including recognizable cartilage rings; however, progressive ischemia of the cartilage was observed pathologically. Moreover, artificial prostheses have also been used for tracheal replacement; however, they have been associated with material migration, rupture, infection and disintegration [6]. Hence, no satisfying methods in the aspect of multilayered structure and its function in tracheal reconstruction have been identified [7–9].

Three-dimensional (3D) bioprinting serves as an additional manufacturing method that uses a 3D bioprinter and enables three-dimensional stacking of certain biomaterials layer by layer according to the intended design. Recent advances in tissue engineering enabled 3D bioprinting using various biocompatible materials including living cells, thereby making the product clinically applicable [10]. Additionally, mesenchymal stem cells (MSC), originally isolated from the bone marrow, are a promising cell type for tissue engineering applications because of their proliferation ability and their potential to differentiate in vitro and in vivo into multiple functional tissue-specific cell types, such as adipocytes, chondrocytes, osteoblasts and skeletal myocytes [11,12].

In this study, we compared the glycosaminoglycan (GAG) synthesis and chondrogenic differentiation level between two different culture methods: bMSC cultured in general medium (MSC group) and chondrogenic-differentiated bMSC cultured in chondrogenic medium (d-MSC group). Using a 3D bioprinter, we developed a novel biocompatible artificial trachea with epithelial cells + bMSC (MSC group) and with epithelial cells + chondrogenic-differentiated bMSC (d-MSC group). The structure of the artificial trachea was considered to be similar to that of the normal trachea. We evaluated the results obtained from the animal study according to the chondrogenic differentiation level of bMSC.

2. Results

2.1. Relative Quantification of Glycosaminoglycan (GAG) in bMSC and Chondrogenic-Differentiated bMSC

Relative quantification of GAG between bMSC and chondrogenic-differentiated bMSC on days 14 and 28 was compared. The relative values of chondrogenic-differentiated bMSC on days 14 and 28 were, respectively, 1.67 ± 0.10 and 2.62 ± 0.11 times higher than those of bMSC (Figure 1).

2.2. Chondrogenic Gene Expression

Gene expression levels were normalized to glyceraldehyde-3-phosphate dehydrogenase (GAPDH) and calculated as the expression relative to that of rabbit ear cartilage cell as control. Quantitative real-time PCR (qRT-PCR) analysis demonstrated that the expression levels of all chondrogenic genes (*ACAN*, Col1α1, Col2α1 and *SOX9*) in d-MSC were higher than those in MSC. The expression level of Col2α1 significantly increased 22.54 fold ± 4.24 in d-MSC. *Aggrecan (ACAN)*, Col1α1 and SOX9 gene expression levels were also increased (7.71 ± 0.75, 5.93 ± 1.38 and 3.10 fold ± 0.41, respectively) in d-MSC compared with those in MSC (Figure 1).

2.3. Structure of the 3D Bioprinted Artificial Trachea

The 3D bioprinted artificial trachea was well manufactured (Figure 2). The SEM image revealed micropores in the innermost and outermost PCL layers and showed that the middle non-pore PCL layer separates the two alginate hydrogel layers (Figure 2). Observation of the artificial trachea stained with the CellTracker™ (Thermo Fisher Scientific, Waltham, MA, USA) under a light microscope with optical filters showed a complete separation of inner layer epithelial cells (stained red) and outer layer bMSC (stained green) (Figure 3). All cells were evenly scattered in the alginate hydrogel.

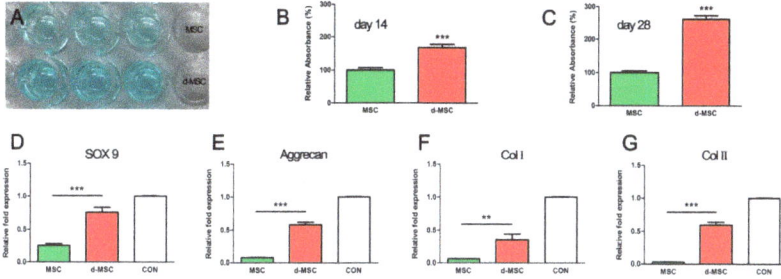

Figure 1. Relative GAG accumulation and chondrogenic marker gene expression level of MSC and d-MSC. (**A**) Alcian blue-stained dissolvents in a 96-well assay plate; the d-MSC group showed a deeper blue color than the MSC group; (**B,C**) Alcian blue absorbance on days 14 and 28. The d-MSC group showed 1.67 ± 0.10 (day 14) and 2.62 ± 0.11 (day 28) times higher values than the MSC group; The relative gene expression levels of SOX9 (**D**); aggrecan (**E**); collagen type 1 (**F**) and collagen type 2 (**G**) were increased in d-MSC. CON = expression level of chondrocyte as positive control. ** $p < 0.01$, *** $p < 0.001$.

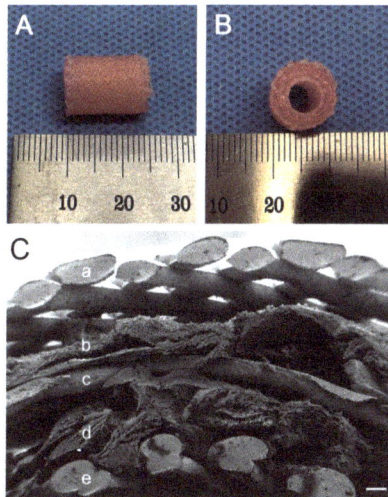

Figure 2. 3D bioprinted artificial trachea. (**A**) Longitudinal view, 15 mm in length. (**B**) Vertical view, five-layered structure with a 5-mm inner diameter and 10-mm outer diameter. (**C**) Magnified structure of the scaffold (SEM image); (a,c,e) layers are composed of PCL; (b) is alginate layer with MSC or d-MSC and (d) is alginate layer with epithelial cells. The innermost (e) and outermost (a) layers showed a microporous feature. The non-porous third layer (c) separates the epithelial cell layer (d) and the MSC or d-MSC layer (b). The scale bar indicates 200 µm.

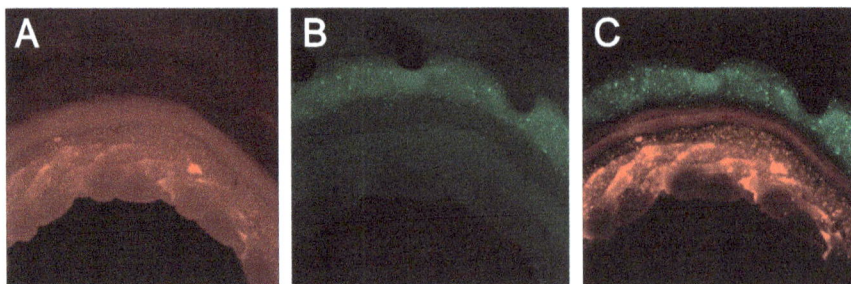

Figure 3. Cell distribution in alginate hydrogel with CellTracker™. (**A**) Epithelial cell in the second alginate hydrogel layer (red); (**B**) MSC in fourth alginate hydrogel layer (green); (**C**) A and B (merged). Each cell was well-printed layer by layer and separated completely.

2.4. Application of the 3D Bioprinted Artificial Trachea

All rabbits that underwent surgery showed no signs of respiratory distress, graft failure, or infection until sacrifice at 12 weeks postoperatively. Plain thoracic radiographs of the MSC and d-MSC groups demonstrated properly sustained tracheal contour without obstruction or stenosis and the 3D-reconstructed CT images indicated good luminal contour of the trachea and no signs of stenosis (Figure 4). The bronchoscopic examination revealed that the tracheal defect was fully re-covered with epithelial mucosa during the experimental period (Figure 5).

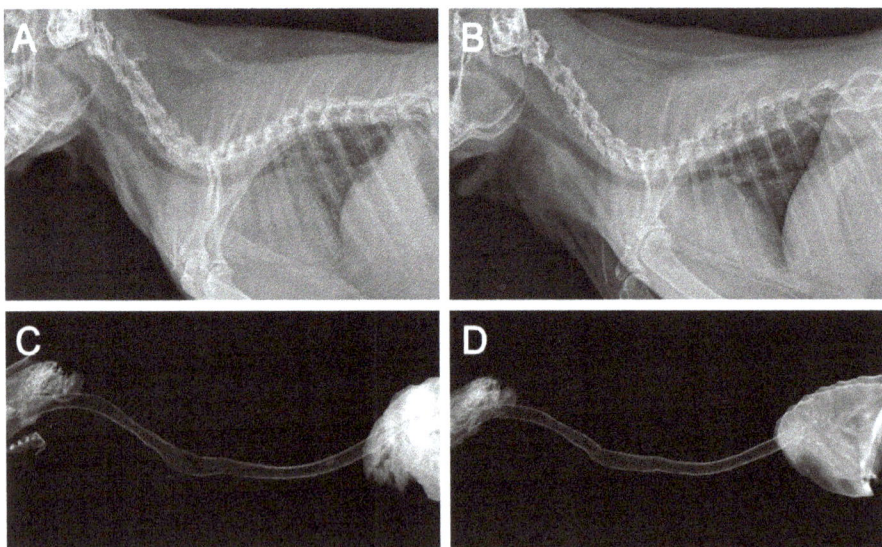

Figure 4. Radiographic findings. Plain lateral thoracic view of the MSC (**A**) and d-MSC (**B**) groups and 3D reconstructed CT image of the MSC (**C**) and d-MSC (**D**) groups at 12 weeks after 3D bioprinted artificial trachea transplantation. Well-sustained tracheal contour was observed with no signs of stenosis or obstruction.

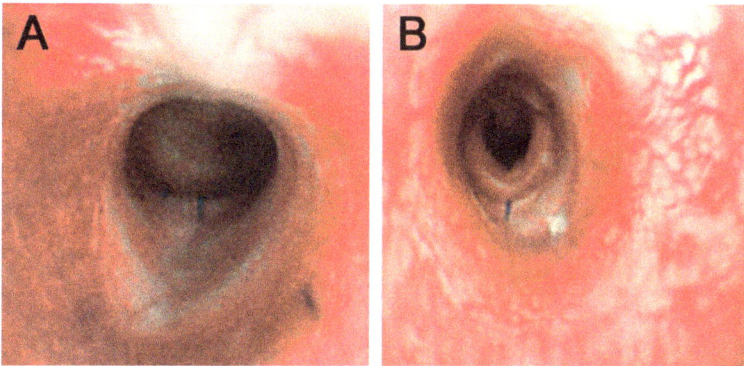

Figure 5. Bronchoscopic findings. Bronchoscopic images of MSC group (**A**) and d-MSC group (**B**) were obtained 12 weeks after the surgery. Tracheal lumen fully covered with epithelial mucosa was observed. Remaining suture materials indicate the implant fixation site.

2.5. In Vivo Epithelialization and Neo-Cartilage Formation

On histopathologic analysis, newly formed respiratory epithelium at the implant regions was observed in all groups. Hematoxylin and eosin (H&E) staining showed that the newly formed epithelium appeared a little rough compared with the normal respiratory epithelium; nevertheless, it has a quite organized structure with cilia (Figure 6). Abundant neovascularization was also found around the implants in all groups. However, neo-cartilage formation was limited to d-MSC group only. On safranin-O/fast green staining, the newly formed cartilage had higher density in cellularity and lighter proteoglycan staining than the normal mature cartilage. The newly formed cartilage was not lining the whole contour of the trachea and was thus limited to a localized area (Figure 7).

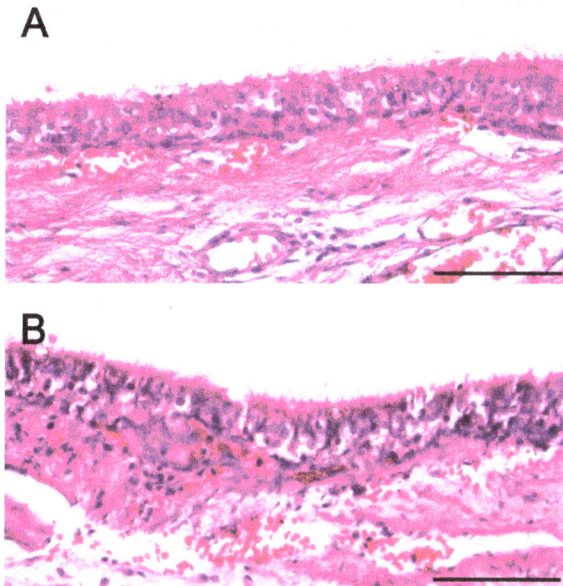

Figure 6. *Cont.*

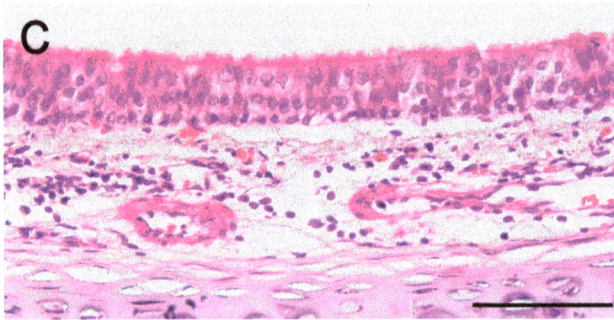

Figure 6. Histopathologic findings. Microscopic images of MSC group (**A**) and d-MSC group (**B**) showed newly formed respiratory epithelium in both groups. Regenerated epithelium in A and B had a rough cell arrangement compared with that of the normal trachea (**C**) but ciliated columnar epithelium was confirmed (Hematoxylin and eosin staining; all scale bars indicate 50 μm).

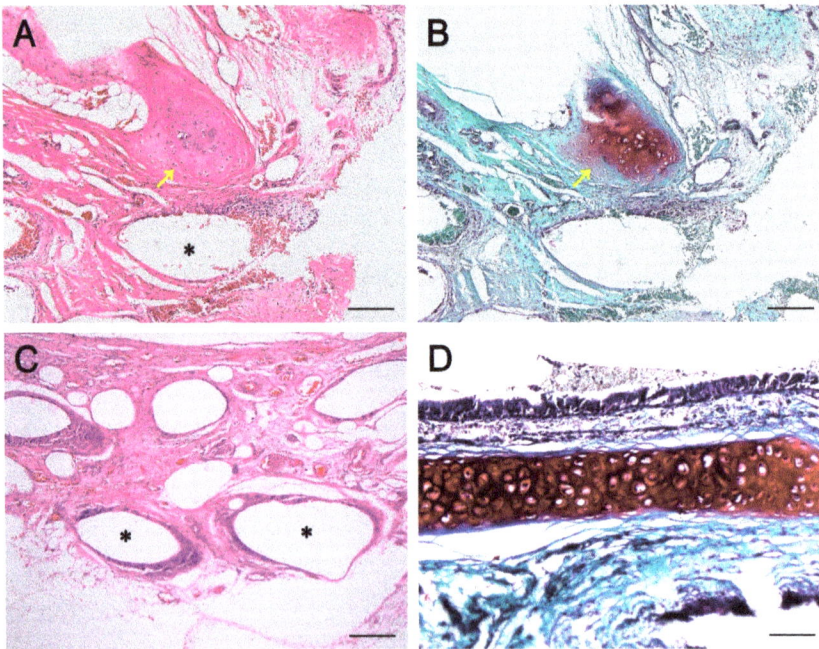

Figure 7. Neo-cartilage formation and neo-vascularization. Neo-vascularization and neo-cartilage formation was observed in the d-MSC group (**A,B**); However, in the MSC group, neo-vascularization was seen but no cartilaginous islet was observed (**C**); Newly formed immature cartilage islet (yellow arrows) had higher cellular density and lighter proteoglycan staining compared with the cartilage of the normal trachea (**D**). A, C and D = hematoxylin and eosin staining; B = safranin O staining. Asterisks (*) indicate fifth polycaprolactone (PCL) layer and all scale bars indicates 50 μm).

3. Discussion

In this study, we cultured rabbit bMSC in a chondrogenic medium and their chondrocyte-like characteristics were identified by modified alcian blue absorbance test and qRT-PCR. Epithelialization and neo-cartilage formation were observed in the animal transplant experiment using a 3D bioprinted artificial trachea, which has autologous epithelial cells and chondrogenic-differentiated bMSC.

Aggrecan, Col2α1 and *SOX-9* are known as chondrogenic differentiation markers; *SOX-9* is pre-chondrogenic marker [13]. Lefebvre et al. identified that *SOX-9* activate aggrecan and Col2α1 genes in cultured cells; hence, it plays an essential role in chondrogenic differentiation [14]. In our study, *SOX-9*, aggrecan, Col1α1 and Col2α1 gene expressions were upregulated in the d-MSC group compared with those in the MSC group. This result is similar to that reported in Bo Wei et al.'s in vitro study of bone marrow MSC-derived extracellular matrix. They revealed that a greater increase in cartilage-like gene expression was observed in the group cultured with transforming growth factor (TGF)-β [15]. Moreover, the qRT-PCR gene expression results indicate that d-bMSC have a greater possibility to give rise to chondrogenic tissue formation than bMSC, which is supported by Barry et al. who demonstrated that TGF could induce chondrogenic differentiation in mesenchymal stem cells [16]. Kojima et al. also demonstrated neo-cartilage formation with supplemental TGF-β in bMSC polymer tissue-engineered trachea and its GAG content levels were also similar to those of a normal cartilage [17]. In our study, the relative accumulation of GAG, assessed by modified alcian blue staining, in cultured cells also supports the qRT-PCR result that chondrogenic-differentiated bMSC have more potency in synthesizing GAG contents than bMSC.

Various materials, that is, from synthetic products to biological tissues, were used for tracheal tissue reconstruction; however, previous reports with sufficiently successful outcomes are limited. Behrend et al. reported in a study of homogenic tracheal transplantation in sheep that the grafts were completely absorbed and replaced by inflammatory scar tissue; thus, the stability of the trachea could not be secured [18]. A study of tissue-engineered allograft, using fibrin-hyaluronan composite gel and chondrocyte, by Kim et al. showed partial success, with fine luminal contour of the regenerated site but insufficient neo-cartilage formation [19]. For a successful tracheal tissue transplantation, the following requirements for the grafts are inevitable: appropriate mechanical properties, adequate blood supply to maintain the characteristic structure and lining with ciliated epithelium [6,7,20]. Several studies showed that bMSC promotes neo-angiogenesis. Han et al. demonstrated that bMSC may enhance neo-vascularization in cryopreserved trachea allograft by upregulating vascular endothelial growth factor expression [21]. Hence, in our study, the abundant neo-vascularization was associated with the angiogenic potential of bMSC.

Patrício et al. demonstrated that PCL/PLA (poly lactic acid) scaffolds produced by solvent casting showed a better result in reduced pore size, mechanical properties and cell adhesion than PCL/PLA scaffolds produced by melt blending [22]. In another recent study, PCL/HA (hydroxyapatite) scaffolds produced using an extrusion-based system revealed successful composite scaffold with fully interconnected pores [23]. Biodegradable materials such as polyglycolic acid (PGA), polylactic acid (PLA), poly(lactic-co-glycolic) acid (PLGA) and polycaprolactone (PCL) used for 3D printing have strength corresponding to the tracheal cartilage and therefore various attempts are being made to apply 3D bioprinting technology to tracheal transplant research [1]. Moreover, recent advances in 3D bioprinting techniques resulted in hybridization of scaffolds with various cells, including mesenchymal stem cells. Zopf et al. reported the use of a customized 3D-printed, biodegradable tracheal prosthesis made with PCL in a patient with tracheobronchomalacia [24]. PCL is a synthetic bioresorbable polymer with potential applications in tissue engineering and thus can be used in a 3D bioprinter without deleterious solvents. It also has excellent mechanical properties and slow degradation in vivo via enzymatic hydrolysis [9]. Moreover, Son et al. demonstrated that a PCL/poly(methyl methacrylate) scaffold is appropriate for neo-bone formation in vivo and cell growth in vitro [25]. Costantini et al. showed that 3D bioprinted scaffolds with bone marrow-derived human mesenchymal stem cells in alginate, as a templating agent for stability during 3D printing, exhibit enhanced chondrogenic differentiation of bMSC in a chondrogenic medium. Additionally, mixing cells with alginate enabled the formulation of biomimetic inks for 3D printing, which in turn could be used in cartilage tissue engineering [26].

In an animal experiment by Go et al., decellularized matrix tracheal tissue transplantation was performed in pigs [27]. They reported that the experimental group, which was transplanted with autologous cells (inner autologous epithelial cells and outer autologous bMSC-derived chondrocytes), had a significantly higher survival rate with no signs of airway collapse or ischemia than the other groups. Furthermore, biological tissues are composed of more than two types of cells. Hence, if scaffolds contain two or more cells, they become more similar to biological tissues and could be more effective in vivo. Currently, having multiple cells in one scaffold is challenging and several investigators have been attempting to overcome this limitation. The advancement in 3D bioprinting techniques with hydrogels is considered an extremely effective solution for various kinds of cells to be applied to a scaffold [28].

As previously mentioned, the trachea is a hollow cylindrical organ with outer C-shaped cartilages covered with respiratory epithelium inside. For tracheal reconstruction, the scaffold should be a favorable environment for respiratory epithelium and neo-angiogenesis and its shape and strength must be appropriate [1]. The artificial trachea used in this study was cylindrical in shape, which is similar to the normal trachea, with five multilayers composed of biocompatible materials (i.e., PCL and alginate hydrogel). The three major aspects in tissue regeneration using tissue engineering include the scaffold, injection of cells and cells seeded or within the scaffold [29] and the last item was involved in our study. We believe that the artificial trachea we manufactured is novel; it contains two different types of cell in the alginate hydrogel separated by middle non-porous PCL layer for mechanical strength and has micropores in the innermost and outermost PCL layer for a smooth communication between cells and adjacent tissues. Furthermore, in our study, we observed newly formed respiratory epithelium in all groups; however, neo-cartilage formation was detected only in the d-MSC group and was limited in a localized area. During tracheal regeneration, cartilage formation is vital as well as epithelialization to maintain mechanical strength and function of the airway. Based on our experiment, chondrogenic differentiation of bMSC was more effective for tracheal cartilage regeneration than non-differentiated bMSC. Nonetheless, further study with a long-term observation and different cell application appears necessary in the near future. Moreover, mechanical test of the tracheal scaffold and dynamic air flow analysis throughout the respiratory system should be conducted prior to clinical application of tissue engineered tracheal replacement.

4. Materials and Methods

This animal study was approved by the Institutional Animal Care and Use Committee of Yonsei University Health System (publication no. 2015-0361, 2015). This study was performed according to the ARRIVE guidelines and the National Institutes of Health Guide for the Care and Use of Laboratory Animals. Approval date: 4 December 2015.

4.1. Primary Cell Culture

4.1.1. Isolation and Culture of Autologous bMSC

Bone marrow-derived mesenchymal stem cells (bMSC) were isolated from each rabbit (New Zealand White rabbits, male, 3 months old; $n = 6$). Briefly, premedication with 5 mg/kg xylazine and 10 mg/kg Zoletil® (Virbac Korea, Seoul, Korea) were administered intramuscularly and anesthesia was maintained by isoflurane inhalation. Bone marrow was harvested from the femur using a 13 G bone biopsy needle and stored in a 50-mL pre-heparinized conical tube (SPL Life Sciences, Gyeonggi-do, Korea). The bone marrow was filtered through a 40-μm cell strainer (Life Sciences, New York, NY, USA) and mixed with phosphate-buffered saline (PBS) of up to 8 mL. The mixture was centrifuged at 1500 rpm for 5 min, the supernatant was discarded and the remaining precipitate was suspended with 8 mL serum-free Dulbecco's Modified Eagle's Medium (low glucose) (Welgene, Daegu, Korea). Subsequently, the mixture was transferred to a 15-mL conical tube (SPL Life Sciences) containing 6 mL of Ficoll-Paque® (Sigma-Aldrich, St. Louis, MO, USA) and centrifuged at 1840 rpm for 30 min. After centrifugation, the interphase was harvested and mixed to the medium (up to 10 mL) in a 15-mL conical tube and centrifuged at 1500 rpm for 5 min and the supernatant was removed. The cell pellet was mixed with the medium supplemented with 10% fetal bovine serum (FBS) (GE Healthcare Life Sciences, Pittsburgh, PA, USA) and 1% penicillin-streptomycin (Thermo Fisher Scientific, Waltham, MA, USA). The culture medium was carefully changed after 3 days and every 2 days thereafter. The culture was maintained at 37 °C in a 5% CO_2 incubator. The bMSC were passaged twice before the experiments.

4.1.2. Isolation and Chondrogenic Differentiation of Autologous bMSC (d-MSC)

The isolation procedure of bMSC was the same as that described above (New Zealand White rabbits, male, 3 months old; $n = 6$). For chondrogenic differentiation, 100 nM dexamethasone, 10% insulin-transferrin-selenium (ITS-premix), 1 μg/mL ascorbic acid, 1% sodium pyruvate, 10 ng/mL human transforming growth factor-β1 were added in the medium [12]. The culture medium was carefully changed after 3 days and every 2 days thereafter. The culture was maintained at 37 °C in a 5% CO_2 incubator. The bMSC were passaged twice before the experiments.

4.1.3. Isolation and Culture of Autologous Epithelial Cells

Epithelial cells were isolated from the rabbits described previously (New Zealand White rabbits, male, 3 months old; $n = 12$). A 4-mm skin biopsy punch was performed under general anesthesia and medial nasal mucosa was harvested from the nostril and stored in PBS containing 1% penicillin-streptomycin for 30 min. Subsequently, submucosal tissue was manually eliminated as much as possible on a sterilized petri dish. Remaining tissue explant was harvested, incubated with 0.2% (w/v) collagenase type II (Thermo Fisher Scientific) in Ham's F-12 medium for 24 h, filtered through a 100-μm nylon cell strainer (BD Biosciences, Franklin Lakes, NJ, USA) and centrifuged at 1500 rpm for 5 min. After the supernatant was discarded, the cells were seeded on a 100-mm cell culture dish (SPL Life Sciences) with Ham's F-12 medium (Welgene) supplemented with 10% FBS (GE Healthcare, Salt Lake City, UT, USA), 1% penicillin-streptomycin (Thermo Fisher Scientific), 10 μg/mL amphotericin B (Enzo Life Sciences, Farmingdale, NY, USA), 50 μg/mL gentamicin (Daesung Microbiological Labs, Gyeonggi-do, Korea), 0.5 μg/mL hydrocortisol (Sigma-Aldrich), 5 ng/mL epidermal growth factor (ProSpec, East Brunswick, NJ, USA), 1.5 μg/mL bovine serum albumin (MP Biomedicals, Santa Ana, CA, USA) and 1×ITS+3 solution (Sigma-Aldrich). Non-adherent cells were removed by washing in PBS and the culture medium was changed every 3 days. The culture was maintained at 37 °C in a 5% CO_2 incubator. The epithelial cells were passaged twice before the experiments.

4.2. In Vitro Study

4.2.1. Modified Alcian Blue Absorbance Test

The relative quantity of GAG contents was determined using modified alcian blue absorbance test. bMSC and chondrogenic-differentiated bMSC were prepared at 5×10^4 cells/mL in a 6-well cell culture plate. Non-adherent cells were removed by washing in PBS and the culture medium was changed every 3 days. The plates were maintained at 37 °C in a 5% CO_2 incubator. On days 14 and 28, each well was washed with PBS twice and 20 μL of alcian blue solution was added into each well. After 15 min, the solutions were discarded and each well was washed with distilled water (DW) thrice. Subsequently, 350 μL of dimethyl sulfoxide (DMSO) was added into each well and the plates were placed on a cell culture rocker system for 10 min. Thereafter, 100 μL of the solution in each well was transferred to a 96-well assay plate (SPL Life Sciences). The absorbance at 610 nm was measured using a VersaMax ELISA Microplate Reader (Molecular Devices, San Jose, CA, USA) [13].

4.2.2. RNA Extraction and Quantitative Reverse Transcription Polymerase Chain Reaction (qRT-PCR)

After 14 days of culture, total RNA of MSC and d-MSC was extracted using the MiniBest™ Universal RNA Extraction kit (TaKaRa Biomedicals, Otsu, Japan) according to the manufacturer's protocol. The quantity and purity of the RNA from each sample were determined by the ratio of the optical density at 260 nm to that at 280 nm using Nanodrop™ (Thermo Fisher Scientific). Using PrimeScript™ RT reagent kit (TaKaRa Biomedicals), we prepared 1 µL of cDNA according to the manufacturer's protocol. The PCR primers for aggrecan (ACAN), collagen type 1 (Col1α1), collagen type 2 (Col2α1) and SOX-9 (SOX9) gene are described in Table 1. qRT-PCR was performed to detect quantitative real-time PCR products from cDNA using SYBR Premix Ex Taq™ (TaKaRa Biomedicals). The qRT-PCR conditions were as follows: 35 cycles of denaturation at 95 °C for 30 s, annealing at 60 °C for 1 min and extension at 72 °C for 1 min in a qRT-PCR detection system (Thermo Fisher Scientific).

Table 1. Primer sequences used in qRT-PCR.

Gene	Primer Nucleotide Sequence
ACAN	Forward: 5-TCGAGGACAGCGAGGCC-3 Reverse: 3-AGAGATGTGCGATGTGGGAGCT-5
Col1α1	Forward: 5-GCGGTGGTTACGACTTTGGTT-3 Reverse: 3-AGTGAGGAGGGTCTCAATCTG-5
Col2α1	Forward: 5-GGCAATAGCAGGTTCACGTACA-3 Reverse: 3-TTCACCCCGTTCTGACAATAGC-5
SOX9	Forward: 5-CACACAGCTCACTCGACCTTG-3 Reverse: 3-GCTCTACTAGGATTTTATTGGCTT-5
GAPDH	Forward: 5-ATGGGGAAGGTGAAGGTCG-3 Reverse: 3-CCAGTGGTCCCGACGAAAAT-5

The threshold cycle (Ct) values were normalized using a Ct value derived from the following: $\Delta Ct = Ct_{target} - Ct_{GAPDH}$; the expression of each RNA in the d-MSC group relative to that in the MSC group (fold change) was described using the following: $2^{-\Delta\Delta Ct}$, where $\Delta\Delta Ct = \Delta Ct_{d\text{-}MSC} - \Delta Ct_{MSC}$.

4.3. In Vivo Study

4.3.1. Manufacturing Artificial Trachea Using a 3D Bioprinter

The artificial trachea was cylindrical in shape and had five layers, an inner and outer diameter of 5 and 10 mm, respectively and a length of 15 mm. Polycaprolactone (PCL; MW = 45,000) (Sigma-Aldrich) was used as the supporting layer in the first, third and fifth layers and sodium alginate (Sigma-Aldrich) was used as the middle viscosity layer in the second and fourth layers from the inside. The PCL in the first and fifth layers were fabricated in diagonal grid patterns with micropores for ease of exchange of growth factors between cells and adjacent tissues, while the PCL in the third layer was fabricated in a helical form without pores to separate different cells in the second and fourth viscosity layers. PCL was placed in a 3D bioprinter (KIMM & Protek Korea, Daejeon, Korea) and dispensed through a 300 µm nozzle at a temperature of around 100 °C and a pneumatic pressure of 400 kPa. Moreover, the middle viscosity layer was fabricated in a helical form (epithelial cells in the second layer and MSC or d-MSC in the fourth layer). 1% calcium chloride solution was added to achieve the appropriate viscosity of sodium alginate hydrogel. Each hydrogel contained 1×10^7 cells/10 mL. The cells encapsulated with alginate hydrogel was loaded into a disposable syringe of the same 3D bioprinter and dispensed through a 400 µm nozzle with a pneumatic pressure of 100 kPa at room temperature. The printing speed of all layer were 200 mm/min. The resulting artificial trachea was soaked in 5% calcium chloride solution for 30 min for gelation of bioprinted sodium alginate hydrogel and was subsequently transferred to a medium for cell preservation. All artificial tracheas were manufactured a day prior to surgical transplantation.

4.3.2. Cell Distribution in the 3D Bioprinted Artificial Trachea

Cell distribution in alginate hydrogel was identified using CellTracker™ Fluorescent Probes (Thermo Fisher Scientific) according to the manufacturer's protocol. Briefly, the dye vial was thawed at room temperature before opening and the dye product was dissolved in DMSO to a final concentration of 10 mM. Serum-free medium was added to the dissolved dye vial and the solution was warmed to 37 °C. The prepared cells were harvested by centrifugation and aspiration of the supernatant. The cells were re-suspended in a pre-warmed solution (epithelial cells in the red-dye vial and bMSC in the green-dye vial) and incubated for 30 min at 37 °C in a 5% CO_2 incubator. Thereafter, centrifugation and removal of the supernatant were performed again and the cells obtained were used in the 3D bioprinting as previously described. The manufactured artificial tracheas were observed under a light microscope with appropriate optical filters according to the color of the dye.

4.3.3. Scanning Electron Microscope (SEM)

The manufactured 3D bioprinted artificial trachea was imaged using a field emission scanning electron microscope with a backscattered electron image detector and an environmental secondary electron detector (JEOL. Ltd., Tokyo, Japan).

4.3.4. Surgical Procedures for Tracheal Replacement Using 3D Bioprinted Artificial Trachea and Post Experimental Observation

After the administration of 5 mg/kg xylazine and 10 mg/kg Zoletil® intramuscularly as pre-medications, 3.0 Fr tracheal tube intubation was performed and general anesthesia was maintained by isoflurane inhalation. The rabbits ($n = 12$) were placed in dorsal recumbent position with the neck slightly extended. The surgical region was shaved and disinfected. A vertical midline skin incision was made and the underlying trachea was carefully dissected from the cervical muscles. After cervical trachea exposure, a half-pipe-shaped partial tracheal resection (approximately 10×10 mm-size) was performed with a no. 15 scalpel. The defect was gently replaced with the pre-manufactured 3D bioprinted artificial trachea, which contains autologous epithelial cells and bMSC ($n = 6$, MSC group) or epithelial cells and chondrogenic-differentiated bMSC ($n = 6$, d-MSC group) (Figure 8). The artificial trachea was 15 mm in length (longitudinally sectioned and half-pipe-shaped to match the resected defect) and securely sutured with 5–0 Vicryl (Ethicon, Somerville, NJ, USA). Standard closure of the skin was performed and postoperative antibiotics were administered subcutaneously once a day for the first week. The animals were placed in cages at 20–24 °C and 40–60% humidity with a 12-h light/dark cycle and were fed standard laboratory rabbit food and water ad libitum. To evaluate the diameter of the airway, plain thoracic radiographs were obtained at after extubation, 4 and 12 weeks postoperatively with VXR-9M radiography system (DRGEM, Gwangmyeong-si, Korea). Bronchoscopic examinations with CV-260SL (Olympus, Tokyo, Japan) and computed tomography (CT) with Brivo 385 CT scanner system (GE Healthcare, Little Chalfont, UK) were performed at postoperative 12 weeks before sacrifice.

4.3.5. Histopathology

All rabbits were euthanized at 12 weeks after transplantation. Tracheal segments including 5 mm of healthy tracheal tissue, both proximal and distal to the transplants, were harvested and fixed in 10% formalin. After fixation, the samples were embedded in paraffin blocks and cut (7-μm thick). All sections were obtained from the middlemost part of the samples including the transplants. The sections were deparaffinized, rehydrated and stained with hematoxylin and eosin (H&E) and safranin-O/fast green. Thereafter, the slides were observed under a light microscope.

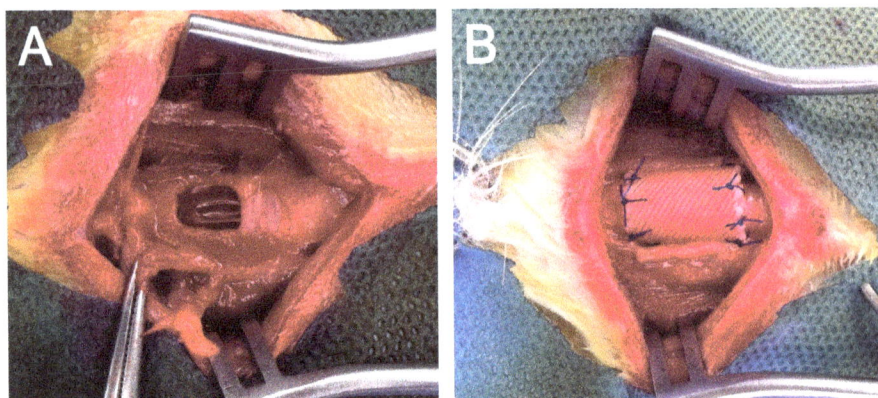

Figure 8. Application of 3D bioprinted artificial trachea. (**A**) Approximately 10 × 10-mm half-pipe-shaped tracheal defect on the ventral part of the trachea; (**B**) The defect was replaced with 3D bioprinted artificial trachea and sutured with 5–0 absorbable suture material.

4.4. Statistical Analysis

All data are expressed as mean ± standard deviation. Statistical analyses were performed using GraphPad Prism 5.0 software (GraphPad Software Inc., San Diego, CA, USA). Normal data distribution was determined using the Shapiro-Wilk test. A two-tailed Student's unpaired *t*-test was used to compare the mean values of all study parameters. A *p* value < 0.05 was considered statistically significant.

5. Conclusions

In this study, we cultured and differentiated rabbit autologous bone marrow-derived mesenchymal stem cells into chondrocyte-like cells and confirmed its chondrogenic features by modified alcian blue absorbance test and qRT-PCR. The relative GAG accumulation level was increased as time goes by in chondrogenic differentiated mesenchymal stem cells. Also, chondrogenic-differentiation marker genes such as SOX-9, aggrecan, Col1α1 and Col2α1 expressions were upregulated in the d-MSC group compared with those in the MSC group.

Moreover, we developed artificial trachea in a novel design by a 3d bioprinting technique, which contains two different kinds of cells (respiratory epithelial cells and chondrogenic-differentiated bone marrow-derived stem cells) in one scaffold. In our animal experiment on tracheal replacement, neo-cartilage formation was observed in a quite short period in d-MSC group, neo-epithelialization and neo-vascularization was identified as well. According to our study, differentiation of the bone-marrow derived mesenchymal stem cells considered to have more potential for tracheal cartilage regeneration.

Author Contributions: S.-W.B., J.L. and D.-H.K. conceived the project, design the experiments and analysis the data. S.-W.B., K.-W.L., J.-H.P. and D.-H.K. performed the experiments. J.L., C.-R.J. and J.Y. performed 3D bio-printing. S.-W.B. wrote the manuscript and D.-H.K. and H.-Y.K. reviewed and edited the manuscript.

Acknowledgments: This research was supported by the Basic Science Research Program through the National Research Foundation of Korea (NRF) funded by the Ministry of Education (NRF-2015R1D1A1A01060374) and the Korea Research Institute of Bioscience and Biotechnology (KRIBB) Research Initiative Program.

Conflicts of Interest: The authors declare no conflict of interest.

References

1. Chang, J.W.; Park, S.A.; Park, J.K.; Choi, J.W.; Kim, Y.S.; Shin, Y.S.; Kim, C.H. Tissue-engineered tracheal reconstruction using three-dimensionally printed artificial tracheal graft: Preliminary report. *Artif. Organs* **2014**, *38*, E95–E105. [CrossRef] [PubMed]

2. Villegas-Álvarez, F.; González-Zamora, J.F.; González-Maciel, A.; Soriano-Rosales, R.; Pérez-Guille, B.; Padilla-Sánchez, L.; Reynoso-Robles, R.; Ramos-Morales, A.; Zenteno-Galindo, E.; Pérez-Torres, A.; et al. Fibrocollagen-covered prosthesis for a noncircumferential segmental tracheal replacement. *J. Thorac. Cardiovasc. Surg.* **2010**, *139*, 32–37. [CrossRef] [PubMed]

3. Doss, A.E.; Dunn, S.S.; Kucera, K.A.; Clemson, L.A.; Zwischenberger, J.B. Tracheal replacements: Part 2. *ASAIO J.* **2007**, *53*, 631–639. [CrossRef] [PubMed]

4. Rose, K.; Sesterhenn, K.; Wustrow, F. Tracheal allotransplantation in man. *Lancet* **1979**, *1*, 433. [CrossRef]

5. Wurtz, A.; Hysi, I.; Kipnis, E.; Zawadzki, C.; Hubert, T.; Jashari, R.; Copin, M.C.; Jude, B. Tracheal reconstruction with a composite graft: Fascial flap-wrapped allogenic aorta with external cartilage-ring support. *Interact. Cardiovasc. Thorac. Surg.* **2013**, *16*, 37–43. [CrossRef] [PubMed]

6. Jungebluth, P.; Moll, G.; Baiguera, S.; Macchiarini, P. Tissue-engineered airway: A regenerative solution. *Clin. Pharmacol. Ther.* **2012**, *91*, 81–93. [CrossRef] [PubMed]

7. Hamilton, N.; Bullock, A.J.; Macneil, S.; Janes, S.M.; Birchall, M. Tissue engineering airway mucosa: A systematic review. *Laryngoscope* **2014**, *124*, 961–968. [CrossRef] [PubMed]

8. Vacanti, C.A.; Paige, K.T.; Kim, W.S.; Sakata, J.; Upton, J.; Vacanti, J.P. Experimental tracheal replacement using tissue-engineered cartilage. *J. Pediatr. Surg.* **1994**, *29*, 201–205. [CrossRef]

9. Sakata, J.; Vacanti, C.; Schloo, B.; Healy, G.; Langer, R.; Vacanti, J. Tracheal composites tissue engineered from chondrocytes, tracheal epithelial cells and synthetic degradable scaffolding. *Transplant. Proc.* **1994**, *26*, 3309–3310. [PubMed]

10. Zein, I.; Hutmacher, W.; Tan, K.; Teoh, S. Fused Deposition Modeling of Novel Scaffold Architectures for Tissue Engineering Applications. *Biomaterials* **2002**, *23*, 1169–1185. [CrossRef]

11. Murphy, S.V.; Atala, A. 3D bioprinting of tissues and organs. *Nat. Biotechnol.* **2014**, *32*, 773–785. [CrossRef] [PubMed]

12. Ciuffreda, M.C.; Malpasso, G.; Musarò, P.; Turco, V.; Gnecchi, M. Protocols for in vitro differentiation of human mesenchymal stem cells into osteogenic, chondrogenic and adipogenic lineages. *Mesenchymal Stem Cells: Methods and Protoc.* **2016**, *1416*, 149–158. [CrossRef]

13. Tanthaisong, P.; Imsoonthornruksa, S.; Ngernsoungnern, A.; Ngernsoungnern, P.; Ketudat-Cairns, M.; Parnpai, R. Enhanced chondrogenic differentiation of human umbilical cord wharton's jelly derived mesenchymal stem cells by GSK-3 Inhibitors. *PLoS ONE* **2017**, *12*, 1–15. [CrossRef] [PubMed]

14. Lefebvre, V.; Behringer, R.R.; De Crombrugghe, B. L-Sox5, Sox6 and SOx9 control essential steps of the chondrocyte differentiation pathway. *Osteoarthr. Cartil.* **2001**, *9*, 69–75. [CrossRef]

15. Wei, B.; Jin, C.; Xu, Y.; Du, X.; Yan, C.; Tang, C.; Ansari, M.; Wang, L. Chondrogenic Differentiation of Marrow Clots after Microfracture with BMSC-Derived ECM Scaffold in Vitro. *Tissue Eng. Part A* **2014**, *20*, 2646–2655. [CrossRef] [PubMed]

16. Barry, F.; Boynton, R.E.; Liu, B.; Murphy, J.M. Chondrogenic differentiation of mesenchymal stem cells from bone marrow: Differentiation-dependent gene expression of matrix components. *Exp. Cell Res.* **2001**, *268*, 189–200. [CrossRef] [PubMed]

17. Kojima, K.; Ignotz, R.A.; Kushibiki, T.; Tinsley, K.W.; Tabata, Y.; Vacanti, C.A. Tissue-engineered trachea from sheep marrow stromal cells with transforming growth factor β2 released from biodegradable microspheres in a nude rat recipient. *J. Thorac. Cardiovasc. Surg.* **2004**, *128*, 147–153. [CrossRef] [PubMed]

18. Behrend, M.; Kluge, E. The fate of homograft tracheal transplants in sheep. *World J. Surg.* **2008**, *32*, 1669–1675. [CrossRef] [PubMed]

19. Kim, D.Y.; Pyun, J.H.; Choi, J.W.; Kim, J.H.; Lee, J.S.; Shin, H.A.; Kim, H.J.; Lee, H.N.; Min, B.H.; Cha, H.E.; et al. Tissue-engineered allograft tracheal cartilage using fibrin/hyaluronan composite gel and its in vivo implantation. *Laryngoscope* **2010**, *120*, 30–38. [CrossRef] [PubMed]

20. Ott, L.M.; Weatherly, R.A.; Detamore, M.S. Overview of tracheal tissue engineering: Clinical need drives the laboratory approach. *Ann. Biomed. Eng.* **2011**, *39*, 2091–2113. [CrossRef] [PubMed]

21. Han, Y.; Lan, N.; Pang, C.; Tong, X. Bone marrow-derived mesenchymal stem cells enhance cryopreserved trachea allograft epithelium regeneration and vascular endothelial growth factor expression. *Transplantation* **2011**, *92*, 620–626. [CrossRef] [PubMed]

22. Patrício, T.; Domingos, M.; Gloria, A.; D'Amora, U.; Coelho, J.F.; Bártolo, P.J. Fabrication and characterisation of PCL and PCL/PLA scaffolds for tissue engineering. *Rapid Prototyp. J.* **2014**, *20*, 145–156. [CrossRef]

23. Domingos, M.; Gloria, A.; Coelho, J.; Bartolo, P.; Ciurana, J. Three-dimensional printed bone scaffolds: The role of nano/micro-hydroxyapatite particles on the adhesion and differentiation of human mesenchymal stem cells. *Proc. Inst. Mech. Eng. Part H J. Eng. Med.* **2017**, *231*, 555–564. [CrossRef] [PubMed]

24. Zopf, D.; Hollister, S.; Nelson, M.; Ohye, R.; Green, G. Bioresorbable airway splint created with a three-dimensional printer. *N. Engl. J. Med.* **2013**, *368*, 2043–2045. [CrossRef] [PubMed]

25. Son, S.R.; Linh, N.T.B.; Yang, H.M.; Lee, B.T. In vitro and in vivo evaluation of electrospun PCL/PMMA fibrous scaffolds for bone regeneration. *Sci. Technol. Adv. Mater.* **2013**, *14*. [CrossRef] [PubMed]

26. Costantini, M.; Idaszek, J.; Szöke, K.; Jaroszewicz, J.; Dentini, M.; Barbetta, A.; Brinchmann, J.E.; Święszkowski, W. 3D bioprinting of BM-MSCs-loaded ECM biomimetic hydrogels for in vitro neocartilage formation. *Biofabrication* **2016**, *8*. [CrossRef] [PubMed]

27. Go, T.; Jungebluth, P.; Baiguero, S.; Asnaghi, A.; Martorell, J.; Ostertag, H.; Mantero, S.; Birchall, M.; Bader, A.; Macchiarini, P. Both epithelial cells and mesenchymal stem cell-derived chondrocytes contribute to the survival of tissue-engineered airway transplants in pigs. *J. Thorac. Cardiovasc. Surg.* **2010**, *139*, 437–443. [CrossRef] [PubMed]

28. Huang, Y.; Zhang, X.; Gao, G.; Yonezawa, T.; Cui, X. 3D bioprinting and the current applications in tissue engineering. *Biotechnol. J.* **2017**. [CrossRef] [PubMed]

29. Fishman, J.M.; Lowdell, M.; Birchall, M.A. Stem cell-based organ replacements-Airway and lung tissue engineering. *Semin. Pediatr. Surg.* **2014**, *23*, 119–126. [CrossRef] [PubMed]

International Journal of
Molecular Sciences

MDPI

Article

3D Printed, Microgroove Pattern-Driven Generation of Oriented Ligamentous Architectures

Chan Ho Park [1,*], Kyoung-Hwa Kim [2], Yong-Moo Lee [2], William V. Giannobile [3] and Yang-Jo Seol [2,*]

[1] Dental Research Institute, School of Dentistry, Seoul National University, 1 Gwanak-ro, Gwanak-gu, Seoul 08826, Korea
[2] Department of Periodontology and Dental Research Institute, School of Dentistry, Seoul National University, 28 Yongon-dong, Chongno-gu, Seoul 110-749, Korea; perilab@snu.ac.kr (K.-H.K.); ymlee@snu.ac.kr (Y.-M.L.)
[3] Department of Periodontics and Oral Medicine, School of Dentistry and Department of Biomedical Engineering, College of Engineering, University of Michigan, 1011 North University Ave., Ann Arbor, MI 48109-1078, USA; wgiannob@umich.edu
*   Correspondence: perioengineer@snu.ac.kr (C.H.P.); yjseol@snu.ac.kr (Y.-J.S.); Tel.: +82-2-880-2323 (C.H.P.); +82-2-744-0051 (Y.-J.S.)

Received: 21 July 2017; Accepted: 4 September 2017; Published: 8 September 2017

Abstract: Specific orientations of regenerated ligaments are crucially required for mechanoresponsive properties and various biomechanical adaptations, which are the key interplay to support mineralized tissues. Although various 2D platforms or 3D printing systems can guide cellular activities or aligned organizations, it remains a challenge to develop ligament-guided, 3D architectures with the angular controllability for parallel, oblique or perpendicular orientations of cells required for biomechanical support of organs. Here, we show the use of scaffold design by additive manufacturing for specific topographies or angulated microgroove patterns to control cell orientations such as parallel (0°), oblique (45°) and perpendicular (90°) angulations. These results demonstrate that ligament cells displayed highly predictable and controllable orientations along microgroove patterns on 3D biopolymeric scaffolds. Our findings demonstrate that 3D printed topographical approaches can regulate spatiotemporal cell organizations that offer strong potential for adaptation to complex tissue defects to regenerate ligament-bone complexes.

Keywords: 3D printing; ligament; microgroove patterns; biopolymer; tissue engineering

1. Introduction

Fibrous connective tissues in musculoskeletal systems require highly specialized spatiotemporal organizations with physical integration and mineralized tissues for physiological responsiveness under biomechanical stimulations [1]. Specific orientations of fibrous tissues on highly organized collagenous constructs have crucial roles in optimizing various biomechanical and biophysical responses like absorption, transmission or the generation of forces [1,2]. However, diseases or traumatic injuries of the musculoskeletal systems could induce instabilities in multiple tissue interfaces or the loss of their skeletal-supportive functions [1]. In the cases of craniofacial and dental complexes, multiple tissue integrities with coordinated ligaments, the periodontal complex (bone-periodontal ligament (PDL)-cementum), facilitate the optimization of various systematic functional responses to support and position the teeth [3–6]. In particular, angulated PDLs with spatiotemporal organizations between the teeth and the alveolar bone significantly contribute to masticatory/occlusal stress absorptions and distributions [3,7], as well as the optimization of mineralized tissue remodeling for tooth-periodontium complexes [8]. Therefore, perpendicular/oblique PDL orientations to the tooth-root surfaces add to the functionalization and revitalization of tooth-supportive biofunctional structures.

Various state-of-the-art approaches with micro-/nano-topographical characteristics on 2D substrates have been developed to generate various cell-material interactions [9–11] and to regulate cell behaviors, such as cell adhesion, migration, proliferation, differentiation or specific cell organizations [12–14]. Beyond 2D perspectives for biomedical applications, additive manufacturing or 3D printing techniques permit spatial designs of specific geometries [15], and many efforts have contributed to manufacturing 3D scaffolding systems for preclinical and clinical scenarios [16–19]. In our example, fiber-guiding scaffolds particularly promoted periodontal regeneration with tissue compartmentalization and limited PDL organizations to the tooth root surface [20]. In addition to preclinical studies, our first human case study using a patient-specific scaffold manufactured by 3D printing attempted to treat a large periodontal defect and to regenerate periodontal complexes (bone-PDL-cementum) [21]. However, there remain challenging limitations for the spatiotemporally control of perpendicular/oblique angulations of ligamentous bundles for physiological functioning restorations in the ligament-bone complexes.

Herein, we investigated a simple, but precisely controllable method to create 3D printed architectures with cell-responsive, micro-topographies with angulated patterns on spatial scaffolds (Figures 1 and 2). The additive manufacturing system creates micron-scaled, layer-by-layer artifacts, which are generally removed for smooth surface qualities. In this study, the angular-controllable microgrooves (known as artifacts) were actively considered for specific surface patterns, which can be programmed in the digital slicing step of the additive manufacturing procedure (Figures 1B and 2A). The manufacturing strategy for biomimetic microenvironments can enable integrated angulated microgroove patterns on ligament-guiding scaffolds to promote spatiotemporally-directional and anisotropic cellular consolidations.

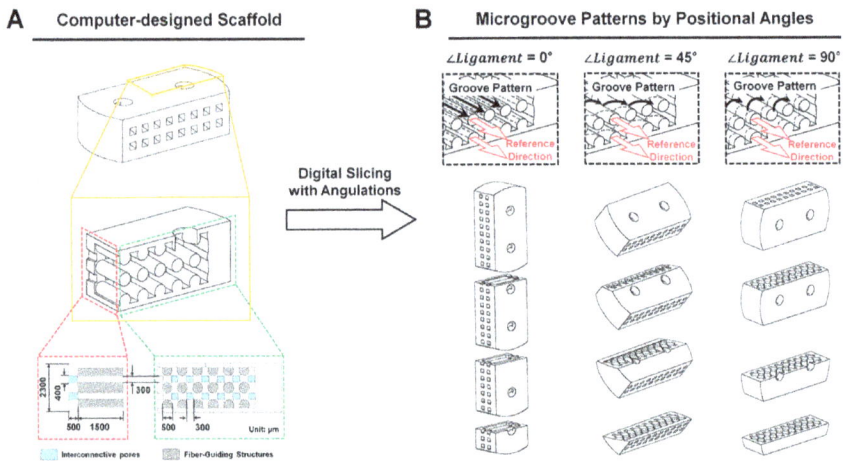

Figure 1. 3D designs of engineered ligament scaffolds. (**A**) After designing 3D scaffolds with three-layered ligament-guiding architectures; (**B**) different angles were set for the additive manufacturing. The yellow part was the quarter of the scaffold to show the designed internal architecture. Green and red dash-lined boxes presented the side and the front views of the scaffold with the design parameters.

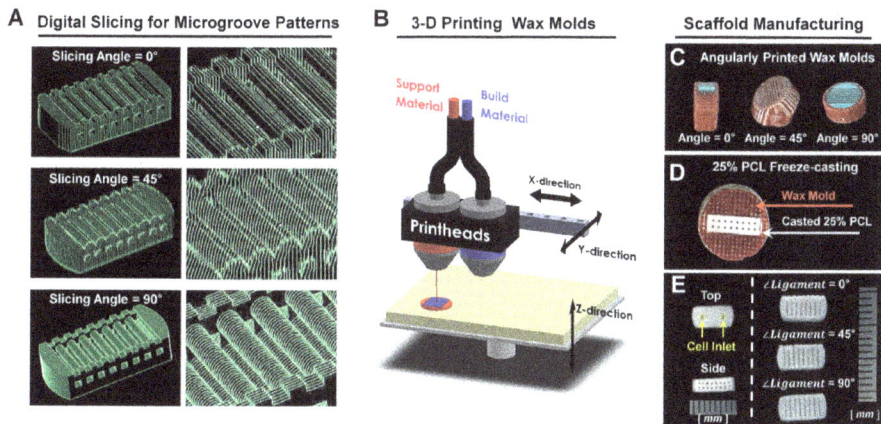

Figure 2. Fabrication of engineered ligament scaffolds. (**A**) The digital slicing procedure created angulated microgroove patterns on ligament architectures; (**B–D**) the additive 3D printing system produced dual wax constructs and 25% PCL (poly-ε-caprolactone) was casted after removing blue wax parts; (**E**) PCL scaffolds were shown with similar external architectures.

2. Results

Scanning electron microscope (SEM) showed each angulated ligament-guiding, microgroove patterns on the PCL scaffolds at a low magnification and rough surfaces on the microgroove wall-sides at a high magnification (Figure 3A–C). In particular, the surface roughness can be generated by the freeze-casting method with various concentrations of PCL in 1,4-dioxane, and different surface properties of the scaffolds can be significantly characterized for the biological responses of the cells. At this point, due to the difficulty of quantitative measurements of roughness on spatial structures of microscopic topographies, the surface roughness was analyzed using the freeze-casted PCL constructs on flat wax molds (Figure S1). Among the different PCL solutions, the surface roughness by the 20% and 25% PCL solutions showed statistical similarities (Ra (20%) = 2.237 ± 0.727 µm; Ra (25%) = 2.00 ± 0.700 µm; Figure S1), and the 25% PCL solution was significantly considered for the scaffolds to enhance cell attachment and viability [22]. In addition to indirect characterizations of scaffold topographies (Figure S1), noisy section-profiles of microgroove patterns on scaffolds can qualitatively show the surface roughness (Figure 3D–F). Based on geometric assessments, the precise additive manufacturing resulted in high reliability and reproducibility to generate microgroove patterns with angulated slicing procedures. From the micro-CT (micro-computed tomography) results, the microgroove patterns on the referential ligament structures were spatially qualitative in the 3D reconstruction, and the microgroove depths of all of the groups were measured with 2D segmented scaffold images (Figure 4). Based on various structural characterizations of the 3D ligament scaffolds (∠*Ligament* = 0°, 45° and 90°), the additive manufacturing system created highly reproducible and adjustable microgroove structures by the angular slicing procedures. Furthermore, the freeze-casting method of the biopolymeric solution (25% PCL solution in 1,4-dioxane) facilitated roughness to enhance cellular attachments to the biomaterial surface.

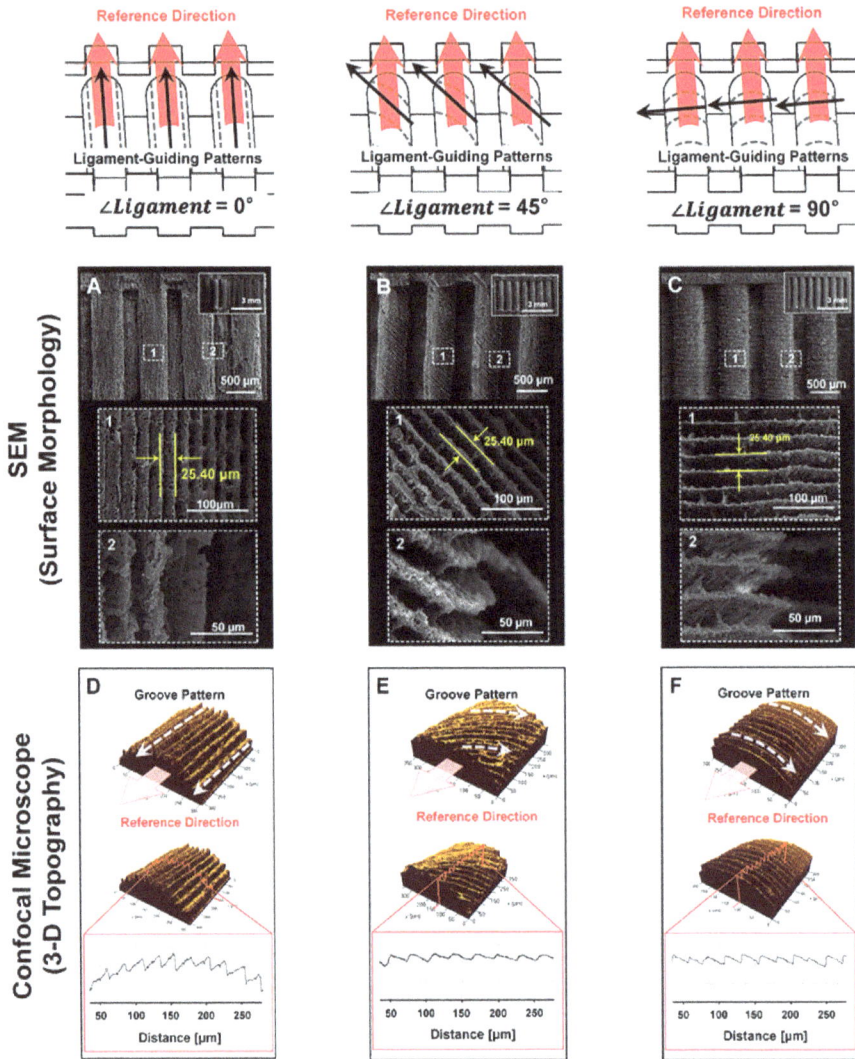

Figure 3. Image analyses of microgroove patterns on ligament architectures of scaffolds. (**A–C**) Scanning electron microscope (SEM) qualitatively demonstrated surface morphologies, roughness and angulated microgroove patterns on ligament scaffolds with the distance between microgroove patterns by the digital slicing procedure (25.40 µm); (**D–F**) the confocal microscope facilitated topographical investigations for different angulated microgroove patterns to the reference direction. In the red boxes of **D–F**, surface topographies and roughness of three different groups ($\angle Ligament = 0°$, 45° and 90°) were characterized and profiled using crossed red-lines on the scaffolds. Based on the surface profiles, intervals of microgrooves can be determined with approximately 25.40 µm.

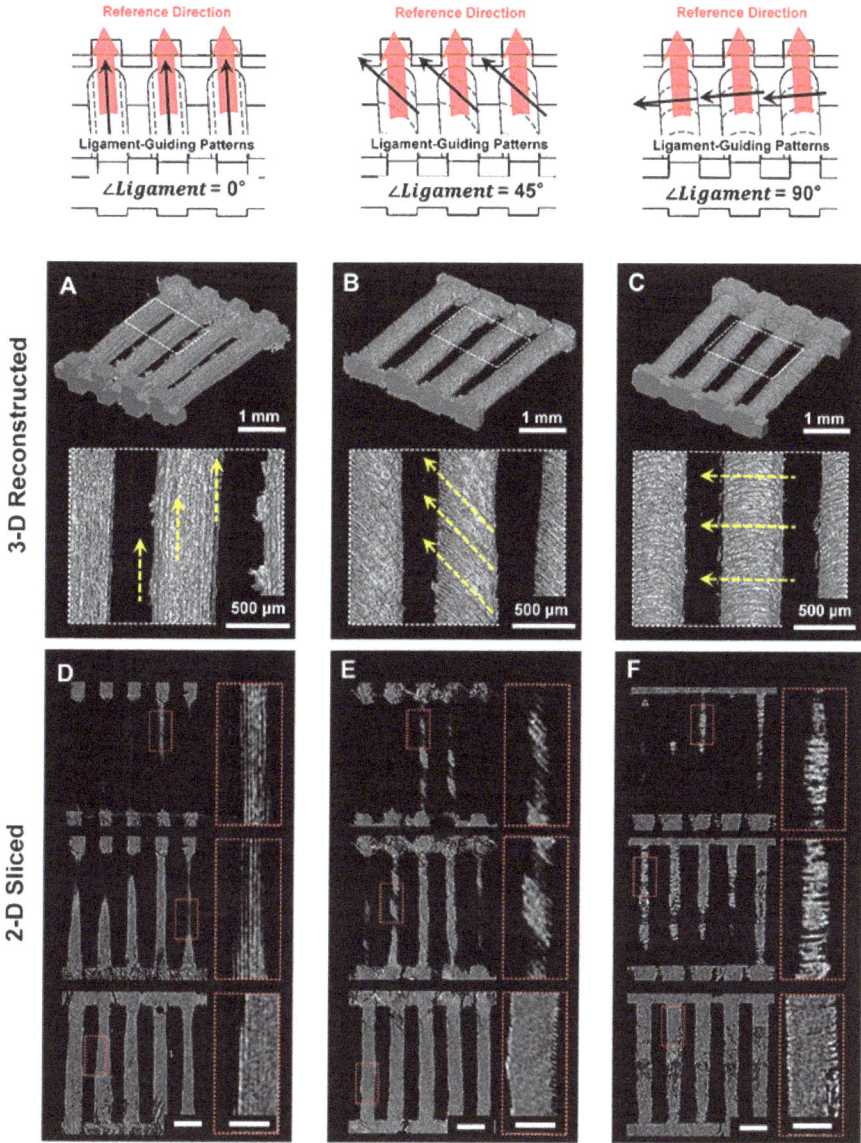

Figure 4. Micro-computed tomographic (micro-CT) images for 3D reconstructed and 2D sliced image analyses. (**A–C**) 3D reconstructed spatial patterns were qualitatively analyzed on scaffold surfaces and (**D–F**) 2D sliced images demonstrated microgroove directionalities after fabrication of engineered ligament scaffolds. The scale bars in **D–F**: 1 mm. The scale bars in the red-dash boxes in **D–F**: 250 μm.

Prior to the investigations of directional cell orientations, the interactions between the microgroove patterns and cell shapes (or nuclear deformations) were required for simple identification of different microgroove patterns, which the 3D printing system can create with a parallel topography to the reference direction ($\angle Ligament = 0°$; Figure S2). Slicing thicknesses of 12.70 μm and 25.40 μm were programmed to manufacture wax molds, and identical patterns were found on the spatially-designed

PDL architectures after the PCL casting (Figure S2). After hPDL cell culturing with 1.0×10^3 cells per scaffold at 7 days, 12.70-μm microgroove patterns exhibited random organizations of cells with their nuclei; however, the 25.40-μm microgrooves had significantly oriented cell collectivities with typical nuclear angles (frequency of nuclear angle (0–10°) = 73.26 ± 14.85%; Figure S2). In addition, the NAR (nuclear aspect ratio), which is calculated with the long-to-short axis ratio and characterized as one (NAR = 1) for the same long and short axes [23,24], can provide statistical identification of cell shapes and associations with cellular orientations. Based on NSI (nuclear shape index) and NAR, two typical microgroove patterns can markedly guide different morphological characteristics on the same microgroove directionalities. Quantification analyses of the nuclear shapes demonstrated that highly organized cell collectivities on the 25-μm microgrooves had a statistically lower circularity (NSI (12.70 μm) = 0.82 ± 0.039 and NSI (25.40 μm) = 0.65 ± 0.037), which indicates between zero for linear and one for circular shapes (Figure S2). The organization of developed F-actin bundles on spatial structures possessing microgroove patterns was qualitatively observed, and actin cytoskeleton orientations were clearly identified (Figure 5A–C) following the angulated patterns ($\angle Ligament = 0°$, 45° and 90°). In addition, the orientations of hPDL cell collectivities were statistically analyzed with the angulations of the anisotropic hPDL nuclei, which strongly corresponded to the microgroove topographies (Figure 5D,E). Approximately 70% or more of the cells were highly aligned on the microgroove patterns on the created topographical-guiding platforms ($\angle Ligament = 0°$, 45° and 90°) in the 7- and 21-day cultures (Figure 5A–C and Figure S3). 3D printed microgroove patterns enabled a significant promotion of cell proliferation, as well as directional organization along three different guidable topographies (Figure S3). Based on the results of the increased hPDL populations after the 21-day cultures, the microgroove topographies with rough surfaces by the freeze-casting method facilitated the promotion of cell attachments and proliferation, as well as angulations (Figure 5).

In addition to nuclear angulations, cell shapes consistently correlated with nuclear morphologies, which characterized the microgroove patterns using the analysis methods, NAR and NSI [23,24]. Angulated nuclei were distinctly identified on individual microgroove patterns; however, the NARs and NSIs (or circularities) of $\angle Ligament$ of 0°, 45° and 90° had no statistical differences in the 7- and 21-day cultures ($p > 0.05$; Figure 6). As shown in Figures 3 and 4, the nuclei were highly elongated (NAR ~2.34 at 7 days and 2.41 at 21 days) and deformed (NSI ~0.64 at 7 days and 0.59 at 21 days) when the hPDL cells were cultured on angulated microgrooves (Figure 6).

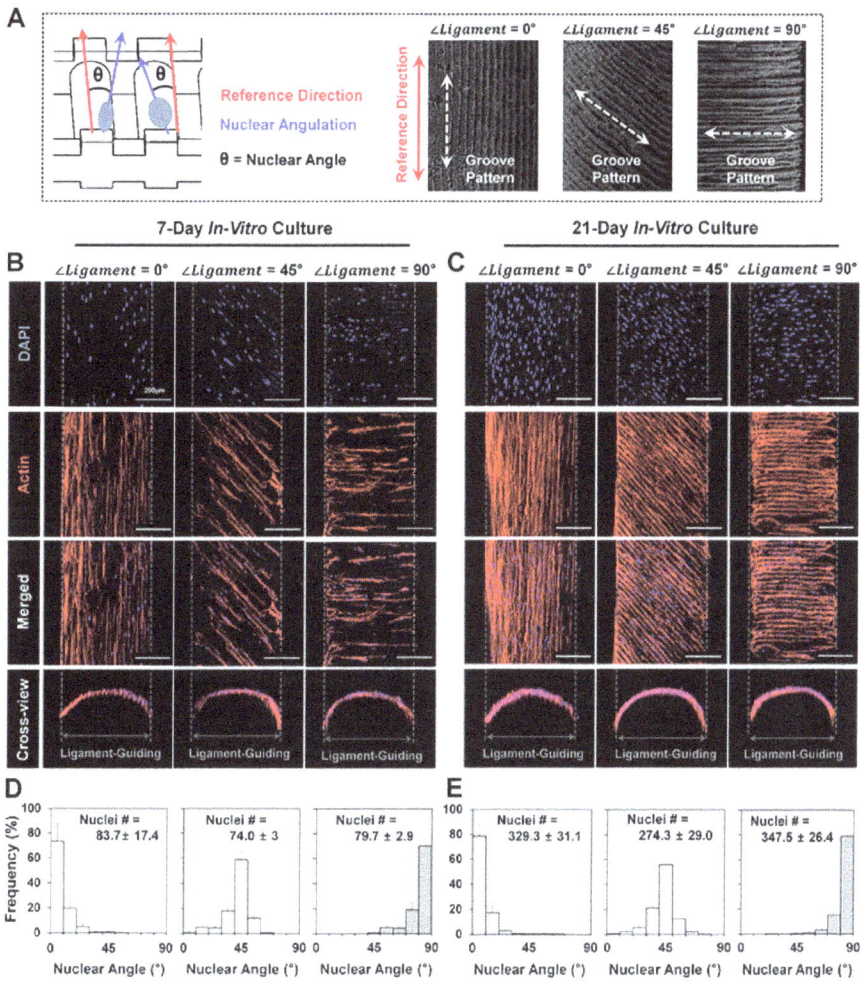

Figure 5. Fluorescence staining method to qualitatively and quantitatively analyze cell and nuclear orientations along microgroove patterns. (**A**) Nuclear angulations were analyzed against the reference direction by 4′,6-diamidino-2-phenylindole (DAPI) staining; (**B**,**C**) the orientations of cellular bundles can be determined by phalloidin-stained actin filaments at 7- and 21-day cultures. Individual microgroove patterns facilitated to angularly organize cells as the in vitro culture period went by ∠*Ligament* = 0°, 45° and 90° with (**D**,**E**) statistical significances of the nuclear angulations. White dash-lines represent the ligament architecture border lines with a 250-μm distance. Nuclei #: the number of analyzed nuclei. Scale bars: 200 μm.

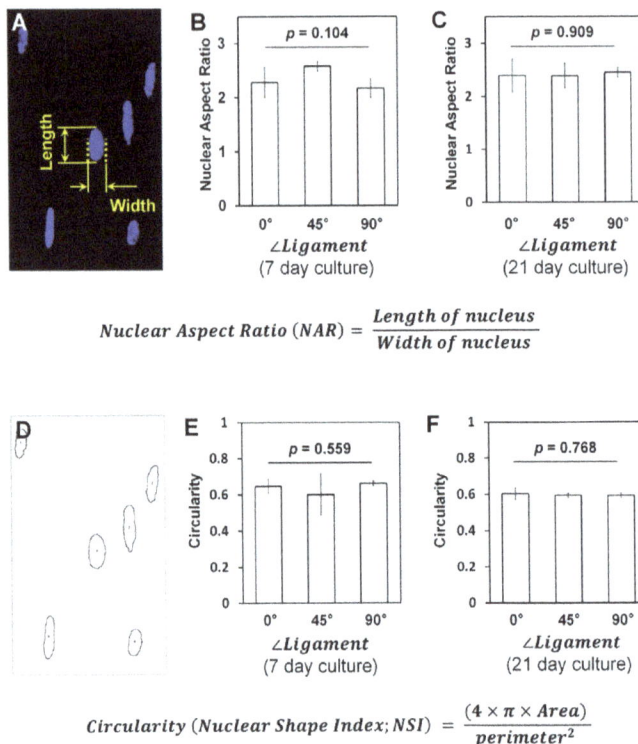

$$Nuclear\ Aspect\ Ratio\ (NAR) = \frac{Length\ of\ nucleus}{Width\ of\ nucleus}$$

$$Circularity\ (Nuclear\ Shape\ Index;\ NSI) = \frac{(4 \times \pi \times Area)}{perimeter^2}$$

Figure 6. Quantification assessments of nuclear deformations using the nuclear aspect ratio and the nuclear shape index (circularity). (**A**) Nuclei were quantitatively assessed for nuclear deformation using NAR, and (**B,C**) three different groups ($\angle Ligament = 0°$, $45°$ and $90°$) had no statistically significant differences; (**D**) the NSI can be quantitatively determined for the cell nuclear elongation based on the perimeters of nucleus, and (**E,F**) all groups for circularity assessments had statistical similarity (*p*-value > 0.05).

3. Discussion

Considering the spatial or dimensional limitations to create scaffolds for large tissue regeneration [25], our investigation could be a possible strategy for manufacturing spatiotemporal architectures with a multitude of microgroove patterns and for creating surface roughness for cell attachments on engineered scaffold surfaces using the freeze-casting method. The appropriate surface properties of 3D structures could improve cell attachment and viability without biochemical modifications given that the surface roughness can regulate cellular activities [26,27]. The roughness on microgroove walls could be acquired by the solvent exchange method with ethanol ($T_f = -114\ °C$) and 1,4-dioxane ($T_f = 11.8\ °C$) at $-20\ °C$ and controlled by using the appropriate concentration of PCL solution (Figure S1). In the quantitative analysis, the 20% and 25% PCL scaffolds had topographical similarities to a typical surface roughness, which Faia-Torres et al. demonstrated to facilitate enhancement of fibroblastic cell attachments and viability without biochemical modifications of the PCL surface [22].

Moreover, the patterned microarchitectures by angulated 3D printing techniques (Figures 3 and 4) can be a key moderator to significantly control cell organizations and tissue morphologies (Figure 5) even though three different types of scaffolds had structural similarities macroscopically (Figure 2E). To evaluate the orientations of scaffold-seeded cell collectivities, significant efforts have contributed to measuring the spatial orientations of a single cell using the cytoskeletal polarity, alignments of actin filaments or deformed nuclear shapes on specifically characterized substrates [23,28,29]. Recently, Versaevel et al. showed an orchestrated correlation between fibroblastic cell orientations and nuclear shapes by the nuclear shape index (NSI; circularity) and the cell shape index on micro-patterned substrates under various mechanistic-regulating microenvironments [24].

Although our model did not apply mechanistic stimulations [24], microgroove patterns can regulate cell orientations by programmed slices and fully-guided nuclear elongations and anisotropic deformations, which are associated with cell orientations [24]. Therefore, the 25-μm microgrooved topography can precisely control cellular orientations as optimal patterns for orientations of cell collectivities. The localized cell adhesion and nuclear elongation/deformation are specifically correlated with F-actin developments in physical-characterized microenvironments, which can also regulate cytoskeleton architectures for cell behaviors [30,31]. In our static condition study, NSI and the nuclear aspect ratio (NAR) were used to measure nuclear orientations corresponding to cell alignments for statistical quantification assessments (Figure 6). Interestingly, the collective cell orientations with high populations were predictably and precisely controlled by the microgroove patterns on the 3D directional ligament scaffolds in the 21-day cultures. At this point, various nano-topographical approaches have demonstrated cell orientations [9,32] or regulations of stem cell responses by anisotropic morphologies or by re-arrangements of cytoskeletal components [33,34]. However, most studies have mainly evaluated individual cell morphologies prior to the increases in cell populations or the formations of strong cell-cell interactions like tissues [32,33].

Based on a simple, but predictable investigation, the technique in this study has strong potential to manufacture spatiotemporal fiber-guiding platforms for 3D fibrous connective tissue formations with specific orientations, which are responsible for functioning restorations in multiple tissue complexes. As examples of the versatilities of 3D printing microgroove patterns, spatial architectures with customized, defect-adapted geometries can be designed for the neogenesis of multiple tissue complexes, which have mineralized structures and fibrous connective tissues in various dimensional interfaces (Figure S4). Based on 3D reconstructed CT-image datasets (Figure S4A), fiber-guiding structures were designed to encompass the tooth-root surfaces for PDL interface regeneration in vivo in the one-wall periodontal osseous defect model (Figure S4B–E). The micro-CT displayed that the customized scaffolds had high adaptability to the created defect geometry (88.30 ± 14.89%; Figure S5), and the SEM showed highly organized and angulated microgroove patterns on the 3D scaffold surfaces (Figure S6).

In general, layer-by-layer artifacts, which are defined as stair stepping errors in the programmed slicing procedure, are commonly removed to increase the surface finish quality and to decrease the critical geometric tolerance for manufacturing accuracy. Recently, 3D printing techniques for biomedical applications have mainly considered cell viability in manufacturing cell-laden spatial constructs with single or multiple cell types [15,35,36]. However, spatiotemporal tissue re-organizations or fibrous re-orientations in engineered microenvironments are the major challenge and are significantly required to restore functionalities in tissue complexes along with cell viabilities [1,5,10]. Our strategy utilized artifactual topographies by the additive manufacturing technique to optimize spatiotemporal cell organizations along engineered directionalities of 3D scaffolds. Microgroove patterns by programmed slicing procedures on the scaffold surface enabled the formation of structural similarities to natural fibrous tissue bundles and fibrous connective tissue orientations with high predictability. In addition, the slice thickness by the additive manufacturing system significantly affected cell organization, and the positional directions to digitally slice designed models were critically influential to create angulated microgroove patterns and to orient cell collectivities with anisotropic nuclear morphologies. Moreover, micron-intervals between the microgrooves that were programmed during the design slices can be significantly correlated to the nuclear deformability regardless of the angulated patterns (Figures S2 and 6). From the perspective of additive manufacturing and 3D printing, spatiotemporal microgroove patterns on 3D printed scaffolds are a promising platform to form hierarchical and functional structures of fibrous connective tissues for tissue engineering and regenerative medicine applications.

4. Materials and Methods

4.1. Computer Design and Polymeric Fabrication of PDL Scaffolds

For the referential direction of PDL, cylindrical structures (0.5-mm diameter and 3.0-mm length) were designed in a scaffold to comprise multi-layered PDL architectures, opened pores on both sides of a scaffold and two cell-seeding inlets on top (Figure 1) in the CAD (computer-aided design) program, Solidworks 2013 software (Dassault Systems SOLIDWORKS Corp., Waltham, MA, USA). The layer thickness of designed molds was set to 25.40 µm during the digital slicing process, and the designed molds were positioned with three different angles (0°, 45° and 90°) to print wax molds (Figure 2). The additive manufacturing procedure created microgroove patterns with consistent thickness (25.40 µm), which was determined in digital slicing steps (Figure S1). Twenty five percent poly-ε-caprolactone (PCL) solution in 1,4-dioxane was freeze-casted into wax molds, and the solvent was extracted by 99% ethanol and double distilled water at −20 °C for 2 days and 4 °C for 3 days, respectively. Then, wax molds were removed using 35–37 °C cyclohexane for 1 day and, after the PCL scaffolds were cooled down at room temperature, 99% ethanol removed cyclohexane. The PCL scaffolds were stored in 70% ethanol at 4 °C until cell seeding.

4.2. Morphological and Topographical Characterizations of Microgroove Patterns on 3D Ligament Architectures

For the qualitative analyses of angulated microgrooves and surface roughness, microgroove patterns on longitudinal architectures in the PCL scaffold were morphologically characterized using the SEM (Figure 2) at 15 kV (S-4700 FE-SEM, Hitachi, Japan). The spatial topographical evaluations were performed for measurements of groove distance (25.40 µm; Figure 3D–F) and angular patterns against the reference direction, which was designed by the CAD for ligament architectures using the topography analysis of the confocal laser scanning microscope (CLSM; Carl Zeiss MicroImaging GmbH, Jena, Germany). The PCL-casted ligament scaffolds were volumetrically and cross-sectionally characterized using micro-CT (SkyScan 1172, Bruker-microCT, Knotich, Belgium), which provide 3D reconstructive and digitally-sectioned images and set to scan with an 11.55 µm³ voxel size under a 40 kV source voltage and a 200 µA source current.

4.3. In Vitro Cell Culture and Fluorescence Staining for Cell Orientation Analyses

For evaluations of cellular orientations or angulations, human PDL cells (hPDLs) were cultivated in Passages 4–5 with minimum essential medium alpha (α-MEM) including 10% fetal bovine serum (FBS) and antibiotics (100 units/mL penicillin). After seeding 1×10^3 cells into a scaffold containing approximately 30.0 mm³, cell-loadable void volume, hPDL-seeded PCL scaffolds were incubated for 7 and 21 days. For nucleus angulation analyses for three different orientations (∠*Ligament* = 0°, 45° and 90°), immunofluorescence was performed with cell nucleus staining with DAPI in blue (4′,6-diamidino-2-phenylindole, Life Technologies (Thermo Fischer Scientific, Waltham, MA, USA) and F-actin staining with phalloidin in red (Alexa Fluoro® 546 Phalloidin, Life Technologies, Carlsbad, CA, USA) after fixing cultured cells at different time points. Using ImageJ software (National Institutes of Health (NIH), Bethesda, MD, USA), angulations of DAPI-stained cell nucleus were measured against the reference direction, which was defined by the designed architecture (Figure 5A–C).

4.4. Nuclear Shape and Deformation Analyses

To quantitatively analyze nuclear deformations by microgroove patterns and microgroove angulations, projected DAPI-stained images were utilized to calculate nuclear aspect ratio (NAR), which was calculated with measured long and short axes of each nucleus and circularity (nuclear shape index (NSI)), which was analyzed with perimeters and area by the ImageJ software.

4.5. 3D Customized Scaffold Developments with Geometric Adaptation to the 1-Wall Periodontal Defect

After scanning the dissected cadaveric mandible by micro-CT (SkyScan 1172, Bruker microCT, Kontich, Belgium), we digitally created the 1-wall periodontal defect. The Solidworks 2013 software was utilized to design fiber-guiding scaffolds with PDL and bone architectures. The Magics 19 software (Materialise Inc., Leuven, Belgium) was utilized to generate defect-fit geometries of scaffolds by booleaning two-image data, which were the 3D reconstructed 1-wall defect and a computer-designed scaffold. 3D printed wax molds having designed scaffold architectures had the freeze-casting procedure using 25% PCL in 1,4-dioxane. Ninety nine percent ethanol and double-distilled water-extracted frozen 1,4-dioxane solvent at −20 °C for 2 days and 4 °C for 2 days, respectively, were used. After removing wax molds by cyclohexane at 35–37 °C, 99% ethanol was used to remove residual cyclohexane in PCL scaffolds at room temperature, and scaffolds were stored in 70% ethanol at 4 °C. For SEM scanning, the solvent was changed to double-distilled water, and the PCL scaffolds were freeze-dried.

4.6. Statistical Analysis

All data were analyzed using the mean ± standard deviation (SD). For comparisons of the nuclear aspect ratio (NAR) and circularity (nuclear shape index (NSI) with three different groups ($\angle Ligament = 0°$, 45° and 90°), the one-way analysis of variance (one-way ANOVA) test with Bonferroni correction was utilized with the α-value set at the 0.05 level of significance.

5. Conclusions

3D printing technologies have been rapidly developed to create various cell-viable hierarchical architectures for tissue engineering, but it is still challenging to spatiotemporally control the orientations of ligamentous bundles for physiological functioning restorations in musculoskeletal complexes. We demonstrate that the different angulated microgroove patterns on 3D printed scaffolds can control the orientation of ligamentous cell bundles with high manufacturing reproducibility. This simple strategy provides the topographical platform to precisely form functional architectures for 3D organizations of fibrous connective tissues. The ligament-guiding architectures can integrate designed bone compartments to create ligament-bone constructs, and the multi-compartmentalized constructs can lead to tissue-functioning restoration with multiple tissue neogenesis in in vivo biomedical applications.

Supplementary Materials: Supplementary materials can be found at www.mdpi.com/1422-0067/18/9/1927/s1.

Acknowledgments: Chan Ho Park was supported by the National Research Foundation of Korea (NRF-2014R1A1A2059301 and NRF-2016R1D1A1B03935686). Yang-Jo Seol was supported by the Bio & Medical Technology Development Program of the NRF funded by the Korean government, MSIP (NRF-2014M3A9E3064466), and William V. Giannobile was supported by NIH/NIDCR DE 13397.

Author Contributions: Chan Ho Park and Yang-Jo Seol developed the 3D printing strategy for angulated organizations of ligament cells and designed all experiments; Chan Ho Park manufactured ligament-guiding scaffolds, performed all in vitro experiments, characterized scaffolds using micro-CT, SEM and confocal micrcopy for topographies and performed the statistical quantifications; Kyoung-Hwa Kim analyzed the data and performed the data statistics; Yong-Moo Lee, Yang-Jo Seol and William V. Giannobile supervised the project and provided the study direction; Chan Ho Park, Yang-Jo Seol and William V. Giannobile wrote and edited the manuscript.

Conflicts of Interest: The authors declare no conflict of interest.

Abbreviations

hPDL	Human periodontal ligament
CAD	Computer-aided design
PCL	Poly-ε-caprolactone

SEM	Scanning electron microscopy
Micro-CT	Micro-computed tomography
CLSM	Confocal laser scanning microscopy
DAPI	4′,6-Diamidino-2-phenylindole
Ra	Arithmetic roughness average
T_f	Freezing temperature
NSI	Nuclear shape index
NAR	Nuclear aspect ratio

References

1. Lu, H.H.; Thomopoulos, S. Functional attachment of soft tissues to bone: Development, healing, and tissue engineering. *Annu. Rev. Biomed. Eng.* **2013**, *15*, 201–226. [CrossRef] [PubMed]
2. Wang, L.; Wu, Y.; Guo, B.; Ma, P.X. Nanofiber yarn/hydrogel core-shell scaffolds mimicking native skeletal muscle tissue for guiding 3D myoblast alignment, elongation, and differentiation. *ACS Nano* **2015**, *9*, 9167–9179. [CrossRef] [PubMed]
3. Park, C.H.; Kim, K.H.; Rios, H.F.; Lee, Y.M.; Giannobile, W.V.; Seol, Y.J. Spatiotemporally controlled microchannels of periodontal mimic scaffolds. *J. Dent. Res.* **2014**, *93*, 1304–1312. [CrossRef] [PubMed]
4. Pagni, G.; Kaigler, D.; Rasperini, G.; Avila-Ortiz, G.; Bartel, R.; Giannobile, W.V. Bone repair cells for craniofacial regeneration. *Adv. Drug. Deliv. Rev.* **2012**, *64*, 1310–1319. [CrossRef] [PubMed]
5. Pilipchuk, S.P.; Monje, A.; Jiao, Y.; Hao, J.; Kruger, L.; Flanagan, C.L.; Hollister, S.J.; Giannobile, W.V. Integration of 3D printed and micropatterned polycaprolactone scaffolds for guidance of oriented collagenous tissue formation in vivo. *Adv. Healthc. Mater.* **2016**, *5*, 676–687. [CrossRef] [PubMed]
6. Park, C.H.; Rios, H.F.; Taut, A.D.; Padial-Molina, M.; Flanagan, C.L.; Pilipchuk, S.P.; Hollister, S.J.; Giannobile, W.V. Image-based, fiber guiding scaffolds: A platform for regenerating tissue interfaces. *Tissue Eng. Part C Methods* **2014**, *20*, 533–542. [CrossRef] [PubMed]
7. Ho, S.P.; Kurylo, M.P.; Fong, T.K.; Lee, S.S.; Wagner, H.D.; Ryder, M.I.; Marshall, G.W. The biomechanical characteristics of the bone-periodontal ligament-cementum complex. *Biomaterials* **2010**, *31*, 6635–6646. [CrossRef] [PubMed]
8. Hurng, J.M.; Kurylo, M.P.; Marshall, G.W.; Webb, S.M.; Ryder, M.I.; Ho, S.P. Discontinuities in the human bone-PDL-cementum complex. *Biomaterials* **2011**, *32*, 7106–7117. [CrossRef] [PubMed]
9. Bae, W.G.; Kim, J.; Choung, Y.H.; Chung, Y.; Suh, K.Y.; Pang, C.; Chung, J.H.; Jeong, H.E. Bio-inspired configurable multiscale extracellular matrix-like structures for functional alignment and guided orientation of cells. *Biomaterials* **2015**, *69*, 158–164. [CrossRef] [PubMed]
10. Sun, Y.; Jallerat, Q.; Szymanski, J.M.; Feinberg, A.W. Conformal nanopatterning of extracellular matrix proteins onto topographically complex surfaces. *Nat. Methods* **2015**, *12*, 134–136. [CrossRef] [PubMed]
11. Kim, D.H.; Lipke, E.A.; Kim, P.; Cheong, R.; Thompson, S.; Delannoy, M.; Suh, K.Y.; Tung, L.; Levchenko, A. Nanoscale cues regulate the structure and function of macroscopic cardiac tissue constructs. *Proc. Natl. Acad. Sci. USA* **2010**, *107*, 565–570. [CrossRef] [PubMed]
12. Downing, T.L.; Soto, J.; Morez, C.; Houssin, T.; Fritz, A.; Yuan, F.; Chu, J.; Patel, S.; Schaffer, D.V.; Li, S. Biophysical regulation of epigenetic state and cell reprogramming. *Nat. Mater.* **2013**, *12*, 1154–1162. [CrossRef] [PubMed]
13. Hu, J.; Hardy, C.; Chen, C.M.; Yang, S.; Voloshin, A.S.; Liu, Y. Enhanced cell adhesion and alignment on micro-wavy patterned surfaces. *PLoS ONE* **2014**, *9*, e104502. [CrossRef] [PubMed]
14. Jiao, A.; Trosper, N.E.; Yang, H.S.; Kim, J.; Tsui, J.H.; Frankel, S.D.; Murry, C.E.; Kim, D.H. Thermoresponsive nanofabricated substratum for the engineering of three-dimensional tissues with layer-by-layer architectural control. *ACS Nano* **2014**, *8*, 4430–4439. [CrossRef] [PubMed]
15. Kolesky, D.B.; Homan, K.A.; Skylar-Scott, M.A.; Lewis, J.A. Three-dimensional bioprinting of thick vascularized tissues. *Proc. Natl. Acad. Sci. USA* **2016**, *113*, 3179–3184. [CrossRef] [PubMed]
16. Kang, H.W.; Lee, S.J.; Ko, I.K.; Kengla, C.; Yoo, J.J.; Atala, A. A 3D bioprinting system to produce human-scale tissue constructs with structural integrity. *Nat. Biotechnol.* **2016**, *34*, 312–319. [CrossRef] [PubMed]
17. Murphy, S.V.; Atala, A. 3D bioprinting of tissues and organs. *Nat. Biotechnol.* **2014**, *32*, 773–785. [CrossRef] [PubMed]

18. Pati, F.; Gantelius, J.; Svahn, H.A. 3D bioprinting of tissue/organ models. *Angew. Chem.* **2016**, *55*, 4650–4665. [CrossRef] [PubMed]

19. Park, C.H.; Rios, H.F.; Jin, Q.; Bland, M.E.; Flanagan, C.L.; Hollister, S.J.; Giannobile, W.V. Biomimetic hybrid scaffolds for engineering human tooth-ligament interfaces. *Biomaterials* **2010**, *31*, 5945–5952. [CrossRef] [PubMed]

20. Park, C.H.; Rios, H.F.; Jin, Q.; Sugai, J.V.; Padial-Molina, M.; Taut, A.D.; Flanagan, C.L.; Hollister, S.J.; Giannobile, W.V. Tissue engineering bone-ligament complexes using fiber-guiding scaffolds. *Biomaterials* **2012**, *33*, 137–145. [CrossRef] [PubMed]

21. Rasperini, G.; Pilipchuk, S.P.; Flanagan, C.L.; Park, C.H.; Pagni, G.; Hollister, S.J.; Giannobile, W.V. 3D-printed bioresorbable scaffold for periodontal repair. *J. Dent. Res.* **2015**, *94*, 153S–157S. [CrossRef] [PubMed]

22. Faia-Torres, A.B.; Guimond-Lischer, S.; Rottmar, M.; Charnley, M.; Goren, T.; Maniura-Weber, K.; Spencer, N.D.; Reis, R.L.; Textor, M.; Neves, N.M. Differential regulation of osteogenic differentiation of stem cells on surface roughness gradients. *Biomaterials* **2014**, *35*, 9023–9032. [CrossRef] [PubMed]

23. Chen, B.; Co, C.; Ho, C.C. Cell shape dependent regulation of nuclear morphology. *Biomaterials* **2015**, *67*, 129–136. [CrossRef] [PubMed]

24. Versaevel, M.; Grevesse, T.; Gabriele, S. Spatial coordination between cell and nuclear shape within micropatterned endothelial cells. *Nat. Commun.* **2012**, *3*, 671. [CrossRef] [PubMed]

25. Rogers, J.A.; Nuzzo, R.G. Recent progress in soft lithography. *Mater. Today* **2005**, *8*, 50–56. [CrossRef]

26. Chen, W.; Shao, Y.; Li, X.; Zhao, G.; Fu, J. Nanotopographical surfaces for stem cell fate control: Engineering mechanobiology from the bottom. *Nano Today* **2014**, *9*, 759–784. [CrossRef] [PubMed]

27. Teo, B.K.; Wong, S.T.; Lim, C.K.; Kung, T.Y.; Yap, C.H.; Ramagopal, Y.; Romer, L.H.; Yim, E.K. Nanotopography modulates mechanotransduction of stem cells and induces differentiation through focal adhesion kinase. *ACS Nano* **2013**, *7*, 4785–4798. [CrossRef] [PubMed]

28. Chen, S.; Nakamoto, T.; Kawazoe, N.; Chen, G. Engineering multi-layered skeletal muscle tissue by using 3D microgrooved collagen scaffolds. *Biomaterials* **2015**, *73*, 23–31. [CrossRef] [PubMed]

29. Gilchrist, C.L.; Ruch, D.S.; Little, D.; Guilak, F. Micro-scale and meso-scale architectural cues cooperate and compete to direct aligned tissue formation. *Biomaterials* **2014**, *35*, 10015–10024. [CrossRef] [PubMed]

30. Fletcher, D.A.; Mullins, R.D. Cell mechanics and the cytoskeleton. *Nature* **2010**, *463*, 485–492. [CrossRef] [PubMed]

31. Gardel, M.L.; Schneider, I.C.; Aratyn-Schaus, Y.; Waterman, C.M. Mechanical integration of actin and adhesion dynamics in cell migration. *Annu. Rev. Cell Dev. Biol.* **2010**, *26*, 315–333. [CrossRef] [PubMed]

32. Ray, A.; Lee, O.; Win, Z.; Edwards, R.M.; Alford, P.W.; Kim, D.H.; Provenzano, P.P. Anisotropic forces from spatially constrained focal adhesions mediate contact guidance directed cell migration. *Nat. Commun.* **2017**, *8*, 14923. [CrossRef] [PubMed]

33. Liu, X.; Wang, S. Three-dimensional nano-biointerface as a new platform for guiding cell fate. *Chem. Soc. Rev.* **2014**, *43*, 2385–2401. [CrossRef] [PubMed]

34. Mosqueira, D.; Pagliari, S.; Uto, K.; Ebara, M.; Romanazzo, S.; Escobedo-Lucea, C.; Nakanishi, J.; Taniguchi, A.; Franzese, O.; Di Nardo, P.; et al. Hippo pathway effectors control cardiac progenitor cell fate by acting as dynamic sensors of substrate mechanics and nanostructure. *ACS Nano* **2014**, *8*, 2033–2047. [CrossRef] [PubMed]

35. Pati, F.; Jang, J.; Ha, D.H.; Won Kim, S.; Rhie, J.W.; Shim, J.H.; Kim, D.H.; Cho, D.W. Printing three-dimensional tissue analogues with decellularized extracellular matrix bioink. *Nat. Commun.* **2014**, *5*, 3935. [CrossRef] [PubMed]

36. Kolesky, D.B.; Truby, R.L.; Gladman, A.S.; Busbee, T.A.; Homan, K.A.; Lewis, J.A. 3D bioprinting of vascularized, heterogeneous cell-laden tissue constructs. *Adv. Mater.* **2014**, *26*, 3124–3130. [CrossRef] [PubMed]

International Journal of
Molecular Sciences

MDPI

Article

A New Bone Substitute Developed from 3D-Prints of Polylactide (PLA) Loaded with Collagen I: An In Vitro Study

Ulrike Ritz [1],*, Rebekka Gerke [1], Hermann Götz [2], Stefan Stein [3] and Pol Maria Rommens [1]

[1] Department of Orthopaedics and Traumatology, BiomaTiCS, University Medical Center, Johannes Gutenberg University, 55131 Mainz, Germany; rebekka.gerke@gmx.de (R.G.); pol.rommens@unimedizin-mainz.de (P.M.R.)

[2] Platform for Biomaterial Research, University Medical Center, BiomaTiCS, Johannes Gutenberg University, 55131 Mainz, Germany; hgoetz@uni-mainz.de

[3] Georg-Speyer-Haus—Institute for Tumor Biology and Experimental Therapy, 60659 Frankfurt, Germany; s.stein@gsh.uni-frankfurt.de

* Correspondence: ritz@uni-mainz.de; Tel.: +49-6131-172-359

Received: 23 October 2017; Accepted: 27 November 2017; Published: 29 November 2017

Abstract: Although a lot of research has been performed, large segmental bone defects caused by trauma, infection, bone tumors or revision surgeries still represent big challenges for trauma surgeons. New and innovative bone substitutes are needed. Three-dimensional (3D) printing is a novel procedure to create 3D porous scaffolds that can be used for bone tissue engineering. In the present study, solid discs as well as porous cage-like 3D prints made of polylactide (PLA) are coated or filled with collagen, respectively, and tested for biocompatibility and endotoxin contamination. Microscopic analyses as well as proliferation assays were performed using various cell types on PLA discs. Stromal-derived factor (SDF-1) release from cages filled with collagen was analyzed and the effect on endothelial cells tested. This study confirms the biocompatibility of PLA and demonstrates an endotoxin contamination clearly below the FDA (Food and Drug Administration) limit. Cells of various cell types (osteoblasts, osteoblast-like cells, fibroblasts and endothelial cells) grow, spread and proliferate on PLA-printed discs. PLA cages loaded with SDF-1 collagen display a steady SDF-1 release, support cell growth of endothelial cells and induce neo-vessel formation. These results demonstrate the potential for PLA scaffolds printed with an inexpensive desktop printer in medical applications, for example, in bone tissue engineering.

Keywords: 3D-printing; polylactide; collagen; biocompatibility; osteogenesis; angiogenesis

---

1. Introduction

During the last decade, much knowledge has been acquired in trauma management. However, large segmental bone defects caused by trauma, infection, bone tumors or revision surgeries still represent a challenge for trauma surgeons all over the world [1,2]. Until now, the gold standard for treatment of large bone defects is autologous bone grafting, which, unfortunately, is associated with severe problems; one or several additional interventions are needed and the transplant material is limited [3]. Although several materials and various implant options have been developed, the perfect solution for filling up critical size defects still remains to be found [4].

New and innovative bone substitutes, which should be non-toxic, endotoxin-free, biocompatible and optimally biodegradable as well as osteoinductive, are needed. One problem concerning bone substitutes is the fact, that they need high mechanical stability and should simultaneously express a porous structure for ingrowth of bone, tissue and cells. Moreover, bone substitutes should stimulate osseointegration. This could be achieved by modifications of the material with active compounds displaying the requested features. One possible solution are materials consisting of a mechanical stable structure (e.g., hydroxyapatite) [5] filled, coated or modified with polymers representing soft part, e.g., hydrogels [6], which can be altered with cells or bioactive molecules [7]. Many studies in the last years demonstrated that the critical point for bone tissue engineering is not the induction of bone growth but the supply of the implant with nutrients and oxygen, showing that vascularization of the implant and its surroundings is the main requirement [8,9]. Therefore, new bone substitutes should also elicit angiogenic effects. One option to create the requested effects together with osteogenesis is the immobilization of bioactive molecules, for example incorporation of SDF-1 (stromal-derived factor), which is known to act chemotactically on endothelial (progenitor) cells and thereby to induce angiogenesis [10,11].

Three-dimensional (3D) printing is a simple procedure to create 3D porous scaffolds used for bone tissue engineering [12]. A promising material is polylactide, a polymer which is known to be degraded by hydrolysis to harmless and non-toxic monomers [13]. Moreover, it displays mechanical stability, which is why it has been studied and employed in various medical studies and applications [14,15].

One soft material being used together with mechanically stable materials is collagen. Collagens are extracellular matrix (ECM) molecules and are widely employed as biomaterials due to their excellent behaviour [16], even in bone tissue engineering [17]. It is inexpensive, well-tolerated and easy to handle and modify. Moreover it is degraded in the body by collagenases, releasing non-toxic and non-immunogenic peptides [18]. Collagens from various sources exist [19], but rat tail or bovine sources are the most commonly used. Collagen has been used for tissue engineering [20,21] and delivery of various bioactive molecules [22] for decades, emphasizing its potential as a biomaterial for various medical applications. Besides bone tissue engineering studies, collagen-based (hydro)gels or scaffolds, have been suggested, inter alia, for use in cartilage repair [23], wound-infection [24], tendon tissue engineering [25], neurodegenerative diseases [26] and nucleus pulposus regeneration [27]. Moreover, Bersini et al. combined computational tools with an experimental approach including hydrogels consisting of fibrin and collagen to engineer vascularized bone-mimicking tissues [28].

In the present study, solid discs as well as porous cage-like 3D structures made of PLA were coated and filled with collagen. These materials were employed for biocompatibility evaluation and testing of endotoxin contamination. The effects on various cell types were tested in vitro and interpreted with regard to their potential usefulness in medical applications.

2. Results and Discussion

2.1. Scaffold Characterization

PLA scaffolds were fabricated using a 3D-printer, Ultimaker 2+, with a 0.25 mm nozzle. Images taken by scanning electron microscopy shown in Figure 1 demonstrate the rough and inherent surface of the printed 3D models due to the limited resolution of the 3D-printer. The three-dimensional reconstruction of the cage-like model as well as the inside views taken by a microcomputed tomography technique demonstrate a scaffold structure with pores large enough for cells to invade from the top as well as from side walls. The comparison of the designed 3D model and the printed results demonstrates some structural imperfections due to the limiting factors of the 3D print technology. Threads and loose particles as well as geometric deviations caused by the fabrication process are estimated as non-relevant and consciously accepted. The hole in the center of the scaffold was designed to facilitate gel injection into the scaffold.

Polylactide is a promising starting material for scaffolds used for biomedical application. However, not many studies utilized low-cost and "easy-to-handle" desktop printers, which can be afforded and handled by everyone [29,30]. The structure of the printed scaffolds hints to positive properties for cell growth as well as modification options; both facts being prerequisites for a material usable in bone tissue engineering.

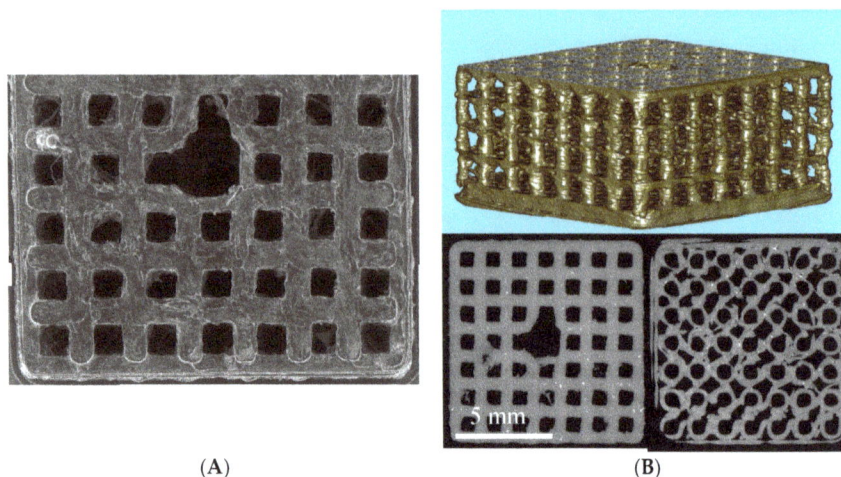

Figure 1. (**A**) Scanning electron microscope (SEM) image gallery of the PLA scaffolds. Single Images (100×) are stacked together manually; (**B**) µCT 3D volume rendering of polylactide (PLA) cage and µCT-slices as multiplanar reconstruction (MPR) images from the top and bottom site of the PLA cage.

2.2. Endotoxin Contamination

Endotoxins are lipopolysaccharides (LPS), located on the outer cell membrane of Gram-negative bacteria. Ubiquitously present, they can enter the blood stream of humans and induce immunological responses resulting in inflammation and infection, possibly leading to multiple organ failure and eventually to death [31,32]. Once contaminated, removal of endotoxins from infected materials is almost impossible as they are resistant to high temperature and even acidic solutions. Only extreme high temperatures or extreme acidic solutions can destroy endotoxins—conditions that would destroy most biomaterials. Therefore, avoiding endotoxin contamination during syntheses processes is highly recommended [33]. To our knowledge, no studies have been performed to determine endotoxin contamination in PLA scaffolds after 3D printing.

In order to ensure, that polylactide scaffolds do not contain endotoxins after printing, we performed limulus amebocyte lysate-assays (LAL-assays) to determine endotoxin contamination before and after printing in the supernatant of PLA incubated in PBS (phosphate buffered saline, Figure 2). PLA was used directly after taking it out of the package (red circle 5) as well as after being in the air for a few days (red circle 6) and analyzed for endotoxin contamination; LAL-assay demonstrates an endotoxin contamination well above the FDA-limit (0.5 EU/mL). In contrast, LAL-assay demonstrated an endotoxin contamination of our PLA printed scaffolds from 0.1–0.25 EU/mL (Figure 2, green circles), which is clearly below the FDA-limit (0.5 EU/mL).

Most likely the high temperatures (up to 240 °C) during printing processes are sufficiently high to destroy any present endotoxins. Similar findings were reported by Neches et al. who describe the intrinsic sterility of 3D printing due to, among other factors, the high fusion modeling temperatures [34]. After printing, the PLA disks or -scaffolds are transferred directly to ethanol-containing solutions and handled under sterile conditions to minimize further endotoxin contaminations.

2.3. Biocompatibility

In order to ensure biocompatibility of our PLA scaffolds, printed cages were incubated in medium for 12, 24 and 48 h. The supernatants were used for incubation of L929 cells seeded in 96 wells for 24 h and viability was tested with a MTT assay. This protocol is in accordance to ISO-10993-5:

Biological evaluation of medical devices. Figure 3 demonstrates the viability of cells for all three time points. Cell viability of 100% corresponds to cells growing on tissue-culture-treated polystyrene substrates (medium control). According to ISO 10993-5, cell viabilities > 70% indicate no cytotoxic effects, whereas cell viability between 0% and 40% represents high cytotoxicity as seen in the positive controls A and B (ZDEC and ZDBC: polyurethane matrices stabilized with organic zinc) with values below 25% viability.

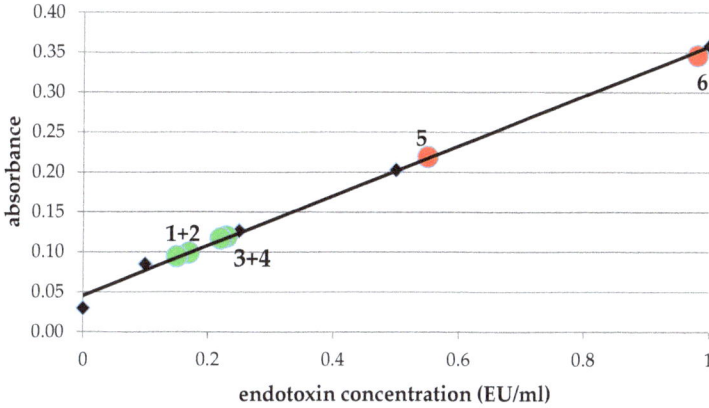

Figure 2. Determination of endotoxin concentration in PLA disks. Endotoxin concentration was determined in supernatants of PLA strings before printing directly after unpacking (red circle 5) and after a few days at the air (red circle 6) and in PLA disks after printing incubated for 24 h (1 + 2) or 48 h (3 + 4): green circles. The trendline corresponds to standards with 0.1, 0.25, 0.5 and 1.0 EU/mL (black quarders). The diagram shows that endotoxin concentrations are well below the FDA limit of 0.5 EU/mL.

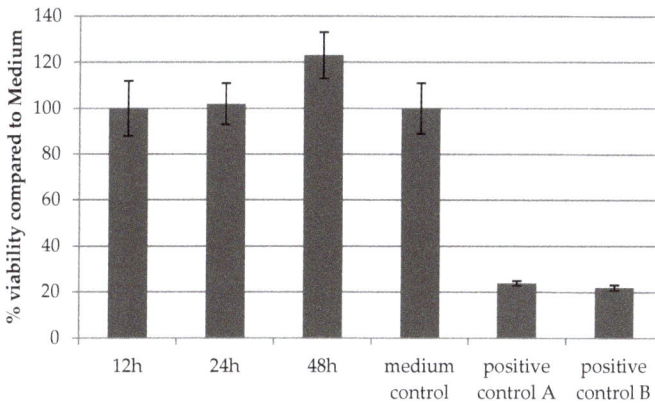

Figure 3. Confirmation of biocompatibility of PLA disks. MTT-tests were performed analogue to ISO 10993-5 and confirmed the biocompatibility of PLA disks as demonstrated by viability of cells after incubation for 12, 24 and 48 h in PLA medium, respectively, compared to standard cell medium. Medium control: standard cultivation medium; positive controls: ZDEC and ZDBC: polyurethane matrices stabilized with organic zinc.

2.4. Microscopic Analyses

Different cell types stably expressing eGFP (human primary osteoblasts: hOB, osteosarcoma cells: SaOS-2) or mCherry (human umbilical vein endothelial cells: HUVEC, normal human dermal fibroblasts: NHDF) were seeded on collagen-coated PLA discs to analyze cell growth. Figure 4 shows an equal distribution of all tested cells seeded on PLA discs.

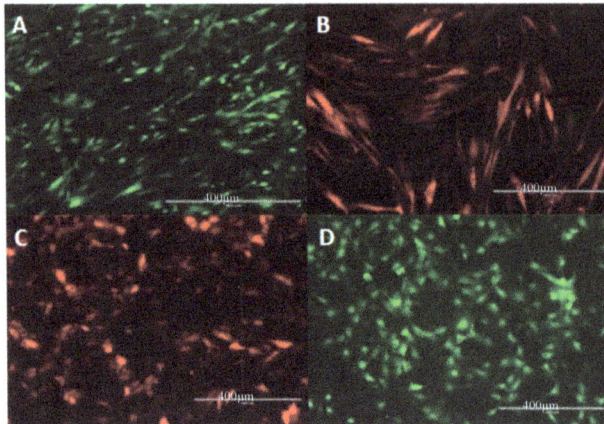

Figure 4. Microscopic analysis of different cell types seeded on PLA discs. (**A**) hOB (human primary osteoblasts; (**B**) normal human dermal fibroblasts: NHDF; (**C**) human umbilical vein endothelial cells: HUVEC; and (**D**) osteosarcoma cells: SaOS-2. Scale bars represent 400 μm.

2.5. Proliferation

For proliferation tests, we seeded cells on PLA scaffolds coated with or without collagen solutions. The results demonstrated statistically significant better proliferation of various cell types on PLA discs coated with rat tail or bovine collagen (Figure 5B,C), compared to non-coated discs (Figure 5A, $p \leq 0.05$). The shape of these growth curves is similar, and, except for hOB cells the overall proliferation of the cells is slightly better on discs coated with bovine collagen, however, these differences are not statistically significant ($p > 0.05$).

It has been demonstrated that PLA is a biocompatible and biodegradable material [13,35,36]. However, only few in vitro studies have been performed concerning 3D-printed scaffolds employing PLA. One study was performed with murine MC3T3 cells, confirming the biocompatibility of PLA scaffolds. This study used different coatings (hyaluron and pullulan) and found different effects of the coatings on cells [30]. Rosenzweig et al. tested chondrocytes and nucleus pulposus cells on PLA scaffolds and demonstrated their increase in proliferation [37]. Yang et al. combined polylactide-co-glycolide (PGLA) with hydroxyapatite and found an osteoinductive effect with human bone marrow derived stem cells (hBMSCs) [38]. A similar study was performed by Senatov [39]. Wurm et al. manufactured PLA samples by fused deposition modeling and tested their biocompatibility with human fetal osteoblasts (hFOB) [40]. They demonstrated that their PLA samples showed no cytotoxicity, however, hFOB demonstrated reduced cell growth compared to polystyrene control, probably due to differences in surface roughness. These effects might be masked in our approach by collagen coating. Rodina tested migration and proliferation of murine mesenchymal stem cells on electrospun PLA scaffolds coated with collagen [41,42]—completely different to our scaffolds, but their study demonstrates possible combination of PLA and collagen. Concerning biocompatibility of PLA, our study is in accordance with the described literature, however, it is the first study analyzing 3D-printed PLA discs coated with collagen with different human cells in vitro. We could demonstrate that the cells adhere to collagen-coated PLA discs and exhibit their typical morphology. Moreover, all cell types proliferate on the discs over a period of ten days. This is especially interesting with regards to endothelial cells as these cells represent an angiogenic cell type and many studies showed that angiogenesis is the critical point of tissue regeneration in almost every tissue. Until now, only few groups demonstrated a positive effect on angiogenic factors of human stem cells in combination with polylactide [43–45].

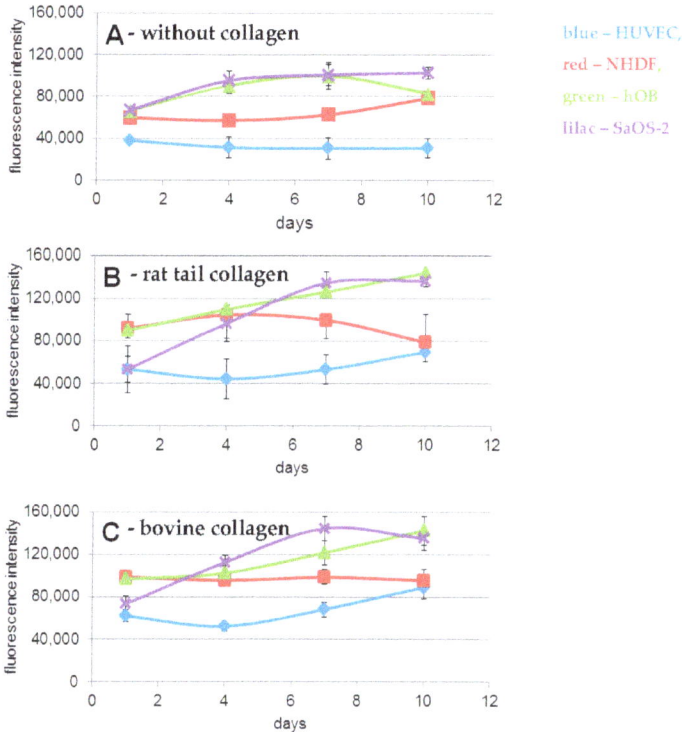

Figure 5. Proliferation assays on PLA discs with different cell types. (**A**) without collagen; (**B**) PLA discs coated with rat tail collagen; (**C**) PLA discs coated with bovine collagen. Blue: HUVEC, red: NHDF, green: hOB and lilac: SaOS-2.

2.6. SDF-1 Immobilization and Release

In order to construct a material which can be applied as bone substitute, a cage-like structure was printed with pores large enough for cells, vessels and/or bone to grow inside. To construct a material that enhances angiogenesis in the first place, we immobilized SDF-1 in the PLA collagen cage and measured the release kinetics of this factor from the cage.

For release kinetics, 500 ng SDF-1 were immobilized in collagen gel in PLA cages and its release was measured over 48 h. Interestingly, we could not observe a high initial burst release, but a relative steady release. After 48 h, 50% of the initially immobilized SDF-1 is still in the cage (Figure 6).

In former studies, we could demonstrated that SDF-1 bound to collagen or hydrogels keeps its functional bioactivity, induces angiogenetic effects and, as a consequence, supports bone tissue regeneration [46–48]. In these studies, we observed a high initial release of immobilized SDF-1, which was sufficient to induce angiogenetic effects; however, a steady release over a longer time period as observed in the present study with the PLA collagen cages seems to be preferable [11,49,50].

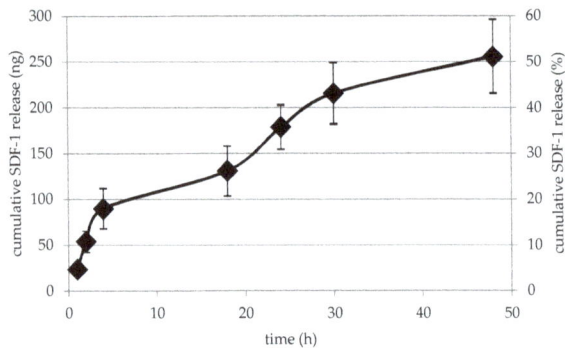

Figure 6. SDF-1 release from PLA collagen cages. SDF-1 is released from collagen inside the 3D-printed cages in a steady manner and after 48 h approximately 50% are still inside the cage.

2.7. Angiogenic Potential

mCherry-expressing HUVECs were seeded on PLA cages loaded with SDF-1 immobilized in bovine collagen (Figure 7). The cells adhere to the cage and start to grow inside it. Thereby they start to form neo-vessel-like structures comparable to cells grown in Matrigel [51].

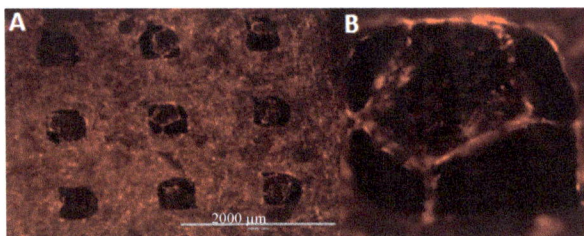

Figure 7. HUVECs seeded on PLA collagen–SDF-1 scaffolds. The cells adhere to the cage and seem to grow from the outside of the cage into the collagen gel (**A**). Thereby they start to form neo-vessel-like structures comparable to cells grown in Matrigel (**B**). Scale bar represents 2000 μm.

These results are promising concerning the application of mechanically stable PLA collagen–SDF-1 constructs as bone substitutes. The produced scaffolds are biocompatible, demonstrate a steady release of the immobilized factors and induce neo-vessel formation in endothelial cells. Until now, no such approach has been tested. Pinese et al. combined PLA polymers as knitted patches with collagen/chondroitin sulfate for ligament regeneration [52]. Heo et al. combined PLA with gelatin hydrogels and showed that human adipose-derived stem cells can be drifted to osteogenic differentiation and suggest this material for bone tissue engineering [53]. Yin used 3D-printed cages for cervical diseases [54] and Kao et al. coated PLA with poly(dopamine) for bone tissue engineering. They demonstrated a higher expression of ALP and osteocalcin in adipose-derived stem cells after seeding on these constructs. Moreover, they could show that some proteins associated with angiogenic differentiation were upregulated [43]. Concerning angiogenesis, Sekula et al. reported a positive effect of PLA on human umbilical cord-derived mesenchymal stem cells on gene expression of endothelial markers; however they did not analyze vessel formation in an angiogenesis assay [45]. Using an approach comparable to Kao [43], Yeh et al. demonstrated an upregulation of osteogenic and angiogenic markers of bone marrow stem cells on 3D-printed PLA scaffolds after immobilization of poly-dopamine [44]. To our knowledge, the present work is the only study testing angiogenic associated aspects with PLA scaffolds printed on a simple desktop 3D-printer coated or filled with collagen I employing inter alia endothelial cells and performing proliferation as well as angiogenesis assays.

Concerning in vivo evaluation of PLA constructs, Chou et al. used a PLA cage as a carrier for bone chips that induced bone regeneration in a rabbit model [13]. They demonstrated that morselized corticocancellous bone chips were converted into a structured cortical bone graft and that the PLA cage was already completely degraded 12 weeks after implantation. This demonstrates that the transfer of our in vitro results to in vivo studies is generally possible.

Our approach displays an inexpensive, easily constructed scaffold with the necessary mechanical stability and a soft material inside that can be modified with various cytokines or even cells to induce angiogenesis, and as a consequence bone regeneration. Proof of concept in a rat in vivo model is part of the follow-up-study. Before analyzing the potential of our scaffold to enhance bone regeneration in an in vivo femur defect model in the rat, further tests to characterize the mechanical stability of our 3D-prints loaded with collagen I will be performed. Mechanical tests as well as in vivo experiments will clarify the potential of our approach for bone tissue engineering.

3. Materials and Methods

3.1. 3D-Printing

For 3D-printing we used commercially available PLA filament (Ultimaker silver metallic PLA, iGo3D, Hannover, Germany) with a diameter of 2.85 ± 0.10 mm. Mechanical, thermal and other properties are listed in the technical data sheet from Ultimaker. Sanchez et al. give a good overview about material characterization methods for PLA filaments in 3D printing [55]. All PLA discs and cages were designed with a 3D modeling software (Autodesk® Inventor Professional 2013, Autodesk, San Rafael, CA, USA) so that they fit to the 24 well of an ultra-low attachment plate (Corning, Wiesbaden, Germany). The 3D model file generated by the Autodesk Inventor software was exported to a convenient file format (STL).

For the 3D print pre-processing this file was imported into the CURA 2.5 software (Ultimaker B.V., Geldermasen, The Netherlands), previewed, scaled and adjusted as necessary. Cura slices the model ready for print with the 3D-printer Ultimaker 2+ (Ultimaker B.V., Geldermasen, The Netherlands) employing the smallest available nozzle size (0.25 mm). A slightly modified high quality profile in Cura 2.5 with a layer height of 0.06 mm, 100% infill, 200 °C nozzle temperature and 60 °C build plate temperature has shown the best results.

Figure 8 demonstrates the difference between the designed 3D model and the 3D printed objects.

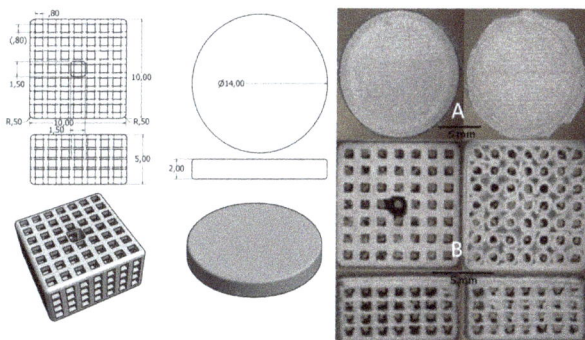

Figure 8. Left side: Technical sketches with dimensions in [mm] for both the PLA disc and the PLA cage; **Right side**: light microscopic pictures of the original 3D-printed objects (**A**) PLA disc; (**B**) PLA cage.

3.2. Quality Control

For quality control purposes, all the printed objects underwent detailed examination by means of stereo light microcopy (Leica MZ 16A) as mentioned above, scanning electron microscopy (FEI Quanta 200FEG) and micro computational tomography (Scanco µCT40). SEM investigation was done under low vacuum condition to avoid disturbing artefacts due to the electrical isolating characteristic of the PLA material. The SEM investigation under low vacuum condition is recommended especially for non-conductive samples with undergoing cavities were the available sputter coating techniques fails to cover the whole sample surfaces.

Parts of the PLA cage surfaces are shown in Figure 9 to demonstrate the limits of precision of the applied 3D printing technique. To make sure that a high degree of connectivity was achieved for the printed material the PLA cages were scanned with a micro-CT system. Micro-CT reconstructions as shown in Figure 10 demonstrate the internal designed structures with a sufficient porosity.

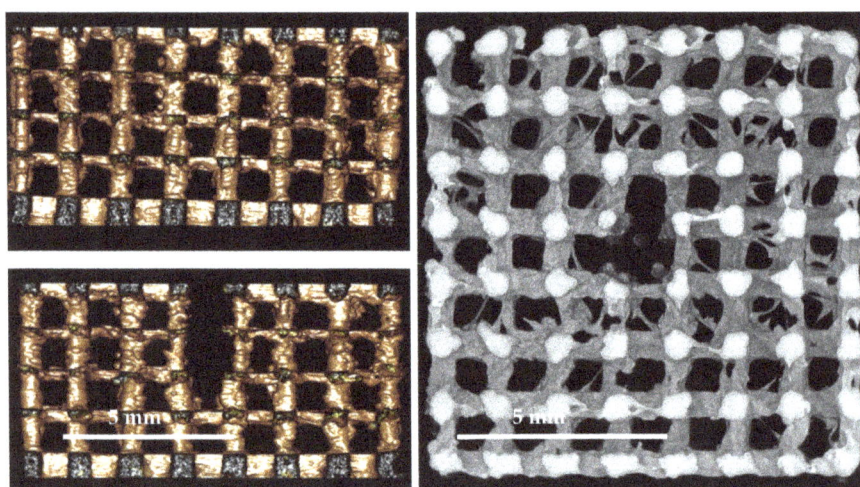

Figure 9. SEM quality control. Magnification: 100×.

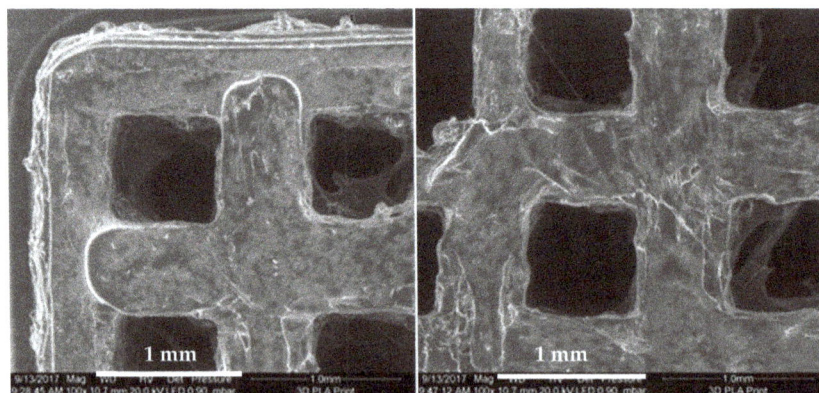

Figure 10. µCT quality control. (**Left**): surface rendered images (view from both sides); (**Right**): MPR image with volume rendering from the internal site of the PLA cage.

3.3. Biocompatibility

In vitro cytotoxicity was analyzed employing the MTT ([3-(4,5-dimethylthiazol-2-yl)-2,5-diphenyltetrazoliumbromid)]) assay analogous to ISO 10993-5. Mouse L929 cells (20.000 cells/well) were seeded in a 96-well tissue culture polystyrene (TCPS) plate for 24 h. Directly after printing PLA constructs were incubated in 500 µL cell media for 24 h. 100 µL of this extract were given to L929 cells in the 96 well plate. After an incubation time of 24 h the MTT assay was performed according to ISO 10993-5. The colorimetric readout was performed at a wavelength of 570 nm (reference wavelength 650 nm). Polyurethane membranes stabilized with organic zinc derivatives ZDEC (zinc diethyl dithiocarbamate) and ZDBC (zinc dibutyl dithiocarbamate) (Food and Drug Safety Center, Hatano Research Institute, Hadano, Japan), were used as positive controls. These controls induce a reproducible cytotoxic reaction.

3.4. Endotoxin Contamination

Prior to the cell experiments, the 3D-printed PLA constructs were tested for endotoxin contamination employing the endpoint chromogenic LAL (limulus amebocyte lysate) assay (Lonza, Basel, Switzerland). The assay was performed according to the manufacturer's recommendations. For preparation of the endotoxin analyte solution, PLA discs and cages were incubated in 1 mL water directly after printing without preceding washing-steps for 24 h at 37 °C (conditions were transferred from ISO 10993-5: 2009; Biological evaluation of medical devices) in order to extract any potential endotoxin from the PLA matrix. the supernatant (50 µL) was employed in the LAL test. Parallel to sampling of the analyte solutions, a standard curve was established and the analysis results compared to positive controls (provided by the manufacturer (Lonza, Basel, Switzerland) within the LAL kit) and negative control (pure endotoxin free water).

3.5. Coating/Filling of Discs/Cages with Collagen I

Solid PLA discs were coated with bovine (Viscofan, Weinheim, Germany) or rat tail collagen (Invitrogen, Karlsruhe, Germany). Collagens were diluted 1:100 with PBS and discs were incubated for one hour to assure even coating.

3D cages were filled with a collagen gel solution following an established protocol [56,57]. Briefly, collagen type I (3 mg/mL bovine, Viscofan, Weinheim, Germany), aqua dest, M199 (10×), NaHCO$_3$ and NaOH and SDF-1 (500 ng, Miltenyi, Bergisch Gladbach, Germany) were combined within an ice bath to prevent polymerization of the solution. Next, 300 µL of the liquid collagen solution was pipetted into a PLA cage sitting in a 24-well ultra-low attachment plate (Corning, Wiesbaden, Germany) and allowed to polymerize.

3.6. Cells

Four cell types were used to analyze cell growth and viability on 3D-printed PLA discs coated with bovine or rat tail collagen. Normal human dermal fibroblasts (NHDF, Promega, Karlsruhe, Germany), human primary osteoblasts (hOB [48,58]), osteoblast-like cells (SaOS, ATCC, Manassas, VA, USA) and endothelial cells (HUVEC, Promega, Karlsruhe, Germany) were seeded onto discs, into or onto cages, respectively, sitting in a 24-well ultra-low attachment plate. Cells (100,000) were seeded in all cases.

3.7. Viability

Proliferation was measured over a 10-day period with alamarBlue-assay (Invitrogen, Karlsruhe, Germany) according to the manufacturer's protocol.

To allow analysis by fluorescent microscopy cells were transduced with lentiviral vectors encoding mCherry or enhanced green fluorescent protein. Vector supernatants were collected and concentrated from transfected 293T producer cells as previously described [59]. For gene transfer, 15.000 human osteoblasts (hOB) or HUVEC were seeded into 24-well tissue culture plates (Greiner,

Frickenhausen, Germany) in 500 μL media supplemented with 5 μg/mL protamine sulfate. Two rounds of transduction on day 1 and 3 were performed at a cumulative multiplicity of infection (MOI) of ~100 to achieve >98% gene marking. Transduction efficiency was confirmed by fluorescence microscopy (Wilovert AFL30, Hundt, Wetzlar, Germany) and flow cytometry FACSCalibur (BD Biosciences, San Jose, CA, USA) using the CellQuestPro Software (BD Biosciences, San Jose, CA, USA). The different cell types were seeded as monocultures on PLA discs or on the cages and the spreading, morphology and distribution of the cells analyzed microscopically with a fluorescent microscope.

3.8. SDF-1 Release Assay

Three-dimensional cages were filled with bovine or rat tail collagen immobilized with fluorescein-linked SDF-1. Release kinetics of SDF-1-FITC from the PLA collagen–cages were measured via fluorescence reading in the supernatant (Glomax-Multidetection System, Promega, Karlsruhe, Germany).

3.9. Statistical Analysis

Statistical analysis was performed using SPSS 22.0 software (SPSS Inc., Chicago, IL, USA). Results are presented as means ± standard deviation. At least triplicate measurements for each time point and experimental condition were performed. Differences corresponding to $p < 0.05$ were considered statistically significant.

4. Conclusions

Polylactide is an interesting material for 3D-printing in biomaterial research. This study confirms its biocompatibility and demonstrates an endotoxin contamination clearly below the FDA limit. Cells of various cell types (osteoblasts, fibroblasts and endothelial cells) grow, spread and proliferate on PLA-printed discs. PLA cages loaded with SDF-1–collagen support cell growth of endothelial cells and induce neo-vessel formation. These results demonstrate the potential for PLA scaffolds in medical applications, for example, in bone tissue engineering. Tests of mechanical stability as well as in vivo tests employing a femur defect model in the rat will define the potential to induce angiogenesis and bone regeneration of the described scaffolds.

Acknowledgments: This project was funded by BiomaTiCS, Mainz. This work represents parts of the doctoral thesis of Rebekka Gerke.

Author Contributions: Ulrike Ritz and Hermann Götz conceived and designed the experiments; Rebekka Gerke and Hermann Götz performed the experiments; Ulrike Ritz and Hermann Götz analyzed the data; Stefan Stein produced the lentiviral vectors; Ulrike Ritz and Pol Maria Rommens wrote the paper.

Conflicts of Interest: The authors declare no conflict of interest.

References

1. Marsell, R.; Einhorn, T.A. The biology of fracture healing. *Injury* **2011**, *42*, 551–555. [CrossRef] [PubMed]
2. Zalavras, C.G. Prevention of Infection in Open Fractures. *Infect. Dis. Clin. N. Am.* **2017**, *31*, 339–352. [CrossRef] [PubMed]
3. Giannoudis, P.V.; Atkins, R. Management of long-bone non-unions. *Injury* **2007**, *38* (Suppl. S2), S1–S2. [CrossRef]
4. Giannoudis, P.V.; Stengel, D. Clinical research in trauma and orthopaedic surgery—Call for action. *Injury* **2008**, *39*, 627–630. [CrossRef] [PubMed]
5. Shi, P.; Wang, Q.; Yu, C.; Fan, F.; Liu, M.; Tu, M.; Lu, W.; Du, M. Hydroxyapatite nanorod and microsphere functionalized with bioactive lactoferrin as a new biomaterial for enhancement bone regeneration. *Colloids Surf. B Biointerfaces* **2017**, *155*, 477–486. [CrossRef] [PubMed]
6. Tozzi, G.; De Mori, A.; Oliveira, A.; Roldo, M. Composite Hydrogels for Bone Regeneration. *Materials* **2016**, *9*, 267. [CrossRef] [PubMed]

7. Zou, Q.; Li, J.; Niu, L.; Zuo, Y.; Li, J.; Li, Y. Modified n-HA/PA66 scaffolds with chitosan coating for bone tissue engineering: Cell stimulation and drug release. *J. Biomater. Sci. Polym. Ed.* **2017**, *28*, 1271–1285. [CrossRef] [PubMed]

8. Filipowska, J.; Tomaszewski, K.A.; Niedźwiedzki, Ł.; Walocha, J.A.; Niedźwiedzki, T. The role of vasculature in bone development, regeneration and proper systemic functioning. *Angiogenesis* **2017**, *20*, 291–302. [CrossRef] [PubMed]

9. Lu, H.; Liu, Y.; Guo, J.; Wu, H.; Wang, J.; Wu, G. Biomaterials with Antibacterial and Osteoinductive Properties to Repair Infected Bone Defects. *Int. J. Mol. Sci.* **2016**, *17*, 334. [CrossRef] [PubMed]

10. Eman, R.M.; Hoorntje, E.T.; Oner, F.C.; Kruyt, M.C.; Dhert, W.J.A.; Alblas, J. CXCL12/Stromal-Cell-Derived Factor-1 Effectively Replaces Endothelial Progenitor Cells to Induce Vascularized Ectopic Bone. *Stem Cells Dev.* **2014**, *23*, 2950–2958. [CrossRef] [PubMed]

11. Kimura, Y.; Tabata, Y. Controlled release of stromal-cell-derived factor-1 from gelatin hydrogels enhances angiogenesis. *J. Biomater. Sci. Polym. Ed.* **2010**, *21*, 37–51. [CrossRef] [PubMed]

12. Roseti, L.; Parisi, V.; Petretta, M.; Cavallo, C.; Desando, G.; Bartolotti, I.; Grigolo, B. Scaffolds for Bone Tissue Engineering: State of the art and new perspectives. *Mater. Sci. Eng. C Mater. Biol. Appl.* **2017**, *78*, 1246–1262. [CrossRef] [PubMed]

13. Chou, Y.C.; Lee, D.; Chang, T.M.; Hsu, Y.H.; Yu, Y.H.; Liu, S.J.; Ueng, S.W. Development of a Three-Dimensional (3D) Printed Biodegradable Cage to Convert Morselized Corticocancellous Bone Chips into a Structured Cortical Bone Graft. *Int. J. Mol. Sci.* **2016**, *17*, 595. [CrossRef] [PubMed]

14. Zhou, R.; Xu, W.; Chen, F.; Qi, C.; Lu, B.Q.; Zhang, H.; Wu, J.; Qian, Q.R.; Zhu, Y.J. Amorphous calcium phosphate nanospheres/polylactide composite coated tantalum scaffold: Facile preparation, fast biomineralization and subchondral bone defect repair application. *Colloids Surf. B Biointerfaces* **2014**, *123*, 236–245. [CrossRef] [PubMed]

15. Naito, Y.; Terukina, T.; Galli, S.; Kozai, Y.; Vandeweghe, S.; Tagami, T.; Ozeki, T.; Ichikawa, T.; Coelho, P.G.; Jimbo, R. The effect of simvastatin-loaded polymeric microspheres in a critical size bone defect in the rabbit calvaria. *Int. J. Pharm.* **2014**, *461*, 157–162. [CrossRef] [PubMed]

16. Bierbaum, S.; Hintze, V.; Scharnweber, D. Functionalization of biomaterial surfaces using artificial extracellular matrices. *Biomatter* **2012**, *2*, 132–141. [CrossRef] [PubMed]

17. Bhuiyan, D.B.; Middleton, J.C.; Tannenbaum, R.; Wick, T.M. Mechanical properties and osteogenic potential of hydroxyapatite-PLGA-collagen biomaterial for bone regeneration. *J. Biomater. Sci. Polym. Ed.* **2016**, *27*, 1139–1154. [CrossRef] [PubMed]

18. Wallace, D.G.; Rosenblatt, J. Collagen gel systems for sustained delivery and tissue engineering. *Adv. Drug Deliv. Rev.* **2003**, *55*, 1631–1649. [CrossRef] [PubMed]

19. Shanmugasundaram, N.; Ravikumar, T.; Babu, M. Comparative physico-chemical and in vitro properties of fibrillated collagen scaffolds from different sources. *J. Biomater. Appl.* **2004**, *18*, 247–264. [CrossRef] [PubMed]

20. Sarker, B.; Hum, J.; Nazhat, S.N.; Boccaccini, A.R. Combining collagen and bioactive glasses for bone tissue engineering: A review. *Adv. Healthc. Mater.* **2015**, *4*, 176–194. [CrossRef] [PubMed]

21. Kuttappan, S.; Mathew, D.; Nair, M.B. Biomimetic composite scaffolds containing bioceramics and collagen/gelatin for bone tissue engineering—A mini review. *Int. J. Biol. Macromol.* **2016**, *93*, 1390–1401. [CrossRef] [PubMed]

22. Motamedian, S.R.; Hosseinpour, S.; Ahsaie, M.G.; Khojasteh, A. Smart scaffolds in bone tissue engineering: A systematic review of literature. *World J. Stem Cells* **2015**, *7*, 657–668. [CrossRef] [PubMed]

23. Rothdiener, M.; Uynuk-Ool, T.; Sudkamp, N.; Aurich, M.; Grodzinsky, A.J.; Kurz, B.; Rolauffs, B. Human osteoarthritic chondrons outnumber patient- and joint-matched chondrocytes in hydrogel culture-Future application in autologous cell-based OA cartilage repair? *J. Tissue Eng. Regen. Med.* **2017**. [CrossRef] [PubMed]

24. Ghica, M.V.; Albu, M.G.; Leca, M.; Popa, L.; Moisescu, S.T. Design and optimization of some collagen-minocycline based hydrogels potentially applicable for the treatment of cutaneous wound infections. *Pharmazie* **2011**, *66*, 853–861. [PubMed]

25. Theiss, F.; Mirsaidi, A.; Mhanna, R.; Kummerle, J.; Glanz, S.; Bahrenberg, G.; Tiaden, A.N.; Richards, P.J. Use of biomimetic microtissue spheroids and specific growth factor supplementation to improve tenocyte differentiation and adaptation to a collagen-based scaffold in vitro. *Biomaterials* **2015**, *69*, 99–109. [CrossRef] [PubMed]

26. Giordano, C.; Albani, D.; Gloria, A.; Tunesi, M.; Rodilossi, S.; Russo, T.; Forloni, G.; Ambrosio, L.; Cigada, A. Nanocomposites for neurodegenerative diseases: Hydrogel-nanoparticle combinations for a challenging drug delivery. *Int. J. Artif. Organs* **2011**, *34*, 1115–1127. [CrossRef] [PubMed]

27. Tsaryk, R.; Gloria, A.; Russo, T.; Anspach, L.; De Santis, R.; Ghanaati, S.; Unger, R.E.; Ambrosio, L.; Kirkpatrick, C.J. Collagen-low molecular weight hyaluronic acid semi-interpenetrating network loaded with gelatin microspheres for cell and growth factor delivery for nucleus pulposus regeneration. *Acta Biomater.* **2015**, *20*, 10–21. [CrossRef] [PubMed]

28. Bersini, S.; Gilardi, M.; Arrigoni, C.; Talo, G.; Zamai, M.; Zagra, L.; Caiolfa, V.; Moretti, M. Human in vitro 3D co-culture model to engineer vascularized bone-mimicking tissues combining computational tools and statistical experimental approach. *Biomaterials* **2016**, *76*, 157–172. [CrossRef] [PubMed]

29. Abed, A.; Assoul, N.; Ba, M.; Derkaoui, S.M.; Portes, P.; Louedec, L.; Flaud, P.; Bataille, I.; Letourneur, D.; Meddahi-Pelle, A. Influence of polysaccharide composition on the biocompatibility of pullulan/dextran-based hydrogels. *J. Biomed. Mater. Res. A* **2011**, *96*, 535–542. [CrossRef] [PubMed]

30. Souness, A.; Zamboni, F.; Walker, G.M.; Collins, M.N. Influence of scaffold design on 3D printed cell constructs. *J. Biomed. Mater. Res. Part B Appl. Biomater.* **2017**. [CrossRef] [PubMed]

31. Gorbet, M.B.; Sefton, M.V. Endotoxin: The uninvited guest. *Biomaterials* **2005**, *26*, 6811–6817. [CrossRef] [PubMed]

32. Lieder, R.; Petersen, P.H.; Sigurjonsson, O.E. Endotoxins-the invisible companion in biomaterials research. *Tissue Eng. Part B Rev.* **2013**, *19*, 391–402. [CrossRef] [PubMed]

33. Ronco, C. Endotoxin Removal: History of a Mission. *Blood Purif.* **2014**, *37*, 5–8. [CrossRef] [PubMed]

34. Neches, R.Y.; Flynn, K.J.; Zaman, L.; Tung, E.; Pudlo, N. On the intrinsic sterility of 3D printing. *PeerJ* **2016**, *4*, e2661. [CrossRef] [PubMed]

35. Goffin, A.L.; Raquez, J.M.; Duquesne, E.; Siqueira, G.; Habibi, Y.; Dufresne, A.; Dubois, P. From interfacial ring-opening polymerization to melt processing of cellulose nanowhisker-filled polylactide-based nanocomposites. *Biomacromolecules* **2011**, *12*, 2456–2465. [CrossRef] [PubMed]

36. Nasonova, M.V.; Glushkova, T.V.; Borisov, V.V.; Velikanova, E.A.; Burago, A.Y.; Kudryavtseva, Y.A. Biocompatibility and Structural Features of Biodegradable Polymer Scaffolds. *Bull. Exp. Biol. Med.* **2015**, *160*, 134–140. [CrossRef] [PubMed]

37. Rosenzweig, D.H.; Carelli, E.; Steffen, T.; Jarzem, P.; Haglund, L. 3D-Printed ABS and PLA Scaffolds for Cartilage and Nucleus Pulposus Tissue Regeneration. *Int. J. Mol. Sci.* **2015**, *16*, 15118–15135. [CrossRef] [PubMed]

38. Yang, Y.; Yang, S.; Wang, Y.; Yu, Z.; Ao, H.; Zhang, H.; Qin, L.; Guillaume, O.; Eglin, D.; Richards, R.G.; et al. Anti-infective efficacy, cytocompatibility and biocompatibility of a 3D-printed osteoconductive composite scaffold functionalized with quaternized chitosan. *Acta Biomater.* **2016**, *46*, 112–128. [CrossRef] [PubMed]

39. Senatov, F.S.; Niaza, K.V.; Zadorozhnyy, M.Y.; Maksimkin, A.V.; Kaloshkin, S.D.; Estrin, Y.Z. Mechanical properties and shape memory effect of 3D-printed PLA based porous scaffolds. *J. Mech. Behav. Biomed. Mater.* **2016**, *57*, 139–148. [CrossRef] [PubMed]

40. Wurm, M.C.; Most, T.; Bergauer, B.; Rietzel, D.; Neukam, F.W.; Cifuentes, S.C.; Wilmowsky, C.V. In-vitro evaluation of Polylactic acid (PLA) manufactured by fused deposition modeling. *J. Biol. Eng.* **2017**, *11*, 29. [CrossRef] [PubMed]

41. Rodina, A.V.; Tenchurin, T.K.; Saprykin, V.P.; Shepelev, A.D.; Mamagulashvili, V.G.; Grigor'ev, T.E.; Lukanina, K.I.; Orekhov, A.S.; Moskaleva, E.Y.; Chvalun, S.N. Migration and Proliferative Activity of Mesenchymal Stem Cells in 3D Polylactide Scaffolds Depends on Cell Seeding Technique and Collagen Modification. *Bull. Exp. Biol. Med.* **2016**, *162*, 120–126. [CrossRef] [PubMed]

42. Rodina, A.V.; Tenchurin, T.K.; Saprykin, V.P.; Shepelev, A.D.; Mamagulashvili, V.G.; Grigor'ev, T.E.; Moskaleva, E.Y.; Chvalun, S.N.; Severin, S.E. Proliferative and Differentiation Potential of Multipotent Mesenchymal Stem Cells Cultured on Biocompatible Polymer Scaffolds with Various Physicochemical Characteristics. *Bull. Exp. Biol. Med.* **2017**, *162*, 488–495. [CrossRef] [PubMed]

43. Kao, C.T.; Lin, C.C.; Chen, Y.W.; Yeh, C.H.; Fang, H.Y.; Shie, M.Y. Poly(dopamine) coating of 3D printed poly(lactic acid) scaffolds for bone tissue engineering. *Mater. Sci. Eng. C Mater. Biol. Appl.* **2015**, *56*, 165–173. [CrossRef] [PubMed]

44. Yeh, C.H.; Chen, Y.W.; Shie, M.Y.; Fang, H.Y. Poly(Dopamine)-Assisted Immobilization of Xu Duan on 3D Printed Poly(Lactic Acid) Scaffolds to Up-Regulate Osteogenic and Angiogenic Markers of Bone Marrow Stem Cells. *Materials* **2015**, *8*, 4299–4315. [CrossRef] [PubMed]

45. Sekula, M.; Domalik-Pyzik, P.; Morawska-Chochol, A.; Bobis-Wozowicz, S.; Karnas, E.; Noga, S.; Boruczkowski, D.; Adamiak, M.; Madeja, Z.; Chlopek, J.; et al. Polylactide- and polycaprolactone-based substrates enhance angiogenic potential of human umbilical cord-derived mesenchymal stem cells in vitro - implications for cardiovascular repair. *Mater. Sci. Eng. C Mater. Biol. Appl.* **2017**, *77*, 521–533. [CrossRef] [PubMed]

46. Brunsen, A.; Ritz, U.; Mateescu, A.; Hofer, I.; Frank, P.; Menges, B.; Hofmann, A.; Rommens, P.M.; Knoll, W.; Jonas, U. Photocrosslinkable dextran hydrogel films as substrates for osteoblast and endothelial cell growth. *J. Mater. Chem.* **2012**, *22*, 19590–19604. [CrossRef]

47. Hertweck, J.; Ritz, U.; Gotz, H.; Schottel, P.C.; Rommens, P.M.; Hofmann, A. CD34+ cells seeded in collagen scaffolds promote bone formation in a mouse calvarial defect model. *J. Biomed. Mater. Res. Part B Appl. Biomater.* **2017**. [CrossRef] [PubMed]

48. Ritz, U.; Kogler, P.; Hofer, I.; Frank, P.; Klees, S.; Gebhard, S.; Brendel, C.; Kaufmann, K.; Hofmann, A.; Rommens, P.M.; et al. Photocrosslinkable polysaccharide hydrogel composites based on dextran or pullulan-amylose blends with cytokines for a human co-culture model of human osteoblasts and endothelial cells. *J. Mater. Chem. B* **2016**, *4*, 6552–6564. [CrossRef]

49. Naderi-Meshkin, H.; Matin, M.M.; Heirani-Tabasi, A.; Mirahmadi, M.; Irfan-Maqsood, M.; Edalatmanesh, M.A.; Shahriyari, M.; Ahmadiankia, N.; Moussavi, N.S.; Bidkhori, H.R.; et al. Injectable hydrogel delivery plus preconditioning of mesenchymal stem cells: Exploitation of SDF-1/CXCR4 axis towards enhancing the efficacy of stem cells' homing. *Cell Biol. Int.* **2016**. [CrossRef] [PubMed]

50. Prokoph, S.; Chavakis, E.; Levental, K.R.; Zieris, A.; Freudenberg, U.; Dimmeler, S.; Werner, C. Sustained delivery of SDF-1alpha from heparin-based hydrogels to attract circulating pro-angiogenic cells. *Biomaterials* **2012**, *33*, 4792–4800. [CrossRef] [PubMed]

51. DeCicco-Skinner, K.L.; Henry, G.H.; Cataisson, C.; Tabib, T.; Gwilliam, J.C.; Watson, N.J.; Bullwinkle, E.M.; Falkenburg, L.; O'Neill, R.C.; Morin, A.; et al. Endothelial cell tube formation assay for the in vitro study of angiogenesis. *J. Vis. Exp.* **2014**, e51312. [CrossRef] [PubMed]

52. Pinese, C.; Gagnieu, C.; Nottelet, B.; Rondot-Couzin, C.; Hunger, S.; Coudane, J.; Garric, X. In vivo evaluation of hybrid patches composed of PLA based copolymers and collagen/chondroitin sulfate for ligament tissue regeneration. *J. Biomed. Mater. Res. Part B Appl. Biomater.* **2017**, *105*, 1778–1788. [CrossRef] [PubMed]

53. Heo, D.N.; Castro, N.J.; Lee, S.J.; Noh, H.; Zhu, W.; Zhang, L.G. Enhanced bone tissue regeneration using a 3D printed microstructure incorporated with a hybrid nano hydrogel. *Nanoscale* **2017**, *9*, 5055–5062. [CrossRef] [PubMed]

54. Yin, X.; Jiang, L.; Yang, J.; Cao, L.; Dong, J. Application of biodegradable 3D-printed cage for cervical diseases via anterior cervical discectomy and fusion (ACDF): An in vitro biomechanical study. *Biotechnol. Lett.* **2017**, *39*, 1433–1439. [CrossRef] [PubMed]

55. Cruz Sanchez, A.; Lanza, S.; Boudaoud, H.; Hoppe, S.; Camargo, M. Polymer Recycling and Additive Manufacturing in an Open Source context: Optimization of processes and methods. In Proceedings of the Annual International Solid Freeform Fabrication Symposium, Austin, TX, USA, 10–12 August 2015; pp. 1591–1600.

56. Heiss, M.; Hellstrom, M.; Kalen, M.; May, T.; Weber, H.; Hecker, M.; Augustin, H.G.; Korff, T. Endothelial cell spheroids as a versatile tool to study angiogenesis in vitro. *J. Off. Publ. Fed. Am. Soc. Exp. Biol.* **2015**, *29*, 3076–3084. [CrossRef] [PubMed]

57. Korff, T.; Augustin, H.G. Integration of endothelial cells in multicellular spheroids prevents apoptosis and induces differentiation. *J. Cell Biol.* **1998**, *143*, 1341–1352. [CrossRef] [PubMed]

58. Hofmann, A.; Ritz, U.; Hessmann, M.H.; Schmid, C.; Tresch, A.; Rompe, J.D.; Meurer, A.; Rommens, P.M. Cell viability, osteoblast differentiation, and gene expression are altered in human osteoblasts from hypertrophic fracture non-unions. *Bone* **2008**, *42*, 894–906. [CrossRef] [PubMed]

59. Brendel, C.; Muller-Kuller, U.; Schultze-Strasser, S.; Stein, S.; Chen-Wichmann, L.; Krattenmacher, A.; Kunkel, H.; Dillmann, A.; Antoniou, M.N.; Grez, M. Physiological regulation of transgene expression by a lentiviral vector containing the A2UCOE linked to a myeloid promoter. *Gene Ther.* **2012**, *19*, 1018–1029. [CrossRef] [PubMed]

International Journal of
Molecular Sciences

MDPI

Review

Electrospun Fibrous Scaffolds for Tissue Engineering: Viewpoints on Architecture and Fabrication

Indong Jun [1], Hyung-Seop Han [1,2], James R. Edwards [1] and Hojeong Jeon [2,3,*]

[1] Botnar Research Centre, Nuffield Department of Orthopaedics, Rheumatology and Musculoskeletal Sciences (NDORMS), University of Oxford, Oxford OX3 7LD, UK; indong.jun@ndorms.ox.ac.uk (I.J.); hyungseop.han@ndorms.ox.ac.uk (H.-S.H.); james.edwards@ndorms.ox.ac.uk (J.R.E.)
[2] Center for Biomaterials, Korea Institute of Science & Technology (KIST), Seoul 02792, Korea
[3] Division of Bio-Medical Science & Technology, KIST School, Korea University of Science and Technology, Seoul 02792, Korea
*   Correspondence: jeonhj@kist.re.kr; Tel.: +82-2-958-5140

Received: 15 November 2017; Accepted: 3 March 2018; Published: 6 March 2018

Abstract: Electrospinning has been used for the fabrication of extracellular matrix (ECM)-mimicking fibrous scaffolds for several decades. Electrospun fibrous scaffolds provide nanoscale/microscale fibrous structures with interconnecting pores, resembling natural ECM in tissues, and showing a high potential to facilitate the formation of artificial functional tissues. In this review, we summarize the fundamental principles of electrospinning processes for generating complex fibrous scaffold geometries that are similar in structural complexity to the ECM of living tissues. Moreover, several approaches for the formation of three-dimensional fibrous scaffolds arranged in hierarchical structures for tissue engineering are also presented.

Keywords: electrospinning;   nanofiber   scaffolds;   tissue   engineering;   extracellular matrix-mimicking geometries

1. Introduction

Tissue engineering is an emerging multidisciplinary field that aims to regenerate damaged or lost tissues/organs of living organisms using a combination of cells and scaffolds [1,2]. These engineering techniques start with scaffolds, providing environments for cells/tissues to grow in an orderly manner, and become functionalized into new tissues/organs [3]. Given the importance of intercellular interactions between scaffolds and implanted surrounding cells/tissues, considerable efforts have been made to design an artificial extracellular matrix (ECM) composed of complex fibrous structures, including glycosaminoglycans, collagen, elastin, and reticular fibers [4–6]. These components are known to provide mechanical and biochemical support to the surrounding cells, and these effects are dependent on tissue type [7–9]. The biomimetic approaches taken in the field of biomaterials seek out innovation in technology from the phenomenon in nature. The relationship between architecture and function that governs normal physiology is equally instrumental in tissue regeneration. Therefore, biomaterials should be designed and engineered with the optimized structural function for the target tissue in mind.

Electrospinning has been used for the fabrication of ECM-mimicking fibrous scaffolds for several decades [10–12]. Electrospinning is a spinning technique that uses electrostatic forces to produce fibrous scaffolds from biocompatible polymers. A simple equipment setup makes electrospinning a versatile way to process all of the different biocompatible polymers into fibrous scaffolds. Several studies have been conducted to apply tissue engineering through controlling parameters of the electrospinning process, including electrospinning process parameters (i.e., the application of an electric field, flow rate, distance between the needle and collector, and diameter of the metallic needle) and solution parameters (i.e., concentration, viscosity, and solution conductivity) [13–17]. These processes can be suitable for the large-scale production of scaffolds with a controllable single fiber diameter, motivating us to attempt to manipulate the process for tissue-specific applications.

As with other biocompatible materials used to mimic native tissues, electrospun fibrous scaffolds provide nanoscale/microscale fibrous structures with interconnecting pores, resembling the natural ECM in tissues, and showing a high potential to facilitate the formation of artificial functional tissues (Figure 1). For example, porosity is observed in skin and bone tissues, which have a large volume fraction of interconnected pores to facilitate cell migration and the transport of nutrients during tissue regeneration [18,19]. However, conventional electrospun fibrous scaffolds are composed entirely of closed-packed fibers, which provide a superficial porous structure, and these porous structures became smaller when the fiber diameter is decreased from the microscale to the nanoscale. Such poor cell infiltration into the fibrous scaffolds results in the formation of two-dimensional (2D) surfaces rather than a three-dimensional (3D) environment, which more closely mimics the ECM. Sisson et al. demonstrated that the small pore size of electrospun fibrous scaffolds could hamper cellular migration into a fibrous scaffold, restricting tissue ingrowth [20].

To date, the scientific community has investigated a considerable number of fabricated structures of electrospun fibers and their corresponding functions for tissue engineering, including the generation of skin, bone, muscle, cartilage, and blood vessels [21–29]. Although these introduced fibrous scaffolds have been successfully applied to mimic native tissues, more research is needed in order to fully understand the cellular responses to sophisticated structures with different scales, dimensions, and spatial arrangements. Therefore, it is important to analyze the sophisticated structures of fibrous scaffolds, including its spatial geometry and 3D form. In this review article, we summarize the fundamental principles of electrospinning processes for generating complex fibrous scaffolds geometries that are similar in structural complexity to the ECM of living tissues. Moreover, several approaches for the formation of 3D fibrous scaffolds arranged in hierarchical structures for tissue engineering are also presented.

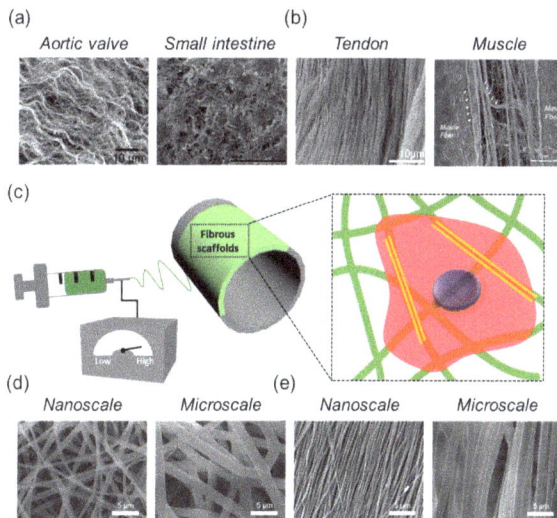

Figure 1. Scanning electron microscope (SEM) images of a natural extracellular matrix (ECM) in distinct types of tissues with (**a**) isotropic direction and (**b**) anisotropic direction [30–33]; (**c**) Schematic illustration of the electrospinning process; (**d**) Representative SEM images of fibrous scaffolds with a controllable fibrous scale with (**d**) randomly and (**e**) aligned fibrous deposition via electrospinning.

2. Fibrous Scaffolds with Hierarchical Structures

Improving our understanding of the structural features of tissues is critical for selecting appropriate materials for reconstructing damaged tissues. Within the last several decades, many studies have aimed to fabricate fibrous nanoscale/microscale structures by applying different electrospinning process parameters (Figure 1). Electrospinning works on a simple principle; a charged polymer jet is collected on a metallic collector when the applied electrostatic charge overcomes the surface tension of the polymer solution. Electrospun fibrous scaffolds are typically assembled into nonwoven networks, which are deposited in an anisotropic (random) fiber orientation. These randomly deposited fibrous scaffolds have potential applications as temporary substitutes for skin and bone tissue engineering, because they replicate the microstructure of natural tissue [21,24]. Moreover, the high surface area of electrospun fibrous scaffolds allows oxygen permeability at the wound site, making these scaffolds suitable substrates for wound dressings.

Tissues in the human body also have unique ECM structures that are found in specific tissues (i.e., the heart, nerve, and blood vessel) exhibiting anisotropic fibrous structures. From a structural viewpoint, structures that can mimic these features are believed to be necessary for precisely guiding cell growth and tissue regeneration, which lead to the recovery of their natural tissue function. For this purpose, electrospun fibrous scaffolds with various alignments have shown an outstanding ability to guide cell morphology and affect cell function when compared with other types of random fibrous scaffolds, both in vitro and in vivo. For example, Gnavi et al. demonstrated that aligned fibrous scaffolds provide contact guidance to cultured nerve cells, resulting in the alignment and elongation of cells along the contacted fiber direction [34]. Lee et al. found that aligned fibrous scaffolds could serve as ideal materials for bone tissue engineering, and that the orientation of fibers plays an important role in the guidance of new bone formation [35]. In addition, other studies have shown that myoblasts and endothelial cells are also affected by cellular morphological changes in aligned fibrous scaffolds [36,37].

Although numerous fibrous scaffolds with ECM-mimicking structures have been introduced to date, most studies of fibrous scaffolds have verified the effects of unitary structures, such as fiber scale, porosity, and orientation. However, most of the native tissues in humans are composed of hierarchical structures and are not unitary structures. For example, bone tissue has a hierarchical organization over various scales ranging from microscale to nanoscale-structured components [38]. Tendons are the connective tissues that bridge muscles to bone, allowing the maximal transmission of forces to produce movement at joints [39]. For these reasons, studies based on fibrous scaffolds with unitary structures may provide us with incomplete knowledge of the unitary structure, and do not consider hierarchical structures. Therefore, hierarchically patterned fibrous structures are required to design scaffolds for tissue engineering, which would allow us to better understand cell behaviors and functions on ECM-mimicking scaffolds. In this section, we discuss recent progress in the production of electrospun fibrous scaffolds with hierarchical structures via a variety of techniques (Figure 2).

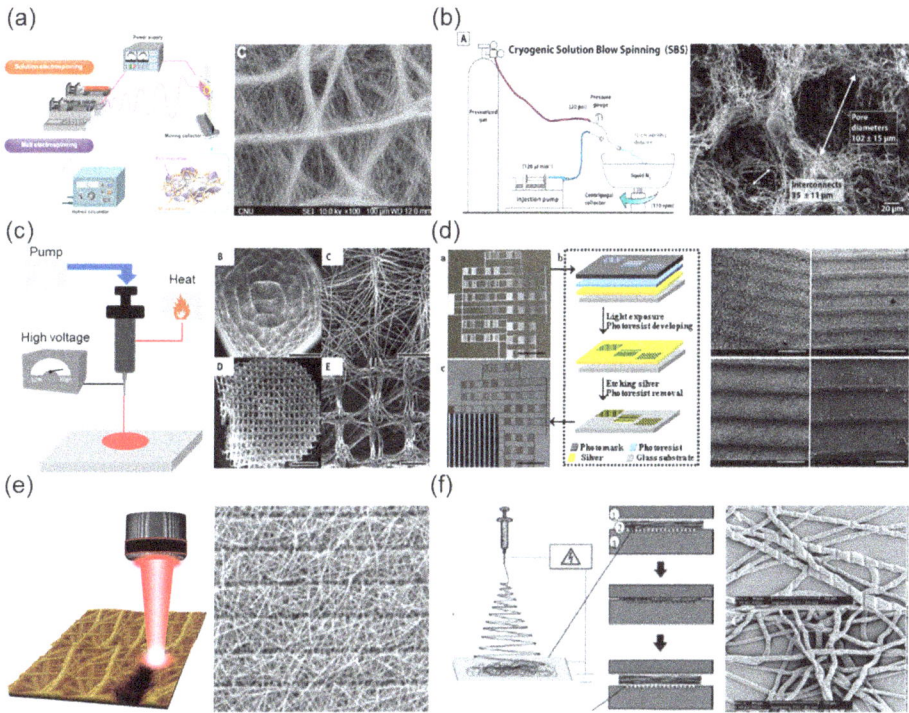

Figure 2. Techniques to control geometry in fibrous scaffolds and representative images. Schematic setup for (**a**) dual extrusion electrospinning [40]: (**b**) cryogenic electrospinning [41]; (**c**) melt electrospinning [42]; and (**d**) micropatterned collector-based electrospinning [43]. Schematic illustration of post-processing techniques using (**e**) laser-based ablation [44] and (**f**) nanoimprinting lithography [45].

2.1. Geometric Control on Fibrous Scaffolds

2.1.1. Dual Extrusion Electrospinning

Dual extrusion electrospinning employs two extrusion processes that function simultaneously. This technique enables the differential control of spatial geometry depending on extrusion conditions, e.g., polymeric concentrations and independent solvents. This permits nanofibrous/microfibrous structures to be combined in a single scaffold and the distribution of electrospun fibers to be controlled. Leverson et al. investigated the cellular response using hybrid scaffolds of mixed nanoscale/microscale fibers fabricated by dual extrusion electrospinning, in which one syringe creates nanoscale fibers, and the other generates microscale fibers [46]. Park et al. prepared layer-on-layer stacks that included alternating layers of random fibers and aligned fibers [47]. These proposed hybrid fibrous structures provide more stable mechanical support than single fibrous scaffolds. Furthermore, the presence of nanoscale fibers in the hybrid-scale scaffolds influences cell behaviors. For example, Kim et al. demonstrated that the osteoblast differentiation ability increased in hybrid scaffolds composed of nanoscale silk fibroin fibers and microscale poly(3-caprolactone) (PCL) fibers [40]. The authors reported that hybrid-scale scaffolds have the advantage of forming ECM-mimicking structures to support cell migration for microscale fibers, whereas nanoscale fibers were used to mimic the structure of the ECM for cell adhesion. Hybrid-scale scaffolds with multiple scales can provide synergistic effects, thereby improving the function of osteoblasts. Therefore, dual extrusion electrospinning is a suitable method for generating scaffolds with different fibrous scales and structures, which is critical for controlling the behavior of cells.

2.1.2. Temperature-Assisted Electrospinning (Cryogenic/Melt)

Fibrous scaffolds can also be fabricated by temperature-assisted electrospinning, using additional temperature-controlled equipment to improve cell permeability. Cryogenic electrospinning uses ice crystals throughout the process of depositing fibrous layers in order to prevent the formation of highly compacted fibrous scaffolds during the electrospinning process. Scaffold porosity can be adjusted from 10–500 μm, depending on various controllable factors such as size and the amount of ice crystals. The ice crystals are then removed through the freeze-drying of the fibrous scaffolds, leaving large void spaces. This technique is a promising technique for developing fibrous scaffolds, permitting cell infiltration into these void spaces. Bulysheva et al. have used cryogenic electrospinning to generate fibrous scaffolds that are capable of inducing an in-growth of fibroblasts and epithelial cells into scaffolds [48]. Similarly, cryogenic solution blow spinning uses a system of concentric nozzles and pressurised gas to blow a polymer solution into a cryogenic liquid. Medeiros et al. have developed a method to generate fibrous scaffolds with controlled porous structures using a combination of thermally induced phase separation and solution blow spinning [41]. Ice microspheres create these fibrous scaffolds, forming interconnected networks with the fibers directly on the surface of the liquid nitrogen. Fibrous scaffolds made using this method have 3D scaffolds that are formed of porous fibers with interconnected macroscale pores.

Alternatively, melt electrospinning is another temperature-assisted electrospinning approach using a higher temperature. Melt electrospinning uses a polymer melt instead of a polymer solution, permitting the controlled fibrous deposition of 3D scaffolds with programmable porosity and alignment. In principle, the polymer is placed in a syringe that can be heated to a desirable temperature (~400 °C) and extruded using air pressure. This technique overcomes the disadvantages of conventional electrospinning, e.g., toxic solvents are not required [49]. This lack of solvent has implications for a wide range of biomedical materials, because the residual solvent does not need to be removed in order to use the scaffolds on cells or tissues. Zaiss et al. used melt electrospinning in order to produce 3D fibrous scaffolds in which the average pore size and fiber diameter on the fibrous scaffolds were 250–300 and 15 μm, respectively, by using an X–Y programed collector stage [42]. Melt blowing technology also uses a higher temperature to generate fibrous scaffolds. In melt blowing electrospinning, the polymer is melted and extruded through a jet, while heated air is blown through an air nozzle. Jenkins et al. reported that microarchitectural fibrous scaffolds were produced using a melted polymer with controlled airflow velocities (500–1400 m^3 air/h/m scaffolds) for rotator cuff tendon tissue engineering [50]. They showed that these fibrous scaffolds have a fiber diameter of 3–5000 nm, which can provide a relevant physiological structure for tendon-like ECM environments.

2.1.3. Micropatterned Collector-Based Electrospinning

Fibrous scaffolds generated using electrospinning are deposited on a conductive metallic collector in a randomly oriented fibrous direction due to their chaotic whipping nature in the electrified liquid jet. These 2D features may be problematic to the broad application of such scaffolds for tissue engineering applications. Zhang et al. utilised a unique patterned collector and fabricated different microscale architectures and macroscale 3D tubular structures [51]. The resulting fibrous scaffolds had hierarchically patterned structures similar to those of the used patterned metallic collector. Another study demonstrated the use of an array of stainless steel beads as the collector, and yielded hierarchically patterned nanoscale fibrous scaffolds with arrayed microscale wells [52]. However, the patterned collectors, which were modulated by weaving or engraving, caused some shortcomings in microscale precision patterning and intricate patterns. In this regard, Liu et al. showed the potential use of a glass template collector patterned with an electrically conductive circuit [43]. They designed a micropatterned glass template on a collector prepared by lithography in order to obtain a micropatterned silver circuit for the selective deposition of electrospun fibers. They demonstrated that lithography could provide a flexible collector surface with microscale precision patterns, and that the cells could be modulated to a precise location and into specific shapes using hierarchically patterned fibrous scaffolds. Furthermore, such strategies using lithographic collectors have also been used for the preparation of fibrous scaffolds for tissue regeneration. For example, Lie et al. fabricated fibrous scaffolds using a lithographic collector with various types of fibrous patterns, such as honeycomb, rectangle, and square shaped, and confirmed that co-cultured cells (cardiomyoblasts, cardiac fibroblasts, and endothelial cells) on honeycomb-patterned fibrous scaffolds showed spontaneous beating similar to that of native cardiomyoblasts [53]. These approaches have yielded materials that mimic the in vivo microenvironment of native tissues, such as spatial arrangement and cell–cell communication, in order to elevate the effects of cell function. Another approach is replacing 2D flat collectors with 3D collecting templates in order to obtain 3D fibrous scaffolds. Zhang et al. introduced methods to fabricate 3D macrofibrous scaffolds with controlled architectures using various 3D collecting templates [51]. They confirmed that 3D fibrous scaffolds with different macroscopic configurations, e.g., length, diameter, and shape, could be fabricated by designing 3D collecting templates.

2.1.4. Post-Processing after Electrospinning

For many years, researchers have attempted to fabricate scaffolds with a hierarchical fibrous structure by generating a nanoscale/microscale structure on prefabricated fibrous scaffolds using chemical modification methods, such as ultraviolet (UV) irradiation [54,55]. However, such conventional fabrication methods for the formation of hierarchical fibrous structures require complicated processes, which tend to result in the inevitable collapse of the fibrous structure and changes in their mechanical properties. To overcome this hurdle, laser processing has been utilised. Laser ablation is a well-known precise fabrication technique that induces negligible thermal stress or collateral damage on the target materials owing to the very short time scales involved in the laser/material interaction (10^{-12} and 10^{-15} s for picosecond and femtosecond lasers, respectively) [56,57]. The laser-based processing of fibrous scaffolds offers several advantages, such as a delicate patterning technique through a rapid and uncomplicated process of direct ablation of the material. Kong et al. used a picosecond laser on aligned nanofibrous scaffolds to create microscale holes (100–200 µm) with a spacing of 50 µm and 200 µm between patterns in engineered corneal tissues [58]. In addition, femtosecond laser patterning can be used to fabricate various patterns on fibrous scaffolds, such as line, well, square, triangle, and pentagon shapes [59,60]. Recent studies have shown that femtosecond laser ablation on fibrous scaffolds may provide an effective in vitro fibrous platform to modulate cell behaviors. Lee et al. employed a femtosecond laser ablation system to create microscale porosity on electrospun nanofibrous scaffolds with diameters of 50 µm, 100 µm, and 200 µm, and spacing of 50 µm and 200 µm between patterns [59]. They demonstrated that femtosecond laser-ablated electrospun fibrous scaffolds not only affected adhesive cell morphology in vitro, but also enabled better cell ingrowth in vivo. Moreover, Jun et al. showed that the orientation of myoblasts could be efficiently modulated by femtosecond laser-ablated microscale grooves on the surface of fibrous scaffolds [44]. They demonstrated that cells initially grew according to the nanoscale random fibrous structure, but eventually reorganised to match the adhesive cellular morphology and orientation of myotube assembly on femtosecond-laser ablated microscale grooves. More importantly, cells showed better ingrowth in laser-ablated fibrous scaffolds than in untreated fibrous scaffolds, which are major drawbacks for conventional electrospinning. Shin et al. evaluated these femtosecond laser-ablated fibrous scaffolds with nanoscale/microscale structures, which were developed to promote the biological function of endothelial cells by resembling the native endothelium [61].

Another simple way to create additional patterns on electrospun fibers is nanoimprinting lithography (NIL). Nandakumar et al. utilised nanoimprinting lithography on electrospun fibers at physiological temperatures [45]. They prepared microscale electrospun fibers (single fibers with a diameter of approximately 6 µm), and then imprinted the fibers within 1–5 min under a high-pressure vacuum at less than 42 °C. Patterns ranging from line to geometric shapes (circles and triangles) were imprinted on prepared fibrous scaffolds.

2.2. Three-Dimensional Fibrous Scaffolds

The fibrous scaffolds produced by conventional electrospinning are usually two-dimensional rather than three-dimensional, hampering cell infiltration. Although several approaches have been explored to increase the porosity of electrospun fibrous scaffolds to overcome cell ingrowth, these approaches do not produce 3D scaffolds with respect to thickness. Several approaches to form 3D fibrous scaffolds using different electrospinning processes have been developed (Figure 3).

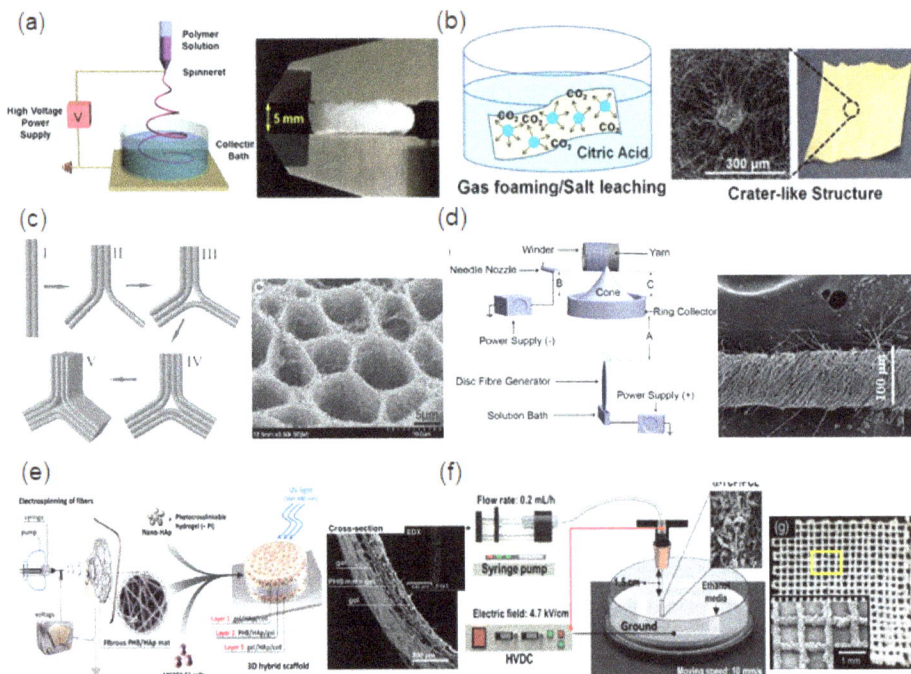

Figure 3. Several approaches to the formation of three-dimensional (3D) fibrous scaffolds using different electrospinning process: (**a**) liquid-collecting electrospinning [62]; (**b**) gas foaming [63]; (**c**) self-assembly [64]; and (**d**) fibrous yarn scaffolds [65]; (**e**) schematic illustration of a hydrogel-integrated fibrous scaffold [66]; (**f**) a hybrid system using 3D printing and electrospinning [67].

2.2.1. Liquid-Collecting Electrospinning

A technique using liquid reservoirs as collectors has attracted attention as a method for preparing 3D fibrous scaffolds. The utilization of liquids with low surface tension, such as water, ethanol, and methanol, causes the extruded fibers to sink during electrospinning, and overcomes the effects of fiber bonding. This results in the loosening of fibrous layers of scaffolds with higher internal porosity. The choice of the liquid to use in the reservoir should be considered according to surface tension and hydrophilicity. For example, Chen et al. fabricated 3D fibrous scaffolds incorporating fibrous morphologies and interconnected pore structures using liquid-collecting electrospinning. In their study, a bath containing diluted alcohol solution was used as the liquid-based collector [62]. During this process, electrospun fibers continuously accumulated in relatively fluffy stacks. After the removal of alcohol through a freezing process at a low temperature, foamed 3D fibrous scaffolds were prepared.

The porosity of the formed 3D fibrous structures was much larger than that of conventional fabricated 2D electrospun fibrous scaffolds. The thickness of the fibrous scaffolds prepared by this method was measured as 5 mm. Recently, Kasuga et al. successfully fabricated macroscale 3D fibrous scaffolds (with a thickness of 40–50 mm) with a cotton-like structure using this method [68]. They claimed that these 3D fibrous scaffolds could improve the localization of neighboring cells at the initial stage after implantation.

2.2.2. Gas Foaming

Three-dimensional fibrous scaffolds using gas foaming provide viable alternatives to open the pores between fibrous networks with a minimal application of force. The gas foaming technique for prepared 2D fibrous scaffolds usually involves three steps: 2D fibrous scaffold/gas solution formation, gas bubble nucleation/growth, and 2D fibrous scaffold expansion to 3D fibrous scaffolds. Jiang et al. reported on the feasibility of this process using prepared 2D fibrous scaffolds and $NaBH_4$ as a gas foaming agent [69]. The prepared 2D fibrous scaffolds were soaked in $NaBH_4$ solution, and bubbles were then formed/generated on the 2D fibrous scaffolds, creating a more open scaffold. Researchers demonstrated that a higher concentration of gas foaming agent significantly increased the thickness of the prepared 2D fibrous scaffolds. In addition, cells successfully infiltrated and grew throughout the gas foamed 3D fibrous scaffolds. A similar technique was reported by Hwang et al. involving the use of a gas foaming/salt leaching technique [63]. Briefly, Hwang et al. fabricated fibrous scaffolds with crater-like structures, enabling them to replicate the 3D ECM fibrous environment. Notably, this scaffold permitted human mesenchymal stem cells to penetrate through the prepared 3D fibrous scaffold (up to 250 μm), whereas most human mesenchymal stem cells were not able to penetrate through the conventional electrospun fibrous scaffolds within seven days.

2.2.3. Self-Assembly

Self-assembly is a process in which a disordered arrangement forms an ordered system as a consequence of specific, local interactions among the components themselves. Self-assembly of the fibrous scaffolds was investigated for several types of polymers, including polyethylene oxide (PEO), polyacrylonitrile (PAN), PCL, and polyvinyl alcohol. Liang et al. utilised precursor polymer solutions (PEO dissolved in water) and fabricated self-assembled fibrous structures with honeycomb patterns [70]. They explained the relationship between the concentration of the polymer solution and the self-assembled formation of honeycomb patterns. In addition, Ahirwal et al. fabricated self-organised fibrous structures in honeycomb patterns using a prepared polymer solution (PCL dissolved in dimethylformamide (DMF)) [71]. Yan et al. studied the mechanisms underlying this phenomenon, and found that the polymer solution concentration, collected substrates, collection distance, and humidity played critical roles in the formation of the self-assembled honeycomb-patterned structures in fibrous scaffolds [64]. For example, self-assembled honeycomb patterns were not observed at higher concentration (i.e., 7% PAN dissolved in DMF). The well-defined 3D honeycomb structures in fibrous scaffolds could only form when the humidity decreased below 60% for the PAN/DMF solution. Moreover, the self-assembled pore size increased as the distance between the polymer jet and collector decreased.

2.2.4. Fibrous Yarn Scaffolds

Based on the hierarchical architecture of soft tissues, such as tendon and ligaments, the fabrication of multifilament yarn scaffolds has been highlighted for tissue engineering applications [72]. Yarns are linear assemblies of single fibers with improved mechanical properties generated through twisting, weaving, and knitting. Recently, considerable efforts have been made to fabricate fibrous yarn scaffolds directly through electrospinning. For example, Ali et al. fabricated continuous and twisted fibrous yarns using a multi-nozzle and rotary funnel collector [65]. They demonstrated that the flow rate of the polymer solution and the speed of the rotating collector could control fibrous yarn production and twist rates. Additionally, Mouthuy et al. used a textile production line to create multifilament yarn, which could mimic the hierarchical architecture of the tendon [73]. Fibrous yarn scaffolds prepared using this method improved the functions of primary human tenocytes, including adherent cell numbers and proliferation. Furthermore, they directly implanted fibrous yarn scaffolds into transected infraspinatus tendons, and confirmed the good safety profile of this method, with only a mild foreign body reaction. Chang et al. reported on twisted fibrous yarns consisting of microfiber and nanofiber yarns that were prepared using a high-speed spinneret tip [74]. With the high rotational speed of the spinneret, charged fibers could be deposited on the ground collector with rotation during jetting to form a twisted fibrous rope. The formation of twisted fibrous yarn is dependent on the distance between the spinneret tip and the ground collector, and such fibrous yarn scaffold systems can be utilised not only for enhancing mechanical strength, but also for generating medical scaffolds with improved therapeutic effects.

2.2.5. Hydrogel-Integrated Fibrous Scaffolds

Hydrogel-integrated fibrous scaffolds may also be used to mimic the native tissue environment. An advantage of combining fibrous scaffolds with hydrogels is that cells can easily migrate through 3D environments, such as the 3D environment of native ECM. Indeed, hydrogel-integrated fibrous scaffolds allow for cellular contact guidance within 3D environments, which cannot be accomplished with separate hydrogel or electrospun fiber systems. Sadat-Shojai et al. attempted to reconstruct the structure using electrospun fibers as the inner layer, and a hydrogel as the 3D structure [66]. Briefly, they prepared an electrospun fibrous layer that was soaked in the precursor gel mixture with a photoinitiator. The 3D hydrogel integrated fibrous scaffold was formed by UV light for 10 s. The inner electrospun fibers were designed to provide mechanical support, while the embedded hydrogels were designed to facilitate cell proliferation and spread. Similarly, Wu et al. prepared 3D hybrid scaffolds based on a fibrous yarn network within a hydrogel shell to mimic the native cardiac tissue structure [75]. They prepared an interwoven aligned fibrous structure via a weaving technique, and then encapsulated the structure within photocurable hydrogels. These 3D hybrid scaffolds promoted the alignment of cardiomyoblasts on each fibrous layer, and individually controlled the cellular orientation of different layers in a 3D hydrogel environment.

2.2.6. Near-Field Electrospinning (NFE) with 3D Printing Technology

Despite extensive studies, it is still not possible to fabricate highly organised fibrous scaffolds with controlled uniformity and architecture using conventional electrospinning generated by the chaotic whipping of liquid jets. NFE is a relatively new method in the field of electrospinning, and has recently been actively applied by researchers. NFE uses a short distance between the spinneret and collector (less than 3 mm) to prevent the bending instability and split in ejected fibers. This method stabilizes the region of the ejected polymer jet to control fiber deposition and enable the production of 3D structures. For example, Lu et al. introduced an NFE method to fabricate 3D microstructures of fibrous scaffolds [76] by balancing the distance between the spinneret and the collector (0.5–3 mm). In addition, a programmable X–Y motion stage was also utilised to deposit the fibrous structures in a predesigned path.

Alternatively, an electrohydrodynamic printing process was developed to deposit the electrohydrodynamically printed fibers into customised patterns by controlling the X–Y motion stage [67]. Briefly, this system enabled microscale fibrous bundles to form from the charged single jet by replacing the solvent of the polymeric solution with an alcohol-based solvent. They determined the optimal processing conditions, including electric field, distance, flow rate, and needle gauge, in order to fabricate 3D microscale fibrous scaffolds. Fibrous scaffolds prepared using this technique promoted osteoblast behaviors and functions.

Fattahi et al. designed 3D fibrous scaffolds in combination with 3D printing technology and NFE [77]. This approach yielded highly organised fibrous architectures with the desired form via an X–Y–Z moving stage. Additionally, He et al. combined melt electrospinning with 3D printing.

Int. J. Mol. Sci. **2018**, *19*, 745

The ejected melted fibers could be precisely stacked in a layer-by-layer manner to form 3D fibrous scaffolds [78]. Three-dimensional printing and electrospinning are relatively new in the field of electrospinning and have recently been actively applied by researchers. A hybrid system using 3D printing and electrospinning can help create 3D fibrous scaffolds with similar complex architectures and the ability to regulate cellular behaviors.

3. Conclusions and Future Perspectives

The native tissue has complex structures with somewhat unique arrangements and architectures of fibrous shapes, and ECM-mimicking materials have been of great interest to scientists, particularly in the field of tissue engineering. Electrospinning offers advantages for the preparation of fibrous scaffolds resembling the fibrillar architecture of the ECM in native tissues, yielding materials with major advantages in tissue engineering. There are many methods for fabricating 2D and 3D fibrous scaffolds, which show structural characteristics similar to those of the native ECM. In this review, we discussed recent advances in the fabrication of fibrous scaffolds with desired geometries and architectures using several electrospinning techniques. This review provided an overview of the fundamental principles of the electrospinning process for generating complex fibrous scaffold geometries similar to the structural complexity of the ECM in living tissue. As our understanding of the origins of these features increases, we can begin to design, fabricate, and apply this knowledge in the biomedical field.

Although there are still many challenges to overcome, electrospinning shows enormous potential in the fabrication of fibrous scaffolds with controllable geometric/architectural structures, enabling researchers to design and develop novel fibrous scaffolds that more closely mimic the structural environment of the native ECM. In the development of "ECM-mimicking materials," the objective is to improve our understanding of native complex structures and prepare highly efficient biomedical scaffolds for tissue engineering. Further studies of fibrous scaffolds are ongoing, and will be useful for achieving efficient tissue engineering.

Acknowledgments: James R. Edwards is grateful for the continued support of the Arthritis Research UK Fellowship (20631). This study was also supported by the Korea Health Industry Development Institute (KHIDI), funded by the Ministry of Health and Welfare (HI16C0133), and Korea Institute of Science & Technology (KIST) project (2Z05190).

Author Contributions: Indong Jun performed the literature review and wrote the manuscript. Hyung-Seop Han participated in the review of the literature. James R. Edwards was responsible for critical review of the manuscript. Hojeong Jeon was responsible for critical review of the manuscript and the discussion.

Conflicts of Interest: The authors declare no conflicts of interest.

References

1. Khademhosseini, A.; Langer, R. A decade of progress in tissue engineering. *Nat. Protoc.* **2016**, *11*, 1775–1781. [CrossRef] [PubMed]
2. Chen, F.M.; Liu, X.H. Advancing biomaterials of human origin for tissue engineering. *Prog. Polym. Sci.* **2016**, *53*, 86–168. [CrossRef] [PubMed]
3. Akbari, M.; Tamayol, A.; Bagherifard, S.; Serex, L.; Mostafalu, P.; Faramarzi, N.; Mohammadi, M.H.; Khademhosseini, A. Textile Technologies and Tissue Engineering: A Path Toward Organ Weaving. *Adv. Healthc. Mater.* **2016**, *5*, 751–766. [CrossRef] [PubMed]
4. Thomas, K.; Engler, A.J.; Meyer, G.A. Extracellular matrix regulation in the muscle satellite cell niche. *Connect. Tissue Res.* **2015**, *56*, 1–8. [CrossRef] [PubMed]
5. Wen, S.L.; Feng, S.; Tang, S.H.; Gao, J.H.; Zhang, L.H.; Tong, H.; Yan, Z.P.; Fang, D.Z. Collapsed Reticular Network and its Possible Mechanism during the Initiation and/or Progression of Hepatic Fibrosis. *Sci. Rep.* **2016**, *6*, 35426. [CrossRef] [PubMed]
6. Mouw, J.K.; Ou, G.Q.; Weaver, V.M. Extracellular matrix assembly: A multiscale deconstruction. *Nat. Rev. Mol. Cell Biol.* **2014**, *15*, 771–785. [CrossRef] [PubMed]

7. Caralt, M.; Uzarski, J.S.; Iacob, S.; Obergfell, K.P.; Berg, N.; Bijonowski, B.M.; Kiefer, K.M.; Ward, H.H.; Wandinger-Ness, A.; Miller, W.M.; et al. Optimization and Critical Evaluation of Decellularization Strategies to Develop Renal Extracellular Matrix Scaffolds as Biological Templates for Organ Engineering and Transplantation. *Am. J. Transplant.* **2015**, *15*, 64–75. [CrossRef] [PubMed]

8. Bonnans, C.; Chou, J.; Werb, Z. Remodelling the extracellular matrix in development and disease. *Nat. Rev. Mol. Cell Biol.* **2014**, *15*, 786–801. [CrossRef] [PubMed]

9. Xu, T.; Zhao, W.; Zhu, J.M.; Albanna, M.Z.; Yoo, J.J.; Atala, A. Complex heterogeneous tissue constructs containing multiple cell types prepared by inkjet printing technology. *Biomaterials* **2013**, *34*, 130–139. [CrossRef] [PubMed]

10. Jiang, T.; Carbone, E.J.; Lo, K.W.H.; Laurencin, C.T. Electrospinning of polymer nanofibers for tissue regeneration. *Prog. Polym. Sci.* **2015**, *46*, 1–24. [CrossRef]

11. Kai, D.; Liow, S.S.; Loh, X.J. Biodegradable polymers for electrospinning: Towards biomedical applications. *Mater. Sci. Eng. C-Mater.* **2014**, *45*, 659–670. [CrossRef] [PubMed]

12. Karuppuswamy, P.; Venugopal, J.R.; Navaneethan, B.; Laiva, A.L.; Sridhar, S.; Ramakrishna, S. Functionalized hybrid nanofibers to mimic native ECM for tissue engineering applications. *Appl. Surf. Sci.* **2014**, *322*, 162–168. [CrossRef]

13. Pillay, V.; Dott, C.; Choonara, Y.E.; Tyagi, C.; Tomar, L.; Kumar, P.; du Toit, L.C.; Ndesendo, V.M.K. A Review of the Effect of Processing Variables on the Fabrication of Electrospun Nanofibers for Drug Delivery Applications. *J. Nanomater.* 2013. [CrossRef]

14. Mirjalili, M.; Zohoori, S. Review for application of electrospinning and electrospun nanofibers technology in textile industry. *J. Nanostruct. Chem.* **2016**, *6*, 207–213. [CrossRef]

15. Zafar, M.; Najeeb, S.; Khurshid, Z.; Vazirzadeh, M.; Zohaib, S.; Najeeb, B.; Sefat, F. Potential of Electrospun Nanofibers for Biomedical and Dental Applications. *Materials* **2016**, *9*. [CrossRef] [PubMed]

16. Correia, D.M.; Ribeiro, C.; Ferreira, J.C.C.; Botelho, G.; Ribelles, J.L.G.; Lanceros-Mendez, S.; Sencadas, V. Influence of Electrospinning Parameters on Poly(hydroxybutyrate) Electrospun Membranes Fiber Size and Distribution. *Polym. Eng. Sci.* **2014**, *54*, 1608–1617. [CrossRef]

17. Motamedi, A.S.; Mirzadeh, H.; Hajiesmaeilbaigi, F.; Bagheri-Khoulenjani, S.; Shokrgozar, M. Effect of electrospinning parameters on morphological properties of PVDF nanofibrous scaffolds. *Prog. Biomater.* **2017**, *6*, 113–123. [CrossRef] [PubMed]

18. Wu, S.L.; Liu, X.M.; Yeung, K.W.K.; Liu, C.S.; Yang, X.J. Biomimetic porous scaffolds for bone tissue engineering. *Mat. Sci. Eng. R* **2014**, *80*, 1–36. [CrossRef]

19. Chaudhari, A.A.; Vig, K.; Baganizi, D.R.; Sahu, R.; Dixit, S.; Dennis, V.; Singh, S.R.; Pillai, S.R. Future Prospects for Scaffolding Methods and Biomaterials in Skin Tissue Engineering: A Review. *Int. J. Mol. Sci.* **2016**, *17*. [CrossRef] [PubMed]

20. Sisson, K.; Zhang, C.; Farach-Carson, M.C.; Chase, D.B.; Rabolt, J.F. Fiber diameters control osteoblastic cell migration and differentiation in electrospun gelatin. *J. Biomed. Mater. Res. A* **2010**, *94A*, 1312–1320. [CrossRef] [PubMed]

21. Chen, S.X.; Liu, B.; Carlson, M.A.; Gombart, A.F.; Reilly, D.A.; Xie, J.W. Recent advances in electrospun nanofibers for wound healing. *Nanomedicine* **2017**, *12*, 1335–1352. [CrossRef] [PubMed]

22. Dias, J.R.; Granja, P.L.; Bartolo, P.J. Advances in electrospun skin substitutes. *Prog. Mater. Sci.* **2016**, *84*, 314–334. [CrossRef]

23. Wang, C.; Wang, M. Electrospun multicomponent and multifunctional nanofibrous bone tissue engineering scaffolds. *J. Mater. Chem. B* **2017**, *5*, 1388–1399. [CrossRef]

24. Xu, T.; Miszuk, J.M.; Zhao, Y.; Sun, H.L.; Fong, H. Electrospun Polycaprolactone 3D Nanofibrous Scaffold with Interconnected and Hierarchically Structured Pores for Bone Tissue Engineering. *Adv. Healthc. Mater.* **2015**, *4*, 2238–2246. [CrossRef] [PubMed]

25. Kheradmandi, M.; Vasheghani-Farahani, E.; Ghiaseddin, A.; Ganji, F. Skeletal muscle regeneration via engineered tissue culture over electrospun nanofibrous chitosan/PVA scaffold. *J. Biomed. Mater. Res. A* **2016**, *104*, 1720–1727. [CrossRef] [PubMed]

26. Elsayed, Y.; Lekakou, C.; Labeed, F.; Tomlins, P. Smooth muscle tissue engineering in crosslinked electrospun gelatin scaffolds. *J. Biomed. Mater. Res. A* **2016**, *104*, 313–321. [CrossRef] [PubMed]

27. Liu, W.W.; Li, Z.Q.; Zheng, L.; Zhang, X.Y.; Liu, P.; Yang, T.; Han, B. Electrospun fibrous silk fibroin/poly(L-lactic acid) scaffold for cartilage tissue engineering. *Tissue Eng. Regen. Med.* **2016**, *13*, 516–526. [CrossRef]

28. Krishnan, K.V.; Columbus, S.; Krishnan, L.K. Alteration of Electrospun Scaffold Properties by Silver Nanoparticle Incorporation: Evaluation for Blood Vessel Tissue Engineering. *Tissue Eng. Part A* **2015**, *21*, S239–S240.

29. Vatankhah, E.; Prabhakaran, M.P.; Semnani, D.; Razavi, S.; Morshed, M.; Ramakrishna, S. Electrospun Tecophilic/Gelatin Nanofibers with Potential for Small Diameter Blood Vessel Tissue Engineering. *Biopolymers* **2014**, *101*, 1165–1180. [CrossRef] [PubMed]

30. Ye, X.; Wang, H.; Zhou, J.; Li, H.; Liu, J.; Wang, Z.; Chen, A.; Zhao, Q. The effect of Heparin-VEGF multilayer on the biocompatibility of decellularized aortic valve with platelet and endothelial progenitor cells. *PLoS ONE* **2013**, *8*, e54622. [CrossRef] [PubMed]

31. Oliveira, A.C.; Garzón, I.; Ionescu, A.M.; Carriel, V.; Cardona Jde, L.; González-Andrades, M.; Pérez Mdel, M.; Alaminos, M.; Campos, A. Evaluation of small intestine grafts decellularization methods for corneal tissue engineering. *PLoS ONE* **2013**, *8*, e66538. [CrossRef] [PubMed]

32. Youngstrom, D.W.; Barrett, J.G.; Jose, R.R.; Kaplan, D.L. Functional Characterization of Detergent-Decellularized Equine Tendon Extracellular Matrix for Tissue Engineering Applications. *PLoS ONE* **2013**, *8*. [CrossRef] [PubMed]

33. Lieber, R.L.; Ward, S.R. Cellular Mechanisms of Tissue Fibrosis. 4. Structural and functional consequences of skeletal muscle fibrosis. *Am. J. Physiol.-Cell Physiol.* **2013**, *305*, C241–C252. [CrossRef] [PubMed]

34. Gnavi, S.; Fornasari, B.E.; Tonda-Turo, C.; Laurano, R.; Zanetti, M.; Ciardelli, G.; Geuna, S. The Effect of Electrospun Gelatin Fibers Alignment on Schwann Cell and Axon Behavior and Organization in the Perspective of Artificial Nerve Design. *Int. J. Mol. Sci.* **2015**, *16*, 12925–12942. [CrossRef] [PubMed]

35. Lee, J.H.; Lee, Y.J.; Cho, H.J.; Shin, H. Guidance of In Vitro Migration of Human Mesenchymal Stem Cells and In Vivo Guided Bone Regeneration Using Aligned Electrospun Fibers. *Tissue Eng. Part A* **2014**, *20*, 2031–2042. [CrossRef] [PubMed]

36. Han, J.J.; Wu, Q.L.; Xia, Y.N.; Wagner, M.B.; Xu, C.H. Cell alignment induced by anisotropic electrospun fibrous scaffolds alone has limited effect on cardiomyocyte maturation. *Stem Cell Res.* **2016**, *16*, 735–745. [CrossRef] [PubMed]

37. Gaharwar, A.K.; Nikkhah, M.; Sant, S.; Khademhosseini, A. Anisotropic poly (glycerol sebacate)-poly (epsilon-caprolactone) electrospun fibers promote endothelial cell guidance. *Biofabrication* **2015**, *7*. [CrossRef]

38. Reznikov, N.; Shahar, R.; Weiner, S. Bone hierarchical structure in three dimensions. *Acta Biomater.* **2014**, *10*, 3815–3826. [CrossRef] [PubMed]

39. Baldino, L.; Cardea, S.; Maffulli, N.; Reverchon, E. Regeneration techniques for bone-to-tendon and muscle-to-tendon interfaces reconstruction. *Br. Med. Bull.* **2016**, *117*, 25–37. [CrossRef] [PubMed]

40. Kim, B.S.; Park, K.E.; Kim, M.H.; You, H.K.; Lee, J.; Park, W.H. Effect of nanofiber content on bone regeneration of silk fibroin/poly(epsilon-caprolactone) nano/microfibrous composite scaffolds. *Int. J. Nanomed.* **2015**, *10*, 485–502.

41. Medeiros, E.L.G.; Braz, A.L.; Porto, I.J.; Menner, A.; Bismarck, A.; Boccaccini, A.R.; Lepry, W.C.; Nazhat, S.N.; Medeiros, E.S.; Blaker, J.J. Porous Bioactive Nanofibers via Cryogenic Solution Blow Spinning and Their Formation into 3D Macroporous Scaffolds. *ACS Biomater. Sci. Eng.* **2016**, *2*, 1442–1449. [CrossRef]

42. Zaiss, S.; Brown, T.D.; Reichert, J.C.; Berner, A. Poly(epsilon-caprolactone) Scaffolds Fabricated by Melt Electrospinning for Bone Tissue Engineering. *Materials* **2016**, *9*. [CrossRef] [PubMed]

43. Liu, Y.W.; Zhang, L.; Li, H.N.; Yan, S.L.; Yu, J.S.; Weng, J.; Li, X.H. Electrospun Fibrous Mats on Lithographically Micropatterned Collectors to Control Cellular Behaviors. *Langmuir* **2012**, *28*, 17134–17142. [CrossRef] [PubMed]

44. Jun, I.; Chung, Y.W.; Heo, Y.H.; Han, H.S.; Park, J.; Jeong, H.; Lee, H.; Lee, Y.B.; Kim, Y.C.; Seok, H.K.; Shin, H.; Jeon, H. Creating Hierarchical Topographies on Fibrous Platforms Using Femtosecond Laser Ablation for Directing Myoblasts Behavior. *ACS Appl. Mater. Interfaces* **2016**, *8*, 3407–3417. [CrossRef] [PubMed]

45. Nandakumar, A.; Truckenmuller, R.; Ahmed, M.; Damanik, F.; Santos, D.R.; Auffermann, N.; de Boer, J.; Habibovic, P.; van Blitterswijk, C.; Moroni, L. A Fast Process for Imprinting Micro and Nano Patterns on Electrospun Fiber Meshes at Physiological Temperatures. *Small* **2013**, *9*, 3405–3409. [CrossRef] [PubMed]

46. Levorson, E.J.; Raman Sreerekha, P.; Chennazhi, K.P.; Kasper, F.K.; Nair, S.V.; Mikos, A.G. Fabrication and characterization of multiscale electrospun scaffolds for cartilage regeneration. *Biomed. Mater.* **2013**, *8*. [CrossRef] [PubMed]

47. Park, S.H.; Kim, M.S.; Lee, B.; Park, J.H.; Lee, H.J.; Lee, N.K.; Jeon, N.L.; Suh, K.Y. Creation of a Hybrid Scaffold with Dual Configuration of Aligned and Random Electrospun Fibers. *ACS Appl. Mater. Interfaces* **2016**, *8*, 2826–2832. [CrossRef] [PubMed]

48. Bulysheva, A.A.; Bowlin, G.L.; Klingelhutz, A.J.; Yeudall, W.A. Low-temperature electrospun silk scaffold for in vitro mucosal modeling. *J. Biomed. Mater. Res. A* **2012**, *100A*, 757–767. [CrossRef] [PubMed]

49. Tourlomousis, F.; Ding, H.; Kalyon, D.M.; Chang, R.C. Melt Electrospinning Writing Process Guided by a "Printability Number". *J. Manuf. Sci. Eng.* **2017**, *139*. [CrossRef]

50. Jenkins, T.L.; Meehan, S.; Pourdeyhimi, B.; Little, D. Meltblown Polymer Fabrics as Candidate Scaffolds for Rotator Cuff Tendon Tissue Engineering. *Tissue Eng. Part A* **2017**, *23*, 958–967. [CrossRef] [PubMed]

51. Zhang, D.; Chang, J. Electrospinning of three-dimensional nanofibrous tubes with controllable architectures. *Nano Lett.* **2008**, *8*, 3283–3287. [CrossRef] [PubMed]

52. Xie, J.W.; Liu, W.Y.; MacEwan, M.R.; Yeh, Y.C.; Thomopoulos, S.; Xia, Y.N. Nanofiber Membranes with Controllable Microwells and Structural Cues and Their Use in Forming Cell Microarrays and Neuronal Networks. *Small* **2011**, *7*, 293–297. [CrossRef] [PubMed]

53. Liu, Y.W.; Xu, G.S.; Wei, J.J.; Wu, Q.; Li, X.H. Cardiomyocyte coculture on layered fibrous scaffolds assembled from micropatterned electrospun mats. *Mater. Sci. Eng. C-Mater.* **2017**, *81*, 500–510. [CrossRef] [PubMed]

54. Dong, Y.X.; Yong, T.; Liao, S.; Chan, C.K.; Ramakrishna, S. Degradation of electrospun nanofiber scaffold by short wave length ultraviolet radiation treatment and its potential applications in tissue engineering. *Tissue Eng. Part A* **2008**, *14*, 1321–1329.

55. Jao, P.F.; Franca, E.W.; Fang, S.P.; Wheeler, B.C.; Yoon, Y.K. Immersion Lithographic Patterning of Electrospun Nanofibers for Carbon Nanofibrous Microelectrode Arrays. *J. Microelectromech. Syst.* **2015**, *24*, 703–715. [CrossRef]

56. Jeon, H.; Koo, S.; Reese, W.M.; Loskill, P.; Grigoropoulos, C.P.; Healy, K.E. Directing cell migration and organization via nanocrater-patterned cell-repellent interfaces. *Nat. Mater.* **2015**, *14*, 918–923. [CrossRef] [PubMed]

57. Malinauskas, M.; Zukauskas, A.; Hasegawa, S.; Hayasaki, Y.; Mizeikis, V.; Buividas, R.; Juodkazis, S. Ultrafast laser processing of materials: From science to industry. *Light-Sci. Appl.* **2016**, *5*. [CrossRef]

58. Kong, B.; Sun, W.; Chen, G.S.; Tang, S.; Li, M.; Shao, Z.W.; Mi, S.L. Tissue-engineered cornea constructed with compressed collagen and laser-perforated electrospun mat. *Sci. Rep.* **2017**, *7*. [CrossRef] [PubMed]

59. Lee, B.L.P.; Jeon, H.; Wang, A.J.; Yan, Z.Q.; Yu, J.; Grigoropoulos, C.; Li, S. Femtosecond laser ablation enhances cell infiltration into three-dimensional electrospun scaffolds. *Acta Biomater.* **2012**, *8*, 2648–2658. [CrossRef] [PubMed]

60. Jenness, N.J.; Wu, Y.Q.; Clark, R.L. Fabrication of three-dimensional electrospun microstructures using phase modulated femtosecond laser pulses. *Mater. Lett.* **2012**, *66*, 360–363. [CrossRef]

61. Shin, Y.M.; Shin, H.J.; Heo, Y.; Jun, I.; Chung, Y.W.; Kim, K.; Lim, Y.M.; Jeon, H.; Shin, H. Engineering an aligned endothelial monolayer on a topologically modified nanofibrous platform with a micropatterned structure produced by femtosecond laser ablation. *J. Mater. Chem. B* **2017**, *5*, 318–328. [CrossRef]

62. Chen, H.; Peng, Y.; Wu, S.; Tan, L.P. Electrospun 3D Fibrous Scaffolds for Chronic Wound Repair. *Materials* **2016**, *9*. [CrossRef] [PubMed]

63. Hwang, P.T.J.; Murdock, K.; Alexander, G.C.; Salaam, A.D.; Ng, J.I.; Lim, D.J.; Dean, D.; Jun, H.W. Poly(-caprolactone)/gelatin composite electrospun scaffolds with porous crater-like structures for tissue engineering. *J. Biomed. Mater. Res. A* **2016**, *104*, 1017–1029. [CrossRef] [PubMed]

64. Yan, G.; Yu, J.; Qiu, Y.; Yi, X.; Lu, J.; Zhou, X.; Bai, X. Self-assembly of electrospun polymer nanofibers: A general phenomenon generating honeycomb-patterned nanofibrous structures. *Langmuir* **2011**, *27*, 4285–4289. [CrossRef] [PubMed]

65. Ali, U.; Zhou, Y.Q.; Wang, X.G.; Lin, T. Direct electrospinning of highly twisted, continuous nanofiber yarns. *J. Text. Inst.* **2012**, *103*, 80–88. [CrossRef]

66. Sadat-Shojai, M.; Khorasani, M.T.; Jamshidi, A. A new strategy for fabrication of bone scaffolds using electrospun nano-HAp/PHB fibers and protein hydrogels. *Chem. Eng. J.* **2016**, *289*, 38–47. [CrossRef]

67. Kim, M.; Yun, H.S.; Kim, G.H. Electric-field assisted 3D-fibrous bioceramic-based scaffolds for bone tissue regeneration: Fabrication, characterization, and in vitro cellular activities. *Sci. Rep.* **2017**, *7*. [CrossRef] [PubMed]

68. Kasuga, T.; Obata, A.; Maeda, H.; Ota, Y.; Yao, X.F.; Oribe, K. Siloxane-poly(lactic acid)-vaterite composites with 3D cotton-like structure. *J. Mater. Sci.-Mater. Med.* **2012**, *23*, 2349–2357. [CrossRef] [PubMed]

69. Jiang, J.; Carlson, M.A.; Teusink, M.J.; Wang, H.J.; MacEwan, M.R.; Xie, J.W. Expanding Two-Dimensional Electrospun Nanofiber Membranes in the Third Dimension By a Modified Gas-Foaming Technique. *ACS Biomater. Sci. Eng.* **2015**, *1*, 991–1001. [CrossRef]

70. Liang, T.; Mahalingam, S.; Edirisinghe, M. Creating "hotels" for cells by electrospinning honeycomb-like polymeric structures. *Mater. Sci. Eng. C Mater. Biol. Appl.* **2013**, *33*, 4384–4391. [CrossRef] [PubMed]

71. Ahirwal, D.; Hebraud, A.; Kadar, R.; Wilhelm, M.; Schlatter, G. From self-assembly of electrospun nanofibers to 3D cm thick hierarchical foams. *Soft Matter* **2013**, *9*, 3164–3172. [CrossRef]

72. Chainani, A.; Hippensteel, K.J.; Kishan, A.; Garrigues, N.W.; Ruch, D.S.; Guilak, F.; Little, D. Multilayered Electrospun Scaffolds for Tendon Tissue Engineering. *Tissue Eng. Part A* **2013**, *19*, 2594–2604. [CrossRef] [PubMed]

73. Mouthuy, P.A.; Zargar, N.; Hakimi, O.; Lostis, E.; Carr, A. Fabrication of continuous electrospun filaments with potential for use as medical fibres. *Biofabrication* **2015**, *7*. [CrossRef] [PubMed]

74. Chang, G.Q.; Li, A.K.; Xu, X.Q.; Wang, X.L.; Xue, G. Twisted Polymer Microfiber/Nanofiber Yarns Prepared via Direct Fabrication. *Ind. Eng. Chem. Res.* **2016**, *55*, 7048–7051. [CrossRef]

75. Wu, Y.B.; Wang, L.; Guo, B.L.; Ma, P.X. Interwoven Aligned Conductive Nanofiber Yarn/Hydrogel Composite Scaffolds for Engineered 3D Cardiac Anisotropy. *ACS Nano* **2017**, *11*, 5646–5659. [CrossRef] [PubMed]

76. Luo, G.X.; Teh, K.S.; Zang, X.N.; Wu, D.Z.; Wen, Z.Y.; Lin, L.W. High Aspect-Ratio 3d Microstructures Via near-Field Electrospinning for Energy Storage Applications. In Proceedings of the 2016 IEEE 29th International Conference on Micro Electro Mechanical Systems (MEMS), Shanghai, China, 24–28 January 2016; pp. 29–32.

77. Fattahi, P.; Dover, J.T.; Brown, J.L. 3D Near-Field Electrospinning of Biomaterial Microfibers with Potential for Blended Microfiber-Cell-Loaded Gel Composite Structures. *Adv. Healthc. Mater.* **2017**, *6*. [CrossRef] [PubMed]

78. He, J.K.; Xia, P.; Li, D.C. Development of melt electrohydrodynamic 3D printing for complex microscale poly (epsilon-caprolactone) scaffolds. *Biofabrication* **2016**, *8*. [CrossRef] [PubMed]

International Journal of
Molecular Sciences

MDPI

Review

Electrospinning of Chitosan-Based Solutions for Tissue Engineering and Regenerative Medicine

Saad B. Qasim [1], Muhammad S. Zafar [2,3,*], Shariq Najeeb [4], Zohaib Khurshid [5], Altaf H. Shah [6], Shehriar Husain [7] and Ihtesham Ur Rehman [8]

[1] Department of Restorative and Prosthetic Dental Sciences, College of Dentistry, Dar Al Uloom University, P.O. Box 45142, Riyadh 11512, Saudi Arabia; s.qasim@dau.edu.sa
[2] Department of Restorative Dentistry, College of Dentistry, Taibah University, Al Madinah, Al Munawwarah 41311, Saudi Arabia
[3] Department of Dental Materials, Islamic International Dental College, Riphah International University, Islamabad 44000, Pakistan
[4] Restorative Dental Sciences, Al-Farabi Colleges, Riyadh 361724, Saudi Arabia; shariqnajeeb@gmail.com
[5] College of Dentistry, King Faisal University, P.O. Box 380, Al-Hofuf, Al-Ahsa 31982, Saudi Arabia; drzohaibkhurshid@gmail.com
[6] Department of Preventive Dental Sciences, College of Dentistry, Dar Al Uloom University, Riyadh 11512, Saudi Arabia; a.shah@dau.edu.sa
[7] Department of Dental Materials, College of Dentistry, Jinnah Sindh Medical University, Karachi 75110, Pakistan; shehriarhusain@gmail.com
[8] Materials Science and Engineering Department, Kroto Research Institute, University of Sheffield, Sheffield S3 7HQ, UK; i.u.rehman@sheffield.ac.uk
* Correspondence: MZAFAR@taibahu.edu.sa or drsohail_78@hotmail.com; Tel.: +966-50-754-4691

Received: 6 December 2017; Accepted: 24 January 2018; Published: 30 January 2018

Abstract: Electrospinning has been used for decades to generate nano-fibres via an electrically charged jet of polymer solution. This process is established on a spinning technique, using electrostatic forces to produce fine fibres from polymer solutions. Amongst, the electrospinning of available biopolymers (silk, cellulose, collagen, gelatine and hyaluronic acid), chitosan (CH) has shown a favourable outcome for tissue regeneration applications. The aim of the current review is to assess the current literature about electrospinning chitosan and its composite formulations for creating fibres in combination with other natural polymers to be employed in tissue engineering. In addition, various polymers blended with chitosan for electrospinning have been discussed in terms of their potential biomedical applications. The review shows that evidence exists in support of the favourable properties and biocompatibility of chitosan electrospun composite biomaterials for a range of applications. However, further research and in vivo studies are required to translate these materials from the laboratory to clinical applications.

Keywords: chitosan; composite solutions; electrospinning; regeneration; tissue engineering

1. Introduction

Electrospinning (ES) is a process that utilises an electric field to control the deposition of polymer fibres onto target substrates. Originally, the process was devised for the textile industry [1]. It is now lauded for its capabilities to economically and efficiently fabricate non-woven meshes of fibres specifically in the field of tissue engineering (TE) where these fibrous scaffolds can mimic both the natural form and function of the extracellular matrix (ECM) [2–4]. Historically, ES was first observed by Rayleigh in 1897, and was further explored by Zeleny in 1914. Later in 1934, Formhals patented the process of ES for the production of collagen acetate fibres using a strong voltage of 57 kV [5]. Since 1980s, ES has been used to create submicron to nano-meters (nm) sized fibres. Fibres of varying characteristics can be acquired by altering different processing parameters. Currently, the ES is not being used solely in the healthcare industry [6,7], but also has a wide range of applications in other fields including energy [8], waste water [9], textile [10] and security domains [6,11–13]. According to the European Patent Office, until 2013, more than 1891 patents had been filed using the term "ES" and 2960 with the term "nano-fibres" in title and abstract [1,14]. More than 200 institutions and universities worldwide have explored ES as a means of producing different types of nano-fibres [5,7]. In terms of dental applications of ES, more than 45 scientific research papers have been published since 2005 [7].

ES has been adapted to obtain natural and synthetic polymer fibrous mats that mimic the extra cellular matrix (ECM). Amongst available biopolymers, CH and its naturally derived composites have been widely adapted for TE applications. Its versatility with respect to variations in the molecular weights and degree of deacetylation in order to target a clinical condition in the field of regenerative medicine is adapted very efficiently. For ease of understanding, there is a brief description of the ES process, various parameters affecting the fabrication and properties of electrospun nano-fibres. The aim of the current review is to assess the current literature about the electrospinning of chitosan and its composites with other natural polymers to be utilized in TE and regenerative applications.

2. Solution Electrospinning Process

The process of ES is based on obtaining fine fibres from polymer solutions via electrostatic forces. The electrospun fibres have small diameters and significantly larger surface area compared to conventional fibres. The essential components of ES are: 1-high voltage power supply, 2-spinneret and a grounded collecting plate such as a metallic screen, and 3-plate or rotating mandrel (Figure 1).

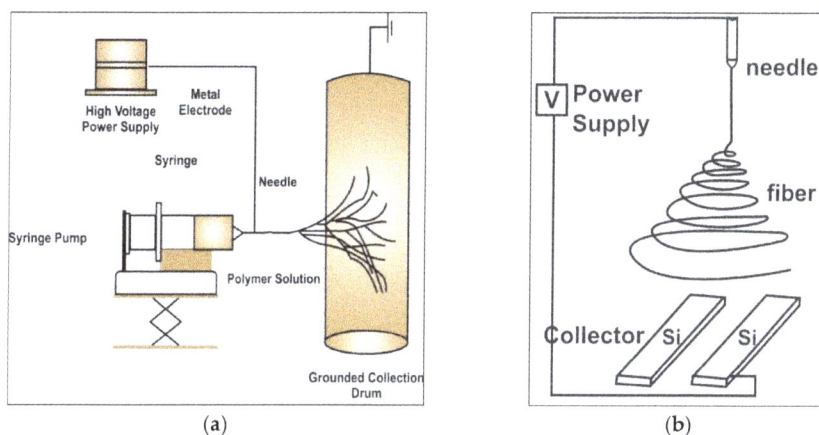

(a)   (b)

Figure 1. The electrospinning process shown schematically (a) electrospinning equipment plate or rotating mandrel (b) aligned collection plates for electrospun nano-fibres [7].

A direct current (DC) voltage is used to generate a potential difference between two terminals within the range of 1–30 kVs that injects a charge of a certain polarity into the polymer solution to accelerate a jet towards a collector of opposite polarity [15]. A syringe pump is used to pump the completely dissolved polymer solutions into the metallic capillary tube [16,17]. The ES equipment can be setup in vertical or horizontal orientations (Figure 1a,b).

In brief, a polymer solution is held by its surface tension at the capillary tip that is subjected to the potential difference created between the spinneret and anode surface (collector). Due to this electric field, a charge is triggered on the liquid surface, changing the pendant drop to a falling jet. When this field reaches a critical value, the repulsive electric forces overcome the surface tension forces. Finally, an unstable charged jet of polymer solution is extruded from the tip of the Taylor cone (conical profile of solvent which bends stretches and thins). Consequently, an unstable and a rapid whipping jet ensues between the capillary tip and collector. The resultant evaporation of the solvent leaves the polymer nano-fibres behind on the collector [18,19].

2.1. Processing Parameters

The characteristics of electrospun fibres are governed by various factors related to the ES dope such as concentration, molecular weight, surface tension, viscosity and conductivity/surface charge density of the polymer solution. In addition, equipment factors such as voltage, feed or flow rate of polymer loaded in the syringe, type of collector and the distance between the collector and needle tip can affect the morphology of electrospun fibres [7]. For instance, the shape of the polymer solution drop is affected by the applied voltage, viscosity of the solution and the flow rate of the syringe [20]. It has also been reported that the higher electrostatic forces cause more stretching of solution due to columbic forces in the jet, thereby generating a stronger electric field ultimately reducing the fibre diameter and higher rate of evaporation of solvent [16,21]. On the other hand, a few researchers have observed that the fibre diameter is proportional to the voltage [22,23]. Velocity of the jet and the material transfer rate are affected by the rate at which the polymer solution is pushed out of a syringe [24]. Slower feed rate facilitates enough time for evaporation of the solvent and is likely to reduce the fibres' diameter at the expense of prolonged processing time. The flow rate affects the pores and surface area as well. Fong et al. have reported the fabrication and detailed characterization of electrospun beaded fibres and suggested that parameters such as charge density, solution viscosity and surface tension are the main factors for controlling bead formation [23].

In terms of collectors, several materials can be used to collect the electrospun nano-fibres [25,26]. Aluminium foils are the most commonly used collectors since, the material must be conductive and remain isolated from the axel [27]. Other alternate options included conductive paper/cloth, parallel or guided bar, wire mesh, rotating wheel or rods. Wang et al. [25] studied the difference in the characteristics of fibres collected on an aluminium foil and a wire screen. They observed that due to the different conducting areas on the wire screen, there was more bead formation compared to an aluminium foil. Fibre alignment and diameter can be changed by altering the speed of a rotating collector. High speed rotating collectors tend to produce more aligned and narrower fibres compared to collectors rotating at a slower rate [28]. Further modification of collectors using different attachments such as pins, [29] mandrels [26] and rods [30] can be achieved to collect electrospun fibres. In order to control fibre characteristics and diameter, the tip to collector distance is also a critical factor. According to Buchko et al. a larger distance between the tip and the collector produces round fibres while a shorter distance produces flat fibres [31]. Reducing the tip to collector distance does not allow sufficient time for the evaporation of the ES solvent prior to hitting the collector. Hence, the moist/wet fibres hitting the collector change their morphology to flat. Hence, the optimal distance is essential to facilitate the evaporation of solvent from the nano-fibres prior to hitting the collector [16].

2.2. Solution Parameters

Solution parameters such as concentration and viscosity influence the electrospun fibre morphology. For instance, lower concentrations result in finer fibres and increased number of beads, while concentrated solutions result in thicker bead free fibres or a reduced number of beads [32]. As the concentration of solution increases, fibres form a more spindle like structure and display uniformly thicker fibres due to higher viscosity [21,33]. Highly viscous solutions can be electrospun to form fibres because the flow rate is unable to be maintained at the tip of the solution which leads to formation of large fibres [33]. As the viscosity is directly related to the concentration, each polymer solution has an optimal concentration for ES.

Solution viscosity is another vital parameter that affects the electrospun fibres in terms of size and morphology. ES of uniform bead-free nano-fibres requires the polymer solution to have an optimal viscosity. Very thick fibres can result if the viscosity is high enough to hinder the ejection of the polymer solution. At a viscosity lower than optimal, the continuous fibre formation is less likely and results in droplets or excessive bead formation. The viscous polymers result in extending the stress relaxation times and limiting the breakage of polymer jets. Consequently, fibres acquired at relatively higher viscosity showed more uniform fibres. Similarly, low surface tension of the solvent facilitates production of nano-fibres with less or no beads [23]. Polymer solutions with greater surface tension obscure ES because of jet instability and generation of sprayed droplets [5,19,34]. Charge density is another important factor defining the outcome of the fibres. Narrower diameter fibres can be obtained if electrical conductivity is increased. In contrast, if the solvent is less conductive this results in insufficient elongation of the jet by electrical force producing uniform fibres and bead formation. The majority of the polymers are usually conductive. The electrospinning polymer solutions have charged ions that are highly efficient in jet formation [16].

The molecular weight (Mw) affects the solution properties such as dielectric constant, conductivity, electrical, viscosity, surface tension, and rheological properties [7,16]. The high Mw solutions are opted corresponding to the intended viscosity for generating fibres. The Mw determines the extent and number of polymer chain entanglements in the solution [5,18,22].

2.3. Effect of Nanoinclusions on CH (Chitosan) Electrospinning

Nanoparticles of various natural and synthetically derived materials have a significant impact on the process of electrospinning [35]. Additions of various natural or synthetic nano biomaterials affects the electrospun fibre morphology, size and diameter [36,37]. The addition of nano hydroxyapatite (HAp) has been explored to obtain fibres within the range of 300 nm. Liverani et al. reported a critical finding about Hap-containing fibres that showed a decrease in average diameter with respect to pure CH fibres [38]. Few other studies have also investigated the effect of nano HAp addition to CH in order to obtain fibres for TE applications. Zhang and co-workers were able to obtain fibres with diameters of 214 nm mimicking naturally mineralized counterparts [39]. In a similar study, Thein-Han et al. compared the addition of micro and nano HAp to CH solutions [40]. Results were suggestive that electrospun fibres obtained whilst using nano HAp had better stem cell attachment compared to micro HAp [40,41].

3. Chitosan-Based Polymers for Solution Electrospinning

A wide variety of natural and synthetic polymers and their composite blends have been used for ES. Examples of electrospun synthetic polymers include poly glycolic acid (PGA), poly lactide co-glycolide (PLGA), polycaprolactone (PCL), polyurethane (PU), poly lactic acid (PLA), polystyrene (PS) and poly vinyl alcohol (PVA). Some of polymers available naturally for ES are, silk, cellulose, collagen, gelatine, hyaluronic acid and CH. Natural and synthetic polymers are also used in combination to manipulate the materials properties (such as thermal stability, mechanical strength and barrier properties) depending on the specific application. Other properties such as cellular affinity, morphological, structural, pore size and degradation can also be altered by copolymers [5].

3.1. Chitosan

Chitosan (CH) is a polysaccharide biomaterial composed of (1–4) acetamido 2 deoxy-β-D glucan, (*N*-acetyl D glucosamine and 2 amino 2 deoxy-β-D glucans. The structure of the biopolymer is shown in Figure 2. CH is obtained by partial deacetylation of chitin. The presence of amino groups in the CH structure differentiates CH from chitin. The degree of deacetylation of CH is indicative of the number of amino groups [42].

The deacetylation of CH is performed either by chemical hydrolysis under harsh alkaline conditions or enzymatic hydrolysis [42]. The alkali extracts the protein and deacetylated chitin at the same time. The processing of crustacean shells involves the extraction of proteins and the dissolution of calcium carbonates, that is accumulated in the crab shell at high concentrations [43]. The deacetylation of chitin is rarely complete; when the degree of deacetylation above 60%, chitin becomes CH (Figure 3).

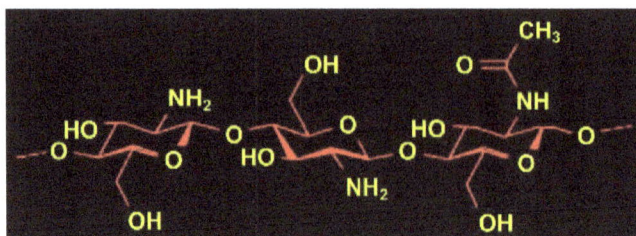

Figure 2. Chemical structure of chitosan showing amide and hydroxyl group that can react and readily form bonds with other natural or synthetic polymers/biomolecules [44].

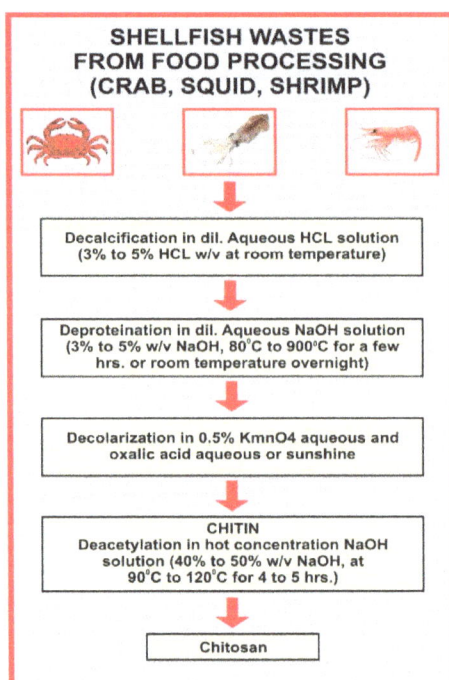

Figure 3. Illustration depicting the deacetylation process adapted to extract CH from chitin [45].

Fabrication of electrospun CH micro or nano-fibres is not an easy task. Common solvents used for dissolving CH are triflouroacetic acid (TFA) or composite solutions of diluted acetic acid copolymerized with polyethylene oxide (PEO) [46]. Different customised approaches performed are either neutralization using the alkaline compounds or cross linkers such as gluteraldehyde and genipin. Hence, the neutralization may ultimately lead to partial or complete loss of features [46].

The majority of the acidic solvents can easily solubilise CH. Its protonation changes CH into a polyelectrolyte in acidic solutions. CH is not able to produce continuous fibres due to the consistent formation of droplets. The high electric field during ES results in repulsive forces among ionic groups within the polymer backbone, hence inhibiting the formation of continuous fibres [47,48]. Although the dual needle setup for CH electrospinning has been rarely reported, it results in repelling of fibres. In addition, adjusting the ideal viscosity of CH electrospinning dope is also challenging. Its rigid D-glucosamine structures, high crystallinity and ability to form hydrogen bonding is responsible for the poor solubility in commonly available organic solvents [49].

Being a cationic polymer, CH in aqueous solution has poly-electrolytic effects. The presence of charged groups results in the significant expansion of polymer coils. In case of electrolyte free polymer solution, the polymer coils shrink and concentrate up. The use of very toxic organic solvents such as hexa-fluoro isopropanol (HFIP) or triflouro acetic acid (TFA) denatures the properties and structure of natural CH [50] and further breaks the inter-chain interactions. Although the number of studies reporting CH fibres for various applications is increasing, each year, very little information is available about the obstacles encountered in obtaining desired fibre properties such as diameter and size. Due to the highly crystalline nature of CH, once dissolved in organic solvents, the formation of hydrogen bonds further complicates the spin-ability [51,52]. In order to overcome this issue, a very low concentration of diluted acetic acid and a fibre-forming agent (PEO) (95:5, CH:PEO) can be used. Sun and Li have mentioned about the problems associated with high viscosity CH limiting its ability to form fibres. It was suggested to perform alkali treatment to hydrolyse CH chains and use a lower molecular weight of CH to overcome this problem [53].

Previously, a number of studies have reported amide bond (NH) formation of CH with organic solvents [54,55]. Yao et al. used lactic acid to generate lactamidated CH in the form of films. CH films were tested for their biocompatibility using fibroblasts [56]. Qu et al. [57] and Toffey et al. [58] used acetic and propionic acids for the regeneration of chitin through amide bond formation. Scanning electron microscopy (SEM) revealed that increase in the polymer concentration resulted in thicker fibres. PEO and CH solutions showed phase separation over time; hence, blended solutions must be electrospun at the earliest preferably within 24 h of blending to prevent any chemical denaturation. The addition of salt before and during ES to these solutions stabilizes them for an extended period of time (Figure 4) [46,59].

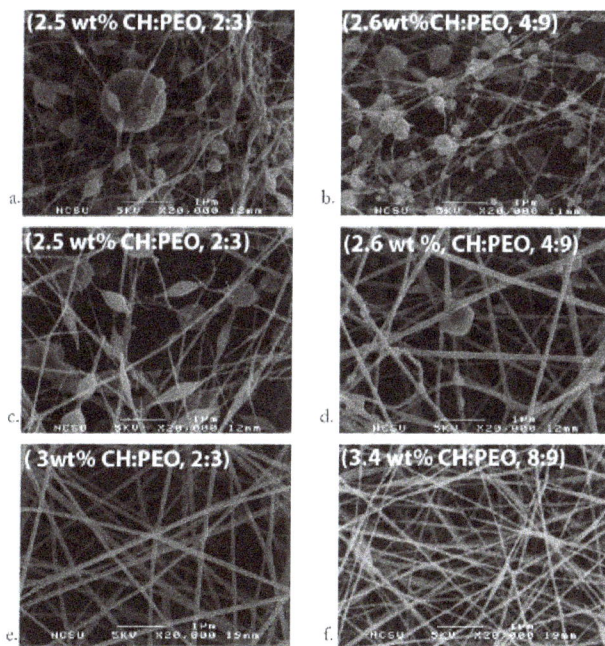

Figure 4. Chitosan PEO nano-fibres depicting the effects of acetic acid concentration, (**a**) 2:3 CH:PEO in 45% (**b**) 4:9 CH:PEO in 36%; (**c**) 2:3 CH:PEO with 2.5 wt.% total polymer 40%; (**d**) 4:9 CH:PEO with 2.6 wt.% total polymer 32%; (**e**) 2:3 CH:PEO with 3 wt.% total polymer blend and 32%; (**f**) 8:9 CH:PEO with 3.4 wt.% total polymer 32% of total acetic acid concentration [59]; scale bar represents 1 μm (Adapted with permission from publisher).

3.2. Chitin/Silk Composite Nano-fibres

Silks are natural polymers produced by insects such as spiders and silkworms. Natural silk fibres have shown favourable features (biocompatibility, non-toxicity) needed for biomaterial applications [60,61]. In addition, the mechanical properties are excellent [62]. Therefore, it has the ability to function under a range of temperatures and humidity [63]. Silk boasts an extensive track record, spanning a period of decades, as a surgical suture material [64]. Silk has also been explored for various biomaterial applications such as TE scaffolds [65,66], and drug delivery [67,68] and bio-dental applications [69–72]. There are many silk-producing animal organisms, however, the main source remains silk worms [73].

Silk harvested from the silkworm (*Bombyx mori*) has the advantage of economic importance as can be domesticated in farms [74]. The *Bombyx mori* (BM) silk has two protein components, a water soluble sericin and water insoluble hydrophobic silk fibroin [74,75]. Glue-like sericin is amorphous in nature and is rich in serine (Mw ~10–300 kDa). It makes approximately 20–30 wt.% of BM silk [76]. Sericin acts as protective coating of silk filaments and cocoons [77] that is permeable to moisture and resistant to oxidation and UV [78]. The sericin has been reported to be associated with the allergic and immunological reactions [79,80], and is hence important to remove sericin completely from fibroin before any biological application can be considered [75,81,82].

The structural component of BM silk is silk fibroin protein (~75 wt.% of total silk) that is a large macromolecule comprised of ~5000 amino acid units [83,84]. The silk fibroin (SF) has crystalline (~66%) and amorphous (~33%) components [85]. The crystalline SF has repeating amino acid units mainly alanine (A), glycine (G) and serine (S) in a typical sequence [G-A-G-A-G-S]$_n$. It forms a β-sheet structure in the spun fibres which is responsible for good mechanical properties [85,86]. In contrast, the amorphous part is mainly composed of phenylalanine (F) and tyrosine (Y). The large side chains of these amino acids lead to hygroscopic properties [87]. SF is further divided into heavy and light chains (H-fibroin and L-fibroin) bonded to each other through disulfide bridges [88,89]. In addition, a glycoprotein (P25) is attached to the SF molecules by non-covalent interactions [89,90]. Considering the unique properties of nanocomposite materials [91] and natural silk, a number of researchers [92–94] have electrospun CH/chitin and silk fibroin (SF) blends using various combinations and solvents (Table 1). Park et al. [93] reported the fabrication of electrospun SF/CH composite nano-fibres using formic acid as an ES solvent. Formic acid is an organic solvent that is highly volatile and has been successfully used for silk fibroin ES [71,95]. The average fibre diameter was reduced with a narrow diameter distribution compared to silk-only nano-fibres. The ionic component of CH results in the increased conductivity of the ES solution, hence, a stronger jet. In addition, intermolecular interactions for example, hydrogen bonding between SF and CH solutions may affect the final properties [93]. The SF nano-fibres are treated with alcoholic solution to induce β-sheet conformation that in turn improves the mechanical properties [96,97]. The CH has a rigid backbone, hence accelerating the conformational changes in SF electrospun nano-fibres [93].

Another study [98] reported ES of chitin/SF-blended fibres using 1,1,1,3,3,3-hexafluoro-2-propanol (HFIP) as an ES solvent. The electrospun fibres were characterized for morphology, dimensional stability and cyto-compatibility. Due to the immiscible nature of SF and chitin, electrospun fibres showed phase-separated structures. The average fibre diameter was reduced by increasing the chitin contents. The solvent induced crystallization and improved the dimensional stability of chitin/SF nano-fibres. In addition, biological properties such as biocompatibility, cell attachment and spreading were evaluated. In terms of potential TE scaffold applications, these electrospun blends exhibited promising characteristics including excellent biocompatibility, cell attachment and spreading for epidermal keratinocytes and fibroblasts [94,98]. Similar findings have been reported by Yoo et al. [94]; chitin and SF solutions were electrospun simultaneously using a hybrid ES technique and nano-fibres were collected at a rotating target. The composition of hybrid materials was controlled by adjusting the solution flow rates. The average fibre diameter was decreased with increasing proportions of CH.

Increasing the proportion of SF improved the mechanical properties whereas higher proportions of CH improved the antibacterial activity of the electrospun fibres. Authors suggested that CH/SF electrospun membranes are a promising candidate for wound healing applications. Chen et al. [99] reported the fabrication of bead-free electrospun nano-fibres of CH/SF blends. The composite nano-fibres were characterized for cellular response using human foetal osteoblasts. SF/CH nano-fibres encouraged the proliferation and differentiation of human foetal osteoblasts. Authors reported the suitability of these materials for bone regeneration applications by choosing a suitable composition. Recently, composite nano-fibres [N-carboxyethyl CH/PVA/SF] have been electrospun using aqueous solutions with suspended SF nanoparticles [92]. These materials had a benefit of being electrospun from aqueous solutions instead of organic solvents during cytotoxicity testing of mouse fibroblasts (L929). These nanomaterials demonstrated good biocompatibility, and hence, can be considered for potential TE applications such as skin regeneration wound dressings [92]. It is clear from previous studies that CH and silk fibroin are compatible with each other for the purpose of ES. A wide range of proportions coupled with ES techniques have been tested with promising outcomes for tissue regeneration applications. However, all reported studies (Table 1) have been conducted in the laboratory environment focusing mainly on physical, mechanical and biological properties.

Table 1. Studies reporting ES of chitosan and silk fibroin composite materials for tissue regeneration application.

Researcher	Solvent	Materials	Key Findings and Significance
Park et al., 2004 [93]	Formic acid	CH/SF blends of variable proportions	Reported the ES of CH/SF blended nano-fibres. The average fibre diameter was reduced with a narrow diameter distribution compared to SF nano-fibres.
Park 2006 [98]	HFIP	Chitin/SF blends of variable proportions	Chitin/SF remains immiscible in nano-fibres The average diameters decreased by increasing chitin contents. Biocompatible and good response for cell attachment and spreading, hence suitable for tissue regeneration applications
Yoo et al., 2008 [94]	HFIP	Chitin/SF blends of variable proportions chitin (5 wt.% in HFIP) and SF (7 wt.% in HFIP)	Confirmed all findings reported by Park et al., 2006 [76]. Chitin/SF solutions were electrospun simultaneously using a hybrid ES technique and nano-fibres were collected on a rotating target. Chitin/SF proportion was controlled by adjusting the flow rates. A narrow fibre diameter distribution (340–920 nm) was observed for chitin/SF nano-fibres compared to SF fibres (140–1260 nm).
Cai 2010 [100]	HFIP, TFE	CH/SF blends; CH contents (0%, 20%, 50%, and 80%)	CH/SF nano-fibrous membranes were successfully electrospun. The average fibre diameter was decreased with the increasing percentage of chitosan. CH/SF composites have better mechanical properties than CS. Electrospun materials were characterized for biocompatibility and antibacterial activity. Authors suggested these membranes as a promising candidate for wound healing applications.
Chen et al., 2012 [101]	mixed solvent [TFA], dichloromethane	CH/SF blends; CH contents (0%, 25%, 50%, 75% and 100%)	Electrospun bead-free CH/SF nano-fibres The composite nano-fibres supported the growth and differentiation of human foetal osteoblasts. Authors reported that a suitable composition of these materials is suitable for bone TE applications.
Zhou et al., 2013 [92]	water	ES dope contained 2.5% (w/v) CH 9% (w/v) PVA in an aqueous solution. SF nanoparticles (4–8 wt.%) were added	Electrospun composite nanofibre membranes using water-soluble N-carboxyethyl CH/PVA/SF nanoparticles The morphology and diameter of the nano-fibres were affected by silk fibroin nanoparticles contents. Presence of intermolecular hydrogen bonding among the molecules of carboxyethyl CH, SF and PVA. Electrospun nanomaterials demonstrated good biocompatibility and can be considered for potential tissue regeneration applications such as skin regeneration wound dressings.

CH (chitosan); SF (silk fibroin); HFIP (1,1,1,3,3,3-hexafluoro-2-propanol); TFE (2,2,2-trifluoroethanol); TFA (trifluoroacetic acid) dichloromethane; PVA (polyvinyl alcohol).

Chitosan/silk fibroin composite materials showed satisfactory properties and biocompatibility essentially required for tissue regeneration and biomedical applications. However, further research including in vivo studies is required to prove these claims for more practical and clinical applications.

3.3. Collagen Chitosan (CC)

Collagen is another natural polymer that has excellent properties (such as biodegradability and biocompatibility) and has been widely explored for TE applications [101,102]. Collagen has been electrospun with and without blending with other polymers such as natural silk [57,58,103], PCL [104], PEO [105] and chitosan [44,101,106]. The purpose of blending other materials with collagen is obvious in terms of modifying the final properties according to potential applications. For example, collagen-CH (CC) blends can modify the mechanical and biological properties to mimic the natural extracellular matrix [107,108]. Previous literature [108–110] has reported the effects of added CH to properties of collagen-based biomaterials. For instance, addition of CH is known to modify the collagen helical characteristics by introducing additional hydrogen bonding that ultimately changes the physical properties. Fourier Transform Infrared (FTIR) spectroscopic analysis has confirmed such molecular interactions between collagen and CH blends [107]. In terms of fibre morphology of CC blends, the fibre diameter was decreased by increasing the CH contents [111]. The various factors affecting the tensile behaviour of electrospun CC materials (single fibres as well as fibrous membrane) have been investigated in detail [44]. Higher tensile strength was observed in case of smaller diameter fibres. In case of electrospun CC membranes, the increase in the ultimate tensile strength was observed by decreasing CH content [44].

Tan et al. fabricated CC composite materials of variable proportions and evaluated the cell viability and proliferation using cells from a human hemopoietic cell line (K562) [112]. The addition of CH (up to 50%) altered the crosslinking pattern of collagen and cellular proliferation. Further increase in the CH content was linked to reduced porosity and cellular proliferation capacity of scaffolds. In addition, CH improved the physical properties such as better stability of fibrous structure and resisting deformation [112]. The behaviours of CC composite materials were also investigated using human periodontal ligament (PDL) cells. The overall adherence and growth capability of periodontal cells was better while using CC scaffolds compared to collagen or CH-only scaffolds. In terms of adherence and growth of cultured PDL cells, the CC scaffolds were better than either CH or collagen scaffolds. It can therefore be suggested that CC composite scaffolds are promising candidates for PDL tissue regeneration [113]. Considering results of using CC composites for TE applications, a number of investigators attempted electrospinning of collagen CH blends [44,101].

In order to fabricate biocompatible wound healing dressings for promoting regeneration of dermal and epidermal layers; nanocomposite fibrous membranes using collagen and CH were electrospun [114]. Characterization of these materials revealed an appreciable degree of crosslinking that resulted in improved mechanical properties (elastic modulus, strength) coupled with a decreased water sorption capability. In terms of biological properties for specific applications; these electrospun membranes were biocompatible and non-toxic to fibroblasts and promoted wound healing. Considering the wound healing potential, the authors declared these electrospun nano-fibre membranes superior to collagen sponge and gauze [114]. Chen and coworkers [101] electrospun CC blends for TE applications and characterized physical, mechanical, thermal properties and biocompatibility using endothelial and smooth muscle cells. The CC scaffolds showed excellent biocompatibility and proliferation for both endothelial and smooth muscle cells suggesting CC as a promising scaffold material for TE applications for regeneration of nerves and blood vessels [101]. Another study [115] has reported the fabrication of electrospun CC targeting the regeneration of nerves and blood vessels. In addition to natural polymers, synthetic polymers (such as thermoplastic PU) were added to enhance the mechanical properties of TE scaffolds and mimic extracellular matrices. The authors reported promising results of in vitro experiments and suggested that in vivo studies are required to validate these results for vascular and nerve regeneration [115].

A potential application of electrospun CC nanofibrous membranes is for corneal TE [116]. The composite CC membranes showed better optical and mechanical properties compared to CH alone. The addition of collagen has been reported to improve the mechanical and physical properties (such as hydrophilicity, optical clarity) without compromising biocompatibility. In addition, CC membranes showed promising characteristics to promote the attachment, spread and viability of cells and has been suggested as a potential candidate for corneal tissue regeneration applications [116]. Recently, CC electrospun scaffolds have been evaluated for guided bone regeneration applications [117]. The CC nanofibre membranes resulted in enhanced cellular proliferation and expression of osteogenic genes in mesenchymal stem cells. There were no apparent signs of inflammation or tissue reaction in the vicinity of the CC membranes suggesting good biocompatibility and potential for guided bone regeneration applications [117]. Although these materials have been evaluated for a range of TE applications such as skin, nerves, vessels, periodontal and bone regeneration, more in vivo and clinical trials are required to validate their properties prior to clinical applications.

3.4. Agarose Chitosan

Agarose is another natural polysaccharide that has been widely used for pharmaceutical and cosmetics applications [118]. Chemically, agarose is a linear polymer composed of repeating units of disaccharide agarobiose. In recent years, many researchers have attempted to fabricate nano-fibres of agarose and CH using various solvents and techniques [119–121]. The mixture of trifluoroacetic acid (TFA) and dichloromethane (DCM) has been reported as a suitable ES solvent for agarose and CH. The agarose added to CH lowered the viscosity of the dope remarkably and resulted in a better compatibility and interaction between agarose and CS molecules [120].

The agarose/CH electrospun fibres (from 30–50% agarose) resulted in smooth cylindrical nano-fibres. However, increasing the agarose content further reduces the viscosity leading to relatively thinner fibres and bead formation [120]. The ES of pure CH results in thicker fibres (in the range of few hundred nanometers to microns), hence agarose can be incorporated to reduce and control the average fibre diameter [120]. Besides chitosan, agarose has been blended with other polymers such as polyacrylamide [119] and polyacrylic acid for ES [122].

There are no biocompatibility or cytotoxicity issues for using agarose-CH blends for biomedical applications. Miguel et al. fabricated agarose-CH thermos-responsive hydrogels for wound healing and skin regeneration [123]. The minimal tissue inflammation and improved healing addressed the excellent biocompatibility and supported cellular proliferation [123].

3.5. Chitosan PEO Composite Electrospinning

The first documented study focusing on the production of electrospun CH-PEO was conducted by Duan et al. [26] using a combination of various ES parameters [concentration (2–8%), CH-PEO ratio (5:1, 2:1 and 1:1) in 2 wt.% acetic acid], voltage of 15 kV, a flow rate of 0.1 mL/h and a stationary collector. While 2% solution only produced beads with no significant fibre formation and 8% solution was too viscous to produce fibres, optimal results were obtained with solutions containing CH and PEO in the ratios of 2:1 and 1:1. One problematic observation of this study was the inconsistent thickness of the fibre diameter. It was seen that thick fibres in the micrometer range were collected directly under the capillary and thinner fibres in the nanometer range were collected elsewhere. This can be attributed to the repulsion of the CH fibres due to their cationic nature. Increasing the molecular mass of PEO had little effect on the ES capabilities of the solution and fibre morphology [26]. Subsequent studies have aimed to improve the ES characteristics of CH-PEO solutions and achieve better control over the quality and fineness of fibres [124–126]. Bhattarai et al. [127] observed that 2% CH and 3% PEO [CH to PEO ratios of 90:10] dissolved in 0.5 M acetic acid produced aligned nano-fibres in the range of 40 nm to 50 μm and claimed improved results compared to Duan et al. [26]. Furthermore, dimethyl sulfoxide was added as a co-solvent to the solution before ES to ease the chain entanglements of CH. An additional advantage of a low PEO is the low swelling of the fibres which increases the structural integrity of the scaffold while in water [127]. In vitro analysis showed that nano-sized CH-PEO fibres favoured cellular attachment and proliferation.

Klossner et al. observed that increasing the total CH-PEO concentration decreases bead formation; however, highly viscous solutions cannot be electrospun [59]. Bead formation was also reduced by decreasing the acetic acid concentration [116]. Additionally, increasing the CH concentration resulted in thicker fibres. Moreover, decreasing the Mw of CH increased the ease of ES. Klossner et al. observed that there was phase separation between CH and PEO after 24 h, hence inhibiting ES [59]. It can be assumed that the CH-PEO solutions in acetic acid have a very short shelf life (<24 h) and must be electrospun within 24 h. The ES capabilities of CH-PEO solutions can be increased by ultra-high molecular weight PEO (UHMWPEO) (Figure 4) [59].

Recently, Qasim et al. reported on processing of ES CH-UHMWPEO solutions that contain UHMWPEO as low as 5 wt.% [3,126] (Figure 5). The main advantage of these fibres is the high CH and lower PEO content that can lead to lesser swelling upon immersion in water. Furthermore, increasing the CH proportion can yield enhanced benefits in terms of antibacterial and osseo-conductive properties. Another way of producing CH and PEO is using coaxial ES of two different blends of polymer solutions [128,129]. Ojha et al. electrospun PEO-coated CH fibres that can be exposed by washing away water-soluble PEO [129]. Conversely, Pakarvan et al. have produced similar fibres albeit in the opposite arrangement i.e., a PEO core coated by CH sheath [128]. Upon washing away the inner core of PEO with water, hollow CH nano-fibres were obtained. Hollow fibres can facilitate cellular attachments and proliferation by providing a larger surface area. The potential area of applications may include haemodialysis and wound-dressings.

Plenty of research has been conducted to improve antimicrobial, regenerative properties and stability of electrospun CH-PEO fibres in combination with poly(hexamethylenebiguanide) hydrochloride (PHMB) or silver nitrate nanoparticles to induce antibacterial properties against *Staphylococcus aureus* and *Escherichia coli* [124,130]. These scaffolds can be advantageous for wound dressings for preventing infections and accelerating the healing. In order to improve the surface properties of the fibres, arginylglycylaspartic acid (RGD) peptides can be crosslinked to the fibres via poly(ethylene glycol) following ES [131]. Compared to unmodified CH-PEO fibres, RGD-modified fibres have superior bioactivity and lead to accelerated tissue regeneration. Recently, incorporation of graphene oxide as a carrier for doxorubicin, an anti-cancer drug, to CH-PEO fibres has made these scaffolds useful as a drug delivery medium to target cancerous tissues directly rather than systemic delivery and avoiding numerous adverse effects [132].

The PEO (as a copolymer) led to the disruption of the CH chain self-association due to hydrogen bonding between $^-$OH and $^+$H ions originating from water molecules [133]. Subsequently, it diminishes repulsion between CH polycationic groups and triggers chain entanglements to cause fibre formation [124,127]. Pakravan et al. used PEO in different percentages and observed the absorption peak at 1112 cm^{-1} in FTIR spectroscopic analysis [125]. This can be assigned to the ether band shifting to a lower wavenumber. Furthermore, PEO reduces the viscosity by breaching intra and/or intermolecular interactions of CH chains. In addition, the flexible PEO chains form around the rigid CH structures [125]. The interaction of CH-PEO is established as a result of solid hydrogen bonding among OH, CH amino groups (Figure 6) and PEO ether groups [125].

Another way of further improving on the regenerative and mechanical properties of CH-PEO nano-fibrous scaffolds is adding another natural or synthetic polymer such as collagen or poly (ε-Caprolactone) (PCL) to the ES solution. However, production of such scaffolds usually involves the use of chemicals or cross-linking agents such as glutaraldehyde and 1,6-diisocyanatohexane (HMDI) that may cause concern for use in clinical settings. Sarkar et al. used tripolyphosphate (TPP), a cross-linking agent, to successfully produce biocompatible cross-linked CH-PEO (5:1 ratio) fibres having diameters as small as 50 nm in 15 M acetic acid [134]. TPP has previously been used to produce biocompatible and non-toxic CH beads for drug delivery applications [135]. It can be a viable alternative to potentially toxic cross-linking agents such as gluteraldehyde and HMDI.

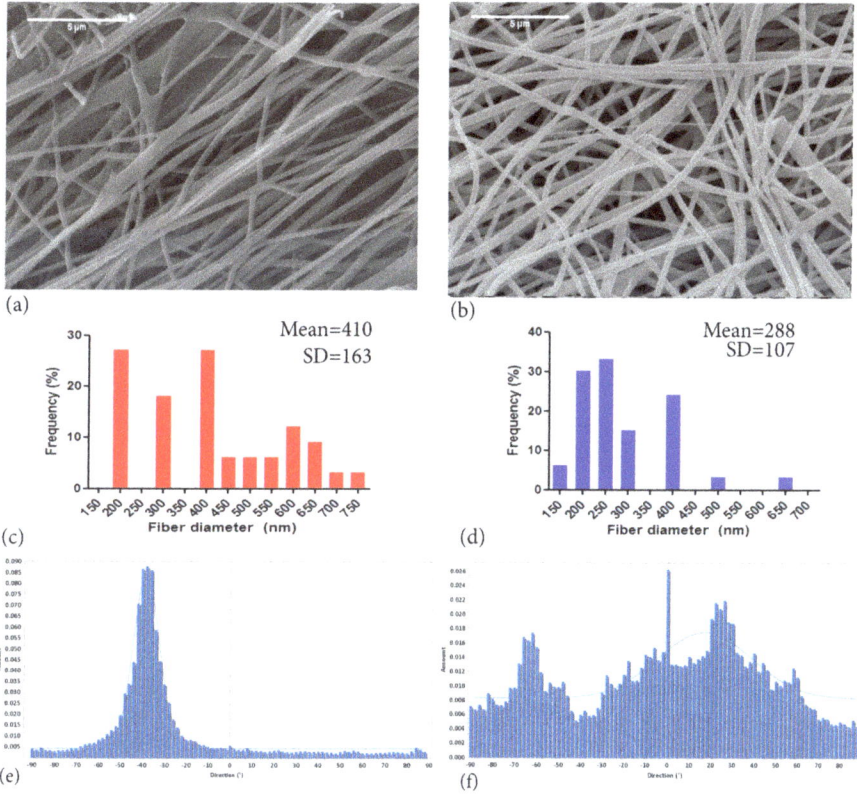

Figure 5. SEM images of electrospun CH-PEO blends; 4.5 wt.% CH:PEO 95:5, and 10:1, 3 wt.% acetic acid in Dimethyl Sulphoxide (DMSO), chitosan electrospun fibres spun using PEO as co-polymer. (**a**) Overly aligned fibres (**b**) random fibres (**c,d**) fibre distribution frequency calculated from 100 fibres (**e,f**) orientation histograms showing distribution of aligned and random fibres. Image adapted with permission from publisher (scale bar = 5 μm) [3].

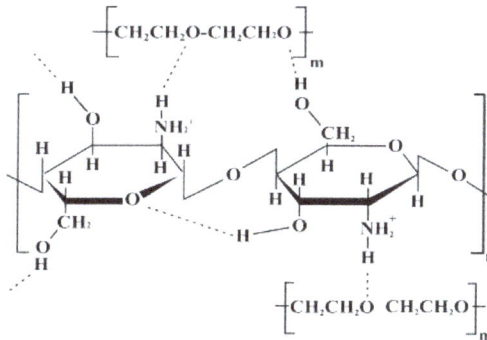

Figure 6. Proposed hydrogen bonding interactions of PEO and CH [125]; (Adapted with permission of publisher).

4. Tissue Engineering and Regenerative Applications of Chitosan-Based Solution Electrospun Fibres

4.1. Neural Tissue Regeneration

Amongst the available methods for scaffold fabrication, biomedical engineers have utilized electrospinning to aid nerve regeneration by synthesizing nerve guidance conduits or other non-porous templates [136]. The electrospun CH scaffolds have also been studied for their neural regenerative potential. Prabhakaran et al. have shown that rats Schwann cells cultured on PCL/CH fibres exhibit significantly higher biocompatibility compared to PCL fibres [137]. Another exciting prospect in neural tissue regeneration is the possibility of constructing electrospun fibrous nanotubes. Electrospun fibrous collagen/CH/thermoplastic polyurethane nanotubes have shown promising results for cultured Schwann cells [115]. Similarly, in vivo studies conducted on sciatic nerve defects in rats suggested that composite nanotubes consisting of electrospun CH/PVA fibres could function as scaffolds [138]. Additionally, the CH/PVA have superior mechanical properties compared to PVA scaffolds [139]. Generally, electrospun CH fibres having aligned morphology induce higher Schwann cell proliferation compared to random fibres [138].

Although using CH composite scaffolds for neural tissue regeneration seems promising, little is known about their long-term in vivo inflammatory effects. Recently, an in vivo study conducted on multi-layered 3D CH fibres enclosed by a PCL shell has exhibited extensive foreign body reactions while implanted in nerve defects [140]. Hence, more studies are pertinent to develop scaffolds that are considered safe for use in human subjects before they are employed in surgical practice.

4.2. Bone Regeneration

Perhaps CH fibres have been most extensively studied as scaffolds for bone regeneration. A typical periodontal defect involves irreversible resorption of alveolar bone. Electrospun CH/PEO scaffolds not only exhibit higher biocompatibility than cast CH membranes, but also possess superior mechanical properties (Table 2) [141].

Table 2. Studies conducted on CH and PEO reporting the orientation, mechanical properties, fibre diameters and aiming at clinical tissue engineering (TE) applications.

Application	Solution (Ratio, %)	Fibre Diameter	Young's Modulus	Orientation	References
Wound dressing	CH:PEO: 0.5 M ACa Triton X or DMSO, 60/40 90/10	few micron down to 40 nm	N/A	Aligned/random	Bhattarai et al. [127]
	4–6 wt.% CH:PEO (2:1, 1:1)	80 to 180 nm	N/A	Random	Duan et al. [26]
	HA/CH (30:70, w/w) 3 wt.% ACa:DMSO 10:1, 15 wt.% Col, 15 wt.% PEO	190 to 230 nm	N/A	Random	Xie et al. [142] (Figure 7)
	7 wt.% CH: TFA: nHA (0.8%, 1%, 2%)	227 nm ± 154 nm 335 nm ± 119 nm (after CRX Genipin)	142 Mpa ± 13 MPa	Random	Frohbergh et al. [143] (Figure 8)
	CH:PEO (3 wt.% ACa, DMSO, 10:1 UHMWPEO, 5%, 10%, 20%)	114 nm ± 19 nm 138 nm ± 15 nm 102 nm ± 1 nm	N/A	Aligned, Random	Zhang et al. [126]
Skin TE	CH grafted PCL, 25 wt.% PCL (DMF, CLF)	423 to 575 nm	N/A	Random	Chen et al. [144]
	CH/PCL/GEL	890 nm ± 364 nm	N/A	Random	Gomes et al. [145]
	CH/PEO/Henna extract (3/4 wt.%)	89 to 64 nm		Random	Yousefi et al. [146]
	5 wt.% CH: TFA, 10 wt.% PCL (40:60, CH:PCL)	175.82 55.95 (A) 215.79 nm ± 44.2 nm	51.54 MPa (A) 8.85 MPA (R)	Aligned/Random	Cooper et al. [147]
Nerve TE	CH:PEO, 4 wt.% in 50 wt.% ACa (50:50, 70:30, 80:20, 90:10)	60–120 nm	N/A	Random	Pakarwan et al. [41]
	CH:PEO, 1.6% (50 to 90% ACa)	10–240 nm	N/a	Random	Kriegel et al. [33,48]
	CH:PEO, 90% ACa	80 nm ± 35 nm	N/A	Random	Desai et al. [149]
	Ag: 5 wt.% CH:PEO 2 wt.% ACa	100 nm (Ag:CH:PEO) 5 nm (CH:PEO)	(YM) 59.2 ± 22.9 (CH:PEO) 322 ± 36.2 (CH:PEO:Ag)	Random	An et al. [150]
Cartilage tissue regeneration	10 mL of 1% CH sol with x mL 5% PEO	NA	2.25 MPa (YM)	Aligned	Subramanian et al. [139]
	CH (PEO):PCL: HAp (15 wt.%)	200 nm	215 MPa (YM)	Aligned & Random	Wu et al. [151]
Periodontal tissue regeneration	CH:PEO (95:5)	410 nm (A) 288 nm (R)	(YM) 357 ± 136 (A) 259 ± 192 (R)	Random & Aligned	Qasim et al. [5]

(N/A) Not applicable, (TE) Tissue engineering, (YM) Young's Modulus, (A) Aligned, (R) Random, (CRX) Cross-linking.

Bioactive ceramics such as HAp can also be incorporated in CH solutions prior to ES to produce bioactive scaffolds capable of accelerating osteoblast proliferation and bone formation (Figure 7 [152,153]). Various combinations of CH with natural polymers such as silk fibroin, collagen and chitin have also been found to induce accelerated proliferation of osteoblasts and mesenchymal cells in vitro [99,153,154]. Recently, in vivo studies conducted on CC fibrous membranes implanted in bone defects in rabbits have exhibited similar efficacy to commercially available collagen-guided tissue regeneration (GTR) membranes [117].

As discussed earlier, the mechanical properties and controlled degradation of GTR membranes have been a concern that can be overcome by adding certain non-toxic cross-linking agents such as genipin (Figure 8). Genipin is a natural cross-linking agent that can be used to reinforce CH and extend the degradation period up to 4–6 months as needed for complete bone healing [155]. Mechanical testing has revealed improved ultimate tensile strength (32 MPa) of such scaffolds that is substantially higher than currently available GTR membranes [156,157].

Recent research has focused on developing CH-based GTR scaffolds that can concurrently be used for bone regeneration and drug delivery to the implantation site. Ferrand et al. reported the possibility to immobilize bone morphogenic protein-2 (BMP-2) on electrospun PCL/CH fibrous scaffolds and enhancing the bone regeneration in vivo [158]. More recently, BMP-7 immobilized on PCL-CH fibres has shown superior osteogenic potential compared to fibres without any growth factors when human mesenchymal stem cells (hMSCs) were cultured [159]. Incorporation of growth factors into CH scaffolds have made it possible to 'kick-start' bone regeneration rather than function solely as barrier membranes. Coupled with the inherent osteogenic potential of CH, such scaffolds are likely to offer an excellent alternative to conventional GTR membranes.

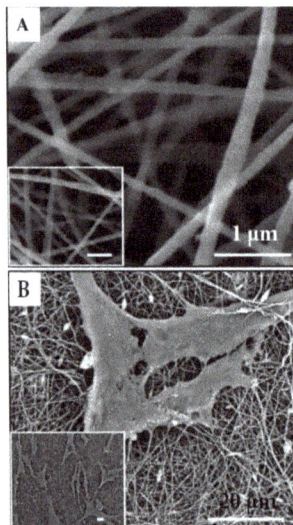

Figure 7. (**A**) SEM micrographs of nanohydroxyapatite/collagen/chitosan fibres; scale bar: 1 μm (**B**) Induced pluripotent stem-cell-derived mesenchymal stem cells (iPSC-MSCs) seeded on HA/chitosan fibres after culturing for 4 days, scale bar: 20 μm (Xie et al., 2016) [142]; (Adapted with permission of publisher).

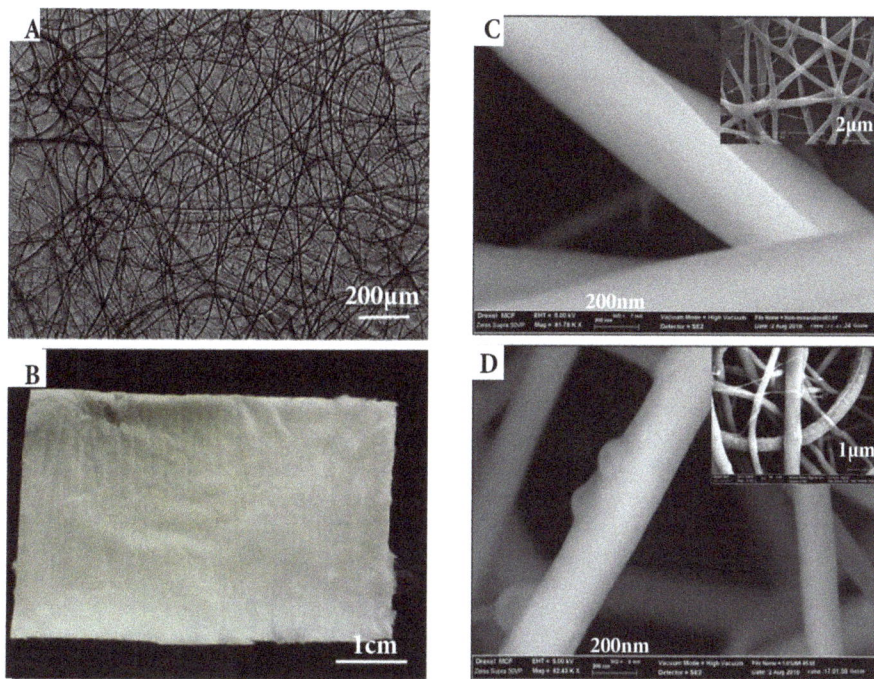

Figure 8. (**A**) Macroscopic image of chitosan fibre and (**B**) fibrous mat; (**C**) Morphology of fibre evaluated by SEM and atomic force microscopy of 0.1% genipin crosslinked and 1% HA loaded; (**D**) 7% chitosan fibres, typical morphology seen inset images [143]; (Adapted with permission of publisher).

4.3. Drug Delivery

Although CH is primarily used as quaternized form to deliver drugs to the implantation sites, the use of fibrous CH scaffolds as delivery media for various drugs has also been reported [160–162]. For instance, electrospun PCL/CH fibres can be used to deliver growth factors for bone regeneration [158,159]. CH fibrous mats impregnated with heparin-bound fibroblast growth factor-2 (FGF-2) stimulated cellular activities of sheep mesenchymal cells indicating a possible mechanism to deliver drugs [163]. Gentamicin immobilized on liposome can be released from CH fibres and has exhibited antimicrobial activity for up to 24 h against *Escherichia coli*, *Pseudomonas aeruginosa* and *Staphylococcus aureus* [164] indicating its potential for wound healing applications. Carbon-based drug carriers such as nano graphene-oxide have also been electrospun along with CH-PEO to produce scaffolds that can release doxorubicin. The primary amino group of CH facilitates cross-linking and ligand attachment for targeted drug delivery. Nanoparticles are negatively charged, and CH is cationic hence promoting electrostatic interaction [132].

4.4. Wound Dressings

Considering the excellent porosity and drug-carrying ability of CH fibres, another major application is production of wound dressings [165]. CH can be electrospun along with synthetic and natural polymers such as PVA, silk fibroin and PLLA to produce dressings [100,165,166]. Antimicrobial enzyme lysozyme can be added to CH-PVA fibrous membranes to prevent wound infections [167]. In addition to drugs, nanoparticles can also be co-electrospun with CH. A dual layered membrane of electrospun CH and adipose-derived human extracellular membrane containing nano-titania (TiO$_2$) particles exhibits higher healing properties in rats [167]. Similarly, nano-silver particles incorporated into electrospun CH/PEO fibres exhibited antibacterial activity against *S. aureus* and *E. coli*, which are both organisms implicated in wound infections [168]. More recently, electrospun CH/arginine fibres exhibited faster wound healing and anti-bacterial properties [169]. Moreover, CH-PVA fibres containing mafenide acetate have shown antibacterial activity against *S. aureus* and *P. aeruginosa* [170].

4.5. Anti-Carious Mucoadhesive Mats

Recently, anti-carious mats constructed from electrospun CH fibres containing antimicrobial agents have been studied for anti-cariogenic potential. CH/thiolated CH mats blended with PVA can be used to deliver anti-caries agents such as *Garcinia mangostana* extract in form of mucoadhesive mats which can be used by patients who may be unable to administer conventional oral hygiene measures to prevent dental caries [171,172].

4.6. Other Applications

The diversity of electrospun CH fibres have led to their use as templates for hepatocyte, chondrogenic and myogenic differentiation. Feng et al. reported CH nano-fibre mesh liver TE applications and tested the biocompatibility using hepatocytes [173]. In another study by Noriega et al. CH nano-fibres were used for culturing chondrocytes. Reported results were suggestive that the matrix geometry was able to regulate and promote the retention of the chondrocyte genotype [174]. A number of investigations have been conducted to study cellular interactions and stem cell fate [147,175,176]. Newman et al. studied the effect of topography by synthesizing aligned and randomly oriented fibres on cell shape and cell differentiation towards osteogenic and myogenic lineages [177].

5. Conclusions and Future Aspects

The present review shows that there is a wealth of scientific evidence available in support of the favourable properties and biocompatibility of chitosan electrospun composite biomaterials for a range of TE and regenerative medicine applications. However, further research including in vivo studies are required to translate these materials from laboratory to clinical applications. Although investigators have been able to alter the instrumentation and solution parameters to mimic natural tissue structure and morphology, the continual process of reporting various possibilities needs further characterisation and clinical trials before their applications for treating medical diseases with predictability. Using electrospinning and augmenting this technique with additives, manufacturers can have further control of the final template. Moreover, clinicians and bioengineers, whilst working together, can solve unexplored regenerative therapies by harnessing the fibre diameter, size, morphology and orientations according to the desired clinical applications. Mimicking structural and functional aspects of natural tissues will have a significant impact on the future of electrospinning of these materials.

Author Contributions: Saad B. Qasim and Shariq Najeeb conceived and wrote the review article focusing on chitosan; Muhammad S. Zafar and Shehriar Husain contributed to the polymer composite part; Zohaib Khurshid and Altaf H. Shah contributed towards collecting the data. Ihtesham Ur Rehman critically reviewed the manuscript.

Conflicts of Interest: The authors declare no conflict of interest.

Abbreviations

CH	Chitosan
DMSO	Dimethyl Sulphoxide
Hap	Hydroxyapatite
ES	Electrospinning
nm	Nanometres
PLA	Poly lactic acid
PLGA	Poly lactide co-glycolide
PU	Polyurethane

Int. J. Mol. Sci. **2018**, *19*, 407

PS	Polystyrene
PVA	Poly vinyl alcohol
SF	Silk fibroin
TE	Tissue Engineering
TFA	Triflouro acetic acid
PEO	Polyethylene oxide
TPP	Tripolyphosphate
CC	Collagen chitosan

References

1. Subbiah, T.; Bhat, G.S.; Tock, R.W.; Parameswaran, S.; Ramkumar, S.S. Electrospinning of nanofibers. *J. Appl. Polym. Sci.* **2005**, *96*, 557–569. [CrossRef]
2. Saito, M.; Tsuji, T. Extracellular matrix administration as a potential therapeutic strategy for periodontal ligament regeneration. *Expert Opin. Biol. Ther.* **2012**, *12*, 299–309. [CrossRef] [PubMed]
3. Qasim, S.B.; Najeeb, S.; Delaine-Smith, R.M.; Rawlinson, A.; Ur Rehman, I. Potential of electrospun chitosan fibers as a surface layer in functionally graded GTR membrane for periodontal regeneration. *Dent. Mater.* **2016**, *33*, 71–83. [CrossRef] [PubMed]
4. Sell, S.A.; Wolfe, P.S.; Garg, K.; McCool, J.M.; Rodriguez, I.A.; Bowlin, G.L. The use of natural polymers in tissue engineering: A focus on electrospun extracellular matrix analogues. *Polymers* **2010**, *2*, 522–553. [CrossRef]
5. Bhardwaj, N.; Kundu, S.C. Electrospinning: A fascinating fiber fabrication technique. *Biotechnol. Adv.* **2010**, *28*, 325–347. [CrossRef] [PubMed]
6. Ramakrishna, S.; Fujihara, K.; Teo, W.-E.; Yong, T.; Ma, Z.; Ramaseshan, R. Electrospun nanofibers: Solving global issues. *Mater. Today* **2006**, *9*, 40–50. [CrossRef]
7. Zafar, M.; Najeeb, S.; Khurshid, Z.; Vazirzadeh, M.; Zohaib, S.; Najeeb, B.; Sefat, F. Potential of electrospun nanofibers for biomedical and dental applications. *Materials* **2016**, *9*, 73. [CrossRef] [PubMed]
8. Sun, G.; Sun, L.; Xie, H.; Liu, J. Electrospinning of nanofibers for energy applications. *Nanomater* **2016**, *6*, 129. [CrossRef] [PubMed]
9. Gibson, P.; Schreuder-Gibson, H.; Rivin, D. Transport properties of porous membranes based on electrospun nanofibers. *Colloids Surf. A Physicochem. Eng. Aspects* **2001**, *187–188*, 469–481. [CrossRef]
10. White, J.; Foley, M.; Rowley, A. A Novel Approach to 3D-Printed Fabrics and Garments. *3D Print Addit. Manufac.* **2015**, *2*, 145–149. [CrossRef]
11. Villarreal-Gómez, L.J.; Cornejo-Bravo, J.M.; Vera-Graziano, R.; Grande, D. Electrospinning as a powerful technique for biomedical applications: A critically selected survey. *J. Biomater. Sci. Polym. Ed.* **2016**, *27*, 157–176. [CrossRef] [PubMed]
12. Anton, F.; Gastell, R.S. Process and Apparatus for Preparing Artificial Threads. U.S. Patent 1,975,504 A, 2 October 1934.
13. Seeram Ramakrishna, K.F.; Teo, W.; Lim, T.; Ma, Z. *An Introduction to Electrospinning and Nanofibers*; World Sci. Pub. Co.: Hackensack, NJ, USA, 2005.
14. Nascimento, M.L.; Araújo, E.S.; Cordeiro, E.R.; de Oliveira, A.H.; de Oliveira, H.P. A Literature Investigation about Electrospinning and Nanofibers: Historical. *Recent Pat. Nanotechnol.* **2015**, *9*, 76–85. [CrossRef] [PubMed]
15. Liang, D.; Hsiao, B.S.; Chu, B. Functional electrospun nanofibrous scaffolds for biomedical applications. *Adv. Drug Deliv. Rev.* **2007**, *59*, 1392–1412. [CrossRef] [PubMed]
16. Agarwal, S.; Wendorff, J.H.; Greiner, A. Use of electrospinning technique for biomedical applications. *Polymer* **2008**, *49*, 5603–5621. [CrossRef]
17. Khan, A.S. *A Novel Bioactive Nano-Composite: Synthesis and Characterisation with Potential Use as Dental Restorative Material*; Queen Mary University of London: London, UK, 2009.
18. Dalton, P.D.; Grafahrend, D.; Klinkhammer, K.; Klee, D.; Möller, M. Electrospinning of polymer melts: Phenomenological observations. *Polymer* **2007**, *48*, 6823–6833. [CrossRef]
19. Huang, Z.-M.; Zhang, Y.Z.; Kotaki, M.; Ramakrishna, S. A review on polymer nanofibers by electrospinning and their applications in nanocomposites. *Compos. Sci. Technol.* **2003**, *63*, 2223–2253. [CrossRef]

20. Baumgart, P.K. Electrostatic Spinning of Acrylic Microfibers. *J. Coll. Interface Sci.* **1971**, *36*, 71–79.

21. Haghi, A.K.; Akbari, M. Trends in electrospinning of natural nanofibers. *Phys. Status Solidi (A)* **2007**, *204*, 1830–1834. [CrossRef]

22. Beachley, V.; Wen, X. Effect of electrospinning parameters on the nanofiber diameter and length. *Mater. Sci. Eng. C Mater. Biol. Appl.* **2009**, *29*, 663–668. [CrossRef] [PubMed]

23. Fong, H.; Chun, I.; Reneker, D. Beaded nanofibers formed during electrospinning. *Polymer* **1999**, *40*, 4585–4592. [CrossRef]

24. Yuan, X.Y.; Zhang, Y.Y.; Dong, C.H.; Sheng, J. Morphology of ultrafine polysulfone fibers prepared by electrospinning. *Polym. Int.* **2004**, *53*, 1704–1710. [CrossRef]

25. Wang, X.F.; Um, I.C.; Fang, D.F.; Okamoto, A.; Hsiao, B.S.; Chu, B. Formation of water-resistant hyaluronic acid nanofibers by blowing-assisted electro-spinning and non-toxic post treatments. *Polymer* **2005**, *46*, 4853–4867. [CrossRef]

26. Duan, B.; Dong, C.H.; Yuan, X.Y.; Yao, K.D. Electrospinning of chitosan solutions in acetic acid with poly(ethylene oxide). *J. Biomater. Sci. Polym. Ed.* **2004**, *15*, 797–811. [CrossRef] [PubMed]

27. Leach, M.K.; Feng, Z.Q.; Tuck, S.J.; Corey, J.M. Electrospinning fundamentals: Optimizing solution and apparatus parameters. *J. Vis. Exp.* **2011**. [CrossRef] [PubMed]

28. Meng, Z.X.; Wang, Y.S.; Ma, C.; Zheng, W.; Li, L.; Zheng, Y.F. Electrospinning of PLGA/gelatin randomly-oriented and aligned nanofibers as potential scaffold in tissue engineering. *Mater. Sci. Eng. C Mater. Biol. Appl.* **2010**, *30*, 1204–1210. [CrossRef]

29. Sundaray, B.; Subramanian, V.; Natarajan, T.S.; Xiang, R.Z.; Chang, C.C.; Fann, W.S. Electrospinning of continuous aligned polymer fibers. *Appl. Phys. Lett.* **2004**, *84*, 1222–1224. [CrossRef]

30. Xu, C.Y.; Inai, R.; Kotaki, M.; Ramakrishna, S. Aligned biodegradable nanotibrous structure: A potential scaffold for blood vessel engineering. *Biomaterials* **2004**, *25*, 877–886. [CrossRef]

31. Buchko, C.J.; Chen, L.C.; Shen, Y.; Martin, D.C. Processing and microstructural characterization of porous biocompatible protein polymer thin films. *Polymer* **1999**, *40*, 7397–7407. [CrossRef]

32. Liu, H.Q.; Hsieh, Y.L. Ultrafine fibrous cellulose membranes from electrospinning of cellulose acetate. *J. Polym. Sci. Part B Polym. Phys.* **2002**, *40*, 2119–2129. [CrossRef]

33. Sukigara, S.; Gandhi, M.; Ayutsede, J.; Micklus, M.; Ko, F. Regeneration of *Bombyx mori* silk by electrospinning—Part 1: Processing parameters and geometric properties. *Polymer* **2003**, *44*, 5721–5727. [CrossRef]

34. Jankovic, B.; Pelipenko, J.; Skarabot, M.; Musevic, I.; Kristl, J. The design trend in tissue-engineering scaffolds based on nanomechanical properties of individual electrospun nanofibers. *Int. J. Pharma* **2013**, *455*, 338–347. [CrossRef] [PubMed]

35. Al-Kattan, A.; Nirwan, V.P.; Munnier, E.; Chourpa, I.; Fahmi, A.; Kabashin, A.V. Toward multifunctional hybrid platforms for tissue engineering based on chitosan (PEO) nanofibers functionalized by bare laser-synthesized Au and Si nanoparticles. *RSC Adv.* **2017**, *7*, 31759–31766. [CrossRef]

36. Vasita, R.; Katti, D.S. Nanofibers and their applications in tissue engineering. *Int. J. Nanomed.* **2006**, *1*, 15–30. [CrossRef]

37. Qasim, S.B.; Rehman, I.U. Application of Nanomaterials in Dentistry. In *Micro and Nanomanufacturing Volume II*; Springer: Berlin, Germany, 2018; Volume 2, pp. 319–336.

38. Liverani, L.; Abbruzzese, F.; Mozetic, P.; Basoli, F.; Rainer, A.; Trombetta, M. Electrospinning of hydroxyapatite–chitosan nanofibers for tissue engineering applications. *Asia-Pac. J. Chem. Eng.* **2014**, *9*, 407–414. [CrossRef]

39. Zhang, Y.; Venugopal, J.R.; El-Turki, A.; Ramakrishna, S.; Su, B.; Lim, C.T. Electrospun biomimetic nanocomposite nanofibers of hydroxyapatite/chitosan for bone tissue engineering. *Biomaterials* **2008**, *29*, 4314–4322. [CrossRef] [PubMed]

40. Thein-Han, W.W.; Kitiyanant, Y. Chitosan scaffolds for in vitro buffalo embryonic stem-like cell culture: An approach to tissue engineering. *J. Biomed. Mater. Res. Part B Appl. Biomater.* **2007**, *80B*, 92–101. [CrossRef] [PubMed]

41. Thein-Han, W.W.; Misra, R.D.K. Biomimetic chitosan–nanohydroxyapatite composite scaffolds for bone tissue engineering. *Acta Biomater.* **2009**, *5*, 1182–1197. [CrossRef] [PubMed]

42. Croisier, F.; Jerome, C. Chitosan-based biomaterials for tissue engineering. *Eur. Polym. J.* **2013**, *49*, 780–792. [CrossRef]

43. Kumar, M.N.R. A review of chitin and chitosan applications. *React. Funct. Polym.* **2000**, *46*, 1–27. [CrossRef]

44. Chen, Z.; Wei, B.; Mo, X.; Lim, C.; Ramakrishna, S.; Cui, F. Mechanical properties of electrospun collagen–chitosan complex single fibers and membrane. *Mater. Sci. Eng. C* **2009**, *29*, 2428–2435. [CrossRef]

45. Nwe, N.; Furuike, T.; Tamura, H. Production, Properties and Applications of Fungal Cell Wall Polysaccharides: Chitosan and Glucan. In *Chitosan for Biomaterials II*; Jayakumar, R., Prabaharan, M., Muzzarelli, R.A.A., Eds.; Springer: Berlin/Heidelberg, Germany, 2011; pp. 187–207.

46. Muzzarelli, R.A. New Techniques for Optimization of Surface Area and Porosity in Nanochitins and Nanochitosans. In *Chitosan for Biomaterials II*; Jayakumar, R., Prabaharan, M., Muzzarelli, R.A.A., Eds.; Springer: Berlin/Heidelberg, Germany, 2011; Volume 244, pp. 167–186.

47. Jayakumar, R.; Prabaharan, M.; Nair, S.V.; Tamura, H. Novel chitin and chitosan nanofibers in biomedical applications. *Biotechnol. Adv.* **2010**, *28*, 142–150. [CrossRef] [PubMed]

48. McKee, M.G.; Hunley, M.T.; Layman, J.M.; Long, T.E. Solution Rheological Behavior and Electrospinning of Cationic Polyelectrolytes. *Macromolecules* **2005**, *39*, 575–583. [CrossRef]

49. Duck Weon Lee, H.L.; Chong, H.N.; Shim, W.S. Advances in Chitosan Material and its Hybrid Derivatives: A Review. *Open Biomater. J.* **2009**, *1*, 10–20.

50. Ohkawa, K.; Cha, D.; Kim, H.; Nishida, A.; Yamamoto, H. Electrospinning of Chitosan. *Macromol. Rap. Commun.* **2004**, *25*, 1600–1605. [CrossRef]

51. Qasim, S.B.; Delaine-Smith, R.M.; Fey, T.; Rawlinson, A.; Rehman, I.U. Freeze gelated porous membranes for periodontal tissue regeneration. *Acta Bimater.* **2015**, *23*, 317–328. [CrossRef] [PubMed]

52. Qasim, S.B.; Husain, S.; Huang, Y.; Pogorielov, M.; Deineka, V.; Lyndin, M.; Rawlinson, A.; Rehman, I.U. In-vitro and in-vivo degradation studies of freeze gelated porous chitosan composite scaffolds for tissue engineering applications. *Polym. Degrad. Stab.* **2016**, *136*, 31–38. [CrossRef]

53. Sun, K.; Li, Z. Preparations, properties and applications of chitosan based nanofibers fabricated by electrospinning. *Express Polym. Lett.* **2011**, *5*, 342–361. [CrossRef]

54. Toffey, A.; Glasser, W.G. Chitin derivatives. II. Time–temperature–transformation cure diagrams of the chitosan amidization process. *J. Appl. Polym. Sci.* **1999**, *73*, 1879–1889. [CrossRef]

55. Qu, X.; Wirsén, A.; Albertsson, A.-C. Structural change and swelling mechanism of pH-sensitive hydrogels based on chitosan and D,L-lactic acid. *J. Appl. Polym. Sci.* **1999**, *74*, 3186–3192. [CrossRef]

56. Yao, F.; Chen, W.; Wang, H.; Liu, H.; Yao, K.; Sun, P.; Lin, H. A study on cytocompatible poly (chitosan-g-L-lactic acid). *Polymer* **2003**, *44*, 6435–6441. [CrossRef]

57. Zhou, J.; Cao, C.; Ma, X.; Lin, J. Electrospinning of silk fibroin and collagen for vascular tissue engineering. *Int. J. Biol. Macromol.* **2010**, *47*, 514–519. [CrossRef] [PubMed]

58. Wang, G.; Hu, X.; Lin, W.; Dong, C.; Wu, H. Electrospun PLGA-silk fibroin-collagen nanofibrous scaffolds for nerve tissue engineering. *In Vitro Cell. Dev. Biol. Anim.* **2011**, *47*, 234–240. [CrossRef] [PubMed]

59. Klossner, R.R.; Queen, H.A.; Coughlin, A.J.; Krause, W.E. Correlation of Chitosan's Rheological Properties and Its Ability to Electrospin. *Biomacromolecules* **2008**, *9*, 2947–2953. [CrossRef] [PubMed]

60. Zuo, B.; Dai, L.; Wu, Z. Analysis of structure and properties of biodegradable regenerated silk fibroin fibers. *J. Mater. Sci.* **2006**, *41*, 3357–3361. [CrossRef]

61. Xu, Y.; Shao, H.; Zhang, Y.; Hu, X. Studies on spinning and rheological behaviors of regenerated silk fibroin/N-methylmorpholine-N-oxide· H₂O solutions. *J. Mater. Sci.* **2005**, *40*, 5355–5358. [CrossRef]

62. Gosline, J.; Guerette, P.; Ortlepp, C.; Savage, K. The mechanical design of spider silks: From fibroin sequence to mechanical function. *J. Exp. Biol.* **1999**, *202*, 3295–3303. [PubMed]

63. Sheu, H.-S.; Phyu, K.W.; Jean, Y.-C.; Chiang, Y.-P.; Tso, I.-M.; Wu, H.-C.; Yang, J.-C.; Ferng, S.-L. Lattice deformation and thermal stability of crystals in spider silk. *Int. J. Biol. Macromol.* **2004**, *34*, 267–273. [CrossRef] [PubMed]

64. Furuzono, T.; Ishihara, K.; Nakabayashi, N.; Tamada, Y. Chemical modification of silk fibroin with 2-methacryloyloxyethyl phosphorylcholine. II. Graft-polymerization onto fabric through 2-methacryloyloxyethyl isocyanate and interaction between fabric and platelets. *Biomaterials* **2000**, *21*, 327–333. [CrossRef]

65. Gellynck, K.; Verdonk, P.C.; Van Nimmen, E.; Almqvist, K.F.; Gheysens, T.; Schoukens, G.; Van Langenhove, L.; Kiekens, P.; Mertens, J.; Verbruggen, G. Silkworm and spider silk scaffolds for chondrocyte support. *J. Mater. Sci. Mater. Med.* **2008**, *19*, 3399–3409. [CrossRef] [PubMed]

66. Nair, L.S.; Bhattacharyya, S.; Laurencin, C.T. Development of novel tissue engineering scaffolds via electrospinning. *Expert Opin. Biol. Ther.* **2004**, *4*, 659–668. [CrossRef] [PubMed]
67. Wenk, E.; Wandrey, A.J.; Merkle, H.P.; Meinel, L. Silk fibroin spheres as a platform for controlled drug delivery. *J. Control. Release* **2008**, *132*, 26–34. [CrossRef] [PubMed]
68. Li, W.-J.; Mauck, R.L.; Tuan, R.S. Electrospun nanofibrous scaffolds: Production, characterization, and applications for tissue engineering and drug delivery. *J. Biomed. Nanotechnol.* **2005**, *1*, 259–275. [CrossRef]
69. Zafar, M.S.; Al-Samadani, K.H. Potential use of natural silk for bio-dental applications. *J. Taibah Univ. Med. Sci.* **2014**, *9*, 171–177. [CrossRef]
70. Kweon, H.; Lee, S.-W.; Hahn, B.-D.; Lee, Y.-C.; Kim, S.-G. Hydroxyapatite and Silk Combination-Coated Dental Implants Result in Superior Bone Formation in the Peri-Implant Area Compared with Hydroxyapatite and Collagen Combination-Coated Implants. *J. Oral Maxillofac. Surv.* **2014**, *72*, 1928–1936. [CrossRef] [PubMed]
71. Zafar, M.S.; Khurshid, Z.; Almas, K. Oral tissue engineering progress and challenges. *Tissue Eng. Regen. Med.* **2015**, *12*, 387–397. [CrossRef]
72. Maniglio, D.; Bonani, W.; Bortoluzzi, G.; Servoli, E.; Motta, A.; Migliaresi, C. Electrodeposition of silk fibroin on metal substrates. *J. Bioact. Compat. Polym.* **2010**, *25*, 441–454. [CrossRef]
73. Zarkoob, S.; Eby, R.; Reneker, D.H.; Hudson, S.D.; Ertley, D.; Adams, W.W. Structure and morphology of electrospun silk nanofibers. *Polymer* **2004**, *45*, 3973–3977. [CrossRef]
74. Hardy, J.G.; Römer, L.M.; Scheibel, T.R. Polymeric materials based on silk proteins. *Polymer* **2008**, *49*, 4309–4327. [CrossRef]
75. Kodama, K. The preparation and physico-chemical properties of sericin. *Biochem. J.* **1926**, *20*, 1208. [CrossRef] [PubMed]
76. Somashekarappa, H.; Annadurai, V.; Subramanya, G.; Somashekar, R. Structure–property relation in varieties of acid dye processed silk fibers. *Mater. Lett.* **2002**, *53*, 415–420. [CrossRef]
77. Kaplan, D.; Adams, W.W.; Farmer, B.; Viney, C. *Silk: Biology, Structure, Properties, and Genetics*; ACS Symposium Series: Washington, DC, USA, 1994.
78. Zhang, Y.-Q. Applications of natural silk protein sericin in biomaterials. *Biotechnol. Adv.* **2002**, *20*, 91–100. [CrossRef]
79. Dewair, M.; Baur, X.; Ziegler, K. Use of immunoblot technique for detection of human IgE and IgG antibodies to individual silk proteins. *J. Allergy Clin. Immunol.* **1985**, *76*, 537–542. [CrossRef]
80. Panilaitis, B.; Altman, G.H.; Chen, J.; Jin, H.-J.; Karageorgiou, V.; Kaplan, D.L. Macrophage responses to silk. *Biomaterials* **2003**, *24*, 3079–3085. [CrossRef]
81. Vaithanomsat, P.; Kitpreechavanich, V. Sericin separation from silk degumming wastewater. *Sep. Purif. Technol.* **2008**, *59*, 129–133. [CrossRef]
82. Yamada, H.; Nakao, H.; Takasu, Y.; Tsubouchi, K. Preparation of undegraded native molecular fibroin solution from silkworm cocoons. *Mater. Sci. Eng. C* **2001**, *14*, 41–46. [CrossRef]
83. Yamaguchi, K.; Kikuchi, Y.; Takagi, T.; Kikuchi, A.; Oyama, F.; Shimura, K.; Mizuno, S. Primary structure of the silk fibroin light chain determined by cDNA sequencing and peptide analysis. *J. Mol. Biol.* **1989**, *210*, 127–139. [CrossRef]
84. Zhou, C.Z.; Confalonieri, F.; Jacquet, M.; Perasso, R.; Li, Z.G.; Janin, J. Silk fibroin: Structural implications of a remarkable amino acid sequence. *Proteins Struct. Funct. Bioinform.* **2001**, *44*, 119–122. [CrossRef] [PubMed]
85. Zhang, Y.-Q.; Shen, W.-D.; Xiang, R.-L.; Zhuge, L.-J.; Gao, W.-J.; Wang, W.-B. Formation of silk fibroin nanoparticles in water-miscible organic solvent and their characterization. *J. Nanopart. Res.* **2007**, *9*, 885–900. [CrossRef]
86. Mita, K.; Ichimura, S.; James, T.C. Highly repetitive structure and its organization of the silk fibroin gene. *J. Mol. Evol.* **1994**, *38*, 583–592. [CrossRef] [PubMed]
87. Zhang, Y.; Lim, C.T.; Ramakrishna, S.; Huang, Z.-M. Recent development of polymer nanofibers for biomedical and biotechnological applications. *J. Mater. Sci. Mater. Med.* **2005**, *16*, 933–946. [CrossRef] [PubMed]
88. Zafar, M.S.; Belton, D.J.; Hanby, B.; Kaplan, D.L.; Perry, C.C. Functional Material Features of *Bombyx mori* Silk Light versus Heavy Chain Proteins. *Biomacromolecules* **2015**, *16*, 606–614. [CrossRef] [PubMed]
89. Inoue, S.; Tanaka, K.; Arisaka, F.; Kimura, S.; Ohtomo, K.; Mizuno, S. Silk fibroin of *Bombyx mori* is secreted, assembling a high molecular mass elementary unit consisting of H-chain, L-chain, and P25, with a 6:6:1 molar ratio. *J. Biol. Chem.* **2000**, *275*, 40517–40528. [CrossRef] [PubMed]

90. Tanaka, K.; Inoue, S.; Mizuno, S. Hydrophobic interaction of P25, containing Asn-linked oligosaccharide chains, with the HL complex of silk fibroin produced by *Bombyx mori*. *Insect Biochem. Mol. Biol.* **1999**, *29*, 269–276. [CrossRef]

91. Khurshid, Z.; Zafar, M.; Qasim, S.; Shahab, S.; Naseem, M.; AbuReqaiba, A. Advances in Nanotechnology for Restorative Dentistry. *Materials* **2015**, *8*, 717–731. [CrossRef] [PubMed]

92. Zhou, Y.; Yang, H.; Liu, X.; Mao, J.; Gu, S.; Xu, W. Electrospinning of carboxyethyl chitosan/poly (vinyl alcohol)/silk fibroin nanoparticles for wound dressings. *Int. J. Biol. Macromol.* **2013**, *53*, 88–92. [CrossRef] [PubMed]

93. Park, W.H.; Jeong, L.; Yoo, D.I.; Hudson, S. Effect of chitosan on morphology and conformation of electrospun silk fibroin nanofibers. *Polymer* **2004**, *45*, 7151–7157. [CrossRef]

94. Yoo, C.R.; Yeo, I.-S.; Park, K.E.; Park, J.H.; Lee, S.J.; Park, W.H.; Min, B.-M. Effect of chitin/silk fibroin nanofibrous bicomponent structures on interaction with human epidermal keratinocytes. *Int. J. Biol. Macromol.* **2008**, *42*, 324–334. [CrossRef] [PubMed]

95. Um, I.C.; Kweon, H.Y.; Lee, K.G.; Park, Y.H. The role of formic acid in solution stability and crystallization of silk protein polymer. *Int. J. Biol. Macromol.* **2003**, *33*, 203–213. [CrossRef] [PubMed]

96. Chen, X.; Zhou, L.; Shao, Z.-Z.; Zhou, P.; Knight, D.P.; Vollrath, F. Conformation Transition of Silk Protein Membranes Monitored by Time-resolved FT-IR Spectroscopy-Conformation Transition Behavior of Regenerated Silk Fibroin Membranes in Alcohol Solution at High Concentration. *Acta Chim. Sin. Chin. Ed.* **2003**, *61*, 625–629.

97. Wang, H.; Zhang, Y.; Shao, H.; Hu, X. A study on the flow stability of regenerated silk fibroin aqueous solution. *Int. J. Biol. Macromol.* **2005**, *36*, 66–70. [CrossRef] [PubMed]

98. Park, K.E.; Jung, S.Y.; Lee, S.J.; Min, B.-M.; Park, W.H. Biomimetic nanofibrous scaffolds: Preparation and characterization of chitin/silk fibroin blend nanofibers. *Int. J. Biol. Macromol.* **2006**, *38*, 165–173. [CrossRef] [PubMed]

99. Chen, J.-P.; Chen, S.-H.; Lai, G.-J. Preparation and characterization of biomimetic silk fibroin/chitosan composite nanofibers by electrospinning for osteoblasts culture. *Nanoscale Res. Lett.* **2012**, *7*, 1–11. [CrossRef] [PubMed]

100. Cai, Z.X.; Mo, X.M.; Zhang, K.H.; Fan, L.P.; Yin, A.L.; He, C.L.; Wang, H.S. Fabrication of Chitosan/Silk Fibroin Composite Nanofibers for Wound-dressing Applications. *Int. J. Mol. Sci.* **2010**, *11*, 3529–3539. [CrossRef] [PubMed]

101. Chen, Z.; Wang, P.; Wei, B.; Mo, X.; Cui, F. Electrospun collagen–chitosan nanofiber: A biomimetic extracellular matrix for endothelial cell and smooth muscle cell. *Acta Bimater.* **2010**, *6*, 372–382. [CrossRef] [PubMed]

102. Helary, C.; Bataille, I.; Abed, A.; Illoul, C.; Anglo, A.; Louedec, L.; Letourneur, D.; Meddahi-Pelle, A.; Giraud-Guille, M.M. Concentrated collagen hydrogels as dermal substitutes. *Biomaterials* **2010**, *31*, 481–490. [CrossRef] [PubMed]

103. Yeo, I.S.; Oh, J.E.; Jeong, L.; Lee, T.S.; Lee, S.J.; Park, W.H.; Min, B.M. Collagen-based biomimetic nanofibrous scaffolds: Preparation and characterization of collagen/silk fibroin bicomponent nanofibrous structures. *Biomacromolecules* **2008**, *9*, 1106–1116. [CrossRef] [PubMed]

104. Mahjour, S.; Sefat, F.; Polunin, Y.; Wang, L.; Wang, H. *Modulation of Fiber Organization within Electrospun Polycaprolactone/Collagen Nanofiber Matrices to Facilitate the Integration of Nanofiber/Cell Multilayered Constructs*; Tissue Engineering Part A, 2015; Mary Ann Liebert, Inc.: New Rochelle, NY, USA, 2015; p. S264.

105. Huang, L.; Nagapudi, K.; Apkarian, R.; Chaikof, E.L. Engineered collagen–PEO nanofibers and fabrics. *J. Biomater. Sci. Polym. Ed.* **2001**, *12*, 979–993. [CrossRef] [PubMed]

106. Kim, N.R.; Lee, D.H.; Chung, P.-H.; Yang, H.-C. Distinct differentiation properties of human dental pulp cells on collagen, gelatin, and chitosan scaffolds. *Oral Surg. Oral Med. Oral Pathol. Oral Radiol. Endodontol.* **2009**, *108*, e94–e100. [CrossRef] [PubMed]

107. Chen, Z.; Mo, X.; Qing, F. Electrospinning of collagen–chitosan complex. *Mater. Lett.* **2007**, *61*, 3490–3494. [CrossRef]

108. Sionkowska, A.; Wisniewski, M.; Skopinska, J.; Kennedy, C.; Wess, T. Molecular interactions in collagen and chitosan blends. *Biomaterials* **2004**, *25*, 795–801. [CrossRef]

109. Taravel, M.; Domard, A. Collagen and its interactions with chitosan: III. Some biological and mechanical properties. *Biomaterials* **1996**, *17*, 451–455. [CrossRef]

110. Taravel, M.; Domard, A. Collagen and its interaction with chitosan: II. Influence of the physicochemical characteristics of collagen. *Biomaterials* **1995**, *16*, 865–871. [CrossRef]

111. Zeng, J.; Chen, X.; Xu, X.; Liang, Q.; Bian, X.; Yang, L.; Jing, X. Ultrafine fibers electrospun from biodegradable polymers. *J. Appl. Polym. Sci.* **2003**, *89*, 1085–1092. [CrossRef]

112. Tan, W.; Krishnaraj, R.; Desai, T.A. Evaluation of nanostructured composite collagen-chitosan matrices for tissue engineering. *Tissue Eng.* **2001**, *7*, 203–210. [CrossRef] [PubMed]

113. Peng, L.; Cheng, X.R.; Wang, J.W.; Xu, D.X.; Wang, G. Preparation and evaluation of porous chitosan/collagen scaffolds for periodontal tissue engineering. *J. Bioact. Compat. Polym.* **2006**, *21*, 207–220. [CrossRef]

114. Chen, J.-P.; Chang, G.-Y.; Chen, J.-K. Electrospun collagen/chitosan nanofibrous membrane as wound dressing. *Colloids Surf. A Physicochem. Eng. Aspects* **2008**, *313*, 183–188. [CrossRef]

115. Huang, C.; Chen, R.; Ke, Q.; Morsi, Y.; Zhang, K.; Mo, X. Electrospun collagen-chitosan-TPU nanofibrous scaffolds for tissue engineered tubular grafts. *Colloids Surf. B Biointerfaces* **2011**, *82*, 307–315. [CrossRef] [PubMed]

116. Li, W.; Long, Y.; Liu, Y.; Long, K.; Liu, S.; Wang, Z.; Wang, Y.; Ren, L. Fabrication and characterization of chitosan–collagen crosslinked membranes for corneal tissue engineering. *J. Biomater. Sci. Polym. Ed.* **2014**, *25*, 1962–1972. [CrossRef] [PubMed]

117. Lotfi, G.; Shokrgozar, M.A.; Mofid, R.; Abbas, F.M.; Ghanavati, F.; Baghban, A.A.; Yavari, S.K.; Pajoumshariati, S. Biological Evaluation (In Vitro and In Vivo) of Bilayered Collagenous Coated (Nano Electrospun and Solid Wall) Chitosan Membrane for Periodontal Guided Bone Regeneration. *Ann. Biomed. Eng.* **2016**, *44*, 2132–2144. [CrossRef] [PubMed]

118. Marinho-Soriano, E.; Bourret, E. Effects of season on the yield and quality of agar from Gracilaria species (Gracilariaceae, Rhodophyta). *Bioresour. Technol.* **2003**, *90*, 329–333. [CrossRef]

119. Cho, M.K.; Singu, B.S.; Na, Y.H.; Yoon, K.R. Fabrication and characterization of double-network agarose/polyacrylamide nanofibers by electrospinning. *J. Appl. Polym. Sci.* **2016**, *133*. [CrossRef]

120. Teng, S.-H.; Wang, P.; Kim, H.-E. Blend fibers of chitosan–agarose by electrospinning. *Mater. Lett.* **2009**, *63*, 2510–2512. [CrossRef]

121. Bao, X.; Hayashi, K.; Li, Y.; Teramoto, A.; Abe, K. Novel agarose and agar fibers: Fabrication and characterization. *Mater. Lett.* **2010**, *64*, 2435–2437. [CrossRef]

122. Kim, J.-M.; Kim, C.; Yoo, S.; Kim, J.-H.; Kim, J.-H.; Lim, J.-M.; Park, S.; Lee, S.-Y. Agarose-biofunctionalized, dual-electrospun heteronanofiber mats: Toward metal-ion chelating battery separator membranes. *J. Mater. Chem. A* **2015**, *3*, 10687–10692. [CrossRef]

123. Miguel, S.P.; Ribeiro, M.P.; Brancal, H.; Coutinho, P.; Correia, I.J. Thermoresponsive chitosan–agarose hydrogel for skin regeneration. *Carbohydr. Polym.* **2014**, *111*, 366–373. [CrossRef] [PubMed]

124. Dilamian, M.; Montazer, M.; Masoumi, J. Antimicrobial electrospun membranes of chitosan/poly(ethylene oxide) incorporating poly(hexamethylene biguanide) hydrochloride. *Carbohydr. Polym.* **2013**, *94*, 364–371. [CrossRef] [PubMed]

125. Pakravan, M.; Heuzey, M.C.; Ajji, A. A fundamental study of chitosan/PEO electrospinning. *Polymer* **2011**, *52*, 4813–4824. [CrossRef]

126. Zhang, Y.Z.; Su, B.; Ramakrishna, S.; Lim, C.T. Chitosan Nanofibers from an Easily Electrospinnable UHMWPEO-Doped Chitosan Solution System. *Biomacromolecules* **2008**, *9*, 136–141. [CrossRef] [PubMed]

127. Bhattarai, N.; Edmondson, D.; Veiseh, O.; Matsen, F.A.; Zhang, M.Q. Electrospun chitosan-based nanofibers and their cellular compatibility. *Biomaterials* **2005**, *26*, 6176–6184. [CrossRef] [PubMed]

128. Pakravan, M.; Heuzey, M.-C.; Ajji, A. Core-shell structured PEO-chitosan nanofibers by coaxial electrospinning. *Biomacromolecules* **2012**, *13*, 412–421. [CrossRef] [PubMed]

129. Ojha, S.S.; Stevens, D.R.; Hoffman, T.J.; Stano, K.; Klossner, R.; Scott, M.C.; Krause, W.; Clarke, L.I.; Gorga, R.E. Fabrication and characterization of electrospun chitosan nanofibers formed via templating with polyethylene oxide. *Biomacromolecules* **2008**, *9*, 2523–2529. [CrossRef] [PubMed]

130. Penchev, H.; Paneva, D.; Manolova, N.; Rashkov, I. Hybrid nanofibrous yarns based on N-carboxyethylchitosan and silver nanoparticles with antibacterial activity prepared by self-bundling electrospinning. *Carbohydr. Res.* **2010**, *345*, 2374–2380. [CrossRef] [PubMed]

131. Wang, Y.-Y.; Lü, L.-X.; Feng, Z.-Q.; Xiao, Z.-D.; Huang, N.-P. Cellular compatibility of RGD-modified chitosan nanofibers with aligned or random orientation. *Biomed. Mater.* **2010**, *5*, 054112. [CrossRef] [PubMed]

132. Ardeshirzadeh, B.; Anaraki, N.A.; Irani, M.; Rad, L.R.; Shamshiri, S. Controlled release of doxorubicin from electrospun PEO/chitosan/graphene oxide nanocomposite nanofibrous scaffolds. *Mater. Sci. Eng. C* **2015**, *48*, 384–390. [CrossRef] [PubMed]

133. Kriegel, C.; Kit, K.M.; McClements, D.J.; Weiss, J. Electrospinning of chitosan-poly(ethylene oxide) blend nanofibers in the presence of micellar surfactant solutions. *Polymer* **2009**, *50*, 189–200. [CrossRef]

134. Sarkar, S.D.; Farrugia, B.L.; Dargaville, T.R.; Dhara, S. Physico-chemical/biological properties of tripolyphosphate cross-linked chitosan based nanofibers. *Mater. Sci. Eng. C* **2013**, *33*, 1446–1454. [CrossRef] [PubMed]

135. Shu, X.Z.; Zhu, K.J. A novel approach to prepare tripolyphosphate/chitosan complex beads for controlled release drug delivery. *Int. J. Pharma* **2000**, *201*, 51–58. [CrossRef]

136. Schmidt, C.E.; Leach, J.B. Neural tissue engineering: Strategies for repair and regeneration. *Annu. Rev. Biomed. Eng.* **2003**, *5*, 293–347. [CrossRef] [PubMed]

137. Prabhakaran, M.P.; Venugopal, J.R.; Chyan, T.T.; Hai, L.B.; Chan, C.K.; Lim, A.Y.; Ramakrishna, S. Electrospun biocomposite nanofibrous scaffolds for neural tissue engineering. *Tissue Eng. Part A* **2008**, *14*, 1787–1797. [CrossRef] [PubMed]

138. Wang, W.; Itoh, S.; Konno, K.; Kikkawa, T.; Ichinose, S.; Sakai, K.; Ohkuma, T.; Watabe, K. Effects of Schwann cell alignment along the oriented electrospun chitosan nanofibers on nerve regeneration. *J. Biomed. Mater. Res. A* **2009**, *91*, 994–1005. [CrossRef] [PubMed]

139. Alhosseini, S.N.; Moztarzadeh, F.; Mozafari, M.; Asgari, S.; Dodel, M.; Samadikuchaksaraei, A.; Kargozar, S.; Jalali, N. Synthesis and characterization of electrospun polyvinyl alcohol nanofibrous scaffolds modified by blending with chitosan for neural tissue engineering. *Int. J. Nanomed.* **2012**, *7*, 25–34.

140. Duda, S.; Dreyer, L.; Behrens, P.; Wienecke, S.; Chakradeo, T.; Glasmacher, B.; Haastert-Talini, K. Outer Electrospun Polycaprolactone Shell Induces Massive Foreign Body Reaction and Impairs Axonal Regeneration through 3D Multichannel Chitosan Nerve Guides. *Biomed. Res. Int.* **2014**, *2014*, 835269. [CrossRef] [PubMed]

141. Subramanian, A.; Vu, D.; Larsen, G.F.; Lin, H.Y. Preparation and evaluation of the electrospun chitosan/PEO fibers for potential applications in cartilage tissue engineering. *J. Biomater. Sci. Polym. E* **2005**, *16*, 861–873. [CrossRef]

142. Xie, J.; Peng, C.; Zhao, Q.; Wang, X.; Yuan, H.; Yang, L.; Li, K.; Lou, X.; Zhang, Y. Osteogenic differentiation and bone regeneration of the iPSC-MSCs supported by a biomimetic nanofibrous scaffold. *Acta Bimater.* **2015**, *29*, 365–379. [CrossRef] [PubMed]

143. Frohbergh, M.E.; Katsman, A.; Botta, G.P.; Lazarovici, P.; Schauer, C.L.; Wegst, U.G.; Lelkes, P.I. Electrospun hydroxyapatite-containing chitosan nanofibers crosslinked with genipin for bone tissue engineering. *Biomaterials* **2012**, *33*, 9167–9178. [CrossRef] [PubMed]

144. Chen, H.; Huang, J.; Yu, J.; Liu, S.; Gu, P. Electrospun chitosan-graft-poly (ε-caprolactone)/poly (ε-caprolactone) cationic nanofibrous mats as potential scaffolds for skin tissue engineering. *Int. J. Biol. Macromol.* **2011**, *48*, 13–19. [CrossRef] [PubMed]

145. Gomes, S.; Rodrigues, G.; Martins, G.; Henriques, C.; Silva, J.C. Evaluation of nanofibrous scaffolds obtained from blends of chitosan, gelatin and polycaprolactone for skin tissue engineering. *Int. J. Biol. Macromol.* **2017**, *102*, 1174–1185. [CrossRef] [PubMed]

146. Yousefi, I.; Pakravan, M.; Rahimi, H.; Bahador, A.; Farshadzadeh, Z.; Haririan, I. An investigation of electrospun Henna leaves extract-loaded chitosan based nanofibrous mats for skin tissue engineering. *Mater. Sci. Eng. C* **2017**, *75*, 433–444. [CrossRef] [PubMed]

147. Cooper, A.; Jana, S.; Bhattarai, N.; Zhang, M. Aligned chitosan-based nanofibers for enhanced myogenesis. *J. Mater. Chem.* **2010**, *20*, 8904–8911. [CrossRef]

148. Kriegel, C.; Kit, K.M.; McClements, D.J.; Weiss, J. Influence of Surfactant Type and Concentration on Electrospinning of Chitosan-Poly(Ethylene Oxide) Blend Nanofibers. *Food Biophys.* **2009**, *4*, 213–228. [CrossRef]

149. Desai, K.; Kit, K.; Li, J.; Zivanovic, S. Morphological and surface properties of electrospun chitosan nanofibers. *Biomacromolecules* **2008**, *9*, 1000–1006. [CrossRef] [PubMed]

150. An, J.; Zhang, H.; Zhang, J.T.; Zhao, Y.H.; Yuan, X.Y. Preparation and antibacterial activity of electrospun chitosan/poly(ethylene oxide) membranes containing silver nanoparticles. *Colloid Polym. Sci.* **2009**, *287*, 1425–1434. [CrossRef]

151. Wu, G.; Deng, X.; Song, J.; Chen, F. Enhanced biological properties of biomimetic apatite fabricated polycaprolactone/chitosan nanofibrous bio-composite for tendon and ligament regeneration. *J. Photochem. Photobiol. B Biol.* **2018**, *178*, 27–32. [CrossRef] [PubMed]

152. Venugopal, J.; Prabhakaran, M.P.; Zhang, Y.Z.; Low, S.; Choon, A.T.; Ramakrishna, S. Biomimetic hydroxyapatite-containing composite nanofibrous substrates for bone tissue engineering. *Philos. Trans. R. Soc. A* **2010**, *368*, 2065–2081. [CrossRef] [PubMed]

153. Zhang, Y.Z.; Reddy, V.J.; Wong, S.Y.; Li, X.; Su, B.; Ramakrishna, S.; Lim, C.T. Enhanced Biomineralization in Osteoblasts on a Novel Electrospun Biocomposite Nanofibrous Substrate of Hydroxyapatite/Collagen/ Chitosan. *Tissue Eng. Part A* **2010**, *16*, 1949–1960. [CrossRef] [PubMed]

154. Lai, G.J.; Shalumon, K.T.; Chen, S.H.; Chen, J.P. Composite chitosan/silk fibroin nanofibers for modulation of osteogenic differentiation and proliferation of human mesenchymal stem cells. *Carbohydr. Polym.* **2014**, *111*, 288–297. [CrossRef] [PubMed]

155. Norowski, P.A., Jr.; Fujiwara, T.; Clem, W.C.; Adatrow, P.C.; Eckstein, E.C.; Haggard, W.O.; Bumgardner, J.D. Novel naturally crosslinked electrospun nanofibrous chitosan mats for guided bone regeneration membranes: Material characterization and cytocompatibility. *J. Tissue Eng. Regen. Med.* **2015**, *9*, 577–583. [CrossRef] [PubMed]

156. Bottino, M.C.; Thomas, V.; Janowski, G.M. A novel spatially designed and functionally graded electrospun membrane for periodontal regeneration. *Acta Bimater.* **2011**, *7*, 216–224. [CrossRef] [PubMed]

157. Bottino, M.C.; Thomas, V.; Schmidt, G.; Vohra, Y.K.; Chu, T.-M.G.; Kowolik, M.J.; Janowski, G.M. Recent advances in the development of GTR/GBR membranes for periodontal regeneration—A materials perspective. *Dent. Mater.* **2012**, *28*, 703–721. [CrossRef] [PubMed]

158. Ferrand, A.; Eap, S.; Richert, L.; Lemoine, S.; Kalaskar, D.; Demoustier-Champagne, S.; Atmani, H.; Mely, Y.; Fioretti, F.; Schlatter, G.; et al. Osteogenetic properties of electrospun nanofibrous PCL scaffolds equipped with chitosan-based nanoreservoirs of growth factors. *Macromol. Biosci.* **2014**, *14*, 45–55. [CrossRef] [PubMed]

159. Eap, S.; Keller, L.; Schiavi, J.; Huck, O.; Jacomine, L.; Fioretti, F.; Gauthier, C.; Sebastian, V.; Schwinte, P.; Benkirane-Jessel, N. A living thick nanofibrous implant bifunctionalized with active growth factor and stem cells for bone regeneration. *Int. J. Nanomed.* **2015**, *10*, 1061–1075.

160. Ignatova, M.G.; Manolova, N.E.; Toshkova, R.A.; Rashkov, I.B.; Gardeva, E.G.; Yossifova, L.S.; Alexandrov, M.T. Electrospun nanofibrous mats containing quaternized chitosan and polylactide with in vitro antitumor activity against HeLa cells. *Biomacromolecules* **2010**, *11*, 1633–1645. [CrossRef] [PubMed]

161. Toshkova, R.; Manolova, N.; Gardeva, E.; Ignatova, M.; Yossifova, L.; Rashkov, I.; Alexandrov, M. Antitumor activity of quaternized chitosan-based electrospun implants against Graffi myeloid tumor. *Int. J. Pharm.* **2010**, *400*, 221–233. [CrossRef] [PubMed]

162. Muzzarelli, R.A. Biomedical exploitation of chitin and chitosan via mechano-chemical disassembly, electrospinning, dissolution in imidazolium ionic liquids, and supercritical drying. *Mar. Drugs* **2011**, *9*, 1510–1533. [CrossRef] [PubMed]

163. Volpato, F.Z.; Almodovar, J.; Erickson, K.; Popat, K.C.; Migliaresi, C.; Kipper, M.J. Preservation of FGF-2 bioactivity using heparin-based nanoparticles, and their delivery from electrospun chitosan fibers. *Acta Bimater.* **2012**, *8*, 1551–1559. [CrossRef] [PubMed]

164. Monteiro, N.; Martins, M.; Martins, A.; Fonseca, N.A.; Moreira, J.N.; Reis, R.L.; Neves, N.M. Antibacterial activity of chitosan nanofiber meshes with liposomes immobilized releasing gentamicin. *Acta Bimater.* **2015**, *18*, 196–205. [CrossRef] [PubMed]

165. Zhou, Y.S.; Yang, D.Z.; Chen, X.M.; Xu, Q.; Lu, F.M.; Nie, J. Electrospun water-soluble carboxyethyl chitosan/poly(vinyl alcohol) nanofibrous membrane as potential wound dressing for skin regeneration. *Biomacromolecules* **2008**, *9*, 349–354. [CrossRef] [PubMed]

166. Zhang, K.H.; Qian, Y.F.; Wang, H.S.; Fan, L.P.; Huang, C.; Yin, A.L.; Mo, X.M. Genipin-crosslinked silk fibroin/hydroxybutyl chitosan nanofibrous scaffolds for tissue-engineering application. *J. Biomed. Mater. Res. Part A* **2010**, *95A*, 870–881. [CrossRef] [PubMed]

167. Charernsriwilaiwat, N.; Opanasopit, P.; Rojanarata, T.; Ngawhirunpat, T. Lysozyme-loaded, electrospun chitosan-based nanofiber mats for wound healing. *Int. J. Pharm.* **2012**, *427*, 379–384. [CrossRef] [PubMed]

168. Wang, X.L.; Cheng, F.; Gao, J.; Wang, L. Antibacterial wound dressing from chitosan/polyethylene oxide nanofibers mats embedded with silver nanoparticles. *J. Biomater. Appl.* **2015**, *29*, 1086–1095. [CrossRef] [PubMed]

169. Antunes, B.P.; Moreira, A.F.; Gaspar, V.M.; Correia, I.J. Chitosan/arginine-chitosan polymer blends for assembly of nanofibrous membranes for wound regeneration. *Carbohydr. Polym.* **2015**, *130*, 104–112. [CrossRef] [PubMed]

170. Abbaspour, M.; Sharif Makhmalzadeh, B.; Rezaee, B.; Shoja, S.; Ahangari, Z. Evaluation of the Antimicrobial Effect of Chitosan/Polyvinyl Alcohol Electrospun Nanofibers Containing Mafenide Acetate. *Jundishapur. J. Microbiol.* **2015**, *8*, e24239. [CrossRef] [PubMed]

171. Samprasit, W.; Kaomongkolgit, R.; Sukma, M.; Rojanarata, T.; Ngawhirunpat, T.; Opanasopit, P. Mucoadhesive electrospun chitosan-based nanofibre mats for dental caries prevention. *Carbohydr. Polym.* **2015**, *117*, 933–940. [CrossRef] [PubMed]

172. Samprasit, W.; Rojanarata, T.; Akkaramongkolporn, P.; Ngawhirunpat, T.; Kaomongkolgit, R.; Opanasopit, P. Fabrication and In Vitro/In Vivo Performance of Mucoadhesive Electrospun Nanofiber Mats Containing alpha-Mangostin. *AAPS PharmSciTech* **2015**, *16*, 1140–1152. [CrossRef] [PubMed]

173. Feng, Z.Q.; Leach, M.K.; Chu, X.H.; Wang, Y.C.; Tian, T.; Shi, X.L.; Ding, Y.T.; Gu, Z.Z. Electrospun chitosan nanofibers for hepatocyte culture. *J. Biomed. Nanotechnol.* **2010**, *6*, 658–666. [CrossRef] [PubMed]

174. Noriega, S.E.; Hasanova, G.I.; Schneider, M.J.; Larsen, G.F.; Subramanian, A. Effect of fiber diameter on the spreading, proliferation and differentiation of chondrocytes on electrospun chitosan matrices. *Cells Tissues Organs* **2012**, *195*, 207–221. [CrossRef] [PubMed]

175. Dang, J.M.; Leong, K.W. Myogenic Induction of Aligned Mesenchymal Stem Cell Sheets by Culture on Thermally Responsive Electrospun Nanofibers. *Adv. Mater.* **2007**, *19*, 2775–2779. [CrossRef] [PubMed]

176. Yang, H.S.; Lee, B.; Tsui, J.H.; Macadangdang, J.; Jang, S.Y.; Im, S.G.; Kim, D.H. Electroconductive Nanopatterned Substrates for Enhanced Myogenic Differentiation and Maturation. *Adv. Healthc. Mater.* **2016**, *5*, 137–145. [CrossRef] [PubMed]

177. Newman, P.; Galenano-Niño, J.L.; Graney, P.; Razal, J.M.; Minett, A.I.; Ribas, J.; Ovalle-Robles, R.; Biro, M.; Zreiqat, H. Relationship between nanotopographical alignment and stem cell fate with live imaging and shape analysis. *Sci. Rep.* **2016**, *6*, 37909. [CrossRef] [PubMed]

MDPI

St. Alban-Anlage 66

4052 Basel

Switzerland

Tel. +41 61 683 77 34

Fax +41 61 302 89 18

www.mdpi.com

International Journal of Molecular Sciences Editorial Office

E-mail: ijms@mdpi.com

www.mdpi.com/journal/ijms